Lecture Notes in Computer Science 4182

Hwee Tou Ng Mun-Kew Leong
Min-Yen Kan Donghong Ji (Eds.)

Information Retrieval Technology

Third Asia Information Retrieval Symposium, AIRS 2006
Singapore, October 16-18, 2006
Proceedings

Volume Editors

Hwee Tou Ng
Min-Yen Kan
National University of Singapore, Department of Computer Science
3 Science Drive 2, Singapore 117543
E-mail: {nght, kanmy} @comp.nus.edu.sg

Mun-Kew Leong
Donghong Ji
Institute for Infocomm Research
21 Heng Mui Keng Terrace, Singapore 119613
E-mail: {mkleong, dhji}@i2r.a-star.edu.sg

Library of Congress Control Number: 2006932723

CR Subject Classification (1998): H.3, H.4, F.2.2, E.1, E.2

LNCS Sublibrary: SL 3 – Information Systems and Application, incl. Internet/Web and HCI

ISSN 0302-9743
ISBN-10 3-540-45780-1 Springer Berlin Heidelberg New York
ISBN-13 978-3-540-45780-0 Springer Berlin Heidelberg New York

Springer is a part of Springer Science+Business Media

springer.com

Typesetting: Camera-ready by author, data conversion by Scientific Publishing Services, Chennai, India
Printed on acid-free paper SPIN: 11880592 06/3142 5 4 3 2 1 0

Preface

Asia Information Retrieval Symposium (AIRS) 2006 was the third AIRS conference in the series established in 2004. The first AIRS was held in Beijing, China, and the 2nd AIRS was held in Cheju, Korea. The AIRS conference series traces its roots to the successful Information Retrieval with Asian Languages (IRAL) workshop series which started in 1996.

The AIRS series aims to bring together international researchers and developers to exchange new ideas and the latest results in information retrieval. The scope of the conference encompassed the theory and practice of all aspects of information retrieval in text, audio, image, video, and multimedia data.

We are happy to report that AIRS 2006 received 148 submissions, the highest number since the conference series started in 2004. Submissions came from Asia and Australasia, Europe, and North America. We accepted 34 submissions as regular papers (23%) and 24 as poster papers (16%).

We would like to thank all the authors who submitted papers to the conference, the seven area chairs, who worked tirelessly to recruit the program committee members and oversaw the review process, and the program committee members and their secondary reviewers who reviewed all the submissions.

We also thank the Publications Chair, Min-Yen Kan, who liaised with the publisher Springer, compiled the camera ready papers and turned them into the beautiful proceedings you are reading now. We also thank Donghong Ji, who chaired the local organization efforts and maintained the AIRS 2006 website, Kanagasabai Rajaraman for publicity efforts, Yee Seng Chan for helping with the START conference management software, and Daniel Racoceanu for organizing and chairing the special session on medical image retrieval.

Lastly, we thank the sponsoring organizations for their support, the Institute for Infocomm Research and the National University of Singapore for hosting the conference, Springer for publishing our proceedings, and AVC for local organization and secretariat support.

August 2006

Mun-Kew Leong
Hwee Tou Ng

Preface

Organization

AIRS 2006 was organized by the Institute for Infocomm Research and co-organized by the National University of Singapore.

Executive Committee

Program Chair	Hwee Tou Ng (National University of Singapore)
General Chair	Mun-Kew Leong (Institute for Infocomm Research, Singapore)
Publication Chair	Min-Yen Kan (National University of Singapore)
Organizing Chair	Donghong Ji (Institute for Infocomm Research, Singapore)
Organizing Committee	GuoDong Zhou (Institute for Infocomm Research, Singapore) Nie Yu (Institute for Infocomm Research, Singapore) Kanagasabai Rajaraman (Institute for Infocomm Research, Singapore)

Area Chairs

Lee-Feng Chien	Academia Sinica
Noriko Kando	National Institute of Informatics
Kui-Lam Kwok	Queens College, City University of New York
Hang Li	Microsoft Research Asia
Ee-Peng Lim	Nanyang Technological University
Jian-Yun Nie	University of Montreal
Hae-Chang Rim	Korea University

Program Committee Members

Jun Adachi	National Institute of Informatics
Robert Allen	Drexel University
Vo Ngoc Anh	University of Melbourne
Thomas Baker	Goettingen State and University Library
Timothy Baldwin	University of Melbourne
Yunbo Cao	Microsoft Research Asia
Aitao Chen	Yahoo!
Hsinchun Chen	University of Arizona

Hsin-Hsi Chen	National Taiwan University
Kuang-hua Chen	National Taiwan University
Lei Chen	Hong Kong University of Science and Technology
Zheng Chen	Microsoft Research Asia
Jean-Pierre Chevallet	IPAL CNRS, Institute for Infocomm Research
Liang-Tien Chia	Nanyang Technological University
Lee-Feng Chien	Academia Sinica
Tat-Seng Chua	National University of Singapore
Hang Cui	National University of Singapore
Koji Eguchi	National Institute of Informatics
Patrick Fan	Virginia Tech
Jianfeng Gao	Microsoft Research
Fredric C. Gey	University of California, Berkeley
Dion Hoe-Lian Goh	Nanyang Technological University
Donna Harman	NIST
Wen-Lian Hsu	Academia Sinica
Jimmy Huang	York University
Makoto Iwayama	Hitachi Ltd.
Myung-Gil Jang	ETRI
Donghong Ji	Institute for Infocomm Research
Rong Jin	Michigan State University
Hideo Joho	Glasgow University
Gareth Jones	City University of Dublin
Min-Yen Kan	National University of Singapore
Ji-Hoon Kang	Chungnam National University
Tsuneaki Kato	Tokyo University
Sang-Bum Kim	Tokyo Institute of Technology
Kazuaki Kishida	Keio University
Youngjoong Ko	Dong-A University
Kui-Lam Kwok	Queens College, City University of New York
Mounia Lalmas	Queen Mary, University of London
Wai Lam	Chinese University of Hong Kong
Gary Geunbae Lee	Pohang University of Science & Technology
Jong-Hyeok Lee	Pohang University of Science & Technology
Gina-Anne Levow	University of Chicago
Haizhou Li	Institute for Infocomm Research
Hang Li	Microsoft Research Asia
Mingjing Li	Microsoft Research Asia
Hee Suk Lim	Hansin University
Chin-Yew Lin	Microsoft Research Asia
Chuan-Jie Lin	National Taiwan Ocean University
Tie-Yan Liu	Microsoft Research Asia
Ting Liu	Harbin Institute of Technology
Yan Liu	The Hong Kong Polytechnic University
Robert Wing Pong Luk	Hong Kong Polytechnic University

Qing Ma	Ryukoku University
Shaoping Ma	Tsinghua University
Helen Meng	The Chinese University of Hong Kong
Satoshi Morinaga	NEC
Sung Hyon Myaeng	Information & Communications University
Takashi Nagatsuka	Tsurumi University
Hiroshi Nakagawa	University of Tokyo
Hidetsugu Namba	Hiroshima City University
Jian-Yun Nie	University of Montreal
Jon Patrick	University of Sydney
Fuchun Peng	Yahoo!
Carol Peters	Italian National Research Council
Ian Ruthven	University of Strathclyde
Tetsuya Sakai	Toshiba Corporate R&D Center
Mark Sanderson	Sheffield University
Jacques Savoy	University of Neuchatel
Koichi Shinoda	Tokyo Institute of Technology
Dawei Song	The Open University
Shigeo Sugimoto	University of Tsukuba
Hiroya Takamura	Tokyo Institute of Technology
Katsumi Tanaka	Kyoto University
Yin-Leng Theng	Nanyang Technological University
Naohiko Uramoto	IBM
Shalini R. Urs	University of Mysore
Hsin-Min Wang	Academia Sinica
Ji-Rong Wen	Microsoft Research Asia
Christa Womster	University of Hildesheim
Kam-Fai Wong	Chinese University of Hong Kong
Lide Wu	Fudan University
Xing Xie	Microsoft Research Asia
Yiyu Yao	University of Regina
Masaharu Yoshioka	Hokkaido University
Yong Yu	Shanghai Jiaotong University
Marcia Zeng	Kent State University
ChengXiang Zhai	University of Illinois, Urbana Champaign
Dell Zhang	Birkbeck, University of London
Jian Zhang	National ICT Australia
Lei Zhang	Microsoft Research Asia
Min Zhang	Tsinghua University
Ming Zhou	Microsoft Research Asia
Justin Zobel	RMIT

Secondary Reviewers

Bo Chen
Yuanhao Chen
Bin Gao
Ming-Qiang Hou
Yang Hu
Feng Jing
Hiroshi Kanayama
Jiaming Li
Xirong Li
Jing Liu
Song Liu
Yiqun Liu
Yuting Liu
Tao Qin
Hironori Takeuchi
Mingfeng Tsai
Bin Wang
Canhui Wang
Changhu Wang
Huan Wang
Min Xu
Xin Yan
Wujie Zheng

Table of Contents

Session 1A: Text Retrieval

Session 1B: Search and Extraction

Session 1C: Text Classification and Indexing

Session 1D: Text Clustering

Session 1E: Information Retrieval Models

Session 2A: Web Information Retrieval

Session 2B: Cross-Language Information Retrieval

Session 2C: Question Answering and Summarization

Session 2D: Natural Language Processing

Session 2E: Evaluation

Session 3A: Multimedia Information Retrieval

Special Session: Medical Image Retrieval

Poster Session

Query Expansion with ConceptNet and WordNet: An Intrinsic Comparison

Ming-Hung Hsu, Ming-Feng Tsai, and Hsin-Hsi Chen*

Department of Computer Science and Information Engineering
National Taiwan University
Taipei, Taiwan
{mhhsu, mftsai}@nlg.csie.ntu.edu.tw,
hhchen@csie.ntu.edu.tw

Abstract. This paper compares the utilization of ConceptNet and WordNet in query expansion. Spreading activation selects candidate terms for query expansion from these two resources. Three measures including discrimination ability, concept diversity, and retrieval performance are used for comparisons. The topics and document collections in the ad hoc track of TREC-6, TREC-7 and TREC-8 are adopted in the experiments. The results show that ConceptNet and WordNet are complementary. Queries expanded with WordNet have higher discrimination ability. In contrast, queries expanded with ConceptNet have higher concept diversity. The performance of queries expanded by selecting the candidate terms from ConceptNet and WordNet outperforms that of queries without expansion, and queries expanded with a single resource.

1 Introduction

Query expansion has been widely used to deal with paraphrase problem in information retrieval. The expanded terms may come from feedback documents, target document collection, or outside knowledge resources [1]. WordNet [2], an electronic lexical database, has been employed to many applications [9], where query expansion is an important one. Voorhees [14] utilized lexical semantic relations in WordNet to expand queries. Smeaton et al [13] added WordNet synonyms of original query terms with half of their weights. Liu et al [7] used WordNet to disambiguate word senses of query terms, and then considered the synonyms, the hyponyms, and the words from definitions for possible additions to a query. Roberto and Paola [10] utilized WordNet to expand a query and suggested that a good expansion strategy is to add those words that often co-occur with the words of the query. To deal with short queries of web users, Moldovan and Mihalcea [8] applied WordNet to improve Internet searches.

In contrast to WordNet, commonsense knowledge was only explored in retrieval in a few papers. Liu and Lieberman [5] used ConceptNet [4] to expand query with the related concepts. However, the above work did not make formal evaluation, so that we were not sure if the effects of introducing common sense are positive or negative. Hsu and Chen [3] introduced commonsense knowledge into IR by expanding concepts in text descriptions of images with spatially related concepts. Experiments

* Corresponding author.

H.T. Ng et al. (Eds.): AIRS 2006, LNCS 4182, pp. 1 – 13, 2006.

showed that their approach was more suitable for precision-oriented tasks and for "difficult" topics. The expansion of this work was done at document level instead of query level. Document contributes much larger contextual information than query.

In the past, few papers have touched on the comparison of ConceptNet and WordNet in query expansion under the same benchmark. We are interested in what effects of these two resources have. If we know which resource is more useful in a certain condition, we are able to improve the retrieval performance further. In this paper, we design some experiments with evaluation criteria to quantitatively measure WordNet and ConceptNet in the aspect of query expansion. We employ the same algorithm, *spreading activation* [12], to select candidate terms from ConceptNet and WordNet for the TREC topics 301-450, which were used in TREC-6, TREC-7 and TREC-8. To compare the *intrinsic characteristics* of these two resources, we propose three types of quantitative measurements including discrimination ability, concept diversity, and retrieval performance.

This paper is organized as follows. In Section 2, we give brief introduction to WordNet and ConceptNet. The comparison methodology is specified in Section 3. Section 4 introduces the experiment environment and discusses the experimental results. Section 5 concludes the remarks.

2 WordNet and ConceptNet

In this section, we give brief introduction to the two resources to be compared. Frameworks and origins of the two knowledgebase are described. A surface comparison of their similarities and differences is also presented.

2.1 WordNet

WordNet appeared in 1993 and has been developed by linguistic experts at Princeton University's Cognitive Science Laboratory since 1985. It is a general-purpose knowledgebase of words, and it covers most English nouns, adjectives, verbs and adverbs. WordNet's structure is a relational semantic network. Each node in the network is a lexical unit that consists of several synonyms, standing for a specific "sense". Such lexical unit is called as 'synset' in WordNet terminology. Synsets in WordNet are linked by a small set of semantic relations such as 'is-a' hierarchical relations and 'part-of' relations. For its simple structure with words at nodes, WordNet's success comes from its ease of use [2][9].

2.2 ConceptNet

ConceptNet is developed by MIT Media Laboratory and is presently the largest commonsense knowledgebase [6]. ConceptNet is a relational semantic network that is automatically generated from about 700,000 English sentences of the Open Mind Common Sense (OMCS) corpus. Nodes in ConceptNet are compound concepts in the form of natural language fragments (e.g. 'food', 'buy food', 'grocery store', and 'at home'). Because the goal of developing ConceptNet is to cover pieces of commonsense knowledge to describe the real world, there are 20 kinds of relations categorized as causal, spatial, functional, etc. ConceptNet has been adopted in many interactive applications [4]. Hsu and Chen [3] utilized ConceptNet to expand image annotations and got improvement for some difficult topics. As commonsense knowledge is deeply context-sensitive, the suitability of ConceptNet for query expansion is still not clear.

2.3 A Surface Comparison

WordNet and ConceptNet have several similarities: (1) their structures are both relational semantic networks; (2) both of them are general-purpose (that is, not domain-specific) knowledgebase; and (3) concepts in the two resources are both in the form of natural language. On the other hand, WordNet and ConceptNet differ from each other in some aspects: (1) as their processes of development differ (manually handcrafted vs. automatically generated), intuitively WordNet has higher quality and robustness; (2) while WordNet focuses on formal taxonomies of words, ConceptNet focuses on a richer set of semantic relations between compound concepts [6]; and (3) WordNet differentiates ambiguous meanings of a word as synsets, however, ConceptNet bears ambiguity of commonsense knowledge in its concepts and relations.

3 Comparison Methodologies

To compare the two knowledgebase in the aspect of query expansion intrinsically, we perform the same algorithm to expand queries. As WordNet and ConceptNet are both relational semantic networks, i.e., useful concepts for expansion in the network are usually those related to the concepts of the query, spreading activation [12] is adopted.

Figure 1 shows the overall procedure of the comparison. Given an original query, we perform spreading activation on WordNet and ConceptNet, respectively. Then, the two expanded queries are compared with three quantitative measurements. The first measurement computes the discrimination ability in information retrieval. The second measurement calculates the concept diversity in relevant documents. The third directly evaluates the performance of retrieval, including two typical evaluation criteria for ad hoc retrieval. All of these measurements are described in Section 3.4.

When we perform spreading activation in a semantic network to expand a query, the node of *activation origin* represents the concept of the given query. The activation origin is the first to be activated, with an initial activation score (e.g., 1.0). Next, nodes one link away from the activation origin are activated, then two links away, and so on.

Equation (1) shown below determines the activation score of node j by three factors: (i) a constant $Cdd \leqq 1$ (e.g., 0.5), which is called *distance discount* that causes a node closer to the activation origin to get a higher activation score; (ii) the activation score of node i; (iii) $W(i,j)$, the weight of the link from i to j. Different relations in the semantic network are of different weights. Neighbor(j) represents the nodes connected to node j.

$$\text{Activation\_score}(j) = C_{dd} \cdot \sum_{i \in \text{Neighbor}(j)} \text{Activation\_score}(i) \cdot W(i,j) \tag{1}$$

Since most traditional IR systems are of bag-of-words model, we select the top N words with the higher activation scores as the expanded query. For a word w, its activation score is the sum of scores of the nodes (i.e., synsets in WordNet) that contain w.

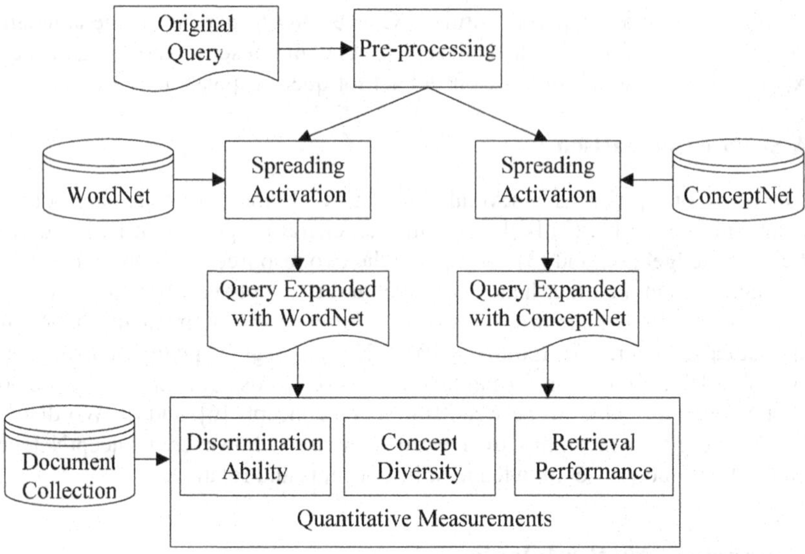

Fig. 1. Overall procedure in our approach

3.1 Pre-processing

Each query was lemmatized and POS-tagged by Brill tagger. Stop-words were removed and each of the remaining words with POS-tags was considered as a *concept* at the following stages.

3.2 Spreading Activation in ConceptNet

In addition to commonsense concepts and relations, ConceptNet also provides a set of tools for reasoning over text [6]. One of these tools, *get_context*(*concepts*), performs spreading activation on all kinds of relations in ConceptNet, to find contextual neighborhood relevant to the concepts as parameters. Different relations are set to different weights in default setting. For example, the weight of 'IsA' relation is 0.9 and the weight of 'DefinedAs' is 1.0, etc. We adopt this tool directly for collecting expanded terms. In our experiments, each word in a compound concept has the same activation score as that of the compound concept. More details about the reasoning tools in ConceptNet please refer to [6].

3.3 Spreading Activation in WordNet

Since each node in WordNet is a synset that contains synonyms of certain sense, spreading activation in WordNet is surely performed on the unit of synset. Because ConceptNet covers most relations in WordNet, we determine the weights of relations in WordNet, shown in Table 1, by referring to the settings in ConceptNet. For each concept (a word with POS-tag) in the query, we choose its most frequent sense (synset) as the activation origin from the corresponding POS. In other words, we do not disambiguate the sense of query terms in this paper for simplicity.

Table 1. Relations in WordNet and their weights for spreading activation

Relation Type	causes	holonyms	hypernyms	hyponyms	meronyms	Pertainyms
Weight	0.8	0.8	0.8	0.9	0.9	0.7

3.4 Quantitative Measurements

The following proposes three types of measurements to investigate the intrinsic differences between WordNet and Concept. They provide different viewpoints for the comparison.

(1) **Discrimination Ability (DA).** Discrimination ability is used to measure how precisely a query describes the information need. In IR, the inverse document frequency (IDF) of a term denotes if it occurs frequently in individual documents but rarely in the remainder of the collection. For IR systems as the state of the art, discrimination ability of a term can be estimated by its IDF value. Hence the discrimination ability of a query is measured and contributed by the IDFs of the terms in the query. For a query q composed of n query terms (q_1, q_2, …, q_n), we define its discrimination ability (DA) as follows.

$$\mathrm{DA}(q) = \frac{1}{n}\sum_{i=1}^{n} \log\left(\frac{Nc}{\mathrm{df}(q_i)}\right) \tag{2}$$

where N_C is the number of documents in a collection, and df(q_i) is the document frequency of query term q_i.

(2) **Concept Diversity (CD).** This measurement helps us observe the concept diversity of an expanded query, relative to the relevant documents. That is, we measure how much an expanded query covers the concepts occurring in the relevant documents. Let tm(.) denote the function that maps the parameter (a document or a query) to the set of its index terms. Let $\{d_{q(1)}, d_{q(2)}, \ldots, d_{q(m)}\}$ denote the set of m documents which are relevant to the query q in the collection. The concept diversity (CD) of a query q is defined as follows.

$$\mathrm{CD}(q) = \frac{1}{m}\sum_{i=1}^{m} \frac{|\,\mathrm{tm}(q) \cap \mathrm{tm}(d_{q(i)})\,|}{|\,\mathrm{tm}(d_{q(i)})\,|} \tag{3}$$

(3) **Retrieval Performance:** This type of measurements includes two typical evaluation criteria for ad hoc retrieval, i.e., average precision (AP) and precision at top 20 documents (P@20).

4 Experiments and Discussion

We adopted the topics and the document collections in the ad hoc track of TREC-6, TREC-7 and TREC-8 as the experimental materials for the comparison. There are 556,077 documents in the collection of TREC-6 and 528,155 documents in TREC-7 and in TREC-8. Only the "title" part was used to simulate short query, since web users often submit short queries to search engines. There are totally 150 topics with identifiers 301-450. However, 4 of them (i.e., topics 312, 379, 392 and 423) are unable to be expanded by spreading activation either in WordNet or in ConceptNet, so that these 4 topics are neglected in the experiments. For each short query, the top 100 words with the higher activation scores form the expanded query. The IR system adopted for the measurement of retrieval performance is Okapi's BM25 [11]. The retrieval performance is measured on the top 1000 documents for each topic.

Figures 2, 3, 4, and 5 show the results of the quantitative measurements, where the x-axis represents topic number, and 301-350, 351-400, and 401-450 are topics of TREC-6, TREC-7 and TREC-8, respectively. To compare the differences between WordNet and ConceptNet, the result presented for each topic is the difference between the two expanded queries, i.e., the measurement of the WordNet-expanded query subtracts that of the ConceptNet-expanded query.

4.1 Preliminary Analysis

Figure 2 shows the differences of two kinds of expansions in discrimination ability (DA). The DA averaged over the 146 experimental topics is 5.676 and 4.191 for the WordNet-expanded and ConceptNet-expanded queries, respectively. From Figure 2, it is obvious that the terms in the WordNet-expanded queries have higher discrimination ability. In other words, they are more specific than those terms in ConceptNet-expanded queries. A specific term highly relevant to a topic can be considered as one of the *kernel words* of that topic. Figure 2 shows that the queries expanded by WordNet are more probable to contain the kernel words.

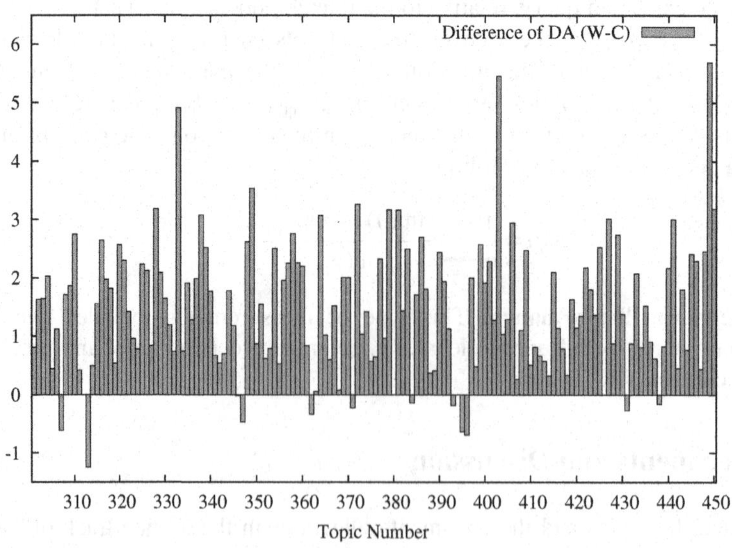

Fig. 2. Differences between WordNet and ConceptNet in discrimination ability (DA)

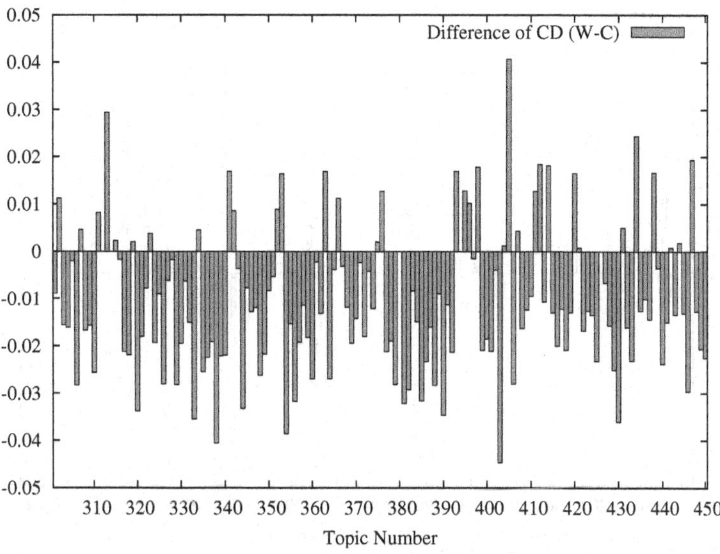

Fig. 3. Differences between WordNet and ConceptNet in concept diversity (CD)

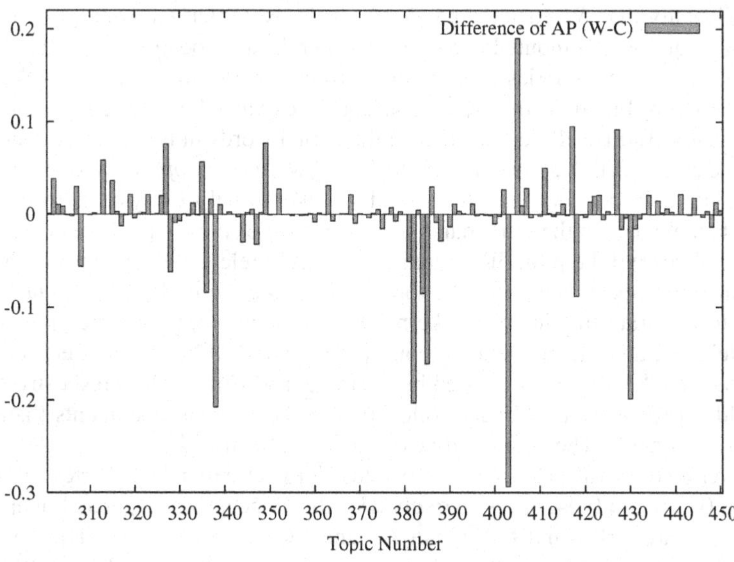

Fig. 4. Topic-by-topic differences of retrieval performance in average precision (AP)

The average concept diversity (CD) is 0.037 and 0.048 for the WordNet-expanded and the ConceptNet-expanded queries, respectively. In contrast to the result of discrimination ability, Figure 3 shows that ConceptNet-expanded queries have higher

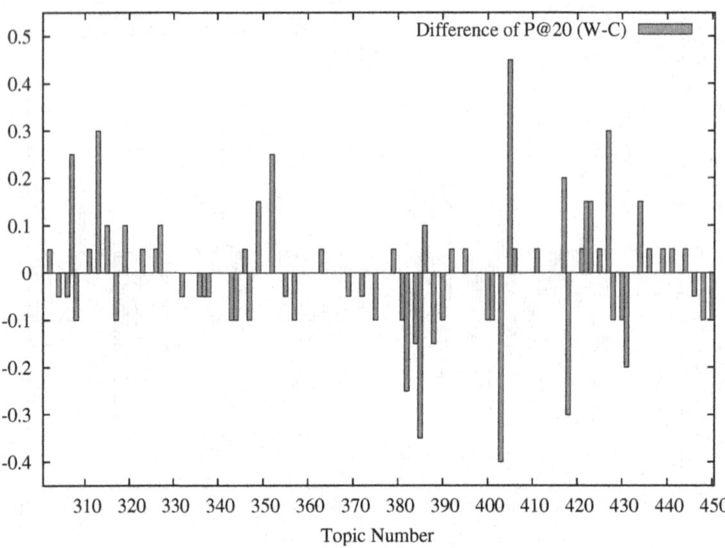

Fig. 5. Topic-by-topic differences of retrieval performance in P@20

concept diversity than WordNet-expanded ones do. Note that the concept diversity of an expanded query is computed according to the relevant documents. As the terms in ConceptNet-expanded queries are usually more general than those in WordNet-expanded queries, Figure 3 shows that ConceptNet-expanded queries cover more of the concepts that would usually co-occur with the kernel words in the relevant documents. We call the concepts that co-occur with the kernel words as *cooperative concepts*.

Expanding a short query with the kernel words will help IR systems to find more relevant documents. On the other hand, co-occurrence of the cooperative concepts and the kernel words will help the IR system to rank truly relevant documents higher than those containing noise. Here we take topic 335 ("adoptive biological parents") as an example to illustrate this idea. The kernel words of this topic may be "pregnancy", "surrogate", etc, and the cooperative concepts may be "child", "bear", etc. Of these, "surrogate" and "child" are suggested by WordNet and ConceptNet, respectively. The pair (child, surrogate) has stronger collocation in the relevant documents than in the irrelevant documents. The detail will be discussed in Section 4.2.

The overall retrieval performances of the WordNet-expanded queries in AP and P@20 are 0.016 and 0.0425, respectively. For the ConceptNet-expanded queries, the performances are 0.019 and 0.0438, in AP and P@20, respectively. These retrieval performances are low because the expanded queries are formed by the top 100 words with the higher activation scores. The simple expansion method introduces too much noise. Figure 4 and Figure 5 show the differences of AP and of P@20 for each topic. We observed that WordNet-expanded queries perform better for some topics, but ConceptNet-expanded queries perform better for some other topics. While WordNet and ConceptNet are different in discrimination ability (Figure 2) and in concept diversity (Figure 3), the two resources can complement each other in the task of ad hoc retrieval. Hence, we made further experiments in the following subsection.

4.2 Further Analysis

In the next experiments, we performed manual query expansion by selecting some of the top 100 words proposed by spreading activation in WordNet or in ConceptNet, to expand the original query. Two research assistants, each of whom dealt with half of the topics of TREC-6, performed the process of manual expansion. They read the topic description in advance, and he/she had no idea about the real content or vocabulary in the relevant documents. This manual selection process was performed separately on the words proposed by WordNet and by ConceptNet. These manually selected words for expansion are called WordNet-consulted (WC) and ConceptNet-consulted (CC) terms, respectively. In this way, we compared four expansion strategies, i.e., original (no expansion), WordNet-consulted, ConceptNet-consulted, and combination of WC and CC. We also increased the weights of the original query terms with different degrees to observe how the performances vary with the degrees. In the experiments, we only used the topics of TREC-6 for analyses.

Figure 6 shows the performances in mean average precision of the four expansion strategies. The x-axis represents the degrees (times) by which the weights of the original query terms are increased. The performance of the original query is 0.221 and doesn't vary with the degrees since there is no expansion. CC slightly performs better than the original when the degree is larger than 3, as well as WC with the degree larger than 6. The slight improvement of CC only or WC only shows that without information about the real content of the relevant documents, effective query expansion for IR is a challenging task even for humans. The best performance is 0.2297, which is obtained with combination of WC and CC at degree 6, and 3.94% increase to the

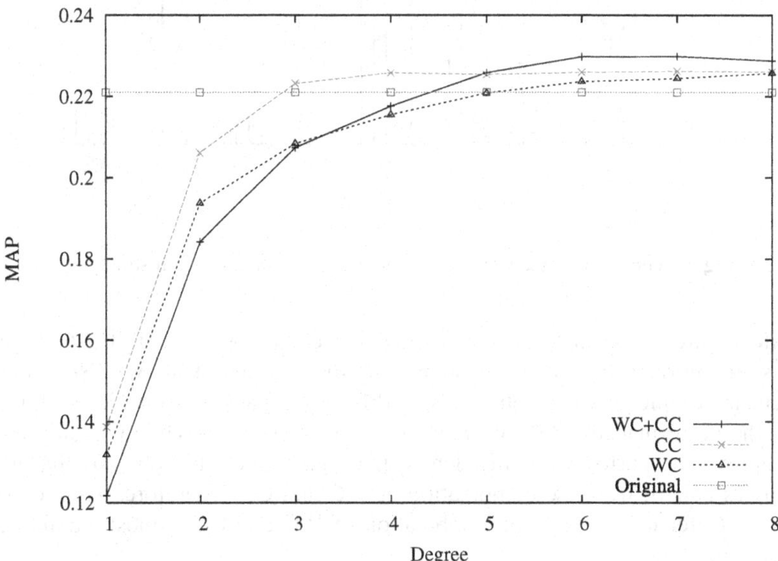

Fig. 6. Performances of the four expansion strategies vs. the weights of the original query term

baseline. This performance improvement on no expansion is examined as significant by a t-test with a confidence level of 95%. The corresponding p-value is 0.034.

Figure 6 also shows that a careful weighting scheme is needed no matter whether WordNet only or ConceptNet only is adopted. With an unsuitable weighting scheme, combination of WC and CC performs even worse than WC only or CC only. In Figure 6, CC performs stably when the degree increases larger than 3, but WC performs stably only after the degree is larger than 6. As the degree stands for how much the weights of original query terms are increased, in the aspect of ranking documents, the degree also stands for how much the weights of CC or WC terms are lightened. While the words proposed by WordNet are usually more specific and influence more heavily on the rankings of retrieved documents, it is shown that an appropriate weighting scheme is more important for WC than for CC.

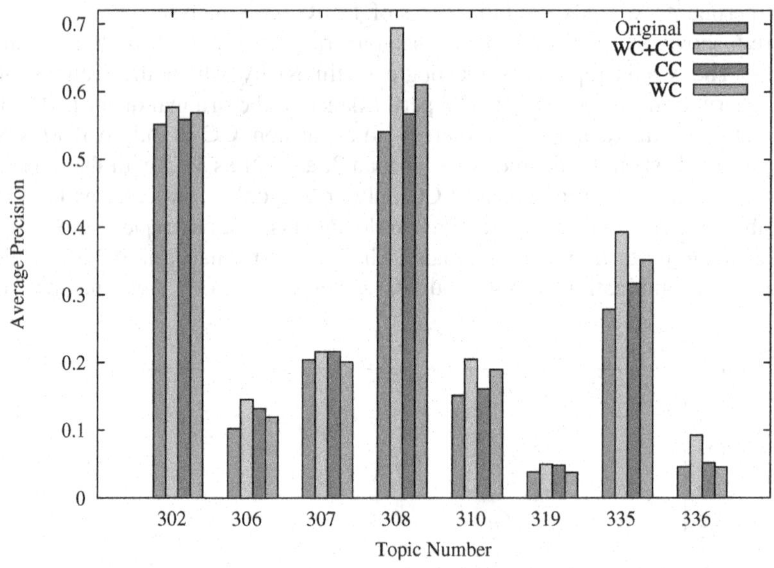

Fig. 7. The performances of 8 topics using different expansion strategies

While Figure 6 shows the result averaged over all topics of TREC-6, Figure 7 shows some more strong evidences supporting the argument that WordNet and ConceptNet can complement each other. Using different expansion strategies with the degree 6, the performances (AP) of eight topics are presented. While CC only or WC only improve the performance of each of the eight topics, it's obvious that all the eight topics benefit from the combination of WC and CC. Therefore, the overall improvement (refer to Figure 6) of combination of WC and CC is mostly exhibited in the eight topics.

We verify the complementary of WordNet and ConceptNet, i.e., frequent co-occurrence of WC terms and CC terms in relevant documents, by the following way. For each pair of CC term t_c and WC term t_w, we calculate LRP(t_c, t_w) using Equation

(4). LRP(t_c, t_w) is a value of *l*ogarithm of the *r*atio of two conditional *p*robabilities: (1) the co-occurrence probability of t_c and t_w in the relevant documents; and (2) the co-occurrence probability of t_c and t_w in the irrelevant documents.

$$\mathrm{LRP}(t_c, t_w) = \log \frac{\mathrm{P}(t_c, t_w \mid R)}{\mathrm{P}(t_c, t_w \mid IRR)} \qquad (4)$$

where *R* and *IRR* represent relevant and irrelevant documents, respectively.

Table 2 shows the title, the CC terms, the WC terms and the term pairs having high LRP values for each topic in Figure 7. The term pairs with high LRP in the eight topics are the major evidences to support that combination of WC and CC is effective as shown in Figures 6 and 7. Note that the pairs (child, surrogate), (life, surrogate) and (human, surrogate) of topic 335 have high LRP values. They also confirm the idea of kernel words and cooperative concepts mentioned in Section 4.1.

Table 2. Illustration of the complementary of WordNet and ConceptNet in 8 topics

Topic	Title of Topic	CC (t_c)	WC (t_w)	(t_c,t_w): LRP(t_c,t_w)
302	poliomyelitis and post-polio	affection flame global	disease place paralysis	(global, paralysis): 5.967 (global, disease): 5.043 (affection, paralysis): 4.983
306	african civilian deaths	war kill nation	megadeath event casualty killing	(war, killing): 3.715 (kill, casualty): 3.598 (nation, killing): 3.408 (nation, casualty): 3.357
307	new hydroelectric projects	state exploration station	proposition risky examination	(station, risky): 3.906 (station, examination): 3.598 (exploration, examination): 3.428
308	implant dentistry	medical reproductive function	artificial prosthesis specialty	(medical, prosthesis): 9.400 (function, prosthesis): 9.009 (medical, artificial): 6.980
310	radio waves and brain cancer	cause carcinogen state produce	corpus radiation	(cause, radiation): 5.234 (produce, radiation): 3.958 (state, radiation): 3.081
319	new fuel sources	find energy natural	material head	(energy, material): 3.744 (energy, head): 3.217 (natural, material): 3.071
335	adoptive biological parents	child human life	surrogate married kinship	(child, surrogate): 8.241 (life, surrogate): 6.840 (human, surrogate): 5.867
336	black bear attacks	animal claw battle	fight strike counterattack	(claw, fight): 6.255 (animal, fight): 3.916

5 Conclusions and Future Works

In this paper, we used the technique of spreading activation to investigate the intrinsic characteristics of WordNet and of ConceptNet. Three types of quantitative measurements, i.e., discrimination ability, concept diversity, and retrieval performance are used to compare the differences between the two resources. With the preliminary analysis and the verification of manual expansion, we have shown that WordNet is good at proposing kernel words and ConceptNet is useful to find cooperative concepts. With an appropriate weighting scheme, the two resources can complement each other to improve IR performance.

In future work, we will investigate an automatic query expansion method to combine the advantages of the two resources. Commonsense knowledge in ConceptNet is deeply context-sensitive so that it needs enough context information for automatic query expansion. Using existing methods such as pseudo relevance feedback or resources such as WordNet to increase the context information can be explored. We also intend to investigate whether complex techniques of word sense disambiguation (WSD) in WordNet are necessary for IR, under the existence of ConceptNet.

Acknowledgements

Research of this paper was partially supported by National Science Council, Taiwan, under the contracts NSC94-2752-E-001-001-PAE and NSC95-2752-E-001-001-PAE.

References

1. Baeza-Yates, Ricardo and Ribeiro-Neto, Berthier: Modern Information Retrieval, Addison-Wesley (1999).
2. Fellbaum, C. (ed.). WordNet: An Electronic Lexical Database. MIT Press (1998).
3. Hsu, Ming-Hung and Chen, Hsin-Hsi: Information Retrieval with Commonsense Knowledge. In: Proceedings of 29th ACM SIGIR International Conference on Research and Development in Information Retrieval (2006).
4. Lieberman, H., Liu, H., Singh, P., and Barry, B. Beating Common Sense into Interactive Applications. AI Magazine 25(4) (2004) 63-76.
5. Liu, H. and Lieberman, H: Robust Photo Retrieval Using World Semantics. In: Proceedings of LREC2002 Workshop: Using Semantics for IR (2002) 15-20.
6. Liu, H. and Singh, P.: ConceptNet: A Practical Commonsense Reasoning Toolkit. BT Technology Journal 22(4) (2004) 211-226.
7. Liu, S., Liu, F., Yu, C.T., and Meng, W.: An Effective Approach to Document Retrieval via Utilizing WordNet and Recognizing Phrases. In: Proceedings of the 27th Annual International ACM SIGIR Conference on Research and Development in Information Retrieval (2004) 266-272.
8. Moldovan, D.I. and Mihalcea, R.: Using WordNet and Lexical Operators to Improve Internet Searches. IEEE Internet Computing 4(1) (2000) 34-43.
9. Morato, Jorge, Marzal, Miguel Ángel, Lloréns, Juan and Moreiro, José.: WordNet Applications. In: Proceedings of International Conference of the Global WordNet Association (2004) 270-278.
10. Navigli, Roberto and Velardi, Paola: An Analysis of Ontology-based Query Expansion Strategies. In: Proceedings of Workshop on Adaptive Text Extraction and Mining at the 14th European Conference on Machine Learning (2003).

11. Robertson, S.E., Walker, S. and Beaulieu, M.: Okapi at TREC-7: Automatic Ad Hoc, Filtering, VLC and Interactive. In: Proceedings of the Seventh Text Retrieval Conference (1998) 253-264.
12. Salton, G. and Buckley, C.: On the Use of Spreading Activation Methods in Automatic Information Retrieval. In: Proceedings of the 11th CM-SIGIR Conference on Research and Development in Information Retrieval (1988) 147-160.
13. Smeaton, Alan F., Kelledy, Fergus, and O'Donell, Ruari: TREC-4 Experiments at Dublin City University: Thresholding Posting Lists, Query Expansion with WordNet and POS Tagging of Spanish. In: Proceedings of TREC-4 (1994) 373-390.
14. Voorhees, Ellen M.: Query Expansion Using Lexical-Semantic Relations. In: Proceedings of the 17th Annual ACM SIGIR Conference on Research and Development in Information Retrieval (1994) 61-69.

Document Similarity Search Based on Manifold-Ranking of TextTiles

Xiaojun Wan, Jianwu Yang, and Jianguo Xiao

Institute of Computer Science and Technology,
Peking University, Beijing 100871, China
{wanxiaojun, yangjianwu, xiaojianguo}@icst.pku.edu.cn

Abstract. Document similarity search aims to find documents similar to a query document in a text corpus and return a ranked list of similar documents. Most existing approaches to document similarity search compute similarity scores between the query and the documents based on a retrieval function (e.g. Cosine) and then rank the documents by their similarity scores. In this paper, we proposed a novel retrieval approach based on manifold-ranking of TextTiles to re-rank the initially retrieved documents. The proposed approach can make full use of the intrinsic global manifold structure for the TextTiles of the documents in the re-ranking process. Experimental results demonstrate that the proposed approach can significantly improve the retrieval performances based on different retrieval functions. TextTile is validated to be a better unit than the whole document in the manifold-ranking process.

1 Introduction

Document similarity search is to find documents similar to a query document in a text corpus and return a ranked list of similar documents to users. The typical kind of similarity search is K nearest neighbor search, namely K-NN search, which is to find K documents most similar to the query document. Similarity search is widely used in recommender systems in library or web applications. For example, *Google*[1] can perform an advanced search with "related" option to find similar web pages with a user-specified web page and *CiteSeer.IST*[2] provides a list of similar papers with the currently browsed paper.

Document similarity search differs from traditional keyword-based text retrieval in that similarity search systems take a full document as query, while current keyword-based search engines usually take several keywords as query. The keyword-based short query is usually constructed by users and can well reflect users' information need, while the document-based long query may contains more redundant and ambiguous information and even greater noise effects stemmed from the presence of a large number of words unrelated to the overall topic in the document.

[1] http://www.google.com

[2] http://citeseer.ist.psu.edu/cs

H.T. Ng et al. (Eds.): AIRS 2006, LNCS 4182, pp. 14–25, 2006.

The popular retrieval functions used in current text retrieval systems include the Cosine function, the Jaccard function, the Dice function [2, 14], the BM25 function in the Okapi system [10, 11] and the vector space model with document length normalization in the Smart system [12, 13], among which the standard Cosine measure is considered as the best model for document similarity search because of its good ability to measure the similarity between two documents. To our knowledge, almost all text search engines find relevant documents to a given query only by the pairwise comparison between the document and the query, thus neglecting the intrinsic global manifold structure of the documents. In order to make up for this limitation, we employ a manifold-ranking process [15, 16] for the initially retrieved documents and make full use of the relationships between the documents. In the manifold-ranking process, documents can spread their ranking scores to their nearby neighbors via a weighted network. Moreover, inspired by passage retrieval [6, 7], the manifold-ranking process is applied at a finer granularity of TextTile [4, 5] instead of the whole document, because a document is usually characterized as a sequence of subtopical discussions (i.e. TextTiles) that occur in the context of a few main topic discussions, and comparison of subtopics is deemed to be more accurate than comparison of the whole document.

Specifically, the proposed retrieval approach consists of two processes: initial ranking and re-ranking. In the initial ranking process, a small number of documents are retrieved based on a popular retrieval function. In the re-ranking process, the query document and the initially retrieved documents are segmented into TextTiles, and then the manifold-ranking algorithm is applied on the TextTiles and each TextTile obtains its ranking score. Lastly, a document gets its final retrieval score by fusing the ranking scores of the TextTiles in the document. The initially retrieved documents are re-ranked and the re-ranked list is returned to users. Experimental results show the encouraging performance of the proposed approach and it can significantly improve the retrieval performances based on different retrieval functions. TextTile is validated to be a better text unit than the whole document in the manifold-ranking process.

The rest of this paper is organized as follows: Popular retrieval functions are introduced in Section 2. The proposed manifold-ranking based approach is described in detail in Section 3. Section 4 gives the experiments and results. Lastly, we present our conclusion in Section 5.

2 Popular Retrieval Functions

2.1 The Cosine Function

The Cosine measure is the most popular measure for document similarity based on the vector space model (VSM). In VSM, a document d is represented by a vector with each dimension referring to a unique term and the weight associated with the term t is calculated by the $tf_{d,t} * idf_t$ formula, where $tf_{d,t}$ is the number of occurrences of term t in document d and $idf_t=1+log(N/n_t)$ is the inverse document frequency, where N is the total number of documents in the collection and n_t is the number of documents containing term *t*. The similarity *sim(q,d)*, between the query document *q* and the document *d*, can be defined as the normalized inner product of the two vectors $\vec{q}$ and $\vec{d}$:

$$\text{sim}_{\text{cosine}}(q,d) = \frac{\vec{q}\cdot\vec{d}}{|\vec{q}||\vec{d}|} = \frac{\sum_{t\in q\cap d}(w_{q,t}\cdot w_{d,t})}{\sqrt{\sum_{t\in q} w_{q,t}^{2} \times \sum_{t\in d} w_{d,t}^{2}}} \quad (1)$$

where *t* represents a term. $q \cap d$ gets the common words between *q* and *d*. Term weight $w_{d,t}$ is computed by $tf_{d,t} * idf_t$.

2.2 The Jaccard Function

The Jaccard function is similar to the Cosine function as follows:

$$\text{sim}_{\text{Jaccard}}(q,d) = \frac{\sum_{t\in q\cap d}(w_{q,t}\cdot w_{d,t})}{\sum_{t\in q} w_{q,t}^{2} + \sum_{t\in d} w_{d,t}^{2} - \sum_{t\in q\cap d}(w_{q,t}\cdot w_{d,t})} \quad (2)$$

2.3 The Dice Function

The Dice function is defined similarly to the Cosine function as follows:

$$\text{sim}_{\text{Dice}}(q,d) = \frac{2\times\sum_{t\in q\cap d}(w_{q,t}\cdot w_{d,t})}{\sum_{t\in q} w_{q,t}^{2} + \sum_{t\in d} w_{d,t}^{2}} \quad (3)$$

2.4 The BM25 Function

The BM25 measure is one of the most popular retrieval models in a probabilistic framework and is widely used in the Okapi system. Given the query document *q*, the similarity score for the document *d* is defined as follows:

$$\text{sim}_{\text{BM25}}(q,d) = \sum_{t\in q} f_{q,t} \times \log\left(\frac{N-n_t+0.5}{n_t+0.5}\right) \times \frac{(K+1)\times f_{d,t}}{K\times\left\{(1-b)+b\frac{dlf_d}{avedlf}\right\}+f_{d,t}} \quad (4)$$

where *t* represents a unique term; *N* is the number of documents in the collection; n_t is the number of documents in which term t exists; $f_{q,t}$ is the frequency of term *t* in *q*; $f_{d,t}$ is the frequency of term *t* in *d*; dlf_d is the sum of term frequencies in *d*; *avedlf* is the average of dlf_d in the collection; *K*=2.0, *b*=0.8 are constants.

2.5 The Vector Space Model with Document Length Normalization

The vector space model with document length normalization (NVSM) is also a popular retrieval model and is used in the Smart system. Given the query document *q*, the similarity score for the document *d* is defined as follows:

$$\mathrm{sim}_{\mathrm{NVSM}}(q,d) = \sum_{t \in q} (1+\log(f_{q,t})) \times idf_t \times \frac{1+\log(f_{d,t})}{1+\log(avef_d)} \times \frac{1}{avedlb + S \times (dlb_d - avedlb)} \quad (5)$$

where idf_t is the inverse document frequency of term t; dlb_d is the number of unique terms in d; $avef_d$ is the average of term frequencies in d (i.e. "dlf_d/dlb_d"); $avedlb$ is the average of dlb_d in the collection; S=0.2 is a constant.

3 The Manifold-Ranking Based Approach

3.1 Overview

The aim of the proposed approach is two-fold: one is to evaluate the similarity between the query and a document by exploring the relationship between all the documents in the feature space, which addresses the limitation of present similarity metrics based only on pairwise comparison; the other is to evaluate document similarity at a finer granularity by segmenting the documents into TextTiles, which addresses the limitation of present similarity metrics based on the whole document.

The proposed approach first segments the query and the documents into TextTiles using the TextTiling algorithm [4, 5], and then applies a manifold-ranking process on the TextTiles. All the TextTiles of a document obtain their ranking scores and the ranking score of the document is obtained by fusing the ranking scores of its TextTiles.

Note that it is of high computational cost to apply the manifold-ranking process to all the documents in the collection, so the above manifold-ranking process is taken as a re-ranking process. First, we use a popular retrieval function (e.g. Cosine, Jaccard, Dice, BM25, NVSM, etc.) to efficiently obtain an initial ranking of the documents, and then the initial k documents are re-ranked by applying the above manifold-ranking process.

Formally, given a query document q and the collection C, the proposed approach consists of the following four steps:

1. **Initial Ranking:** The initial ranking process uses a popular retrieval function to return a set of top documents $D_{\mathrm{init}} \subseteq C$ in response to the query document q, $|D_{\mathrm{init}}|=k$. Each document $d_i \in D_{\mathrm{init}}$ ($1 \le i \le k$) is associated with an initial retrieval score $InitScore(d_i)$.
2. **Text Segmentation:** By using the TextTiling algorithm, the query document q is segmented into a set of TextTiles $\chi_q = \{x_1, x_2, \ldots, x_p\}$ and all documents in D_{init} are segmented respectively and the total set of TextTiles for D_{init} is $\chi_{D_{\mathrm{init}}} = \{x_{p+1}, x_{p+2}, \ldots, x_n\}$.
3. **Manifold-Ranking:** The manifold-ranking process in applied on the whole set of TextTiles: $\chi = \chi_q \cup \chi_{D_{\mathrm{init}}}$, and each TextTile x_j ($p+1 \le j \le n$) in $\chi_{D_{\mathrm{init}}}$ gets its ranking score f_j^*.

4. **Score Fusion:** The final score *FinalScore*(d_i) of a document $d_i \in D_{\text{init}}$ ($1 \le i \le k$) is computed by fusing the ranking scores of its TextTiles. The documents in D_{init} are re-ranked according to their final scores and the re-ranked list is returned.

The steps 2-4 are key steps in the re-ranking process and they will be illustrated in detail in next sections, respectively.

3.2 The Text Segmentation Process

There have been several methods for division of documents according to units such as sections, paragraphs, or fixed length sequences of words, or semantic passages given by inferred shift of topic [6, 7]. In this study, we adopt semantic passages to represent subtopics in a document. As mentioned in [4, 5], the text can be characterized as a sequence of subtopical discussions that occur in the context of a few main topic discussions. For example, a news text about China-US relationship, whose main topic is the good bilateral relationship between China and the United States, can be described as consisting of the following subdiscussions (numbers indicate paragraph numbers):

1 Intro-the establishment of China-US relationships
2-3 The officers exchange visits
4-5 The culture exchange between the two countries
6-7 The booming trade between the two countries
8 Outlook and summary

We expect to acquire the above subtopics in a document and use them in the manifold-ranking process instead of the whole document. The most popular TextTiling algorithm is used to automatically subdivide text into multi-paragraph units that represent subtopics.

The TextTiling algorithm detects subtopic boundaries by analyzing patterns of lexical connectivity and word distribution. The main idea is that terms that describe a subtopic will co-occur locally, and a switch to a new subtopic will be signaled by the ending of co-occurrence of one set of terms and the beginning of the co-occurrence of a different set of terms. The algorithm has the following three steps:

1) Tokenization: The input text is divided into individual lexical units, i.e. pseudo-sentences of a predefined size;
2) Lexical score determination: All pairs of adjacent lexical units are compared and assigned a similarity value;
3) Boundary identification: The resulting sequence of similarity values is graphed and smoothed, and then is examined for peaks and valleys. The subtopic boundaries are assumed to occur at the largest valleys in the graph.

For TextTiling, subtopic discussions are assumed to occur within the scope of one or more overarching main topics, which span the length of the text. Since the segments are adjacent and non-overlapping, they are called TextTiles.

The computational complexity is approximately linear with the document length, and more efficient implementations are available, such as Kaufmann [8] and JTextTile [3].

3.3 The Manifold-Ranking Process

Manifold-ranking [15, 16] is a universal ranking algorithm initially used to rank data points along their underlying manifold structure. The prior assumption of manifold-ranking is: (1) nearby points are likely to have the same ranking scores; (2) points on the same structure (typically referred to as a cluster or a manifold) are likely to have the same ranking scores. An intuitive description of manifold-ranking is as follows: A weighted network is formed on the data, and a positive rank score is assigned to each known relevant point and zero to the remaining points which are to be ranked. All points then spread their ranking score to their nearby neighbors via the weighted network. The spread process is repeated until a global stable state is achieved, and all points obtain their final ranking scores.

In our context, the data points are denoted by the TextTiles in the query document q and the top documents in D_{init}. The manifold-ranking process in our context can be formalized as follows:

Given a set of data points $\chi = \chi_q \cup \chi_{D_{init}} = \{x_1, x_2, \ldots x_p, x_{p+1}, \ldots, x_n\} \subset R^m$, the first p points are the TextTiles in the query document q and the rest n-p points are the TextTiles in the documents in D_{init}. Let $f: \chi \rightarrow R$ denotes a ranking function which assigns to each point x_j ($1 \le j \le n$) a ranking value f_j. We can view f as a vector $f = [f_1, f_2, \ldots, f_n]^T$. We also define a vector $y = [y_1, y_2, \ldots, y_n]^T$, in which y_j=1 ($1 \le j \le p$) for the TextTiles in q and y_j= $InitScore(d_i)$ ($p+1 \le j \le n$) for the TextTiles in any document d_i in D_{init}, where $x_j \in d_i$, $d_i \in D_{init}$, which means that the initial retrieval score of a document is used as the initial ranking scores of the TextTiles in the document. The manifold-ranking algorithm goes as follows:

---

1. *Compute the pairwise similarity among points (TextTiles) and using the standard Cosine function.*
2. *Connect any two points with an edge. We define the affinity matrix W by* $W_{ij}=sim_{cosine}(x_i,x_j)$ *if there is an edge linking* x_i *and* x_j*. Note that we let* $W_{ii}=0$ *to avoid loops in the graph built in next step.*
3. *Symmetrically normalize W by* $S=D^{-1/2}WD^{-1/2}$ *in which D is the diagonal matrix with (i,i)-element equal to the sum of the i-th row of W.*
4. *Iterate* $f(t+1) = \alpha S f(t) + (1-\alpha) y$ *until convergence, where* α *is a parameter in (0,1).*
5. *Let* f_j^* *denote the limit of the sequence* $\{f_j(t)\}$*. Each TextTiles* x_j ($p+1 \le j \le n$) *gets its ranking score* f_j^*.

Fig. 1. The manifold-ranking algorithm

In the above iterative algorithm, the normalization in the third step is necessary to prove the algorithm's convergence. The fourth step is the key step of the algorithm, where all points spread their ranking score to their neighbors via the weighted network. The parameter of manifold-ranking weight α specifies the relative contributions to the ranking scores from neighbors and the initial ranking scores. Note that *self-reinforcement* is avoided since the diagonal elements of the affinity matrix are set to zero.

The theorem in [16] guarantees that the sequence $\{f(t)\}$ converges to

$$f^* = \beta(I - \alpha S)^{-1} y \tag{6}$$

where β=1-α. Although f^* can be expressed in a closed form, for large scale problems, the iteration algorithm is preferable due to computational efficiency. Usually the convergence of the iteration algorithm is achieved when the difference between the scores computed at two successive iterations for any point falls below a given threshold (0.0001 in this study).

Using Taylor expansion, we have

$$\begin{aligned} f^* &= \beta(I - \alpha S)^{-1} y \\ &= \beta(I + \alpha S + \alpha^2 S^2 + K) y \\ &= \beta(y + \alpha S y + \alpha S(\alpha S y) + K) \end{aligned} \tag{7}$$

From the above equation, if we omit the constant coefficient β, f^* can be regarded as the sum of a series of infinite terms. The first term is simply the vector y, and the second term is to spread the ranking scores of the TextTiles to their nearby TextTiles, and the third term is to further spread the ranking scores, etc. Thus the effect of the TextTiles in the documents is gradually incorporated into the ranking score.

3.4 The Score Fusion Process

The final retrieval score of a document $d_i \in D_{\text{init}}$ is computed by fusing the ranking scores of its TextTiles as follows:

$$FinalScore(d_i) = \frac{\sum_{x_j \in d_i} \lambda_j f_j^*}{|d_i|} \tag{8}$$

where $\lambda_j = \text{sim}_{\text{cosine}}(x_j, d_i)$ is the cosine similarity between the TextTile x_j and its associated document d_i, which measures the importance of the TextTile x_j in the document d_i. $|d_i|$ represents the number of TextTiles in the document d_i. This normalization avoids favoring long documents.

Finally, the documents in D_{init} are re-ranked according to their final scores and the re-ranked list is returned[3].

4 Experiments

4.1 Experimental Setup

In the experiments, the manifold-ranking based approach ("MR+TextTile") is compared with two baseline approaches: "Cosine" and "MR+Document". The "Cosine"

[3] Only the top k documents (i.e. the documents in D_{init}) in the original ranked list are re-ranked and the rest documents in the original ranked list still hold their initial ranks.

baseline does not apply the manifold-ranking process and directly ranks the documents by their Cosine similarity with the query document, which is the popular way for document similarity search. The "MR+Document" baseline is adopted in [16], which uses the whole document instead of TextTile in the manifold-ranking process, and thus it does not need the steps of text segmentation and score fusion. The manifold-ranking process is also applied on top documents retrieved by other retrieval functions.

To perform the experiments, a ground truth data set is required. We build the ground truth data set from the TDT-3 corpus, which has been used for evaluation of the task of topic detection and tracking [1] in 1999 and 2000. TDT-3 corpus is annotated by Linguistic Data Consortium (LDC) from 8 English sources and 3 Mandarin sources for the period of October through December 1998. 120 topics are defined and about 9000 stories are annotated over these topics with an "on-topic" table presenting all stories explicitly marked as relevant to a given topic. According to the specification of TDT, the on-topic stories within the same topic are similar and relevant. After removing the stories written in Chinese, we randomly chose 40 topics as a test set, while the others were used as a training set.

Sentence tokenization was firstly applied to all documents. Stop words were removed and Porter's stemmer [9] was used for word stemming. The JTextTile tool with default setting was employed to segment each document into TextTiles. The total stories are considered as the document collection for search, the first document within the topic is considered as the query document and all the other documents within the same topic are the relevant (similar) documents, while all the documents within other topics are considered irrelevant (dissimilar) to the query document. A ranked list of 500 documents was required to be returned for each query document based on each retrieval approach. The higher the document is in the ranked list, the more similar it is with the query document. For the proposed manifold-ranking process, the number of initially retrieved documents is typically set to 50, i.e. $|D_{init}|=k=50$.

As in TREC[4] experiments, we use the average precisions at top N results, i.e. *P@5* and *P@10*, as evaluation metrics. The precision at top N results for a query is calculated as follows:

$$P@N = \frac{|C \cap R|}{|R|}, \tag{9}$$

where R is the set of top N retrieved documents, and C is the set of similar documents defined above for a given query document. The precision is calculated for each query and then the values are averaged across all queries.

Note that the number of documents within each topic is different and some topics contain even less than 5 documents or 10 documents, so its corresponding *P@5* or *P@10* may be low.

4.2 Experimental Results

The precision values of our proposed approach ("MR+TextTile") and two baseline approaches (i.e. "Cosine" and "MR+Document") are compared in Table 1, when the

[4] http://trec.nist.gov

manifold-ranking weight α is set to 0.3, which is tuned on the training set. The upper bounds are the ideal values under the assumption that all the relevant (similar) documents are retrieved and ranked higher than those irrelevant (dissimilar) documents in the ranked list. Seen from Table 1, the proposed approach significantly outperforms the two baseline systems. We can also see that the "MR+Document" baseline achieves almost the same *P@5* value with the "Cosine" baseline and the higher *P@10* value than the "Cosine" baseline, which demonstrates that manifold-ranking process can benefit document ranking.

Table 1. Performance comparison of the proposed approach and baseline approaches (* indicates that the performance change over baseline1-"Cosine" is statistically significant, and # indicates that the performance change over baseline2-"MR+Document" is statistically significant, i.e. p-value<0.05 for t-test)

	Baseline1 (Cosine)	Baseline2: (MR+Document)	Our Approach: (MR+TextTile)	Upperbound
P@5	0.830	0.825	0.855*#	0.935
P@10	0.720	0.738*	0.763*#	0.863

The performances of two MR-based approaches (i.e. "MR+TextTile" & "MR+Document") with different manifold-ranking weight α are shown and compared in Figures 2 and 3. Seen from the figures, with appropriate values of the manifold-ranking weight (i.e. α<0.5), the proposed approach (i.e. "MR+TextTile") can significantly outperform the approach of "MR+Document", which demonstrates that TextTile is a more appropriate unit than the whole document for the manifold-ranking process. This result can be explained by that a document is usually characterized as a sequence of subtopical discussions that occur in the context of a few main topic discussions and each TextTile can represent a subtopic with coherent text, and thus the manifold-ranking process can work at a finer granularity.

Figure 4 explores the influence of the number of initially retrieved documents (i.e. *k*) on the performance of the proposed approach (i.e. "MR+TextTile"). Seen from the figure, when *k* is larger than 75, the system performances almost do not alter any more. This shows that a small number of initially retrieved documents work well in the re-ranking process and it will not improve the retrieval performance by increasing the number of initially retrieved documents.

In addition to the Cosine function, other popular retrieval functions are explored in the experiments, including the Jaccard function, the Dice function, the BM25 function and the VSM with document length normalization (NVSM). We use these functions for initial ranking and get the initial retrieved documents associated with their initial retrieval scores, and then the proposed re-ranking process (steps 2-4) is applied. The performances of the systems based only on the retrieval functions and the performances ofthe systems using the re-ranking process are compared in Table 2. For example, "Jaccard" denotes the system using the Jaccard function to retrieve the documents, while "Jaccard+MR" denotes the system using the Jaccard function to retrieve the initial *k*

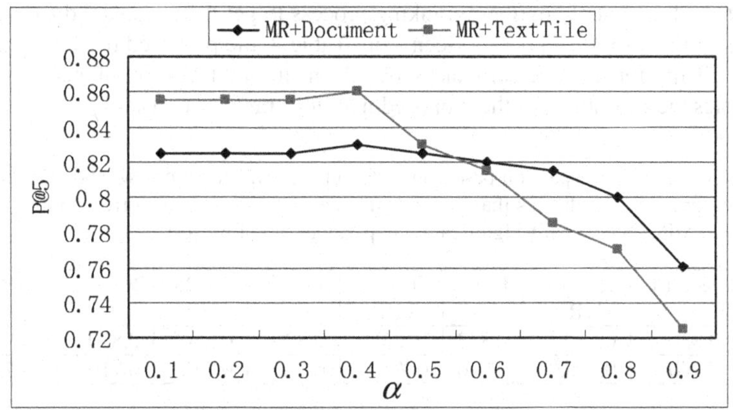

Fig. 2. P@5 comparison of MR-based approaches with different α

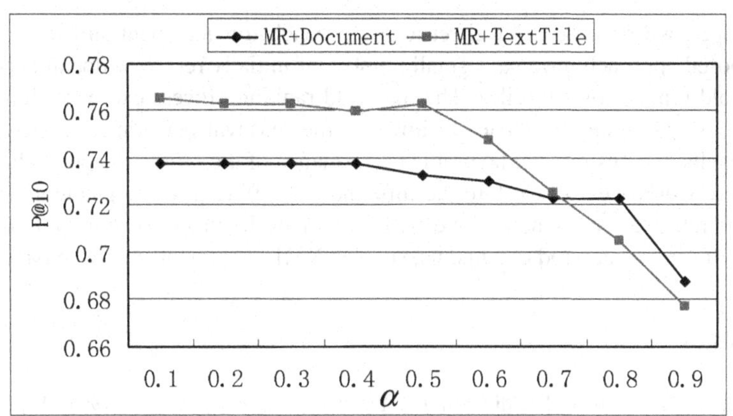

Fig. 3. P@10 comparison of MR-based approaches with different α

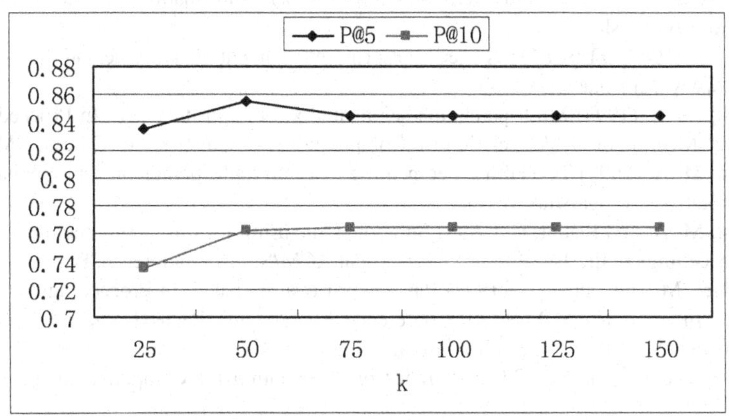

Fig. 4. Performance comparison of the proposed approach with different k

documents and then applying the re-ranking process to get the re-ranked document list. Here we still have k=50 and α=0.3. Seen from Table 2, the proposed re-ranking process improves all the retrieval performances based on different retrieval functions, which demonstrates the robustness of the proposed manifold-ranking process.

Table 2. Performance comparison based on different retrieval functions w/ and w/o the manifold-ranking process (* indicates that the performance change over the corresponding retrieval approach w/o MR is statistically significant, i.e. p-value<0.05 for t-test)

	Jaccard	Jaccard +MR	Dice	Dice +MR	BM25	BM25 +MR	NVSM	NVSM +MR
P@5	0.835	0.850	0.832	0.850*	0.820	0.850*	0.810	0.845*
P@10	0.740	0.758*	0.740	0.753*	0.720	0.740*	0.710	0.723

5 Conclusion

In this paper, we propose a novel retrieval approach for document similarity search. The proposed approach re-ranks a small number of initially retrieved documents based on manifold-ranking of TextTiles. The manifold-ranking process can make full use of the relationships among TextTiles to improve the retrieval performance. The experimental results demonstrate the favorable performance of the proposed approach.

In future work, we will explore the influence of different text segmentation methods on the retrieval performance. We will also adapt the proposed retrieval approach to search of semi-structured documents, such as XML documents and web pages.

References

1. Allan, J., Carbonell, J., Doddington, G., Yamron, J. P. and Yang, Y. (1998) Topic detection and tracking pilot study: final report. In Proceedings of DARPA Broadcast News Transcription and Understanding Workshop, 194-218
2. Baeza-Yates, R., and Ribeiro-Neto, B. (1999) Modern Information Retrival. ACM Press and Addison Wesley
3. Choi, F. (1999) JTextTile: A free platform independent text segmentation algorithm. http://www.cs.man.ac.uk/~choif
4. Hearst, M. A. (1994) Multi-paragraph segmentation of expository text. In Proceedings of the 32nd Meeting of the Association for Computational Linguistics, Los Cruces, NM
5. Hearst, M. A. (1997) TextTiling: Segmenting text into multi-paragraph subtopic passages. Computational Linguistics, 23(1): 33-64
6. Hearst, M. A. and Plaunt, C. (1993) Subtopic structuring for full-length document access. In Proceedings of the 16th Annual International ACM/SIGIR Conference, Pittsburgh, PA
7. Kaszkiel, M. and Zobel, J. (1997) Passage retrieval revisited. In Proceedings of the 20th Annual International ACM/SIGIR Conference, Philadelphia, Pennsylvania
8. Kaufmann, S. (1999) Cohesion and collocation: using context vectors in text segmentation, Proceedings of the 37th conference on Association for Computational Linguistics, 591-595
9. Porter, M. F. (1980) An algorithm for suffix stripping. Program, 14(3): 130-137

10. Robertson, S., Walker, S. (1994) Some simple effective approximations to the 2-poisson model for probabilistic weighted retrieval. In Proc. of the 17th International ACM/SIGIR Conference on Research and Development in Information Retrieval , 232-241
11. Robertson, S., Walker, S. and Beaulieu, M. (1999) Okapi at TREC–7: automatic ad hoc, filtering, VLC and filtering tracks. In Proceedings of TREC'99
12. Salton, G. (1991) The SMART retrieval system: experiments in automatic document processing. Prentice-Hall
13. Singhal, A., Buckley, C. and Mitra, M. (1996) Pivoted document length normalization. In Proceedings of SIGIR'96.
14. van Rijsbergen, C. J. (1979) Information Retrieval. Butterworths, London
15. Zhou, D., Bousquet, O., Lal, T. N., Weston, J. and SchÖlkopf, B. (2003) Learning with local and global consistency. In Proceedings of NIPS-2003
16. Zhou, D., Weston, J., Gretton, A., Bousquet, O. and SchÖlkopf, B. (2003). Ranking on data manifolds. In Proceedings of NIPS-2003

Adapting Document Ranking to Users' Preferences Using Click-Through Data

Min Zhao[1,*], Hang Li[2], Adwait Ratnaparkhi[3], Hsiao-Wuen Hon[2], and Jue Wang[1]

[1] Institute of Automation, Chinese Academy of Sciences, Beijing, China
[2] Microsoft Research Asia, Beijing, China
[3] Microsoft Corporation, Redmond, WA, USA

Abstract. This paper proposes a new approach to ranking the documents retrieved by a search engine using click-through data. The goal is to make the final ranked list of documents accurately represent users' preferences reflected in the click-through data. Our approach combines the ranking result of a traditional IR algorithm (BM25) with that given by a machine learning algorithm (Naïve Bayes). The machine learning algorithm is trained on click-through data (queries and their associated documents), while the IR algorithm runs over the document collection. We consider several alternative strategies for combining the result of using click-through data and that of using document data. Experimental results confirm that any method of using click-through data greatly improves the preference ranking, over the method of using BM25 alone. We found that a linear combination of scores of Naïve Bayes and scores of BM25 performs the best for the task. At the same time, we found that the preference ranking methods can preserve relevance ranking, i.e., the preference ranking methods can perform as well as BM25 for relevance ranking.

1 Introduction

This paper is concerned with the problem of improving document ranking in information retrieval by using *click-through data.* Suppose that we have a search engine. Given a query, it can return a list of documents ranked based on their relevance to the query. During the use of the search engine we have collected users' click-through data. The click-through data can represent the *general trends of users' preferences* on documents with respect to queries. Our goal here is to re-rank the documents in future search using the click-through data so that the most likely to be clicked documents are re-ranked on the top from among the relevant documents. That is to say, we are to re-rank documents on the basis of users' preferences among the relevant documents.

Usually users only look at the top ranked documents in search results (cf., [22] [23] [24]), and thus if we can rank users' preferred documents on the top, then we will be able to improve users' search experiences, for example, save users' time in search.

To the best of our knowledge, no investigation has been conducted on the problem, although there is some related work as described in Section 2.

---

[*] Min Zhao is currently researcher at NEC Lab China, Beijing.

H.T. Ng et al. (Eds.): AIRS 2006, LNCS 4182, pp. 26–42, 2006.

In this paper, we employ a machine learning approach to address the above problem and investigate a number of simple, but in our view, fundamental methods.

We employ BM25 for the relevance ranking in the initial search engine (In our experiments, the search engines used for collecting click-through data were also based on BM25). We use the click-through data to train a classifier which can categorize queries into documents with scores. Specifically, queries are viewed as instances, documents are viewed as categories, and queries invoking clicks of a document are regarded as instances 'classified' into the category. As classifier, we make use of Naïve Bayes. Given a new query, BM25 and Naïve Bayes can respectively construct a ranked list of documents, on the basis of relevance and preference respectively. We consider a number of methods to combine the two lists. Linear combination is the main method among them.

Two questions may arise here. First, do the preference ranking methods preserve relevance? In other words, are the documents ranked top by the methods still relevant? Second, are the preference ranking methods indeed effective? Which method is more effective in performing the task?

We conducted experiments in order to answer the above questions, using data from two search engines: Encarta and Microsoft Word Online Help. First, experimental results indicate that our preference ranking methods can preserve *relevance ranking*. Second, experimental results indicate that the preference ranking methods using click-through data significantly perform better than BM25 for *preference ranking*. Furthermore, among the methods, the linear combination of scores from BM25 and Naïve Bayes performs the best. (We evaluated preference ranking results in terms of *click distribution*).

In conclusion, our methods can significantly improve preference ranking and at the same time preserve relevance ranking.

2 Related Work

2.1 Use of Click-Through Data

Document retrieval can be described as the following problem. A search engine receives a query from a user, and then returns a ranked list of retrieved documents based on their *relevance* to the query. The list contains the links to the documents, as well as the titles and snippets of the documents. The user clicks the links of the documents which he considers *relevant and interesting*, after reading the titles and snippets.

Click-through data can be recorded in operation of the search engine, as described in [11]. Here, click-through data refers to <query, clicks> pairs recorded during document retrieval. The clicks are the set of documents clicked by the user after sending the query.

Some work has been done on using query logs in information retrieval and web data mining (e.g., [1]), because query logs contain a large amount of useful information. Here, by query logs we mean all the information which can be obtained during search of documents. Query log data, thus, contains click-through data. (Note that some researchers do not distinguish the two terms.)

Methods of using click-through data have been proposed for meta-search function training, term expansion, query clustering, implicit feedback, etc.

For example, Joachims [11] proposes a method of utilizing click-through data in learning of a meta-search function. Specifically, he employs a method called Ranking SVM. His method is unique in that it takes the relative positions of the clicks in a ranked list as training data. He has applied the method to the adaptation of a meta-search engine to a particular group of users.

Oztekin et al [17] propose several methods for meta-search. They have compared the methods by using a new metric called 'average position of clicks' calculated based on click-through data. The authors claim that the implicit evaluation method based on users' clicks can offer more statistically reliable results than an explicit evaluation method based on users' relevance judgments.

Cui et al [6] try to use click-through data to solve the problem of mismatch between query terms and document terms. With click-through data, the probabilistic correlations between query terms and documents terms can be extracted and high quality term expansion can be conducted.

Furthermore, Beeferman etc. [4] propose methods for harvesting clusters of queries and web pages by using click-through data. Ling et al [14] investigate a similar problem, but focus on the issue of finding generalized query patterns and using them in the cache of a search engine.

In [12] [18] and [21], click-through data is regarded as implicit feedback and used to improve relevance ranking.

DirectHit was a commercial internet search engine (http://www.directhit.com). It is claimed that DirectHit uses click-through data in ranking of retrieved documents (http://www.searchengines.com/directhit.html). Their problem setting is similar to that in the current research; however, details of the technologies used are not known.

2.2 Meta Search and Relevance Feedback

Meta-search attempts to improve ranking by combining ranking results returned by several search engines, while relevance feedback manages to improve ranking by using feedbacks on relevance from users. Both of them consider performing ranking based on *relevance*, not based on users' *preference*.

Meta-search and related problems such as result fusion have been intensively studied [2] [3] [8] [9] [11] [13] [15] [17] [25]. One of the key issues in meta-search is to construct a meta-ranking function which combines the scores or ranks returned by ranking functions. Many meta-ranking functions have been proposed (cf., [2] [16]), which are based on averaging, sum, linear combination, voting, interleaves, logistic regression, etc.

Relevance feedback is an iterative process as follows. Given documents returned by the search engine, the user is asked to mark them as relevant or irrelevant. The query is then expanded with terms extracted from the marked relevant documents.

Next, documents are retrieved with the new query. The process is repeated (cf., [20]). Since relevance judgments can be burdens to users, automatic ways of getting feedbacks, such as pseudo-relevance feedback [5] and implicit relevance feedback [21] [26], have been proposed. The main purpose of relevance feedback is to refine user's query so that user's information needs can be expressed more accurately.

3 Problem and Evaluation

3.1 Preference Ranking Using Click-Through Data

The problem we address in this paper is whether and how we can use click-through data to improve preference ranking. More specifically, we consider a re-ranking (or adaptation) problem as follows: a search engine first receives a query from a user, and then presents a ranked list of retrieved documents based on their relevance to the query. (We assume here that the document collection and the relevance ranking mechanism (e.g., BM25) are fixed). Next, a re-ranking engine re-orders the ranked list of documents based on users' preferences, using click-through data. The top ranked documents should be relevant and preferred by users.

In one extreme case, the re-ranking engine completely ignores the initial relevance ranking and creates a new preference ranking. In the other extreme case, the re-ranking engine only uses the initial relevance ranking.

Preference and relevance are different notions, although closely related to each other. If users think that some documents are more worth-reading (important and interesting) than the others among the relevant documents, then we say that they prefer those documents. In this paper, preferences are assumed to be from a group of users, not a single user. Preference is more subjective and dynamic, while relevance is more objective and static. When a query is fixed, its relevant documents are almost fixed, while users' preferences may change depending on user groups and time periods.

In this paper, we assume that click-through data can reflect the general trends of users' preferences. (Note that users' preferences can be measured by other factors like 'dwell time', cf., [7]). It is likely that a user clicks a document which he find uninteresting or even irrelevant later. The percentage of such 'noisy' clicks over total clicks should be low, however. (Otherwise, the search engine will not be used by users.) Therefore, click-through data can and should be used in preference ranking. We also assume here that there is no 'click spam'.

Figure 1 shows the differences between the current problem and other information retrieval problems: conventional search, meta-search, relevance feedback. The current problem focuses on preference ranking, while the other problems focus on relevance ranking.

In practice, usually users only look at the top ranked documents in search results (cf., [22] [23] [24]), and thus putting users' preferred documents on the top is necessary and important for document retrieval.

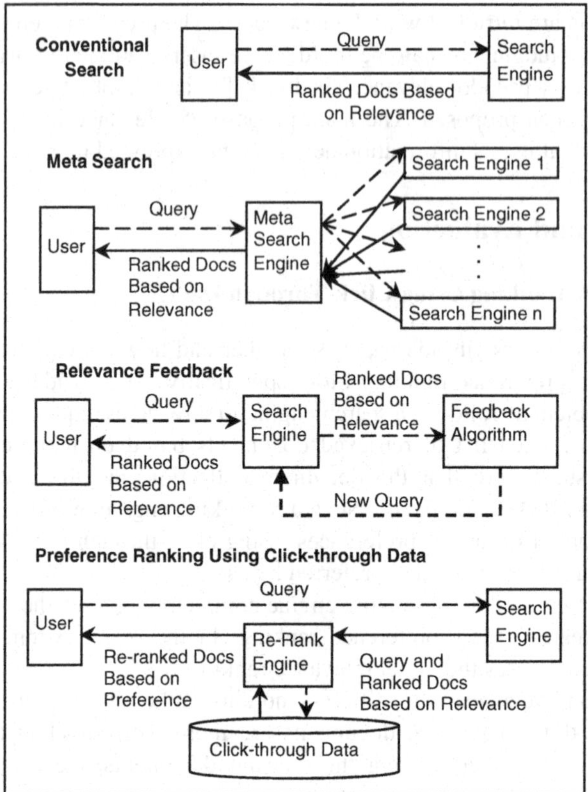

Fig. 1. Comparison of retrieval problems

3.2 Click Distribution

An ideal way of evaluating the performance of a preference ranking method might be to select a number of users and a set of queries, and let every user to explicitly judge preference on each of the documents to each of the queries. Furthermore, for each query, counting the number of preference judgments on each document, ranking the documents in descending order of their preference counts, and taking the order as the correct 'answer'. This evaluation can give accurate results. However, it is costly, and thus is not practical.

We consider here the use of '*click distribution*' for the evaluation. Click distribution is the distribution of document positions clicked by users in the ranked lists. If a user clicks *k* documents, a method that places the *k* documents in higher ranks (ideally the first k ranks) is better than a method that does not.

The advantage of using the measure is that a large amount of click-through data can be used. The disadvantage is that click-through data may contain noise. However, as is explained above, the percentage of noise should be low. Thus, click distribution is a good enough measure for evaluation of preference ranking methods.

4 Methods

We propose a machine learning approach to address the problem of preference ranking using click-through data. It turns out to be a problem of using click-through data to re-rank documents so that the most likely to be clicked documents are re-ranked on the top from among the relevant documents.

We first use click-through data to train a classifier which categorizes queries into documents with scores. It is actually a problem of text classification. Specifically, we view queries as instances, documents as categories, and queries invoking clicks of a document as instances 'classified' into the category. We next combine the results given by the classifier and the results given by the initial search engine. It is actually a problem of result fusion. Specifically, we combine the information from two data sources: the classifier and the search engine, in order to achieve better performances than using information from either of the sources alone.

As the initial search engine, we employ the ranking method used in the Okapi system [19], which is called BM25 (*BM* hereafter). (This is because in the experiments described below click-through data were collected from search engines based on BM25.)

For click-through data classification, we employ Naïve Bayes (hereafter *NB*).

For combination of the two methods, we consider a number of strategies: (1) data combination (*DC*), (2) ranking with NB first and then BM (*NB+BM*), and (3) linear combination of BM and NB scores (*LC-S*) and linear combination of BM and NB ranks (*LC-R*). BM and NB themselves are the combination methods in two extreme cases: the former does not make use of click-through data and the latter makes only use of click-through data. DC first combines click-through data with document data, and then uses BM.

In practice, some search engines only return a ranked list of documents without scores. NB, NB+BM, and LC-R can still work under such circumstances.

4.1 Notation

$D = \{d\}$ is a document collection.
$Q = \{q\}$ is a set of queries, and a query q consists of words.
$T = \{t\}$ is a set of words.
$R(q) = <d_1, d_2, \dots d_m>$ is a ranked list of retrieved documents with respect to query q, returned by a method. Some methods only return a ranked list, while other methods associate each d_i $(1 \le i \le m)$ with a score $S(q, d)$, satisfying $S(q, d_1) \ge S(q, d_2) \ge \dots \ge S(q, d_m)$.
$C = \{<q, c>\}$ is a set of click-through data, where $c = <d_1, d_2, \dots d_k>$ is a set of documents clicked by a user.
$Qd = \{q\}$ is the subset of Q related to document d in click-through data.

4.2 BM – Initial Ranking Function

In BM, given query q, we first calculate the following score (using the default values of parameters in the formula of BM25) of each document d with respect to it

$$S_{BM}(q,d)=\sum_{t\in q}\left\{\frac{tf_{t,d}\cdot\log(\frac{N-df_t+0.5}{df_t+0.5})}{0.5+1.5\frac{dl_d}{dl_{avg}}+tf_{t,d}}\right\},$$

where $tf_{t,d}$ is term frequency of word t in document d, df_t is document frequency of word t in collection D, N is number of documents in D, dl_d is length of document d, and dl_{avg} is average length of documents in D.

We next rank documents according to their scores $S_{BM}(q,d)$, and then obtain a ranked list of documents $R_{BM}(q)$.

4.3 NB – Click-Through Classification

We can view each document in the given collection as a category, and the queries associated with each of the documents as instances classified into the category. The association can be found in click-through data (cf., Figure 2). Thus, we can construct a classifier on the basis of the click-through data and formalize the problem of document retrieval as that of text classification.

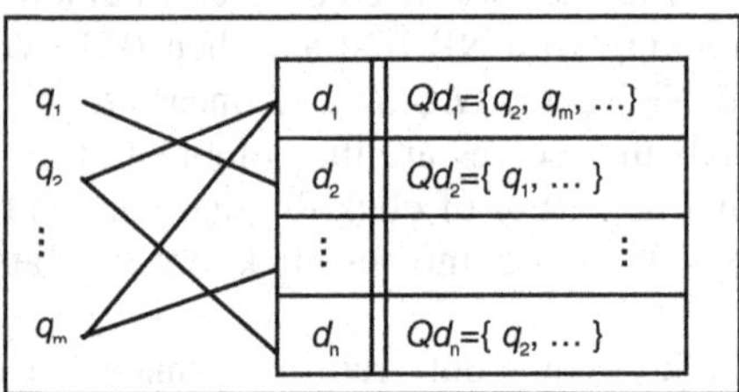

Fig. 2. Relationship between queries and documents

In NB, we employ Naïve Bayes as classifier [16]. Given query q (a sequence of words t_1, t_2, …, t_l) and category d, the conditional probability of $p(d|q)$ is calculated as follows.

$$p(d\,|\,q)=\frac{p(d)\cdot p(t_1,t_2,\ldots,t_l\,|\,d)}{p(t_1,t_2,\ldots,t_l)}\propto p(d)\cdot\prod_{t\in d}p(t\,|\,d)\cdot$$

We calculate the probabilities using click-through data.

$$p(d)\cdot\prod_{t\in q}p(t\,|\,d)=p(d)\cdot\prod_{t\in q}\left\{\frac{tf_{t,Qd}+1}{tf_{Qd}+|V_Q|}\right\}.$$

Similarly we have

$$p(\bar{d}\,|\,q)=p(\bar{d})\cdot\prod_{t\in q}p(t\,|\,\bar{d})=p(\bar{d})\cdot\prod_{t\in q}\left\{\frac{tf_t-tf_{t,Qd}+1}{tf_Q-tf_{Qd}+|V_Q|}\right\}.$$

Here, Qd is subset of queries in Q associated with document d, $tf_{t,Qd}$ is term frequency of word t in Qd, tf_t is term frequency of t in Q, tf_{Qd} is total term frequency in Qd, tf_Q is total term frequency in Q, and $|V_Q|$ is number of unique words in Q. Furthermore, $p(d)$ and $p(\bar{d})$ are prior probabilities of category d and its converse category $\bar{d}$, and we assume that they are equal for all documents. The converse category $\bar{d}$ consists of all documents except d.

We use log of ratio of the two probabilities as score of document d with respect to query q.

$$S_{NB}(q,d) = \log\left\{\frac{p(d \mid q)}{p(\bar{d} \mid q)}\right\}.$$

In NB, we rank documents by $S_{NB}(q,d)$ and then obtain $R_{NB}(q)$.

4.4 DC – Data Combination

The above two methods rank documents based upon information from only one data source: either document or click-through. In the sequel, we consider a number of methods which combine the information from both data sources.

The simplest method of combination might be to combine the queries to their associated documents, view them as pseudo-documents, and run BM with the pseudo-documents.

More precisely, for each document d and its associated query set Qd, we create a pseudo-document d' by combing d and Qd. Given a new query q, apply BM to the collection $D'=\{d'\}$ and get $S_{BM}(q,d')$ for each document. Let $S_{DC}(q,d) = S_{BM}(q,d')$, we can rank the documents with the scores and obtain $R_{DC}(q)$.

4.5 NB+BM – NB First and BM Next

In NB+BM, given the initial ranked list by BM, we pick up from the list those documents whose NB scores are larger than a predetermined threshold, rank them by NB, and put them on the top of the list. The assumption is that click-through data reflects users' preferences and thus it is better to use the information first for preference ranking.

Let $R_{BM}(q)$ be the ranked list of documents returned by BM. Some of them can be ranked by NB with the ranked list denoted as $R_{NB}(q)$. We put the documents in $R_{NB}(q)$ at the top, and the remaining documents behind according to $R_{BM}(q)$. We then obtain $R_{NB+BM}(q)$. For example, if $R_{BM}(q) = \langle d_1, d_2, d_3, d_4 \rangle$ and $R_{NB}(q) = \langle d_3, d_2 \rangle$, then $R_{NB+BM}(q) = \langle d_3, d_2, d_1, d_4 \rangle$.

4.6 LC-S and LC-R – Linear Combination

We can employ a fusion strategy to combine the results of NB and BM. The most widely used method for such fusion is linear combination.

We consider two sorts of linear combination methods. In *Linear Combination of Scores* (LC-S), we rank documents by linearly combining the normalized scores returned by BM and NB:

$$S_{LC-S}(q,d) = S_{BM}(q,d) + \alpha \cdot S_{NB}(q,d) \ ,$$

where $S_{BM}(q,d)$ and $S_{NB}(q,d)$ are normalized to [0, 1], α is the weight in (0, 1).

In *Linear Combination of Ranks* (LC-R) we rank documents by linearly combining the normalized ranks returned by BM and NB. Suppose that document d is ranked at i^{th} position in $R_{BM}(q)$ and at j^{th} position in $R_{NB}(q)$, and there are m documents in $R_{BM}(q)$ and n documents in $R_{NB}(q)$, then we have

$$S_{LC-R}(q,d) = \frac{m-i}{m} + \alpha \cdot \frac{n-j}{n} \ ,$$

where α is weight.

We determine the weight α in the linear combination by training on click-through data.

As measure for training, we use *Percentage of Clicks on Top N* (*PC-TopN* for short). PC-TopN represents the percentage of clicked documents on top-N positions among all clicked documents. Actually, PC-TopN is the result at point N in a *click distribution*. Since we use click distribution for evaluation, it is natural to use the same measure for training (For training, we can use a single value instead of a distribution as the measure).

Given a specific value of α, we can create a linear combination model and calculate its corresponding PC-TopN value with the *training* data. From a number of possible values of α, we select the one that maximizes PC-TopN with respect to the training data. That is to say, we choose the α that achieves the best performance in document preference ranking. Since there is only one parameter α in our linear combination model, there is no need to employ a complicated parameter optimization method.

5 Experiments and Results

We conducted two experiments on two data sets. One was to investigate whether our preference ranking methods can preserve relevance ranking. The other was to examine whether our methods outperform the baseline BM25 for preference ranking, and among them which method works best.

5.1 Data Sets

The two data sets used in our experiments are 'Encarta' and 'WordHelp'. Encarta is an electronic encyclopedia on the web (http://encarta.msn.com), which contains 44,723 documents. Its search engine (based on BM) has recorded click-through data. We obtained data collected over two months. WordHelp is from Microsoft Word Online Help, which contains 1,340 documents. In an experimental setting, its search engine (also based on BM) collected click-through data during four months. Details of the two data sets are shown in Table 1.

Table 2 shows randomly selected queries in the click-through data of each data set.

The titles and snippets in both WordHelp and Encarta are created by humans, and thus are very accurate. Thus, for the two data sets it is possible for users to make preference judgments based on title and snippet information.

Table 1. Details of WordHelp and Encarta click-through data

	Word	Encarta
Time span	4 months	2 months
Total number of queries	~34,000	~4,600,000
Total number of unique queries	~13,000	~780,000
Average query length	~1.9	~1.8
Average clicks per query	~1.4	~1.2

Table 2. Example queries in WordHelp and Encarta click-through data

WordHelp	xml, table contents, watermark, signature, speech, password, template, shortcut bar, fax, office assistant, equation editor, office shortcut bar, highlight, language, bookmark
Encarta	electoral college, China, Egypt, Encarta, sex, George Washington, Christmas, Vietnam war, dogs, world war II, solar system, abortion, Europe

5.2 Training and Testing Data

The click-through data in Encarta and WordHelp were recorded in chronological order. We separated them into two parts according to time periods, and used the first parts for training and the second parts for testing.

For Encarta, we had 4,500,000 $<q, c>$ click-through pairs for training, and 10,000 pairs for testing. For WordHelp, we had 30,000 $<q, c>$ click-through pairs for training and 2,000 pairs for testing.

For NB and NB+BM, we made use of the training data (click-through data) in the construction of classifiers. For LC-R and LC-S, we made use of the training data in the construction of classifiers and a randomly selected subset of training data (10,000 instances in Encarta and 2,000 instances in WordHelp) in the tuning of linear combination weights. For DC, we utilized the training data and the document data in the construction of models.

When training the weights in LC-R and LC-S, we tried different values of N for PC-TopN: PC-Top1, PC-Top2, PC-Top5, PC-Top10, and PC-Top20. It seems that N only slightly affects the weights and the final ranking results have little differences with different values of N used in training. Thus, in the experiments reported in this paper, N was fixed at 5.

In order to conduct precise analysis, we further created three data sets on the basis of the same click-through data for both Encarta and WordHelp. They are referred to as one_click, first_click, and all_clicks, respectively. In one_click, we used the $<q, c>$ pairs only having single clicks as instances. In first_click, we extracted the first clicks in all click-through pairs to create instances. In all_clicks, we used all click-through pairs as instances. For example, given two click-through instances $<q_1, \{d_1, d_2\}>$ and $<q_2, \{d_3\}>$, one_click = $\{<q_2, d_3>\}$, first_click = $\{<q_1, d_1>, <q_2, d_3>\}$, and all_clicks = $\{<q_1, d_1, d_2>, <q_2, d_3>\}$. Due to limitation of space, we only report the results with first_click in this paper. The results with one_click and all_clicks have the same tendencies.

5.3 Experiment 1

We randomly selected 40 queries from the testing (click-through) data of Encarta and that of WordHelp respectively. More precisely, for each data set, we sorted all the queries in terms of frequency, then equally divided the sorted set into four subsets, and randomly picked up 10 queries from each of the subsets. In this way, we were able to select queries from different frequency ranges.

For each query, the six ranking methods described in Section 4 were employed to produce a ranked list, and the top-5 documents returned by the methods were merged together. Next, humans made judgments on the relevance of the documents. We then used the measure of top 1, 2, and 5 precisions to evaluate the performances of the six methods for *relevance ranking*.

We report here the results when 10% of Encarta and 20% of WordHelp training data were respectively used for training of the ranking methods, in accordance with the results reported for Experiment 2 in Section 5.4.

The results in Table 3 indicate that all the *preference* ranking methods using click-through data improve upon or at least work as well as the initial relevance ranking method – BM. (Some of the improvements are statistically significant according to our statistical testing). Thus, it is safe to say that the preference ranking methods can preserve relevance ranking.

It should be noted that NB, the method which ranks documents by using click-through data alone, can still achieve better results than BM. The result indicates that preference is strongly related to relevance.

Table 3. Performance of six methods with respect to WordHelp and Encarta data in terms of top 1, 2 and 5 precisions (Pre@1, Pre@2, Pre@5)

Collection	Method	Pre@1	Pre@2	Pre@5
WordHelp	LC-S	**0.850**	**0.700**	**0.455**
	NB+BM	0.700	0.625	0.450
	NB	0.700	0.613	0.445
	DC	0.700	0.600	0.435
	LC-R	0.675	0.575	0.390
	BM	0.650	0.537	0.390
Encarta	LC-S	**0.825**	**0.563**	**0.300**
	NB+BM	0.750	0.525	0.280
	NB	0.750	0.525	0.280
	DC	0.750	0.525	0.270
	LC-R	0.725	0.500	0.260
	BM	0.450	0.388	0.250

5.4 Experiment 2

We applied the methods of BM, NB, DC, NB+BM, LC-R, and LC-S to both Encarta and WordHelp data for preference ranking.

All the results with Encarta and WordHelp show almost the same tendencies.

We used different proportions of training data in creating NB, DC, NB+BM, LC-R, and LC-R. We only show the results here when 20% and 100% of WordHelp training data, and 10% and 100% of Encarta training data are available. This is because they are representative of the general trends in the entire results.

We found that the use of click-through data is effective for enhancing preference ranking. Nearly in all cases, the methods using click-through (i.e., LC-S, LC-R, NB, NB+BM, DC) perform better than the method without using it (i.e., BM).

We also found that when the size of training (click-through) data is large, LC-S, NB+BM, and NB perform nearly equally best, and when the size of training (click-through) data is small, LC-S performs best. Therefore, it is better to always employ LC-S.

The results were obtained on the basis of the measure: click distribution as described in Section 3.2. Again, click distribution represents the percentage of clicked documents on top *N* positions among clicked documents. The higher percentage of clicks on top, the better a method is. For each method, we evaluated the percentages of clicks on top 1, top 2, top 10, top 20, and all. We also evaluated the percentages of clicks in intervals between rank 1-1, rank 2-2, rank 3-5, rank 6-10, rank 11-20, and rank 21-.

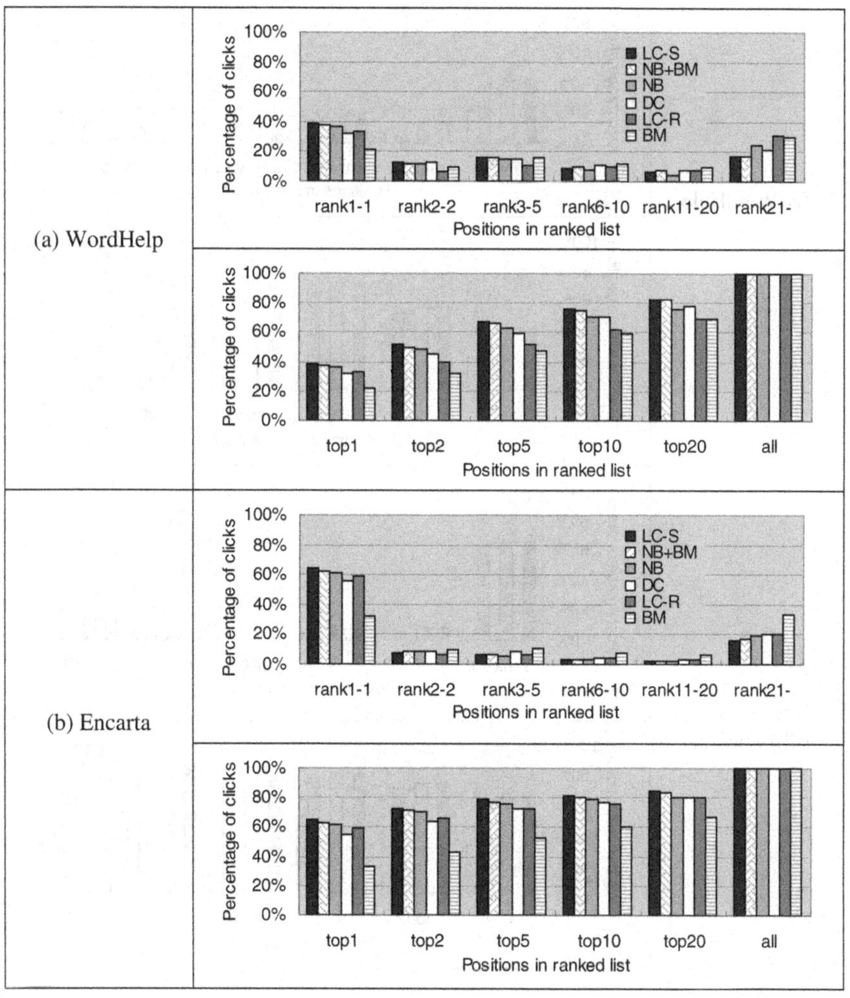

Fig. 3. Click distributions with 20% WordHelp training data and 10% Encarta training data

Figure 3 shows the click distributions when 10% of Encarta and 20% of WordHelp training (click-through) data are respectively available. Among the methods, LC-S performs the best.

Figure 4 shows the click distributions when 100% training data is available. Among the methods, LC-S, NB+BM, and BM perform best.

We conducted sign test [10] to see the significance of differences between methods on percentage of clicks on top5 (significance level was set to 0.005).

We use '{A, B}' to represent 'methods A and B do not have significant difference in performance', and use 'A >> B' to represent 'method A is significantly better than method B'.

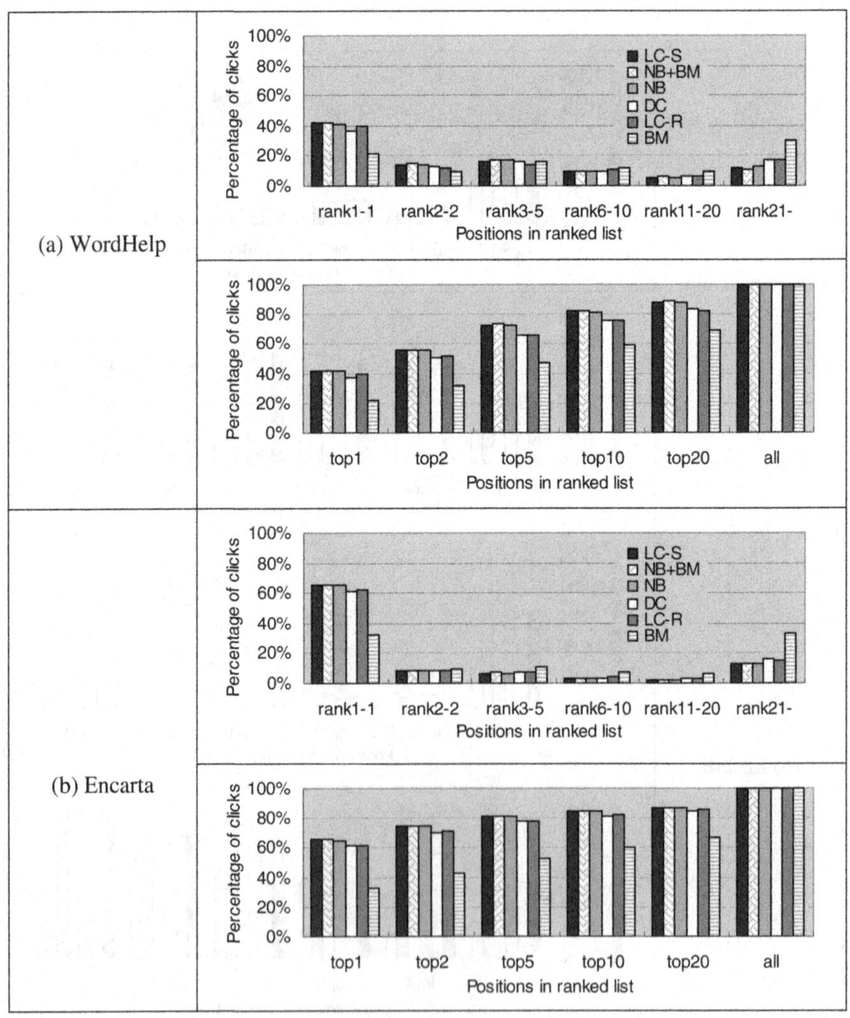

Fig. 4. Click distributions with 100% WordHelp and Encarta training data

With 20% training data for WordHelp and 10% training data for Encarta, we have:

(1) WordHelp: {LC-S, NB+BM} >> NB >> DC >> LC-R >> BM
(2) Encarta: LC-S >> NB+BM >> NB >> {DC, LC-R} >> BM

With 100% training data, we get the same results for WordHelp and Encarta:

(3) {LC-S, NB+BM, NB} >> {DC, LC-R} >> BM.

5.5 Discussions

We discuss why the use of click-through data can help improve preference ranking and why LC-S can work best when different sizes of click through data are available.

We examined how the proportion of 'unseen' queries in testing data changes when training (click-through) data gets accumulated. Figure 5 shows the results. We see that the number of previously unseen queries in the testing data decreases when training data increases no matter whether it is in time order or reverse order. For WordHelp, eventually only 34% of testing queries is previously unseen, and for Encarta, only 15% of testing queries is previously unseen. Therefore, using click-through data (already observed click records) can help improve prediction of users' future clicks, i.e., preference ranking.

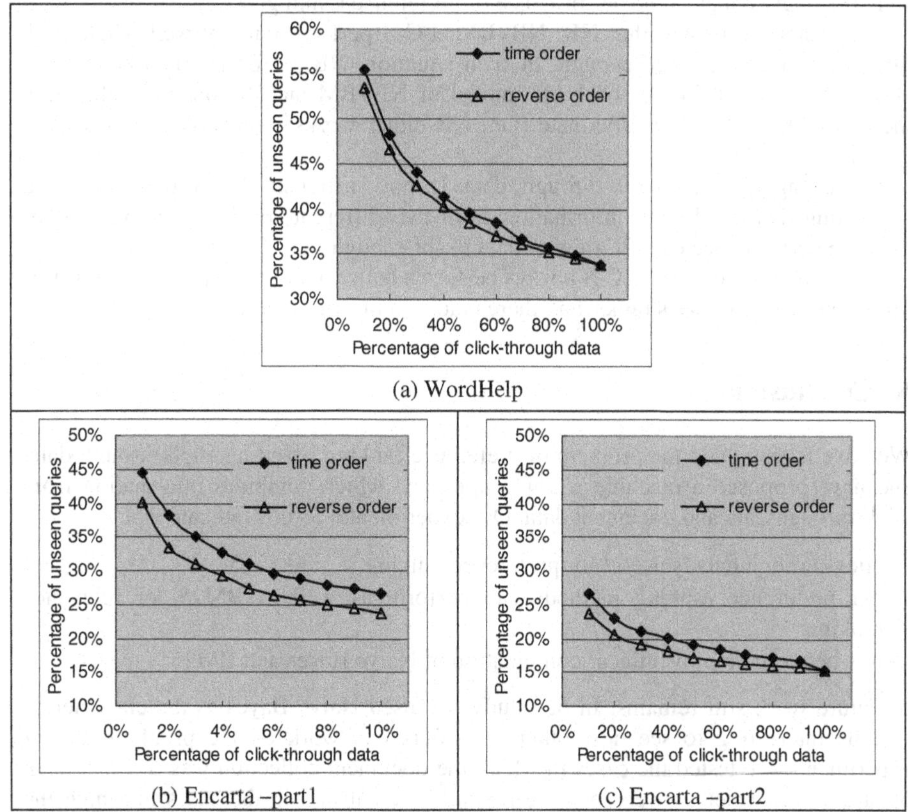

Fig. 5. Percentages of unseen queries in testing data versus sizes of WordHelp and Encarta training data

Table 4. Distribution of queries and clicked documents in click-through data (Type represents unique queries / clicked documents, and Token represents all queries / clicked documents)

Frequency		1	2-3	4-10	11-50	51-
WordHelp query	Type	76.5%	14.1%	6.3%	2.6%	0.5%
	Token	29.9%	12.5%	14.4%	21.5%	21.7%
WordHelp clicked document	Type	3.2%	5.7%	22.6%	49.5%	19.0%
	Token	0.1%	0.4%	4.6%	35.8%	59.1%
Encarta query	Type	68.3%	19.7%	7.4%	3.1%	1.5%
	Token	11.4%	7.4%	7.2%	11.9%	62.1%
Encarta clicked document	Type	8.0%	11.5%	20.4%	29.1%	31.0%
	Token	0.1%	0.2%	0.9%	4.8%	94.0%

Table 4 shows the frequency distributions of queries and clicked documents in the click-through data. For both data sets, query frequency is heavily distributed on a small number of common terms, and clicked document frequency is heavily distributed on a small number of common documents. The statistics indicate that prediction of users' clicks based on click-through data is at least possible for common queries.

From the results, we see that the way of using click-through data, as we propose, can effectively enhance the performance of preference ranking.

It appears reasonable that NB, NB+BM, LC-S perform equally well when click through data size is large enough. In such situation, NB performs much better than BM, and thus both in the 'back-off' model of NB+BM and the linear combination model of LC-S, NB does dominate (i.e., NB either works frequently or has a large weight).

Interestingly, when click-through data is not sufficient, LC-S performs best, suggesting that combining information from two different sources effectively helps improve performance even if training data is not enough.

It is easy to understand LC-S always performs better than LC-R, as the former uses scores and the latter uses ranks, and there is loss in information when using LC-R.

6 Conclusions

We have investigated the problem of preference ranking based on click-through data, and have proposed a machine learning approach which combines information from click-through data and document data. Our experimental results indicate that

- –click-through data is useful for preference ranking,
- –our preference ranking methods can perform as well as BM25 for relevance ranking,
- –it is better to employ a linear combination of Naïve Bayes and BM25.

Future work still remains. In our study, we used Naïve Bayes as the classifier. It will be interesting to see how other classifiers can work on the problem. In our experiments, we tested the cases in which the document collection was fixed. It is an open question whether we can make a similar conclusion in the cases in which the document collection dynamically changes during recording of click-through data.

References

1. Anick, P. Using terminological feedback for web search refinement – a log-based study. In Proceedings of SIGIR'03 (2003) 88-95.
2. Aslam, J.A., and Montague, M.H. Models for Metasearch. In Proceedings of SIGIR'01 (2001) 275-284.
3. Bartell, B.T., Cottrell, G.W., and Belew, R.K. Automatic combination of multiple ranked retrieval systems. In Proceedings of SIGIR'94 (1994) 173-181.
4. Beeferman, D., and Berger, A. Agglomerative clustering of a search engine query log. In Proceedings of SIGKDD'00 (2000) 407-416.
5. Buckley, C., Salton, G., Allan J., and Singhal A. Automatic Query Expansion Using SMART: TREC 3. TREC-1994 (1994) 69-80.
6. Cui, H., Wen, J.R., Nie, J.Y., and Ma, W.Y. Probabilistic query expansion using query logs. In Proceedings of WWW'02 (2002) 325-332.
7. Dumais, S., Joachims, T., Bharat, K, and Weigend, A. SIGIR 2003 Workshop Report: Implicit Measures of User Interests and Preferences. SIGIR Forum, 37(2), Fall 2003. (2003) 50-54.
8. Fox, E.A. and Shaw, J.A. Combination of multiple searches. In Proceedings of TREC-2 (1994) 243-249.
9. Greengrass, E. Information Retrieval: a Survey (2000). http://www.cs.umbc.edu/cadip/readings/IR.report.120600.book.pdf
10. Hull, D.A. Using Statistical Testing in the Evaluation of Retrieval Experiments. In Proceedings of SIGIR'93 (1993) 329-338.
11. Joachims, T. Optimizing search engines using clickthrough data. In Proceedings of SIGKDD'02 (2002) 133-142.
12. Joachims, T., Granka, L., Pan, B., Hembrooke, H., Gay, G. Accurately interpreting clickthrough data as implicit feedback. SIGIR'2005 (2005) 154-161
13. Lee, J.H. Analyses of multiple evidence combination. In Proceedings of SIGIR'97 (1997) 267-276.
14. Ling, C.X., Gao, J.F., Zhang, H.J., Qian, W.N., and Zhang, H.J. Improving Encarta search engine performance by mining user logs. International Journal of Pattern Recognition and Artificial Intelligence. 16(8) (2002) 1101-1116.
15. Manmatha, R., Rath, T., and Feng, F. Modeling score distributions for combining the outputs of search engines. In Proceedings of SIGIR'01 (2001) 267-275.
16. Mitchell, T.M. Machine Learning. McGraw Hill (1997).
17. Oztekin, B., Karypis, G., and Kumar, V. Expert agreement and content based reranking in a meta search environment using mearf. In Proceedings of WWW'02 (2002) 333-344.
18. Radlinski, F., Joachims, T. Query chains: learning to rank from implicit feedback. KDD'2005 (2005) 239-248
19. Robertson, S.E., Walker, S., Jones, S., Hancock-Beaulieu, M.M. and Gatford, M., Okapi at TREC-3. In Overview of the Third Text REtrieval Conference (TREC-3) (1995) 109-126.
20. Salton, G., and Buckley, C. Improving retrieval performance by relevance feedback. Journal of American Society for Information Sciences. (1990) 41:288--297.
21. Shen, X., Tan, B., Zhai, C. Context-sensitive information retrieval using implicit feedback. SIGIR'2005 (2005) 43-50.
22. Silverstein, C., Henzinger, M., Marais, H., and Moricz, M. Analysis of a Very Large AltaVista Query Log. Technical Report SRC 1998-014, Digital Systems Research Center (1998).

23. Spink, A., Jansen B.J., Wolfram, D., and Saracevic, T. From e-sex to e-commerce: web search changes. IEEE Computer, 35(3) (2002) 107-109.
24. Spink, A., Wolfram, D., Jansen B.J., and Saracevic, T. Searching the web: the public and their queries. Journal of the American Society of Information Science and Technology, 52(3) (2001) 226-234.
25. Vogt, C.C., and Cottrell, G.W. Predicting the performance of linearly combined IR systems. In Proceedings of SIGIR'98 (1998) 190-196.
26. White, R.W., Ruthven, I., and Jose J.M. The use of implicit evidence for relevance feedback in web retrieval. In Proceedings of ECIR'02 (2002) 93-109.

A PDD-Based Searching Approach for Expert Finding in Intranet Information Management

Yupeng Fu, Rongjing Xiang, Min Zhang, Yiqun Liu, and Shaoping Ma

State Key Lab of Intelligent Technology and Systems,
Tsinghua University, Beijing, 100084
richbird@gmail.com

Abstract. Expert finding is a frequently faced problem in Intranet information management, which aims at locating certain employees in large organizations. A Person Description Document (PDD)-based retrieval model is proposed in this paper for effective expert finding. At first, features and context about an expert are extracted to form a profile which is called the expert's PDD. A retrieval strategy based on BM2500 algorithm and bi-gram weighting is then used to rank experts which are represented by their PDDs. This model proves effective and the method based on this model achieved the best performance in TREC2005 expert finding task.Comparative studies with traditional non-PDD methods indicate that the proposed model improves the system performance by over 45%. [1]

1 Introduction

Driven by the information explosion, organizations are giving more emphasis to the management of the increasing mass of knowledge they accumulate and search on it. However, as reported by Susan Feldman [1], knowledge workers still spend from 15% to 35% of their time searching for information and 40% corporate users can't find the information they need on intranets. Among the corporate information retrieval tasks users perform, expert finding is a frequently faced problem and a variety of practical scenarios of organizational situations that lead to that problem have been extensively presented in the literature [2], [3], [4].

Because of the target shifting from document to expert, simply using searching engine to trace an expert over the documents in organization is not an effective approach. The reason lies that there is a variety of data structures (Web-pages, Emails and Database et al) and formats (e.g. HTML, PDF, WORD) in an enterprise information environment, and it's difficult to extract and combine useful information from those different resources. Furthermore, it's also a challenge to build up a relationship between query and expert via documents because experts' information is not all accurately and explicitly documented. Traditional retrieval algorithm which is generally based on keyword matching may not lead to the relevant experts.

[1] Supported by the Chinese National Key Foundation Research & Development Plan (2004CB318108), Natural Science Foundation (60223004, 60321002, 60303005, 60503064) and the Key Project of Chinese Ministry of Education (No. 104236).

H.T. Ng et al. (Eds.): AIRS 2006, LNCS 4182, pp. 43–53, 2006.

The main contributions of our work are that we propose a person-centered model which builds a direct relationship between query and expert by searching among Person Description Document (PDD)s for each expert. This model also provides a convenient approach to handle the problem of mining and integrating expertise evidence in multi-resource. We also adopt a term dependence model for ranking retrieved PDDs, which helps to improve searching precision and accuracy.

The rest of paper is organized as follows. Section 2 gives a brief review of related work on expert finding. Section 3 introduces details of the PDD-based search model and term dependency retrieval method. Experiment evaluation is presented in Section 5 to assess the performance of the proposed algorithm. Finally come the discussions and conclusion.

2 Related Works

Since it has not been sufficiently addressed in research and is of immense practical importance in real organizations, expert finding attracts lots of researchers' interest in recent years. In the beginning, expert databases (aka "knowledge directories", "knowledge maps") through manual entry of expertise data are many organization used as automated assistance to expert finding. *Microsoft's SPUD* [5], *Hewlett-Packard's CONNEX* [6] and the *SAGE People Finder* [7] are examples of this approach. However, expert databases are known to suffer from numerous shortcomings [8]: manually developing the databases is a labor intensive and expensive task and databases are usually infected by the propensity of the experts to devote time to offer detailed description of their expertise.

Nowadays, people attempt to develop systems that exploit experts by retrieving information from corporate documents in organizations. Generally speaking, there are two alternative expert search models: the query-time generated and the aggregated search models.

In query-time generated models, source databases are searched employing information retrieval systems, and the matching sources are mined for experts. This approach allows using most recent information without maintaining additional information internally. The system *ExpertFinder* created by Mattox and his colleague [9] is a typical query-time generated system. It works by linking documents found through queries to the search engine then gathers the evidence and combines to a single score for one person. The *ContactFinder system* [10] and some commercial systems like *Autonomy's IDOL Server* [11] also employ query-time generated models to find expert.

In other approaches of associating experts to aggregated expertise models, experts are linked to a pre-constructed or dynamically generated central representation of expertise which can be a kind of knowledge model (ontology), organizational structure, etc. Several studies have been conducted on this model, for example, *Tacit's KnowledgeMail* [12] and *Lotus's Discovery Server* [13]. A typical system built by Nick Craswell, and David Hawking is *P@NOPTIC Expert* [14]. This system maintains a list of experts, gathered from the corporate documents which mention those people to build "employee documents", and exploit experts in those documents.

Query-time generated model has one shortcoming that it is hard to manipulate expertise information and to allow for an open and multi-purpose exploit of the expertise data from all types of documents [15], but organizations often have a mixture of data types and each data type has particular characteristics, including inter and intra document structure. To overcome this shortcoming, the proposed PDD-based search model in this paper is a typical aggregated expert search model which intuitively integrates expertise evidence from multi-source. However, unlike previous aggregated search models, the search model in this paper consists of a refined person description document constructed from web pages and a novel PDD ranking model using bi-gram retrieval.

3 Person Description Document (PDD) Model for Expert Search

3.1 PDD-Based Search Model

As mentioned, expert finding suffers from two tough problems. First, expertise data are distributed across the organization in heterogeneous sources and in various formats (Web sites, databases, Emails and the like). How well expertise indicators like terms and phrases reflect expertise is mainly a factor of how the source in which these indicators occur relates to the expert, irrespective of the indicators' direct statistical (or otherwise) correlation with the expert [5]. Second, most of the times the expertise information is not fully documented, and usually only a part of a document is related to the expert whom it mentions. Hence how to solve these problems is our research focus.

Definition 1: There is potential expertise evidence distributed in heterogeneous sources, which can be recognized and extracted to reform a document that describes the particular expert. This refined document is called **Person Description Document (PDD)**. In an organization, all these PDDs compose a PDD collection.

A PDD represents the expertise data associated with an expert. This PDD-based ranking method represents whether or not a PDD is relevant to a specific area where the user wants to find the experts.

The model is based on the assumption that each expert can be represented by a list of keywords which are related to him, and in a limited document collection expert evidence are extracted as PDD to approximate the ideal description of him. As a whole, PDD-based search model estimates $p(e \mid q)$, the probability of the expert e given the query q. specifically. The model is defined as:

$$p(e \mid q) = p(\theta_e \mid q) \approx p(\overline{\theta}_e \mid q) \tag{1}$$

where θ_e denotes the ideal description of expert e, and $\overline{\theta}_e$ stands for the PDD of person e to approach θ_e. The PDD-based ranking method will be described in section 3.3.

This PDD approach gains benefits from two processes: information clustering and data cleansing. This model aggregates expertise information to tackle the problem of heterogeneity and distribution of sources and implements information extraction process as filters to documents avoiding searching for documents altogether. PDD approach centralizes expertise data which can fortify the description of the person while in query-time generated model expertise data is dispersed in piles of documents.

Another advantage is that the PDD approach could provide flexible accurate expertise evidence extraction. Each data type in the corporation has particular characteristics and variant expertise evident extraction strategies are designed for different data type. What's more, PDDs could be human readable. In the returned PDD list, the user could choose to look at a PDD for a candidate expert and from that PDD link to the various sources that contribute to this PDD. That is to say, PDD is a convenient expertise information index for user to efficiently access the historical documents related to that expert.

Intuitively, variant weights are given to expertise data from different source so as to aggregate information effectively, which has already been discussed by some previous work [14]. How to collect not-documented expertise information, especially from web pages, becomes the key point that concerned instead. In section 3.2 we will discuss extracting expertise information from Web Pages as component of PDD.

3.2 Construction of PDD from Web Pages

Three kinds of features have been proposed to construct the PDD from web-pages, namely *context information*, *distance-weighted functional information*, and *group information.*

Context information for an expert is the words around the expert identifier which have been proved to be supportive to him. This information is a window of text which supplies detailed and descriptive information.

Distance-weighted functional information is generated using Markup terms in HTML documents, such as the *Title, Heading*, *Bold*, etc, which have significant contribution in describing web pages, containing summary words and have abundantly expressive force [16]. However, it's not accurate and appropriate to simply extract all the Markup information as component of PDD. A document may have more than one topic and an author usually mentions a person within one topic, hence the indiscriminate Markup information extraction may cause the topic-drift problem. A distance factor is considered in the process of extraction. In the text blocks with the occurrence of a person, the nearest Markup information ahead of the person are considered to be reliably potential expertise information. For example, the caption of a paragraph that mentioned a person may supply recapitulative expertise evidence of him.

Group information is extracted considering both plain text information and Markup terms in the documents. Based on the phenomena that persons often co-occur in the context as a group, we examine a mixture of patterns that group information expressed. For example, a list of person identifier may represent members of a group, and in a table, the left column list person names and in the right column give the corresponding description information like the professional title. Correctly recognizing the group patterns bring on a precise expertise mining that group description information and individual information of group members are respectively treat and extracted as group-based context. A link relationship is built within group members in our PDD. When an expert to the given topic is affirmed, his or her group mates are provided as an expanded answer.

In the experiment section, the effects of all the above PDD-features will be discussed.

3.3 Term Dependency Ranking Model for PDD-Based Search

Expert related queries are very well-defined and specific [17]. Generally they contain more than one word and replete with professional keywords which are correlated with each other [18]. Hence an assumption is made in our model that in these queries, adjacent words have strong relationship with each other. Based on our previous work [19], a ranking method for PDD-based search is proposed. The basic idea is that if adjacent words in the query are also adjacent in the document, then the document would be more likely to be relevant.

In implementation, this method is combined with BM2500 term weighting by treating word pairs in a query as additional terms for query, then calculating its information and adding to the original query term weights as a summation. The traditional BM2500 formula given by Robertson [20] is that:

$$W = \sum_{t \in query\_terms} W^{(1)} \frac{(k_1+1)tf_t(k_3+1)qtf_t}{(K+tf_t)(k_3+qtf_t)}, \tag{2}$$

where $W^{(1)} = \log \frac{N - df_t + 0.5}{df_t + 0.5}$

Where k_1 and k_3 are constant, df denotes document frequency of query term, tf_t and qtf_t denote term frequency in description document and in query. N is the total document number of collection. Our modified BM2500 considering word pair is as follow:

$$W = (1-\lambda) \sum_{t \in query\_terms} W^{(1)} \frac{(k_1+1)tf_t(k_3+1)qtf_t}{(K+tf_t)(k_3+qtf_t)} + \lambda \sum_{wp \in query\_wordpairs} W_2, \tag{3}$$

where $W^{(1)} = \log \frac{N - df_t + 0.5}{df_t + 0.5}, W_2 = W_2^{(1)} \frac{(k_1+1)tf_{wp}(k_3+1)qtf_{wp}}{(K+tf_{wp})(k_3+qtf_{wp})},$

$$W_2^{(1)} = \log \frac{N - df_{wp} + 0.5}{df_{wp} + 0.5}.$$

Where df_{wp} stands for exact document frequency of word pair in the collection and λ is multiplied to the word pair part of the total weights to constrain the impact of the word pair weight on the final documents rank. It should be noted that generally speaking, W_2 is smaller than original weights for DF of word pair is not high relative to df_t. The intuition behind this formula is that as λ augmented more, the stronger relationship between word pair in queries is emphasized. Now we describe our empirical results.

4 Experiments

4.1 Experimental Setting

The PDD-based search model approach is evaluated on web page part of dataset in TREC (Text REtrieval Conference, http://trec.nist.gov/) 2005 expert finding task. The collection is a crawl of the public W3C (*.w3c.org) sites and comprises 331,037 documents, in which 132643 are Web pages. Participants could use all the documents in order to rank a list of 1092 candidate experts. The 50 queries are consisting of professional short terms like "Semantic Web Coordination". In TREC 2005 enterprise track, the search topics were so-called "working groups" of the W3C, and the experts were members of these groups. These ground-truth lists were not part of the collection but were located after the crawl was performed. This enabled TREC to dry-run this task with minimal effort in creating relevance judgments. The official results are constructed manually by TREC.

4.2 Effect of Component in PDD

The effect of the expertise data that were extracted is examined, and the results are shown in figure 1 and figure 2. In these experiments, λ in word pair ranking method is set to 1.

Figure 1 illustrates the effect of the component used solely as PDD. Features like <title>, <bold>, <heading1&2>, anchor text, context are selected as experimental target. The results show that <title> and <heading12> field achieve good performance and context field also provide about 0.2 mean average precision, though not as much as former two. The bold text and anchor text provide low MAP correspondingly. Then the fields are merged together and Figure 2 shows the effect of their combination. We take the best-performed feature <title> as baseline for comparison. The figure shows that combination of <title> and anchor text give minor increase to MAP while combination of <title> with <bold> and with <heading12> decrease it a bit. The combination of <title>, <heading12> and context achieves the best performance in these experiments. Anchor text field is added to the former combination but it doesn't bring more enhancements.

The fact that <title> and <heading> fields are reliable and effective expertise evidence is reasonable. After observation on many organizational HTML documents,

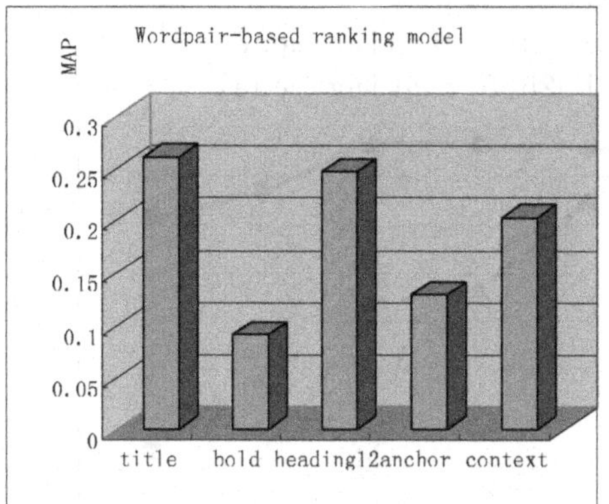

Fig. 1. Performance with different features used solely as PDD

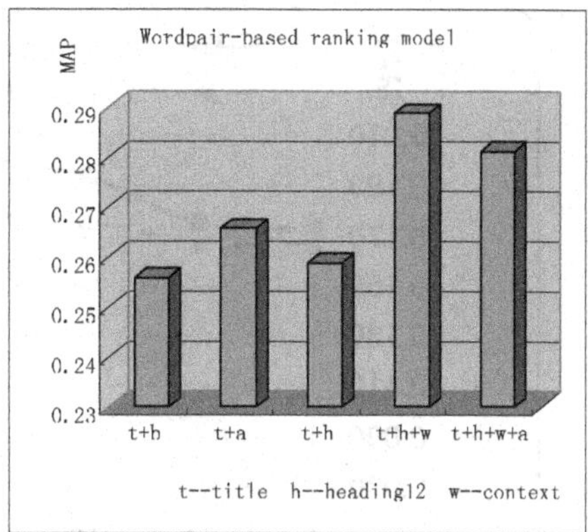

Fig. 2. Performance with combination of various features as PDD

it's found that the titles of lots of documents and heading information of sections contain descriptive information of the content. Especially some documents on introducing technology and a group of experts, they often put the keywords and abstractions in title and heading field. These fields represent experts who document mentions well by containing summary information.

Context information and group information are combined in these experiments as person-based and group-based context information. The results prove that it is important and dependable expertise data.

At last, anchor text and bold text don't provide great increase to MAP when they are combined with other effective fields. Anchor text may contain repeated expertise information that title field represents and may contain some noise of no use. For bold text, the results are explained from two aspects: in one hand, the amount of bold text extracted is not large; in the other hand, it's frustrate to determine the effectiveness of bold text because the author use bold text to emphasize may not be have relevance with expert it mentions.

4.3 Effect of Word Pair Based Ranking Model

In this section, the effectiveness of our ranking method based on word pair is examined.

First the impact that the method has on size of context length is illustrated in Figure3 which has two parabolas. The upper curve shows that using our ranking method the average precision increase as window-size rises and achieves best performance at 60-80 words, then MAP falls when window-size becomes larger. The lower curve is same as upper one in shape; however, it falls rapidly as window-size grows. Using word pair can achieve better performance than original BM2500 method on context field, and at the peak of upper curve, the proposed PDD-based approach has 47.9% improvement.

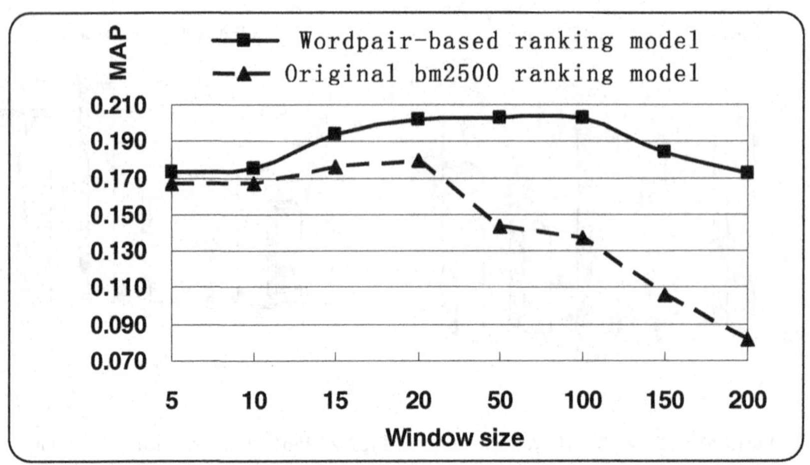

Fig. 3. Comparison of MAP on context as PDD with different ranking model

Table 1. Comparison of wordpair-based ranking model effect on features of PDD

Feature	MAP		R-prec		p@10	
	With wp	Without wp	With wp	Without wp	With wp	Without wp
Title	*0.2620*	0.2479	*0.3060*	0.2996	*0.4280*	0.4140
Bold	*0.0911*	0.0906	*0.1343*	0.1297	*0.1915*	0.1872
Heading12	*0.2476*	0.2389	*0.2890*	0.2835	*0.4220*	0.4060
Anchor text	*0.1290*	0.1261	*0.1736*	0.1748	*0.2688*	0.2667

Second, Table 1 illustrates a consistent improvement on other features of PDD after word pair based ranking method applied. Notice that although comparing with the impact on context, other features don't show significantly effectiveness, an all-pervading improvement in all main evaluation is demonstrated.

The basic idea of the word pair based ranking model is that if adjacent words in the query are also adjacent in the document, then the document is relevant. When the text area is not long enough, this method doesn't express its usefulness much. The reason is that when extraction is limit to a small range of text, the probability of extracting adjacent words that are also adjacent in queries is small. The experiments prove this by the fact that using features of little amount and small window-sized context, the improvement of this ranking method is trivial. When the window enlarged, the precision increase because more adjacent query words are included in PDD. However, when the window-size is too large, in an extreme case, the entire document is viewed as a window, the performance is not good. Too much noise are covered in the context at this situation.

4.4 Comparison Experiment on Search Model

In this section, the effectiveness of PDD-based search model is examined by comparing to experiments of query-time generated search model.

Firstly, a contrastive experiment is conducted which involves two steps. In step 1, documents relevant to the query are retrieved using traditional IR model; and in step 2 candidates are explored in the returned documents. While the previous step takes full advantage of well-studied IR models, in the second step true experts are selected within the relevant documents using voting method. Three voting ways are attempted, name-occurrence counting, document-similarity-based and document-ranking-based schemes. The latter two calculate a person's score according to the similarity score or the ranking of the documents where he or she appears. In all these schemes, the features considered in PDD are also involved in contrastive experiment to clarify whether or not the feature information is factor that makes effort on the gap and the results shows that there's trivial difference. Additionally, word pair based ranking model is used in the first phrase.

Table 2 shows the improvement of results of PDD approach on the result of the scheme based on documents ranking. The results illustrate a considerable gap between the results of the two models. In table 2, the official top results using query-time generate model in TREC2005 expert finding task is also presented and we achieved the best performance in the evaluation as participant.

Table 2. Comparative best results between PDD-based search model and our contrastive model

Model	MAP	p@10
PDD-based search model	0.2890	0.4500
Best result in the contrastive model	0.1986	0.2450
Improvement	***+45.5%***	***+83.7***
Best result using query-time generated model in TREC2005	0.2668	0.3700

The results show that the PDD-based search model is more effective by a more than 45% improvement to contrastive experiment in search precision. The seasons can be analyzed in several aspects.

Potential expertise information is extracted from the documents at mean time this process cleanses the data and abandons useless information. Then this data is aggregated as PDD which means that a relationship is built up between documents and persons. PDD approach centralizes expertise data which can fortify the description of the person while in query-time generated model expertise data is dispersed in piles of documents. Notice that BM2500 normalized the PDD length at retrieval step acting as a person normalization which avoids that in such cases as certain irrelevant people whose names appear in almost every document and query-time generated model will assign high scores to them.

Query-time generated model first retrieve documents relevant to the query, however, it's not reasonable to say that because the document has the biggest similarity to the query the persons appear in the document are experts to the query. The PDD is non query-related and the expertise data in PDD just represents the person, so it's intuit to say that more similar the PDD to the query, more relative the person to query.

Table 3. Comparative details of Web pages corpus between official collection and PDD collection

SOURCE	Docs	PDD Docs	Size (MB)	PDD Size (MB)
WEB	45,975	***872***	1,043	***43***
PEOPLE	1,016	***58***	3.0	***0.066***
Wiki	19,605	***158***	181	***3.5***
Other	3,538	***558***	47	***1.7***

Meanwhile, the PDD-based search is efficient. Table 3 illustrates that the PDD collection is much smaller than original document collection both in document number and in collection size. The PDD documents number is 2.3% of original source and the total PDD collection size is 3.78 % of original collection size. PDD generation is an offline job, therefore the benefit that providing more efficiency online service is gained with small index. Oppositely, query-generated model use the original document collection to index and to retrieve, so it put all the expert exploitation at online time.

5 Conclusion and Future Work

This paper proposes a PDD based search model to solve the expert finding problem. This model develops a strategy to extract the features and context from semi-structured web pages as filter to non-expertise data then combine these expertise data together as PDD to fortify the evidence of expert and to reduce number of documents in collection. A word pair based ranking method is adopted to rank the retrieved PDDs effectively.

The contributions of the paper are in the following two aspects. First, a person-centered model is proposed, which builds a direct relationship between query and expert by searching among PDD for each expert. This model also provides a convenient approach to handle the problem of mining and integrating expertise evidence in multi-resource. Especially, this paper proposes how to mine expertise information in web pages. Second, a term dependence ranking model is proposed which helps to improve search precision.

This research is mainly focus on finding expert using web pages resource and one of the future works is to extract expertise information from multi-sources, especially from emails which is another import part of corporate data and to found a effective way to integrate this information as one PDD.

Another future work is to make more application on other vertical search fields. Regarding the characteristic of PDD-based search model, this approach can be extended to object-centralized search such as software search, MP3 search and so on.

References

1. Susan Feldman "The high cost of not finding information" In Content, Document and Knowledge Management KMWorld-Volume 13, Issue 3, March 2004
2. D. Mattox, "Expert Finder", The Edge: The MITRE Advanced Technology Newsletter, vol. 2, no. 1, 1998. (at http://www.mitre.org/pubs/edge/June_98).
3. D. W. McDonald and M. S. Ackerman, "Just Talk to Me: A Field Study of Expertise Location", Proceedings of the 1998 ACM Conference on Computer Supported Cooperative Work (CSCW'98), Seattle, WA: ACM Press, pp. 315-324, 1998.
4. A. S. Vivacqua, "Agents for Expertise Location". Proceedings of the AAAI Spring Symposium on Intelligent Agents in Cyberspace, Stanford, CA, pp. 9 – 13, March 1999.
5. T. H. Davenport and L. Prusak, "Working Knowledge : How Organizations Manage What They Know". Boston, MA: Harvard Business School Press, 1998.
6. http://www.carrozza.com/connex/
7. http://sage.fiu.edu/Mega-Source.htm
8. D. Yimam. Expert Finding systems for organizations: Domain analysis and the demoir approach. In *Beyond Knowledge Management: Sharing Expertise*. MIT Press, 2000.
9. D. Mattox, M. Maybury and D. Morey, "Enterprise Expert and Knowledge Discovery". In Proceedings of the 8th International Conference on Human-Computer Interaction (HCI International'99), Munich, Germany, pp. 303-307, August 1999.
10. B. Krulwich and C. Burkey. The ContactFinder agent:Answering bulletin board questions with referrals. InAAAI-96, 1996.
11. www.autonomy.com
12. www.tacit.com
13. www.lotus.com/kd
14. Nick Craswell, David Hawking, Anne-Marie Vercoustre,Peter Wilkins, "P@NOPTIC Expert: Searching for Experts not just for Documents", CSIRO, Australia 2003
15. D. Yimam Seid and A. Kobsa, "Centralization vs. Decentralization Issues in Internet-based Knowledge Management Systems: Experiences from Expert Recommender Systems". Workshop on Organizational and Technical Issues in the Tension Between Centralized and Decentralized Applications on the Internet (TWIST2000). Institute for Software Research, University of California, Irvine, CA, 2000.
16. Glover E, Tsioutsiouliklis K, Lawrence S, Pennock D, Flake G. Using Web structure for classifying and describing Web pages. In: Proc.of the Int'l World Wide Web Conf. (www 2002). Hawaii: ACM Press, 2002. 562-569. http://www2002.org/CDROM/refereed/504/index.html
17. H. Kautz, B. Selman and A. Milewski, "Agent Amplified Communication". In The Proceedings of the Thirteenth National Conference on Artificial Intelligence (AAAI-96), Portland, OR, 1996, pp. 3 – 9
18. David Hawking, "Challenges in Enterprise Search". CSIRO ICT Centre, Fifteenth Australasian Database Conference, ADC2004
19. M. Zhang, C.Lin, YQ.Liu, L.Zhao and SP.Ma "THUIR at TREC 2003: Novelty, Robust and Web" State Key Lab of Intelligent Tech. & Sys., CST Dept, Tsinghua University, 2003
20. S. E. Robertson, S. Walker, Okapi/Keenbow at TREC-8, In The Eighth Text REtrieval Conference (TREC 8), 1999, pp. 151-162

A Supervised Learning Approach to Entity Search

Guoping Hu[1], Jingjing Liu[2], Hang Li, Yunbo Cao, Jian-Yun Nie[3], and Jianfeng Gao

Microsoft Research Asia, Beijing, China
[1] iFly Speech Lab, University of Science and Technology of China, Hefei, China
[2] College of Information Science & Technology, Nankai University, Tianjin, China
[3] Département d'informatique et de recherche opérationnelle Université de Montréal

Abstract. In this paper we address the problem of entity search. Expert search and time search are used as examples. In entity search, given a query and an entity type, a search system returns a ranked list of entities in the type (e.g., person name, time expression) relevant to the query. Ranking is a key issue in entity search. In the literature, only expert search was studied and the use of co-occurrence was proposed. In general, many features may be useful for ranking in entity search. We propose using a linear model to combine the uses of different features and employing a supervised learning approach in training of the model. Experimental results on several data sets indicate that our method significantly outperforms the baseline method based solely on co-occurrences.

Keywords: Information Retrieval, Entity Search, Expert Search, Time Search, Supervised Learning.

1 Introduction

Most of the research on Information Retrieval (IR) focuses on search of documents. Despite the progress on document search, many specialized searches are still not well studied. This paper tries to address entity search, which provides search of specific types of information. More precisely, the user types a search query and designates the type of entity, and the system returns a ranked list of entities in the type (persons, times, places, organizations, or URLs) that are likely to be associated with the query. Entity search includes expert search, time search, place search, organization search, and URL search.

Entity search will be particularly useful at enterprise. People at enterprise are often interested in obtaining information of specific types. For example, an employee of a company may want to know the time of the next event concerning a product, the right person to contact for a problem, and so on.

Traditional search approach does not support entity search. For example, for a query looking for "date of IT Exhibition," each of the words will be used as a keyword, including "date." As a consequence, documents containing the keywords "date," "IT" (if not removed as a stop word) and "exhibition" will be retrieved. However, the documents or passages retrieved do not necessarily contain the required information, i.e. the date of the exhibition in this example.

There is an increasing interest in entity search in the research community. For example, expert search was investigated in several search engines and studies [2, 6, 7, 13, 14, 15, 21 and 22]. There was also a task on expert search at TREC 2005.

H.T. Ng et al. (Eds.): AIRS 2006, LNCS 4182, pp. 54–66, 2006.

However, we observe that all the proposed methods relied on traditional IR techniques or simple features for ranking entities. For example, Craswell et al. [3] proposed using the co-occurrences of people and keywords in documents as evidence of association between them, and ranking people according to the strength of association.

One may consider dealing with the problem with Question Answering (QA). However, there is a striking difference between QA and entity search. Well formed questions are assumed to be inputted in QA, while most queries in entity search are just keywords. Therefore, a great number of the techniques used for QA cannot be adequately applied here.

In this paper, we aim to develop an appropriate approach for entity search. Ranking is a key issue in entity search. We propose a new ranking method that utilizes a variety of features in a linear combination model. Supervised learning is employed to train the model.

Two specific types of search are considered as examples in this paper: time search and people search. For time search, experiments have been carried out on two different collections: one from the intranet of Microsoft and the other from TREC Web and QA Tracks. For expert search, experiments have been performed with the TREC expert search data. The experimental results indicate our method performs significantly better than the baseline methods.

The rest of the paper is organized as follows. In Section 2 we introduce related work, and in Section 3 we explain the problem of entity search. In Section 4, we describe our proposed approach to entity search. Section 5 gives our experimental results. We conclude in Section 6.

2 Related Work

2.1 Traditional Search

Conventional Information Retrieval aims to identify documents or passages that may be 'relevant' to the query. The evidence used for the relevance judgment is the appearance of the query terms in the documents or passages. The query terms are usually weighted according to their occurrences in the documents or passages using models such as TF-IDF (e.g., [20]), BM25 (e.g., [19]), and Language Model (e.g., [17]). Usually these term weighting methods do not require labeled data for training. In that sense, the methods are unsupervised.

There is also a new trend in IR recently that manages to employ supervised learning methods for training ranking functions. Herbrich et al. [9] cast the IR problem as ordinal regression. They proposed a method for training the ordinal regression model on the basis of SVM. Gao et al. [8] proposed conducting discriminate training on a linear model for IR and they observed a significant improvement of accuracy on document retrieval by using the method. The success of the approach is mainly due to the following two factors: (1) The model provides the flexibility of incorporating an arbitrary number of features; (2) The model is trained in a supervised manner and thus has a better adaptation capability.

2.2 Entity Search

Entity search tries to identify entities strongly associated with query terms. There were several studies on entity search. The most studied type of entity was people (or expert).

Expert search was investigated in [2, 6, 7, 13, 14, 15, 21 and 22]. There was also a task on expert search at TREC 2005 [25]. All the existing methods for entity search only exploited simple features or traditional IR techniques for ranking. For example, Craswell et al. [3] proposed using co-occurrences between people and query words as evidence to rank people.

Many features may be useful for entity search, including new features that are not used in traditional IR. Therefore, an appropriate approach to entity search should easily incorporate new features. In this paper, we propose employing supervised learning to train an entity search model. This approach is chosen because of its flexibility and its capability of adaptation. To our knowledge, no previous study has explored the same approach for entity search.

2.3 Factoid Question Answering

Question Answering (QA) is a task that aims to provide the user with the correct answer to a question. Many QA systems were developed, including Webclopedia [10], NSIR [18], MultiText [4], MULDER [12], AskMSR [1], and the statistical QA system of IBM [11].

Factoid QA is the most studied subtask, which tries to answer questions about facts. It usually contains the following steps: (1) question type identification, (2) question expansion, (3) passage retrieval, (4) answer ranking, and (5) answer generation. In question type identification, the type of the question (that is also the type of the answer) is identified. In question expansion, the synonymous expressions of the question are created. In passage retrieval, relevant passages are retrieved with the content words in the question and its expansions. In answer ranking, the retrieved passages are ranked, and the potential answers within the passages are marked (in the identified question type). Finally, in answer generation, a single answer is generated from the top ranked passages.

People may take QA as an appropriate means for entity search. However, the assumptions for QA may not hold for entity search. First, QA heavily relies on NLP techniques to analyze the question. The basic assumption is that the question is a well formulated question. For entity search, we have to deal with queries instead of complete questions. That means that NLP would not help entity search. Second, users may be satisfied with one or several passages as search results as in entity search; getting a unified answer as in QA is more desirable, but not necessarily needed. Therefore, for entity search we do not necessarily need to employ the same methodologies used in QA.

3 Entity Search

The problem of entity search such as expert search and time search can be described as follows. The system maintains a collection of documents. When the user inputs a query (just like people usually do in search of documents) and designates a type, the system returns a ranked list of entities in the type based on the information in the documents. The entities are ranked based on the likelihood of being associated with the query. An entity is considered being associated with the query if there exists strong relationship between the query and the entity (e.g., a person is an expert on the

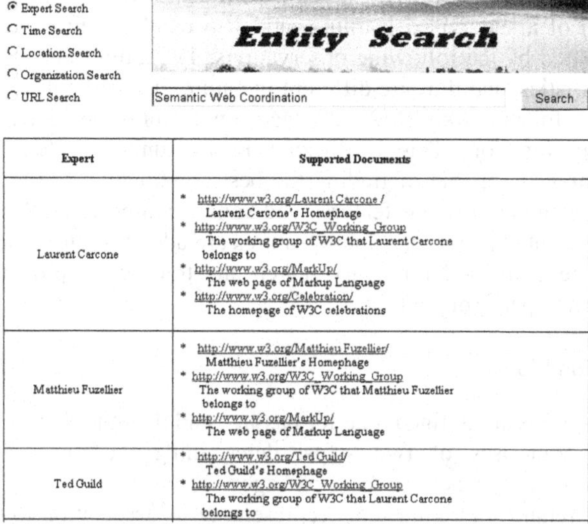

Fig. 1. An illustrative interface of entity search system

topic of the query). In practice, an association presented in a document can be bogus; however, we do not consider the problem in this paper.

A single user interface can be considered for entity search. There are check boxes on the UI, and each represents one type of entity (as shown in Figure 1). The user can designate the types of information to search by checking the corresponding boxes. The user can then input queries in the search box.

Figure 1 shows an example of expert search. When the user checks the box "expert search" and inputs the query "semantic web coordination," the system returns a list of people that are likely to be experts on the area. Each answer (person) is associated with supporting documents. The documents can help the user to figure out whether the answers are correct.

4 Our Method for Entity Search

In entity search, we first need to recognize the targeted entities in the documents. A simple method is used here: We utilize heuristic rules to conduct the entity recognition. For example, in time search, rules are created to identify time expressions such as "Nov. 17, 2004." Here, we do not consider conducting anaphoric resolution, for example, identifying the date of "yesterday" in a document.

For each identified entity, a passage is constructed. A passages is a window of fixed size around an identified entity; but it can also be delimited by the natural boundaries (tags in HTML documents such as <table>, <ol>, <p>, <ul>, <pre>, <li>, <dl>, <dt>, <tr>, <hr>). The entity in a passage is called the *center* of the passage. Around the center, features are extracted and used for the ranking of the passage with regard to the query.

The approach used in our study, which makes it different from most existing ones, is the utilization of a supervised learning model to combine different features. Our method is motivated by the following observations: 1) Features are usually extracted according to heuristics, and thus are different in nature. It is difficult to combine them in a traditional IR function like BM25. 2) Once new features are added, one has to re-tune the ranking function. This is difficult if the tuning is done manually. 3) Supervised learning is capable of finding the best combination of the features in an automatic manner, given a set of features and some training examples. The training process can be executed again, once new features are added. In this study, our goal is to develop a general method for entity search and thus we employ the supervised learning approach to perform the task.

4.1 Ranking Function

Two ranking functions are defined: one for passage and the other for entity. Our entity search method comprises of two steps: determining the top K passages and determining the top N entities from these passages.

Each passage retains a ranking score representing its association to the query. Let us denote it as $s(q,p)$. Suppose that the query q contains several keywords (content words). Then, $s(q,p)$ is determined according to the following equation:

$$s(q,p) = \sum_{w \in q \cap p} \varphi(w,p) \tag{1}$$

where w stands for a keyword appearing in both the query q and the passage p, and $\varphi(w,p)$ denotes the weight of w. We will explain the function $\varphi(w,p)$ in more details in the next subsection. We select the top K passages based on the $s(q,p)$ scores.

Once the K top passages are identified, a *weighted voting* is conducted to determine the top N entities, in which each passage votes for its own entity (in its center). Specifically, the ranking score of an entity e with respect to the query q is calculated as follows:

$$s(q,e) = \sum_{p \in P(e)} s(q,p) \tag{2}$$

where $s(q,e)$ is a score representing the association between query q and entity e, and $P(e)$ is the set of passages supporting e. The top N entities are those with the highest $s(q,e)$ scores.

4.2 Learning the Weighting for a Keyword

In conventional IR, the weight of a keyword is determined according to its frequency. For entity search, we believe that this weighting schema is not appropriate. Another mechanism should be adopted. For example, a keyword may be more important if it is closer to the entity (center) in the passage. Here, we define the general weighting schema of $\varphi(w,p)$ - the weight of query word w in a passage p - as a linear function of features. Each keyword w corresponds to a vector of m features denoted as $(x_{w,p,1},\ldots,x_{w,p,m})$. Then $\varphi(w,p)$ is calculated as follows:

$$\varphi(w,p)=\sum_{j=1}^{m} c_j x_{w,p,j} \quad (3)$$

where $c_j, j=1,\ldots,m$ are weights of different features, which are trained by using Hill Climbing algorithm as described later.

4.3 Feature Sets

As features, we utilize information on the distance between center and keyword, the characteristics of keyword, the characteristics of passage, etc. We describe the features in this section.

Time Search. For time search, we define 35 features and Table 1 shows some examples of them. Here, we denote the entity as center, and the keyword as word. Some features take on numerical values, while some features are Boolean features.

Expert Search. For expert search, we define 15 features and Table 2 shows some examples of them. Here, we denote the entity in a passage as center, and the keyword as word too.

Table 1. Example of features for Time Search

Group	Type	Feature Description
1	Features of *word*	IDF of *word*
		Term Frequency (TF) of *word* in original document
		TF * IDF of *word*
		Distribution of *word* in original document, measured in entropy (c.f. [23])
		Part of speech of *word*
2	Features of passage	Bag of words in passage
		Size of passage
3	Features on position	Distance function $\exp(-\alpha \cdot d)$ where d is minimum distance between *center* and *word*, and α is parameter.
		Does *word* occur immediately preceding or following center?
		Does *word* occur in the same sentence or paragraph as *center*?
		Does *word* occur in the previous or next sentence of *center*?
4	Features on relation with other entity	Is there any other entity in the same sentence as *center*?
		Is there any other entity close to *word*?

4.4 Training

The key problem in ranking is to make a correct utilization of the features, i.e. a correct setting of the weights $(c_1,\ldots,c_m)$ in equation (3). We resort to supervised learning to train the weights in the linear combination function. To do this, we need to utilize a set of

Table 2. Example of features for Expert Search

Group	Type	Feature Description
1	Features of *word*	Term Frequency (TF) of *word*
		TF * IDF of *word*
2	Features on position	Distance function $\exp(-\alpha \cdot d)$ where *d* is minimum distance between *center* and *word* and α is parameter.
		Does *word* occur immediately preceding or following *center*?
3	Features on metadata	Does *word* occur in the title of document?
4	Features on structure of document	Does *word* occur in the same section with *center?* (We parse the HTML document to get the section information)

training data. The training data contains a set of queries, a set of passages obtained as described above, and the relevance judgments provided by human judges.

Different algorithms can be employed in training of the weights $(c_1,\ldots,c_m)$. In this study, we use the Hill-Climbing algorithm (e.g., [16]) because of its simplicity. The Hill-climbing algorithm is slightly modified so as to deal with over-tuning. This algorithm works as follows:

(1) Initialize each weight c_j as 0.0;
(2) For $j = 1$ to m
   (a) Set $t = c_j$
   (b) Try different weight values for t, conduct ranking using the training data and evaluate the performance of *entity ranking* in terms of Mean Average Precision
   (c) Record the highest performance achieved in b) and the corresponding weight value
   (d) Reset c_j as t;
(3) Record the highest performance achieved in step (2) and the corresponding weight, if the improvement of step (2) is larger than the threshold σ (this is to avoid over-tuning), then adopt the new weight (note that only one weight is updated here), and go to step (2). Otherwise, terminate the algorithm.

5 Experimental Results

5.1 Baseline Methods

Our method is compared with two baseline methods:

(1) **BM25:** In this method we calculate the BM25 score of the passage with respect to the query.
(2) **Distance:** In this method we calculate the following distance score:

$$\exp(-\alpha \cdot d(w,t)) \quad (4)$$

where w denotes a keyword and t denotes the *center*, d is the *minimum* distance between t and w, and α is a parameter.

The first baseline method is representative of the traditional IR approach. These baseline methods are chosen because they are commonly used in the previous studies. For example, Craswell et al.'s method based on co-occurrence [3] is similar to the BM25 method. The second one is similar to a typical QA approach [18].

For all the two methods, the top K passages are identified according to their passage ranking functions. Then two voting methods are used to further rank entities: simple voting or weighted voting. Simple voting means voting entities by the number of supported passages, and weighted voting is similar to that in equation (2).

5.2 Evaluation Measures

We use Mean Average Precision (MAP) and R-Precision [5] as the measures for evaluations of entity search.

5.3 Time Search

Data Sets. Our experiments on time search were carried out on two data sets:

1) The first one was created from the query logs of a search engine on the intranet of Microsoft in a time period (9/2004). First the queries containing the clue words "when," "schedule," "day," and "time" were collected, and then the clue words were removed from the queries. The remaining parts were used as pseudo queries time search. This set contains 100 queries (referred to as MS hereafter). The documents are from the same intranet.

2) The other query set was created from the temporal questions in the TREC QA Track (i.e., questions with "when," "in what time," etc). Again, stop words and time clues were removed from the queries. This query set contains 226 queries (referred to as TREC hereafter). The documents are those used in TREC Web Track.

For each query, each of the methods tested returned 100 answers. These answers were judged manually by 2 human evaluators. For 23% of the MS queries correct answer could not be found in the retrieved documents, and the number was 51% for TREC.

Time Expression Identification Experiment. As our method depends on the quality of time expression identification, we first conducted experiments on the time expression identification. 300 documents were randomly selected. A human annotator was asked to annotate all the time expressions within them. This enabled us to evaluate the time identification method. Table 3 shows the evaluation results. We see that in general our identification method obtains high accuracies. The lower precision for the MS data is due to the occurrences of confusing product names and programming codes in the data set (e.g., "Money 2002").

Time Expression Ranking Experiment. We conducted time search using our method and the two baselines. The top 100 answers with each method were used. We performed *4-fold cross validation*, i.e. 3/4 of the queries were used in training and 1/4

Table 3. Results of time expression identification method

Dataset	Annotated	Identified	Matched	Precision	Recall
MS	2,018	2,447	1,987	0.8120	0.9846
TREC	3,589	3,783	3,544	0.9368	0.9875

Table 4. Results of time expression ranking

MS	MAP	R-Precision
BM25 (Simple Voting)	0.1483	0.1241
BM25 (Weighted Voting)	0.1559	0.1377
Distance (Simple Voting)	0.3291	0.2880
Distance (Weighted Voting)	0.3440	0.3070
Our method	**0.3952**	**0.3659**
TREC	**MAP**	**R-Precision**
BM25 (Simple Voting)	0.1232	0.1012
BM25 (Weighted Voting)	0.1277	0.1158
Distance (Simple Voting)	0.1951	0.1760
Distance (Weighted Voting)	0.2070	0.1863
Our method	**0.2454**	**0.2234**

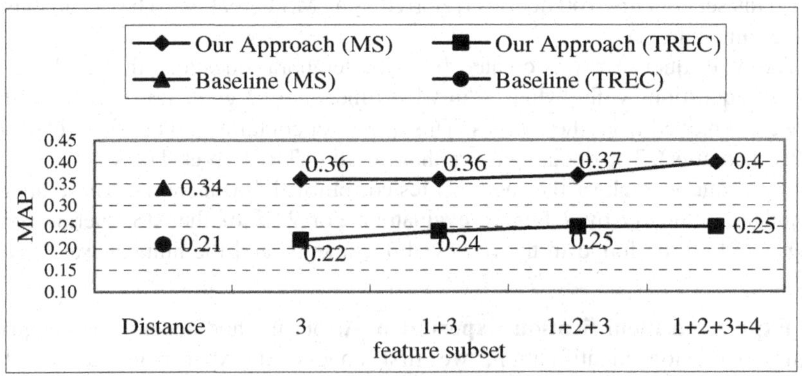

Fig. 2. Ranking results with different feature subsets for passage scoring learning

for testing. Table 4 shows the time search results of the different methods in terms of MAP and R-Precision.

We can observe that Distance performs better than BM25. This shows that the traditional IR method that does not incorporate a specific treatment of entities is not sufficient for entity search. We can also observe that weighted voting performs better than simple voting. However, the differences are not large. Comparing our method with the baseline methods, we can see that our learning method outperforms both baseline methods with quite large margins. One may notice that the overall accuracies in terms of MAP and R-Precision are not very high. An important reason is that 23% of the MS queries and 51% of the TREC queries do not have correct answers.

In order to analyze the contribution of each feature subset (representing different aspects) in our learning based method, we used different subsets of the features in our ranking functions. The feature subsets 1, 2, 3 and 4 in Figure 2 correspond to the four feature aspects described in Section 4.3. From Figure 2, it can be seen that when more features are used the search results can be improved. This result validates our hypothesis that more features should be used in entity search. It also shows that the supervised learning method we propose is capable of combine different features so as to take advantage of each of them.

5.4 Expert Search

Data Set. In our expert search experiments, we used the data set in the expert search task of enterprise search track at TREC 2005. The document collection was crawled from the public W3C [24] sites in June 2004, which contains 331,307 web pages, including specifications, email discussion, wiki pages and logs of source control. The ground truth on expert search was obtained from an existing database of W3C working groups. For each query (the name of the working group), a list of people (the group members) is associated as experts on the topic. There are in total 1,092 members in all the groups. (For details of the TREC data, see [25]).

Personal Name Identification. Heuristic rules were used to identify personal names in the documents. In order to evaluate the accuracy of this process, 500 documents from the W3C document collection were randomly selected. All the personal names from these documents were manually checked by a human annotator. The evaluation result shows that our name identification method can work well: the precision and recall are 100% and 90%, respectively. The process missed some names mainly because some irregular variants of names exist in the corpus (e.g. “danc“ for ”Dan Connolly”). In general, our name identification process is satisfactory.

Expert Ranking Experiment. In the official tests in TREC, two sets of queries were provided: a set of 10 training topics and another set of 50 topics for testing. However, we notice that quite large differences between the training and testing queries exist: All the queries in the training set are single phrases, while some queries in the test data are combinations of several phrases. Therefore, we only used the 50 test queries in our experiment, and conducted 10-fold cross-validation in our evaluation. The results reported below are the averaged results over 10 trials.

Table 5 shows the results on expert search for our method and the baseline methods. Different from time search, we do not observe a large difference between the BM25 method and the Distance method: both baseline methods perform quite poorly. In comparison, our method based on supervised learning outperforms them significantly. This confirms again that our method is more suitable for entity search.

Table 5. Results of the baseline methods and our method

Methods	MAP	R-Precision
BM25 (Simple Voting)	0.1516	0.1623
BM25 (Weighted Voting)	0.1231	0.1352
Distance (Simple Voting)	0.1430	0.1625
Distance (Weighted Voting)	0.1330	0.1644
Our method	0.2684	0.3190

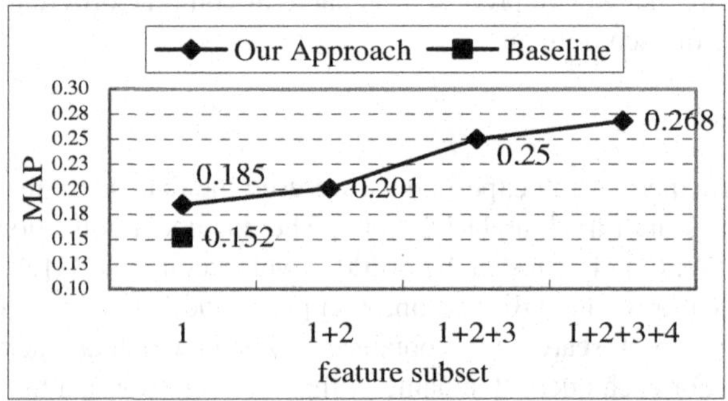

Fig. 3. Ranking results with different feature subsets for passage scoring learning in expert search

In order to see the contributions of different features, we ran our method with different subsets of features. From Figure 3, we can see that the more the features used, the better the performance achieved. This is similar to the observation on time search. Therefore, we have more evidence to say that it is advantageous to incorporate more features in entity search and employ a supervised learning method to appropriately combine the uses of the features.

6 Conclusion

Entity search such as expert search and time search is useful in many search scenarios, particularly in enterprise. Methods based on the use of co-occurrence have been proposed in the literature for expert search. This paper explores a new approach for entity search on the basis of supervised learning. Specially, it makes use of a linear combination model for ranking of entities, which incorporates many different useful features.

We have applied our method to two of the entity searches – time search and expert search in this paper. Experimental results show that the proposed method performs significantly better than the baseline methods solely using co-occurrence or distance.

The main contributions of this work are (1) a proposal of a supervised learning approach to entity search and an empirical verification of the effectiveness of the approach; and (2) identification of some useful features for time search and expert search.

Time search and expert search are just the first entity searches we have investigated. Our long term goal is to construct a search system in which all the commonly required entity searches are developed effectively. Such a system would offer great facilities for people to find their required types of information.

References

1. Eric Brill, Susan Dumais and Michele Banko, *An Analysis of the AskMSR Question-Answering System*, EMNLP 2002
2. Christopher S. Campbell, Paul P. Maglio, Alex Cozzi, Byron Dom, *Expertise Identification using Email Communications*, CILM'03, 2003
3. N. Craswell, D. Hawking, A. M. Vercoustre and P. Wilkins, *P@NOPTIC Expert: Searching for Experts not just for Documents*. In Ausweb, 2001.
4. Gordon V. Cormack, Charles L. A. Clarke, D. I. E. Kisman, Christopher R. Palmer, *Fast Automatic Passage Scoring (MultiText Experiments for TREC-8)*. TREC 1999
5. Gordon V. Cormack and Thomas R. Lynam, *Statistical Precision of Information Retrieval Evaluation*, SIGIR'06, August 6–11, 2006, Seattle, Washington, USA.
6. S. Deerwester, S. T. Dumais, G. W. Fumas, T. K. Landauer, and R. Harshman, *Indexing by Latent Structure Analysis*, Journal of the American Society for Information Sciences. ACM Press 391-407, 1990.
7. B. Dom, I. Eiron, A. Cozzi and Z. Yi, *Graph-Based Ranking Algorithms for E-mail Expertise Analysis*, in Proc. of the 8th ACM SIGMOD workshop on Research issues in data mining and knowledge discovery, 2003.
8. Jianfeng Gao, Haoliang Qi, Xinsong Xia, and Jian-Yun Nie. 2005. *Linear discriminant model for information retrieval*. In: SIGIR2005.
9. R. Herbrich, T. Graepel and K. Obermayer. 2000. *Large margin rank boundaries for ordinal regression*. Advances in Large Margin Classifiers, pp. 115-132. MIT Press, Cambridge, MA.
10. Eduard H. Hovy, Laurie Gerber, Ulf Hermjakob, Michael Junk, Chin-Yew Lin, *Question Answering in Webclopedia*. TREC 2000
11. Abraham Ittycheriah, Salim Roukos, *IBM's Statistical Question Answering System-TREC 11*. TREC 2002
12. Cody C. T. Kwok, Oren Etzioni, Daniel S. Weld, *Scaling question answering to the Web*. WWW-2001: 150-161
13. M. E. Maron, S. Curry, and P. Thompson. *An inductive search system: Theory, design and implementation*. IEEE Transaction on Systems, Man and Cybernetics, 16(1):21–28, 1986.
14. Dave Mattox, Mark Maybury, Daryl Morey, *Enterprise Expert and Knowledge Discovery*, Proceedings of the HCI International '99
15. D. W. McDonald, *Evaluating Expertise Recommendations*. In Proc. of the ACM 2001 international conference on Supporting Group Work (GROUP'01), Boulder, CO, 2001.
16. William Morgan, Warren Greiff, John Henderson, *Direct Maximization of Average Precision by Hill-Climbing, with a Comparison to a Maximum Entropy Approach*, HLT-NAACL 2004, pp. 93-96
17. Jay M. Ponte, W. Bruce Croft, *A Language Modeling Approach to Information Retrieval*. SIGIR1998: 275-281

18. Dragomir R. Radev, Weiguo Fan, Hong Qi, Harris Wu, Amardeep Grewal, *Probabilistic question answering on the web*. WWW 2002: 408-419
19. Stephen E. Robertson, Steve Walker, Micheline Hancock-Beaulieu, Mike Gatford, A. Payne, *Okapi at TREC-4*. TREC 1995
20. Gerard Salton, James Allan, Chris Buckley, *Approaches to Passage Retrieval in Full Text Information Systems*. SIGIR 1993: 49-58
21. Mayssam Sayyadian, Azadeh Shakery, AnHai Doan, ChengXiang Zhai, *Toward Entity Retrieval over Structured and Text Data*, WIRD'04, the first Workshop on the Integration of Information Retrieval and Databases (WIRD'04), 2004
22. L.A. Steer and K.E. Lochbaum, *An Expert/Expert Locating System Based on Automatic Representation of Semantic Structure*, in Proc. of the Fourth IEEE Conference on Artificial Intelligence Applications, 1988.
23. Li Zhang, Yue Pan, Tong Zhang, *Focused named entity recognition using machine learning*. SIGIR 2004: 281-288
24. World Web Consortium (W3C), http://w3.org
25. TREC 2005, http://trec.nist.gov/tracks.html

Hierarchical Learning Strategy in Relation Extraction Using Support Vector Machines

GuoDong Zhou[1,2], Min Zhang[2], and Guohong Fu[3]

[1] School of Computer Science and Technology, Suzhou University, China 215006
gdzhou@suda.edu.cn
[2] Institute for Infocomm Research, Singapore 119613
mzhang@i2r.a-star.edu.sg
[3] Department of Linguistics, The University of Hong Kong, Hong Kong
ghfu@hotmail.com

Abstract. This paper proposes a novel hierarchical learning strategy to deal with the data sparseness problem in relation extraction by modeling the commonality among related classes. For each class in the hierarchy either manually predefined or automatically clustered, a discriminative function is determined in a top-down way. As the upper-level class normally has much more positive training examples than the lower-level class, the corresponding discriminative function can be determined more reliably and effectively, and thus guide the discriminative function learning in the lower-level, which otherwise might suffer from limited training data. In this paper, the state-of-the-art Support Vector Machines is applied as the basic classifier learning approach using the hierarchical learning strategy. Evaluation on the ACE RDC 2003 and 2004 corpora shows that the hierarchical learning strategy much improves the performance on least- and medium- frequent relations.

1 Introduction

With the dramatic increase in the amount of textual information available in digital archives, there has been growing interest in Information Extraction (IE), which identifies relevant information (usually of pre-defined types) from text documents in a certain domain and put them in a structured format.

According to the scope of the ACE program (ACE 2000-2005), IE has three main objectives: Entity Detection and Tracking (EDT), Relation Detection and Characterization (RDC), and Event Detection and Characterization (EDC). This paper will focus on the ACE RDC task, which detects and classifies various semantic relations between two entities. For example, we want to determine whether a person is a citizen of a country, based on the evidence in the context. Extraction of semantic relationships between entities can be very useful for applications such as question answering, e.g. to answer the query "Who is the president of the United States?", and information retrieval, e.g. to expand the query "George W. Bush" with "the president of the United States" via his relationships with the country "the United States".

One major challenge in relation extraction is due to the data sparseness problem (Zhou et al 2005). As the arguably largest annotated corpus in relation extraction, the ACE RDC 2003 corpus shows that different relation subtypes are much unevenly

H.T. Ng et al. (Eds.): AIRS 2006, LNCS 4182, pp. 67 – 78, 2006.

distributed and a few subtypes, such as the subtype "Founder" under the type "ROLE", suffers from a small amount of annotated data. While various machine learning approaches, such as generative modeling (Miller et al 2000), maximum entropy (Kambhatla 2004) and support vector machines (Zhao et al 2005; Zhou et al 2005), have been applied in the relation extraction task, no explicit learning strategy is proposed to deal with the data sparseness problem.

This paper proposes a novel hierarchical learning strategy to deal with the data sparseness problem by modeling the commonality among related classes. Through organizing various classes hierarchically according to their relatedness, a discriminative function is determined for a given class in a top-down way by capturing the commonality among related classes to guide the training of the given class. Here, support vector machines (SVM) is applied as the basic classifier learning approach. The reason that we choose SVM is that SVM represents the state-of-the-art in the machine learning community and support both feature-based and kernel-based learning. For SVM, the guidance is done by, when training a lower-level class, discounting those negative training instances which do not belong to the support vectors of the upper-level class. By doing so, the data sparseness problem can be well dealt with and much better performance can be achieved, especially for those relations with small or medium amounts of annotated examples. Evaluation on the ACE RDC 2003 and 2004 corpora shows that the hierarchical learning strategy achieves much better performance than the flat learning strategy on least- and medium-frequent relations. It also shows that our system much outperforms the previous best-reported system by addressing the data sparseness issue using the hierarchical learning strategy.

The rest of this paper is organized as follows. Section 2 presents related work. Section 3 describes the hierarchical learning strategy using SVM. Finally, we present experimentation in Section 4 and conclude this paper in Section 5.

2 Related Work

The relation extraction task was formulated at MUC-7(1998) and extended at ACE (2000-2005). With the increasing popularity of ACE, it is starting to attract more and more researchers within the NLP community. Representative related works can be classified into three categories according to different approaches they used: generative models (Miller et al 2000), tree kernel-based approaches (Zelenko et al 2003; Culotta et al 2004; Bunescu et al 2005; Zhang et al 2005), and feature-based approaches (Kambhatla 2004; Zhao et al 2005[1]; Zhou et al 2005).

Miller et al (2000) augmented syntactic full parse trees with semantic information corresponding to entities and relations, and built generative models to integrate various tasks such as POS tagging, named entity recognition and relation extraction. The problem is that such integration using a generative approach may impose big challenges such as the need of a large annotated corpus. To overcome the data sparseness problem, generative models typically apply some smoothing techniques to integrate different scales of contexts in parameter estimation, e.g. the back-off approach in Miller et al (2000).

---

[1] Here, we classify this paper into feature-based approaches since the feature space in the kernels of Zhao et al (2005) can be easily represented by an explicit feature vector.

Zelenko et al (2003) proposed extracting relations by computing kernel functions between parse trees. Culotta et al (2004) extended this work to estimate kernel functions between augmented dependency trees and achieved the F-measure of 45.8 on the 5 relation types in the ACE RDC 2003 corpus[2]. Bunescu et al (2005) proposed a shortest path dependency kernel. They argued that the information to model a relationship between two entities can be typically captured by the shortest path between them in the dependency graph. It achieved the F-measure of 52.5 on the 5 relation types in the ACE RDC 2003 corpus. Zhang et al (2005) adopted a clustering algorithm in unsupervised relation extraction using tree kernels. To overcome the data sparseness problem, various scales of sub-trees are applied in the tree kernel computation. Although tree kernel-based approaches are able to explore the huge implicit feature space without much feature engineering, further research work is necessary to make them effective and efficient.

Comparably, feature-based approaches achieved much success recently. Kambhatla (2004) employed maximum entropy models with various features derived from word, entity type, mention level, overlap, dependency tree, parse tree and achieved the F-measure of 52.8 on the 24 relation subtypes in the ACE RDC 2003 corpus. Zhao et al (2005) combined various kinds of knowledge from tokenization, sentence parsing and deep dependency analysis through SVM and achieved the F-measure of 70.1 on the 7 relation types of the ACE RDC 2004 corpus[3]. Zhou et al (2005) further systematically explored diverse lexical, syntactic and semantic features through SVM and achieved F-measure of 68.1 and 55.5 on the 5 relation types and the 24 relation subtypes in the ACE RDC 2003 corpus respectively. To overcome the data sparseness problem, feature-based approaches normally incorporate various scales of contexts into the feature vector extensively. These approaches then depend on adopted learning algorithms to weight and combine each feature effectively. For example, an exponential model and a linear model are applied in the maximum entropy models and support vector machines respectively to combine each feature via the learned weight vector.

As discussed above, although various approaches have been employed in relation extraction, they only apply some implicit approaches to overcome the data sparseness problem. Until now, there are no explicit ways to resolve the data sparseness problem in relation extraction. Currently, all the current approaches apply the flat learning strategy which equally treats training examples in different relations independently and ignore the commonality among different relations. This paper proposes a novel hierarchical learning strategy to resolve this problem by considering the relatedness among different relations and capturing the commonality among related relations. By doing so, the data sparseness problem can be well dealt with and much better performance can be achieved, especially for those relations with small or medium amounts of annotated examples.

3 Hierarchical Learning Strategy

Traditional classifier learning approaches apply the flat learning strategy. That is, they equally treat training examples in different classes independently and ignore the

[2] The ACE RDC 2003 corpus defines 5/24 relation types/subtypes between 4 entity types.

[3] The ACE RDC 2004 corpus defines 7/23 relation types/subtypes between 7 entity types.

commonality among related classes. The flat strategy will not cause any problem when there are a large amount of training examples for each class, since, in this case, a classifier learning approach can always learn a nearly optimal discriminative function for each class against the remaining classes. However, such flat strategy may cause big problems when there is only a small amount of training examples for some of the classes. In this case, a classifier learning approach may fail to learn a reliable (or nearly optimal) discriminative function for a class with a small amount of training examples, and, as a result, may significantly affect the performance of the class and even the overall performance.

To overcome the inherent problems in the flat learning strategy, this paper proposes a hierarchical learning strategy which explores the inherent commonality among related classes through a class hierarchy. In this way, the training examples of related classes can help in learning a reliable discriminative function for a class with only a small amount of training examples in a top-down way. Here, the state-of-the-art support vector machines (SVM) is applied as the basic classifier learning approach. For SVM, the guidance is done by, when training a lower-level class, discounting those negative training instances which do not belong to the support vectors of the upper-level class.

In the following, we will first describe SVM, followed by the hierarchical learning strategy using SVM. Finally, we will consider several ways in building the class hierarchy.

3.1 Support Vector Machines

Support Vector Machines (SVM) is a supervised machine learning technique motivated by the statistical learning theory (Vapnik 1998). Based on the structural risk minimization of the statistical learning theory, SVM seeks an optimal separating hyper-plane to divide the training examples into two classes and make decisions based on support vectors which are selected as the only effective instances in the training set.

Basically, SVM is only for binary classification. Therefore, we must extend it to multi-class classification, such as the ACE RDC task. For efficiency, we apply the *one vs. others* strategy, which builds K classifiers so as to separate one class from all others. Moreover, we map the SVM output into the probability by using an additional sigmoid model:

$$p(y=1 \mid f) = \frac{1}{1+\exp(Af+B)} \tag{1}$$

where f is the SVM output and the coefficients A & B are to be trained using the model trust alorithm as described in Platt (1999). The final decision of an instance in multi-class classification is determined by the class which has the maximal probability from the corresponding SVM.

The reason why we choose SVM is that SVM represents the state-of–the-art in the machine learning community and supports both feature-based learning and kernel-based learning. Moreover, there are good implementations of the algorithm available.

In this paper, we use the binary-class SVMLight[4] deleveloped by Joachims (1998). By default, the simple linear kernel is applied unless specified.

3.2 Hierarchical Learning Strategy Using Support Vector Machines

Assume we have a class hierarchy for a task, e.g. the one in the ACE RDC 2003 corpus as shown in Table 1 of Section 4.1. The hierarchical learning strategy explores the inherent commonality among related classes in a top-down way. For SVM, the guidance of related classes in training a discriminative function for a given class is done by discounting those negative training instances which do not belong to the support vectors of the upper-level class. That is, the lower-level discriminative function has the preference toward the discriminative function of its upper-level class. Here, all the positive relation examples (i.e. those under the "YES" relation) are always kept since their number is always much less that of the negative relation examples (i.e. those under the "NON" relation) in most classification tasks, such as the ACE RDC task using the ACE RDC 2003 and 2004 corpora.

For an example, let's look at the training of the "Spouse" relation subtype in the class hierarchy as shown in Table 1:

- Train a SVM for the "YES" relation vs. the "NON" relation.
- Train a SVM for the "SOCIAL" relation type vs. all the remaining relation types (including the "NON" relation) by discounting the negative relation examples, which do not belong to the support vectors of the upper level SVM classifier for the "YES" relation vs. the "NON" relation, and keep the remaining examples.
- Train a SVM for the "Spouse" relation subtype vs. all the remaining relation subtypes under all the relation types (including the "NON" relation) by discounting the negative relation examples, which do not belong to the support vectors of the upper level SVM classifier for the "SOCIAL" relation vs. all the remaining relation types (including the "NON" relation) , and keep the remaining examples.
- Return the above trained SVM as the classifier for the "Spouse" relation subtype.

Please note: considering the argument order of the two mentions, one more step is necessary to differentiate the two different argument orders for a non-symmetric relation subtype. For this issue, please see Section 4.1 for more details.

In the case of using SVM as the basic classifier learning approach, the support vectors of a given class are much biased to the ones of the upper class which covers the given class and its related classes due to discounting negative relation examples which do not belong to the support vectors of the upper level SVM classifier. In some sense, related classes help by discounting the effect of un-related classes when determining the support vectors for a given class. As a result, this can largely reduce the noisy effect. Moreover, this can largely reduce the search space in training and make SVM training converge much faster. Our experimentation shows that similar

---

4 Joachims has just released a new version of SVMLight for multi-class classification. However, this paper only uses the binary-class version. For details about SVMLight, please see http://svmlight.joachims.org/

performance is achieved when the discounting weight ranges from 0.1 to 0.3. Therefore, the discounting weight is set to 0.2 throughout the paper. That is, the cost of a discounted training example is multiplied by 0.2.

In this way, the training examples in different classes are not treated independently any more, and the commonality among related classes can be captured via the hierarchical learning strategy. The intuition behind this strategy is that the upper-level class normally has more positive training examples than the lower-level class so that the corresponding discriminative function can be determined more reliably. In this way, the training examples of related classes can help in learning a reliable discriminative function for a class with only a small amount of training examples in a top-down way and thus alleviate its data sparseness problem.

3.3 Building the Class Hierarchy

We have just described the hierarchical learning strategy using a given class hierarchy. Normally, a rough class hierarchy can be given manually according to human intuition, such as the one in the ACE RDC 2003 corpus. In order to explore more commonality among sibling classes, we make use of binary hierarchical clustering for sibling classes at both lowest and all levels. This can be done by first using the flat learning strategy to learn the discriminative functions for individual classes and then iteratively combining the two most related classes using the cosine similarity function between their discriminative functions, e.g. using the support vector sets in SVM, in a bottom-up way. The intuition is that related classes should have similar hyper-planes to separate from other classes and thus have similar sets of support vectors in SVM.

- **Lowest-level hybrid:** Binary hierarchical clustering is only done at the lowest level (i.e. the relation subtype level) while keeping the upper-level class hierarchy. That is, only sibling classes at the lowest level are hierarchically clustered.
- **All-level hybrid:** Binary hierarchical clustering is done at all levels in a bottom-up way. That is, sibling classes at the lowest level are hierarchically clustered first and then sibling classes at the upper-level. In this way, the binary class hierarchy can be built iteratively in a bottom-up way.

4 Experimentation

This paper focuses on the ACE RDC task with evaluation on the ACE RDC 2003 and 2004 corpora.

4.1 Experimental Setting

Evaluation is mainly done on the ACE RDC 2003 corpus. The training data consists of 674 documents (~300k words) with 9683 relation examples while the held-out testing data consists of 97 documents (~50k words) with 1386 relation examples. Table 1 lists various types and subtypes of relations for the corpus, along with their occurrence frequencies in the training data. It shows that this corpus suffers from a small amount of annotated data for a few subtypes such as the subtype "Founder" under the type "ROLE". Here, we also divide various relation subtypes into three bins: large/middle/small, according to their training data sizes. For this corpus, 400 is used as as the lower threshold for the large bin and 200 as the upper threshold for the small bin[5]. As a result, the large/medium/small bin includes 5/8/11 relation subtypes in the corpus, respectively.

Table 1. Relation statistics in the training data of the ACE RDC 2003 corpus

Type	Subtype	Freq	Bin Type
AT	Based-In	347	Medium
	Located	2126	Large
	Residence	308	Medium
NEAR	Relative-Location	201	Medium
PART	Part-Of	947	Large
	Subsidiary	355	Medium
	Other	6	Small
ROLE	Affiliate-Partner	204	Medium
	Citizen-Of	328	Medium
	Client	144	Small
	Founder	26	Small
	General-Staff	1331	Large
	Management	1242	Large
	Member	1091	Large
	Owner	232	Medium
	Other	158	Small
SOCIAL	Associate	91	Small
	Grandparent	12	Small
	Other-Personal	85	Small
	Other-Professional	339	Medium
	Other-Relative	78	Small
	Parent	127	Small
	Sibling	18	Small
	Spouse	77	Small

Detailed evaluation has been also done on the ACE RDC 2004 corpus, which contains 451 documents and 5702 relation instances. For comparison with Zhao et al (2005), we only use the same 348 nwire and bnews documents, which contain 125k words and 4400 relation instances. Evaluation is done using 5-fold cross-validation and shows similar tendency with the ACE RDC 2003 corpus. To avoid redundancy, we will only report the final performance on the ACE RDC 2004 corpus.

[5] A few minor relation subtypes only have very few examples in the testing set. The reason to choose this threshold is to guarantee a reasonable number of testing examples in the small bin. For the ACE RC 2003 corpus, using 200 as the upper threshold will fill the small bin with about 100 testing examples while using 100 will include too few testing examples for reasonable performance evaluation.

In this paper, we adopt the same feature set as applied in Zhou et al (2005): word, entity type, mention level, overlap, base phrase chunking, dependency tree, parse tree and semantic information. By default, we use the SVM with the simple linear kernel unless specified. Moreover, we also explicitly model the argument order of the two mentions involved. For example, when comparing mentions m1 and m2, we distinguish between m1-ROLE.Citizen-Of-m2 and m2-ROLE.Citizen-Of-m1. Note that, in the ACE RDC 2003 task, 6 of these 24 relation subtypes are symmetric: "NEAR.Relative-Location", "SOCIAL.Associate", "SOCIAL.Other-Relative", "SOCIAL.Other-Professional", "SOCIAL.Sibling", and "SOCIAL.Spouse". In this way, we model relation extraction in the ACE RDC 2003 task as a multi-class classification task with 43 (24X2-6+1) classes, two for each relation subtype (except the above 6 symmetric subtypes) and a "NONE" class for the case where the two mentions are not related. For the ACE RDC 2004 task, 6 of these 23 relation subtypes are symmetric: "PHYS.Near", "PER-SOC.Business", "PER-SOC.Family", "PER-SOC.Other", "EMP-ORG.Partner", and "EMP-ORG.Other". In this way, we model relation extraction in the ACE RDC 2004 task as a multi-class classification task with 41 (23X2-6+1) classes, two for each relation subtype (except the above 6 symmetric subtypes) and a "NONE" class for the case where the two mentions are not related.

4.2 Experimental Results

Table 2 compares the hierarchical learning strategy and the flat learning strategy with the existing class hierarchy on the ACE RDC 2003 corpus using SVM. It shows that the hierarchical strategy outperforms the flat strategy by 1.9 (57.4 vs. 55.5) in F-measure, largely due to its gain in recall. As a result, the hierarchical strategy achieves the F-measure of 57.4 using SVM.

Table 2. Comparison of the hierarchical learning strategy vs. the flat learning strategy using the existing class hierarchy on the ACE RDC 2003 corpus

Strategy	P%	R%	F
Flat	63.1	49.5	55.5
Hierarchical	63.7	52.3	57.4

Table 3. Comparison of the hierarchical learning strategy using different class hierarchies on the ACE RDC 2003 corpus

Class Hierarchy	P%	R%	F
Existing	63.7	52.3	57.4
Entirely Automatic	63.5	52.5	57.4
Lowest-level Hybrid (based on the existing one)	64.1	52.8	57.9
All-level Hybrid (based on the existing one)	64.3	52.9	58.0

Table 3 compares the performance of the hierarchical learning strategy using different class hierarchies on the ACE RDC 2003 corpus. It shows that, the lowest-level hybrid approach, which only automatically updates the existing class hierarchy at the lowest level, slightly improves the performance by 0.5 (57.9 vs. 57.4) in F-measure while further updating the class hierarchy at upper levels in the all-level hybrid approach only has very slight effect (58.0 vs. 57.9) in F-measure. This is largely due to the fact that the major data sparseness problem occurs at the lowest level, i.e. the relation subtype level in the ACE RDC 2003 corpus. As a result, the hierarchical learning strategy using the class hierarchy built with the all-level hybrid approach achieves the F-measure of 58.0 using SVM, which outperforms the flat learning strategy by 2.5 (58.0 vs. 55.5) in F-measure. In order to justify the usefulness of our hierarchical learning strategy when a rough class hierarchy is not available and difficult to determine manually, we also experiment using entirely automatically built class hierarchy without considering the existing class hierarchy. Table 3 shows that using automatically built class hierarchy performs comparably (63.7%/52.3%/57.4 vs. 63.5%/52.5%/57.4) with using only the existing one.

Table 4. Comparison of the hierarchical and flat learning strategies on the relation subtypes of different training data sizes using SVM. Notes: the figures in the parentheses indicate the cosine similarities between the support vectors sets of the discriminative functions learned using the two strategies.

Bin Type	Large Bin (0.95)			Middle Bin (0.90)			Small Bin (0.76)		
(Similarity)	P%	R%	F	P%	R%	F	P%	R%	F
Flat	68.6	57.3	62.4	67.8	34.7	45.9	35.0	24.7	29.0
Hierarchical	69.4	58.5	63.4	68.2	41.8	51.8	42.3	31.2	35.9

With the major goal of resolving the data sparseness problem for the classes with a small amount of training examples, Table 4 compares the hierarchical and flat learning strategies on the relation subtypes of different training data sizes using SVM on the ACE RDC 2003 corpus. Here, we divide various relation subtypes into three bins: large/middle/small, according to their available training data sizes. Please see Table 1 for details. Table 4 shows that the hierarchical learning strategy outperforms the flat learning strategy by 1.0(63.4 vs. 62.4)/5.9(51.8 vs. 45.9)/6.9(35.9 vs. 29.0) in F-measure on the large/middle/small bin respectively. This indicates that the hierarchical learning strategy performs much better than the flat learning strategy for those classes with a small or medium amount of annotated examples although the hierarchical learning strategy only performs slightly better by 1.0 than the flat learning strategy on those classes with a large size of annotated corpus. This suggests that the proposed hierarchical learning strategy can well deal with the data sparseness problem in the ACE RDC 2003 corpus using SVM.

An interesting question is about the similarity between the discriminative functions learned using the hierarchical and flat learning strategies. Table 4 compares the similarities between the weighted support vector sets of the discriminative functions for the two strategies and different bins. Since different relation subtypes contain quite different numbers of training examples, the similarities are weighted by the training example numbers of different relation subtypes. Table 4 shows that the discriminative functions learned using the two strategies are very similar (with the similarity about 0.95) for the relation subtypes belonging to the large bin, while there exist quite differences between the discriminative functions learned using the two strategies for the relation subtypes belonging to the medium/small bin with the similarity about 0.90/0.76 respectively. This means that the use of the hierarchical learning strategy over the flat learning strategy only has very slight change on the discriminative functions for those classes with a large amount of annotated examples while its effect on those with a small amount of annotated examples is obvious. This contributes to and explains (the degree of) the performance difference between the two strategies on the different training data sizes as shown in Table 4.

Due to the difficulty of building a large annotated corpus, another interesting question is about the adaptability of the hierarchical learning strategy and its comparison with the flat learning strategy. Figure 1 shows the effect of different training data sizes for some major relation subtypes while keeping all the training examples of remaining relation subtypes on the ACE RDC 2003 corpus using SVM. It shows that the hierarchical learning strategy performs much better than the flat learning strategy when only a small amount of training examples is available. It also shows that the hierarchical learning strategy can achieve stable performance much faster than the flat learning strategy. Finally, it shows that the ACE RDC 2003 corpus suffer from the lack of training examples. Among the three major relation subtypes, only the subtype "Located" achieves steady performance.

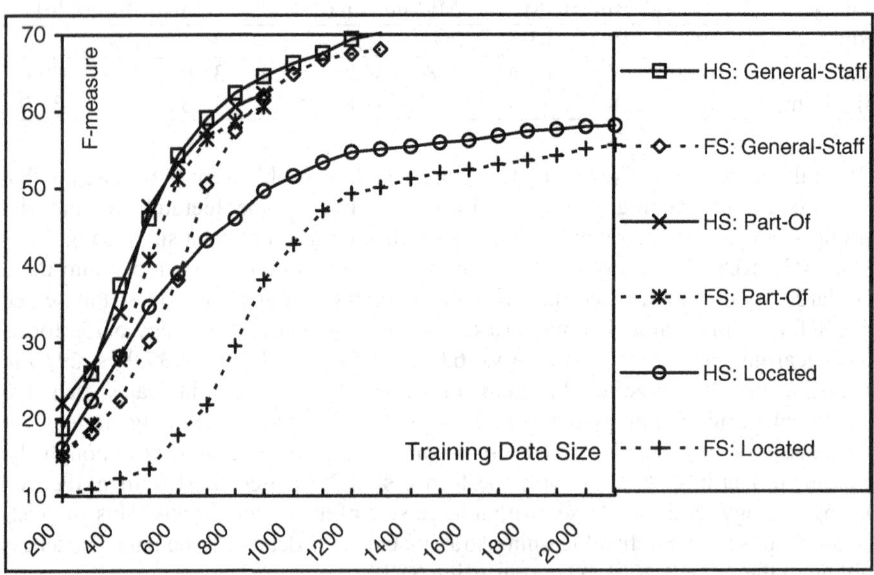

Fig. 1. Comparison on adaptability of the hierarchical learning strategy vs. the flat learning strategy for some major relation subtypes on the ACE RDC 2003 corpus (Note: FS for the flat learning strategy and HS for the hierarchical learning strategy)

Finally, we also compare our system with the previous best-reported systems, such as Zhou et al (2005) and Zhao et al (2005) on the ACE RDC 2003 and 2004 corpora respectively. For the ACE RDC 2003 corpus, Table 6 shows that our system with the hierarchical learning strategy outperforms the previous best-reported system by 2.5

(58.0 vs. 55.5) in F-measure using SVM with linear kernel, in the extraction of the 24 relation subtypes. Moreover, we also evaluate our system using the polynomial kernel with degree d=2 in SVM. It shows that this can further improve the F-measure by 2.2 (60.2 vs. 58.0) in the extraction of 24 relation subtypes. As a result, our system significantly outperforms Zhou et al (2005) by 4.7 (60.2 vs. 55.5) in F-measure on the 24 relation subtypes. For the ACE RDC 2004 corpus, Table 6 shows that our system with the hierarchical learning strategy outperforms the previous best-reported system by 0.5 (70.8 vs. 70.3) in F-measure using SVM with linear kernel on the 7 relation types. Moreover, we also evaluate our system using the polynomial kernel with degree d=2 in SVM. It shows that this can further improve the F-measure by 2.1 (64.9 vs. 62.8) on the 23 relation subtypes. As a result, our system significantly outperforms Zhao et al (2005) by 2.3 (72.6 vs. 70.3) in F-measure on the 7 relation types.

Table 6. Comparison of our system with other best-reported systems (The figures outside/inside the parentheses are for extraction of the relation types/subtype)

System	2003			2004		
	P%	R%	F	P%	R%	F
Ours: Hierarchical+SVM (linear)	64.3 (82.2)	52.9 (58.2)	58.0 (68.1)	74.2 (82.6)	54.5 (62.0)	62.8 (70.8)
Ours: Hierarchical+SVM (polynomial)	70.8 (82.8)	52.3 (59.3)	60.2 (69.1)	74.8 (83.4)	57.3 (64.1)	64.9 (72.6)
Zhou et al(2005): Flat+SVM(linear)	63.1 (77.2)	49.5 (60.7)	55.5 (68.0)	- (-)	- (-)	- (-)
Zhao et al(2005):Flat+SVM (composite polynomial)	- (-)	- (-)	- (-)	- (69.2)	- (70.5)	- (70.3)

5 Conclusion

This paper proposes a novel hierarchical learning strategy to deal with the data sparseness problem in relation extraction by modeling the commonality among related classes. For each class in a class hierarchy, a discriminative function is determined in a top-down way. In this way, the upper-level discriminative function can effectively guide the lower-level discriminative function learning. In this paper, the state-of-the-art Support Vector Machines is applied as the basic classifier learning approach using the hierarchical learning strategy and the guidance is done by, when training a lower-level class, discounting those negative training instances which do not belong to the support vectors of the upper-level class. Evaluation on the ACE RDC 2003 and 2004 corpora shows that the hierarchical learning strategy performs much better than the flat learning strategy in resolving the critical data sparseness problem in relation extraction.

In the future work, we will explore the hierarchical learning strategy using other guidance principle. Moreover, just as indicated in Figure 1, most relation subtypes in the ACE RDC 2003 corpus (arguably the largest annotated corpus in relation extraction) suffer from the lack of training examples. Therefore, a critical research in relation extraction is how to rely on semi-supervised learning approaches (e.g. bootstrap) to alleviate its dependency on a large amount of annotated training examples and achieve better and steadier performance.

References

ACE. (2000-2005). Automatic Content Extraction. http://www.ldc.upenn.edu/Projects/ACE/

Bunescu R. & Mooney R.J. (2005). A shortest path dependency kernel for relation extraction. *HLT/EMNLP'2005*: 724-731. 6-8 Oct 2005. Vancover, B.C.

Collins M. (1999). Head-driven statistical models for natural language parsing. *Ph.D. Dissertation*, University of Pennsylvania.

Culotta A. and Sorensen J. (2004). Dependency tree kernels for relation extraction. *ACL'2004*. 423-429. 21-26 July 2004. Barcelona, Spain.

Miller G.A. (1990). WordNet: An online lexical database. *International Journal of Lexicography*. 3(4):235-312.

Miller S., Fox H., Ramshaw L. and Weischedel R. (2000). A novel use of statistical parsing to extract information from text. *ANLP'2000*. 226-233. 29 April - 4 May 2000, Seattle, USA

MUC-7. (1998). *Proceedings of the 7th Message Understanding Conference (MUC-7)*. Morgan Kaufmann, San Mateo, CA.

Kambhatla N. (2004). Combining lexical, syntactic and semantic features with Maximum Entropy models for extracting relations. *ACL'2004(Poster)*. 178-181. 21-26 July 2004. Barcelona, Spain.

Platt J. (1999). Probabilistic Outputs for Support Vector Machines and Comparisions to regularized Likelihood Methods. In *Advances in Large Margin Classifiers*. Edited by Smola .J., Bartlett P., Scholkopf B. and Schuurmans D. MIT Press.

Zelenko D., Aone C. and Richardella. (2003). Kernel methods for relation extraction. *Journal of Machine Learning Research*. 3(Feb):1083-1106.

Zhang M., Su J., Wang D.M., Zhou G.D. and Tan C.L. (2005). Discovering Relations from a Large Raw Corpus Using Tree Similarity-based Clustering, *IJCNLP'2005, Lecture Notes in Computer Science (LNCS 3651)*. 378-389. 11-16 Oct 2005. Jeju Island, South Korea.

Zhao S.B. and Grisman R. (2005). Extracting relations with integrated information using kernel methods. ACL'2005: 419-426. Univ of Michgan-Ann Arbor, USA, 25-30 June 2005.

Zhou G.D., Su J. Zhang J. and Zhang M. (2005). Exploring various knowledge in relation extraction. *ACL'2005*. 427-434. 25-30 June, Ann Arbor, Michgan, USA.

Learning to Separate Text Content and Style for Classification

Dell Zhang[1] and Wee Sun Lee[2]

[1] School of Computer Science and Information Systems
Birkbeck, University of London
London WC1E 7HX, UK
dell.z@ieee.org

[2] Department of Computer Science and Singapore-MIT Alliance
National University of Singapore
Singapore 117543
leews@comp.nus.edu.sg

Abstract. Many text documents naturally have two kinds of labels. For example, we may label web pages from universities according to their categories, such as "student" or "faculty", or according the source universities, such as "Cornell" or "Texas". We call one kind of labels the content and the other kind the style. Given a set of documents, each with both content and style labels, we seek to effectively learn to classify a set of documents in a new style with no content labels into its content classes. Assuming that every document is generated using words drawn from a mixture of two multinomial component models, one content model and one style model, we propose a method named *Cartesian EM* that constructs content models and style models through Expectation Maximization and performs classification of the unknown content classes transductively. Our experiments on real-world datasets show the proposed method to be effective for *style independent text content classification.*

1 Introduction

Text classification [1] is a well established area of machine learning. A text classifier is first trained using documents with pre-assigned class labels and then offered test documents for which it must guess the best class labels.

We identify and address a special kind of text classification problem where every document is associated with a pair of independent labels (c_i, s_j) where $c_i \in C = \{c_1, ..., c_m\}$ and $s_j \in S = \{s_1, ..., s_n\}$. In other words, the label space is the Cartesian product of two independent sets of labels, $C \times S$, as shown in Figure 1. This problem setting extends the standard one-dimensional (1D) label space to two-dimensions (2D). Following the terminology of computational cognitive science and pattern recognition [2], we call C content labels and S style labels. Given a set of labeled training documents in $n-1$ styles $\{s_1, ..., s_{n-1}\}$ and a set of test documents in a new style s_n, how should computers learn a classifier to predict the content class for each test document?

H.T. Ng et al. (Eds.): AIRS 2006, LNCS 4182, pp. 79–91, 2006.

This machine learning problem is less explored, yet occurs frequently in practice. For example, consider a task of classifying academic web pages into several categories (such as "faculty" and "student"): one may have labeled pages from several universities (such as "Cornell", "Texas", and "Washington") and need to classify pages from another university (such as "Wisconsin"), where the categories can be considered as content classes and the universities can be regarded as style types. Other examples include: learning to classify articles from a new journal; learning to classify papers from a new author; learning to classify customer comments for a new product; learning to classify news or messages in a new period, and so on. The general problem is the same whenever we have a two (or more) dimensional label for each instance.

	s_1	s_2	...	s_{n-1}	s_n
c_1					
c_2					
$\vdots$					
c_{m-1}					
c_m					

Fig. 1. The Cartesian product label space

Since we care about the difference among content classes but not the difference among style types, it could be beneficial to separate the text content and style so that the classifier can focus on the discriminative content information but ignore the distractive style information. However, existing approaches to this problem, whether inductive or transductive, simply discard the style labels. Assuming that every document is generated using words drawn from a mixture of two multinomial component models, one content model and one style model, we propose a new method named *Cartesian EM* that constructs content models and style models through Expectation-Maximization [3] and performs classification transductively [4]. Our experiments on real-world datasets show that the proposed method can not only improve classification accuracy but also provide deeper insights about the data.

2 Learning with 1D Labels

2.1 Naïve Bayes

Naïve Bayes (NB) is a popular supervised learning algorithm for probabilistic text classification [5]. Though very simple, NB is competitively effective and highly efficient [6].

Given a set of labeled training documents $D = \{d_1, ..., d_{|D|}\}$, NB fits a generative model which is parameterized by θ and then applies it to classify unlabeled test documents. The generative model of NB assumes that a document d is generated by first choosing its class $c_i \in C = \{c_1, \ldots, c_m\}$ according to a prior distribution of classes, and then producing its words independently according to a multinomial distribution of terms conditioned on the chosen class [7].

The prior class probability $\Pr[c_i]$ can be estimated by

$$\Pr[c_i] = \left(\sum_{d \in D} \Pr[c_i \mid d]\right) \Big/ |D|, \tag{1}$$

where $\Pr[c_i \mid d]$ for a labeled document d is 1 if d is in class c_i or 0 otherwise.

With the "naïve" assumption that all the words in the document occur independently given the class, the conditional document probability $\Pr[d \mid c_i]$ can be estimated by

$$\Pr[d \mid c_i] = \prod_{w_k \in d} \left(\Pr[w_k \mid c_i]\right)^{n(w_k, d)}, \tag{2}$$

where $n(w_k, d)$ is the number of times word w_k occurs in document d. The conditional word probability $\Pr[w_k \mid c_i]$ can be estimated by

$$\Pr[w_k \mid c_i] = \frac{\eta + \sum_{d \in D} n(w_k, d) \Pr[c_i \mid d]}{\sum_{w \in V} \left(\eta + \sum_{d \in D} n(w, d) \Pr[c_i \mid d]\right)}, \tag{3}$$

where V is the vocabulary, and $0 < \eta \leq 1$ is the Lidstone's smoothing parameter [6].

Given a test document d, NB classifies it into class

$$\hat{c}(d) = \arg\max_{c_i \in C} \Pr[c_i \mid d]. \tag{4}$$

The posterior class probability $\Pr[c_i \mid d]$ for an unlabeled document d can be computed via Bayes's rule:

$$\Pr[c_i \mid d] = \frac{\Pr[d \mid c_i] \Pr[c_i]}{\Pr[d]} = \frac{\Pr[d \mid c_i] \Pr[c_i]}{\sum_{c \in C} \Pr[d \mid c] \Pr[c]}. \tag{5}$$

2.2 1D EM

The parameter set θ of the above 1D generative model includes $\Pr[c_i]$ for all $c_i \in C$ and $\Pr[w_k \mid c_i]$ for all $w_k \in V$, $c_i \in C$. If the set of labeled documents is not large enough, NB may not be able to give a good estimation of θ therefore the classification accuracy may suffer. However, it has been shown that the model estimation could be improved by making use of additional unlabeled documents (such as the test documents). This is the idea of semi-supervised learning, i.e., learning from both labeled data and unlabeled data.

NB can be extended to semi-supervised learning by incorporating a set of unlabeled documents through Expectation-Maximization [8, 9]. The principle of

```
Initialization: estimate θ from the labeled documents only,
using equations (1) and (3).
while ( θ has not converged ) {
        E-step: calculate the probabilistic class labels for
        the unlabeled documents based on the current θ, using
        equation (5).
        M-step: re-estimate θ from both the labeled documents
        and the unlabeled documents that have been assigned
        probabilistic class labels, using equations (1) and
        (3).
}
Classify the unlabeled documents using equation (4).
```

Fig. 2. The 1D EM algorithm

Expectation-Maximization (EM) will be further explained later in the next section. Since this EM based method is designed for the standard 1D label space, we call it *1D EM*. Please refer to [8, 9] for its detailed derivation.

3 Learning with 2D Labels

3.1 Cartesian Mixture Model

Let's consider the Cartesian product label space $C \times S$, where $C = \{c_1, ..., c_m\}$ and $S = \{s_1, ..., s_n\}$. In this 2D label space, every document $d \in D$ is associated with a pair of independent labels (c_i, s_j) where $c_i \in C$ and $s_j \in S$.

We introduce a 2D generative model, *Cartesian mixture model*, for this problem of learning with 2D labels. It naturally assumes that each content class $c_i \in C$ or style type $s_j \in S$ corresponds to a multinomial model. A document d is generated by first choosing its content class c_i and style type s_j according to a prior distribution of label-pairs, and then producing its words independently according to a mixture of two component models that correspond to c_i and s_j respectively. That is to say, every specific occurrence of word $w_k \in V$ in a (c_i, s_j) document is generated from either the content model $\Pr[w_k | c_i]$ or the style model $\Pr[w_k | s_j]$, though we do not know which one is actually responsible. Therefore the probability of a word $w_k \in V$ to occur in (c_i, s_j) documents can be calculated by

$$\Pr[w_k | c_i, s_j] = \lambda \Pr[w_k | c_i] + (1 - \lambda) \Pr[w_k | s_j], \tag{6}$$

where $\lambda \in [0,1]$ is a parameter used for weighting the component models. In this paper, the same weighting parameter λ is used for all label-pairs, but our method can be easily extended to allow every label-pair have its own weighting parameter.

Since the content label and the style label are independent, we have

$$\Pr[c_i, s_j] = \Pr[c_i]\Pr[s_j]. \tag{7}$$

The prior content class probability $\Pr[c_i]$ can still be estimated using equation (1). The prior style type probability $\Pr[s_j]$ can be estimated similarly by

$$\Pr[s_j] = \left(\sum\nolimits_{d \in D} \Pr[s_j \mid d]\right) \Big/ |D|, \tag{8}$$

where $\Pr[s_j \mid d]$ for a labeled document d is 1 if d is in style s_j or 0 otherwise.

Let $|d|$ denote the length of document d, and $o_{p,d}$ denote the word that occurs in the *p*-th position of document d. By imposing the extended "naïve" assumption that all the words in the document occur independently given the content class and the style type, the conditional document probability $\Pr[d \mid c_i, s_j]$ can be estimated by

$$\Pr[d \mid c_i, s_j] = \prod\nolimits_{p=1}^{|d|} \Pr[o_{p,d} \mid c_i, s_j] = \prod\nolimits_{w_k \in d} \left(\Pr[w_k \mid c_i, s_j]\right)^{n(w_k, d)}, \tag{9}$$

where $n(w_k, d)$ is the number of times word w_k occurs in document d. The conditional word probability $\Pr[w_k \mid c_i, s_j]$ given by equation (6) involves λ, $\Pr[w_k \mid c_i]$ and $\Pr[w_k \mid s_j]$ whose estimation will be discussed later.

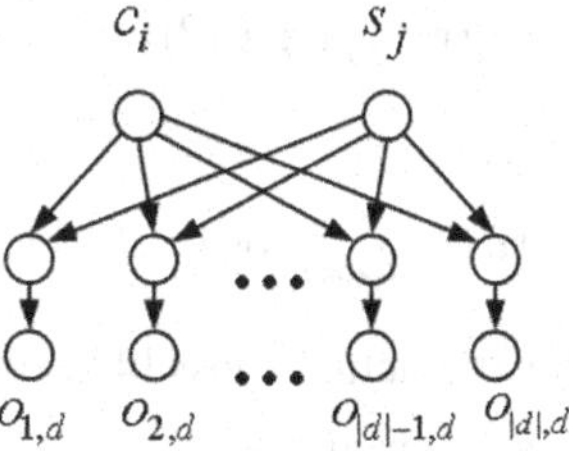

Fig. 3. The graphical model representation of a document as a Bayesian network

In our problem setting (§1), the training documents are fully labeled, but the test documents are only half-labeled (we only know that the test document is in style s_n but we do not know which content class it belongs to). Given a test document d whose style is known to be s_n, we can predict its content class to be

$$\hat{c}(d) = \arg\max_{c_i \in C} \Pr[c_i \mid d, s_n]. \tag{10}$$

The posterior class probability $\Pr[c_i \mid d, s_n]$ for an half-labeled document d in style s_n can be calculated using Bayes's rule and equation (7):

$$\Pr[c_i \mid d, s_n] = \frac{\Pr[d \mid c_i, s_n]\Pr[c_i, s_n]}{\Pr[d \mid s_n]\Pr[s_n]} = \frac{\Pr[d \mid c_i, s_n]\Pr[c_i]}{\sum_{c \in C} \Pr[d \mid c, s_n]\Pr[c]}. \tag{11}$$

3.2 Cartesian EM

The parameter set θ of the above Cartesian mixture model includes $\Pr[c_i]$ for all $c_i \in C$, $\Pr[s_j]$ for all $s_j \in S$, $\Pr[w_k \mid c_i]$ for all $w_k \in V$, $c_i \in C$, $\Pr[w_k \mid s_j]$ for all $w_k \in V$, $s_j \in S$, and λ. One difficulty to estimate θ is that we would not be able to get the values of λ, $\Pr[w_k \mid c_i]$ and $\Pr[w_k \mid s_j]$ by using maximum likelihood estimation in a straightforward manner, because for every observed word occurrence we do not know exactly whether the content model or the style model generated it. Furthermore, we need to take the test documents into consideration while estimating θ (at least $\Pr[w_k \mid s_j]$), but for every test document we do not know exactly which content class it comes from.

The Expectation-Maximization (EM) algorithm [3] is a general algorithm for maximum likelihood estimation when the data is "incomplete". In this paper, we propose a new EM based method named *Cartesian EM* that constructs the Cartesian mixture model and applies it to predict the content classes of the test documents in a new style. Cartesian EM actually belongs to the family of transductive learning [4], a special kind of semi-supervised learning that the learning algorithm can see the set of test examples and make use of them to improve the classification accuracy on them.

A common method for estimating the model θ is maximum likelihood estimation in which we choose a θ that maximizes its likelihood (or equivalently log-likelihood) for the observed data (in our case the set of documents D and their associated labels):

$$\hat{\theta} = \arg\max_\theta L(\theta) = \arg\max_\theta \log L(\theta) = \arg\max_\theta \log \Pr[D \mid \theta] . \tag{12}$$

Given the model θ, we have

$$\log \Pr[D \mid \theta] = \sum_{d \in D} \log \Pr[d, c(d), s(d)] , \tag{13}$$

where $c(d)$ and $s(d)$ stand for the actual (possibly unknown) content label and the actual style label of document d respectively.

For a fully-labeled document d, we know that $\Pr[c_i \mid d]$ is 1 if d is in content class c_i or 0 otherwise and $\Pr[s_j \mid d]$ is 1 if d is in style type s_j or 0 otherwise, therefore we can calculate the log-likelihood of d given the model θ using Bayes's rule and equation (7):

$$\begin{aligned}\log \Pr[d, c(d), s(d)] &= \sum_{i=1}^{m} \sum_{j=1}^{n} \Pr[c_i \mid d] \Pr[s_j \mid d] \log \Pr[d \mid c_i, s_j] + \\ &\sum_{i=1}^{m} \sum_{j=1}^{n} \Pr[c_i \mid d] \Pr[s_j \mid d] \log \Pr[c_i] + \sum_{i=1}^{m} \sum_{j=1}^{n} \Pr[c_i \mid d] \Pr[s_j \mid d] \log \Pr[s_j]\end{aligned} . \tag{14}$$

Furthermore, using equation (6) and equation (9), we get

$$\log \Pr[d \mid c_i, s_j] = \sum_{p=1}^{|d|} \log\left(\lambda \Pr[o_{p,d} \mid c_i] + (1-\lambda) \Pr[o_{p,d} \mid s_j]\right) . \tag{15}$$

Now we see that there are two obstacles to estimating θ: one is that the test documents are only half-labeled hence equation (14) is not applicable to their log-likelihood computation, the other is that equation (15) contains logarithms of sums hence hard to maximize.

The basic idea of the EM algorithm is to augment our "incomplete" observed data with some latent/hidden variables so that the "complete" data has a much simpler likelihood function to maximize [10]. In our case, we introduce a binary latent variable for each content class $c_i \in C$ and each (half-labeled) test document d to indicate whether document d is in the content class c_i, i.e.,

$$y(c_i, d) = \begin{cases} 1 & \text{if document } d \text{ is in content class } c_i \\ 0 & \text{otherwise} \end{cases}. \tag{16}$$

Moreover, we introduce a binary latent variable for each word occurrence $o_{p,d}$ to indicate whether the word has been generated from the content model or the style mode,

$$z(p, d) = \begin{cases} 1 & \text{if } o_{p,d} \text{ is from the content model} \\ 0 & \text{otherwise} \end{cases}. \tag{17}$$

If the two types of latent variables are observed, the data is complete and consequently the log-likelihood function becomes much easier to maximize. For a half-labeled test document d, we know that the latent variable $y(c_i, d)$ is 1 if d is in content class c_i or 0 otherwise and $\Pr[s_j \mid d]$ is 1 if d is in style type s_j or 0 otherwise, therefore as in equation (14) we can calculate its log-likelihood given the model θ and the latent variables $y(c_i, d)$:

$$\begin{aligned} \log \Pr[d, c(d), s(d)] &= \sum_{i=1}^{m} \sum_{j=1}^{n} y(c_i, d) \Pr[s_j \mid d] \log \Pr[d \mid c_i, s_j] + \\ &\sum_{i=1}^{m} \sum_{j=1}^{n} y(c_i, d) \Pr[s_j \mid d] \log \Pr[c_i] + \sum_{i=1}^{m} \sum_{j=1}^{n} y(c_i, d) \Pr[s_j \mid d] \log \Pr[s_j] \end{aligned}. \tag{18}$$

In addition, with the help of the latent variables $z(p, d)$, equation (15) for computing $\log \Pr[d \mid c_i, s_j]$ can be re-written as

$$\log \Pr[d \mid c_i, s_j] = \sum_{p=1}^{|d|} \begin{pmatrix} z(p, d) \log\left(\lambda \Pr[o_p(d) \mid c_i]\right) + \\ (1 - z(p, d)) \log\left((1-\lambda) \Pr[o_p(d) \mid s_j]\right) \end{pmatrix}, \tag{19}$$

because we assume that we know which component model has been used to generate each word occurrence.

The EM algorithm starts with some initial guess of the model $\theta^{(0)}$, then iteratively alternates between two steps, called the "E-step" (expectation step) and the "M-step" (maximization step) respectively [10]. In the E-step, it computes the expected log-likelihood for the complete data, or the so-called "Q-function" denoted by $Q(\theta; \theta^{(t)})$, where the expectation is taken over the computed conditional distribution of the latent variables given the current setting of model $\theta^{(t)}$ and the observed data. In the M-step, it re-estimates the model to be $\theta^{(t+1)}$ by maximizing the Q-function. Once we have a new generation of model parameters, we repeat the E-step and the M-step. This process continues until the likelihood converges to a local maximum.

The major computation to be carried out in the E-step is to estimate the distributions of latent variables, in our case, $y(c_i,d)$ and $z(p,d)$. For a half-labeled test document d in style s_n, we have

$$\Pr[y(c_i,d)=1]=\Pr[c_i \mid d,s_n]=\frac{\Pr[d \mid c_i,s_n]\Pr[c_i]}{\sum_{c\in C}\Pr[d \mid c,s_n]\Pr[c]} \quad (20)$$

via using equation (11), and obviously $\Pr[y(c_i,d)=0] = 1-\Pr[y(c_i,d)=1]$. For a word occurrence $o_{p,d}=w_k$ in a document d with label-pair (c_i,s_j), we have

$$\Pr[z(p,d)=1]=\frac{\lambda\Pr[w_k \mid c_i]}{\lambda\Pr[w_k \mid c_i]+(1-\lambda)\Pr[w_k \mid s_j]}, \quad (21)$$

and $\Pr[z(p,d)=0] = 1-\Pr[z(p,d)=1]$. Since the value of $\Pr[z(p,d)=1]$ given by the above equation is same for every word occurrence $o_{p,d}=w_k\in V$ in (c_i,s_j) documents, we introduce a new notation z_{ijk} to represent it.

The M-step involves maximizing the Q-function,

$$\theta^{(t+1)}=\arg\max_\theta Q(\theta;\theta^{(t)}). \quad (22)$$

In our case, the Q function can be obtained by combining the equations (13), (14), (18), (19), (20) and (21), and taking expectation over latent variables:

$$\begin{aligned}Q(\theta;\theta^{(t)}) &= \sum_{d\in D}\sum_{i=1}^{m}\sum_{j=1}^{n}\Pr[c_i \mid d]\Pr[s_j \mid d]E(d,c_i,s_j)\\ &+\sum_{d\in D}\sum_{i=1}^{m}\sum_{j=1}^{n}\Pr[c_i \mid d]\Pr[s_j \mid d]\log\Pr[c_i]\\ &+\sum_{d\in D}\sum_{i=1}^{m}\sum_{j=1}^{n}\Pr[c_i \mid d]\Pr[s_j \mid d]\log\Pr[s_j]\end{aligned}, \quad (23)$$

where

$$E(d,c_i,s_j)=\sum_{w_k\in d}\left(n(w_k,d)\left(E(w_k,c_i)+E(w_k,s_j)\right)\right), \quad (24)$$

$$E(w_k,c_i)=z_{ijk}\log\left(\lambda\Pr[w_k \mid c_i]\right), \quad (25)$$

$$E(w_k,s_j)=(1-z_{ijk})\log\left((1-\lambda)\Pr[w_k \mid s_j]\right). \quad (26)$$

The next model estimation $\theta^{(t+1)}$ should maximize $Q(\theta;\theta^{(t)})$, meanwhile the model parameters need to obey some inherent constraints such as

$$\sum_{w \in V} \Pr[w \mid c_i] = 1 . \quad (27)$$

The M-step turns out to be a constrained optimization problem. Using the Lagrange multiplier method, we can get an analytical solution to this problem. The derived update rules for the M-step are as follows. The prior content class probabilities $\Pr[c_i]$ are updated using equation (1), while the prior style type probabilities $\Pr[s_j]$ are kept unchanged. The conditional word probabilities should be adjusted using the following equations:

$$\Pr[w_k \mid c_i] = \left(\eta + \sum_{d \in D} \sum_{s \in S} Z_1(w_k, d, c_i, s) \right) \Bigg/ \left(\sum_{w \in V} \left(\eta + \sum_{d \in D} \sum_{s \in S} Z_1(w, d, c_i, s) \right) \right) \quad (28)$$

where

$$Z_1(w_k, d, c_i, s_j) = n(w_k, d) \Pr[c_i \mid d] \Pr[s_j \mid d] z_{ijk} , \quad (29)$$

and similarly,

$$\Pr[w_k \mid s_j] = \left(\eta + \sum_{d \in D} \sum_{c \in C} Z_0(w_k, d, c, s_j) \right) \Bigg/ \left(\sum_{w \in V} \left(\eta + \sum_{d \in D} \sum_{c \in C} Z_0(w, d, c, s_j) \right) \right) \quad (30)$$

where

$$Z_0(w_k, d, c_i, s_j) = n(w_k, d) \Pr[c_i \mid d] \Pr[s_j \mid d] (1 - z_{ijk}) . \quad (31)$$

Besides, the weight parameter should be re-estimated by

$$\lambda = \left(\sum_{w \in V} \sum_{d \in D} \sum_{c \in C} \sum_{s \in S} Z_1(w, d, c, s) \right) \Bigg/ \left(\sum_{w \in V} \sum_{d \in D} \sum_{c \in C} \sum_{s \in S} (Z_1(w, d, c, s) + Z_0(w, d, c, s)) \right) . \quad (32)$$

The EM algorithm is essentially a hill-climbing approach, thus it can only be guaranteed to reach a local maximum. When there are multiple local maximums, whether we will actually reach the global maximum depends on where we start: if we start at the "right hill", we will be able to find a global maximum. In our case, we ignore the style labels and use the 1D estimation equations (1) and (3) to initialize $\Pr[c_i]$ and $\Pr[w_k \mid c_i]$ from the training documents, just as in the standard NB algorithm. The initial values of $\Pr[s_j]$ and $\Pr[w_k \mid s_j]$ can be estimated similarly by ignoring the class labels, but from both the training documents and the test documents, using equation (8) and

$$\Pr[w_k \mid s_j] = \left(\eta + \sum_{d \in D} n(w_k, d) \Pr[s_j \mid d] \right) \Bigg/ \left(\sum_{w \in V} \left(\eta + \sum_{d \in D} n(w, d) \Pr[s_j \mid d] \right) \right) . \quad (33)$$

The initial value of the weighting parameter λ is simply set to $1/2$ that puts equal weights to the component models.

```
Initialization: set λ=1/2 , and estimate the probabilities
Pr[c_i] , Pr[w_k | c_i] , Pr[s_j] , Pr[w_k | s_j] just as in the 1D situation,
using equations (1), (3), (8), (33).
while ( θ has not converged ) {
        E-step: calculate Pr[c_i | d] for the half-labeled test
        documents using equation (20), and calculate z_ijk using
        equation (21), based on the current θ .
        M-step: re-estimate θ using equations (28), (30),
        (32), with the help of the latent variables.
}
Classify the half-labeled test documents using equation (10)
.
```

Fig. 4. The Cartesian EM algorithm

4 Experiments

We have conducted experiments on three real-world datasets to evaluate the effectiveness of the proposed Cartesian EM method for text classification in the 2D problem setting (stated in §1). Three methods, NB, 1D EM and Cartesian EM (C. EM), were compared in terms of classification accuracy. The Lidstone smoothing parameter was set to a good value $\eta = 0.1$ [6], and document frequency (df) based feature selection [11] were performed to let NB achieve optimal average performance on each dataset. The document frequency has been shown to have similar effect as information gain or chi-square test in feature selection for text classification.

The WebKB dataset (http://www.cs.cmu.edu/afs/cs.cmu.edu/project/theo-20/www/data/) contains manually classified Web pages that were collected from the computer science departments of four universities ("Cornell", "Texas", "Washington" and "Wisconsin") and some other universities ("misc."). In our experiments, each category is considered as a content class, and each source university is considered as a style type because each university's pages have their own idiosyncrasies. Only pages in the top four largest classes were used, namely "student", "faculty", "course" and "project". Furthermore, unlike the popular setting, the pages from the "misc." universities were not used either, because we thought it made little sense to train a style model for them. All pages were pre-processed using the Rainbow toolkit [12] with the options "--skip-header --skip-html --lex-pipe-command=tag-digits --no-stoplist --prune-vocab-by-doc-count=3". The experimental results on this dataset are shown in Table 1. The weighting parameter values found by Cartesian EM were consistently around 0.75.

The SRAA dataset (http://www.cs.umass.edu/~mccallum/code-data.html) is a collection of 73,218 articles from four newsgroups (simulated-auto, simulated-aviation, real-auto and real-aviation). In our experiments, "auto" and "aviation" are considered as content classes and "simulated" and "real" are considered as style types. All articles were pre-processed using the Rainbow toolkit [12] with the option "--skip-header --skip-html --prune-vocab-by-doc-count=200". The experimental results on this dataset are shown in Table 1. The weighting parameter values found by Cartesian EM were consistently around 0.95. It is also possible to take "simulated" and "real" as classes while "auto" and "aviation" as styles, but in our experiments, Cartesian EM did not work well in that situation. We think the reason is that the actual weight of "auto" and "aviation" models are overwhelming (around 0.95) so that Cartesian EM would drift away and consequently fail to get good discrimination between "simulated" and "real". This is a known pitfall of the EM algorithm. Therefore Cartesian EM may be not suitable when the text content is significantly outweighed by text style. It may be possible to detect this situation by observing λ and back off to a simpler method like NB when λ is small.

The 20NG dataset (http://people.csail.mit.edu/people/jrennie/20Newsgroups) is a collection of approximately 20,000 articles that were collected from 20 different newsgroups [13]. The "bydate" version of this dataset along with its train/test split was used. In our experiments, each newsgroup is considered as a content class, and the time period before and after the split point are considered as two style types. It is realistic and common to train a classifier on documents before a time point and then test it on documents thereafter. All articles were pre-processed using the Rainbow toolkit [12] with the option "--prune-vocab-by-occur-count=2 --prune-vocab-by-doc-count=0". The experimental results on this dataset are shown in Table 1. The weighting parameter value found by Cartesian EM was about 0.95.

To sum up, Cartesian EM compared favorably with NB and 1D EM in our experiments. In the case where Cartesian EM improved performance, the improvements over the standard 1D-EM and Naive Bayes on these three datasets are all statistically significant (P-value < 0.05), using the McNemar's test. It is able to yield better understanding (such as the weight of content/style) as well as increase classification accuracy.

Table 1. Experimental results

Dataset	Test Style	NB	1D EM	C. EM
WebKB	Cornell	81.42%	**84.96%**	**84.96%**
	Texas	81.75%	82.94%	**83.33%**
	Washington	77.25%	79.22%	**80.00%**
	Wisconsin	82.79%	81.17%	**84.42%**
SRAA	simulated	80.11%	91.85%	**94.39%**
	real	91.99%	87.00%	**92.68%**
20NG	new articles	79.55%	80.92%	**81.78%**

5 Related Works

The study of separating content and style has a long history in computational cognitive science and pattern recognition. One typical work in this area is [2]. Our work exploits the characteristic of text data, and the proposed Cartesian EM method is relatively more efficient than the existing methods.

Various mixture models have been used in different text classification problems. However, most of them assume that a document is generated by only one component model [14], while our Cartesian mixture model assumes that a document is generated by two multinomial component models. In [15], a mixture model is proposed for multi-label text classification, where each document is assumed to be generated by multiple multinomial component models, one per document label. In [16], a mixture model is proposed for relevance feedback, where the feedback documents are assumed to be generated by two multinomial component models, one known background model and one unknown topic model, combined via a fixed weight. The Cartesian mixture model is different with these existing works as it is designed for documents with 2D labels and it works in the context of transductive learning.

One straightforward way to extend 1D EM to 2D scenario is to simply regard each label-pair (c,s) as a pseudo 1D label. In other words, each cell in the Cartesian product label matrix (as shown in Figure 1) is associated with a multinomial model. That method named EM2D has been proposed for the problem of learning to integrate web taxonomies [17]. However, EM2D is not able to generalize content classifiers to new styles thus not applicable to our problem. In contrast, Cartesian EM assumes that each row or column in the Cartesian product label matrix corresponds to a multinomial model and the observations (cells) are generated by the interaction between content models (rows) and style models (columns). This is to simulate the situation where some words in the document are used for style purposes while other words in the same document are used to describe the content.

6 Conclusion

The ability of human being to separate content and style is amazing. This paper focuses on the problem of *style-independent text content classification*, and presents an EM based approach, Cartesian EM, that has been shown to be effective by experiments on real-world datasets.

References

1. Sebastiani, F.: Machine Learning in Automated Text Categorization. ACM Computing Surveys **34** (2002) 1-47
2. Tenenbaum, J.B., Freeman, W.T.: Separating Style and Content with Bilinear Models. Neural Computation **12** (2000) 1247-1283
3. Dempster, A.P., Laird, N.M., Rubin, D.B.: Maximum Likelihood from Incomplete Data via the EM Algorithm. Journal of the Royal Statistical Society, Series B **39** (1977) 1-38
4. Vapnik, V.N.: Statistical Learning Theory. Wiley (1998)
5. Mitchell, T.: Machine Learning. McGraw Hill (1997)
6. Agrawal, R., Bayardo, R., Srikant, R.: Athena: Mining-based Interactive Management of Text Databases. Proceedings of the 7th International Conference on Extending Database Technology (EDBT), Konstanz, Germany (2000) 365-379

7. McCallum, A., Nigam, K.: A Comparison of Event Models for Naive Bayes Text Classification. AAAI-98 Workshop on Learning for Text Categorization, Madison, WI (1998) 41-48
8. Nigam, K., McCallum, A., Thrun, S., Mitchell, T.: Learning to Classify Text from Labeled and Unlabeled Documents. Proceedings of the 15th Conference of the American Association for Artificial Intelligence (AAAI), Madison, WI (1998) 792-799
9. Nigam, K., McCallum, A., Thrun, S., Mitchell, T.: Text Classification from Labeled and Unlabeled Documents using EM. Machine Learning **39** (2000) 103-134
10. Zhai, C.: A Note on the Expectation-Maximization (EM) Algorithm. (2004)
11. Yang, Y., Pedersen, J.O.: A Comparative Study on Feature Selection in Text Categorization. Proceedings of the 14th International Conference on Machine Learning (ICML), Nashville, TN (1997) 412-420
12. McCallum, A.: Bow: A Toolkit for Statistical Language Modeling, Text Retrieval, Classification and Clustering. (1996)
13. Lang, K.: NewsWeeder: Learning to Filter Netnews. Proceedings of the 12th International Conference on Machine Learning (ICML), Tahoe City, CA (1995) 331-339
14. Pavlov, D., Popescul, A., Pennock, D.M., Ungar, L.H.: Mixtures of Conditional Maximum Entropy Models. Proceedings of the 20th International Conference on Machine Learning (ICML), Washington DC, USA (2003) 584-591
15. McCallum, A.: Multi-Label Text Classification with a Mixture Model Trained by EM. AAAI'99 Workshop on Text Learning (1999)
16. Zhai, C., Lafferty, J.D.: Model-based Feedback in the Language Modeling Approach to Information Retrieval. Proceedings of the 10th ACM International Conference on Information and Knowledge Management (CIKM), Atlanta, GA (2001) 403-410
17. Sarawagi, S., Chakrabarti, S., Godbole, S.: Cross-Training: Learning Probabilistic Mappings between Topics. Proceedings of the 9th ACM SIGKDD International Conference on Knowledge Discovery and Data Mining (KDD), Washington DC, USA (2003) 177-186

Using Relative Entropy for Authorship Attribution

Ying Zhao, Justin Zobel, and Phil Vines

School of Computer Science and Information Technology, RMIT University
GPO Box 2476V, Melbourne, Australia
{yizhao, jz, phil}@cs.rmit.edu.au

Abstract. Authorship attribution is the task of deciding who wrote a particular document. Several attribution approaches have been proposed in recent research, but none of these approaches is particularly satisfactory; some of them are ad hoc and most have defects in terms of scalability, effectiveness, and efficiency. In this paper, we propose a principled approach motivated from information theory to identify authors based on elements of writing style. We make use of the Kullback-Leibler divergence, a measure of how different two distributions are, and explore several different approaches to tokenizing documents to extract style markers. We use several data collections to examine the performance of our approach. We have found that our proposed approach is as effective as the best existing attribution methods for two class attribution, and is superior for multi-class attribution. It has lower computational cost and is cheaper to train. Finally, our results suggest this approach is a promising alternative for other categorization problems.

1 Introduction

Authorship attribution (AA) is the problem of identifying who wrote a particular document. AA techniques, which are a form of document classification, rely on collections of documents of known authorship for training, and consist of three stages: preprocessing of documents, extraction of style markers, and classification based on the style markers. Applications of AA include plagiarism detection, document tracking, and forensic and literary investigations. Researchers have used attribution to analyse anonymous or disputed documents [6, 14]. The question of who wrote Shakespeare's plays is an AA problem. It could also be applied to verify the authorship of e-mails and newsgroup messages, or to identify the source of a piece of intelligence.

Broadly, there are three kinds of AA problem: binary, multi-class, and one-class attribution. In binary classification, all the documents are written by one of two authors and the task is to identify who of the two authors wrote unattributed documents. Several approaches to this problem have been described [4, 6, 10]. Multi-class classification [1,5,12], in which documents by more than two authors are provided, is empirically less effective than binary classification. In one-class

H.T. Ng et al. (Eds.): AIRS 2006, LNCS 4182, pp. 92–105, 2006.

classification, also referred to as authorship verification, some documents training are written by a particular author while the authorship of the remainder is different but unknown [15]. The task is to determine whether a given document is produced by the target author. This is more difficult again, as it is easier to characterize documents as belonging to a certain class rather than to any class except the specified one.

Existing approaches use a range of methods for extracting features, most commonly style markers such as function words [1, 5, 10, 12] and grammatical elements such as part of speech [2, 21, 22]. Given these markers, AA requires use of a classification method such as support vector machines (SVMs) [5,15] or principal component analysis [1,10,12]. However, much previous work in the area is marred by lack of use of shared benchmark data, verification on multiple data sets, or comparison between methods—each paper differs in both style markers and classification method, making it difficult to determine which element led to success of the AA method, or indeed whether AA was successful at all. It is not clear whether these methods scale, and some are ad hoc. In previous work, we compared some of these methods using common data collections [25] drawn from the TREC data [8] and other readily available sources, and found Bayesian networks and SVMs to be superior to the other approaches given function words as tokens. A secondary contribution of this new paper is extension of this previous work to grammatical style markers.

Our primary contribution is that we propose a new AA approach based on relative entropy, measured using the Kullback-Leibler divergence [17]. Language models have been successfully used in information retrieval [24], by, in effect, finding the documents whose models give the least relative entropy for the query. Here we explore whether relative entropy can provide a reliable method of categorization, where the collection of documents known to be in a category are used to derive a language model. A strong motivation for exploring such an approach is efficiency: the training process is extremely simple, consisting of identifying the distinct terms in the documents and counting their occurrences. In contrast, existing categorization methods are quadratic or exponential.

To test the proposed method, we apply it to binary and multi-class AA, using several kinds of style marker. For consistency we use the same data collections as in our previous work [25]. We observe that our method is at least as effective for binary classification as the best previous approaches, Bayesian networks and SVMs, and is more effective for multi-class classification.

In addition, we apply our method to the standard problem of categorization of documents drawn from the Reuters newswire [16]. AA is a special case of text categorization, but it does not necessarily follow that a method that is effective for AA will be effective for categorization in general, and *vice versa*. However, these preliminary experiments have found that KLD is indeed an effective general categorization technique, with effectiveness comparable to that of SVMs. We infer that, given appropriate feature extraction methods, the same categorization techniques can be used for either problem.

2 Background

The basic processes of AA consists of three stages: text preprocessing, feature extraction, and categorization. In the first stage, the text is standardized and may be annoted with lexical information. In the second stage, features are identified in the transformed text. In the third stage, feature sets are compared to determine likely authorship. A variety of AA approaches have been proposed, differing in all three stages.

In style-based classification, both lexical and grammatical markers have been used. Function words are a lexical style marker that has been widely used [1,5, 10,12], on the basis that these words carry little content: a typical author writes on many topics, but may be consistent in the use of the function words used to structure sentences. Some researchers have included punctuation symbols, while others have experimented with n-grams [13,18,19]. Grammatical style markers have also been used for AA [2,21,22], with natural language processing techniques are used to extract features from the documents. However, the AA performance is subject to the performance of the corresponding natural-language tools that are used.

Once stylistic features have been extracted, they must be used in the way to classify documents. Several researchers have applied machine-learning techniques to AA. Diederich et al. [5] and Koppel et al. [15] have used SVMs in their experiments. Diederich et al. used a collection of newspaper articles in German, with seven authors and between 82 and 118 texts for each author. Documents with fewer than 200 words were not used, as they were considered to not have enough authorial information. Accuracies of 60% to 80% were reported. The data used by Koppel et al. consists of 21 English books by a total of 10 authors. An overall accuracy of 95.7% was reported; due to the small size of data collection, the high accuracy may not be statistically significant. In our previous work [25], we used a large data collection and tested five well-known machine learning methods. We concluded that machine learning methods are promising approaches to AA. Amongst the five methods, Bayesian networks were the most effective.

Principle component analysis (PCA) is a statistical technique that several researchers have employed for AA [1,10,12]. Baayen et al. [2] used PCA on a small data collection, consisting of material from two books. Holmes et al. [10] applied PCA to identify the authorship of unknown articles that have been tentatively attributed to Stephen Crane. The data consisted of only fourteen articles known to have been written by Crane and seventeen articles of unknown authorship. PCA has largely been used for binary classification. In our initial investigation [25], PCA appears ineffective for multi-class classification. Additionally, PCA is not easily scalable; it is based on linear algebra and uses eigenvectors to determine the principle components for measuring similarity between documents. In most cases, only the first two principle components are used for classification and other components are simply discarded. Although other components may contain less information compared to the first two components, discarding will cause information loss, which may reduce the classification effectiveness.

Compression techniques and language models are another approach to AA, including Markov chains [14, 21, 22] and n-gram models [13, 18, 19]. Khmelev and Tweedie [14] used Markov chains to identify authorship for documents in Russian. Character level n-grams are used as style markers. An accuracy of 73% was reported as the best result in multi-class classification, but in most cases there were generally only two instances of each authors' work, raising doubts as to the reliability of the results. Peng et al. [19] applied character level n-gram language models to a data collection of newswire articles in Greek. The collection contains documents by 10 authors, with 20 documents for each author. Although an average of 82% accuracy was reported, the size of collection is probably too small to draw any representative conclusions. The question of whether character-level n-grams are useful as style markers is, therefore, unclear. As the full text of the documents was retained in these experiments, it is possible that the effectiveness of topic markers rather than style markers was being measured.

Another compression-based approach is to measure the change in compressed file size when an unknown document is added to a set of documents from a single author. Benedetto et al. [3] used the standard LZ77 compression program and reported an overall accuracy of 93%. In their experiment, each unknown text is appended to every other known text and the compression program is applied to each composite file as well as to the original text. The increase in size due to the unknown text can be calculated for each case, and the author of the file with smallest increase is assumed to be the target. However, Goodman [7] failed to reproduce the original results, instead achieving accuracy of only 53%.

More fundamentally, the approach is based on two poor premises. One is that the full text of the data is used, so that topic as well as style information is contributing to the outcomes; document formatting is a further confounding factor. The other premiss is that compression is an unreliable substitute for modelling. Compression techniques build a model of the data, then a coding technique uses the model to produce a compact representation. Typical coding techniques used in practice have ad hoc compromises and heuristics to allow coding to proceed at a reasonable speed, and thus may not provide a good indication of properties of the underlying model. By using off-the-shelf compression rather than examining properties of the underlying model, much accuracy may be lost, and nothing is learnt about which aspects of the modelling are successful in AA. In the next section we explore how models can be directly applied to AA in a principled manner.

For classification tasks in general, two of the most effective methods are SVMs and Bayesian networks. SVMs [20] have been successfully used in applications such as categorization and handwriting recognition. The basic principle is to find values for parameters α_i for data points that maximize

$$\sum_i \alpha_i - \frac{1}{2} \sum \alpha_i \alpha_j y_i y_j (x_i \cdot x_j)$$

These values define a hyperplane, where the dimensions correspond to features. Whether an item is in or out of a class depends on which side of the hyperplane it lies. However, the computational complexity of SVM is a drawback. Even the best algorithm gives $O(n^2)$ computational cost, for n training samples.

A Bayesian network structure [9] is an acyclic directed graph in which there is one node in the graph for each feature and each node has a table of transition probabilities for estimating probabilistic relationships between nodes based on conditional probabilities. There are two learning steps in Bayesian networks, learning of the network structure and learning of the conditional probability tables. The structure is determined by identifying which attributes have the strongest dependencies between them. The nodes, links, and probability distributions are the structure of the network, which describe the conditional dependencies. However a major drawback of this approach is that asymptotic cost is exponential, prohibiting use of Bayesian networks in many applications.

3 Entropy and Divergence

Entropy measures the average uncertainty of a random variable X. In the case of English texts, each $x \in X$ could be a token such as a character or word. The entropy is given by:

$$H(X) = -\sum_{x \in X} p(x) \log_2 p(x)$$

where $p(x)$ is the probability mass function of a random character or word. $H(X)$ represents the average number of bits required to represent each symbol in X. The better the model, the smaller the number of bits.

For example, we could build a model for a collection of documents by identifying the set W of distinct words w, the frequency f_w with which each w occurs, and the total number $n = \sum_w f_w$ of word occurrences. This model is context free, as no use is made of word order. The probability $p(w) = f_w/n$ is the maximum likelihood for w, and

$$n \times \left(-\sum_w p(w) \log_2 p(w) \right) = -\sum_w f_w \log_2 \frac{f_w}{n}$$

is the minimum number of bits required to represent the collection under this model. The compression-based AA techniques considered above can be regarded as attempting to identify the collections whose models yield the lowest entropy for a new document, where however the precise modelling technique is unknown and the model is arbitrarily altered to achieve faster processing.

A difficulty in using direct entropy measurements on new documents is that the document may contain a new word w that is absent from the original model, leading to $p(w) = 0$ and undefined $\log_2 p(w)$. We examine this issue below.

Another way to use entropy is to compare two models, that is, to measure the difference between two random variables. A mechanism for this measurement of relative entropy is the *Kullback-Leibler divergence* (KLD) [17], given by:

$$KLD(p||q) = \Sigma_{x \in X} \; p(x) \log_2 \frac{p(x)}{q(x)}$$

where $p(x)$ and $q(x)$ are two probability mass functions.

In this paper, we propose the use of KLD as a categorization technique. If a document with probability mass function p is closer to q than to q'—that is, has a smaller relative entropy—then, we hypothesise, the document belongs in the category corresponding to q. The method is presented in detail later.

We use simple language modelling techniques to estimate the probability mass function for each document and category. Language models provide a principle for quantifying properties of natural language. In the context of using language models for AA, we assume that the act of writing is a process of generating natural language. The author can be considered as having a model generating documents of a certain style. Therefore, the problem is to quantify how different the authors' models are.

Given a token sequence $c_1c_2 \ldots c_n$ representing a document we need to estimate a language model for the document. In an ideal model, we would have enough data to use context to estimate a high $p(c_i|c_1 \ldots c_{i-1})$ should be obtained for each token occurrence. However, in common with most use of language models in information retrieval, we use a unigram model; for example, if the tokens are words, there are simply not enough word sequences to estimate multigram probabilities, and thus we wish only to estimate each $p(c_i)$ independently.

Therefore, the task is to find out a probability function to measure the probability of each component that occurs in the document. The most straightforward estimation in language modelling is the maximum likelihood estimate, in which the probability of each component is given by the frequency normalized by the total number of components in that document d (or, equivalently, category C):

$$p_d\left(c\right) = \frac{f_{c,d}}{|d|}$$

where $f_{c,d}$ is the frequency of c in d and $|d| = \sum_{c' \in d} f_{c',d}$. We then can determine the KLD between a document d and category C as

$$\begin{aligned} KLD\left(p_d||p_C\right) &= \sum_{c \in C \cup d} p_d\left(c\right) \log_2 \frac{p_d\left(c\right)}{p_C\left(c\right)} \\ &= \sum_{c \in C} \frac{f_{c,d}}{|d|} \log_2 \frac{f_{c,d} \cdot |C|}{f_{c,d} \cdot |d|} \end{aligned} \tag{1}$$

KLD as a Classifier for Authorship Attribution

Given author candidates $A = \{a_1 \ldots a_j\}$, it is straightforward to build a model for each author by aggregating the training documents. We can build a model for an unattributed document in the same way. We can then determine the author model that is most similar to the model of the unknown document, by calculating KLD values between author models and unknown documents to identify the target author for which the KLD value is the smallest.

However, it is usually the case that some components are missing in either the training documents or the documents to be attributed. This generates an undefined value in equation 1, and thus a KLD value cannot be computed. This is a standard problem with such models, and other researchers have explored a variety of smoothing techniques [24] to calculate the probability of missing components.

The Dirichlet prior is an effective smoothing technique for text-based applications, in particular information retrieval. We use Dirichlet smoothing to remove these zero probabilities, under which the probability of component c in document d (or equivalently, category C) is:

$$\begin{aligned} p'_d(c) &= \frac{|d|}{\lambda + |d|}\frac{f_{c,d}}{|d|} + \frac{\lambda}{\lambda + |d|} p_B(c) \\ &= \frac{f_{c,d}}{\lambda + |d|} + \frac{\lambda}{\lambda + |d|} p_B(c) \end{aligned}$$

where λ is a smoothing parameter and $p_B(c)$ is the probability of component c in a *background model*. For short documents, the background probabilities dominate, on the principle that the evidence for the in-document probabilities is weak. As document length grows, the influence of the background model diminishes. Choice of an appropriate value for λ is a tuning stage in the use of language models.

In principle the background model could be any source of typical statistics for components. Intuitively it makes sense to derive the model from other documents of similar type; in attributing newswire articles, for example, a background model derived from poetry seems unlikely to be appropriate. As background model, we use the aggregate of all known documents, including training and test, as this gives the largest available sample of material. There is no reason why a background model could not be formed this way in practice.

In estimating KLD, the same background model is used for documents and categories, so KLD is computed as

$$\begin{aligned} KLD(p_d \| p_C) &= \\ \sum_{c \in C \cup d} &\left[\left(\frac{f_{c,d}}{\lambda + |d|} + \frac{\lambda}{\lambda + |d|} p_B(c) \right) \times \log_2 \frac{\frac{f_{c,d}}{\lambda+|d|} + \frac{\lambda}{\lambda+|d|} p_B(c)}{\frac{f_{c,C}}{\lambda+|C|} + \frac{\lambda}{\lambda+|C|} p_B(c)} \right] \end{aligned} \tag{2}$$

By construction of the background model, $d \subset C$, so there are no zeroes in the computation.

4 Feature Types

Function words are an obvious choice of feature for authorship attribution, as they are independent of the content but do represent style. A related choice of feature is punctuation, though the limited number of punctuation symbols mean that their discrimination power must be low.

An alternative is to use lexical elements. We explored the use of parts of speech, that is, lexical categories. Linguists recognize four major categories of words in English: nouns, verbs, adjectives, and adverbs. Each of these types can be further classified according to morphology. Most part-of-speech tag sets make use of the same basic categories; however, tag sets differ in how finely words are divided into categories, and in how categories are defined.

In this paper, we propose the following approach to use of parts of speech in authorship attribution. We applied NLTK (a Natural Language ToolKit)[1] to extract the part-of-speech tags from each original document. The part-of-speech tag set we used to tag our data collection in text preprocessing is the "brown" tag set. For simplicity, and to ensure that our feature space was not too sparse, we condensed the number of distinct tags from 116 to 27, giving basic word classes whose statistical properties could be analysed.

A further refinement is to combine the classes. We explore combinations of function words, parts of speech, and punctuation as features in our experiments.

5 Experiments

We used experiments on a range of data sources to examine effectiveness and scalability of KLD for attribution. In preliminary experiments, we also examined the effectiveness of KLD for other types of classification problems. Several data collections were used in our experiments, including newswire articles from the Associated Press (AP) collection [8], English literature from the Gutenberg Project, and the Reuters-21578 test collection [16]. The first two data collections are used for AA. The Reuters-21578 test collection was used to examine the applicability of KLD for general categorization.

AP. From the AP newswire collection we have selected seven authors who each contributed over 800 documents. The average document length is 724 words. These documents are splitted into training and testing groups. The number of documents used for training was varied to examine the scalability of the methods. This collection was used in our previous work [25].

Gutenberg project. We wanted to test our technique on literary works, and thus selected the works of five well known authors from the Gutenberg project[2]: *Haggard*, *Hardy*, *Tolstoy*, *Trollope*, and *Twain*. Each book is divided into chapters and splitted for training and testing. Our collection consists of 137 books containing 4335 chapters. The number of chapters from each author ranges from 492 to 1174, and the average chapter length is 3177 words. In our experiments, the number of chapters used for training is randomly selected and varied.

Reuters-21578. These documents are from the Reuters newswire in 1987, and have been used as a benchmark for general text categorization tasks. There are 21578 documents. We use the Modapte split [16] to group documents for training and testing. The top eight categories are selected as the target classes; these are *acq*, *crude*, *earn*, *grain*, *interest*, *money-fx*, *ship*, and *trade*.

[1] Available from `http://nltk.sourceforge.net/index.html`.

[2] `www.gutenberg.org`

Table 1. Effectiveness (percentage of test documents correctly attributed) for Bayesian networks, SVMs, and KLD attribution on two-class classification. The data is the AP collection, with function words as features. Best results in each case are shown in **bold**.

Docs per author	Bayes network	KLD $\lambda = 10$	KLD $\lambda = 10^2$	KLD $\lambda = 10^3$	KLD $\lambda = 10^4$	SVM
50	78.90	89.24	**89.98**	89.67	77.83	85.81
100	81.55	90.93	**91.19**	91.17	82.10	89.38
200	84.18	91.74	**91.81**	91.67	87.38	91.12
400	84.82	92.05	92.19	92.19	89.86	**92.40**
600	84.46	92.17	92.14	92.24	90.74	**92.86**

We used the KLD method in a variety of ways to examine robustness and scalability of classification. We first conducted experiments for *two-class classification*, that is, to discriminate between two known authors. In this context, all the documents used for training and testing are written by either one of these two candidates. *Multi-class classification*, also called n-class classification for any $n \geq 2$, is the extension of two-class classification to arbitrary numbers of authors.

We applied KLD classification to all three data collections for both binary classification and n-class classification. In all experiments, we compared our proposed KLD language model method to Bayesian networks, which was the most effective and scalable classification method in our previous work [25]. In addition, we have made the first comparison between a KLD classifier with SVM, a successful machine learning method for classification. We used leave-one-out validation method to avoid the overfitting problem and estimate the true error rate for classification. The linear kernel was selected as most text categorization problems are linear separable [11]. More complex kernel functions have not been shown to significantly increase the classification rate [20, 23]. The package used in our experiments is *SVM-light.*[3]

We also investigated the significance of different types of features that can be used to mark authorial structure of a particular document. As discussed above, we have used function words, parts of speech, and punctuation as features; these were used both separately and in combination.

Two-class experiments. Our experiments were for the two-class classification task. The results were reported in Table 1, where outcomes are averaged across all 21 pairs of authors, because significant inconsistencies were observed from one pair of authors to another in our previous reported experiments [25]. We tested different values of λ: 10, 10^2, 10^3, and 10^4.

We observed that the best results were obtained for value of $\lambda = 10^2$ and $\lambda = 10^3$. To examine the scalability of KLD attribution, we have increased the number of documents used for training and maintained the same set of test documents. As can be seen, the accuracy of classification increases as the number of documents for training is increased, but appears to plateau. The KLD

[3] Available from `http://svmlight.joachims.org`.

Table 2. Effectiveness (percentage of test documents correctly attributed) of KLD attribution with $\lambda = 10^2$ on AP, using different feature types, for two-class classification

Docs per author	func word	POS tags	POS(punc)	combined
50	**89.98**	83.00	83.38	88.38
100	**91.19**	82.90	83.21	88.79
200	**91.67**	82.90	83.79	89.62
400	**92.19**	83.29	83.67	89.36
600	**92.14**	83.07	83.52	89.17

Table 3. Effectiveness (percentage of test documents correctly attributed) for Bayesian networks, SVMs, and KLD attribution on two-class classification. The data is the Gutenberg collection, with function words as features.

Docs per author	Bayes network	KLD $\lambda = 10^2$	KLD $\lambda = 10^3$	KLD $\lambda = 10^4$	SVM
50	93.50	94.70	**94.80**	84.30	91.40
100	95.10	95.80	**96.00**	88.90	94.85
200	95.10	96.10	**96.50**	93.70	**96.50**
300	95.38	96.50	96.70	95.50	**97.20**

method is markedly more effective than the Bayesian network classifier. With a small number of documents for modelling, the KLD method is more effective than SVM, while with a larger number of documents SVM is slightly superior.

As noted earlier, the computational cost of the SVM and Bayesian network methods is quadratic or exponential, whereas the KLD method is approximately linear in the number of distinct features. It is thus expected to be much more efficient; however, the diversity of the implementations we used made it difficult to meaningfully compare efficiency.

We next examined discrimination power of different feature types, using KLD classification on the two class classification task. As discussed above, we used function words, part-of-speech (POS) tags, POS with punctuation, and a combined feature set containing all previous three types of feature. Results are reported in Table 2, which shows the average effectiveness from the 21 pairs of authors. Function words were best in all cases, and so we concentrated on these in subsequent experiments. With all feature types, effectiveness improved with volume of training data, but only up to a point.

We then tested KLD attribution on the Gutenberg data we had gathered. Average effectiveness is reported in Table 3. The trends were similar to those observed on the AP collection. Again, our proposed KLD method is consistently more effective than Bayesian networks, and SVM is more effective than KLD only when a larger number of training documents is used; when SVM is superior, the difference is slight. In combination these results show that KLD attribution can be successfully used for binary attribution.

We applied the KLD approach to the Gutenberg data to examine the discrimination power of different feature types. Results are shown in Table 4. In

Table 4. Effectiveness (percentage of test documents correctly attributed) of KLD method for Gutenberg attribution, using different feature types, on two-class classification

Docs per author	functions word	POS tags	POS(punc)	combined
50	94.80	86.00	91.10	**96.10**
100	**96.00**	86.35	93.05	95.70
200	**96.50**	85.65	93.40	96.30
300	**96.70**	86.15	93.10	96.34

Table 5. Effectiveness (percentage of test documents correctly attributed) of Bayesian networks and KLD attribution for the AP data, on two- to five-class classification

Number of authors	Bayes network	KLD $\lambda = 10^2$	KLD $\lambda = 10^3$	KLD $\lambda = 10^4$
50 documents per author				
2	89.67	**92.14**	91.41	74.86
3	79.49	**84.21**	83.97	64.55
4	75.83	**81.43**	81.14	52.77
5	71.72	76.15	**76.27**	48.36
300 documents per author				
2	90.46	**94.95**	94.91	91.82
3	85.22	**88.70**	88.61	85.24
4	80.63	87.00	**87.05**	82.05
5	76.33	82.84	**83.11**	77.15

one case, the combined feature set is superior; in the remainder, the best feature type is again the function words.

Multi-class experiments. We next examined the performance of the KLD method when applied to multi-class classification. In the two-class experiments, the function words were the best at discrimination amongst different author styles; in the following experiments, then, we compared Bayesian networks and the KLD classification method using only function words as the feature set. SVMs were not used, as they cannot be directly applied to multi-class classification.

For each test, we used 50 and 300 documents from each author for training. The outcomes were again averaged from all possible author combinations, that is 21 combinations for 2 and 5 authors, and 35 combinations for 3 and 4 authors. As shown in Table 5, with appropriate λ values, the KLD approach consistently and substantially outperforms Bayesian networks. Smaller values of λ are the more effective, demonstrating that the influence of the background model should be kept low.

We then ran the corresponding experiments on the Gutenberg data, as shown in Table 6. The outcomes were the same as that on the AP data, illustrating that the method and parameter settings appear to be consistent between collections.

General text categorization. In order to determine the suitability of KLD classification for other types of classification tasks, we used the Reuters-21578

Table 6. Effectiveness (percentage of test documents correctly attributed) of Bayesian networks and KLD attribution for the Gutenberg data, on two- to five-class classification

Number of authors	Bayes network	KLD $\lambda = 10^2$	KLD $\lambda = 10^3$	KLD $\lambda = 10^4$
50 documents per author				
2	93.50	94.70	**94.80**	84.30
3	88.80	**92.33**	91.87	71.97
4	87.67	**89.80**	89.15	62.75
5	86.00	**87.60**	87.00	54.80
300 documents per author				
2	95.38	96.50	**96.70**	95.50
3	91.13	94.73	**94.90**	92.30
4	88.75	92.80	**93.00**	90.05
5	87.25	91.00	**91.20**	88.20

Table 7. Effectiveness (precision, recall, and accuracy) of KLD classification and SVM for general text categorization on the Reuters-21578 test collection

categories top 8 (1 vs. n)	relevant/irrelevant (same train/test split)	KLD($\lambda = 10^2$) rec/pre/acc	SVM rec/pre/acc
acq	668/1675	**95.81**/93.70/96.97	94.01/96.32/97.27
crude	150/2193	**96.58**/62.95/96.24	69.33/91.23/97.61
earn	1048/1295	97.23/90.02/93.94	**98.19**/98.19/98.38
grain	117/2226	**99.15**/71.17/97.95	84.62/99.00/99.19
interest	80/2263	**92.50**/45.68/95.99	37.50/93.75/97.78
money-fx	123/2224	**95.12**/54.42/95.56	69.11/80.95/97.52
ship	54/2289	**85.19**/33.58/95.78	24.07/86.67/98.16
trade	103/2240	**93.20**/52.17/95.95	67.98/87.50/98.16

collection to test topic-based classification using KLD. In the Reuters-21578 data collection, documents are often assigned to more than one category. (This is a contrast to AA, in which each document has only one class.) In our experiment, we chose the first category as the labelled class, as it is the main category for that document. In common with standard topic classification approaches we used all document terms as the classification features.

In these preliminary experiments—we do not claim to have thoroughly explored the application of KLD to general categorization—we tested n-class classification, where $n = 8$, both with and without stemming. We compared KLD classification and SVM in terms of precision, recall, and overall accuracy. Accuracy measures the number of documents correctly classified. Thus for any given category, it is calculated as the total number of documents correctly classified as belonging to that category, plus the total number of documents correctly classified as not belonging to that category, divided by the total number of documents classified. Results are shown in Table 7. KLD classification consistently achieves higher recall than SVMs, but with worse precision and slightly lower accuracy. We conclude that KLD classification is a plausible method for general text categorization, but that further exploration is required to establish how best it should be used for this problem.

6 Conclusions

We have proposed the use of relative entropy as a method for identifying authorship of unattributed documents. Simple language models have formed the basis of a recent series of developments in information retrieval, and have the advantage of simplicity and efficiency. Following simple information theoretic principles, we have shown that a basic measure of relative entropy, the Kullback-Leibler divergence, is an effective attribution method.

Here and in other work we have explored alternative attribution methods based on machine learning methods. These methods are computationally expensive and, despite their sophistication, at their best can only equal relative entropy. We have also explored other feature extraction methods, but the results show that function words provide a better style marker than do tokens based on parts of speech or patterns of punctuation. Compared to these previous methods, we conclude that relative entropy, based on function word distributions, is efficient and effective for two-class and multi-class authorship attribution.

Acknowledgements. This work was supported by the Australian Research Council.

References

1. H. Baayen, H. V. Halteren, A. Neijt, and F. Tweedie. An experiment in authorship attribution. *6th JADT*, 2002.
2. H. Baayen, H. V. Halteren, and F. Tweedie. Outside the cave of shadows: Using syntactic annotation to enhance authorship attribution. *Literary and Linguistic Computing*, 11(3):121–132, 1996.
3. D. Benedetto, E. Caglioti, and V. Loreto. Language trees and zipping. *The American Physical Society*, 88(4), 2002.
4. J. N. G. Binongo. Who wrote the 15th book of oz? an application of multivariate statistics to authorship attribution. *Computational Linguistics*, 16(2):9–17, 2003.
5. J. Diederich, J. Kindermann, E. Leopold, and G. Paass. Authorship attribution with support vector machines. *Applied Intelligence*, 19(1-2):109–123, 2003.
6. G. Fung. The disputed federalist papers: Svm feature selection via concave minimization. In *Proceedings of the 2003 Conference on Diversity in Computing*, pages 42–46. ACM Press, 2003.
7. J. Goodman. Extended comment on language trees and zipping, 1995.
8. D. Harman. Overview of the second text retrieval conference (TREC-2). *Information Processing & Management*, 31(3):271–289, 1995.
9. D. Heckerman, D. Geiger, and D. Chickering. Learning bayesian networks: the combination of knowledge and statistical data. *Machine Learning*, 20:197–243, 1995.

10. D. I. Holmes, M. Robertson, and R. paez. Stephen crane and the new-york tribune: A case study in traditional and non-traditional authorship attribution. *Computers and the Humanities*, 35(3):315–331, 2001.
11. T. Joachims. Text categorization with support vector machines: Learning with many relevant features. In C. Nédellec and C. Rouveirol, editors, *Proceedings of ECML-98, 10th European Conference on Machine Learning*, number 1398, pages 137–142, Chemnitz, DE, 1998. Springer Verlag, Heidelberg, DE.
12. P. Juola and H. Baayen. A controlled-corpus experiment in authorship identification by cross-entropy. *Literary and Linguistic Computing*, 2003.
13. V. Keselj, F. Peng, N. Cercone, and C. Thomas. N-gram-based author profiles for authorship attribution. In *Pasific Association for Computational Linguistics*, pages 256–264, 2003.
14. D. V. Khmelev and F. J. Tweedie. Using markov chains for identification of writers. *Literary and Linguistic Computing*, 16(4):229–307, 2002.
15. M. Koppel and J. Schler. Authorship verification as a one-class classification problem. In *Twenty-first International Conference on Machine Learning*. ACM Press, 2004.
16. D. D. Lewis, Y. Yang, T. G. Rose, and F. Li. Rcv1: A new benchmark collection for text categorization research. *J. Mach. Learn. Res.*, 5:361–397, 2004.
17. Manning and H. Schze. *Foundations of Statistical Natural Language Processing*. MIT Press, Cambridge, may 1999.
18. F. Peng, D. Schuurmans, V. Keselj, and S. Wang. Language independent authorship attribution using character level language models. In *10th Conference of the European Chapter of the Association for Computational Linguistics, EACL*, 2003.
19. F. Peng, D. Schuurmans, and S. Wang. Language and task independent text categorization with simple language models. In *NAACL '03: Proceedings of the 2003 Conference of the North American Chapter of the Association for Computational Linguistics on Human Language Technology*, pages 110–117, Morristown, NJ, USA, 2003. Association for Computational Linguistics.
20. B. Scholkopf and A. J. Smola. *Learning with Kernels: Support Vector Machines, Regularization, Optimization and Beyond*. The MIT Press, 2002.
21. E. Stamatatos, N. Fakotakis, and G. Kokkinakis. Automatic authorship attribution. In *Proceedings of the 9th Conference of the European Chapter of the Association for Computational Linguistics*, pages 158–164, 1999.
22. E. Stamatatos, N. Fakotakis, and G. Kokkinakis. Computer-based authorship attribution without lexical measures. *Computers and the Humanities*, 35(2):193–214, 2001.
23. V. Vapnik and D. Wu. Support vector machine for text categorization, 1998.
24. C. Zhai and J. Lafferty. A study of smoothing methods for language models applied to information retrieval. *ACM Trans. Inf. Syst.*, 22(2):179–214, 2004.
25. Y. Zhao and J. Zobel. Effective authorship attribution using function word. In *2nd Asian Information Retrieval Symposium*, pages 174–190. Springer, 2005.

Efficient Query Evaluation Through Access-Reordering

Steven Garcia and Andrew Turpin

School of Computer Science and Information Technology
RMIT University, GPO Box 2476V, Melbourne 3001, Australia
{garcias, aht}@cs.rmit.edu.au

Abstract. Reorganising the index of a search engine based on access frequencies can significantly reduce query evaluation time while maintaining search effectiveness. In this paper we extend access-ordering and introduce a variant index organisation technique that we label access-reordering. We show that by access-reordering an inverted index, query evaluation time can be reduced by as much as 62% over the standard approach, while yielding highly similar effectiveness results to those obtained when using a conventional index.

Keywords: Search engines, index organisation, efficiency, access-ordering.

1 Introduction

Web search engines currently process queries over collections composed of several billion documents. With the number of queries submitted by users of the Web, the need to service requests in a timely and efficient manner is of critical importance.

A key component of modern text search engines that allow efficient query processing is the inverted list [18]. Several techniques have been proposed to optimise the organisation of inverted lists to allow efficient query processing. One such technique is the access-ordered index [7]. In this approach, inverted lists are reorganised based on past user queries to allow faster processing at query time.

One of the major drawbacks of the access-ordered index is the difficulty in compressing the index effectively. In this paper we extend the work on access-ordered indexes and present a variant that we label *access-reordered* indexes that improves index compression levels over the previous approach.

A second issue that arises when dealing with an access-ordered index is the durability of the index. That is, if the index ordering is based on past user queries, then how long will the reorganisation be able to process queries efficiently as user requests change over time? In this paper we explore the effect of training set size, and show that an access-ordered index based on a minimal number of queries can be used to produce results that are approximately equivalent to those of an index ordered by queries gathered over a significantly larger period of time.

H.T. Ng et al. (Eds.): AIRS 2006, LNCS 4182, pp. 106–118, 2006.

This paper is organised as follows. In Section 2 we present background on efficient query evaluation in search engines. In Section 3 our variant index organisation technique is introduced, and in Section 4 we present experimental results that demonstrate the effectiveness of our proposed technique. Conclusions and further research issues are discussed in Section 5.

2 Background

In this section we discuss efficient text search. Specifically, we focus on inverted index organisation, and present recent efficiency optimisations that take advantage of index reorganisation.

A well known data structure used in efficient text search is the *inverted index*. For each term t in a collection, an *inverted list* is created that stores information regarding the occurrences of that term. For a ranked search system, an inverted list is composed of *postings*, consisting of $< d, f_{d,t} >$ pairs, where d is a document identifier in which the term occurs and $f_{d,t}$ is the within document term frequency of term t in document d. As an example, consider the inverted list for the term KALIMDOR:

$$< 1,5 >; < 12,3 >; < 14,2 >$$

In this list we see that the term appears five times in, document 1, three times in document 12, and twice in document 14.

An inverted list organised by ordinal document identifiers is known as a *document-ordered* index. Due to the ordinal nature of the *document-ordered* index, inverted list compression gains can be made by storing the difference between adjacent values. This difference is referred to as the *d–gap*. For our example above, the inverted list for KALIMDOR becomes:

$$< 1,5 >; < 11,3 >; < 2,2 >$$

where the first posting refers to document 1, the second to document 12 $(1+11)$, and the third posting to document 14 $(12+2)$. By compacting the values in this manner, variable byte and variable bit encoding techniques can more effectively compress the index [6,8]. Scholer et al. found that when compressing inverted lists, variable byte encoding provides effective compression, with efficient encoding and decoding [13]. In this work we make use of variable byte integer compression for our inverted lists.

Once an index has been constructed, the evaluation of a user query can be performed using one of many well known similarity metrics [12]. We make use of the Okapi BM25 [16] metric that evaluates the predicted similarity between a query q and a document d as:

$$\text{sim}(q,d) = \sum_{t \in q} \log\left(\frac{N - f_t + 0.5}{f_t + 0.5}\right) \times \frac{(k_1+1)f_{d,t}}{K + f_{d,t}}$$

where N is the number of documents in the collection and f_t is the number of distinct documents that contain term t. K is set to $k_1((1-b) + b \times L_d/\text{avl})$

where k_1 and b are tuning parameters; L_d is the length of document d, and avl is the average document length in the collection. Our implementation does not consider query-term frequency, which is not used in this context.

Evaluation of a query proceeds as follows. For each term in the query, the corresponding inverted list is loaded from disk. The list is decoded, and for each posting a partial similarity between the document and query is calculated and stored in an *accumulator* structure. If the posting represents a document that has not been previously seen, a new accumulator is initialised for the document with the calculated partial score. If the posting represents a document that has been previously seen, then the accumulator for that document is updated to include the new partial similarity. When all of the query term lists have been processed, the set of accumulators is partially sorted to produce the top R scoring documents, and the results are returned to the user.

For queries containing terms that are common in the collection, the number of accumulators that need to be initialised can be large. This can increase main memory usage and query evaluation time. Moffat and Zobel have proposed techniques that limit the number of accumulators at query evaluation, and have shown that such techniques can be as effective as having an unlimited number of accumulators [9].

The internal representation of the collection as used by the search engine is critical to service user information needs quickly. Standard query evaluation requires processing the entire inverted list for each query term. Reorganising inverted lists, so that the postings towards the head of each list are the most important postings for that term, has a key benefit: query evaluation time can be reduced by only processing the early portion of each inverted list. The list pruning techniques applied at query evaluation time vary with each reordering scheme and are discussed next.

One well known approach to collection reordering is that of *frequency-ordering* where the postings within the inverted lists are organised by descending within document frequency $f_{d,t}$ value [11]. This approach is based on the assumption that those documents that contain the term most frequently, are the most important in the list. At query evaluation time inverted lists are processed in inverse document frequency order. When a new list is about to be processed, two threshold values a_{ins} and a_{add} are calculated. Postings are decoded sequentially, and as the list is processed the within document frequency of each posting is compared to the a_{ins} threshold. While the within document frequency values remain greater than the a_{ins} threshold, new accumulators are initialised. Then, when the within document frequency values fall below a_{ins}, and while they remain greater than the a_{add} threshold, processing continues with only existing accumulators being updated. Finally, when the within document frequency values fall below the a_{add} threshold, the processing of the list terminates.

A side effect of the list reordering is that d-gaps can no longer be used to compact the inverted lists as adjacent postings are no longer ordered by ascending document identifier. However, Persin and Zobel found that by storing a single

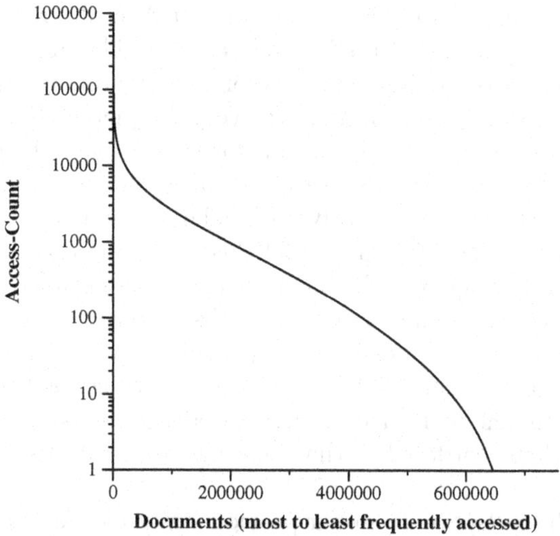

Fig. 1. Frequency of document access over 20 million queries in a 7.5 million document collection

$f_{d,t}$ value for blocks of postings that share the same frequency value, frequency-ordering can achieve comparable compression gains to a standard index that uses d-gap compaction.

Anh and Moffat proposed an alternate inverted list organisation approach where inverted list postings are organised by the contribution of the posting to the similarity metric [1,2]. In an *impact-ordered* inverted list, the postings are ordered by descending impact of the document on the similarity measure. The definition of an impact varies between similarity metrics.

Several pruning strategies were proposed for impact-ordered indexes that dynamically prune lists based on the impact of the current posting being processed, the number of equi-impact blocks of postings processed to date, and the document frequency of the current query term.

Compression of the inverted lists is difficult for impact-ordered lists, as the values in the lists may be real numbers requiring a floating point representation. To overcome this difficulty, Anh and Moffat propose quantising the impacts to a predefined integer range. As in frequency-ordered lists, the impact value need only be stored once per group of postings that share the same quantised impact value.

Examination of Web query logs has shown that term appearance in query logs is non-uniform. In an exploration of Web query logs, Spink et. al. found that the 75 most common search terms accounted for 9% all query terms [17]. Garcia et. al. demonstrated that such a skew in query terms leads to a non-uniform distribution of documents returned to the user at query time [7]. That is, given a query log, the search system will in general return a subset of documents from the collection more frequently than all other documents.

Figure 1 shows the skew in distribution of document access on a collection of 7.5 million documents when 20 million queries from Lycos.de are run against the collection. The analysis is based on the top 1,000 results returned for each query. This query log is discussed further in Section 4. The most frequently occurring document appears in the result set for over 1 million queries, that is, it appears in the results approximately once every 20 queries. Further, the 10% most frequently accessed documents account for over 70% of the documents in the result sets. Based on such trends, Garcia et. al. proposed an index organisation technique where the most frequently accessed documents are placed towards the head of the inverted lists [7]. They label this technique *access-ordering.*

In access-ordering, frequency of document access is determined as follows: for each document in a collection a counter is initialised to zero. A query log is run over the collection, and for each document appearing in the top 1,000 results of a query, the counter is incremented by 1. After the query log has been completely processed, each document will have an associated *access-count.* Each inverted list in the index is then reordered so that postings are organised by descending access-count.

Consider again the collection from our previous example. Assume that we run four queries over our collection, taking the top 3 results per query as follows: Query 1 – D_5, D_3, D_{14}; Query 2 – D_1, D_{14}, D_5; Query 3: – D_{12}, D_1, D_9; and Query 4 – D_{14}, D_2, D_7. After processing the queries, we can determine an access-count for each document. For example, the access-count for document D_1 is 2 as it appears in the result set for two queries. We then store the access-counts in the inverted lists within posting triples $< d, f_{d,t}, a_d >$, where d and $f_{d,t}$ are defined as above, and a_d is the access count for document d. The access-ordered inverted list for the term KALIMDOR becomes:

$$< 14, 2, 3 >; < 1, 5, 2 >; < 12, 3, 1 >$$

Note that document D_{14} now is the first posting in the list, as it appeared most frequently in the result sets.

Garcia et. al. explored several early termination heuristics ranging from simple approaches such as processing a fixed number of postings per list, and only processing postings with a minimum access count, to more elaborate techniques such as tracking the average accumulator contribution of a list as it is processed and terminating when a threshold is passed. They found the most effective pruning technique was an adaption of the frequency-ordered pruning approach, where processing terminates when the contribution of a posting falls below a threshold value that is calculated once per inverted list. In this scheme, a threshold value is calculated for each term prior to processing its inverted list as follows:

$$a_{allow} = (c_{allow} . S_{max})/({w_t}^2 + 1)$$

where w_t is the weight of the query term t, S_{max} is the largest accumulator seen to date while processing this query, and c_{allow} is a global tuning parameter. Terms are processed in descending inverse document frequency order, with list processing terminating when the access-count of a term falls below the a_{allow} value. Accumulator limiting is also used to bound main memory usage as required.

The initial work with access-ordering and list pruning reports savings in processing time of up to 25% over the conventional approach. However, as access-ordered lists do not have ordered document identifiers or ordered document contributions, the advantages of d-gap compaction are minimal. Indeed, without compaction, the index size grows by 60% when ordering lists by access-count.

An alternative to index reorganisation is *collection reordering.* In a reordered collection, instead of reorganising the inverted lists independently, the document identifiers of the entire collection are remapped to meet a specified criterion. Typically, similar documents are assigned document identifiers in the same proximity. The key benefit of this approach is that if similar documents are clustered together, then the inverted lists of the terms that appear in these documents will contain postings that have a reduced average d-gap. This in turn leads to compression gains. Blandford and Blelloch [3] and Silvestri et. al. [14,15] propose collection reordering techniques based on the clustering of similar documents, and show that this can reduce index size by as much as 23% over a conventional approach.

3 Access-Reordering

Access-ordered indexes have been shown to be an efficient technique to reduce the costs of query evaluation. However, the increase in index size offsets part of the potential benefit of the index organisation. Also, as the index reorganisation is based on an ordering determined from a query log, it is possible that the ordering will become less effective over time, as the pattern of user queries change. In this work, we propose a novel extension to access-ordered indexes that overcomes the issue of index size, and explore the effect of temporal changes in user queries.

Unlike other index reorganisation techniques, access-ordered indexes are reordered by a collection–wide ordering. All lists are ordered by the same set of access-counts.

We propose an improvement to compression by reassigning document identifiers. Specifically, after establishing access-counts, documents are assigned an identifier based on their rank in the access-ordering. That is, the document with the highest access count is assigned the document identifier 1, the document with the second highest access count is assigned the document identifier 2, and so on.

Consider the following example. After processing a query log we are left with the following access scores in a collection of 15 documents: $D_1 = 12$, $D_3 = 11$, $D_5 = 2$, $D_7 = 1$, $D_9 = 13$, $D_{12} = 5$, $D_{14} = 33$. All other documents have an access score of 0. Based on these values, D_{14} with an access score of 33 is the most accessed document, so it is assigned document identifier 1. Similarly, D_1 with an access-count of 12 is the third highest scoring document, and is assigned document identifier 3. Finally, D_{12} is ranked fifth and is assigned identifier 5. Our *access-reordered* list for KALIMDOR becomes:

$$< 1, 2, 3 >; < 3, 5, 2 >; < 5, 3, 1 >$$

Note that with ascending document identifiers, we can now make use of d-gap compaction, which should lead to reduced index size.

An advantage of access-ordering is that changes to the access-counts can be reflected in the index by resorting the inverted list of each term. Lists can be updated independently, as needed, without reference to the rest of the collection. On the other hand, remapping the document identifiers makes updates difficult as it requires updating all postings in all lists.

For the access-reordering approach to be effective, remapping of the index must not be required at frequent intervals. We therefore need to explore the stability of a generated ordering. In particular, two questions need to be addressed: first, do changes in user queries over time cause the index to become unstable for list pruning? Second, how many training queries are required to generate an access-reordered index that can efficiently service user queries?

4 Results

In this section we present experimental results that demonstrate the effectiveness of access-reordering. We begin with a discussion of the experimental environment. We then explore the effects of pruning at query time, and show that there is minimal difference between the results returned by a conventional system, and the results returned by our approach. We also show that an index trained on as few as 100,000 queries can be used to process as many as 20 million queries with effective query time pruning. Finally, we demonstrate the efficiency of our system by showing that our approach with pruning can process queries up to 62% faster than the conventional approach.

4.1 Experimental Environment

All experiments were conducted on an Intel Pentium 4 server with 2Gb of RAM. The server made use of the Linux Fedora Core 3 operating system. For all timed experiments, the system cache was flushed between runs.

We made use of the the Zettair search engine[1]; an open source search engine developed by the RMIT Search Engine Group. Queries are bag-of-words or ranked queries, with similarity scores evaluated using the Okapi BM25 measure described in Section 2.

4.2 Collections and Query Logs

Our experiments made use of a subset of the Gov2 collection from the 2004 Text REtrieval Conference (TREC) [4]. The Gov2 collection is composed of Web documents from the .gov domain crawled during 2004. Our subset consisted of the first 7.5 million documents in the collection as distributed by TREC. This subset accounted for the first 100Gb of uncompressed documents.

To explore the effect of training query set size, a large query log was required. We experimented with two distinct query logs with different properties. The first query log from Lycos.de[2] contains over 200 million time–ordered queries from

[1] See: `http://www.seg.rmit.edu.au/zettair/`

[2] See: `www.lycos.de`

the Lycos Web search engine. Unfortunately, a high proportion of these queries contain German terms that do not occur frequently in the Gov2 collection.

The second query log was generated from data provided by Microsoft[3] in 2003. We were provided a list of over 500 million queries ordered from most to least frequently occurring. Queries that occurred less than three times during the period in which the queries were gathered were not included in the list. While more compatible with the collection, this list does not provide a time reference for each query, therefore simulating a stream of queries over time requires randomisation of the query log, and does not reflect trends in query use such as those explored by Ozmutlu et. al. [10] and Diaz and Jones [5]. For the purposes of this work, we generated a simulated time–ordered query log of 10 million queries by sequentially selecting queries from the Microsoft list of queries based on their frequency of occurrence.

4.3 Relevance Framework

To evaluate the effectiveness of our system we make use of the well–known information retrieval measures *recall*, *precision* and *mean average precision* [18]. Recall measures the proportion of known relevant documents that are returned by a search system, and precision measures the proportion of relevant documents returned in a result set. Mean average precision is mean of average precision values for each query, where average precision is the average of precision values for each relevant document in the retrieved results.

To measure the effect of training set size and query drift, a large number of queries and relevance judgments are required. While the query logs that we work with meet the former requirement, relevance judgments are unavailable. However it is still possible to measure change in system performance by measuring the change in the results returned by two search systems when processing the same query log.

In our approach, we process a query log using system A over our collection and record of all returned results in the set R_A. We label this the *oracle* run. A second run with the same query log and collection is processed using search system B and the results are stored in R_B. By treating R_A as the set of relevant documents, and R_B as the returned result set, we can compare the difference between the two runs using precision and recall. Under such an approach precision and recall values of 100% indicate that both systems produce identical results. Lower values of recall indicate dramatically different result sets, and low values of precision at N indicate that system B produces different results ranked within the top N documents to the results returned by system A.

4.4 Effects of Pruning

To explore the effects of pruning at query time, we compared the result sets of our access-reordered index with query time pruning with the results returned by a standard index. Using 1,000 queries from each log, we marked the results

[3] See: `www.msn.com`

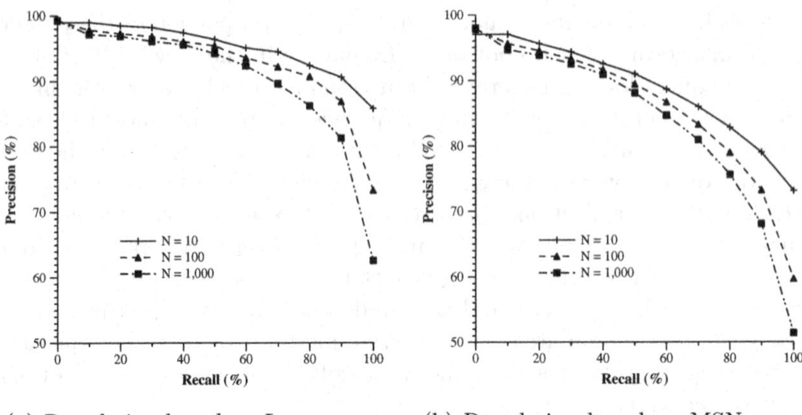

(a) Reordering based on Lycos query log.

(b) Reordering based on MSN query log.

Fig. 2. Difference in results returned by an access-reordered index with query time pruning, to that of a standard index with no pruning. N is the number of results examined per query.

returned by the standard approach as the oracle run, and compared these to the results returned by our access-reordered scheme. The 1,000 queries selected for this experiment were not part of the queries used to train the access-counts. The pruning parameter c_{allow} was trained, for each query log, to process on average 10% of the inverted lists.

Figure 2 shows the difference in result sets between the two runs when we consider the top 10, 100 and 1,000 returned results per query. Figure 2(a) compares an index reordered by access-counts based on the first 20 million Lycos queries, while Figure 2(b) compares an index reordered by access-counts generated by 10 million random Microsoft queries. For the collection reordered by access-counts from the Lycos queries, we can see that, for 10 results per query ($N = 10$), precision at 100% recall drops to only 85.79%. As we increase the number of results considered per query, the similarity between our approach and the standard differ more, however, across all three levels of considered results, for up to 90% recall, the precision never drops below 80%. Figure 2(b) shows that an access-reordered index based on the Microsoft log produces similar results. The pruning scheme therefore gives robust performance across a broad range of recall levels.

4.5 Stability of Access-Reordering

For a search system, the cost of reordering a collection is significant. When the collection is reordered, the entire index must be rebuilt. In this experiment, we show that with as few as 50,000 queries, we can generate a collection ordering that gives search results that do not differ greatly from results obtained using an index reordered on access counts based on 20 million queries.

To examine the effects of query drift over time, we generated indexes based on query logs from 50,000 queries up to 20 million queries from the temporally

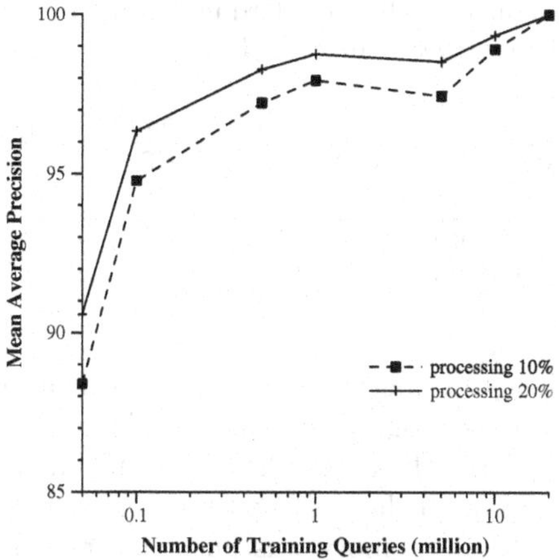

Fig. 3. Comparison of results returned by indexes trained on varying numbers of training queries, compared to an index trained on 20 million queries

ordered Lycos log. For each reordered index, we ran 1,000 test queries and collected the top 1,000 results per query. We treated the results returned by the index trained on 20 million queries as the oracle run, and compared the difference in results returned by indexes trained on the smaller sized query sets. If the ordering produced is stable, then an index trained on a relatively small amount of queries should be able to return similar results to those of an index trained on a large number of queries.

The mean average precision of results returned by indexes trained on query logs of varying sizes are compared to the oracle run in Figure 3. The figure shows that when pruning so that only 20% of the inverted lists are processed on average, the mean average precision obtained when training with only 50,000 queries is already over 90%. Further, with 100,000 training queries the results are 96% similar.

The figure also shows the similarity between the indexes when pruning the inverted lists to process 10% of the postings on average. In this case, little performance is lost compared to that of 20% processing.

The Microsoft log is generated by sampling queries based on their frequency of occurrence. As such, results for this experiment using the Microsoft log are not presented as they do not exhibit the temporal property of query drift that is being investigated for this experiment.

4.6 Efficiency

We have shown that an access-reordered index with pruning produces results comparable to those obtained using a standard index. We now compare the efficiency of query evaluation when using an access-reordered index to the standard approach.

Table 1. Time to process a query with a standard index and an access-reordered index when c_{allow} is tuned to process 10% of inverted list

Index	Milliseconds per query
Baseline	67
Access-reordered	25
Baseline (stop)	36
Access-reordered (stop)	20

To compare the efficiency of query evaluation, we timed both systems using 100,000 queries from the Lycos log. To avoid bias, the timing queries were different to the ones used to train the access-reordered index. An index trained on 20 million queries was used, with the c_{allow} parameter set to process 10% of the inverted lists on average. Main memory was flushed between each run to avoid caching effects.

Table 1 shows the time to process a single query using a standard index and an access-reordered index. The access-reordered approach with pruning is 62% faster.

In our experiments, the access-reordered index was 2% larger than the standard index, as a result of storing a single access-count value for each document in the collection. This is a significant improvement over the results reported by Garcia, et al. with access-ordering where index size grew by as much as 60% [7].

When pruning inverted lists at query time, a large saving comes from not processing postings in terms that occur frequently in the collection. These terms, often referred to as *stop-words*, are commonly disregarded by search systems. To investigate this effect we performed timings on stopped queries using the same c_{allow} parameter. In the case of stopping, the access-reordered approach remains 44% faster than the standard approach.

5 Conclusions and Future Work

We have presented a new index representation that builds on access-ordered indexes. This technique overcomes the compression difficulties raised by the access-ordered approach by generating an index that is equivalent in design to a standard document-ordered index. We have shown that even when inverted lists are pruned at query time by up to 90%, the returned results are comparable to the results returned by a standard index with no pruning.

We have also shown that an access-ordering is robust, where an index based on as little as 100,000 queries can produce similar results to orderings based on larger query training sets. This suggests that an access-reordered index has a durable life span, and that the index need not be rebuilt frequently to allow for query drift.

Finally, we have shown that access-reordered indexes with pruning are significantly faster than the standard approach, both with and without stopping.

Several aspects of this work however remain to be explored. Specifically, given the cost of reordering the collection, the ability to add new documents into an already constructed index would be advantageous. However, determining the access-count to assign to new documents is problematic. One possible approach is to query the collection using each new document, and assigning it the access-count of the highest result. Further, we would like to explore optimisations to the collection reordering technique so that the entire index need not be reordered at once, but instead reordered dynamically as the inverted lists are used.

Acknowledgments

We would like to thank Hugh Williams, Adam Cannane and Falk Scholer for their ideas and input into this work. We would also like to thank Klye Peltonen for access to the Microsoft query log. This work was supported in part by Australian Research Council Grant DP0558916.

References

1. V. N. Anh and A. Moffat. Impact transformation: Effective and efficient web retrieval. In K. Järvelin, M. Beaulieu, R. Baeza-Yates, and S. H. Myaeng, editors, *Proc. ACM SIGIR Conference on Research and Development in Information Retrieval*, pages 3–10, Tampere, Finland, Aug. 2002.
2. V. N. Anh and A. Moffat. Simplified similarity scoring using term ranks. In G. Marchionini, A. Moffat, J. Tait, R. Baeza-Yates, and N. Ziviani, editors, *Proc. ACM SIGIR Conference on Research and Development in Information Retrieval*, pages 226–233, Salvador, Brazil, Aug. 2005.
3. D. Blandford and G. Blelloch. Index compression through document reordering. In J. Storer and M. Cohn, editors, *Data Compression Conference*, pages 342–351, Snowbird, Utah, Apr. 2002.
4. C. Clarke, N. Craswell, and I. Soboroff. Overview of the TREC 2004 terabyte track. In *Proc. TREC Text REtrieval Conference*, 2004.
5. F. Diaz and R. Jones. Using temporal profiles of queries for precision prediction. In M. Sanderson, K. Järvelin, J. Allan, and P. Bruza, editors, *Proc. ACM SIGIR Conference on Research and Development in Information Retrieval*, pages 18–24, Sheffield, United Kingdom, July 2004.
6. P. Elias. Universal codeword sets and representations of the integers. *IEEE Transactions on Information Theory*, IT-21(2):194–203, Mar. 1975.
7. S. Garcia, H. E. Williams, and A. Cannane. Access-ordered indexes. In V. Estivill-Castro, editor, *Proc. ACSC Australasian Computer Science Conference*, pages 7–14, Dunedin, New Zealand, Jan. 2004. Australian Computer Society, Inc.
8. S. W. Golomb. Run-length encodings. *IEEE Transactions on Information Theory*, IT–12(3):399–401, July 1966.
9. A. Moffat and J. Zobel. Fast ranking in limited space. In *Proc. IEEE ICDE Conference on Data Engineering*, pages 428–437, Houston, Texas, Feb. 1994.
10. S. Ozmutlu, A. Spink, and H. C. Ozmutlu. A day in the life of web searching: an exploratory study. *Information Processing & Management*, 40(2):319–345, Mar. 2004.
11. M. Persin, J. Zobel, and R. Sacks-Davis. Filtered document retrieval with frequency-sorted indexes. *Journal of the American Society for Information Science*, 47(10):749–764, Oct. 1996.

12. J. M. Ponte and W. B. Croft. A language modeling approach to information retrieval. In W. B. Croft, A. Moffat, C. J. van Rijsbergen, R. Wilkinson, and J. Zobel, editors, *Proc. ACM SIGIR Conference on Research and Development in Information Retrieval*, pages 275–281, Melbourne, Australia, Aug. 1998.
13. F. Scholer, H. E. Williams, J. Yiannis, and J. Zobel. Compression of inverted indexes for fast query evaluation. In K. Järvelin, M. Beaulieu, R. Baeza-Yates, and S. H. Myaeng, editors, *Proc. ACM SIGIR Conference on Research and Development in Information Retrieval*, pages 222–229, Tampere, Finland, Aug. 2002.
14. F. Silvestri, S. Orlando, and R. Perego. Assigning identifiers to documents to enhance the clustering property of fulltext indexes. In M. Sanderson, K. Järvelin, J. Allan, and P. Bruza, editors, *Proc. ACM SIGIR Conference on Research and Development in Information Retrieval*, pages 305–312, Sheffield, United Kingdom, July 2004.
15. F. Silvestri, R. Perego, and S. Orlando. Assigning document identifiers to enhance compressibility of web search engines indexes. In H. M. Haddad, A. Omicini, R. L. Wainwright, and L. M. Liebrock, editors, *Proc. ACM SAC Symposium on Applied Computing*, pages 600–605, Nicosia, Cyprus, Mar. 2004.
16. K. Sparck-Jones, S. Walker, and S. Robertson. A probabilistic model of information retrieval: Development and comparative experiments. Parts 1&2. *Information Processing & Management*, 36(6):779–840, Nov. 2000.
17. A. Spink, D. Wolfram, B. J. Jansen, and T. Saracevic. Searching the web: the public and their queries. *Journal of the American Society for Information Science*, 52(3):226–234, 2001.
18. I. Witten, A. Moffat, and T. Bell. *Managing Gigabytes: Compressing and Indexing Documents and Images.* Morgan Kaufmann Publishers, Los Altos, CA, second edition, 1999.

Natural Document Clustering by Clique Percolation in Random Graphs

Wei Gao and Kam-Fai Wong

Department of Systems Engineering and Engineering Management,
The Chinese University of Hong Kong,
Shatin, N.T., Hong Kong
{wgao, kfwong}@se.cuhk.edu.hk

Abstract. Document clustering techniques mostly depend on models that impose explicit and/or implicit priori assumptions as to the number, size, disjunction characteristics of clusters, and/or the probability distribution of clustered data. As a result, the clustering effects tend to be unnatural and stray away more or less from the intrinsic grouping nature among the documents in a corpus. We propose a novel graph-theoretic technique called *Clique Percolation Clustering* (CPC). It models clustering as a process of enumerating adjacent maximal cliques in a random graph that unveils inherent structure of the underlying data, in which we unleash the commonly practiced constraints in order to discover natural overlapping clusters. Experiments show that CPC can outperform some typical algorithms on benchmark data sets, and shed light on natural document clustering.

1 Introduction

Document clustering is an important technique that facilitates the navigation, search and analysis of information in large unstructured text collections. It uses an unsupervised process to identify inherent groupings of similar documents as a set of clusters such that the intra-cluster similarity is maximized and the inter-cluster similarity is minimized.

Generally, clustering has three fundamental issues to solve: a data presentation model, a data similarity measure, and a clustering algorithm that builds the clusters using the data model and the similarity measure. Most existing clustering methods are based on vector space model [1,17] and represent document as a feature vector of unique content-bearing words that occur in the document sets, which is also known as "bag-of-words" model. Document similarity is calculated using one of the mathematical association measures, such as Euclidean distance, Cosine, Overlap, or Dice coefficients, etc., formulated with the feature vectors. Many clustering models and algorithms have been proposed. From different perspectives, they can be categorized into agglomerative or divisive, hard or fuzzy, deterministic or stochastic [11].

Most existing clustering algorithms optimize criterion functions with respect to the similarity measure in use over all the documents assigned to each potential partition of the collection [11,22]. They always impose some explicit and/or

H.T. Ng et al. (Eds.): AIRS 2006, LNCS 4182, pp. 119–131, 2006.

implicit constraints with respect to the number, size, shape and/or disjoint characteristics of target clusters. For example, partitional algorithms like k-means assumes the cluster number k and does not allow one document belonging to multiple groups. Although fuzzy clustering, such as fuzzy C-means/medoids algorithm [3,13], does support overlapping clusters by the membership function and fuzzifier parameter, they are still confined by cluster number and can find only spherical shape clusters (due to the assumption like k-means that each cluster can be described by a spherical Gaussian). Some algorithms, e.g. *EM* (Expectation-Maximization) clustering, are model-based, assuming Naive Bayes or Gaussian Mixture model [2,14]. They strongly presume certain probabilistic distributions of clustered documents and try to find the model that maximizes the likelihood of the data. When data cannot fit the hypothetical distribution, poor cluster quality can result. k-way clustering or bisection algorithms [22] adapt all kinds of criterion functions, but require to specify cluster number and force clusters to be equally sized. Recently, spectral clustering [8,9] based on graph partitioning has emerged as one of the most effective clustering tools, whose criterion functions are based on max-flow/min-cut theorem [5]. However, they prohibit overlapping clusters which ought to be important for document clustering.

In this paper, we define natural document clustering as a problem of finding unknown number of overlapping document groups with varied sizes and arbitrary data distributions. We try to obtain the clustering results with these free characteristics by removing as many external restrictions as possible and leaving things to the inherent grouping nature among documents. For this purpose, we propose a document clustering technique using a novel graph-theoretic algorithm, named *Clique Percolation Clustering* (CPC). The general idea is to identify adjacent maximal complete subgraphs (maximal cliques) in the document similarity graph using a threshold clique. Certain adjacent maximal cliques are then merged to form one of document clusters that can be fully explored by the threshold clique. Although CPC does introduce an explicit parameter k, which is the size of threshold clique, our algorithm can automatically settle the critical point, at which the natural clustering of the collection can be achieved. We show that CPC can outperform some representative algorithms with experiments on benchmark data.

The rest of this paper is organized as follows: Section 2 gives related work; Section 3 describes the proposed CPC method and its algorithmic implementation; Section 4 presents experiments and results; Finally, we conclude this paper.

2 Related Work

2.1 Graph-Based Document Representation

Two types of graph-based representations have been proposed for modeling documents in the context of clustering. One is the graph obtained by computing the pairwise similarities between the documents [9], and the other is obtained by viewing the documents and the terms as a bipartite graph (co-clustering) [8]. Our work use the first model.

In general, suppose $V = \{d_1, d_2, \ldots, d_{|V|}\}$ is a collection of documents. We represent the collection by an undirect graph $G = (V, E)$, where V is the vertex set and E is the edge set such that each edge $\{i, j\}$ is a set of two adjacent vertices d_i, d_j in V. The adjacent matrix M of the graph is defined by

$$M_{ij} = \begin{cases} w_{ij} & \text{if there is an edge } \{i,j\} \\ 0 & \text{otherwise} \end{cases} , \tag{1}$$

where each entry M_{ij} is the edge weight, and w_{ij} is the value of similarity metric (in what follows we assume Cosine coefficient) between d_i and d_j. The graph can also be unweighted where an edge exists indicating that the distance of its two vertices is smaller than some threshold, in which case M_{ij} is binary.

A *clique* in G is a subset $S \subseteq V$ of vertices, such that $\{i, j\} \in E$ for all distinct $\{d_i, d_j\} \in S$. Thus, any two vertices are adjacent in a clique that constitutes a complete subgraph of G. A clique is said to be *maximal* if its vertices are not a subset of the vertices of a larger clique. The maximal cliques are considered the strictest definition of a cluster [16]. In graph theory, enumerating all maximal cliques (equivalently, all maximal independent sets or all minimal vertex covers) is a fundamental combinatorial optimization problem and its worst-case complexity is believed NP-hard [4,21].

2.2 Graph-Theoretic Clustering

Traditional hierarchical agglomerative clustering (HAC) are intrinsically graph-based. HAC treats each data point as a singleton cluster and then successively merges pairs of clusters until all clusters have been merged into a single cluster that contains all documents. Single-link, complete-link and average-link are the most popular HAC algorithms.

In single-link algorithm [19], the similarity between clusters is measured by their most similar members (minimum dissimilarity). Generally, agglomerative process are rather computationally intensive because the minimum of inter-cluster distances must be found at each merging step. For single-link clustering, an efficient implementation of Minimum Spanning Tree (MST) algorithms of a weighted graph is often involved. Therefore, single-link produces clusters that are subgraphs of the MST of the data and are also connected components. It is capable of discovering clusters of varying shapes, but often suffers from the so-called chaining effect. Complete-link [12] measures the similarity between two clusters by their least similar members (maximum dissimilarity). From graph-theoretic perspective, complete-link clusters are non-overlapping cliques and are related to the node colorability of graphs. Complete-link is not vulnerable to chaining effect, but generates excessive compact clusters and is thus very sensitive to outliers. Average-link clustering [6] is a compromise between single-link and complete-link: the similarity between one cluster and another is the averaged similarity from any member of one cluster to any member of the other cluster; it is less susceptible to outliers and elongated chains.

3 Clique Percolation Clustering Model

3.1 Preliminaries

Suppose $|V|$ number of documents are given in a measure space with a similarity metric w_{ij}. We define a binary relation $\sim_t$ between documents on $G = \{V, E\}$ with respect to parameter t: $i \sim_t j := w_{ij} \leq t$, which is self-reflexive, symmetric and non-transitive. There is an edge $\{i, j\} \in E$ connecting vertices d_i and d_j whenever $i \sim_t j$ with respect to threshold t. Fig. 1 illustrates that given a matrix reflecting the distances between 7 documents and the t value, a series of graphs for the relation $i \sim_t j$ are produced with different connectivity densities. Clearly, if each maximal clique were considered as a cohesive form of cluster, we could discover different number of clusters from these graphs, where $t = 0.5$, 2.5 and 3.5 results in 7, 5 and 3 number of clusters, respectively. Different from HAC clusters, they are planar and overlapping rather than hierarchical and disjoint. They also display interesting properties of natural clusters except for excessive intra-cluster cohesiveness like complete-link clusters.

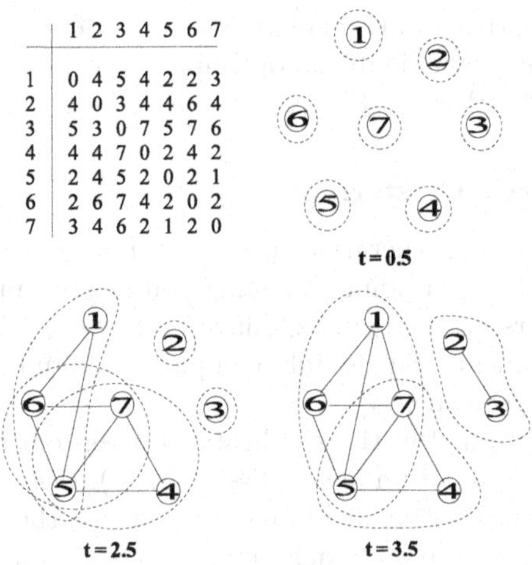

Fig. 1. Graphs with respect to the threshold level t and different cohesive clusters (in dotted regions) resulted from the respective t value

The series of graphs parameterized by t above can be seen as random graphs with constant set of vertices and a changing set of edges generated with some probability p, the probability that two vertices can be connected by an edge. Intuitively, tuning the value of t is somehow equivalent to adding or removing some edges according to p in a monotonic manner. In order for an appropriate t, we first try to determine p_c, the critical value of p, and then derive t given p_c by making use of their interdependency relationship. The critical value p_c is defined as the probability, under which a giant k-clique percolation cluster will emerge in the graph, and is known as the percolation threshold for a random network [7].

At this threshold, the percolation transition takes place (see Section 3.2). For clustering, the assumption behind is that no cluster can be excessively larger than others by commanding $p < p_c$.

3.2 k-Clique Percolation

Concepts. The concept of k-clique percolation for random networks was recently studied in biological physics in [7]. Its successful applications for uncovering community structure of co-authorship networks, protein networks and word association graphs can be found in [15]. Here we briefly describe some related notions.

Definition 1. *k-clique is a complete subgraphs of k vertices.*

Definition 2. *k-clique adjacency: Two k-cliques are adjacent if they share $k-1$ vertices, i.e., if they differ only in a single vertex.*

Definition 3. *k-clique percolation cluster is a maximal k-clique-connected subgraph, i.e., it is the union of all k-cliques that are k-clique adjacent.*

Definition 4. *k-clique adjacency graph is a compressed transformation of the original graph, where the vertices denote the k-cliques of the original graph and there is an edge between two vertices if the corresponding k-cliques are adjacent.*

Moving a particle from one vertex of a k-clique adjacency graph to another along an edge is equivalent to rolling a k-clique template (threshold clique) from one k-clique of the original graph to an adjacent one. A k-clique template can be thought of as an object that is isomorphic to a complete graph of k vertices. It can be placed onto any k-clique of the original graph, and rolled to an adjacent k-clique by relocating one of its vertices and keeping its other $k-1$ vertices fixed. Thus, the k-clique percolation clusters of a graph are all those subgraphs that can be fully explored by rolling a k-clique template in them [7]. Fig. 2 illustrates the effects of one-step rolling of a k-clique template (for $k = 2, 3, 4$ when $t = 3.5$) that produce different topologies of clusters. Note that a k-clique percolation cluster is equivalent with all maximal cliques adjacent by at least $k - 1$ vertices. Thus, compared to the strict maximal clique clusters aforementioned (see Section 3.1), the cohesiveness of a k-clique percolation cluster can be adjusted by the k value. In addition, such clusters are connected components on a k-clique adjacency graph that can be discovered with efficient algorithms. The goal of CPC is to find all k-clique percolation clusters.

Percolation Threshold p_c. How to estimate the threshold probability p_c of k-clique percolation with respect to k? Under such p_c (critical point), a giant k-clique percolation cluster that is excessively larger than other clusters will take place [10,7]. Intuitively, the greater the value of p ($p > p_c$), the more likely the giant cluster appears, and the bigger its size is (which includes most graph nodes), as if using a k-clique can percolate the entire graph.

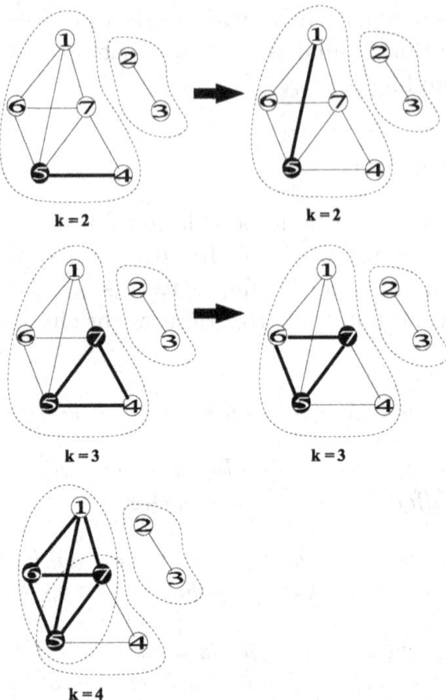

Fig. 2. Effects of k-clique template rolling in a relocation step (black nodes are fixed and bold edges are involved when rolling) with respect to different k that results in different k-clique percolation clusters (in dotted regions)

Consider the heuristic condition of template rolling at the percolation threshold: after rolling a k-clique template from a k-clique to an adjacent one by relocating one of its vertices, the expectation of the number of adjacent k-cliques, where the template can roll further by relocating another of its vertices, be equal to 1. The intuition behind is that a larger expectation value would allow an infinite series of bifurcations for the rolling, ensuring that a giant cluster is present in the graph. The expectation value can be estimated as $(k-1)(|V|-k)p_c^{k-1} = 1$, where $(k-1)$ is the number of template vertices that can be selected for the next relocation, $(|V|-k)$ is the number of potential destinations for this relocation, out of which only the fraction p^{k-1} is acceptable, because each of the new $k-1$ edges must exist in order to reach a new k-clique after relocation. Therefore, we get the percolation threshold function $p_c(k)$ with respect to k and $|V|$:

$$p_c(k) = \frac{1}{[(k-1)(|V|-k)]^{\frac{1}{k-1}}} . \tag{2}$$

For $k = 2$ in particular, $p_c(2) = 1/(|V| - 2)$ gives the percolation threshold of 2-clique connectedness (edge connectedness) of the graph, i.e., most graph nodes can be fully explored by a traversal along the edges.

Generation of Random Graph. By no means, an appropriate graph for clustering can be obtained without prior information regarding the global or local statistics of node connectivity in terms of certain degree distribution. In order to generate such a graph, one commonly specifies a series of hard threshold values of edge weight t, and then determines a good value t_c by trial and error. However, its time cost is generally very expensive due to the complexity of graph-theoretic approaches. Thus, t_c is usually hard to achieve. The concept of clique percolation provides a fundamental probabilistic formalism for determining the critical point, with which we can estimate t_c more directly without prior knowledge on statistics of graph and save the time cost of trial and error.

We examine the co-relation between p and t. Given p_c, we estimate the bound(s) of t_c so that the graph with the equivalent connectivity as that at the percolation threshold can be generated. Because p-t are monotone, an appropriate graph for clustering could be obtained using t slightly less than t_c. This is to prohibit the emergence of a giant cluster at the critical point. For simplification, we derive the upper bound of t_c by an approximation:

$$t_c(k) = p_c(k) \times (w_{max} - w_{min}) \ , \tag{3}$$

where w_{max} and w_{min} are the maximum and minimum values of document similarity in the collection, respectively.

3.3 Algorithmic Implementation

The clustering process is turned out to be a problem of finding all maximal cliques and then merging those with $k - 1$ common nodes into clusters. The proposed CPC method includes 5 major steps:

1. Preprocessing: Eliminate words in the stop list, use Porter's stemmer as the stemming algorithm, build document vectors, and create a $|V| \times |V|$ document similarity matrix $\mathbf{A}$.
2. Given k as parameter, compute $p_c(k)$ using (2), and compute $t_c(k)$ using (3).
3. For each entry in matrix $\mathbf{A}$, if $w_{ij} < t_c(k)$, then reassign $w_{ij} = 1$, otherwise set $w_{ij} = 0$; Create document similarity graph G using the updated $\mathbf{A}$ as adjacent matrix.
4. Enumerate all maximal cliques in G using Algorithm 1.
5. Create a $M \times M$ adjacent matrix $\mathbf{B}$ (where M is the number of maximal cliques identified), find k-clique percolation clusters using Algorithm 2 on $\mathbf{B}$.

Enumerating Maximal Cliques. Algorithms for finding maximal cliques (step 4 above) was studied in [4,21] and achieved processing time bounded by $O(v^2)$ and $O(nm\mu)$[1], respectively. Their algorithms are distinctive because they

[1] v is the number of maximal cliques. n, m and μ are the number of the vertices, edges and all the maximal independent sets, respectively. Note that each maximal independent set of a graph G corresponds one-to-one to each maximal clique of the complementary graph of G [4,21].

can be applied to a graph of comparatively large size. We implement an efficient counterpart of the algorithm using back-tracking method (see Algorithm 1). A maximal click is output at each end of back-track. Thus the running time is $O(v)$. Because v may be exponential with the growth of the number of vertices in worst case, our method is not a polynomial time algorithm either.

Algorithm 1. Enumerate All Maximal Cliques

---

input: Vertex set V and edge set E of graph G.
output: All maximal cliques of G into MC.

procedure EnumMC (MC, U, E)
if $U = \phi$ **then**
    output MC
    return
end if
for every vertex $u \in U$ **do**
    $U \Leftarrow U - \{u\}$
    EnumMC ($MC \cup \{u\}, U \cap \{v|(v,u) \in E\}, E$)
end for
end procedure

$MC \Leftarrow \phi$
EnumMC (MC, V, E)

Finding k-Clique Percolation Clusters. When all the maximal cliques are enumerated, a clique-clique adjacent matrix is prepared. It is symmetric where each row (column) represents a maximal clique and matrix values are equal to the number of common vertices between the two cliques (the diagonal entries are the clique sizes). Note that the intersection of two cliques is always a clique with at least $k-1$ (common) nodes. The k-clique percolation clusters are the one-to-one correspondents to the connected components in the clique-clique adjacency graph, which can be obtained using Algorithm 2 (step 5 above). The algorithm first creates a clique-clique adjacent matrix $\mathbf{B}$, in which every off-diagonal entry smaller than $k-1$ and every diagonal element smaller than k are erased (line 2–12), and then carrying out a depth-first-search (DFS) to find all the connected components. The resulted connected components are used as indices of maximal cliques for outputting k-clique percolation clusters.

4 Experimental Evaluations

4.1 Data Sets

We conduct the performance evaluations based on Reuters-21578[2] corpus, which is popular for document clustering purpose. It contains 21,578 documents that

[2] http://www.daviddlewis.com/resources/testcollections/reuters21578

Algorithm 2. Find All k-Clique Percolation Clusters

input: A set of all maximal cliques MC and k.
output: All k-clique percolation clusters into CPC.

```
procedure Find-k-CPC (CPC, MC, k)
B ⇐ 0 {Initialize entries in B as 0}
for i from 1 to M do
   for j from 1 to M do
      B[i][j] ⇐ |MC_i ∩ MC_j| {Count common nodes of two maximal cliques}
      if (i = j) ∧ (B[i][j] < k) then
         B[i][j] ⇐ 0 {Off-diagonal element < k is replaced by 0}
      else if (i ≠ j) ∧ (B[i][j] < k − 1) then
         B[i][j] ⇐ 0 {Diagonal element < k − 1 replaced by 0}
      end if
   end for
end for
CC ⇐ φ {Initialize connected component set CC}
i ⇐ 1 {Initialize recursion counter i}
CC ⇐ DFS (CC, B, i) {Recursive DFS to find connected components CC in B}
CPC ⇐ OutputCPC (CC, MC) {CC is as index of MC for the output of CPC}
end procedure
```

Find-k-CPC (CPC, MC, k)

are manually grouped into 135 classes. The number of documents for different clusters is very unbalanced, ranging from 1 to 3,945. Many documents have multiple category labels, and documents in each cluster have a broad scope of contents. In our experiments, we remove the clusters with less than 5 documents. We then extract 9,459 documents with unique class labels to form one of our data sets TS1, and rest of 11,084 documents with multiple class labels form TS2. At last, we result in 51 classes in TS1 and 73 classes in TS2. Table 1 shows the statistics of the original Reuters corpus (ORG) and the two resulted data sets.

4.2 Evaluation Metrics

We adopt two quality metrics widely used for document clustering [20], i.e., F-measure and Entropy. The F-measure of a class i is defined as $F(i) = \frac{2PR}{P+R}$. The precision and recall of a cluster j with respect to a class i are defined as: $P = Precision(i,j) = \frac{N_{ij}}{N_j}$ and $R = Recall(i,j) = \frac{N_{ij}}{N_i}$, where N_{ij} is the number of members of class i in cluster j, N_j is the size of cluster j, and N_i is the size of class i. The overall F-measure of the clustering result is the weighted average of $F(i)$:

$$F = \frac{\sum_i (|i| \times F(i))}{\sum_i |i|} \quad , \tag{4}$$

where $|i|$ is the number of documents in class i.

Table 1. Statistics of data sets ORG (Reuters-21578 original corpus), TS1 and TS2

	ORG	TS1	TS2
# of documents	21578	9459	11084
# of clusters	135	51	73
max cluster size	3945	3945	3508
min cluster size	1	5	5
avg. cluster size	186	153	167

Entropy provides a measure of homogeneity of a cluster. The higher the homogeneity of a cluster, the lower the entropy, and vice versa. For every cluster j in the clustering result, we compute p_{ij}, the probability that a member of cluster j belongs to class i. The entropy of each cluster j is calculated using $E_j = -\sum_i p_{ij} \log(p_{ij})$, where the sum is taken over all classes. The total entropy for a set of clusters is calculated as the sum of entropies of each cluster weighted by the size of that cluster:

$$E = \sum_{j=1}^{m} (\frac{N_j}{N} \times E_j) \ , \tag{5}$$

where N_j is the size of cluster j, m is the number of clusters, and N is the size of document collection.

4.3 Performance Evaluation

Experiment 1. Table 2 shows the performance of CPC given the different size of threshold clique. Obviously CPC produces more clusters than the number of categories in the benchmark. This is because Reuters corpus are manually classified according to a set of pre-defined keywords (one for each class). Thus the schema for the categorization is rather unifarious. One document with less discriminative features may belong to more groups and the grouping criterion could be more diverse. CPC is less limited by external constraints, which favors multifarious categorization schemes, and thus has more clusters.

The results on TS2 are better than on the other two data sets in terms of both F-measure and Entropy. Because TS2 contains documents all belonging to multiple classes, we think the better results on it can be attributed to CPC favoring overlapping clusters. We originally expected that the results on ORG would be far and few between the performances on TS1 and TS2, but the worst results of the three are observed on it. One possible reason is that we had pruned the classes with less than 5 documents for producing TS1 and TS2, where fewer small clusters are remained. This may indicate that CPC is disadvantageous in identifying excessively small clusters. We can also observe that the algorithm gives the best results when $k = 4$. Note that when $k = 2$, the performance is significantly poorer than other choices of k. This is because at $k = 2$, the procedure of CPC algorithm is degenerated to find connected components, which is regarded as the most relaxed criterion for clustering.

Table 2. Performance of CPC with respect to different sizes of the threshold clique

	# of clusters			F-measure			Entropy		
k	ORG	TS1	TS2	ORG	TS1	TS2	ORG	TS1	TS2
2	454	106	183	0.501	0.529	0.542	0.381	0.319	0.322
3	328	94	112	0.762	0.836	0.863	0.224	0.107	0.109
4	273	85	97	0.748	0.833	0.875	0.205	0.111	0.104
5	306	91	124	0.603	0.771	0.852	0.237	0.135	0.107
6	362	99	138	0.579	0.725	0.852	0.253	0.151	0.107

Experiment 2. In this experiment, we compare CPC with the other three representative clustering algorithms, k-means, single-link and complete-link. Because it is impossible to command CPC to produce exact number of clusters with the benchmark, we use $k = 4$ for the threshold clique, which is the optimal solution based on Table 2 and also brings about the closest number of clusters to the benchmark. To make fair comparisons under this condition, we examine every one of k-means, single-link and complete-link twice: one with the same number of clusters as the benchmark, and the other with the same number of clusters as CPC. The results are denoted by KM-b, KM-c, SL-b, SL-c, CL-b and CL-c (suffixes b and c represent benchmark and CPC, respectively). Furthermore, because k-means is well-known to be sensitive to local optima, we repeat the algorithm 50 times with different initializations (initial centroids) and choose the best outcomes achieved. In order to align with the clustering results of CPC and the benchmark, we stop the HAC process of single-link and complete-link when the specified number of clusters are left.

Table 3 shows that CPC outperforms other algorithms on all three test sets irrespective of the cluster number used. When compared with KM-b, SL-b and CL-b, CPC can only produce proximate number of clusters, but performs better on both measurements. This indicates that CPC clustering, although multifarious, is still more accurate than other clusterings with exact the same number of clusters as benchmark. When using the same number of clusters as CPC, the results of KM-c, SL-c and CL-c are even worse in some extent. In addition, CPC performs better on TS2 than TS1, and all the remaining algorithms demonstrate an opposite outcome, i.e., the results on TS1 are relatively better than TS2. This testifies the advantages of our method over the partitional algorithms that can only produce disjoint clusters.

k-means performs the worst among the three. Its poor performance on TS2 is very obvious because k-means can only produce spherical partitional clusters of the corpus. Single-link results in clusters that are connected components and complete-link clusterings are non-overlapping cliques. It is reasonable for complete-link performing better than single-link. The superiority of CPC stems from several main reasons: First, CPC aims to form natural clusters that should be by all means overlapping for Reuters corpus; Second, the cohesiveness of CPC clusters is moderate in comparison with relaxed single-link and restricted

Table 3. Comparisons of CPC and the other three representative algorithms, k-means (KM), single-link (SL) and complete-link (CL). We use $k = 4$

	# of clusters			F-measure			Entropy		
	ORG	TS1	TS2	ORG	TS1	TS2	ORG	TS1	TS2
CPC	273	85	97	0.748	0.833	0.875	0.205	0.111	0.104
KM-b	135	51	73	0.512	0.631	0.391	0.310	0.286	0.372
KM-c	273	85	97	0.503	0.509	0.344	0.326	0.315	0.360
SL-b	135	51	73	0.547	0.572	0.466	0.307	0.290	0.325
SL-c	273	85	97	0.533	0.559	0.485	0.310	0.302	0.317
CL-b	135	51	73	0.694	0.727	0.715	0.219	0.279	0.195
CL-c	273	85	97	0.670	0.759	0.734	0.235	0.162	0.183

complete-link; Third, because CPC does not assume each cluster described by any distribution, it tends to be more flexible and natural than model-based approaches like spherical k-means.

5 Conclusion and Future Work

We present a novel clustering algorithm CPC by applying clique percolation technique introduced from the area of biological physics. A more generalized framework related to it is the so-called "small-world network" describing many kinds of community structures in nature and society, which is extensively studied in random networks [10]. This is the first time for the clique percolation being applied in document clustering. The preliminary results demonstrate it is feasible and promising for document clustering. We are confident that CPC is interesting and worth of further studies. There are still many issues left to be studied more deeply. One is the determination of threshold values of t_c according to the percolation threshold probability p_c. So far, the mathematical relationship between them is not exact and clear-cut. To generate an appropriate random graph from p_c, an alternative is to make use of the degree distribution of graph vertices. For each vertex, some nearest neighbors associated with its degree distribution are considered to produce the connectivity instead of depending on the harsh cut by t_c (derived from p_c). This will lead to the further exploration on techniques to analyze complex networks. Furthermore, due to the NP-hardness of maximal clique enumeration algorithms, the CPC method is time-consuming. More efficient maximal cliques enumeration algorithm is required. In the future, we will also compare CPC to more advanced clustering algorithms, such as spectral clustering [8,9] and Information Bottleneck method [18].

References

1. Baeza-Yates, R., Ribeiro-Neto, B.: Modern information retrieval. Addison-Wesley, New York
2. Baker, L., McCallum, A.: Distributional clustering of words for text classification. In Proc. of the 21th ACM SIGIR Conference (1998):96–103

3. Bezdek, J.C.: Pattern recognition with fuzzy objective function algorithms. Plenum Press, New York
4. Bron, C., Kerbosch, J.: Finding all cliques of an undirected graph. Communications of the ACM **16** (1971):575–577
5. Cormen, T.H., Leiserson, C.E., Rivest, R.L., Stein C.: Introduction to algorithms, 2nd Edition. McGraw-Hill
6. Cutting, D., Karger, D., Pedersen, J., Tukey, J.W.: Scatter/Gather: A cluster-based approach to browsing large document collections. In Proc. of the 15th ACM SIGIR Conference (1992):318–329
7. Derenyi, I., Palla, G., Vicsek T.: Clique percolation in random networks. Physics Review Letters **95** (2005):160202
8. Dhillon, I.S.: Co-clustering documents and words using bipartite spectral grpah partitioning. In Proc. of the 7th ACM-KDD (2001):269–274
9. Ding, C.H.Q., He, X.F., Zha, H.Y., Gu, M., Simon, H.D.: A min-max cut algorithm for graph partitioning and data clustering. In Proc. of IEEE ICDM (2001):107–114
10. Dorogovtsev, S.N., Mendes, J.F.F.: Evolution of networks. Oxford Press, New York
11. Jain, A.K., Murty, M.N., Flynn, P.J.: Data clustering: a review. ACM Computing Surveys **31** (1999):264–323
12. King, B.: Step-wise clustering procedures. Journal of the American Statistical Association **69** (1967):86–101
13. Krishnapuram, R., Joshi, A., Nasraoui, O., Yi, L.Y.: Low-complexity fuzzy relational clustering algorithms for web mining. IEEE Transactions on Fuzzy Systems **9** (2001):595–607
14. Liu, X., Gong, Y.: Document clustering with clustering refinement and model selection capabilitities. In Proc. of the 25th ACM SIGIR Conference (2002):191–198
15. Palla, G., Derenyi, I., Farkas, I., Vicsek, T.: Uncovering the overlapping community structure of complex netowrks in nature and society. Nature **435** (2005):814–818
16. Raghavan, V.V., Yu, C.T.: A comparison of the stability characteristics of some graph theoretic clustering methods. IEEE Transactions on Pattern Analysis and Machine Intelligence **3** (1981):393–402
17. Salton, G.: Automatic text processing: the transformation, analysis, and retrieval of information by computer. Addison-Wesley, New York
18. Slonim, N., Tishby, N.: Document clustering using word clusters via the information bottleneck method. In Proc. of the 23th ACM SIGIR Conference (2000): 208–215
19. Sneath, P.H.A., Sokal, R.R.: Numerical taxonomy: the principles and practice of numerical classification. Freeman, London, UK
20. Steinbach, M., Karypis, G., Kumar, V.: A comparison of doucment clustering techniques. In Proc. of KDD-2000 Workshop on Text Mining (2000)
21. Tsukiyama, S., Ide, M., Ariyoshi, H., Shirakawa, I.: A new algorithm for generating all the maximal independent sets. SIAM Journal on Computing **6** (1977):505–517
22. Zhao, Y., Karypis, G.: Criterion functions for document clustering. Technical Report #01-40, Department of Computer Science, University of Minnesota

Text Clustering with Limited User Feedback Under Local Metric Learning*

Ruizhang Huang, Zhigang Zhang, and Wai Lam

Department of Systems Engineering and Engineering Management,
The Chinese University of Hong Kong,
Shatin, Hong Kong
{rzhuang, zgzhang, wlam}@se.cuhk.edu.hk

Abstract. This paper investigates the idea of incorporating incremental user feedbacks and a small amount of sample documents for some, not necessarily all, clusters into text clustering. For the modeling of each cluster, we make use of a local weight metric to reflect the importance of the features for a particular cluster. The local weight metric is learned using both the unlabeled data and the constraints generated automatically from user feedbacks and sample documents. The quality of local metric is improved by incorporating more precise constraints. Improving the quality of local metric will in return enhance the clustering performance. We have conducted extensive experiments on real-world news documents. The results demonstrate that user feedback information coupled with local metric learning can dramatically improve the clustering performance.

1 Introduction

Human beings are quite familiar with grouping text documents into categories. However, with the rapid growth of the Internet and the wide availablitity of electronic documents, it is infeasible for a user to browse all the documents and determine the underlying clusters. Therefore, it is very useful if a system can simulate human learning and discover the clusters automatically. Text clustering technique deals with finding a structure in a collection of unlabeled text documents. Often, text clustering is conducted in a completely automated manner. However, a small amount of user provided information often offers good hints on guiding the clustering to a better partition. Our goal is to incorporate limited user provided information into the automatic clustering process which in return improves the clustering performance.

Although browsing the whole document collection is not feasible, it is much easier for users to read a small amount of documents and grasp a preliminary

* This paper is substantially supported by grants from the Research Grant Council of the Hong Kong Special Administrative Region, China (Project Nos: CUHK 4179/03E and CUHK4193/04E), the Direct Grant of the Faculty of Engineering, CUHK (Project Code: 2050363), and CUHK Strategic Grant (No: 4410001). This work is also affiliated with the Microsoft-CUHK Joint Laboratory for Human-centric Computing and Interface Technologies.

H.T. Ng et al. (Eds.): AIRS 2006, LNCS 4182, pp. 132–144, 2006.

understanding on the document set. Users can provide a small amount of sample documents on some, not necessarily all, clusters. These sample documents are helpful to guide the clustering process. This problem can be viewed and handled as a semi-supervised clustering problem. After the clustering process, some unlabeled documents are grouped into existing clusters with samples. Besides, new clusters are also generated.

After the clusters are learned, users may examine the clustering results and give a small amount of feedback information. The user feedback is useful since it can provide the learning bias in the subsequent clustering task. Another round of clustering process is then invoked taking into consideration of the newly user feedback information. The user feedback can be provided in such an incremental manner until a satisfactory partition of the document collection is obtained. Some natural user feedbacks are:

- Document d should belong to cluster S.
- Document d should not belong to cluster S.
- Document d should be switched from cluster S_i to S_j.
- Two documents should be in the same cluster.
- Two documents should be in different clusters.

For the modeling of each cluster, it is common to find that each cluster emphasizes on different features. In our model, a weight metric is associated with each cluster. The weight metric captures a set of weights for the features of each cluster. The more important a feature, the higher weight is assigned to the feature. This local weight metric contributes to the distance calculation and it is learned using both constraints and the unlabeled data via a separate optimization procedure. The quality of metric is improved by incorporating more precise constraints automatically generated from user feedbacks. After the quality of local metric for every cluster is improved, the clustering results will also be improved in the subsequent learning process.

We have conducted extensive experiments comparing the two variants of our proposed approach. One of them is to incrementally consider the constraints generated from user feedbacks. We also compared with the approach without local metric learning. From the experimental results, our approach of learning local metric from user feedbacks is more effective than the approach without local metric learning. User feedbacks can dramatically improve the clustering performance especially when there are a small amount of constraints.

2 Related Work

Semi-supervised clustering is of great interest in recent years. Wagstaff and Cardie proposed instance-level hard constraints to improve the performance of clustering algorithms [9]. Demiriz et al. [4] proposed an unsupervised clustering method which labels each cluster with class membership, and simultaneously optimizes the misclassification error of the resulting clusters. Basu et al. [1] explored the use of labeled data to generate initial seed clusters, as well as the use of constraints generated from labeled data to guide the clustering process. All of the approaches use only user provided labeled documents or constraints to guide the clustering algorithm. However, the distance measure does not consider the user provided information.

Bilenko et al. [3] described an approach that unifies the constraint-based and metric-based semi-supervised clustering methods. It restricts that constraints are needed for every cluster. Sugota et al. [2] investigated a theoretically motivated framework for semi-supervised clustering that employs Hidden Random Markov Fields. It learns a single distance metric for all clusters, forcing them to have similar shapes. A common shortcoming of all of the above semi-supervised clustering algorithms is that they do not consider the user feedback information in an incremental manner.

There are also some recent research works on learning better metric over the input space. Frigui and Nasraoui [6] proposed an approach that performs clustering and feature weighting simultaneously in an unsupervised manner. Jing et al. [7] presented a modification to the original FW-KMeans to solve the sparsity problem that occurs in text data where different sets of words appear in different clusters. All of these approaches do not consider user feedback to guide the metric learning. Only the unlabeled data are considered. Xing et al. [10] presented an algorithm that, given examples of similar pairs of points, can learn a distance metric that respects these relationships. The limitation of this approach is that it only uses the constraints to learn the metric. We propose an approach considering both the unlabeled documents and constraints to learn the local weight metric for each cluster.

3 Our Learning Approach

3.1 Constraint Generation from User Feedback

We formulate the problem as a semi-supervised clustering problem. The sample documents and user feedbacks are handled by a set of constraints. Two kinds of constraints namely, must-link constraints and cannot-link constraints are adopted. If a must-link relationship is specified for two documents d_i and d_j, they must be grouped into the same cluster. In some situations, there is a further specification that they should be in the same cluster as a particular sample document. If a cannot-link relationship is specified for two documents d_i and d_j, they must be grouped into different clusters.

After the clusters are learned, users can examine the clustering result and give feedbacks on the quality of the clusters. The user feedback information is used to guide the clustering process to a better partition in the next round. The user feedback information is transformed automatically into a set of constraints. Suppose cluster S contains some previously provided sample documents. If a user suggests document d should belong to cluster S, d will form must-link constraints with all the sample documents of S. d will also be regarded as a sample document of S. Similarly if a user suggests d should not belong to S, it will form cannot-link constraints with all the sample documents of S. If a user suggests document d

should be switched from cluster S_i to S_j, it will form must-link constraints with all the sample documents of S_j and cannot-link constraints with all the sample documents of S_i. A user can also specify that two documents should belong to the same cluster. In this case, a must-link constraint is generated. Similarly, if a user specifies that two documents should belong to different clusters, the two documents form a cannot-link constraint.

3.2 Metric-Based Distance Measure

We make use of cosine similarity to measure the similarity between two documents or between a document and a cluster centroid. Standard cosine similarity fails to estimate similarity accurately if different features contribute to different extent to the similarity value. We design a local metric A for each cluster. Each feature is assigned a weight to indicate its contribution to the similarity of two documents in that particular cluster. Precisely, the local weight metric $A = \{a_i, i = 1 \ldots m\}$ is a set of weights assigned to each feature of the cluster. The metric-based distance is calculated as follows:

$$\Psi_A(X, Y) = 1 - \frac{\sum_{i=1}^{m} x_i a_i y_i}{||X||_A ||Y||_A} \tag{1}$$

where X and Y are the vector representation of two documents; m is the number of features in A; x_i, a_i, and y_i are the values of feature i in the vector representation of X, metric A, and Y respectively; $||X||_A$ denotes the weighted L_2 norm: $||X||_A = \sqrt{\sum_{i=1}^{m} x_i^2 a_i}$.

3.3 Objective Function

Clusters are learned by minimizing an objective function. The objective function is composed of two parts, namely, the document distance function and the penalty function, which considers the documents and constraints respectively. An optimal partition is obtained when the overall distance of the documents from the cluster centroids is minimized while a minimum number of constraints are violated.

The document distance function Υ measures the distance of the documents from the cluster centroids. The distance is calculated with the help of the local weight metric A_S for cluster S. This function is formulated as follows:

$$\Upsilon = \sum_{S=1}^{K} \sum_{d \in D_S} \Psi_{A_s}(d, \mu_S) \tag{2}$$

where K is the number of clusters; D_S is the set of documents that belong to the cluster S; d and μ_S are the vector representation of a document and the centroid of the cluster S respectively; A_S is the local metric associated with cluster S; $\Psi_{A_s}(d, \mu_S)$ measures a metric-based distance between document d to cluster centroid μ_S. Since the amount of labeled documents is small compared with unlabeled documents, the document distance function Υ is mainly related to the unlabeled documents.

The penalty function Δ is related to the constraints. It measures the cost of violation of the constraints. We denote the set of must-link constraints as M and the set of cannot-link constraints as C. A traditional must-link constraint is violated when two documents are grouped into different clusters. When a must-link constraint M is specified for two documents and the documents are regarded as sample documents for a particular cluster, the following situation is regarded as violating a must-link constraint. A must-link constraint is specified for two documents. One of the documents is grouped into the correct cluster as a particular sample document. The other document is assigned to a wrong cluster. When a cannot-link constraint C is specified for two documents and they are grouped into the same cluster, we regard this situation as a violation of cannon-link constraints. We make use of the distance of the two documents in the constraints as the penalty cost for violating the constraints in M and C. The penalty function Δ is formulated as follows:

$$\Delta = \sum_{(d_i,d_j)\in M} \Psi_{A_S}(d_i,d_j)\beta(S_{d_i},S_{d_j},L_{d_i}) + \sum_{(d_i,d_j)\in C} (1-\Psi_{A_S}(d_i,d_j))\delta(S_{d_i}=S_{d_j}) \quad (3)$$

$$\beta(S_{d_i},S_{d_j},L_{d_i}) = \begin{cases} 1 \ \text{ when } \ S_{d_i}=L_{d_i} \quad S_{d_i}\neq S_{d_j} \\ 1 \ \text{ when } \ S_{d_j}=L_{d_i} \quad S_{d_i}\neq S_{d_j} \\ 0 \ \text{otherwise} \end{cases}$$

where S_d is the cluster that d belongs to; A_S is local weight metric for cluster S; L_d is the actual cluster to which document d should belong; δ is the indicator function. For the traditional must-link constraint, L_{d_i} is not known from the user feedback and sample documents. β will be replaced by the indicator function. Therefore, the penalty function Δ is formulated as follows:

$$\Delta = \sum_{(d_i,d_j)\in M} \Psi_{A_S}(d_i,d_j)\delta(S_{d_i}\neq S_{d_j}) + \sum_{(d_i,d_j)\in C} (1-\Psi_{A_S}(d_i,d_j))\delta(S_{d_i}=S_{d_j}) \quad (4)$$

Consequently, the objective function Ξ is formulated as follows:

$$\Xi = \alpha\Upsilon + (1-\alpha)\Delta \quad (5)$$

where α is a parameter for balancing the contribution of the document distance and constraints in the objective function. Therefore the semi-supervised clustering task is to minimize the following objective function:

$$\begin{aligned} \Xi = {} & \alpha \textstyle\sum_{S=1}^{K} \sum_{d\in D_S} \Psi_{A_S}(d_i,\mu_S) \\ & +(1-\alpha)(\textstyle\sum_{(d_i,d_j)\in M} \Psi_{A_S}(d_i,d_j)\beta(S_{d_i},S_{d_j},L_{d_i}) \\ & +\textstyle\sum_{(d_i,d_j)\in C}(1-\Psi_{A_S}(d_i,d_j))\delta(S_{d_i}=S_{d_j})) \end{aligned} \quad (6)$$

3.4 EM Process

The optimization for finding the clusters minimizing the objective function expressed in Equation 6 is achieved by an iterative EM process. The outline of the algorithm is presented in Figure 1.

```
1 Initialization: the centroid of each cluster is initialized.
2 Repeat until convergence
3   E-step: Given the centroid and the local weight metric for each cluster,
        update the document assignment to clusters.
4   M-step(I): Given the current cluster assignment, re-calculate the centroid
        of each cluster.
5   M-step(II): Given the current cluster assignment, learn the local weight
        metric by a separate optimization process for each cluster.
```

Fig. 1. The EM process in semi-supervised clustering

The user provided samples are used as seeds to initialize the cluster centroids. Let λ be the number of clusters that the user has provided sample documents. We assume that $\lambda \leq K$. In addition to λ sample document centroids, we select $K - \lambda$ centroids using the farthest-first algorithm [2].

In the E-step, the assignment of the documents to the clusters are updated. Each unlabeled document is assigned to the cluster that minimizes the objective function.

The M-step consists of two parts. Every cluster centroid μ_S is re-estimated using the current document assignment as follows:

$$\mu_S = \frac{\sum_{d \in D_S} d}{||\sum_{d \in D_S} d||_{A_S}} \tag{7}$$

where D_S is the set of documents that belong to the cluster S; A_S is the local metric associated with cluster S;

The second part of the M-step performs local weight metric learning for each cluster. The local weight metric is learned by a separated optimization process presented in Section 3.5.

Two different methods are investigated for incorporating the new constraints to the semi-supervised clustering process. The first method is denoted by Method-B. It conducts the next round of clustering process with the updated constraints without considering the current document partition. After the new constraints are obtained from the user feedbacks, the semi-supervised clustering is invoked from Step 1 of the algorithm shown in Figure 1. The clusters are initialized with updated sample documents and farthest-first algorithm. Clustering process is conducted considering the whole set of the updated constraints.

The second method is denoted by Method-I. It conducts the semi-supervised learning with feedback in an incremental manner. The current clustering result is a local optimal partition. User feedback information provides some suggestions on the direction where the learning process should go in the next round. Illustrated by Figure 1, we invoke the clustering process starting from Step 2 with the current clustering result. The semi-supervised clustering is conducted by considering the whole set of the updated constraints.

3.5 Method for Local Metric Learning

Given the current document assignment, the local metric is learned to obtain different weights on the features used by each cluster. We cast the metric learning as a separate optimization problem for each cluster. The objective function is composed of the objective function components Ξ_S of each cluster as follows:

$$\begin{aligned}\Xi &= \sum_{S=1}^{K} \Xi_S \\ &= \sum_{S=1}^{K} (\alpha \Upsilon_S + (1-\alpha)\Delta_S)\end{aligned} \tag{8}$$

where Υ_S is the document function for cluster S; Δ_S is the penalty function for cluster S. Υ_S and Δ_S are formulated as follows:

$$\begin{aligned}\Upsilon_S &= \sum_{d \in D_S} \Psi_{A_S}(d, \mu_S) \\ \Delta_S &= \sum_{\kappa_S} \Psi_{A_S}(d_i, d_j)\beta(S_{d_i}, S_{d_j}, L_{d_i}) \\ &\quad + \sum_{(d_i,d_j) \in C \cap (d_i,d_j) \in D_S} (1 - \Psi_{A_S}(d_i, d_j))\delta(S_{d_i} = S_{d_j}) \\ \kappa_S &= \{(d_i, d_j) | (d_i, d_j) \in M \cap (L_{d_i} = S) \cap ((d_i \in D_S) \cup (d_j \in D_S))\}\end{aligned} \tag{9}$$

where κ_S is the set of must-link constraints involved in cluster S; D_S is the set of documents that belong to the cluster S; d and μ_S are the vector representation of a document and the centroid of the cluster S respectively; A_S is the local metric associated with cluster S; where S_d is the cluster that d belongs to; L_d is the actual cluster to which document d should belong; $\Psi_{A_S}(\cdot,\cdot)$ measures the metric-based distance between two documents or between a document and a cluster centroid; δ is the indicator function. For the traditional must-link constraint, the Δ_S is formulated as follows:

$$\begin{aligned}\Delta_S = &\sum_{(d_i,d_j) \in M \cap ((d_i \in D_S) \cup (d_j \in D_S))} \Psi_{A_S}(d_i, d_j)\delta(S_{d_i} \neq S_{d_j}) \\ &+ \sum_{(d_i,d_j) \in C \cap (d_i,d_j) \in D_S} (1 - \Psi_{A_S}(d_i, d_j))\delta(S_{d_i} = S_{d_j})\end{aligned} \tag{10}$$

The metric is learned by minimizing the objective function Ξ_S. We employ the steepest descent method to conduct the optimization for the metric A_S for each cluster using the objective function component for each cluster. Every weight a_S in the local metric A_S is updated using the formula:

$$a_S^{\iota+1} = a_S^{\iota} + \eta(-\frac{\partial \Xi_S}{\partial a_S^{\iota}}) \tag{11}$$

where a_S^{ι} is the value of a_S in the ι-th iteration; η is the steep length; $-\frac{\partial \Xi_S}{\partial a_S^{\iota}}$ is the gradient direction for the the ι-th iteration which is calculated as follows:

$$\begin{aligned}\frac{\partial \Xi_S}{\partial a_S^{\iota}} &= \alpha \sum_{x_i \in D_S} \frac{\partial \Psi_{A_S}(x_i, \mu_S)}{\partial a_S^{\iota}} \\ &\quad +(1-\alpha)(\sum_{\kappa_S} \frac{\partial \Psi_{A_S}(d_i,d_j)}{\partial a_S^{\iota}} \beta(S_{d_i}, S_{d_j}, L_{d_i}) \\ &\quad + \sum_{(d_i,d_j) \in C \cap (d_i,d_j) \in D_S} (1 - \frac{\partial \Psi_{A_S}(d_i,d_j)}{\partial a_S^{\iota}})\delta(S_{d_i} = S_{d_j})) \\ \kappa_S &= \{(d_i, d_j) | (d_i, d_j) \in M \cap (L_{d_i} = S) \cap ((d_i \in D_S) \cup (d_j \in D_S))\}\end{aligned} \tag{12}$$

4 Experimental Result

4.1 Datasets

Two real-world data sets were used for conducting our experiments. The first dataset was derived from the TDT3 document corpus[1]. We selected the native English newswire news documents with human assigned topics. The human assigned topics are also grouped into 12 general topic categories such as "Election", "Sports", etc. This dataset contains 2,450 news stories organized in 12 classes corresponding to those general topic categories. The second dataset is derived from Reuters RCV1 corpus [8]. The RCV1 corpus is an archive of 800,000 manually categorized newswire stories. The news stories are organized in four hierarchical groups. We made use of part of the RCV1 corpus to form a dataset for our experiments. We did not select those small clusters which contain a small number of news stories. Stories belonging to at most one of clusters in the first level and the second level were randomly selected. There were 10,429 news stories, in total, organized in four classes were chosen. In particular, there are 6037, 1653, 2239, and 500 news stories in each class respectively.

We pre-processed the TDT3 dataset by stop-word removal and TF-IDF weighting. The RCV1 dataset has been pre-processed[2]. High-frequency and low-frequency words removal were conducted for the two datasets, following the methodology presented in [5]. The thresholds for removing high-frequency and low-frequency words for TDT3 dataset were set to 30 and 2 respectively. The thresholds for removing high-frequency and low-frequency words for RCV1 dataset were set to 1000 and 1 respectively.

4.2 Evaluation Methodology

We make use of FScore measure to evaluate the quality of a clustering solution [11]. Each true class L is mapped to an appropriate system output cluster S to which it matches the best. The suitability of the cluster to the class is measured using the F value. The F value combines the standard precision measure P and recall measure R used in information retrieval as follows:

$$\begin{aligned} F(L_i, S_j) &= \tfrac{2*P(L_i,S_j)*R(L_i,S_j)}{P(L_i,S_j)+R(L_i,S_j)} \\ P(L_i, S_j) &= n_{ij}/n_{s_j} \\ R(L_i, S_j) &= n_{ij}/n_{l_i} \end{aligned} \quad (13)$$

where n_{ij} is the number of documents in cluster S_j belonging to L_i; n_{s_j} is the number of documents in cluster S_j; n_{l_i} is the number of documents belonging to L_i; The FScore for the class L_i is the maximum F value for each system generated cluster.

$$FScore(L_i) = \max_{S_j \in T} F(L_i, S_j) \quad (14)$$

[1] The description of the corpus can be found at http://projects.ldc.upenn.edu/TDT3/TDT3_Overview.html

[2] The pre-processed RCV1 dataset can be found at http://www.daviddlewis.com/resources/testcollections/rcv1/

where T is the set of system generated clusters. The FScore of the entire system generated clusters is the summation of $FScore$ for each class L_i weighted by the size of the class.

$$FScore = \sum_{i=1}^{K} \frac{n_i}{n} FScore(L_i) \tag{15}$$

where n is the total number of documents in the dataset.

4.3 Experimental Setup

We make use of the set of true classes as a reference to simulate the user feedbacks. The clustering solution is compared with the true classes. For each cluster with sample documents, some wrongly assigned documents are selected. The correct cluster of the document is provided if the correct cluster has some sample documents. For the clusters without sample documents, we select document pairs to simulate the user feedbacks. One document in such pairs belongs to a cluster with sample documents while the other does not belong to the clusters with sample documents.

We conducted experiments for our proposed two methods, namely, Method-B and Method-I. For comparative investigation, we also conducted experiments for our proposed method but without local metric learning. The number of clusters, K, for TDT3 dataset and RCV1 dataset was set to 12 and 4 respectively. In each experiment, the number of clusters with user provided sample is less than K. For TDT3 dataset, Figure 2, Figure 3, and Figure 4 depict the clustering performance of 9, 10, and 11 clusters with user provided samples respectively. In each experiment, the initial sample documents for each cluster was set to 10. The feedback information was provided incrementally. We repeat such experiment setup for different combinations of clusters with user provided samples. The performance was measured by the average of the FScore of 3 combinations of clusters.

The experiments for RCV1 dataset were conducted in a similar way. The number of clusters with user provided sample was 2 and 3 as shown in Figure 5 and Figure 6. The performance was measured by the average of the FScore of 4 combinations of clusters with sample documents. The initial sample documents for each cluster was 40. In each round, 20 user feedbacks were provided for each cluster.

4.4 Analysis of the Results

From the experimental results, it shows that the Method-B and Method-I outperform the approach without local metric learning in all of the experiments. The local metric learning is effective in improving the clustering performance.

Incorporating user feedbacks is useful especially when there are a few number of constraints. After some iterations, when the number of constraints for each cluster reach to 25 for TDT3 dataset and 100 for RCV1 dataset, the performance

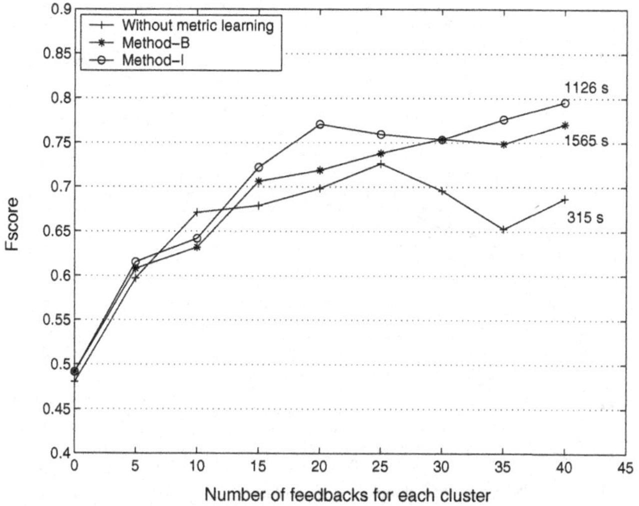

Fig. 2. Clustering performance of TDT3 dataset. The number of clusters with sample documents is 9. Method-B refers to our proposed approach which conducts the clustering with metric learning and considers the constraints generated from user feedbacks in batch. Method-I refers is an incremental version of Method-B. Without metric learning refers to the approach which conducts the clustering without metric learning. The computational time (in second) for each method is also depicted.

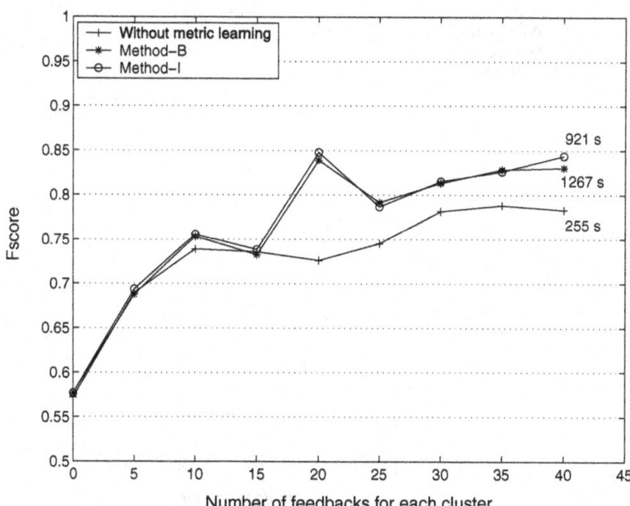

Fig. 3. Clustering performance of TDT3 dataset. The number of clusters with sample documents is 10. Method-B refers to our proposed approach which conducts the clustering with metric learning and considers the constraints generated from user feedbacks in batch. Method-I refers is an incremental version of Method-B. Without metric learning refers to the approach which conducts the clustering without metric learning. The computational time (in second) for each method is also depicted.

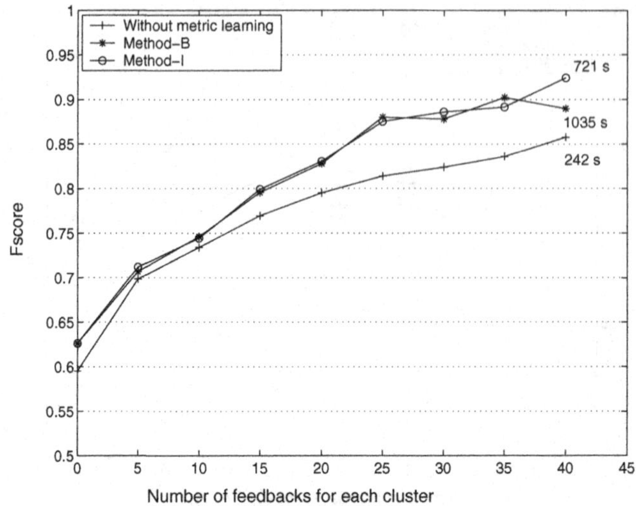

Fig. 4. Clustering performance of TDT3 dataset. The number of clusters with sample documents is 11. Method-B refers to our proposed approach which conducts the clustering with metric learning and considers the constraints generated from user feedbacks in batch. Method-I refers is an incremental version of Method-B. Without metric learning refers to the approach which conducts the clustering without metric learning. The computational time (in second) for each method is also depicted.

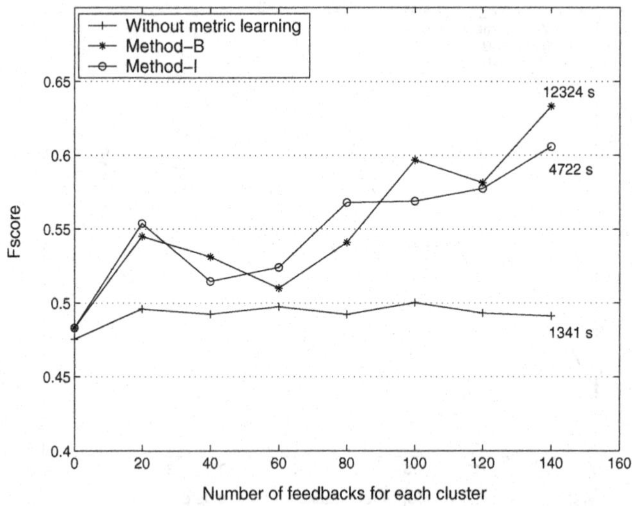

Fig. 5. Clustering performance of RCV1 dataset. The number of clusters with sample documents is 2. Method-B refers to our proposed approach which conducts the clustering with metric learning and considers the constraints generated from user feedbacks in batch. Method-I refers is an incremental version of Method-B. Without metric learning refers to the approach which conducts the clustering without metric learning. The computational time (in second) for each method is also depicted.

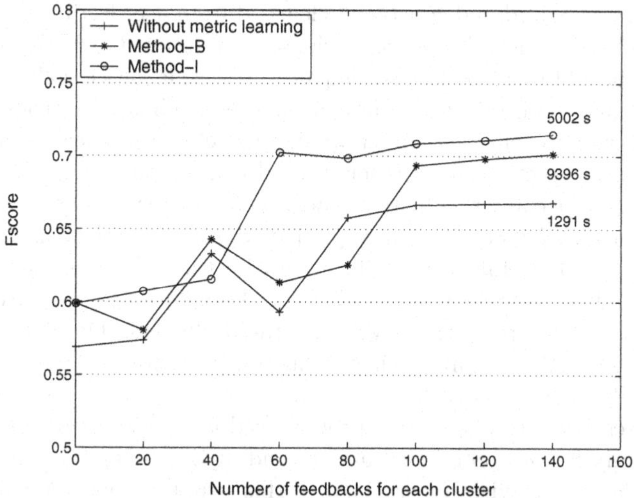

Fig. 6. Clustering performance of RCV1 dataset. The number of clusters with sample documents is 3 Method-B refers to our proposed approach which conducts the clustering with metric learning and considers the constraints generated from user feedbacks in batch. Method-I refers is an incremental version of Method-B. Without metric learning refers to the approach which conducts the clustering without metric learning. The computational time (in second) for each method is also depicted.

of the clustering becomes stable. Providing more user feedbacks does not affect the clustering performance dramatically. Therefore, only a small number of user feedback is needed to guide the clustering process to a better document partition.

From all the experimental results, it shows that the performance of Method-B and Method-I are close to each other. However, the execution time of Method-I is much faster than Method-B. In TDT3 corpus, Method-I uses 28%, 27%, and 30% less time than Method-B for the clustering performance of 9, 10, and 11 clusters with user provided samples respectively. In RCV1 corpus, it takes 62% and 47% less time than Method-B. Incremental semi-supervised clustering is computationally more efficient than conducting the clustering process in batch[3].

5 Conclusions and Future Work

We have presented an approach that incorporates incremental user feedback into text clustering. User feedbacks and a small amount of sample documents are provided for some clusters and are transformed into a set of constraints. A local weight metric is associated with each cluster reflecting the importance of each feature of the cluster. The local weight metric is learned with the help of unlabeled data and constraints. The quality of local weight metric can be

[3] The experiments were conducted on a 3.6 GHz PC with 2 GB memory. The operating system is Linux Fedora Core 4.

improved with newly updated constraints learned from user feedbacks. After the quality of local weight metric for every cluster is improved, a better document partition can be obtained in the subsequent clustering process. Experimental results show that our proposed approach on user feedback incorporation and local metric learning is effective for improving the clustering performance.

There are some directions for further study. Incorporating more natural types of user feedbacks especially for the newly generated clusters can be one direction. Currently, the user feedbacks in our approach capture the relationship between documents. Another useful user feedback can be the term feedback reflecting the relationship between features and clusters. Also, currently we only generate cannot-link constraints from the newly generated clusters. The clustering performance can be further improved if we can consider more user feedbacks and generate more kinds of constraints.

Another direction is to allow users provide feedbacks with a confidence value. Since a user only has a preliminary understanding on the dataset, he/she may provide some feedbacks with uncertainty. In this situation, we can assign a confidence value to the user feedbacks. The confidence value can help to judge how much the clustering process will rely on the user feedbacks.

References

1. S. Basu, A. Banerjee, and R. Mooney. Semi-supervised clustering by seeding. In *Proceedings of the Nineteenth International conference on Machine Learning*, 2002.
2. S. Basu, M. Bilenko, and R. J. Mooney. A probabilistic framework for semi-supervised clustering. In *Proceedings of the tenth ACM SIGKDD International Conference on Knowledge Discovery and Data Mining*, pages 59–68, 2004.
3. M. Bilenko, S. Basu, and R. J. Mooney. Integrating constraints and metric learning in semi-supervised clustering. In *Proceedings of the International Conference on Machine Learning*, 2004.
4. A. Demiriz, K. Bennett, and M. Embrechts. Semi-supervised clustering using genetic algorithms. In *Artificial Neural Networks In Engineering*, 1999.
5. I. S. Dhillon and D. S. Modha. Concept decompositions for large sparse text data using clustering. *Machine Learning*, 42(1):143–175, 2001.
6. H. Frigui and O. Nasraoui. Unsupervised learning of prototypes and attribute weights. *Pattern recognition*, 37(3):567–581, 2004.
7. L. Jing, M. K. Ng, J. Xu, and J.Z. Huang. Subspace clustering of text documents with feature weighting K-Means algorithm. In *Proceedings of the Pacific-Asia Conference on Knowledge Discovery and Data Mining*, pages 802–812, 2005.
8. T. Rose, M. Stevenson, and M. Whitehea. The Reuters Corpus Volume 1– from yesterday 's news to tomorrow's language resources. In *Proceedings of Third International Conference on Language Resources and Evaluation*, 2002.
9. K. Wagstaff and C. Cardie. Clustering with instance-level constraints. In *Proceedings of the Seventeenth International Conference on Machine Learning*, 2000.
10. E. P. Xing, A. Y. Ng, M. Jordan, and S. Russell. Distance metric learning, with application to clustering with side-information. *Advances in NIPS*, 15, 2003.
11. Y. Zhao and G. Karypis. Hierarchical clustering algorithms for document datasets. *Data Mining and Knowledge Discovery*, 10:141–168–231, 2005.

Toward Generic Title Generation for Clustered Documents

Yuen-Hsien Tseng[1], Chi-Jen Lin[2], Hsiu-Han Chen[2], and Yu-I Lin[3]

[1] National Taiwan Normal University,
No. 162, Sec, 1, Heping East Road,Taipei, Taiwan, R.O.C., 106
samtseng@ntnu.edu.tw
[2] WebGenie Information LTD.
B2F., No.207-1, Sec. 3, Beisin Rd.Shindian, Taipei, Taiwan, R.O.C., 231
{dan, sophia}@webgenie.com.tw
[3] Taipei Municipal Univ. of Education
1, Ai-Kuo West RoadTaipei, Taiwan, R.O.C., 100
jg141@mail.jges.tpc.edu.tw

Abstract. A cluster labeling algorithm for creating generic titles based on external resources such as WordNet is proposed. Our method first extracts category-specific terms as cluster descriptors. These descriptors are then mapped to generic terms based on a hypernym search algorithm. The proposed method has been evaluated on a patent document collection and a subset of the Reuters-21578 collection. Experimental results revealed that our method performs as anticipated. Real-case applications of these generic terms show promising in assisting humans in interpreting the clustered topics. Our method is general enough such that it can be easily extended to use other hierarchical resources for adaptable label generation.

Keywords: Hypernym search, WordNet, correlation coefficient.

1 Introduction

Document clustering is a powerful technique to detect topics and their relations for information analysis and organization. However, unlike document categorization where a set of labels or terms is predefined for each category, clustered documents require post-assignment of concise and descriptive titles to help analysts interpret the results. Most existing work selects the title words from the terms contained in the documents themselves. Although this is justifiable, this may not be sufficient. It would be desirable to further suggest generic topic terms for ease of analysis, especially in the applications where documents cover a wide range of domain knowledge. Examples of this need are often found in topic analysis for patent or scientific publications [1-3].

In this work, we attempt to automatically create generic labels which do not necessarily exist in the clustered documents for easier cluster interpretation. As an example, if documents in a cluster were talking about *tables*, *chairs*, and *beds*, then a title labeled "*furniture*" would be perfect for this cluster, especially when this hypernym does not occur in it. This kind of problem was often solved by human

H.T. Ng et al. (Eds.): AIRS 2006, LNCS 4182, pp. 145 – 157, 2006.

experts, such as those in [4-5], where cluster titles were given manually. To make our automatic approach feasible, external resources such as WordNet or other hierarchical knowledge structures are used. Our method first selects content-indicative terms for each cluster. A hypernym search algorithm is then applied to map these terms into generic titles.

The rest of the paper is organized as follows: Section 2 reviews some related work. Section 3 introduces our method for content-indicative term extraction. Section 4 describes the hypernym search algorithm based on WordNet. Section 5 details the experiments that evaluate our method. Section 6 discusses the results and suggests possible improvement. Section 7 concludes this work and shows its implications.

2 Related Work

Labeling a clustered set of documents is an inevitable task in text clustering applications. Automatic labeling methods mainly rely on extracting significant terms from clustered documents, where the term significance can be calculated very differently from clustering algorithms to algorithms.

For example, in the vector space model, where clusters are represented as weighted sums or centroids of the document vectors, terms with heaviest weights in the cluster vectors are extracted as the cluster labels. In [6-7], the term weight in a document vector is the normalized term frequency (*TF*), while in [8], it is a version of the *TFxIDF* (Inverse Document Frequency) weighting scheme. As to its effectiveness, the authors in [9] pointed out (although without experiments) that the simple centroid-based approach outperformed the probabilistic odds scheme which computes the ratio of the conditional probability of a term appearing in a cluster over the sum of the conditional probabilities of the term in other clusters.

In the Self-Organization Map (SOM) method [10], where clusters are organized in a 2-D map, the label of a cluster is the term having the highest goodness measure. This measure is the square of the relative term frequency in the cluster normalized by the sum of the relative term frequencies in other distant clusters.

In an application to group terms for detecting events over time [11], the cluster title consists of the highest ranked named entity followed by the highest ranked noun phrase. The ranks of these terms were obtained by sorting the maximum chi-square values of the terms occurring in a time interval.

In clustering web search results, the longest phrases occurred in most documents in a cluster were used as its title [12].

In other related fields, such as document summarization and translation, there were tasks in the Document Understanding Conference (DUC) [13] to generate very short summaries. These short 10-words summaries have the potential to serve as cluster titles. However, most participants use extraction-based methods [14]. Even there were studies that generate document titles not from the document themselves, a large corpus of documents with human-assigned titles is required to train the "translation model" so as to map document words into human-assigned titles [15]. Besides, these summaries tend to be event-descriptive rather than topic-indicative for a set of documents.

As can be seen, despite there are various techniques to label a set of documents, there are few studies that attempted to deal with the problem that we propose.

3 Cluster Descriptor Selection

The methods to extract cluster descriptors in the above-mentioned studies are mostly related to their clustering algorithms. In our application, we would like to have a general approach that is independent of the clustering process. To this end, we seek solutions from text categorization where selecting the best content-revealing or category-predictive features has been widely studied.

Yang et al [16] had compared five different methods for selecting category-specific terms. They found that *chi-square* is among the best that lead to highest categorization performance. The chi-square method computes the relatedness of term T with respect to category C in the manner:

$$\chi^2(T,C) = \frac{(\text{TP} \times \text{TN - FN} \times \text{FP})^2}{(\text{TP + FN})(\text{FP + TN})(\text{TP + FP})(\text{FN + TN})}. \tag{1}$$

where TP (True Positive), FP (False Positive), FN (False Negative), and TN (True Negative) denote the number of documents that belong or not belong to C while containing or not containing T, respectively. With content terms sorted by chi-square values in descending order, top-ranked terms can be selected as the cluster title.

However, chi-square does not distinguish negatively related terms from positively ones. A remedy to this problem is the use of the *correlation coefficient* (*CC*), which is just the square root of the chi-square:

$$Co(T,C) = \frac{(\text{TP} \times \text{TN - FN} \times \text{FP})}{\sqrt{(\text{TP + FN})(\text{FP + TN})(\text{TP + FP})(\text{FN + TN})}}. \tag{2}$$

As pointed out by Ng et al [17], correlation coefficient selects exactly those words that are highly indicative of membership in a category, whereas the chi-square method not only picks out this set of terms but also those terms that are indicative of non-membership in that category. This is especially true when the selected terms are in small number. As an example, in a small real-world collection of 116 documents with only two exclusive categories: construction vs. non-construction in civil engineering tasks, some of the best and worst terms that are computed by both methods are shown in Table 1. As can be seen, "engineering" is a lowest-ranked term (-0.7880) based on correlation coefficient in the non-construction category, while it is a top-ranked term in both categories based on chi-square (0.6210 is the square of -0.7880). Therefore, instead of chi-square, correlation coefficient is used as our basic word selection method.

Table 1. Some best and worst terms computed by chi-square and correlation coefficient in a collection with two exclusive categories

Chi-square				Correlation Coefficient			
Construction		Non-Construction		Construction		Non-Construction	
engineering	0.6210	**engineering**	0.6210	engineering	0.7880	equipment	0.2854
improvement	0.1004	improvement	0.1004	improvement	0.3169	procurement	0.2231
…				…			
kitchen	0.0009	kitchen	0.0009	communiqué	-0.2062	improvement	-0.3169
update	0.0006	update	0.0006	equipment	-0.2854	**engineering**	-0.7880

A further analysis of the correlation coefficient method reveals that it may be effective for large number of short documents. But without considering the occurring frequency of the term in each document (i.e., *TF*), it tends to select category-specific terms that are not generic enough for long documents. Therefore, we choose only those terms whose document frequency in a cluster exceeds a ratio *r* of the number of documents in that cluster. We denote this revised method as CC_r, where *r* is a tunable parameter and is 0.5 in our implementation. Another remedy is to multiply the term's CC with its total term frequency in the cluster (TFC), denoted as *CC*x*TFC*, where *TFC* is the sum of a term's TF over all documents in the cluster.

4 Generic Title Generation

The cluster descriptors generated in the above may not be topic-indicative enough to well summarize the contents of the clusters. One might need to map the identified clusters into some predefined categories for supporting other data mining tasks (e.g. [18]). If the categories have existing data for training, this mapping can be recast into a standard text categorization problem, to which many solutions can be applied. Another need arises from the situation that there is no suitable classification system at hand, but some generic labels are still desired for quick interpretations. This case is often solved by human experts, where cluster titles are given manually. Below we propose an automatic solution by use of an extra resource, i.e., WordNet.

WordNet is a digital lexical reference system [19]. English nouns, verbs, adjectives and adverbs are organized into synonym sets. Different relations, such as hypernym, hyponym, meronym, or holonym, are defined to link the synonym sets. With these structures, one can look up in WordNet all the hypernyms of a set of given terms and then choose the best among them with some heuristic rules. Since the hypernyms were organized hierarchically, the higher is the level, the more generic are the hypernyms. To maintain the specificity of the set of terms while revealing their general topics, the heuristics have to choose as low-level common hypernyms as possible. When there are multiple choices, ranks should be given to order the hypernyms in priority.

In our implementation, we look up for each given term all its hypernyms alone the path up to the root in the hierarchical tree. The number of occurrence (*f*) and the depth in the hierarchy (*d*) of an encountered hypernym are recorded. With the root being given a depth value of 0, a weight proportional to the normalized f and d is calculated for each hypernym as follows:

$$weight(hypernym) = \frac{f}{nt} \times 2 \times \left(\frac{1}{1+\exp^{-c \times d}} - 0.5 \right) \cdot \tag{3}$$

where *nt* is the number of given terms to normalize the occurrence *f* to a value ranges from 0 to 1 and *c* (0.125 in our implementation) is a constant to control the steepness of the sigmoid function *1/(1+exp(-c x d))* whose value approaches 1 (-1) for large positive (negative) *d*. Since the depth *d* only takes on non-negative value, the actual range of the sigmoid is from 0.5 to 1. It is thus subtracted with 0.5 and then multiplied by 2 to map the value of *d* into the normalized range: 0 to 1. Note that a term having no hypernym or not in WordNet is omitted from being counted in *nt*. Also note that a term can have multiple hypernyms and thus multiple paths to the root. A hypernym is counted only once for each given term, no matter how many times the multiple paths of this term pass this hypernym. This weight is finally used to sort the hypernyms in decreasing order to suggest priority.

In the example mentioned at the beginning, where the 3 terms were given: *table*, *chair*, and *bed*, their hypernym: "*furniture*" did result from the above calculation with a highest weight 0.3584 based on WordNet version 1.6.

5 Experiments

5.1 Document Collections

Two collections were used to evaluate the proposed method. One contains 612 patent documents downloaded on 2005/06/15 from the USPTO's website [20] with "National Science Council" (NSC) as the search term in the assignee field. NSC is the major government agency that sponsors research activities in Taiwan. Institutes, universities, or research centers, public or private, can apply research fundings from NSC. Once the research results have been used to apply for US patents, the intellectual property rights belong to NSC. In other words, NSC becomes the assignee of the patents. (However, this policy has been changed since year 2000 so that the number of NSC patents per year declines since then.) Due to this background, these documents constitute a knowledge-diversified collection with relative long texts (about 2000 words per document) describing various advanced technical details. Analysis of the topics in this collection becomes a non-trivial task as very few analysts know of such diversified technologies and fields. Although each patent has pre-assigned International Patent Classification (IPC) or US Patent Classification (UPC) codes, many of these labels are either too general or too specific such that they do not fit the intended knowledge structures for topic interpretation. Therefore, using them alone do not meet the requirement of the analysis. As such, computer-suggested labels play important roles in helping humans understand this collection. This work is in fact motivated by such a need.

The other collection is a subset of the widely used Reuters-21578 collection for text categorization. It contains brief stories (about 133 words per story) in the financial domain. Each document was assigned at least one category to denote its topics and there are a total of 90 categories based on Yang's preprocessing rules [21]. In this work we used the largest 10 categories of them to evaluate our method.

5.2 Document Clustering

Since it would be hard to have a consensus on the number of clusters to best organize the NSC collection, we clustered it using different options and parameters to get various views on it. Specifically, we analyzed the collection based on document clustering and term clustering as well.

The 612 US patent documents were parsed, filtered, segmented, and summarized. Structured information in the documents such as assignees, inventors, and IPC codes were removed. The remaining textual parts were segmented into several main sections (Abstract, Claims, Field of the Invention, Background of the Invention, Summary of the Invention, and Detailed Descriptions) based on the regularity of the patent document style. Since these sections varies drastically in length, they were summarized using extraction-based methods [22] to select at most 6 best sentences from each section. They were then concatenated to yield the document surrogates for key term extraction, term co-occurrence analysis, indexing, and clustering. These document surrogates did not include the Claims section due to its legal-oriented contents.

From the document surrogates, 19,343 key terms (terms occur at least twice in a document) were extracted. They were used in the term co-occurrence analysis to see which terms are often co-occurred with each other in the same sentence [23]. Only 2714 of the key terms occur in at least two documents and have at least one term co-occurred with them. These 2714 terms were clustered by a complete-link method into 353 small-size clusters, based on how many co-occurred terms they share. These clusters were then repeatedly clustered, again based on the common co-occurred terms (*co-words*), into 101 medium-size clusters, then into 33 large-size clusters, and finally into 10 topic clusters. The reason that we performed this multi-stage clustering is due to the fact that even we set a lowest threshold, we could not obtain as low as 10 clusters in a single step. The 353 small-size clusters are actually obtained with the similarity threshold set to 0.0. In other words, among the 2714 terms, no two terms in different small-size clusters share the same co-words. Through this term clustering, the topics of the collection were revealed as the key terms were merge into concepts, which in turn were merge into topics or domains. This approach has the merit that it can be efficiently applied to very large collections due to the fast co-occurrence analysis [23].

As another way for topic analysis, the 612 NSC patents were clustered directly based on their summary surrogates. As with the above term clustering, they were first clustered into 91 topics, which in turn were grouped into 21 sub-domains, from which 6 major domains were found.

We did not cluster the Reuters collection since it already has been organized by its predefined labels. These labels were used to compare with our cluster titles.

5.3 Evaluation of Descriptor Selection

For comparison, we used three methods to rank cluster terms for descriptor selection, namely *TFC*, $CC_{0.5}$, and *CCxTFC*, as defined in Section 3. These ranking methods were applied to three sets of clustering results. The first is the first-stage document clustering as described above. The second is the second-stage document clustering. The third is the third-stage term clustering based on their co-words. For each cluster, at most 5 best terms were selected as its descriptors.

Two master students majored in library science were invited to compare the relative quality of these descriptors under the same clusters. For each ranking method, the number of cases where it has the best title quality over the other methods was counted. Multiple best choices for a cluster are allowed and those cluster titles that are hard to assess can be omitted from being considered. The ranking methods are coded in such a way that the assessors do not know which method is used to generate the examined titles. The assessment results are shown in Table 2. Note hierarchical clustering structures were shown to the assessors. They were free to examine whatever clusters or sub-clusters they were interested in. This is why the numbers of examined clusters differ between them.

In spite of this difference, this preliminary experiment shows that cluster descriptors selected by $CC_{0.5}$ or CCxTFC are favorable by one assessor, while those generated by TFC are not by either. This is somewhat useful to know, since most past studies use TFC or a variation of it to generate cluster titles.

Table 2. Comparison of three descriptor selection methods among three cluster sets

Ranking method / Cluster set	Assessor	No. of clusters examined	No. of clusters that TFC is the best		No. of clusters that $CC_{0.5}$ is the best		No. of clusters that CCxTFC is the best	
First-stage doc. clusters	1	73	5	7%	52	**71%**	20	27%
	2	19	6	32%	12	63%	13	**68%**
Second-stage doc. clusters	1	13	0	0%	9	**69%**	6	46%
	2	7	1	14%	3	43%	5	**71%**
Third-stage term clusters	1	16	5	31%	12	**75%**	9	56%
	2	10	5	50%	4	40%	8	**80%**

5.4 Evaluation of Generic Title Generation

The proposed title mapping algorithm was applied to the final-stage results of the document and term clustering described above. The first set has 6 clusters and the second has 10. Their best 5 descriptors selected by CCxTFC are shown in the second column of Table 3.

The proposed method was compared to a similar tool called InfoMap [24] which is developed by the Computational Semantics Laboratory at Stanford University. This online tool finds a set of taxonomic classes for a list of given words. It seems that WordNet is also used as its reference system, because the output classes are mostly WordNet's terms. Since no technical details about InfoMap were found on the Web, we cannot implement the InfoMap's algorithm by ourselves. Therefore, an agent program was written to send the descriptors to InfoMap and collect the results that it returns. Only the top-three candidates from both methods are compared. They are listed in the last two columns in Table 3, with their weights appended.

The reasonable classes are marked in bold face in the table. As can be seen, the two methods perform similarly. Both achieve a level of 50% accuracy in either set.

The 10 largest categories in the Reuters collection constitute 6018 stories. Since they are both short and in a large number in each of these categories, the basic CC method was used to select the cluster descriptors. As can be seen from the third column in Table 4, 8 (those in boldface) out of 10 machine-selected descriptors clearly coincide with the human labels, showing that the effectiveness of the CC method in this collection. When compared to the results made by [25], as shown in Table 5, where *mutual information* (MI) was used to rank statistically significant features for text categorization, the CC method yields better topical labels than the MI method.

Table 3. Descriptors and their machine-derived generic titles for the NSC collection

ID	Cluster's Descriptors	WordNet	InfoMap
1	acid, polymer, catalyst, ether, formula	1:substance, matter:0.1853 2:drug:0.0980 **3:chemical compound:0.098**	**1:chemical compound:1.25** 2:substance, matter:1.062 3:object, physical object:0.484
2	silicon, layer, transistor, gate, substrate	1:object, physical object:0.1244 2:device:0.1211 3:artifact, artefact:0.1112	1:object, physical object:0.528 2:substance, matter:0.500 3:region, part:0.361
3	plastic, mechanism, plate, rotate, force	1:device:0.1514 2:base, bag:0.1155 3:cut of beef:0.1155	1:device:0.361 2:entity, something:0.236 3:chemical process:0.0
4	output, signal, circuit, input, frequency	**1:communication:0.1470** **2:signal, signaling, sign:0.1211** 3:relation:0.0995	**1:signal, signaling, sign:1.250** **2:communication:1.000** 3:abstraction:0.268
5	powder, nickel, electrolyte, steel, composite	1:substance, matter:0.1483 **2:metallic element, metal:0.1211** 3:instrumentation:0.0980	**1:metallic element, metal:0.500** 2:substance, matter:0.333 3:entity, something:0.203
6	gene, protein, cell, acid, expression	1:substance, matter:0.1112 2:object, physical object:0.0995 3:chemical compound:0.0980	1:entity, something:0.893 2:chemical compound:0.500 3:object, physical object:0.026
1	resin, group, polymer, compound, methyl	1:substance, matter:0.1853 **2:chemical compound:0.098** 3:whole:0.0717	1:substance, matter:2.472 2:object, physical object:0.736 **3:chemical compound:0.5**
2	circuit, output, input, signal, voltage	**1:communication:0.1470** **2:signal, signaling, sign:0.1211** 3:production:0.0823	**1:signal, signaling, sign:1.250** **2:communication:1.000** 3. round shape:0.000
3	silicon, layer, material, substrate, powder	1:substance, matter:0.1483 2:object, physical object:0.1244 3:artifact, artefact:0.1112	1:substance, matter:2.250 2:artifact, artefact:0.861 3:object, physical object:0.833
4	system, edge, signal, type, device	1:artifact, artefact:0.1483 **2:communication:0.1470** 3:idea, thought:0.0980	1:instrumentality:1.250 **2:communication:0.861** 3:artifact, artefact:0.750
5	solution, polyaniline, derivative, acid, aqueous	1:communication:0.1633 2:legal document,:0.1540 3:calculation, computation:0.1372	1:drug of abuse, street drug:0.000 **2:chemical compound:0.0** 3:set:0.000
6	sensor, magnetic, record, calcium, phosphate	1:device:0.1514 2:object, physical object:0.1244 **3:sound/audio recording:0.12**	1:device:0.312 2:fact:0.000 3:evidence:0.000
7	gene, cell, virus, infection, plant	1:structure, construction:0.1225 2:contrivance, dodge:0.1155 3:compartment:0.1029	1:entity, something:0.790 **2:life form, living thing:0.5** 3:room:0.000
8	density, treatment, strength, control, arrhythmia	1:property:0.1112 2:economic policy:0.1020 3:attribute:0.0995	1:power, potency:1.250 2:property:0.674 3:condition, status:0.625
9	force, bear, rod, plate, member	1:pistol, handgun, side arm:0.1020 2:unit, social unit:0.0980 3:instrumentation:0.0980	1:unit, social unit:1.250 2:causal agent, cause,:0.625 3:organization:0.500
10	transistor, layer, channel, amorphous, effect	1:artifact, artefact:0.1390 2:structure, body structure:0.1225 **3:semiconductor:0.1029**	1:anatomical structure:0.500 2:artifact, artefact:0.040 3:validity, validness:0.000

As to the generic titles shown in Table 4, about 70% of the cases lead to reasonable results for both the WordNet and the InfoMap methods. It is known that the Reuters labels can be organized hierarchically for further analysis [26]. For example, *grain*, *wheat*, and *corn* can be merged into a more generic category, such as *foodstuff*. As can be seen from Table 4, this situation was automatically captured by the proposed method.

Table 4. Descriptors and their generic titles for the Reuters collection

ID	Category	Descriptors	WordNet	InfoMap
1	earn	NET, QTR, Shr, cts Net, Revs	1:goal:0.2744 2:trap:0.2389 3:**income**:0.2389	1:trap:1.000 2:game equipment:1.000 3:fabric, cloth, textile:1.000
2	acq	**acquire, acquisition,** stake, company, share	1:device:0.1514 2:stock certificate, stock:0.1386 3:wedge:0.1386	1:asset:0.500
3	money-fx	currency, **money market**, central banks, TheBank, yen	1:backlog, stockpile:0.0850 2:airplane maneuver:0.0850 3:marketplace, mart:0.0770	1:medium of exchange, **monetary system**:0.750
4	grain	wheat, **grain**, tonnes, agriculture, Corn	1:seed:0.1848 2:cereal, cereal grass:0.1848 3:**foodstuff**, food product:0.182	1:weight unit:1.250 2:grain, **food grain**:1.250 3:cereal, cereal grass:1.250
5	crude	**crude oil**, bpd, OPEC, mln barrels, petroleum	1:lipid, lipide, lipoid:0.2150 2:oil:0.1646 3:**fossil fuel**:0.1211	1:oil:0.750 2:**fossil fuel**:0.750 3:lipid, lipide, lipoid:0.500
6	trade	**trade**, tariffs, Trading surplus, deficit, GATT	1:**business**:0.0823 2:UN agency:0.0823 3:prevailing wind:0.0823	1:liability, financial obligation, indebtedness, pecuniary obligation:0.062
7	interest	rate, money market, BANK, prime, discount	1:charge:0.1372 2:airplane maneuver:0.0850 3:allowance, adjustment:0.0850	1:charge:0.111
8	ship	**ship**, vessels, port, Gulf, TANKERS	1:craft:0.2469 2:instrumentation:0.1959 3:**vessel**, watercraft:0.1848	1:craft:0.812 2:physical object:0.456 3:**vessel**, watercraft:0.361
9	wheat	**wheat**, tonnes, grain agriculture, USDA	1:weight unit:0.1792 2:**foodstuff**, food product:0.151 3:executive department:0.1386	1:weight unit:1.250 2:**foodstuff**, food product:0.500
10	corn	**corn**, maize, tonnes, soybean, grain	1:seed:0.1848 2:cereal, cereal grass:0.1848 3:**foodstuff**, food product:0.182	1:cereal, cereal grass:1.250 2:weight unit:1.250 3:**foodstuff**, food product:0.790

Table 5. Three best words in terms of MI and their categorization *break-even precision* (BEP) rate of the 10 largest categories of Reuters [25]

Category	1st word	2nd word	3rd word	BEP
earn	vs+	cts+	loss+	93.5%
acq	shares+	vs-	inc+	76.3%
money-fx	dollar+	vs-	exchange+	53.8%
grain	wheat+	tonnes+	grain+	77.8%
crude	oil+	bpd+	OPEC+	73.2%
trade	trade+	vs-	cts-	67.1%
interest	rates+	rate+	vs-	57.0%
ship	ships+	vs-	strike+	64.1%
wheat	wheat+	tonnes+	WHEAT+	87.8%
corn	corn+	tonnes+	vs-	70.3%

6 Discussions

The application of our generic title generation algorithm to the NSC patent collection leads to barely 50% of the clusters to have reasonable results. However, this is due to the limit of WordNet in this case. WordNet does not cover all the terms extracted from the patent documents. Also WordNet's hypernym structures may not reflect the knowledge domains desired for analyzing these patents. On one hand, if the texts of the documents better fit the vocabulary of WordNet, the performance improves, as is the case in the Reuters collection. On the other hand, it is thus anticipated that if a suitable classification system can be found for the document collection to be analyzed, the hypernym search algorithm may lead to more desirable labels. We have tried Google's directory for the NSC collection, but the categories returned from searching the descriptors did not meet the intension to analyze the collection. So far we haven't found any better classification systems that improve the generic titles of the NSC clusters. However, even with the current level of performance using WordNet, this method, as well as the method to select the cluster descriptors, are still helpful, as noted by some NSC analysts.

As an example, Figure 1 shows a topic map of the 612 NSC patents. In this figure, each circle denotes an identified topic, the size of the circle denotes the number of patents belonging to the topic, and the number in the circle corresponds to the topic ID. With the *Multi-Dimensional Scaling* (MDS) method [27] to map these clustered topics, the relative distance between each topic in the map reflects their relatedness (while the absolute position and orientation of each topic does not matter). Details of the topic titles and their IDs are shown in Table 6. The 6 major domains circled by dashed lines in the figure were those derived from the final-stage document clustering and were labeled by hand with the help of the proposed method.

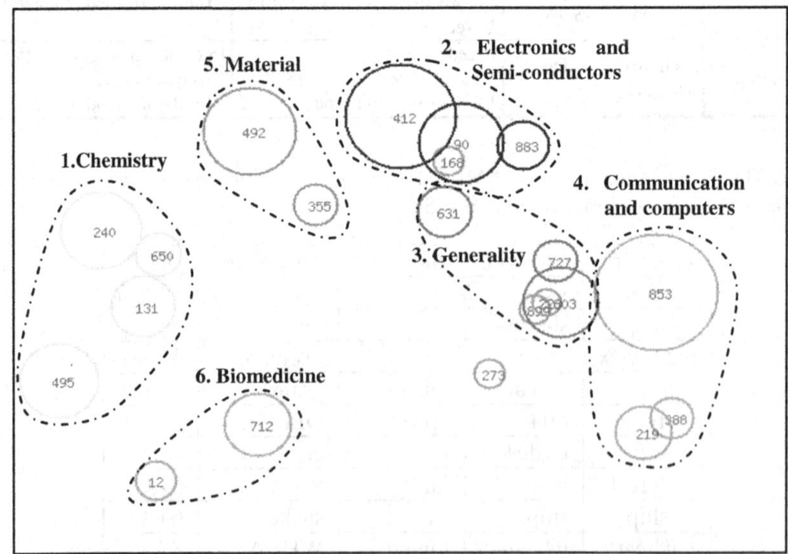

Fig. 1. A top map resulted from analyzing the 612 NSC patent documents

Table 6. Final-stage document clusters from the NSC patents

1: 122 docs. : 0.2013 (acid:174.2, polymer:166.8, catalyst:155.5, ether:142.0, formula:135.9)
 * 108 docs. : 0.4203 (polymer:226.9, acid:135.7, alkyl:125.2, ether:115.2, formula:110.7)
  o 69 docs. : 0.5116 (resin:221.0, polymer:177.0, epoxy:175.3, epoxy resin:162.9, acid: 96.7)
   + ID=131 :26 docs.:0.2211(polymer: 86.1, polyimide: 81.1, aromatic: 45.9, bis: 45.1, ether ...)
   + ID=240 : 43 docs.:0.1896(resin:329.8, acid: 69.9, group: 57.5, polymer: 55.8, monomer: 44.0)
  o ID=495:39 docs.:0.1385(compound: 38.1, alkyl: 37.5, agent: 36.9, derivative: 33.6, formula ...)
 * ID=650 : 14 docs. : 0.1230(catalyst: 88.3, sulfide: 53.6, iron: 21.2, magnesium: 13.7, selective: 13.1)

2: 140 docs. : 0.4068 (silicon:521.4, layer:452.1, transistor:301.2, gate:250.1, substrate:248.5)
 * 123 docs. : 0.5970 (silicon:402.8, layer:343.4, transistor:224.6, gate:194.8, schottky:186.0)
  o ID=412 : 77 docs. : 0.1503(layer:327.6, silicon:271.5, substrate:178.8, oxide:164.5, gate:153.1)
  o ID=90 : 46 docs. : 0.2556(layer:147.1, schottky:125.7, barrier: 89.6, heterojunction: 89.0, ...
 * ID=883 : 17 docs. : 0.1035(film: 73.1, ferroelectric: 69.3, thin film: 48.5, sensor: 27.0, capacitor ...)

3: 66 docs. : 0.2203 (plastic:107.1, mechanism: 83.5, plate: 79.4, rotate: 74.9, force: 73.0)
 * 54 docs. : 0.3086 (plastic:142.0, rotate:104.7, rod: 91.0, screw: 85.0, roller: 80.8)
  o ID=631 : 19 docs.:0.1253(electromagnetic: 32.0, inclin: 20.0, fuel: 17.0, molten: 14.8, side: 14.8)
  o ID=603 : 35 docs. : 0.1275(rotate:100.0, gear: 95.1, bear: 80.0, member: 77.4, shaft: 75.4)
 * ID=727 : 12 docs. : 0.1155(plasma: 26.6, wave: 22.3, measur: 13.3, pid: 13.0, frequency: 11.8)

4: 126 docs. : 0.4572 (output:438.7, signal:415.5, circuit:357.9, input:336.0, frequency:277.0)
 * 113 docs. : 0.4886 (signal:314.0, output:286.8, circuit:259.7, input:225.5, frequency:187.9)
  o ID=853 : 92 docs. : 0.1052(signal:386.8, output:290.8, circuit:249.8, input:224.7, light:209.7)
  o ID=219 : 21 docs. : 0.1934(finite: 41.3, data: 40.7, architecture: 38.8, comput: 37.9, algorithm: ...
 * ID=388 : 13 docs. : 0.1531(register: 38.9, output: 37.1, logic: 32.2, addres: 28.4, input: 26.2)

5: 64 docs. : 0.3131 (powder:152.3, nickel: 78.7, electrolyte: 74.7, steel: 68.6, composite: 64.7)
 * ID=355 : 12 docs. : 0.1586(polymeric electrolyte: 41.5, electroconductive: 36.5, battery: 36.1, ...
 * ID=492 : 52 docs. : 0.1388(powder:233.3, ceramic:137.8, sinter: 98.8, aluminum: 88.7, alloy: 63.2)

6: 40 docs. : 0.2501 (gene:134.9, protein: 77.0, cell: 70.3, acid: 65.1, expression: 60.9)
 * ID=12 : 11 docs. : 0.3919(vessel: 30.0, blood: 25.8, platelet: 25.4, dicentrine: 17.6, inhibit: 16.1)
 * ID=712 : 29 docs. : 0.1163(gene:148.3, dna: 66.5, cell: 65.5, sequence: 65.1, acid: 62.5)

7 Conclusions

Cluster labeling is important for ease of human analysis. In the attempt to produce generic cluster labels, a hypernym search algorithm based on WordNet is devised. Our method takes two steps: the content-indicative terms are first extracted from the documents; these terms are then map to their common hypernyms. Because the algorithm uses only the depth and occurrence information of the hypernyms, it is general enough to be able to adopt other hierarchical knowledge systems without much revision. Real-case experiments using two very different collections justify our anticipation on the performance of the proposed method. Possible improvements were suggested and have been tried. The equality in performance with other similar systems such as InfoMap suggests that our well-described method are among the very up-to-date approaches to solve the problem like this.

Acknowledgments. This work is supported in part by NSC under the grants: NSC 93-2213-E-030-007- and NSC 94-2524-S-003-014-.

References

1. Noyons, E. C. M. & Van Raan, A. F. J. "Advanced Mapping of Science and Technology," Scientometrics, Vol. 41, 1998, pp. 61-67.
2. "The 8th Science and Technology Foresight Survey - Study on Rapidly-developing Research Areas - Interim Report", Science and Technology Foresight Center, National Institute of Science & Technology Policy, Japan, 2004.
3. Hideyuki Uchida, Atsushi Mano and Takashi Yukawa, "Patent Map Generation Using Concept-Based Vector Space Model" Proceedings of the Fourth NTCIR Workshop on Evaluation of Information Access Technologies: Information Retrieval, Question Answering, and Summarization, June 2-4, 2004, Tokyo, Japan.
4. Patrick Glenisson, Wolfgang Gla"nzel, Frizo Janssens and Bart De Moor , "Combining Full Text and Bibliometric Information in Mapping Scientific Disciplines," Information Processing & Management, Vol. 41, No. 6, Dec. 2005, pp. 1548-1572.
5. Kuei-Kuei Lai and Shiao-Jun Wu, "Using the Patent Co-citation Approach to Establish a New Patent Classification System," Information Processing & Management, Vol. 41, No. 2, March 2005, pp.313-330.
6. Cutting, D. R. Karger, J. O. Pedersen, and J. W. Tukey. "Scatter/gather: A cluster-based approach to browsing large document collections," Proceedings of the 15th ACM-SIGIR Conference, 1992, pp. 318-329.
7. Marti A. Hearst and Jan O. Pedersen, "Reexamining the Cluster Hypothesis: Scatter/Gather on Retrieval Results," Proceedings of the 19th ACM-SIGIR Conference, 1996, pp. 76-84.
8. Yiming Yang, Tom Ault, Thomas Pierce and Charles W. Lattimer, "Improving Text Categorization Methods for Event Tracking," Proceedings of the 23rd ACM-SIGIR Conference,, 2000, pp. 65-72.
9. Mehran Sahami, Salim Yusufali, and Michelle Q. W. Baldonaldo, "SONIA: A Service for Organizing Networked Information Autonomously," Proceedings of the 3rd ACM Conference on Digital Libraries, 1998, pp. 200-209.
10. Krista Lagus, Samuel Kaski, and Teuvo Kohonen, "Mining Massive Document Collections by the WEBSOM Method," Information Sciences, Vol 163/1-3, pp. 135-156, 2004.
11. Russell Swan and James Allan, "Automatic Generation of Overview Timelines," Proceedings of the 23rd ACM-SIGIR Conference, 2000, pp. 49-56.
12. Oren Zamir and Oren Etzioni, "Web document clustering: a feasibility demonstration," Proceedings of the 21st ACM-SIGIR Conference, 1998, pp. 46-54.
13. Document Understanding Conferences, http://www-nlpir.nist.gov/projects/duc/.
14. Michele Banko, Vibhu O. Mittal, and Michael J. Witbrock, "Headline Generation Based on Statistical Translation," ACL 2000.
15. Paul E. Kennedy, Alexander G. Hauptmann, "Automatic title generation for EM," Proceedings of the 5th ACM Conference on Digital Libraries, 2000, pp.
16. Yiming Yang and J. Pedersen, "A Comparative Study on Feature Selection in Text Categorization," Proceedings of the International Conference on Machine Learning (ICML'97), 1997, pp. 412-420.
17. Hwee Tou Ng, Wei Boon Goh and Kok Leong Low, "Feature Selection, Perception Learning, and a Usability Case Study for Text Categorization," Proceedings of the 20th ACM-SIGIR Conference, 1997, pp. 67-73.
18. Ronen Feldman, Ido Dagan, Haym Hirsh, "Mining Text Using Keyword Distributions," Journal of Intelligent Information Systems, Vol. 10, No. 3, May 1998, pp. 281-300.

19. WordNet: a lexical database for the English language, Cognitive Science Laboratory Princeton University, http://wordnet.princeton.edu/.
20. "United States Patent and Trademark Office", http://www.uspto.gov/
21. Yiming Yang and Xin Liu, "A Re-Examination of Text Categorization Methods," Proceedings of the 22nd ACM-SIGIR Conference, 1999, pp. 42-49.
22. Yuen-Hsien Tseng, Dai-Wei Juang, Yeong-Ming Wang, and Chi-Jen Lin, "Text Mining for Patent Map Analysis," Proceedings of IACIS Pacific 2005 Conference, May 19-21, 2005, Taipei, Taiwan, pp.1109-1116.
23. Yuen-Hsien Tseng, "Automatic Thesaurus Generation for Chinese Documents", Journal of the American Society for Information Science and Technology, Vol. 53, No. 13, Nov. 2002, pp. 1130-1138.
24. Information Mapping Project, Computational Semantics Laboratory, Standford University, http://infomap.stanford.edu/.
25. Ron Bekkerman, Ran El-Yaniv, Yoad Winter, Naftali Tishby, "On Feature Distributional Clustering for Text Categorization," Proceedings of the 24th ACM-SIGIR Conference, 2001, pp.146-153.
26. Ido Dagan and Ronen Feldman, "Keyword-based browsing and analysis of large document sets," Proceedings of the Symposium on Document Analysis and Information Retrieval (SDAIR-96), Las Vegas, Nevada, 1996.
27. Joseph B. Kruskal, "Multidimensional Scaling and Other Methods for Discovering Structure," pp. 296-339 in "Statistical Methods for Digital Computers" edited by Kurt Enslein, Anthony Ralston, and Herbert S. Wilf, Wiley: New York, 1977.

Word Sense Language Model for Information Retrieval

Liqi Gao, Yu Zhang, Ting Liu, and Guiping Liu

Information Retrieval Laboratory, School of Computer Science and Technology, Box 321, Harbin Institute of Technology, Harbin, P.R. China, 150001,
{lqgao, yzhang, tliu, gpliu}@ir-lab.org
http://www.ir-lab.org/

Abstract. This paper proposes a word sense language model based method for information retrieval. This method, differing from most of traditional ones, combines word senses defined in a thesaurus with a classic statistical model. The word sense language model regards the word sense as a form of linguistic knowledge, which is helpful in handling mismatch caused by synonym and data sparseness due to data limit. Experimental results based on TREC-Mandarin corpus show that this method gains 12.5% improvement on MAP over traditional tf-idf retrieval method but 5.82% decrease on MAP compared to a classic language model. A combination result of this method and the language model yields 8.92% and 7.93% increases over either respectively. We present analysis and discussions on the not-so-exciting results and conclude that a higher performance of word sense language model will owe to high accurate of word sense labeling. We believe that linguistic knowledge such as word sense of a thesaurus will help IR improve ultimately in many ways.

1 Introduction

Natural languages are inherently full of ambiguity and synonymity. One word can be used in various contexts with different meanings, while dissimilar words may represent a same meaning in certain contexts. The variance of word meanings in natural language has posed problems for the applications related to the Natural Language Processing (NLP).

The field of Information Retrieval (IR) is no exception to this problem. For example, if a retrieval system encountered the word 'bank' in a query, should the system regard the word as the meaning of 'either side of rivers' or 'the place where money is deposited'? Besides, a system would perform better if it finds words as 'resume' when a user inputs phrases like 'curriculum vitae'. IR systems based on keywords usually suffer from problems out of variance meanings of words.

For that matter, it seems to be reasonable to assume that an IR system will improve its performance if the documents are represented by words and word senses rather than the former only. Besides, word senses, defined in thesaurus, can be viewed as knowledge representation. It is obvious that every thesaurus

H.T. Ng et al. (Eds.): AIRS 2006, LNCS 4182, pp. 158–171, 2006.

published owes much to expert staffs of lexicographers, who contribute to refining knowledge of inner connections between words in a language. IR systems must benefit from this kind of knowledge if they make the most of resources like thesauri. Research of years ago focused on this point of using word senses in information retrieval. Classical methods were most based on empirical methods. And some of experiments on disambiguated collection showed a drop in retrieval performance (e.g. [1][2]).

In recent years, language modeling approaches for IR has been prevailing, which are based on statistical models with solid mathematics foundation. Their improvement in performance makes them the state-of-the-art methods in IR evaluations.

In this paper we present a word sense language modeling method, taking word senses into account within a language model framework. The word sense model has a close connection with classic language models but the starting point towards retrieval is diverse. Experiments were conducted using TREC-Mandarin collection and TREC-5/6 topics. The collection was labeled with word sense codes defined in Tongyici Cilin, a well-known Chinese thesaurus in NLP.

The remainder of this paper is organized in this way: Section 2 reviews previous methods using word senses in IR; Section 3 presents the word sense language model; Section 4 and Section 5 describe series of experiments and analysis. Our conclusion on the attempt use of word senses in language model and future work come in Section 6.

2 Previous Work

There has been a lot of work dealing with combination of word sense with information retrieval.

The first work where a word sense based algorithm was used with an IR system was done by Weiss [3]. He tested the algorithm on five handpicked words in the Aviation Data Internet (ADI) collection. Resolving ambiguities before information retrieval, Weiss showed that he improved performance of his IR system by 1% for the five test words. Krovetz and Croft [2] established more interesting results, giving information about the relation between Word Sense Disambiguation (WSD) and IR. They concluded that WSD did not have a very important impact on IR, but that disambiguation could be beneficial to IR when there were a few words in common between the query and the document. Voorhees [4] presented a method using WordNet to disambiguate word sense for text retrieval. Retrieval experiments comparing the effectiveness of sense-based vectors which are based on IS-A relations contained in WordNet with stem-based vectors show stem-based vectors to be superior overall. Voorhees concludes that the IS-A relation does not fit to word sense selection for IR. In 1994, Mark Sanderson [1] used a technique that introduces additional sense ambiguity into a collection, doing a research that goes beyond previous work in this field to reveal the influence that ambiguity and disambiguation have on a probabilistic IR system. And in late 1990s, O. Uzuner [5] developed a

WSD algorithm which will be used in disambiguating natural language queries to an IR system. In 2003, Stokoe et al [6] make a study, exploring the using of word sense disambiguation in retrieval effectiveness and subsequent evaluation of a statistical word sense disambiguation system which demonstrates increased precision from a sense based vector space retrieval model over traditional tf-idf techniques.

A large amount of work dealing with language modeling for information retrieval has been done since 1998. The basic approach for using language models for IR assumes that the user has a reasonable ideal of the terms that are likely to appear in the "ideal" document that can satisfy his/her information need, and the query terms the user chooses can distinguish the "ideal" document from the rest of the collection (e.g. [7]). The query is generated as the piece of text representative of the "ideal" document. The task of the system is then to estimate, for each of the documents in the collection, which is most likely to be the ideal document. Liu et al gives a nice review of statistical language modeling for IR [8].

3 Word Sense Language Model for IR

3.1 Word Sense Representation

When concerning with the word sense, we need to two assumptions [9]:

(1) Only one sense of a word is used in each occurrence;
(2) Each word has a fixed number of senses.

It is necessary to find a practical way to represent word senses in IR systems. There are mainly two ways: one is to pick up representative words for word senses and the other is to design a coding system for word sense representation. We follow the latter way. In our system, we use a thesaurus called Tongyici Cilin, which has assigned a unique code to a word sense. For the sake of further discussion, we will describe the details about the thesaurus.

Tongyici Cilin (Cilin for short) is a well-known Chinese thesaurus in Chinese Natural Language Processing (NLP). It was first published approximately in the 1980's. Cilin contains more than 70,000 of words, all of which are arranged into three-level categories according to the word senses of words. The top level of Cilin consists of 12 main classes, denoted by a single capital letter. The second level of Cilin consists of 94 classes which are labeled by two lowercase letters. At the third level, concepts have been classified into 1,428 classes named after two-bit numbers as the labels. Information Retrieval Laboratory (IR-Lab, HIT) extended Cilin from three levels to five levels with more than 30,000 old words replaced or added (see Figure 3.1). The fourth level in the latest version has been extended for concept clusters, labeled by a single capital letter, for words with closer meaning compared to the third level. The deepest level in which words are nearly synonyms stands for atomic concepts. As a whole, every word sense can be coded uniquely within a tree structure. Our work has been based on the extended version of Cilin.

Every word sense has been assigned an eight-bit code for five levels. Look at Table 1. It defines bits and symbols for each level. There are three extra flags at eighth bit of codes for future extension. All the words in the same class of the fifth level category can be regarded as of similar meaning. For example, given the word "小时" (xiaoshi/hour), we can find the synonyms such as "钟点"(zhongdian/hour), "钟头"(zhongtou/hour) and "时"(shi/time or time unit).

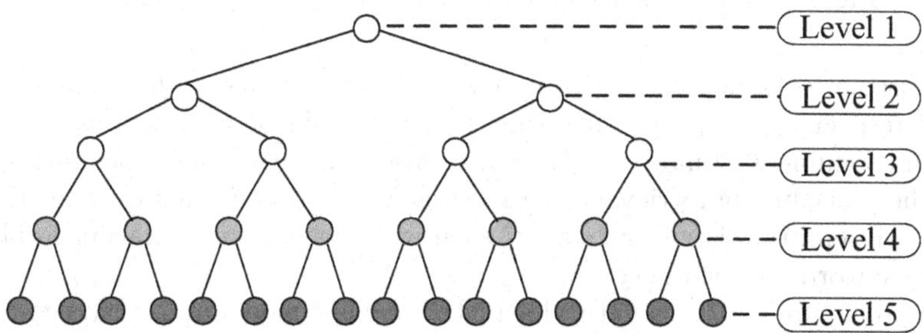

Fig. 1. This figure illustrates the hierarchical structure of "Tongyici Cilin (Extended)"

Table 1. This table shows each bit of the code defined in Cilin

<table>
<tr><th>bit</th><th>property</th><th>level</th><th>example</th></tr>
<tr><td>1</td><td>top</td><td>1</td><td>D</td></tr>
<tr><td>2</td><td>medium</td><td>2</td><td>A</td></tr>
<tr><td>3</td><td rowspan="2">bottom</td><td rowspan="2">3</td><td>1</td></tr>
<tr><td>4</td><td>5</td></tr>
<tr><td>5</td><td>cluster</td><td>4</td><td>B</td></tr>
<tr><td>6</td><td rowspan="2">atom</td><td rowspan="2">5</td><td>0</td></tr>
<tr><td>7</td><td>2</td></tr>
<tr><td>8</td><td>flag</td><td></td><td>-</td></tr>
</table>

Three-level codes of word sense in the thesaurus Cilin are usually used for labeling because this level is a balance of similar meanings and distinguishable characteristics. Figure 2 is an example of a tagged sentence with word sense codes in Cilin:

In other languages, similar forms of word sense codes for synonyms in thesauri are available, such as Roger's Thesaurus, WordNet and the like. Those codes, though differ from code form or definition of Cilin mentioned above, can also work in applications of word sense language model for IR. It is just like that the language models work on both English words and Chinese words.

3.2 The Word Sense Language Model

In the language modeling approaches to information retrieval, it is a common way to estimate multinomial models over terms for each document D in the collection

美国/Di02 国务卿/Di15 贝克/-1 昨天/Ca23 和/Kc01 今天/Ca23 分别/Ka23 会见/ Hi05 来访/Hi02 的/Kd01 英国/Di02 外交大臣/Af08 赫德/-1 和/Kc01 法国/Di02 外交部长/Af10 迪马/-1 ，/-1 双方/Dd05 就/Ka07 海湾/Cb08 战争/Di11 后/Cb04 如何/Ka35 保障/Hi37 地区/Cb08 和平/Ef01 问题/Da01 交换/Hi27 了/Kd05 意见/ Df14 。/-1

Fig. 2. It is an example of a sentence after word sense resolution

C, the whole set of documents to be searched. Given a query Q, documents are ranked respect to the probability that $P(Q|D)$ could be observed as a sample from the document D models. The word sense language model goes further. It takes the procedure of retrieval as a two-stage generation: (1) a sequence of word sense representation is generated; (2) each word sense representation produces respective words in the query.

The procedure goes as a natural way. In language modeling for IR, the query is usually viewed as a specific representation of a user's information need. We assume that the user constructs a query in two steps: he or she forms meanings of the information need in the mind first and then choose proper words and expressions in order to write out exact query phrases or paragraphs. The word sense retrieval model follows as the user's way by calculating the probability of a document generating the query at the word sense level first and then probability from senses to query words.

The model calculates the probability of by covering all possible word sense tagging forms:

$$P(Q|D) = \sum_{S} P(Q, S|D) = \sum_{S} P(S|D)P(Q|S, D) \tag{1}$$

where S stands for any possible word sense resolution schemas.

It is nearly impossible to enumerate all possible schemas in (1) to calculate the probability and difficult to estimate so many parameters. One assumption would simply this model [10]: following the common practice in statistical language modeling, we can assume that there exists a primary word sense resolution which dominates the sum over all possible resolutions. We will just use S representing S^* for simplicity below. The equation (1) can be approximately simplified as follows:

$$P(Q|D) = P(S^*|D)P(Q|S^*, D) \tag{2}$$

such that $S^* = \arg\max_{S} P(S|Q)$.

We can find that equation (2) is consist of two parts: one is the language model of word sense codes $P(S|D)$ and the other $P(Q|S, D)$ is a model describing the generation of query Q given a word sense resolution S and documents D.

Since there has been a large amount of work dealing with language models in information retrieval, any reasonable language model in IR could be applied in modeling on word sense codes. One of the most popular language models for

IR is based on an assumption that each term is dependent on previous terms, which is also called history. Let S be the word sense code sequence for query Q where s_i is the ith word sense code corresponding to the ith word in Q. Then the word sense code language model can be rewritten as:

$$P(S|D) = \prod_i P(s_i|s_1^{i-1}, D) \tag{3}$$

If the history is limited as size N with a Markov assumption, the equations can be approximated as follows:

$$P(S|D) \doteq \prod_i P(s_i|s_{i-N+1}^{i-1}, D) \tag{4}$$

When concerning with the generation model $P(S|D)$, we make an assumption that every word sense generates a term independently. Following that assumption, the model can be rewritten as:

$$P(Q|S, D) = \prod_i P(Q|s_i, D) = \prod_i P(q_i|s_i, D) \tag{5}$$

Thus by combining two models (4) (5), we get the ranking strategy function based on word sense language model:

$$P(Q|D) = \prod_i P(s_i|s_{i-N+1}^{i-1}, D)P(q_i|s_i, D) \tag{6}$$

Usually, a logarithm form is needed:

$$\log P(Q|D) = \sum_i \left(\log P(s_i|s_{i-N+1}^{i-1}, D) + \log P(q_i|s_i, D)\right) \tag{7}$$

In later experiment, we carried out a unigram model of $P(S|D)$, which yields:

$$\log P(Q|D) = \sum_i (\log P(s_i|D) + \log P(q_i|s_i, D)) \tag{8}$$

3.3 Compare with Other Models

In this section, we will discuss differences and similarities between the word sense language model and some of previously proposed retrieval models based on language modeling.

The word sense language model is different from traditional approaches in IR based on word senses. Most of these approaches focus on disambiguates of word senses with a vector space model, retrieving on sense codes compared with those on words. The weighing strategies are usually empirical. Our model, based on language model, provides a combination within a statistical model.

Our model has a close with recent language modeling approaches for IR such as translation model and dependency language model for IR.

If we take the procedure from a word sense to a concrete word as a translation progress, our model is quite similar with translation model for IR proposed by A. Berger in [11]. But the start point is different. Berger viewed the information retrieval as a translation from documents to query, the modeling of which needs to estimate translation probabilities between terms in the same language. The estimation proves to be hard because it is necessary to provide a large enough corpus, which must include sufficient information on synonymies and related words, for training parameters. The word sense language model takes it natural from word senses to words within thesauri. The estimation is relatively easier than translation model because it is easier to label words with sense codes than to find synonymies by hand.

The word sense language model is also close to dependence language model for IR proposed by Gao *et al* [10]. But both of the two models work with different starting points and knowledge. The dependence language model uses structural information, such as dependency links, for better performance. The word sense language model makes the most of word senses for both parameters smoothing at the sense level and clustering of synonymies at the generating procedure. Besides, Gao's model aims to capture long distance dependency relations while our model tries to exploit abstract word senses for IR.

The reason why those models are more or less similar is that they are all derivations or modifications of a classic model – Hidden Markov Model (HMM), which provides a framework for combining language model with knowledge of linguistics.

4 Parameter Estimation

In equation (3), two main categories of parameters are needed to be estimated as follows.

4.1 Estimating S^*

Recall that S^* is the most probable resolution of word senses to given sentences or phrases. To determine such resolution, a module of word sense disambiguation (WSD) could be used. Much attention has been paid to the research of WSD, which studies how to select correct word senses within a context. WSD is quite difficult because word senses have to be determined by considering contexts, which is hard to model and analysis. Borrowing methods, we implement a tagger for word sense resolution using Naïve Bayesian model [12]:

$$S^* = \arg\max_S P(S|V) \tag{9}$$

The model was trained with 11,331 hand-labeled sentences. Those sentences are all Chinese and the sense codes are based on three-level codes in Tongyici Cilin – the thesaurus introduced above. There are 1,791 words with ambiguous senses in the training data, as shown in Table 2.

Table 2. The distributions of word senses in training data

No of sense	count	ratio
1	18824	91.31%
2	1399	6.24%
3	238	0.74%
4	82	0.26%
5	39	0.12%
6	16	0.05%
7	6	0.02%
8	7	0.02%
10	2	0.01%
12	1	0.00%
13	1	0.00%
labeled	20615	-
unlabeled	11073	-

In this implement, most words were labeled as the word sense code with the largest probability. The precision of this implement of word sense resolution is 75% approximately.

4.2 Estimating $P(s_i|D)$ and $P(q_i|s_i, D)$

We use a two-stage smoothing method, one of the state-of-the-art approaches, to estimate the unigram probability [13].

$$\hat{P}(s_i|D) = (1-\alpha)P(s_i|D) + \alpha P(s_i|C) \doteq (1-\alpha)\frac{C_D(s_i) + C_C(s_i)}{\sum\limits_{s_i} C_C(s_i) + \mu} + \alpha\frac{C_C(si) - \delta}{\sum\limits_{s_i} C_C(s_i)} \tag{10}$$

where μ is a parameter of the Dirichlet distribution, and δ is a constant discount.

And the equation for generation query terms can be estimated in a similar way:

$$\hat{P}(q_i|s_i, D) = (1-\beta)\frac{C_D(q_i, s_i)}{C_D(s_i)} + \beta\frac{C_C(q_i, s_i)}{C_C(s_i)} \tag{11}$$

5 Experiment Results

5.1 Experiment Setup

Our experiment is based on a collection of Chinese corpus: TREC Mandarin (LDC2000T52). For Chinese track in TREC-5 and TREC-6, the collection of TREC-Mandarin consists of the People's Daily and Xinhua News articles includes 164,761 documents, the volume of which is about 170 MB. Before indexing, the whole collection will be labeled with word sense codes of Cilin using a tagger trained out of hand-marked word sense corpus, as we mentioned before.

The topics used for evaluation are provided by TREC as well as relevance judgment for test and evaluation of Chinese text retrieval. Topics in Chinese are similar with those in English and other languages in TREC Tracks. Each topic consists of several fields: document number or ID, title, description and narrative. We constructed queries by merging the title, description and part of narrative.

5.2 Results with Labeled Collection

We carried out several methods: tf-idf (with words), tf-idf (with word sense codes and words), the word sense language model method and a general language method (unigram) on the labeled collection.

Table 3. Results on precision of four methods

	tf-idf(word)	tf-idf(WS)	WSLM	Unigram
P@Recall				
0.00	0.6320	0.6370	0.7517	0.7699
0.10	0.5365	0.5591	0.6188	0.6483
0.20	0.4804	0.5004	0.5540	0.5815
0.30	0.4283	0.4499	0.4962	0.5174
0.40	0.3816	0.3989	0.4338	0.4608
0.50	0.3445	0.3618	0.3764	0.4029
0.60	0.2955	0.3116	0.3070	0.3445
0.70	0.2397	0.2517	0.2343	0.2623
0.80	0.1731	0.1894	0.1521	0.1785
0.90	0.0852	0.0933	0.0709	0.0826
1.00	0.0108	0.0125	0.0084	0.0091
Average	0.3280	0.3423	0.3640	0.3871
Prec@N				
5	0.4481	0.4481	0.5815	0.6259
10	0.4574	0.4629	0.5685	0.5888
15	0.4481	0.4506	0.5580	0.5666
20	0.4324	0.4407	0.5333	0.5425
30	0.4154	0.4327	0.5012	0.5049
100	0.3092	0.3312	0.3485	0.3675
200	0.2304	0.2426	0.2503	0.2625
500	0.1275	0.1304	0.1351	0.1368
1000	0.0735	0.0740	0.0735	0.0735
MAP	0.3136	0.3280	0.3530	0.3748
R-Prec	0.3498	0.3769	0.3864	0.4043

Table 3 shows the results of methods for TREC Mandarin topics 1-54. In the figure we see the eleven point recall/precision results as well as the non-interpolated average precision and R-precision.

The results of the first two columns were obtained by using Lemur toolkit. The first column was based on Chinese words only. We use a module for Chinese word segmentation with an accurate higher than 95%. The experiment of the second column is based on a pair of words and sense code like this form "*word/sense code*". From Figure 3.1, two results are shown as "tf-idf(word)" and "tf-idf(WS)", based on words and sense codes with words respectively. They are only baseline results for comparison. It is interesting that the tf-idf based on word sense codes with words performs a little better than results on words only. The reason might be that word sense resolution reduces noisy by labeling a null sense ("-1") to those out-of-vocabulary terms, such as punctuation and some of unknown words.

Column 3 shows the results based on our model. We implemented a unigram word sense language model in the experiment (named WSLM).

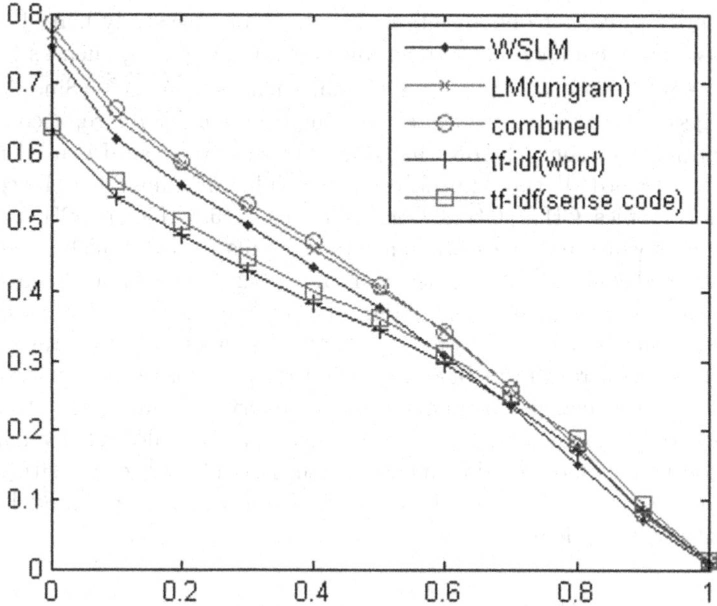

Fig. 3. The Precision-Recall Curves for word sense language model(WSLM), unigram language model, combined results of the two models and results of two tf-idf methods

The fourth column is the result of unigram based on words. A unigram language model for IR [14][15] is used in the experiment for comparison:

$$P(Q|D) = \prod_{q \in Q} (\lambda P(q|C) + (1 - \lambda)P(q|D)) \tag{12}$$

From the result in Table 4 and Figure 5.2, we can see that WSLM performs better than tf-idf approaches with 12.56% and 7.62% increases but a little worse than unigram based on words only with 5.82% decrease. We expected that the word sense language model would help improve IR performance due to synonym knowledge and smoothing effect to some extent. But the results turn out to be opposite. After careful analysis, we find one of main reason leading to this result

Table 4. Performance compared with MAP percent change over WSLM

	MAP	R-Prec	P@10	% change
tf-idf(word)	0.3136	0.3498	0.4574	+12.56
tf-idf(WS)	0.3280	0.3769	0.4629	+7.62
WSLM	0.3530	0.3864	0.5685	–
GLM	0.3748	0.4043	0.5888	-5.82

is that incorrect sense resolution has a much deleterious effort on a retrieval performance, causing mismatches of terms in the documents and the query.

For example, we explored a few ambiguous words in the 54 topics first. The Chinese word '中'seems to be an interesting word for case study here. It carries out at least two meanings: (1) among and within something, such as '大会中' ('during a conference'); (2) short for China, such as '中美关系'(Sino-America relationships). These two meanings are distinguishable. Therefore, according to the assumption mentioned before that only one sense of a word is used in each occurrence, the word'中'should be labeled with different codes due to its context. Among the 54 topics, CH12, CH23, CH32, CH33, CH46, CH50 and CH54 contain the ambiguous word and the word sense tagger fails to distinguish senses of it within restricted contexts. Most of the word meaning 'China' in occurrences were labeled with the former sense as 'among, within and the like'. The case is the same in the collection because we use a same implement of sense tagger. The incorrect resolution results in biased distributions of word senses which cause mismatch between relative documents and the query. It is unrealistic to amend all the incorrectly labeled word senses in such a large collection by hands. If we just modified part of the documents, it can not reflect the real situation in retrieval and yet the real degree of effectiveness of the word sense model. Acute word sense resolution is needed in the word sense language model.

5.3 Results with Combination

Does the word sense based method turn out to hinder the process of retrieval only? Though from the compare with unigram language model it cause a minor due to low precision of word sense resolution, it might be unfair to conclude that the model is totally weak and of no use. It is possible that the sense model describes the documents from a different logical view, which can contribute to boost retrieval results. We carry out an experiment, combining two results of language model based on words and word sense codes to a final result using a linear interpolation:

$$r(D) = \lambda r'_1(D) + (1 - \lambda) r'_2(D) \quad (13)$$

where $r'(D)$ is a normalized rank function of a retrieval model.

Figure 5.2 shows the combined result. Table 5 presents MAP, P@10 and R-Precision of two models and the combined result separately. Figure 4 shows scores with different λ. It is interesting to find out that there is relative improvement

Table 5. Result of combination at λ=0.6

	WSLM	GLM	combined	%over WSLM	%over unigram
MAP	0.3498	0.3530	0.3810	+8.92%	+7.93%
P@10	0.5629	0.5685	0.5960	+5.88%	+4.84%
R-Prec	0.3806	0.3864	0.4110	+7.99%	+6.37%

over both the word sense language model and the unigram language model for retrieval.

From those results, we can see that WSLM provides another logical view towards retrieval, which can be used in boosting retrieval performance rather than hindering. This result shows positive effectiveness of word sense in retrieval. Otherwise, the combination of result must have gone worse.

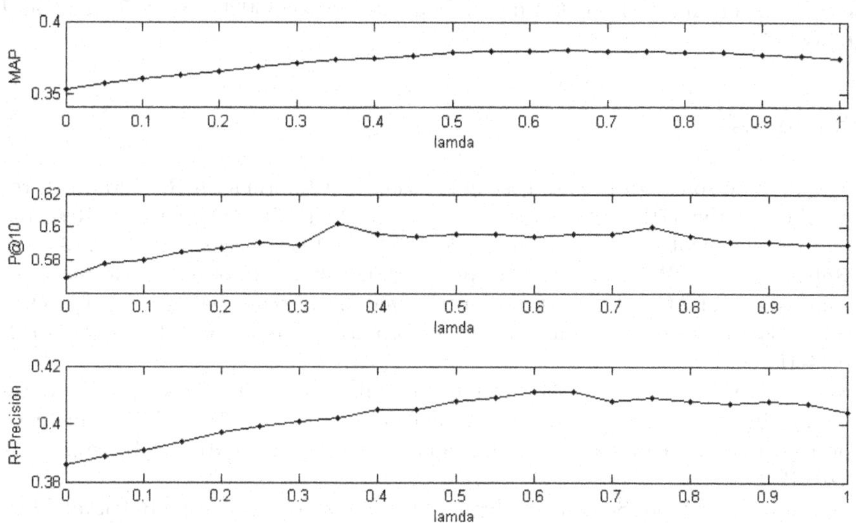

Fig. 4. This figure illustrates the relations between λ and MAP, P@10, R-Precision. From the figure, we can see that when λ is between 0.6 and 0.8, the three evaluation indicators all trend towards acme

6 Conclusion and Future Work

We presented a word sense language model for IR. This language model takes word sense codes as sense representation for computing into account, which is able to handle synonymy and reduce data sparseness because diverse words can be treated as a same sense and the space of sense codes is relatively smaller than the space of words. The experiment results show that word sense language model performs better than traditional approaches such as tf-idf based, though a minor decrease on average scores compared to a unigram language model based on words for IR. Close analysis finds that the accurate of word sense resolution highly concerns with the performance of this model. However by dominating two results from word sense language model and unigram, we obtain improvement over either of results. This result indicates that word sense language model and word-based language model views the documents from different aspects, which can boost the retrieval result together.

We hold a belief that information retrieval needs linguistic knowledge to achieve its ultimate goals. From this start point, we explore a method combining thesaurus-based knowledge into a statistical model for IR, which is quite different from traditional approaches. In the future we conduct more experiments to seek for the effectiveness of this kind of model. This progress owes much to development and research in natural language processing as well as research in statistical approaches in IR, such as language modeling.

Acknowledgements. This research has been supported by the National Natural Science Foundation of China via Grant No.60435020, No.60575042 and No.60503072.

References

1. Mark Sanderson: Word Sense Disambiguation and Information Retrieval. In Proceedings of the 17th Annual International ACM SIGIR Conference on Research and Development in Information Retrieval(SIGIR 1994), pages 142-151, 1994.
2. Robert Krovetz, W.Bruce Croft: Lexical Ambiguity and Information Retrieval. In Proceedings of ACM Transactions on Information Systems, pages 115-141, 1992.
3. S.F.Weiss: Learning to Disambiguate. Information Storage and Retrieval, 1973; 9:33-41.
4. Voorhees, Ellen M: Using WordNet to Disambiguate Word Senses for Text Retrieval. In Proceedings of the 16th Annual International ACM SIGIR Conference on Research and Development in Information Retrieval (SIGIR 1993), pages 171-180, 1993.
5. Ozlem Uzuner: Word Sense Disambiguation Applied to Information Retrieval. Master paper of Engineering in Electrical Engineering and Computer Science at the Massachusetts Institute of Technology.1998:5-20
6. M.P.O. Christopher Stokoe and J. Tait: Word Sense Disambiguation in Information Retrieval Revisited. In Proceedings of the 26th annual international ACM SIGIR conference on Research and development in Information Retrieval(SIGIR2003), pages 159-166, 2003.
7. Ponte J. and Croft W. B.: A Language Modeling Approach to Information Retrieval. In Proceedings of the 21st Annual International ACM-SIGIR Conference on Research and Development in Information Retrieval(SIGIR1998), pages 275 281, 1998.
8. Xiaoyong Liu and W. Bruce Croft: Statistical Language Modeling for Information Retrieval, the Annual Review of Information Science and Technology, vol. 39, 2003.
9. Hinrich Schütze and Jan O.Pedersen: Information Retrieval Based on Word Senses. In Proceedings 4th Annual Symposium on Document Analysis and Information Retrieval (SDAIR1995), pages161-175, 1995.

10. Jianfeng Gao, Jian-Yun Nie, Guangyuan Wu and Guihong Cao: Dependence Language Model for Information Retrieval. In Proceedings of the 27th ACM SIGIR Conference on Research and Development in Information Retrieval (SIGIR2004), pages 170-177, 2004.
11. Adam Berger, John Lafferty: Information Retrieval as Statistical Translation. In Proceedings of the 22nd ACM SIGIR Conference on Research and Development in Information Retrieval (SIGIR1999), pages 222-229, 1999.
12. Ting Liu, Zhimao Lu, Sheng Li: Implement a Full-Text Automatic System for Word Sense Tagging (2004). Journal of Harbin Institute of Technology, Volume 37, No.12, pages 1603-1604, pages 1649.
13. Chengxiang Zhai and John Lafferty: Two-Stage Language Models for Information Retrieval. In Proceedings of the 25th Annual International ACM SIGIR Conference on Research and Development in Information Retrieval (SIGIR 2002), pages 49-56, 2002.
14. D.R.H. Miller, T. Leek, and R.M. Schwartz: A Hidden Markov Model Information Retrieval System. In the Proceedings of the 22nd International Conference on Research and Development in Information Retrieval (SIGIR1999), pages 214-221, 1999.
15. Fei Song, W.Bruce Croft: A General Language Model for Information Retrieval. In Proceedings of the Conference on Information and Knowledge Management (CIKM1999), pages 316-321, 1999.

Statistical Behavior Analysis of Smoothing Methods for Language Models of Mandarin Data Sets

Ming-Shing Yu[1], Feng-Long Huang[2,*], and Piyu Tsai[2]

[1] Department of Information Science, National Chung-Hsing University, Taichung 40227, Taiwan
msyu@nchu.edu.tw

[2] Department of Computer Science and Information Engineering, National United University, MiaoLi 360, Taiwan
{flhuang, pytsai}@nuu.edu.tw

Abstract. In this paper, we discuss the properties of statistical behavior and entropies of three smoothing methods; two well-known and one proposed smoothing method will be used on three language models in Mandarin data sets. Because of the problem of data sparseness, smoothing methods are employed to estimate the probability for each event (including all the seen and unseen events) in a language model. A set of properties used to analyze the statistical behaviors of three smoothing methods are proposed. Our proposed smoothing methods comply with all the properties. We implement three language models in Mandarin data sets and then discuss the entropy. In general, the entropies of proposed smoothing method for three models are lower than that of other two methods.

Keywords: Language models, smoothing methods, statistical behaviors, cross entropy, natural language processing.

I Introduction

Language models (LM) have widely been used in various tasks of natural language processing (NLP), such as speech recognition, machine translation, part-of-speech tagging, spelling correction and word sense disambiguation, etc, [2], [5], [10], [16]. Many research works of language modeling approach, such as [23], [24] and [25], have been used on information retrieval (IR).

An event can be regarded as a possible type of n-gram [1],[3],[13] in LM, $n>=1$. We can calculate the probability for the each occurred event according to its count in training corpora. When the unseen events occur, smoothing methods will be used to re-estimate the probability for each event.

For the possible word's estimation, the word sequence W_{max} with maximum conditional probability $P(W)$ in n-gram model will be expressed as:

---

[*] Correspondence author.

H.T. Ng et al. (Eds.): AIRS 2006, LNCS 4182, pp. 172 – 186, 2006.

$$W_{\max} = \arg\max_w P(w_1^m) = \arg\max_w \prod_{i=1}^{m} P(w_i \mid w_{i-n+1}^{i-1}) \tag{1}$$

As shown in Eq. (1), the probability for each event can be obtained by training the bigram model. Therefore the probability of a word bigram b will be written as:

$$P(w_i \mid w_{i-1}) = \frac{C(w_{i-1}w_i)}{\sum_w C(w_{i-1}w)} \tag{2}$$

where $C(w_i)$ denotes the count of word w_i in training corpus. The probability P of Eq. (2) is the relative frequency and such a method of parameter estimation is called *maximum likelihood estimation* (MLE). In the general case of n-gram models, Eq. (2) can be rewritten as:

$$P(w_i \mid w_{i-n+1}) = \frac{C(w_{i-n+1}^{i-1}w_i)}{\sum_w C(w_{i-n+1}^{i-1}w)}. \tag{3}$$

1.1 Cross Entropy and Perplexity

Two common metrics to evaluate language model is called *cross entropy* and *perplexity*. For a testing data set T which contains a set of events, $e_1,e_2,\ldots,e_m$, the probability for the testing set $P(T)$ can be described as:

$$P(T) = \prod_{i=1}^{m} P(e_i), \tag{3}$$

where m denotes the number of events in testing set T and $P(e_i)$ denotes the probability of event e_i, obtained from n-gram language model, assigning to event e_i. The *entropy* $H(T)$ can be regarded the bit size needed to encode each word in testing set T. $H(T)$ can be shown as:

$$\begin{aligned} H(T) &= -\sum_x P(x)\log_2 P(x) \\ &= -\sum_{i=1}^{m} P(e_i)\log_2 P(e_i). \end{aligned} \tag{4}$$

The perplexity $PP(T)$ is the weighted average number of words in testing set T. Perplexity $PP(T)$ is defined in term of entropy $H(T)$:

$$PP(T) = 2^{H(T)} \tag{5}$$

In general, lower entropy $H(T)$ and perplexity $PP(T)$ for testing set T leads to the better performance of language model. As seen in Eq. (5), smaller entropy $H(T)$ leads to lower perplexity too. In this paper, entropy metric will be used to compare the proposed smoothing methods with the previous methods.

Cross entropy CH is a useful yardstick measuring the ability of a language model to predict a source of data. If the language model is good enough for predicting the future output of the source, the cross entropy should be small. In the general case, $CH >= H$ (where H employs the best possible language model, the source itself).

1.2 Zero Count: The Smoothing Issues

According to the Eqs (2)~(3), we can estimate the most probable word sequence W with MLE. However, such method will lead to the degradation of performance. For a given word w_{i-1} in bigram model, if bigram $w_{i-1}w_i$ never occur in the training corpus, then $C(w_{i-1}w_i)$ is equal to 0. It is apparent that Eqs. (2) and (3) are both equal to zero.

It is not reasonable and good to assign 0 to the unseen events. If we should assign certain probability to such events, how is the probability assigned? The schemes used to resolve the problems are called *smoothing* methods [4],[7],[15],[18]. The probability obtained from MLE will be adjusted and redistributed. Such a process will maintain the total probability to be unity. Usually, the smoothing methods can improve the performance of language models.

Although there are currently several corpora containing more than twenty millions (20M) of training texts (such as WSJ corpus in English texts), the problems related with zero count for novel events always exist in NLP. As illustrated in [13], the events may be either word sequences (like n-grams) or single word (like Mandarin characters or words).

1.3 Overview of Several Smoothing Methods

The principal purpose of smoothing is to alleviate the zero count problems, as described above. In this section we will describe the smoothing methods in detail. There are several well-known smoothing methods [9],[11],[12],[22] in various applications. In the paper, three following methods are discussed: *Additive discounting*, *Witten-Bell* [17], and our proposed Smoothing method.

1.3.1 Additive Discount Method

Additive smoothing method is intuitively simple. A small amount$\delta$is added into the count for all n-grams (including all seen and unseen n-grams). It is apparent that the count for each unseen and seen bigrams is tuned uniformly to δ and c+ δ . The adjusted count c^* for bigrams b_i, which occurred c times (c>=0) in training corpus, is defined as follows:

$$c^* = (c+\delta)\frac{N}{N+B\delta}. \tag{6}$$

Referring to Eq. (7), the smoothed probability P^* for a given bigram b_i can be rewritten as follows:

$$P^*_{i,N} = \frac{c+\delta}{N+B\delta}, \tag{7}$$

where B=A^2 is the number of bigrams in a language. $N = \sum_{w} C(w_{i-1}w)$, A is the number of alphabets for a language, such as the number of Mandarin characters.

Typically, 0<δ<=1. The caseδ=1 is called *add-1* smoothing. In such case, the discounted count $c_i^* = (c_i+1)\frac{N}{N+B}$ for $i \geq 0$ and $\frac{N}{N+B}$ is regarded as the normalization factor (NF) for count c_i. Clearly, the count of each type of bigrams is increased by 1.

According to the previous experiments [3], the performance was usually degraded by using *add-one* smoothing.

1.3.2 Witten-Bell *C*

Here we discuss one of five *W-B* smoothing schemes (called *W-B C*), introduced by Wetten and Bell[1][17].

It is more complex than the additive discount technique. The basic concept is recurring:

> **Key concept:** *use the count of things you've seen just once to help estimate the count of things never occurred.*

The *W-B C* is described as:

$$c_i^* = \begin{cases} \dfrac{S}{U}\dfrac{N}{N+S}, & \text{if } c_i = 0 \\ c_i \dfrac{N}{N+S}, & \text{if } c_i > 0 \end{cases} \tag{8}$$

where *U*, *S* and *N* denote the types of all possible unseen bigrams, seen bigrams in training corpus and the number of all the seen bigrams in training corpus, respectively.

The discounted probability will be expressed for seen bigrams as:

$$P_{i,N}^* = \frac{c_i}{N+S}, \qquad \text{if } c_i > 0 \tag{9}$$

The probability mass P_{mass} for all unseen bigrams assigned by *W-B C* is obtained as:

$$\sum_{i:c_i=0} P_{i,N}^* = \frac{S}{N+S}, \qquad \text{if } c_i{=}0 \tag{10}$$

The probability for each unseen bigram will be derived from Eq. (10) divided uniformly by *U*:

$$P_{0,N}^* = \frac{1}{U}\frac{S}{N+S} \tag{11}$$

As shown in Eq (8), it is obvious that the redistributed count c^* for each bigram which doesn't occurs is equal to *S*/*U*. The size of c^* is subject to the ratio of *S* and *U*. The ratio may be greater or less than 1, depending on the value of *S* and *U*.

1.3.3 Proposed Smoothing Method (Y-H)

In this section, we will propose a novel smoothing methods (so-called Y-H smoothing). The probability mass P_{mass} assigned to unseen events by our method is heuristic. Basic Concept of our proposed smoothing method will be described:

---

[1] There are 5 methods in [17]; method *A*, *B*, *C*, *P* and *X*. We just discuss one of them (*W-B C*) in this paper.

In case for a bigram, our method calculates the smoothed probabilities as:

$$Q(w_{i-1}w_i) = \begin{cases} \dfrac{d}{U(N+1)} & \text{for } c(w_{i-1}^i) = 0, \\ \dfrac{c(w_{i-1}^i)}{N} \dfrac{N+1-d}{N+1} & \text{for } c(w_{i-1}^i) \geq 1, \end{cases} \tag{12}$$

where d denotes a constant ($0<d<1$) and independent of U.

When computing the smoothed probability, the proposed method don't employ interpolating scheme to combine the high order models and lower order models. As shown of Eq. (12), $(N+1-d)/(N+1)$ is the normalization factor for Q^* of seen bigrams. The probabilities for all the seen bigrams will be discounted by the normalization factor and then the accumulated probability then is re-distributed to the unseen bigrams. All the unseen bigrams will share uniformly the distribution mass $d_A/(N+1)$,

$$\sum_{i:c_i=0} P_i^* = \frac{d}{N+1} \qquad \text{for } c_i = 0 \tag{13}$$

2 The Properties to Analyze Statistical Behaviors

In this section, we will propose five properties which can be regarded as statistical features of LMs. These properties will be further used to analyze the statistical behaviors of smoothing methods in next section.

A. Property 1

The smoothed probability for any one bigram b_i should falls between 0 and 1 (0,1), which is described as follows:

$$0 < P_{i,N}^* < 1 \quad \text{,for all bigrams } b_i\ (1<=i<=B) \text{ on any training size } N \tag{14}$$

where $P_{i,N}^*$ is the smoothed probability for a bigram b_i (or word w_i) on training size N, B is the number of types of bigrams.

B. Property 2

The summation of smoothed probability P^* for all the bigrams is necessarily equal to 1 on any training size N. Total smoothed probability P is summed as:

$$P_{1,N}^* + P_{2,N}^* + \ldots + P_{B,N}^* = \sum_{b_i \in seenbigrams} P_{i,N}^* + \sum_{b_j \in unseenbigrams} P_{j,N}^* = 1 \tag{15}$$

,

where B denotes the total number of bigrams.

C. Property 3

The smoothed probability assigned to the bigrams b with different count should satisfy all the following inequality equations[2]:

---

[2] The property was first proposed in [15] and we make a little modification.

$$Q^*_{c,N} < Q^*_{c+1,N}, \qquad \text{for } c=0,1,2,\ldots, \tag{16}$$

where $Q^*_{c,N}$ is the smoothed probability for the bigram b_i with c counts on training corpus of size N.

Inequality Eq. (16) describes the concept that smoothed probability for any bigram with same count should be same on any training size N. Furthermore, the probability for bigram b_{c+1} with c+1 counts should be larger than that of bigrams with c counts.

D. Property 4

Comparing to the probability P prior to smoothing process, the smoothed probability P^* for all bigrams will be changed. Property 4 can be expressed as follows:

$$Q^*_{0,N} > Q_{0,N}, \qquad \text{for } c = 0 \qquad \text{//for all the unseen bigrams with 0 count} \tag{17}$$

$$Q^*_{c,N} < Q_{c,N}, \qquad \text{for } c \geq 1 \qquad \text{//for all the seen bigrams with count >=1} \tag{18}$$

Property 4 shows $Q^*_{0,N}$ for unseen bigrams will be larger than original $Q_{0,N}$ while $Q^*_{c,N}$ will be decreased for all bigrams with more than one count (c>=1). The probability mass P_{mass} discounted from all seen bigrams is distributed uniformly to the smoothed probability for unseen bigrams.

E. Property 5

Three notations B, S and U can be expressed as $B=S+U$ for bigram models. When the number of training size is increased, all the smoothed probability Q^* for bigrams with same counts on training size N+1 should be decreased a bit while comparing to the Q^* on training size N. For instance, when an incoming bigram (say b_{N+1}) occurs, the training size is increase by one (now N=N+1). The smoothed probability Q^* on N+1 training set should be less than the probability Q^* on N for $c \geq 0$, except the P^* for the incoming bigram b_{N+1}:

$$Q^*_{c+1,N+1} = \frac{c(\bullet)+1}{N+1}. \tag{19}$$

In other words, in addition to the Q^* of b_{N+1} at training size N+1, all other smoothed probability Q^* at training size N+1 will be decreased than those at training size N. Although both the numerator and denominator of Eq. (19) are increased by 1, due to N>>c, so the inequality equation $Q^*_{c,N} < Q^*_{c+1,N+1}$ hold. In summary, property 5 can be expressed as:

$$Q^*_{c,N} > Q^*_{c,N+1} \qquad \text{for all bigrams with count } c \text{ >=0, and} \tag{20}$$

$$Q^*_{c,N} < Q^*_{c+1,N+1} \qquad \text{for the new bigram } b_{N+1}. \tag{21}$$

where $Q^*_{c,N}, Q^*_{c+1,N+1}$ denote the smoothed probability for bigram with c counts on training size N and N+1. Note that denotes the smoothed probability for unseen bigrams with 0 count.

3 Properties Analysis for Three Smoothing Methods

From the statistical view, smoothed probability for bigrams computed from various smoothing methods should still comply with these properties. Based on the statistical properties, we will analyze the rationalization of each smoothing models. In this section, several smoothing methods are analyzed for five proposed properties in previous section. Although we focus on the bigram models in this paper, all these properties can be easily expanded into *n*-gram models (n>=3) for all smoothing methods.

3.1 Additive Discount Method

In this paper, we set the additive parameter δ to 1. Therefore, the method refers to *add-1* smoothing. Based on the Eq. (9), the probability $Q_{i,N}$ of bigrams with c counts (c >= 0) is discounted by the normalized factor $N/(N+B)$.

$$Q^*_{i,N} = \frac{c_i^*}{N} = \frac{(c_i+1)N}{N(N+B)} = \frac{c_i+1}{N+B},$$

It is obvious that the property 1 holds. $\sum_c P^*_{c,N} = \sum_c \frac{c_i+1}{N+B} = 1$. Hence, property 2 holds. For the smoothed property of all bigrams, property 3 still holds.

The property 4 for additive method is analyzed as:

As shown of Eq. (18), $Q^*_{c,N}, Q_{c,N}$ on training size N, can be expressed as follows:

$$Q_{c,N} = \frac{c}{N} \text{ and } Q^*_{c,N} = \frac{c+1}{N+B},$$

$$Q^*_{c,N} - Q_{c,N} = \frac{c+1}{N+B} - \frac{c}{N} = \frac{N-cB}{N(N+B)}. \tag{22}$$

According to property 4, Eq. (22) should be negative while numerator (N-cB) of Eq. (22) may be positive (>0), in certain case of three parameters N, c and V. For instance, assuming that N=mB, m>=1 and c<=m-1, therefore N - cB>0. In other words, $Q^*_{c,N} > Q_{c,N}$. Hence, property 4 does not hold. $Q^*_{c,N}$ and $Q^*_{c,N+1}$ can be shown as:

$$Q^*_{c,N} = \frac{c+1}{N+B}, Q^*_{c,N+1} = \frac{c+1}{(N+1)+B}. \quad Q^*_{c,N} - Q^*_{c,N+1} = (c+1)\frac{1}{(N+B)(N+1+B)} > 0$$

Hence Eq. (22) of property 5 holds. We can also prove that for $Q^*_{c,N} < Q^*_{c+1,N+1}$, the new bigram b_{N+1}, Eq. (21) in property 5 still holds. Therefore property 5 holds for this method.

3.2 Witten-Bell Method

As shown in Eq. (15) and (16), smoothed probability P^* for any bigram will be (0,1), property 1 holds. The smoothed probability for all bigrams will be summed up to 1.

Therefore, property 2 still holds. The probability ratio[3] for bigrams with zero and c counts can be expressed as follows:

$$Q^*_{0,N} : Q^*_{c,N} = \frac{S}{B} : c \tag{23}$$

Based on the Eq. (16) of property 3, Eq. (23) should be less than 1. It is obvious that the ratio may be greater than 1, on certain S, U and c (c>=1). For instance, S=9000, U=4000 and c=1 (e.g., for Mandarin character unigram model), S/U>c (9000/4000>1). On the other words, it is possible that $Q^*_{0,N} \geq Q^*_{1,N}$. Hence, property 3 does not hold.

Finally, property 5 will be analyzed as follows. Smoothed probabilities Q^*, derived from Eq. (20), for bigrams b_i with c counts on training size N and N+1 are calculated as:

Case I: When N=N+1, U keep unchanged and U>1.

$$Q^*_{0,N} : Q^*_{0,N+1} = \frac{S}{U(N+S)} : \frac{S}{U(N+1+S)} > 1, \text{ for } c = 0.$$

$$Q^*_{c,N} : Q^*_{c,N+1} = \frac{c}{N+S} : \frac{c}{N+1+S} > 1, \text{ for } c >= 1.$$

Case II: When N=N+1, U=U-1 and U>1.

$$Q^*_{0,N} : Q^*_{0,N+1} = \frac{S}{U(N+S)} : \frac{S}{U(N+1+S)} > 1, \text{ for } c = 0.$$

$$Q^*_{c,N} : Q^*_{c,N+1} = \frac{c}{N+S} : \frac{c}{N+1+S} > 1, \text{ for } c >= 1.$$

Hence, we can also prove that $Q^*_{i,N} > Q^*_{i,N+1}$. Hence, property 5 holds for this method.

3.3 Proposed Smoothing Method (Y-H)

We analyze furthermore the statistical behaviors of the proposed method. As shown in Eq. (13), the smoothed probabilities for any seen and unseen bigram can be (0,1). Therefore, property 1 does hold. The total probabilities P for seen and seen bigrams can be accumulated as:

$$\sum_{i:c_i \geq 1} Q^*_{i,N} + \sum_{i:c_i=0} Q^*_{i,N} = \sum_{i:c_i>0} \frac{c_i}{N} \frac{N+1-d}{N+1} + \sum_{i:c_i=0} \frac{d}{U(N+1)}$$
$$= \frac{N}{N} \frac{N+1-d}{N+1} + \frac{d}{N+1} = 1$$

[3] The notations Q^* here is P^* in Eqs. (21)~(23).

So, property 2 does hold. The smoothed probability Q^* for bigrams with c and c+1 counts on N training data is calculated as follows. For c=0 and 1,

$$
\begin{aligned}
Q^*_{1,N} - Q^*_{0,N} &= \frac{1}{N}\frac{N+1-d}{N+1} - \frac{d}{U(N+1)} \\
&= \frac{N(U-d)+U(1-d)}{UN(N+1)} \\
&= \frac{N}{N}\frac{N+1-d_A}{N+1} + \frac{d_A}{N+1} = 1
\end{aligned} \tag{24}
$$

Due to U>=1 and 0<d<1, numerator $N(U\text{-}d)+U(1\text{-}d)$ of Eq. (24) is positive (>0). For c>1:

$$
\begin{aligned}
Q^*_{c,N} - Q^*_{c+1,N} &= \frac{c(N+1-d)}{N(N+1)} - \frac{(c+1)(N+1-d)}{N(N+1)} \\
&= \frac{-(N+1-d)}{N(N+1)}
\end{aligned} \tag{25}
$$

Because N>=1 and 0<d<1, Eq. (25) will be less than 0. Referring to the results of Eqs. (24) and (25), we can conclude that property 3 does hold. Original and smoothed probability for a bigram with c counts is as follows:

$$
Q^*_{c,N} - Q_{c,N} = \frac{d}{U(N+1)} - 0 > 0 \ \text{ for } c = 0. \tag{26}
$$

$$
Q^*_{c,N} - Q_{c,N} = \frac{c(N+1-d)}{N(N+1)} - \frac{c}{N} = \frac{-cd}{N(N+1)} < 0 \ \text{ for } c >= 1. \tag{27}
$$

As shown of Eq. (26) and (27), we can conclude property 4 does hold. Finally, we analyze property 5. Smoothed probabilities Q^* for bigrams with c >=0 on N and N+1 training data are calculated as:

Case I: When N=N+1, U is unchanged and U>1. In such case, the incoming event ever occurred at least once in current training set.

$$
\begin{aligned}
Q^*_{0,N} - Q^*_{0,N+1} &= \frac{d}{U(N+1)} - \frac{d}{U(N+2)} \\
&= \frac{d}{U(N+1)(N+2)} > 0 \quad \text{for } c = 0.
\end{aligned} \tag{28}
$$

$$
Q^*_{c,N} = \frac{c}{N}\frac{N+1-d}{N+1}, \quad Q^*_{c,N+1} = \frac{c}{N+1}\frac{N+2-d}{N+2}.
$$

$$
\begin{aligned}
Q^*_{c,N} - Q^*_{c,N+1} &= \frac{c(N+1-d)}{N(N+1)} - \frac{c(N+2-d)}{(N+1)(N+2)} \\
&= \frac{c(N+2-2d)}{N(N+1)(N+2)} > 0 \qquad \text{for } c >= 1.
\end{aligned} \tag{29}
$$

It is obvious that Eq. (29) is greater than 0 (N >=1 and 0<d<1). From the results of Eqs. (28) and (29), property 5 does hold in this case.

Case II: When N=N+1, U is decreased by 1 and U>1. In this case, the incoming bigram never occurred in current training set. On the other words, the bigram is a novel and then S will be increased by 1 (S=S+1 and U=U-1.)

$$Q^*_{0,N} - Q^*_{0,N+1} = \frac{d}{U(N+1)} - \frac{d}{(U-1)(N+2)}$$
$$= d_A \frac{U-N-2}{U(U-1)(N+1)(N+2)} \quad \text{for } c = 0. \tag{30}$$

As shown of Eq. (30), numerator (U-N+2) may be less or greater than 0 in terms of N and U:

$$Q^*_{0,N} - Q^*_{0,N+1} \text{ will be} \begin{cases} >0 & \text{if } N < U+2. \\ =0 & \text{if } N = U+2. \\ <0 & \text{if } N > U+2. \end{cases} \tag{31}$$

When N<U+2, property 5 can holds while N>=U+2 the property can't hold yet in such situation. In fact, because of the insufficiency of the training corpus, the training data N is usually much less than the unseen bigrams U in language processing (N<<U).

Table 1 shows the relationship between 3 smoothing methods and five proposed properties. Among these smoothing methods, there isn't any method which complies completely with 5 proposed properties. *Add-1 discounting* and *W-B* don't comply only one of five properties. All other methods do not comply with more than two properties. Notations O and X denote the method does and does not comply with the proposed property, respectively.

Table 1. Four smoothing methods and proposed statistical properties

property / method	1	2	3	4	5
Additive discount	O	O	O	X	O
Witten-Bell (C)	O	O	X	O	O
Our proposed method	O	O	O	O	O

4 Experiment Results

We will first describe the empirical data sets and three Mandarin models in this section. The probability mass P_{mass} assigned to unseen events by various smoothing methods are analyzed. The entropy of all smoothing method discussed for three models are shown. We further discuss the relationship between the P_{mass} and entropy which is a metric for evaluating a LM. Whether the volume of P_{mass} affecting the entropy H of LM or not will be shown.

4.1 Data Sets and Empirical Models

In the following experiments, two text sources are used as data sets; the news texts collected from Internet and ASBC corpus [18]. The HTML tags and all unnecessary symbols are extracted and there are about 7M Mandarin characters in news texts. The Academic Sinica Balanced Corpus version 3.0 (ASBC) includes 316 text files distributed in different fields, 118MB memory size and 5.22 millions of words labeled with a POS tag. Our corpus contains totally up to 12M Mandarin characters.

In the paper, we have constructed three models to evaluate the entropy of smoothing methods discussed; Mandarin character unigrams, character bigrams and word unigrams model. The entropy of each method is calculated on various data size in our experiments, from 1 M to 12M Mandarin characters. The first two models employ up to 12M Mandarin characters (unigrams and bigrams) and the 3rd model use about up to 5M Mandarin words in ASBC corpus.

4.2 Probability Mass Assigned to Unseen Events

Table 2 shows the probability mass P_{mass} to be redistributed to all unseen bigrams and normalized factor for each seen bigram. P_{mass} is varied primarily with the smoothing methods and then with respect to the parameters N, U, S and some constants. It is obvious that the term, normalization factor (*NF*) in each smoothing formula, will affect the probability discounted from the probability P assigned to the events with c counts ($c >= 1$) prior to smoothing process.

Table 2. The probability mass P_{mass} and normalization factor (*NF*) in various smoothing methods

perperty / method	probability mass for all novels	*N. F.* for all seen events
Additive discount	$U/(N+U)$	$N/(N+B)$
Witten-Bell (C)	$S/(N+S)$	$N/(N+S)$
our method	$d/(N+1)$	$(N+1-d)/(N+1)$

4.3 Entropy Evaluation for Three Language Models

In the paper, three language models; Mandarin character unigrams, character bigrams and word unigram, will be evaluated to compare the entropy. These models employ the training data sets depicted in this section to compute the entropy. The entropy with respect to these methods will be shown in term of various training size N. Furthermore, we will observe the relationship between the probability mass P_{mass} assigned to all unseen events and entropy.

As shown in Figure 1, the entropy for various smoothing methods is displayed. Three smoothing methods are evaluated. The constant d in our method are set to 0.8, 0.4, 0.1 and 0.01 to analyze the effectiveness of such constant. For the Mandarin character unigram model in top of Figure 1, the trend of entropy for all methods is similar and the maximum entropy will happen at training size 1M for all the methods.

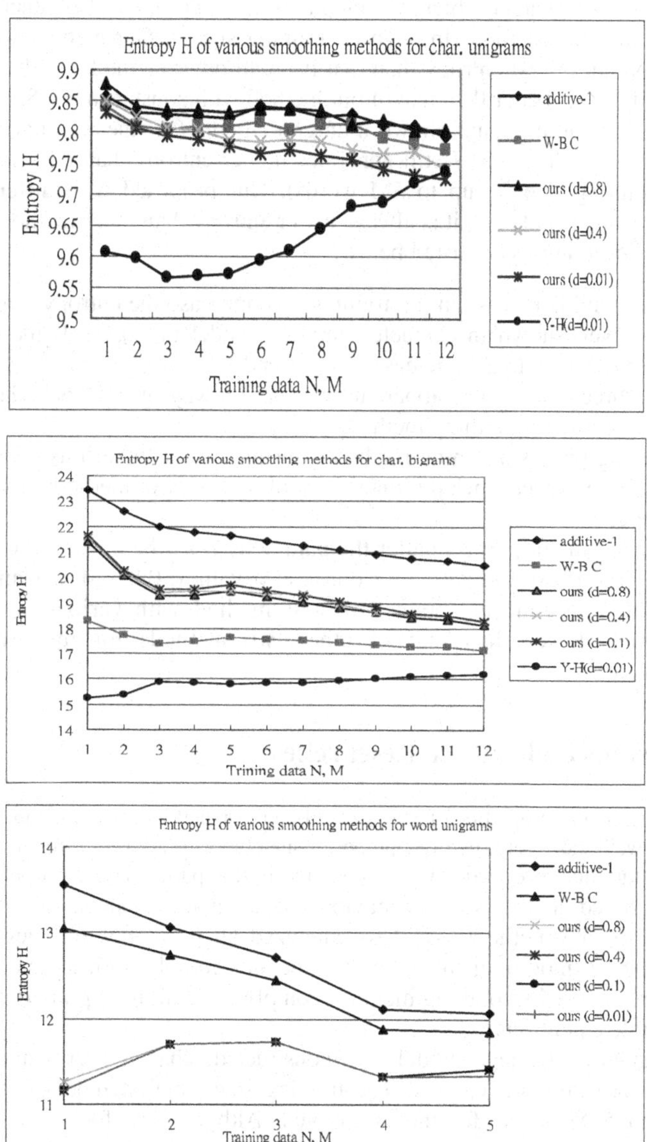

Fig. 1. Entropy H of smoothing methods with various training data N; (top) for Mandarin character unigram model, (middle) for Mandarin character bigram model, (bottom) for Mandarin word unigram model

The entropy of our methods is lower than that of other two methods. Among the methods, our method with parameter d=0.01 will obtain lowest entropy through all training size N. It is obvious that our method with any d<1 can obtain lower entropy than that of *W-B C*, which can be regarded as the special case of our method with d=1 (see Eq. (14)). It is obvious that *Add-1* always has higher entropy than all other methods through all training size N (1M<=N<=12M).

For the Mandarin character bigram, smoothing method *Add-1* also generate highest entropy than all other methods through all training size N. Our method with smaller d =0.01 will obtain lowest entropy than that from all other methods. In the model, the performance *W-B C* is better than that of our method with constant d = 0.8, 0.4 and 0.1.

Finally, we look at the Mandarin word unigram model (see the bottom of Figure 1). First, Smoothing method *Add-1* still generates higher entropy than all other methods through all training size N (up to 5M words). Our proposed with different d will obtain almost same results and it is always lower entropy than other methods.

Several observations are listed below:

1) The trend is that when the training size increase the entropy will decrease. However, shown in character unigram model of Figure 1, the entropy of our method with d=0.01 has reverse trend.
2) For three models, it is apparent that *Add-1* always generates highest entropy among three smoothing methods.
3) Among three smoothing methods, our method with various constant d can generate lower entropy than that of other two methods for all three models.
4) For our method, the smaller the parameter d is, the lower the entropy will be. In our experiments, four constant d values, 0.8, 0.4, 0.1 and 0.01 are used. The entropy obtained by our method with various constant d is always lower than that of other two methods for three models on various N.

5 Conclusions and Future Researches

We have discussed the properties of statistical behavior and entropies of three smoothing methods; two well-known and one our proposed smoothing method used for three language models. Five statistical properties were presented in this paper. These properties, which can be considered as statistical behaviors of language models, are applied for analyzing smoothing methods. We have analyzed all properties for these methods. Among published smoothing methods, they do not comply with all the properties. However, our proposed smoothing methods comply with all these properties based on some reasonable conditions.

Three Mandarin language models are constructed; character unigram, character bigram and word unigram models. The training size for first model is up to 12M characters and 5 M words for the last model. Although we focus on the bigram models in this paper, all these properties can be easily expanded into n-gram models (n>=3) for all smoothing methods. Although only two well-known smoothing methods are discussed in the paper, the set of properties can be expanded into all other methods to analyze the statistical behaviors.

The topics in future include: evaluation for bigram model and trigram models of the smoothing methods; the perplexity between the various size of training data and smoothing methods; the relationship between the constant d in our method and perplexity. In other paper, we will further discuss the perplexity and the performance of real NLP applications in which the smoothing method is applied into POS tagging, Mandarin word semantics or WSD issues.

References

[1] Brown P. F., Pietra V. J., deSouza P. V., Lai J. C., and Mercer R. L., 1992, Class-Based n-gram Models of Natural Language, Computational Linguistics, Vol. 18, pp. 467-479.

[2] Brown P. F., Della Pietra S. A., Della Pietra V. J., Lai J. C., and Mercer R. L., 1992, An Estimate of an Upper Bound for the Entropy of English, Computational Linguistics, Vol. 18, pp. 31-40.

[3] Chen Standy F. and Goodman Joshua, 1999, An Empirical study of smoothing Techniques for Language Modeling, Computer Speech and Language, Vol. 13, pp. 359-394.

[4] Church K. W. and Gale W. A., 1991, A Comparison of the Enhanced Good-Turing and Deleted Estimation Methods for Estimating Probabilies of English Bigrams, Computer Speech and Language, Vol. 5, pp 19-54.

[5] Dagan I., Maucus S and Markovitch, 1995, Contextual Word Similarity and Estimation from Sparse Data, Computer Speech and Language, Vol. 9, pp. 123-152.

[6] Essen U. and Steinbiss, 1992, Cooccurrence Smoothing for Stochastic Language Modelling, IEEE International conference on Acoustic, Speech and Signal Processing, Vol. 1, pp. 161-164.

[7] Good I. J., 1953, The Population Frequencies of Species and the Estimation of Population Parameters, Biometrika, Vol. 40, pp. 237-264.

[8] Jelinek F., 1997, Automatic Speech Recognition-Statistical Methods, M.I.T.

[9] Jelinek F. and Mercer R. L., 1980, Interpolated Estimation of Markov Source Parameters from Spars Data, Proceedings of the Workshop on Pattern Recognition in Practice, North-Holland, Amsterdam, The Northlands, pp. 381-397.

[10] Juraskey D. and Martin James H., 2000, Speech and Language Processing, Prentice Hall.

[11] Katz S. M., March 1987, Estimation of Probabilities from Sparse Data for the Language Models Component of a Speech Recognizer, IEEE Trans. On Acoustic, Speech and Signal Processing, Vol. ASSP-35, pp. 400-401.

[12] Knerser R. and Ney H., 1995, Improved Backing-Off for M-gram Language Modeling, IEEE International conference on Acoustic, Speech and Signal Processing, pp. 181-184.

[13] Nádas A., 1984, Estimation of Probabilities in the Language Model of the IBM Speech Recognition System, IEEE Transactions on Acoustics, Speech, Signals Processing, Vol. 32, No. 4, pp. 859-861.

[14] Nádas A., 1985, On Turing's Formula for Word Probabilities, IEEE Trans. On Acoustic, Speech and Signal Processing, Vol. ASSP-33, pp. 1414-1416.

[15] Ney H. and Essen U., 1991, On Smoothing Techniques for Bigram-Based Natural Language Modeling, IEEE International conference on Acoustic, Speech and Signal Processing, pp. 825-828.

[16] Su K. Y., Chiang T. H., Chang J. S., A Overview of Corpus-Based Statistical-Oriented (CBSO) Techniques for Natural Language Processing, Computational Linguistics and Chinese Language Processing, vol. 1, no. 1, pp.101-157, August 1996.

[17] Witten L. H. and Bell T. C., 1991, The Zero-Frequency Problem: Estimating the Probabilities of Novel Events in Adaptive Text Compression, IEEE Transaction on Information theory, Vol. 37, No. 4, pp. 1085-1094.
[18] Huang C.-R., 1995, Introduction to the Academic Sinica Balance Corpus, Proceeding of ROCLLING VII, pp. 81-99.
[19] Algort P. H. and Cover T. M., 1988, A Sandwich Proof of the Shannon- McMillan-Breiman Theorem, Ahe Annals of Probability, Vol. 16, No. 2, pp. 899-909.
[20] Jurafsky D. and Martin J. H., 2000, Speech and Language Processing, Prentice Hall, Chapter 6.
[21] Juang B. H and Lo S. H., 1994, On the Bias if the Turing-Good Estimate of Probabilities, IEEE Trans. On Signal Processing, Vol. 42, No. 2, pp. 496-498.
[22] Hermann Ney, Ute Essen, Reinhard Kneser, December 1995, On the Estimation of 'Small' Probabilities by Leaving-One-Out, Vol. 17, No. 12, IEEE PAMI, pp. 1202-1212.
[23] Xuehua Shen, ChengXiang Zhai, Active Feedback in Ad Hoc Information Retrieval, Proceedings of ACM SIGIR 2005.
[24] Seung-Hoon Na, In-Su Kang, Ji-Eun Roh, Jong-Hyeok Lee, An Empirical Study of Query Expansion and Cluster-Based Retrieval in Language Modeling Approach, AIRS 2005.
[25] Guihong Cao, Jian-Yun Nie, Jing Bai, Integrating Word Relationships into Language Models Proceedings of ACM SIGIR 2005

No Tag, a Little Nesting, and Great XML Keyword Search*

Lingbo Kong[1], Shiwei Tang[1,2], Dongqing Yang[1], Tengjiao Wang[1], and Jun Gao[1]

[1] Department of Computer Science and Technology, Peking University, Beijing, China, 100871
lbkong@db.pku.edu.cn,
{dqyang, tjwang, gaojun}@pku.edu.cn

[2] National Laboratory on Machine Perception, Peking University,Beijing, China, 100871
tsw@pku.edu.cn

Abstract. Keyword search from Informational Retrieval (IR) can be seen as one most convenient processing mode catering for common users to obtain interesting information. As XML data becomes more and more widespread, the trend of adapting keyword search on XML data also becomes more and more active. In this paper, we first try nesting mechanism for XML keyword search, which just uses a little nesting skill. This attempt has several benefits. For example, it is convenient for common users, because they need not to know any organization knowledge of the target XML data. Secondly, the nesting pattern can be easily transformed into structural hints, which has same mechanism as what XML data model does. Finally, since there is no need of label information, we can retrieve XML fragments from different schemas. Besides, this paper also proposes a new similarity measuring method for retrieved XML fragments which can be from different schemas. Its kernel is KCAM (Keyword Common Ancestor Matrix) structure, which stores the level information of SLCA (Smallest Lowest Common Ancestor) node between two keywords. By mapping XML fragments into KCAMs, the structural similarity can be computed using matrix distance. KCAM distance can go well with the nesting keyword method.

1 Introduction

XML is rapidly emerging as the *de facto* standard for data representation and exchange on Web applications, such as Digital Library, Web service, and Electronic business. XML–fashioned data has become one popular data type. Fig. 1(a) is one instance. Along with the promising future of XML data, the management

* Supported by Project 2005AA4Z307 under the National High-tech Research and Development of China, Project 60503037 under the National Natural Science Foundation of China (NSFC), Project 4062018 under Beijing Natural Science Foundation(BNSF).

H.T. Ng et al. (Eds.): AIRS 2006, LNCS 4182, pp. 187–204, 2006.

research around this kind of data also becomes one popular issue for database community. How to retrieve interesting information from XML data is one important part of this issue. The properties of semi–structured and self–description make this problem challenging, because the processing not only should consider the structural information, but also must not ignore the semantic implied in the labels.

Generally there are two kinds of directions focusing on this issue. The first one inherits the DBMS's habit. It first defines elaborate query languages (Always in regular expression style). Then users should learn their syntaxes and use them to describe their query patterns. The system receives the query, does pattern matching and finally returns the matched results. We mark this as XML Query type. Examples are Lorel, XML–QL, XML–GL, Quilt, XPath [2], XQuery [3]. XPath and XQuery are recommended by W3C organization and are the delegates of this kind of query languages. Regretfully they are inconvenient for common users. In order to use them to describe the query patterns, users not only should learn new mechanism but also should know the data organization information beforehand, such as labels and label relationships. We can see this from Fig. 2 and Fig. 3. Users should understand the meaning of "//", "text()" and "@position". More inconveniently, they should also know the labels of the target XML data, such as "article", "title", even the relationships of labels.

Seeing the disadvantage of elaborate XML query languages, more and more researchers resort to the advantages from Information Retrieval (IR), and try to adapt IR properties into XML data processing. This is the other direction. There are also two kinds of directions during this procedure. The first one is to extend the query languages mentioned above so as to absorb IR properties, marked as XML IR/query type. Regretfully this kind of endeavors cannot cast off the disadvantages from the carriers, and still is inconvenient for common users. The other direction is to integrate keyword search into XML data processing [4,5,6,7,8,9].

When we investigate the researches following XML keyword search, we found there are two deficiencies. The first one is that there is no good method to express structural information in keyword search, however it is known that the structural information means a lot for XML data. The common improvement for keyword expression is to append label or label path information with keywords,such as "author:Botnich title:Bibliography" [6,4]. We can see that this skill still requires users know label information. Further more, the appended labels or label paths also cause one limitation, that is we cannot retrieve the XML fragments which are intimate with "Botnich Bibliography" but labeled in other labels. The second deficiency is that most similarity measuring methods for XML ranking either are dependent on labels and relative positions of nodes [10,11,12,13,14,15,16], or are not sensitive enough for structural discrimination of retrieved XML fragments [6,17,5,7,8]. For example the TF*IDF (TF means Term Frequency; while IDF means Inverse Document Frequency) like similarity measuring methods introduced in [18,5,7] only use node level distribution to sort the retrieved fragments, they do not capture the correlation relationship between keyword nodes which is one important hint for illustrating structural

```
<?xml version="1.0" encoding="UTF-8"?>

<proceedings>
 <issue>
  <articles>
   <article>
    <title>Bibliography on Data Design</title>
    <authors>
     <author position=0>Karen Botnich</author>
    </authors>
   </article>
   <article>
    <title>Face Recognition</title>
    <authors>
     <author position=0>Cola Cohen</author>
    </authors>
   </article>
  </articles>
  <publisher>
   <year>2003</year>
   <address>
    <country>Germany</country>
   </address>
  </publisher>
 </issue>
</proceedings>
```

(a)

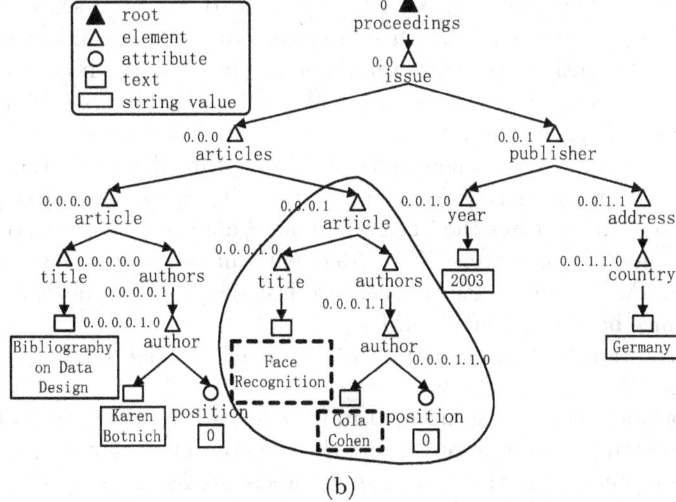

(b)

Fig. 1. The XML document and its tree model. The integer sequence at the left of node is the Dewey code [1]. The two dash line blocks are the keyword nodes for "Cohen" and "Face Recognition" respectively. The tightest XML fragment for keywords "Cohen Face Recognition" is circled. The node with label "article" and Dewey code "0.0.0.1" is the SLCA node for "Cohen Face Recognition".

```
//articles[
.//author[text()= 'Botnich'][ @position= '01']]
//title
```

Fig. 2. XPath query instance

```
LET $titles:= document("proceedings.xml")
    //articles//title

FOR $title in $titles
RETURN <titles>
        <title>
          <name>$title</name>
        </title>
        </titles>
```

Fig. 3. XQuery instance

information. As for the methods in [16,14], their computation heavily depends on the label information and relative positions of nodes. So they are not suitable for XML keyword ranking from different XML schemas. More details about the deficiencies can be found in Section 2.3.

In this paper we reconsider the keyword searching on XML data. We first import nesting mechanism so as to endow keywords with some structural hints. For example we use "((Cohen) (Face Recognition))" to illustrate that the two words belong to two different nodes and they together act as one unit in higher level. Consequently we need not care the string like "Cohen Face Recognition" in XML data at all. The other benefit of this nesting keywords is that we can still retrieve fragments satisfying its structural hints from different XML schemas. This is not easy for the extended keywords in [4] and [6]. By the way, this nesting keyword mechanism itself can also include label information. Since the retrieved XML fragments can be from different XML schemas, XML keyword search processing need new ranking scheme, which should not only be independent of labels, but also be more sensitive to structural difference between the fragments. Most similarity measures are deficient for this kind of situations because of their sensitivity for labels. Triggered from the SLCA (Smallest Lowest Common Ancestor) concept [4,9], we introduce a new structure to capture the structural features in fragments. Its name is KCAM (from Keyword Common Ancestor Matrix), in which each element stores the SLCA node level information corresponding to the keyword nodes. Since every XML fragment can be mapped into its corresponding KCAM, the similarity between two fragments can be measured by the matrix distance between their KCAMs.

The contributions of this paper can be concluded as follows:

1. We reconsider the nesting mechanism for keyword searching on XML data. We use nested parentheses to capture the structural information which users may be interested in. Since there is no labels needed in nesting keywords, we can use it to retrieve fragments even they are from different schemas. Besides, it is easy to absorb labels into nesting keywords like what [4,6] do.
2. Though nesting mechanism can fetch more interesting fragments, the fact, which the retrieved fragments are from different schemas, makes it difficult to measure the structural similarity between fragments. Triggered by SLCA concept, we propose KCAM distance to cope with this trouble. It first transforms every fragment into KCAM, in which each element stores the level information

of SLCA node between two keywords. Then the matrix distance between two KCAMs can capture the structural similarity of the two fragments.

3. We design and implement related algorithms. The extensive experiments verify that the KCAM distance is effective and efficient to discriminate the structural difference between fragments even they are from different schemas.

Section 2 reviews the related work, including three aspects–XML IR query languages, Dewey encoding and XML similarity measures. Section 3 illustrates the nesting keywords. Section 4 discusses the KCAM structure and proposes the computation of KCAM distance. Experiments are illustrated in Section 5. Finally, Section 6 concludes this paper.

2 Related Work

There are mainly three parts of related work associated with this paper. The first is about the IR like query descriptors used to XML data. We briefly retrospect XML query languages in Section 2.1. SLCA problem is illustrated in Section 2.2. The third Section 2.3 shows the researches of similarity measuring techniques for XML fragments.

2.1 XML IR Query Languages

There are two kinds of strategies when adapting IR properties into XML data processing. The first one is to extend XML query languages so as to absorb IR properties [19,20,21,22,23,24,25,26,27,28]. Among them, the extension on XPath and XQuery is the delegate, and W3C recommends one specification, namely XQuery FullText. Just as mentioned in Section 1, this kind of extensions is hard to use for common users. Real XQuery/FT sentence in Fig. 4 illustrates this point. From it we can see that XQueryFT also requires users to learn the syntaxes and data organization information.

```
FOR $i IN document("proceedings.xml")//articles
WHERE $i//authors/author contains Cohen
        AND $i//title contains 'Face'
        AND $i//title contains 'Recognition'
RETURN <result>
        <author>$i//authors/author</author>
        <title>$i//title</title>
       </result>
```

Fig. 4. XQueryFT instance

The other strategy is to use keyword search on XML data. Besides pure keywords only, other consideration is to append labels or label paths with keywords [4,6]. For example, a query in [6] is a list of query terms, where a query term can be a keyword ("Recognition"), a tag ("article:"), or tag–keyword combination ("author:Cohen"). One concrete example is "Face + Recognition article: author:Cohen". While [4] just replaces tag with label path. For example it uses "(//article/title, Recognition)" to confine the target nodes. We can see the essence of this kind of appending is just to help filter the leaf nodes. It is helpless to capture the structural hints the user need indeed. Other direct keyword query extenders [19,29,30] have similar limitations.

2.2 SLCA Problem

Different from the keyword search in traditional Information Retrieval, the targets of XML keyword search always are the XML fragments satisfying given keywords. [4,6,9] transform it as the SLCA problem, defined as Definition 1.

Definition 1 (SLCA problem). *Given one labeled directed tree, G = (V_G , E_G, r, A), and a sequence of keywords W = { k_1, k_2, ..., k_k}, the SLCA problem is to find all nodes which are the roots of all tightest XML fragments S = { S_1, S_2, ..., S_n } corresponding to W from G. The tightest XML fragment S_i ($1 \le i \le n$) has following properties.*

1. *Each S_i must include W;*
2. *There is no any subtree in S_i which includes W;*

The circled part in Fig. 1(b) is the tightest fragment for keywords "Cohen Face Recognition". The "article" node with Dewey code "0.0.0.1" is the SLCA node. The Dewey code for this SLCA node is just the common longest prefix of Dewey codes corresponding to "Cohen" and "Face Recognition" nodes, whose Dewey codes are "0.0.0.1.0" and "0.0.0.1.1.0" respectively. This obviously benefits from that Dewey encoding adopts the node path information into the codes [1].

By the way, we can see from Fig. 1(b) that the SLCA node for keywords has fixed structural relationship with the keyword nodes. We observe that the relative level information of SLCA nodes with the keyword nodes still keeps fixed even after we exchange the positions of keyword nodes. This simple observation triggers us to incorporate SLCA information in similarity measuring technique. More details are illustrated in Section 4.

2.3 Similarity Measures for XML Fragments

After retrieving XML fragments for given keywords, the next important task is to rank them so as to return top targets to users. There are two kinds of techniques for this problem. The first is to use tree edit distance concept [31,32,33]. After proposing three edit operations (Relabel, Delete and Insert) and their operation cost, the distance between two unordered label trees, T_1 and T_2, is defined as the smallest cost of transforming tree T_1 into T_2 just using the three operations mentioned above. We mark it as $EDist(T_1, T_2)$. Since the computation complication of this kind of techniques, more researchers propose many approximate techniques for this problem.

The related approximate similarity measures for this problem always inherit VSM (Vector Space Model) model [34] and $TF * IDF$ concept [35], and can be categorized into four classes. Before the review of approximate similarity measures, we first illustrate three XML fragments for keywords $\{k_1, k_2, k_3, k_4, k_5\}$ in Fig. 5 here. The intention is to help readers intuitively understand the difference of our KCAM and other methods at later discussion by real instances. The three fragments have same label domain, "{a, b, c, d }". The fragment in Fig. 5(a) is the source one. We achieve fragment in Fig. 5(b) by exchanging the position of 'k1' and 'k4' of Fig. 5(a). We can directly infer that two fragments of Fig. 5(a) and Fig. 5(b) are different in structure while having same text distribution similarity. When we exchange the position of 'k2' and 'k3' in Fig. 5(a), we get the Fig. 5(c). The two fragments of Fig. 5(a) and Fig. 5(c) in fact are same in structure and text distribution similarity.

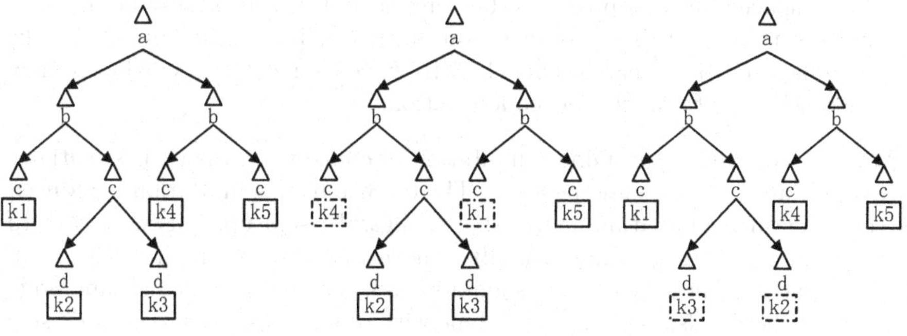

(a) Source XML fragment (b) XML fragment after exchanging K1 and K4 (c) XML fragment after exchanging K2 and K3

Fig. 5. Three XML fragments. The latter two are the variants after exchanging the sequence of some two elements.

The four kinds of approximate techniques are:

- Extended TF*IDF [29,12,36,5,18,7]. This kind of methods model XML fragment similarity problem as multi-variant regression problem based on the text distributions on leaf nodes. They first calculate the text distribution similarity using the concept of $TF * IDF$ on leaf nodes. Then they use the hierarchical information of XML fragment to calculate the final value, which corresponds to the SLCA node of those leaf nodes. This kind of measures only use hierarchical information to realize the regression, and cannot distinguish the structural difference of the fragments. So the three fragments in Figure 5 always are same according to this kind of measures.
- MLP model [13]. It has two parts to simulate the similarity between fragments. One is MLP (Maximum Leaf Path) length vector distance, which is focusing on structural similarity simulation, and the other is label histogram distance. One MLP is the node path corresponding to the longest path among the pathes from one node to all its leaf nodes. We can infer from [13] that MLP length vector has similar drawback as methods in Extended $TF*IDF$.

- Path bag model [17]. This kind of methods uses node label path distribution vector to simulate the distance between XML fragments. They do not consider the correlation structure between leaf nodes, so they take for granted that the two XML fragments in Fig. 5(a) and Figure 5(b) are same. [36] also introduces XPath model with "node position" information trying to absorb the branch information. Nevertheless this also leads to bad situation, i.e., it becomes "node position" sensitive, which will see Fig. 5(a) and Figure 5(c) different. Obviously this method is sensitive with labels, so that it is helpless for fragments with same structure but different labels.
- Structural TF*IDF [37,11,14,15,16]. This kind of methods also absorb the "term" concept in VSM and $TF * IDF$, but the "term" here changes to the "*Twig* unit". So the key task of this method is to determine the *Twig* unit vector space, which is proved to be complicated. [14] itself also admits this and resort to path based approximation, which has similar drawback like Path bag model. Researches in [11,15,16] have similar ideas. Obviously they are sensitive with labels and node position.

We can see that the prominent limitation of current work is their sensitivity with labels and relative node position. The essence of this limitation is induced by the structural simulation using label strings. To sum up, it is necessary to develop new similarity method mainly concentrating on structural feature of XML fragments. The new method should be independent of the labels and node position. It should also be easy to combine with other skills so as to satisfy users when they are interested in fragments with special labels. Our KCAM here is one example, we will discuss it in detail at Section 4.

Before that, we follow the tree edit distance concept and introduce ID edit distance of tree structure as the standard to illustrate the structural similarity between XML fragments, which is independent of node labels.

Definition 2 (ID edit distance). *The edit distance between two XML fragments, T_1 and T_2, satisfying given keywords W, is denoted as $EDist_{I,D}(T_1, T_2)$. It is the minimum cost of all ID edit sequences that transform T_1 to T_2 or vice versa. An ID edit sequence consists only of insertions or deletions of nodes: $EDist_{I,D}(T_1, T_2) = min\{c(S)|S \text{ is an ID edit sequences transforming } T_1 \text{ to } T_2\}$*

Intuitively we have $EDist_{I,D}(T_1, T_2) \leq EDist(T_1, T_2)$, where $EDist(T_1, T_2)$ is the popular edit distance with additional Relabeling. We can see that $ED_{I,D}(T_1, T_2)$ mainly concentrates on the structure information.

3 Nesting Keywords

According to [35], the structural data processing has long been noticed. Regretfully the languages developed for it are not satisfactory. For instance, since the concentrated data type at that time is mainly the Web data, Baeza–Yates, R. investigates the query languages developed for Web data, and concludes at page 392 that they are "meant to be used by programs, not final users". We can see from Section 2.1 that the XML Query (including XML IR/query) processing has similar suspicion. Though keyword search has not this kind of limitation, there is no structural hints in keywords either. This is a pity after all when adapting keyword search for semi–structured XML data, in which structure plays a significant role. During the reading of the book [35], one interesting idea come to my head, that is how about peeling off the labels from the XPath like queries, while at the same time keeping the keywords? This is the inchoation of the nesting keywords here.

In fact the concept of nesting keywords is simple. Its kernel is to organize the keywords also in nested form. For example we can use parentheses to bracket the interested keywords together. The artifice here is that the parentheses can be nested. Obviously this skill has the ability to endow the keywords with structural hints, since the XML specification uses this mechanism in same way. Besides this skill also is easy for common users to use. The most interesting thing about this skill is that it has great potential to describe query pattern, from pure keywords to part of XPath.

1. The simplest form of nesting keywords is that all keywords are bracketed in one parentheses. For example here is one keyword query "(Cohen Face Recognition)". It is clear that this query is just the pure keywords.
2. It is also easy to attach nesting keywords with labels or label paths. "author:(Cohen) title:(Face Recognition)" is one example. This means that nesting keywords method also has same power as keyword pattern descriptor proposed in [4] and [6].
3. From the two skills mentioned above, we can see that though they have comparative description power as those developed for XML keyword search, they do not touch the structural core yet. Fortunately this is quite easy in fact. Here is one instance "((Cohen) (Face Recognition))". It means that Face Recognition and Cohen belong to two different nodes, and the two nodes together are the two leaf nodes of one root node. It is just like the Twig in XPath but without label information here.
4. When we add more complicated auxiliary information, nesting keywords can even cover partial power of XPath or XQuery. Here is another query, "article:(author:(Cohen) title:(Face Recognition))". It represents one labeled tree with three nodes. The root node is "article". Two leaf nodes are "author" and "title" respectively with words "Cohen" and "Face Recognition".

Though we can attach more additional information to expand the description power of nesting keywords, the interesting thing from above illustration is the endowed structural hints proposed in the third item (3) which is meaningful for XML keyword search. Since it implies structural hints in it and need not to know the labels, it not only can help common users to organize their queries in more accurate way, but also can conduct the subsequent processing in more affirmatory style. The latter advantage can be found from the KCAM's usage in Section 4 when measuring the similarity between query keywords and the retrieved XML fragments.

4 KCAM (Keyword Common Ancestor Matrix)

The kernel structure of our new similarity measure is the Keyword Common Ancestor Matrix (KCAM in short). It adopts the position independence of SLCA node with its leaf nodes. By transforming each XML fragment into its corresponding KCAM, the structural difference between XML fragments can be reflected by the matrix distance of their corresponding KCAMs. The KCAM for one XML fragment satisfying given keywords is defined as follows.

Definition 3 (KCAM). *Given one keyword sequence $W = k_1, k_2, \cdots, k_k$, T is one tightest XML fragment satisfying the keywords W. The KCAM A for s is one $k \times k$upper triangle matrix whose element $a_{i,j}(1 \leq i \leq k, 1 \leq j \leq k)$ is determined by following equation:*

$$a_{i,j} = \begin{cases} \text{Level of leaf node corresponding to } k_i & i = j \\ \text{Level of SLCA node of } k_i \text{ and } k_j & i \neq j \end{cases}$$

According to the KCAM definition above, we can see that one KCAM captures correlation relationship between keyword nodes, which is ignored in most approximate techniques mentioned in Section 2.3. After we get the KCAMs corresponding to retrieved XML fragments, the structural difference between two arbitrary XML fragments can be determined by the matrix distance between their KCAMs. This distance is named as KDist$(\cdot,\cdot)$, defined as follows.

Definition 4 (KCAM Distance). *Given one keyword sequence $W = k_1, k_2, \cdots, k_k$, T_1 and T_2 are two tightest XML fragment satisfying the keywords W. According to KCAM definition above, we can construct two corresponding KCAMs, A_{T_1} and A_{T_2}. Then the* KDist *between A_{T_1} and A_{T_2} is defined as follows.*

$$\text{KDist}(T_1, T_2) = \frac{||A_{T_1} - A_{T_2}||_F}{k-1} \tag{1}$$

Where the matrix norm used in Equation (1) is the Frobenius norm. Its calculation is defined as follows.

$$||A||_F = \sqrt{\sum_{i=1}^{k} \sum_{j=i}^{k} a_{i,j}^2} \tag{2}$$

$(k-1)$ *is one experiential factor so as to guarantee that* KDist *here is the lower bound of* $\text{EDist}_{I,D}$, *that is* $\text{KDist} \leq \text{EDist}_{I,D}$. *Its proof is omitted for room reason.*

Based on the two definitions above, we can easily discriminate the XML fragments proposed in Fig. 5. Their KCAMs are illustrated in Fig. 6. Based on the definition of $KDist(T_1, T_2)$, we can see the distance between Fig. 5(a) and Fig. 5(b) is $\sqrt{6}/4 \approx 0.612$. The distance between Fig. 5(a) and Fig. 5(c) is 0.

	K1	K2	K3	K4	K5
K1	3	2	2	1	1
K2		4	3	1	1
K3			4	1	1
K4				3	2
K5					3

(a) KCAM for Source XML fragment in Figure 5(a)

	K1	K2	K3	K4	K5
K1	3	1	1	1	2
K2		4	3	2	1
K3			4	2	1
K4				3	1
K5					3

(b) KCAM for XML fragment in Figure 5(b)

	K1	K2	K3	K4	K5
K1	3	2	2	1	1
K2		4	3	1	1
K3			4	1	1
K4				3	2
K5					3

(c) KCAM for XML fragment in Figure 5(c)

Fig. 6. Three KCAM instances for XML fragments in Figure 5

Clearly we can use KCAM method to distinguish Fig. 5(a) and Fig. 5(b), and at the same time we will not classify Fig. 5(a) and Fig. 5(c) as same.

KCAM mechanism has similar properties with MLP scheme proposed in [13]. MLP also has the ability to simulate the structural similarity, but is different from other methods in Extended $TF * IDF$, which essentially care the weight of keywords in retrieved fragments.Different from level information of our SLCA nodes, it uses the maximum length among all pathes from the node to its leaves. So MLP cannot distinguish the difference of fragments in Fig. 5(a) and Figure 5(b). MLP is also independent of labels and node position. Finally, MLP can also be combined with other label filtering skill when users are interested more at specific domain.

5 Experiments

5.1 System Architecture

Finally the system architecture is illustrated in Fig. 7. Our experimental system mainly has two modules. One is the query processing module, which has two blocks, Query processor and Ranker. The other is the data management module, which is comprised of Dewey Index Management block and Indexer block. The former parses the nesting keywords, calculates SLCA nodes based on the retrieved Dewey codes from data management module, and uses KCAM in Ranker to measure the structural similarity of fragments. The latter is in charge with parsing the XML data in Indexer, and managing the indexed XML data in Dewey Index Management block.

We do not illustrate the computation details in each function block, for minded readers can find them when following the mathematical illustrations of this paper and the related references. The only thing reminded here is that we suggest that it is simpler to compute KDist using its matrix vector. We have following transformation from matrix to its vector according to matrix theory. So the matrix distance becomes the vector distance now. Since the algorithm design is quite straightforward following above mathematical equations, we omit the KCAM construction algorithm from this paper.

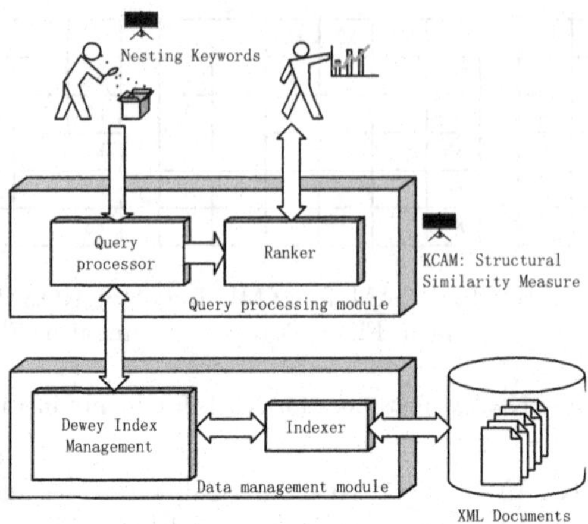

Fig. 7. The architecture of our experimental system

$$\begin{pmatrix} a_{1,1} & a_{1,2} & \cdots & a_{1,k} \\ & a_{2,2} & \cdots & a_{2,k} \\ & & \vdots & \vdots \\ & & & a_{k,k} \end{pmatrix} \Rightarrow (a_{1,1}, \cdots, a_{1,k}, a_{2,2}, \cdots, a_{2,k}, \cdots, a_{k,k}) \tag{3}$$

5.2 Experimental Environment

Synthesized Dataset. Even though there are many ways to get experimental XML data, such as synthesized data from several tools (XMark, XMLGenerator), or some ready benchmark XML datasets (DBLP, Shakespeare, Sigmod, INEX), they are not adequate when to judge the performance of XML keyword searching methods. The reason is that they are difficult to control the node number for specific keyword, such as we need 20,0000 nodes for "Bibliography". So in this paper we verify the efficiency of our KCAM method on synthesized XML data made as follows.

1. We first choose one XML fragment with k leaf nodes, $G = (V_G, E_G, r, A_G)$. *counter* is a counter and its initial value is 1. TS is the template set, its initial is $\varnothing$.
2. We construct the first fragment template by inserting one *Dummy* node as the new root of G, $G_T = (V_{G_T}, E_{G_T}, r_T, A_{G_T})$, and insert it into TS. r_T corresponds to the *Dummy* node. Then we can get the Dewey codes for the k leaf nodes, $D_T = \{d_1, d_2, \ldots, d_k\}$. $maxL_D = max\{L_{d_i} | 1 \leq i \leq k\}$, corresponds to the maximum length of D_T, while L_{d_i} means the length of Dewey code d_i.
3. Randomly select one template from TS, and run one step of following three operations on it. Increase the counter by 1 at each time, and put the new fragment into TS as the new template.

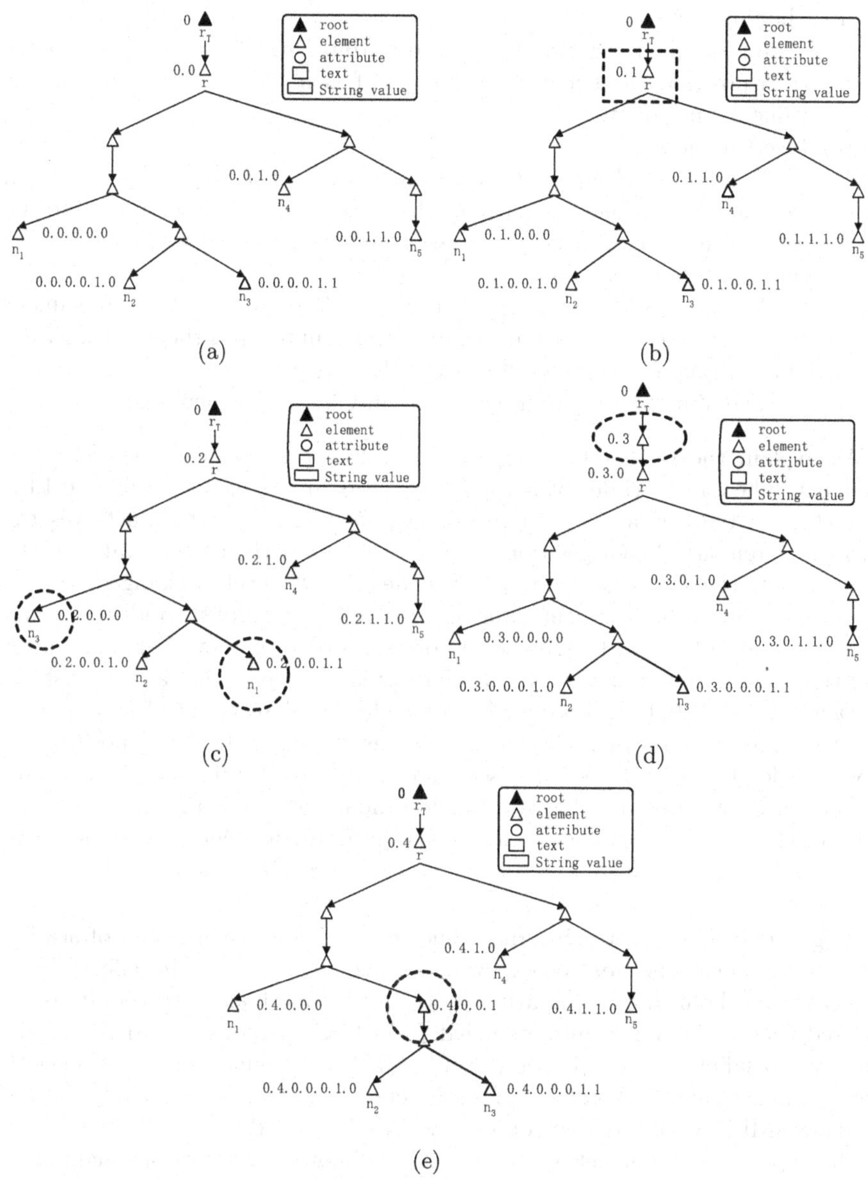

Fig. 8. Illustrations for synthesized dataset. (a) is the source fragment; (b) is the new fragment after one Copying action on (a); (c) is the new fragment after one Shuffling action on (a); (d) and (e) are two new fragments after Inserting action on (a).

(a) Copying action

Change the integer of each d_i at level 2 to the current value of the counter. This means to copy the subtree and insert it into r_T as new subtree.

(b) Shuffling action

Randomly select two Dewey codes from { d_1, d_2, ..., d_k }, exchange their positions, and change the integer of each d_i at level 2 to the current value of the counter.

(c) Inserting action

Randomly select one Dewey code d_i from { d_1, d_2, ..., d_k }. Then randomly select one integer from $[1,L_{d_i}]$,choose the position next to the selected integer to insert one integer and refresh all Dewey codes of the selected template.

Assuming n $(2 \leq n \leq L_{d_i})$ is the candidate position, the refreshment means to insert the current value of the counter into the same position of all Dewey codes whose prefix equals to $p_{d_i}(n)$.

The aim of this action is to randomly insert one new node.

We explain the procedure using Figure 8. Fig. 8(a) is the initial template fragment. The Dewey codes of n_1, n_2, n_3, n_4, n_5 are "0.0.0.0.0", "0.0.0.0.1.0", "0.0.0.0.1.1","0.0.1.0" and "0.0.1.1.0" respectively. Figure 8(b) illustrates the fragment after one Copying action on Fig. 8(b). Since the current value of the counter is 1, the integers of all Dewey codes at 2nd level is changed to "1". Figure 8(c) shows the fragment after one Shuffling action of n_1 and n_3 on Figure 8(a). Figure 8(d) and Figure 8(e) correspond to two new fragments after inserting new nodes respectively. The former is to insert node at 1st position on Dewey code "0.0.0.0". Since all Dewey codes are descendant of Dewey code "0", they are all refreshed. The latter is to insert new node at 5th position on Dewey code "0.1.0.0.1.0". We can see that only "0.0.0.0.1.0" and "0.0.0.0.1.1" are refreshed because only they are the descendant of "0.0.0.0.1".

From the Fig. 8 we can see that the synthesized fragments are random not only on structure, but also on level distribution of SLCA nodes.

Configurations for Experiments. Since the description capability of nesting keywords is clear, the goal of experiments here is to verify the effectiveness of KCAM mechanism by comparing it with other methods. Regretfully as we inferred from Section 2.3, only extended $TF * IDF$ methods and MLP are not sensitive to label or node position. Since MLP has similar computation with methods in extended $TF*IDF$, and it is independent of labels and node position, we choose MLP as the comparison object with our KCAM.

The experimental dataset is constructed by following instructions mentioned above. We first build several template fragments like Fig. 8(a), and then increase the number of fragments by randomly shuffling the keyword position and inserting new node. There are four parameters to control the fragments. One is keyword number (10 in our experiment). The second is the number of template fragments. The third is the times of randomly insertion. We set it as 20. The last one is the exchange times among keywords for each template fragments and their variants after insertion.

After getting synthesized dataset, we do two kinds of experiments. The first one is for structural sensitivity. We randomly select 10 groups, each of which has

30 fragments. We carry out KCAM and MLP on them, record the ratio of the precision values between KCAM and MLP, and illustrate the average statistic. The second kind is for the performance scalability. We separate the dataset into 9 groups, from 5000 to 45000. We run both methods on each group, and illustrate the running time of each method.

All experiments are run on a Dell Dimension 8100 computer with 1.4GHz Pentium IV processor and 256MB of physical main memory. The Operating System is Windows 2000 Professional. We use JAVA as the programming language. The JDK is Java 2 SDK Standard Edition Version 1.4.1. We execute each test 6 times with the performance result of the first run discarded.

5.3 Experimental Result

Fig. 9 is the result for precision comparison of KCAM and MLP. When the number of keywords is specified, we run MLP and KCAM on the sampled XML fragments. After we obtain the precision of KCAM and MLP, we get the ratio of the two precision values corresponding to the two methods. From Figure 9 we can see that the KCAM/MLP precision ratio is adjacent when the keyword number is small. While the ratio become larger when the keyword number increases. This evolution verifies that KCAM method is more sensitive with the structural differences than MLP method.

Fig. 10 illustrates the runtime performance result of the two methods. We first construct two kinds of synthesized data corresponding to 5 and 10 keywords with different number of fragments from 5000 to 45000. Then we run KCAM and MLP respectively on the datasets. "KCAM5Key" and "KCAM10Key" mean that there are 5 and 10 keywords in all fragments when running KCAM. "MLP5Key" and "MLP10Key" mean that there are 5 and 10 keywords in all fragments when running MLP. From Figure 10 we can see that when the number of keywords is small, KCAM has comparatively performance with MLP even there are 45000 fragments. When there is many keywords, the performance of KCAM becomes lower than MLP, and the gap becomes larger and larger as fragment number increases.

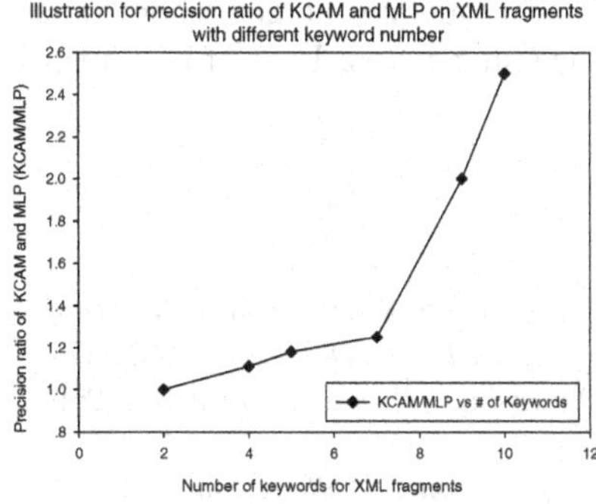

Fig. 9. Precision ratio of KCAM and MLP on different XML fragments

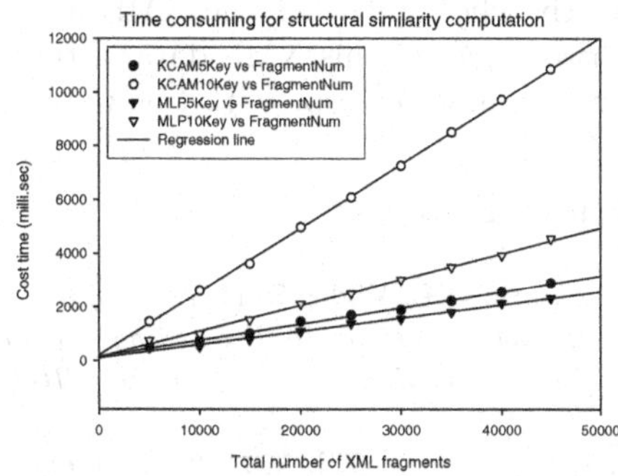

Fig. 10. Time consuming for KCAM and MLP with fragment scale increasing

Though the performance of KCAM is lower than MLP from Figure 10, KCAM's merit is still obvious according to Figure 9, that is KCAM has more power to distinguish the structural difference. Besides, we notice that the largest cost of KCAM performance is lower than 12,000 millisecond. This shows that the computation of KCAM is still efficient and has pragmatic value for real application.

6 Conclusion and Future Work

In this paper we reconsider the keyword processing on XML data. Seeing that the current XML keyword extensions lack structural information, we propose nesting keywords mechanism so as to endow keywords with structural hints. Though this mechanism has simple concept, it is significant not only for common users but also for similarity measures of retrieved XML fragments. When using nesting keywords to retrieve information from XML database, the retrieved XML fragments can be from different schemas (Pure keywords has same property). But current similarity measures are incapable for this situation: they are either insensitive for subtle structural differences between fragments, or too sensitive with labels and node positions. Aiming at this problem, this paper proposes KCAM distance method. Its kernel is keyword common ancestor matrix, in which each element stores the level information of SLCA node corresponding to keyword nodes in XML fragments. Since the level information of SLCA node implies the relative position information between keyword nodes, KCAM can capture structural information of XML fragment. Therefore the matrix distance between two KCAMs can be used to reflect the structural difference of the two corresponding XML fragments.

Our future work will concentrate on the data redundancy problem in current XML data processing systems. During the realization of our experimental system, we investigate the developed XML systems. We find there is data redundancy in current XML data processing scheme. That is the XML Query type processing (including XML IR/query) and XML IR/keyword type processing have their own particular XML data representation in their realization. Consequently when querying one XML data in different query schemes, there must be two copies of this XML data in the two processing types. Part of our future work attempts to cope with this redundancy.

References

1. Tatarinov I., Viglas S. D. *Storing and Querying Ordered XML Using a Relational Database System*. In *Proceedings of the 2002 ACM SIGMOD International Conference on Management of Data (SIGMOD)*. Madison, Wisconsin. June 3-6, 2002. 204–215.
2. Clark.J, DeRose. S. *XML Path Language(XPath) version 1.0 w3c recommendation. World Wide Web Consortium.* 1999, November
3. Chamberlin.D, et al. *XQuery: A Query Language for XML W3C working draft. Technical Report WD-xquery-20010215, World Wide Web Consortium.* 2001, February

4. Schmidt.A, Kersten. L.M, Windhouwer. M. *Querying XML documents made easy: Nearest concept queries.* In: *Proceedings of the 17th International Conference on Data Engineering (ICDE)*, 321–329. 2001, April.
5. Guo.L, et al. *XRANK: Ranked Keyword Search over XML Documents. SIGMOD 2003.* 2003, June 9-12
6. Cohen.S, et al. *Xsearch: A semantic search engine for xml. Proceedings of the 29th VLDB Conference*, 33–44. 2003, September 9-12
7. Weigel.F, et al. *Content and Structure in Indexing and Ranking XML. WebDB.* 2004.
8. Botev.C, Shanmugasundaram. J. *Context-Sensitive Keyword Search and Ranking for XML. Eighth International Workshop on the Web and Databases (WebDB 2005).* 2005, June 16-17
9. Xu.Y, Papakonstantinou. Y. *Efficient Keyword Search for Smallest LCAs in XML Databases. ACM SIGMOD 2005.* 2005, June 14-16
10. Schlieder.T, Meuss. H. *Result ranking for structured queries against xml documents. DELOS Workshop: Information Seeking, Searching and Querying in Digital Libraries.* 2000.
11. Guha.S, et al. *Approximate XML Joins. Proceedings of the 2002 ACM SIGMOD International Conference on Management of Data (SIGMOD).* 2002, June 3-6
12. Yu.C, Qi.H, Jagadish. V. H. *Integration of IR into an XML Database. INEX Workshop*, 162–169. 2002.
13. Kailing.K, et al. *Efficient Similarity Search for Hierarchical Data in Large Databases. Advances in Database Technology - EDBT 2004, 9th International Conference on Extending Database Technology*, **ISBN 3-540-21200-0**, 676–693. 2004, March 14-18
14. Amer-Yahia.S, et al. *Structure and Content Scoring for XML. Proceedings of the 31st International Conference on Very Large Data Bases (VLDB)*, 361–372. 2005, August 30 - September 2.
15. Yang.R, Kalnis.P, Tung. K. A. *Similarity Evaluation on Tree-structured Data. ACM SIGMOD Conference.* 2005, June 13-16
16. Augsten.N, Böhlen. H.M, Gamper. J. *Approximate Matching of Hierarchical Data Using pq-Grams. Proceedings of the 31st International Conference on Very Large Data Bases (VLDB)*, 301–312. 2005, August 30 - September 2.
17. Joshi.S, et al. *A Bag of Paths Model for Measuring Structural Similarity in Web Documents. SIGKDD'03.* 2003, August 24-27
18. Carmel.D, et al. *Searching XML Documents via XML Fragments. SIGIR'2003.* 2003, July 28-August 1.
19. Wolff. E.J, Flörke.H, Cremers. B. A (1999). *XPRES: A ranking approach to retrieval on structured documents.* University of Bonn. Technical Report IAI-TR-99-12.
20. Florescu.D, Kossmann.D, Manolescu. I. *Integrating Keyword Search into XML Query Processing. WWW.* 2000.
21. Fuhr.N, Großjohann. K. *XIRQL: A query language for information retrieval in XML documents. In International Conference on Information Retrieval (SIGIR).* 2001.
22. Bremer. M.J, Gertz. M. *XQuery/IR: Integrating XML Document and Data Retrieval. WebDB.* 2002.
23. Chinenyanga. T.T, Kushmerick. N. *An expressive and efficient language for XML information retrieval. Journal of the American Society for Information Science and Technology (JASIST)*, **53**(6), 438–453. 2002

24. Theobald.A, Weikum. G. *The index-based XXL search engine for querying XML data with relevance ranking. In Proceedings of the 8th Conference on Extending Database Technology (EDBT)*, 477–495. 2002, March.
25. Al-Khalifa.S, Yu.C, Jagadish. V. H. *Querying Structured Text in an XML Database. SIGMOD 2003.* 2003, June 9-12
26. Amer-Yahia.S, Botev.C, Shanmugasundaram. J. *TeXQuery: A FullText Search Extension to XQuery. In Proceedings of the 13th conference on World Wide Web*, 583–594. 2004, May 17-22
27. Amer-Yahia.S, Lakshmanan. V.L, Pandit. S. *FleXPath: Flexible Structure and Full-Text Querying for XML. SIGMOD 2004.* 2004, June 13-18
28. Curtmola.E, et al. *GalaTex: A Conformant Implementation of the XQuery FullText Language. Informal Proceedings of the Second International Workshop on XQuery Implementation, Experience, and Perspectives (XIME-P).* 2005, June 16-17.
29. Wolff. E.J, Flörke.H, Cremers. B. A. *Searching and browsing collections of structural information. In Proceedings of IEEE Advances in Digital Libraries (ADL 2000)*, 141–150. 2000, May.
30. Woodley.A, Geva. S. *NLPX - An XML-IR System with a Natural Language Interface. Proceedings of the 9th Australian Document Computing Symposium.* 2004, December 13.
31. Zhang.K. *On the editing distance between unordered labeled trees. Information Processing Letters*, **42**(3), 133–139. 1992
32. Shasha.D, Zhang. K (1997). *Approximate Tree Pattern Matching.* In A. Apostolico & Z. Galil (Ed.), *Pattern Matching Algorithms.* Oxford University.
33. Bille.P. *A survey on tree edit distance and related problems. Theoretical Computer Science*, **337**(1-3), 217–239. 2005, June
34. Salton.G (1968). *Automatic Information Organization and Retrieval.* New York: McGraw-Hill.
35. Baeza-Yates.R, Ribeiro-Neto. B (1999). *Modern Information Retrieval.* (pp. 19–73). Pearson Education Limited.
36. Kotsakis.E. *Structured Information Retrieval in XML documents. Proceedings of the 2002 ACM symposium on Applied computing*, 663–667. 2002, March.
37. Schlieder.T, Meüss. H. *Querying and ranking XML documents. Journal of the American Society for Information Science and Technology*, **53**(6), 489–503. 2002, May.

Improving Re-ranking of Search Results Using Collaborative Filtering

Rohini U[1] and Vamshi Ambati[2]

[1] Language Technologies Research Center
International Institute of Information Technology
Hyderabad, India
rohini@research.iiit.ac.in
[2] Regional Mega Scanning Center
International Institute of Information Technology,
Hyderabad, India
vamshi@iiit.ac.in

Abstract. Search Engines today often return a large volume of results with possibly a few relevant results. The notion of relevance is subjective and depends on the user and the context of search. Re-ranking of these results to reflect the most relevant results to the user, using a user profile built from the relevance feedback has proved to provide good results. Our approach assumes implicit feedback gathered from a search engine query logs and learn a user profile. The user profile typically runs into sparsity problems due to the sheer volume of the WWW. Sparsity refers to the missing weights of certain words in the user profile. In this paper, we present an effective re-ranking strategy that compensates for the sparsity in a user's profile, by applying collaborative filtering algorithms. Our evaluation results show an improvement in precision over approaches that use only a user's profile.

1 Introduction

In general, interactions with current day web search engines could be characterized as "one size fits all". This means that all queries, posed by different users are treated similarly as simple keywords where the aim is to retrieve web pages matching the keyword. As a result, though the user has a focused information need, due to the excess information on the WWW, the amount of results returned for a particular keyword search is enormous. This places burden on the user to scan and navigate the retrieved material to find the web pages satisfying his actual information need. For example, two different users may use exactly the same query "Java" to search for different pieces of information - 'Java island in Indonesia' or 'Java programming language'. Existing IR systems would return a similar set of results for both these users. Incorporating the user's interests and focus into the search process is quite essential for disambiguating the query and providing personalized search results.

One way to disambiguate the words in a query is to associate a categorical tag with the query. For example, if the category "software" or the category "travel"

H.T. Ng et al. (Eds.): AIRS 2006, LNCS 4182, pp. 205–216, 2006.

is associated with the query "java", then the user's intention becomes clear. By utilizing the selected categories as a context for the query, a search engine is likely to return documents that are more suitable to the user. Current search engines such as Google or Yahoo! have hierarchies of categories to help users to specify his/her categories manually to the query. Unfortunately, such extra effort can not be expected from the user in a web search scenario. Instead it is preferred to automatically obtain a set of categories for a user query directly by a search engine. However, categories returned from a typical search engine are still independent of a particular user and many of the returned document results could belong to categories that may not reflect the intention of the searcher. This demands further personalization of the search results.

In this paper, we propose a two phase strategy to personalize search results over the WWW. We first learn a user profile based on his relevance feedback and use it effectively in a re-ranking phase to provide personalized search results. Since it is difficult to assume that the users will provide the relevant documents ([1], [2], [3] etc) explicitly, we make use of the implicit feedback given by the users which are captured in search engine interactions as "Query logs" or "Click through data". In the rest of the paper, we use the terms click through data, query log data and implicit feedback interchangeably. Such data consists of the queries, clicked documents and the identity of the user say ip address and is invaluable for research in search personalization.

Liu et al [4] successfully built user profiles for re-ranking by incorporating query categories in the learning process. We follow an approach similar to them. We first infer the category of a query using existing search engines and open directory project (ODP). We use the queries, their respective categories and the corresponding clicked documents in the learning of a user profile for the user. The user profile is represented as a matrix containing the pairs (term, category) and their corresponding weights. Machine Learning algorithms are used to automatically learn these term weights in the matrix. Each element represents how important the term is when the user is searching for a query of the given category. Re-ranking of the search results based on the user profile thus built, has shown improvement in performance. Though category helps to disambiguate the query, it adds another extra dimension to the user profile. This typically brings in sparsity in the user profile, which was observed in our case. Sparsity refers to the missing weights of certain words in the user profile.

In this paper, we present an effective re-ranking solution that compensates for the sparsity in a user profile, by collaborative filtering algorithms. A great deal of information overlap exists in web searches among users ([5], [6], [7] [8], etc). This overlap is seen due to users with similar information needs, posing similar queries. To our knowledge, this vast and rich source of information overlap hasn't much been properly exploited for the WWW. Collaborative filtering algorithms work exceptionally well in a community like environment with significant overlap of information needs and interests. The novelty of our re-ranking algorithm lies in addressing the sparsity in the user profile by exploiting the information overlap, using collaborative filtering. Our approach shows an improvement in the overall results when compared to a re-ranking performed based on just the user profile.

The rest of the paper is organized as follows. Section 2 discusses the Related Work, Section 3 discusses the proposed approach of learning user profiles, the re-ranking strategy, and addressing the sparsity in the user profiles. Section 4 describes the experimental setup and evaluation. Section 5 described our conclusions and future work.

2 Related Work

The related work related to the approach proposed in this paper, can broadly be classified as work done in personalized search and work done in, collaborative filtering applied to search.

2.1 Personalized Search

There has been a growing literature available with regard to personalization of search results. In this section, we briefly overview some of the available literature. Page et al [9] proposed personalized PageRank as a modification to the global PageRank algorithm. However, the computation of personalized PageRank in the paper is not addressed beyond the original algorithm. Haveliwala [10] used personalized PageRank scores to enable topic sensitive web searches. However, no experiments based on a user's context such as browsing patterns, bookmarks and so on were reported. Pretschner [11] used ontology to model a users interests, which are studied from users browsed web pages. Speretta and Gauch [12] used users search history to construct user profiles. Liu et. al [4] performed personalized web search by mapping a query to a set of categories using a user profile and a general profile learned from the user's search history and a category hierarchy respectively. Shen et. al [13] proposed a decision theoretic framework for implicit user modeling for personalized search. They consider the short term context in modeling a user. Radlinski and Joachims [[14], [15]] learn a ranking function using Support Vector Machines and using it to improving search results.

2.2 Collaborative Filtering and Search

Chidlovski et al [16] describes the architecture of a system performing collaborative re-ranking of search results. The user and community profiles are built from the documents marked as relevant by the user or community respectively. These profiles essentially contain the terms and their appropriate weights. Re-ranking of the search results is done using the term wights using adapted cosine function. The search process and the ranking of relevant documents are accomplished within the context of a particular user or community point of view. However the paper does not discuss much about the experimental details. Sugiyama et.al [17] performed personalization by adapting to users interests without any effort from users. Further, they modified the traditional memory based collaborative filtering algorithm to suit to the web search scenario and used it to improve the search results. They constructed a user-term weights matrix analogous to user-item matrix in memory based collaborative filtering algorithms and then applied traditional collaborative filtering predictive algorithms to predict a term weight in each user profile. Lin et. al [18] presented an approach to perform personalized web search based on PLSA, Probabilistic Latent Semantic Analysis, a technique which stems from linear algebra. They extracted a co-occurrence triple containing the users, queries, and web pages by mining the web-logs of the users and modeled the latent semantic relationship between them using PLSA. Armin Hust [19] performed query expansion by using previous search queries by one or more users and their relevant documents. This query expansion method reconstructs the query as a linear combination of existing old queries. The terms of the relevant documents of these existing old queries are used for query expansion. However, the approach does not take the user into account. In ([6], [5], [20], [7], [20], [21]) a novel approach to web search - Collaborative Web search was introduced. It combined techniques for exploiting knowledge of the query-space with ideas from social networking to develop a Web search platform capable of adapting to the needs of communities of users. In brief, the queries submitted and the results selected by a community of users are recorded and reused in order to influence the results of future searches for similar queries. Results that have been reliably selected for similar queries in the past are promoted. Rohini and Vamshi [22] proposed an approach for re-ranking of search results in a digital library scenario. The user profiles were constructed from the documents marked as relevant or irrelevant. Re-ranking of the results is done using the user profile and profile of others users in the community. They assumed and assigned a set of static communities for each user which the user has selected while registering with the system. Also, the user also selects the community before posing the query and the re-ranking is done based on the community selected.

Several other works ([23], [24], [8], [25], [19] etc) have made use of past queries mined from the query logs to help the current searcher.

3 Proposed Approach

The proposed approach to search result personalization consists of two phases. The first is a learning phase and the second is a retrieval/re-ranking phase. We use "click through data" from a real world search engine, www.alltheweb.com, to build and test our proposed approach. In this section we discuss in detail our approach of learning user profiles and re-ranking search results for personalization.

3.1 Mapping Query to a Category

User profiles are learned on implicit feedback data, annotated with the category of the query posed. However, the click through data used here, does not consist of an associated category for the query. We therefore enhance the click through data by assigning category information to all the queries using the ODP, Open Directory Project (http://dmoz.org). The DMOZ Open Directory Project (ODP) is the largest, most comprehensive human-edited web page catalog currently available. It covers 4 million sites filed into more than 590,000 categories (16 wide-spread top-categories, such as Arts, Computers, News, Sports, etc.) Currently, there are more than 65,000 volunteering editors maintaining it. ODP's data structure is organized as a tree, where the categories are internal nodes and pages are leaf nodes.

In our work, we consider only the top most ODP categories in the hierarchy to classify the query into categories. The category information of the query can be obtained by posing the query to one or more of the directory services available (directory.google.com, http://dmoz.org). which returns the related categories to the query. Otherwise, based on the categories of the top 10 documents in the search results, the most common document category is chosen and selected as the query category. Other effective solutions for query categorization exist, like training a text classifier on the documents contained in the ODP data. Such a classifier could be used to categorize the clicked documents in the click-through data. Improvements in performance of query categorization are always possible, and will enhance our proposed approach. However, for simplicity we currently focus on the former approach for query categorization that depends on direct ODP lookup.

3.2 Learning User Profiles

We use machine learning algorithms for learning the user profiles from the implicit feedback provided by the user. The input to the learning algorithms is a user's implicit relevance feedback, gathered from the click through data, along with the query and the associated category. Learning the user profile involves learning the weights of certain features extracted from the implicit feedback. The effectiveness of the user profile depends to a large extent on the representation of features. As mentioned earlier, we consider the features (term, category) to effectively represent the context through the category. The weights of the features represent the importance of the term for the respective category.

We considered SVM for learning the weights of the features for its success in various text applications [26], [27]. An SVM is trained using our proposed features and at the end of the training phase, the weights of the features are learned which constitutes our user profile. The procedure of learning weights is similar to Radlinski and Joachims [14]. SVM^{light} [28] has been used for training the SVM.

3.3 Re-ranking

Re-ranking of the results is done by first retrieving a set of documents matching the query using a search engine. Then the top documents returned by the search engine are reranked using the user profile in the following manner. At first the test query category is inferred similar to the learning phase as discussed in section above (Mapping query to a category). Then for each word in the document, the weight of the pair (term, category) is obtained from the user profile.

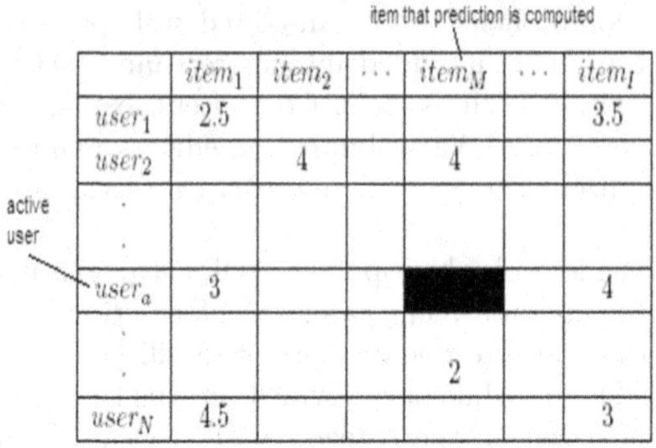

item that prediction is computed

active user

	$item_1$	$item_2$	...	$item_M$	...	$item_I$
$user_1$	2.5					3.5
$user_2$		4		4		
. .						
$user_a$	3			■		4
. .				2		
$user_N$	4.5					3

Fig. 1. Sample user-item matrix

Let c be the identified category of the query, t be a word in the document D_j and w_{t,D_j} be its weight in the document D_j (typically the term frequency TF or TFIDF etc). t_c represent the pair (term, category). UP_{a,t_c} represents the weight of t_c in the user profile of the user a. CP_{a,t_c} represents the predicted weight of t_c using collaborative filtering. Then the document ranks are computed as weighted combination of its term frequency TF in the document and the weight obtained from his user profile as shown in Equation (1).

$$R_{D_j,Q_c} = \sum_{t \in D_j} \alpha w_{t,D_j} + (1-\alpha) W_{a,t_c} \tag{1}$$

$$W_{a,t_c} = \begin{cases} UP_{a,t_c} & \text{if} \quad UP_{a,t_c} \neq 0 \\ CP_{a,t_c} & \text{otherwise} \end{cases} \tag{2}$$

Re-ranking is done by sorting the documents in decreased order of their rank. Based on experimentation, we set the value of α to be 0.7.

3.4 Addressing Sparsity in the User Profile to Improve Re-ranking

Usage of a (term, category) pairs helps to disambiguate the query and act as good contextual information in building a user profile. However, this typically brings in sparsity in the user profile, due to an added dimension to the user profile - category of the query. Sparsity refers to missing weights of certain words in the user profile. We address the sparsity in the userprofile using collaborative filtering to improve the re-ranking of the documents. Certain weights of the pairs (term, category) not occurring the user's profile are predicted using the adapted version of the collaborative filtering which we present below. In the following subsections, we first briefly review the pure collaborative filtering algorithms, especially neighborhood-based algorithms, and then describe the adapted collaborative filtering algorithms to address the sparsity in the userprofiles and then present how we make predictions of the pairs (term, category).

Overview of the Pure Collaborative Filtering Algorithm
Collaborative filtering is one of the most successful recommendation algorithms. They have been popular for recommending news [[29], [30]], audio CDs, movies, music [31], research papers etc. Recommendations are typically computed using the feedback taken from all the users in the community represented in a user-item matrix. The entries in the user-item matrix are the ratings given by the respective users for the respective items. Collaborative filtering can broadly be seen as the problem of predicting missing values in a user-item ratings matrix. Figure 1 shows a simplified example of a user-item ratings matrix. In the neighborhood-based algorithm [32], a subset of users is first chosen based on their similarity to the active user, and a weighted combination of their rating is then used to produce predictions for the active user. The algorithm can be summarized in the following steps: 1. Weight all users with respect to similarity to the active user. This similarity between users is measured as the Pearson correlation coefficient between their rating vectors. 2. Select n users that have the highest similarity with the active user. These users form the neighborhood. 3. Compute a prediction from a weighted combination of the neighbors ratings.

In step 1, $S_{a,u}$, which denotes similarity between users a and u,and is computed using the Pearson correlation coefficient as shown in Equation (3) where $r_{a,i}$ is the rating given to item i by user a, and $\overline{r_a}$ is the mean rating given by user a, and I is the total number of items. In step 2, i.e., neighborhood-based methods, a subset of appropriate users is chosen based on their similarity to the active user computed in the above step, and a weighted aggregate of their ratings is used to generate predictions for the active user in the next step 3. In step 3, predictions are computed as the weighted average of deviations from the neighbors mean as shown in Equation (4)

$$S_{a,u} = \frac{\sum_{i=1}^{I}(r_{a,i} - \overline{r_a})(r_{u,i} - \overline{r_u})}{\sqrt{\sum_{i=1}^{I}(r_{a,i} - \overline{r_a})^2 \sum_{i=1}^{I}(r_{u,i} - \overline{r_u})^2}} \tag{3}$$

$$p_{a,i} = \overline{r_a} + \frac{\sum_{u=1}^{n}(r_{u,i} - \overline{r_u})S_{a,u}}{\sum_{i=1}^{I} S_{a,u}} \tag{4}$$

Adapted Collaborative Filtering Algorithm
In the pure collaborative filtering algorithms described above, we considered a user-item ratings matrix. Similarly, we now consider user-(term,category) matrix (see Figure 2). Each row in the matrix represents the entries in the user profile of the respective user. By representing the user profile in this fashion, collaborative filtering algorithms can directly be applied.

The prediction of the (term,category) weights are computed by first identifying a set of similar users (ie users who has similar (term,category) weights as measured using Equation (5). Then using these users, the predictions are computed analogous to pure collaborative filtering as shown in Equation (6) Then the re-ranking of the document is done as described in Equation (1).

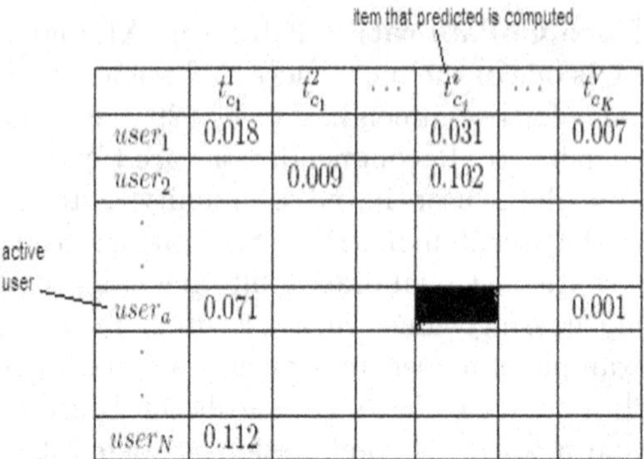

	$t^1_{c_1}$	$t^2_{c_1}$	...	$t^i_{c_j}$	...	$t^V_{c_K}$
$user_1$	0.018			0.031		0.007
$user_2$		0.009		0.102		
.						
$user_a$	0.071					0.001
.						
$user_N$	0.112					

Fig. 2. Sample user-(term,category) matrix

$$S_{a,u} = \frac{\sum_{t_c \in (T,C)} (w_{a,t_c} - \overline{w_a})(w_{u,t_c} - \overline{w_u})}{\sqrt{\sum_{t_c \in (T,C)} (w_{a,t_c} - \overline{w_a})^2 \sum_{t_c \in (T,C)} (w_{u,t_c} - \overline{w_u})^2}} \tag{5}$$

$$p_{a,t_c} = \overline{w_a} + \frac{\sum_{u=1}^{n} (w_{u,t_c} - \overline{w_u}) S_{a,u}}{\sum_{t_c \in (T,C)} S_{a,u}} \tag{6}$$

where $p(a, t_c)$ is the predicted computed for term t in query Category c and is equal to CP_{a,t_c}.

4 Experiments

4.1 Data and Experimental Setup

Query log data used in the experiments consist of the query, the clicked URLs for the query and the user identifier (ip addresses) and the time of click of the document. Such information though invaluable for research on information retrieval, is not released by major search engines. Recently, Alltheweb.com[1] has made available its search logs for research purposes. The data was collected from queries mainly submitted by European users on 6 February 2001. The data set contains approximately a million queries submitted by over 200,000 users and 977,891 unique click URLs. Further information on the data can be found in [33].

We use the query log data released by Alltheweb.com to perform our experiments and to evaluate the proposed approach We first divide the query logs into a large chunk of training clickthrough data, used for learning user profiles and a smaller chunk for testing and evaluating the approach. A direct evaluation experiment of our proposed re-ranking algorithm can not be performed on the present day's search indices of Alltheweb.com or any other search engine for that matter. Document repositories on the WWW have been changing drastically and undergo restructuring. Hence evaluation results can not be based on

[1] http://alltheweb.com

user profiles learnt from the query logs used in the current experiment. Therefore we first obtain all the documents corresponding to the queries in the testing data by crawling the click URLs and storing them as a repository. We were only successful in retrieving about 40% of the actual click URLs due to broken links and restructing of the WWW. These retrieved documents constitute the document repository used in current test experiments. With the volume of query log data we are working with, this repository could be considered as an analog to the WWW that corresponds to the query logs in discussion. For the purposes of these experiments, we name this repository as the mini-WWW, consisting of about 35,000 documents. We also pick queries from the query log data and pose it to Google to fetch and download the top 100 documents. These documents are added to the mini-WWW. This prevents any kind of bias that may have been introduced in the construction of mini-WWW from click URLs in the query log data. With availability of every day query log data we expect the proposed approaches to scale and be useful in the WWW scenario. We used Lucene [2], an open source search engine for indexing this mini-WWW repository. All the evaluations reported below are obtained by performing our experiments through Lucene's search engine.

4.2 Evaluation

The testing data extracted from the clickthrough data is now used for evaluating the performance of our re-ranking approach. The test data consists of 5,000 queries posed by 780 users, with an average repetition of 15.9% in the queries. Repeated experiments have been conducted by using subsets of this training data. Each query from the testing data set is posed to the search engine for the mini-WWW and results obtained are cross-validated with references to the actual clicked documents in the testing data. We follow an evaluation approach similar to the one followed in [5] . We compare three methods of re-ranking. Firstly we consider the ranking provided by search engine, in this case Lucene's default ranking. The second approach tested is the ranking based on only the user profile. The third is the proposed approach for addressing the data sparsity problem using collaborative filtering. We refer to them as 'unranked', 'only user profile' and 'collaborative' respectively. The evaluation metrics used for comparison are minimum accuracy and precision @ N, N=5, 10 and 20. We could not evaluate the standard collaborative filtering measures like MAE etc because we assume boolean relevant judgments as opposed to the former which use ratings typically ranging from 0-5.

Minimum Accuracy. Minimum accuracy has been used in [5] in evaluation of their approach. It measures the ability of a search engine to return at least a single relevant result in returned results. We compare the top 30 results returned by our ranking approaches in calculating the minimum accuracy. The percentage of the queries for which at least one relevant result is returned is computed. The results are presented in Figure 3.

[2] http://lucene.apache.org

Table 1. Precision@N values for the approaches 'uranked' (UP), 'only user profile' (PP) and 'collaborative' (CP)

Method	Precision@5	10	20
unranked (UP)	0.09	0.18	0.22
only user profile (PP)	0.15	0.26	0.34
collaborative (CP)	0.23	0.38	0.55
Improvement(PP over UP)	6%	8%	12%
Improvement(CP over PP)	8%	12%	21%

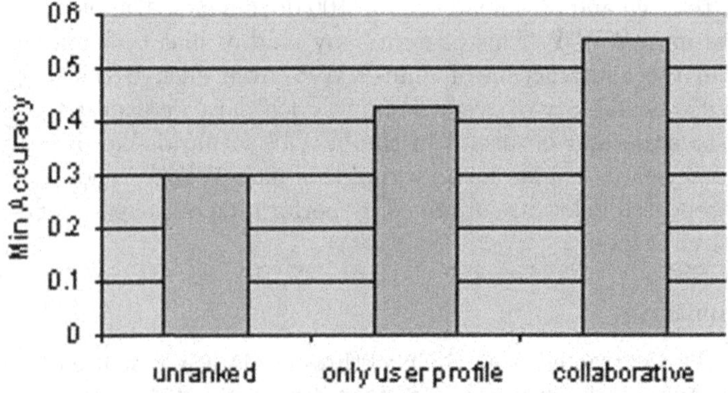

Fig. 3. Minimum accuracy comparison

Precision. We used precision at N (p@N) defined as the number of relevant documents at a given cut-off rank at N. It is a widely used metrics for evaluating approaches performing re-ranking of results. The reported value is averaged over all users and queries. We compared the three approaches mentioned above. The precision values for the three approaches 'unranked' (UP), 'only user profile' (PP) and 'collaborative' (CP). The results are shown in the Table 1. As it can be seen from the table, our approach showed an improvement over other approaches.

5 Conclusions

In this paper, we proposed a two phase strategy to personalize search results over the WWW. We first learn a user profile from his "clickthrough data", collected from a real world search engine. This user profile is then used in a re-ranking phase to personalize the search results. We also used query and its category information to in learning the user profile. Category information helps to disambiguate the query and focus on the information need. However, in the scenario of WWW search, it adds another extra dimension to the user profile, typically bringing in sparsity in the user profile. We propose an effective re-ranking strategy that compensates for the sparsity in a user's profile, using collaborative filtering algorithms. We evaluate our approach using standard information retrieval metrics, to show an improvement in performance over earlier re-ranking strategies based on only user profile.

Acknowledgements

We would like to thank Dr. Vasudeva Varma, for dicussions on collaborative filtering and re ranking of search results.

References

1. Kelly, D., Teevan, J.: Implicit feedback for inferring user preference: a bibliography. SIGIR Forum **37** (2003) 18–28
2. Kelly, D., Belkin, N.J.: Reading time, scrolling and interaction: Exploring implicit sources of user preferences for relevance feedback during interactive information retrieval. In: Proceedings of the 24th Annual International Conference on Research and Development in Information Retrieval (SIGIR '01). (2001) 408–409
3. Kim, J., Oard, D., Romanik, K.: Using implicit feedback for user modeling in internet and intranet searching. Technical report, College of Library and Information Services, University of Maryland at College Park (2000)
4. Liu, F., Yu, C., Meng, W.: Personalized web search by mapping user queries to categories. In: Proceedings of the Eleventh International Conference on Information and Knowledge Management (CIKM '02), ACM Press. (2002) 558–565
5. Balfe, B.S.E., Freyne, J., Briggs, P., Coyle, M., Boydell, O.: Exploiting query repetition & regularity in an adaptive community-based web search engine. User Modeling and User-Adapted Interaction: The Journal of Personalization Research (2004) 383–423
6. Smyth, B., Balfe, Boydell, O., Bradley, K., Briggs, P., Coyle, M., Freyne, J.: A live-user evaluation of collaborative web search. In: Proceedings of the 19th International Joint Conference on Artificial Intelligence (IJCAI'05), Edinburgh, Scotland. (2005) 1419–1424
7. Balfe, E., Smyth, B.: An analysis of query similarity in collaborative web search. In: Proceedings of the European Conference on Information Retrieval, Springer-Verlag (2005) 330–344
8. Fitzpatrick, L., Dent, M.: Automatic feedback using past queries: Social searching? In: In Proceedings of the Annual International ACM SIGIR Conference on Research and Development in Information Retrieval, ACM Press (1997) 306–313
9. Page, L., Brin, S., Motwani, R., Winograd, T.: The pagerank citation ranking: Bringing order to the web. Technical report, Stanford Digital Library Technologies Project (1998)
10. Haveliwala, T.H.: Topic-sensitive pangerank. In: Proceedings of the 11th International World Wide Web Conference (WWW2002). (2002) 517–526
11. Pretschner, A., Gauch, S.: Ontology based personalized search. In: ICTAI. (1999) 391–398
12. Speretta, M., Gauch, S.: Personalized search based on user search histories. In: Web Intelligence. (2005) 622–628
13. Shen., X., Tan., B., Zhai., C.: Context-sensitive information retrieval using implicit feedback. In: Proceedings of SIGIR 2005. (2005.) 43–50

14. Radlinski, F., Joachims, T.: Evaluating the robustness of learning from implicit feedback. In: ICML Workshop on Learning In Web Search. (2005)
15. Radlinski, F., Joachims, T.: Query chains: Learning to rank from implicit feedback. In: Proceedings of the ACM Conference on Knowledge Discovery and Data Mining (KDD), ACM 2005. (2005)
16. Chidlovskii, B., Glance, N., Grasso, A.: Collaborative re-ranking of search results. In: Proceedings of AAAI-2000 Workshop on AI for Web Search. (2000)
17. Sugiyama, K., Hatano, K., Yoshikawa., M.: Adaptive web search based on user profile constructed without any effort from users. In: Proceedings of WWW 2004. (2004.) 675 – 684
18. Lin., H., Xue., G.R., Zeng., H.J., Yu, Y.: Using probabilistic latent semantic analysis for personalized web search. In: Proceedings of APWEB'05. (2005) 707–717
19. Hust, A.: Query expansion methods for collaborative information retrieval. Inform., Forsch. Entwickl. **19** (2005) 224–238
20. Smyth, B., Balfe, E., Briggs, P., Coyle, M., Freyne, J.: Collaborative web search. In: In Proceedings of the 18th International Joint Conference on Artificial Intelligence, IJCAI-03, Morgan Kaufmann (2003) 1417–1419
21. Freyne, J., Smyth, B., Coyle, M., Balfe, E., Briggs, P.: Further experiments on collaborative ranking in community-based web search. Artificial Intelligence Review **21** (2004) 229–252
22. Rohini, U., Vamshi, A.: A collaborative filtering based re-ranking strategy for search in digital libraries. In: proceedings of 8th ICADL - 2005. (2005) 192–203
23. Raghavan, V.V., Sever, H.: On the reuse of past optimal queries. Proceedings of the Annual International ACM SIGIR Conference on Research and Development in Information Retrieval (1995) 344–350
24. Ji-Rong Wen, J.Y., Zhang, H.J.: Query clustering using user logs. ACM Transactions on Information Systems (TOIS) **20** (2002) 59–81
25. Glance, N.S.: Community search assistant. In: In Proceedings of the International Conference on Intelligent User Interfaces, ACM Press (2001) 91–96
26. Vapnik, V.N.: The nature of statistical learning theory. (1995)
27. Yang, Y., Liu, X.: A re-examination of text categorization methods. In: Proceedings of the 22nd Annual International ACM Conference on Research and Development in Information Retrieval (SIGIR'99). (1999) 42–49
28. Joachims, T.: Making large-scale svm learning practical. Advances in Kernel Methods - Support Vector Learning (1999)
29. Resnick, P., Iacovou, N., Suchak, M., Bergstorm, J.R.P.: Grouplens: An open architecture for collaborative filtering of netnews. In: Proc. of the ACM 1994 Conference on Computer Supported Cooperative Work (CSCW '94). (1994) 175–186
30. Konstan, J.A., Miller, B.N., Maltz, D., Herlocker, J.L., Gordon, L.R., Riedl, J.: Grouplens: Applying collaborative filtering to usenet news. In: Communications of the ACM. Volume 40 of 3. (1997) 77–87
31. Cohen, W.W., Fan, W.: Web-collaborative filtering: recommending music by crawling the web. Computer Networks **33** (2000) 685–698
32. Herlocker, J.L., Konstan, J.A., Borchers, A., Riedl, J.: An algorithmic framework for performing collaborative filtering. In: SIGIR. (1999) 230–237
33. Bernard J. Jansen, A.S.: An analysis of web searching by european alltheweb.com users. Information Processing and Management **41** (2005) 361–381

Learning to Integrate Web Catalogs with Conceptual Relationships in Hierarchical Thesaurus

Jui-Chi Ho, Ing-Xiang Chen, and Cheng-Zen Yang

Department of Computer Science and Engineering
Yuan Ze University, Taiwan, R.O.C.
{ricky, sean, czyang}@syslab.cse.yzu.edu.tw

Abstract. Web catalog integration has been addressed as an important issue in current digital content management. Past studies have shown that exploiting a flattened structure with auxiliary information extracted from the source catalog can improve the integration results. Although earlier studies have also shown that exploiting a hierarchical structure in classification may bring better advantages, the effectiveness has not been testified in catalog integration. In this paper, we propose an enhanced catalog integration (ECI) approach to extract the conceptual relationships from the hierarchical Web thesaurus and further improve the accuracy of Web catalog integration. We have conducted experiments of real-world catalog integration with both a flattened structure and a hierarchical structure in the destination catalog. The results show that our ECI scheme effectively boosts the integration accuracy of both the flattened scheme and the hierarchical scheme with the advanced Support Vector Machine (SVM) classifiers.

1 Introduction

With the fast growth of various Web information sources, many applications are in need of integrating different Web portals and many on-line vendors [1]. For example, B2C companies such as Amazon may want to merge catalogs from several on-line vendors into its catalog to provide customers versatile contents. B2B e-commerce is also an application reported in recent research [1,8,9,13]. In these studies, the importance of catalog integration is discussed. That is, an earlier study shows that only about 20% of the categorized sites retrieved from both Yahoo! and Google catalogs are the same, which implies that users may need to spend much effort browsing different Web catalogs to gain the required materials [3]. Therefore, an integrated Web catalog service not only can help users gain more relevant and organized information in one catalog but also can save them much time to surf among different Web catalogs.

In the past few years, several studies have been proposed to enhance catalog integration performance [1,3,9,10,12,13,15,16,19,20]. As noted in [1], catalog integration should not be a simple classification task. When some implicit source information is exploited, the integration accuracy can be highly improved. In their study, a foremost approach is designed by enhancing the Naive Bayes classifier with implicit source information [1]. Recently, other state-of-the-art studies employ Support Vector Machines (SVMs) (e.g. [3,12,15,19]) and the Maximum Entropy model (e.g. [16]) to enhance the

H.T. Ng et al. (Eds.): AIRS 2006, LNCS 4182, pp. 217–229, 2006.

accuracy performance of Web catalog integration. Their results show that the accuracy improvements can be highly achieved.

However, these studies only consider a flattened catalog structure. Moreover, they do not comprehensively study the hierarchical relationships between the categories and subcategories existing in the destination catalog. We are motivated to conduct research on hierarchical catalog integration because early studies have shown that exploiting a hierarchical structure in classification can bring better advantages [2,4,14]. In [16], it is also shown that the source hierarchical information can improve the accuracy. However, the study focuses on a flattened structure for catalog integration and does not discuss the effectiveness of the hierarchical structure of the destination catalog.

In this paper, we propose an enhanced catalog integration (ECI) approach with the conceptual relationships extracted from the hierarchical thesaurus in the source catalog to improve the integration performance. We further apply such an ECI approach to a flattened catalog structure (ECI-F) and a hierarchical catalog structure (ECI-H) and examine the consistent accuracy improvement of catalog integration. For the flattened categories, since the documents of lower level categories are merged upward to the second-level categories as the task definition in [1], the SVM classifier distinguishes a second-level category from all other second-level categories. For the hierarchical categories, the SVM classifier is learned to first distinguish the top-level categories, and then distinguish the second-level categories from other categories within the same top level. The accuracy of both the flattened and hierarchical integration structures is measured by considering only the second-level accuracy where the documents of the lower level categories are merged upward to the second-level ones.

We have conducted several experiments with real-world catalogs from Yahoo! and Google. We have also discussed the accuracy improvements of the ECI-F and ECI-H approaches respectively over the flattened structure and the hierarchical structure. The results show that the improvements of the ECI approach compared with original SVM are very noticeable, and the accuracy is consistently improved in every category on the basis of both the flattened structure and the hierarchical structure. The accuracy improvement of the flattened structure is boosted over 11% on average, and the accuracy improvement of the hierarchical structure is boosted over 16% on average.

The rest of the paper is organized as follows. Section 2 briefly states the problem definitions, assumptions, and limitations. Section 3 reviews the related studies of catalog integration. Section 4 describes the details of the ECI strategy and depicts our integration approach. Section 5 presents the experimental results and discusses the influence factors. Finally, Section 6 concludes the paper.

2 Problem Statement

As the formal definitions in [1], we assume that there are two catalogs participating in the integration process. One is the source catalog S with a set of m categories S_1, S_2, ..., S_m. The other is the destination catalog D with a set of n categories D_1, D_2, ..., D_n. The integration process is performed by merging each document d in S into a correspondent category in D. In addition, we assume that the catalogs are homogeneous and overlapped with some common documents. This means that the catalogs are not orthogonal, so the implicit source information can be exploited. Our real-world data

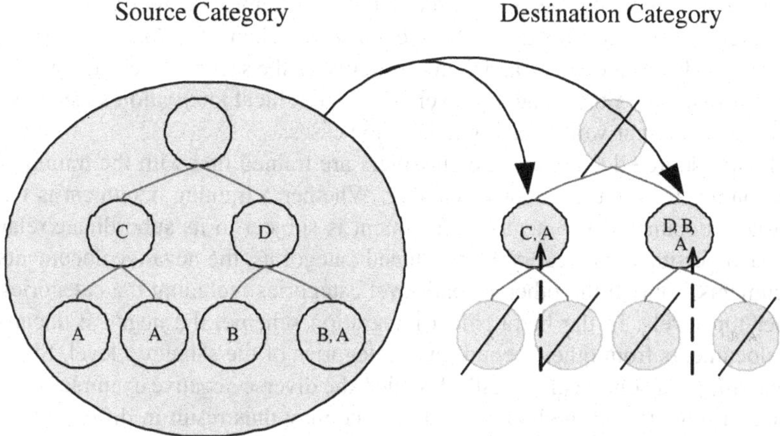

Fig. 1. The process of a flattened catalog integration

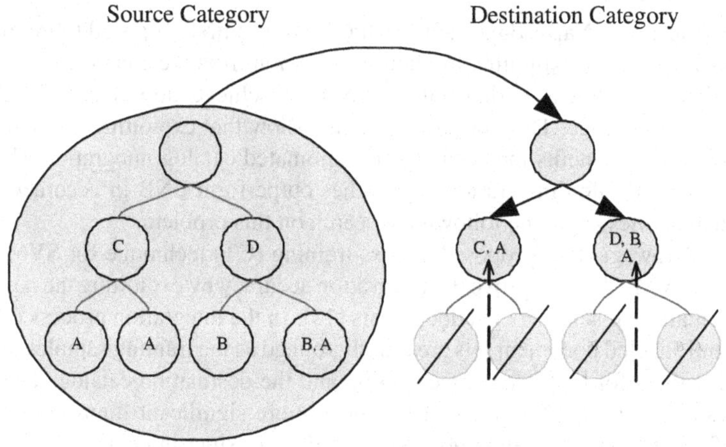

Fig. 2. The process of a hierarchical catalog integration

sets also support this overlapping assumption, and a flattened scheme and a hierarchical scheme are employed in catalog integration.

In the flattened integration scheme, we follow the integration model used in [1] in which the category hierarchies are flattened. Fig. 1 illustrates the process of a flattened integration scheme in which the documents under the second level are merged upward to the second level. Each classifier then distinguishes a second-level category from all other second-level categories. Although this cannot model many real-world cases in which catalogs are hierarchical, the flat catalog assumption is still helpful in investigating the effectiveness of the implicit source information in catalog integration.

In the hierarchical integration scheme, the top-level structure is reserved to help catalog integration. Fig. 2 depicts the process of a hierarchical integration scheme. To compare the hierarchical scheme with the flattened scheme, the documents under the second level are merged upward to the second level. Then, the classifier first distinguishes the top-level categories, and then distinguishes the second-level categories from other categories within the same top level. The hierarchical integration scheme is thus much closer to the real-world catalog integration cases.

To classify source documents, the classifiers are trained first with the training documents coming from the destination catalog. Whether a training document is treated as a positive document or a negative document is subject to its subordinate relationship to each destination category. For flattened categories, the negative documents are the documents from all the other second-level categories including the categories under other top levels. In the hierarchical integration scheme, the negative documents are the documents from other second-level categories of the same top level. Since the SVM classifier is a "one-against-rest" classifier, the diverse negative examples between the flattened scheme and the hierarchical scheme may thus result in different training models.

3 Literature Review

In 2001, an enhanced Naive Bayes approach (ENB) was first proposed to improve the integration accuracy by exploiting implicit information from the source catalog [1]. In their experiments with real-world catalogs, ENB can achieve more than 14% accuracy improvement on average. Their promising results show that exploiting implicit source information indeed benefits the accuracy for automated catalog integration. However, most of the later catalog integration approaches outperform ENB in accuracy performance. Their studies begin the following research on this problem.

In 2003, Sarawagi et al. proposed a cross-training (CT) technique for SVM classifiers, termed SVM-CT, to improve the integration accuracy by exploiting the native category information of semi-labeled documents [12]. In the integration process of SVM-CT, two semi-labeled document sets are crossly trained as the training samples to model the two classifiers for both the source catalog and the destination catalog. The experimental results show that SVM-CT can achieve more significant improvements than NB classifiers in most cases and improves the SVM classifier in nearly half cases. Although the CT technique achieves the improvements for SVM classifiers, their results also reveal that SVM-CT may be unstable in real-world catalog integration.

In 2003, Tsay et al. proposed two techniques to improve the accuracy of classifiers [15]. The first technique is called *probabilistic enhancement* (PE) that uses category information to enhance probabilistic classifiers such as NB classifiers. The second technique is called *topic restriction* (TR) that can be applied to general classifiers such as SVM, which is termed as SVM-TR here. Besides, they tried to combine NB and SVM with different dynamic settings to further improve the integration accuracy. The experimental results of real Web catalogs show that both techniques can significantly improve the accuracy of the NB and SVM classifiers on an average of over 10%. Although both techniques are promising to improve integration accuracy, however, the TR enhancement for SVM can only achieve less than 0.2% improvement.

In 2004, Zhang and Lee proposed two approaches to improve the accuracy: the *cluster shrinkage* (CS-TSVM) approach [19] and the co-bootstrapping approach [20]. The main idea behind CS-TSVM is to shrink all objects in a documents category to the cluster center. Although CS-TSVM overall outperforms ENB in macro and micro F-scores in the experiments, the proposed shrinkage process may faultily shrink the negative objects because it just shrink all objects. The co-bootstrapping approach employs the boosting technique, because it can find the optimal combination of heterogeneous weak hypotheses automatically without manually adjusting feature weights. In contrast, other machine learning algorithms, such as SVM, are required to adjust relative combination weights. Although the co-bootstrapping algorithm can optimize the combination of heterogeneous weak hypotheses automatically, the results show that the improvements are overall lower than CS-TSVM. Moreover, co-bootstrapping is discouraged by its poor efficiency.

In 2005, Chen et al. studied the effectiveness of SVM classifiers in catalog integration by using other embedded auxiliary catalog information [3]. They proposed an iterative SVM-based approach (SVM-IA) to consistently improve the integration performance. The results show that SVM-IA has a prominent accuracy performance, and the performance is more stable than SVM-CT. However, the accuracy improvement of SVM-IA is constrained by the auxiliary information after several iterations, and thus no obvious breakthrough is achieved.

In 2005, Wu et al. first extracted the hierarchical information from the source catalog and employed the Maximum Entropy model to improve the accuracy of catalog integration [16]. Their results show that the accuracy improvement is very promising and is consistently better than ENB. Although this study indicates that the hierarchical information from the source catalog is helpful to improve the integration accuracy, it has not been further proved for a hierarchical catalog.

The surveyed past studies only consider the flattened structure. Although the performance improvements are significant, the integration effectiveness based on a hierarchical structure has not been addressed. Therefore, the integration performance for both a flattened structure and a hierarchical structure needs to be further studied.

4 The Enhanced Catalog Integration Approach

In our study, SVM classifiers are used with linear kernel functions [17], $f : X \in R^n \to R$, to find a hyperplane that can separate the positive examples, $f(x) \geq +1$, from the negative examples, $f(x) \leq -1$. The linear function is in the form of $f(x) = (w, x) + b = \sum_{i=1}^{n} w_i x_i + b$ where $(w, b) \in R^n \times R$. The linear SVM is trained to find the optimal values of w and b such that $||w||$ is minimized. These trained SVM are employed in our enhanced catalog integration (ECI) approach which boosts their performance in catalog integration. In the following, we describe the ECI approach and the enhanced process of a flattened (ECI-F) scheme and the hierarchical (ECI-H) scheme.

4.1 Hierarchical Label Information

To improve the integration accuracy, a weighting formula is designed to extract the semantic concepts existing in the source catalog. In Equation 1, the weight of each thesaurus is exponentially decreased according to the increased levels to represent the semantic concept extracted from the source labels. Equation 1 calculates the feature weight of each document, where L_i is the relevant label weight assigned as $1/2^i$ with an i-level depth, and f_x is the occurrence ratio of feature x in the document. The label weights are exponentially decreased to minimize the influence of the labels which express very general concepts about the documents. Other functions can be used, such as $1/i^2$, to achieve similar results.

$$FeatureWeight = \lambda \times \frac{L_x}{\sum_{i=0}^{n} L_i} + (1-\lambda) \times f_x \qquad (1)$$

Table 1 shows the weights of different-level labels, where L_0 is the document level, L_1 is one level upper, and so on to L_n for n levels upper. A threshold λ is used to accommodate the weights of the source thesauri to fit in with the destination catalog. In the experiments, different values of λ for the source catalogs were examined. The experimental results suggest that a small value for λ is preferable to avoid overweighting the label information. With such a thesaurus weighting scheme, the conceptual relationships of the hierarchical source categories can be transformed and added into the test documents.

To build the enhanced classifiers for destination categories, the same enhancement on hierarchical label information is also applied to the destination catalog to strengthen the discriminative power of the classifiers. Likewise, the weights of the features and native category label information in the destination catalog are calculated in accordance with Equation 1. In this paper, the setting of $\lambda = 0.05$ to the destination catalog is only reported for the sake of space saving. Similar results can be obtained as in the source catalog.

Table 1. The weights assigned for different labels

Hierarchical Level	Label Weight
Document Level (L_0)	$1/2^0$
One Level Upper (L_1)	$1/2^1$
Two Levels Upper (L_2)	$1/2^2$
⋮	⋮
n Levels Upper (L_n)	$1/2^n$

4.2 The ECI-F and ECI-H Integration Schemes

The ECI approach is to extract the conceptual relationships between the hierarchical Web thesaurus from the source catalog. In the source catalog, the source hierarchical thesaurus are added into the test documents with different label weights calculated by the weighting formula. In ECI, the destination catalog is processed differently according to the integration type: flattened or hierarchical. Fig. 3 and Fig. 4 separately illustrate the use of a source hierarchy with the enhanced integration process of the ECI-F scheme for flattened catalog integration and the ECI-H scheme for hierarchical catalog integration.

In Fig. 3, the test documents with the source hierarchical thesaurus information are directly integrated into the second-level categories. In Fig. 4, the test documents with the source hierarchical thesaurus information are first integrated into the top-level categories and then are integrated into the second-level categories. In the hierarchical structure, the ECI-H scheme further removes the features occurring in the top-level categories to avoid misleading integration in the second-level categories. Since some documents may appear in more than one category in the destination catalog, Fig. 3 and Fig. 4 further show that the document A can be integrated into more than one category.

The proposed ECI approach is employed for both the flattened structure and the hierarchical structure as described in Fig. 5. In the source catalog, the hierarchical thesaurus

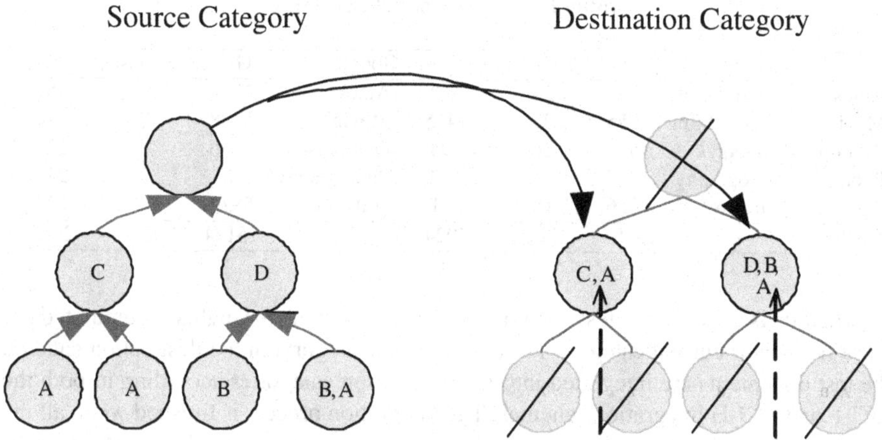

Fig. 3. The process of an ECI-F catalog integration

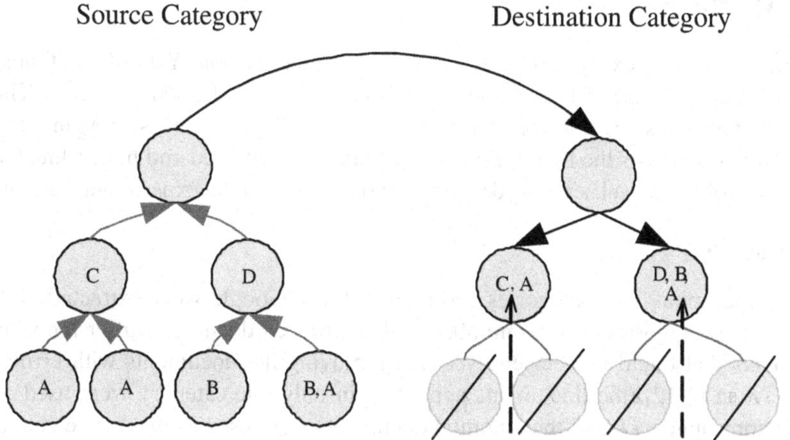

Fig. 4. The process of an ECI-H catalog integration

```
repeat
    repeat /* Add the source information.*/
        S ← {S_x}
    until Doc=∅
    classify test documents from S into D
    repeat /*Integrate all categories.*/
        recursive call to SVM classifiers
    until Class=∅
end repeat
```

Fig. 5. The ECI integration approach

Table 2. The experimental categories

	Yahoo!	| Y-G |	| Y Class |	| Y Test |	Google	| G-Y |	| G Class |	|G Test |
Autos	/Automotive/	1681	24	412	/Autos/	1096	14	451
Movies	/Movies_Film/	7255	27	1415	/Movies/	5188	27	374
Outdoors	/Outdoors/	1579	26	194	/Outdoors/	2396	23	226
Photo	/Photography/	1304	19	212	/Photography/	615	9	222
Software	/Software/	1876	15	691	/Software/	5829	59	723
Total		13695	111	2924		15124	132	2996

information is added to the test documents with different label weights accumulated upward from their current categories to the top-level category. In the destination catalog, the test documents are integrated into the destination categories according to both the ECI-F and ECI-H integration schemes. The integration process is finished when all the test documents from their source categories are integrated into the designated destination categories.

5 Experiments

We have conducted experiments with real-world catalogs from Yahoo! and Google to study the performance of ECI-F and the ECI-H schemes with SVM^{light} [6,7]. The experimental results show that the ECI approach consistently improves SVM in all cases, and effectively boosts the integration accuracy of both flattened and hierarchical structures. The following subsections describe the data sets and the experimental results.

5.1 Data Sets

In the experiment, five categories from Yahoo! and Google were extracted. Table 2 shows these categories and the number of the extracted documents after ignoring the documents that could not be retrieved and removing the documents with error messages. As in [1,12], the documents appearing in only one category were used as the destination catalog D, and the common documents were used as the source catalog S. If the number of the documents of a certain category is less than 10, the category would be merged upward to the top-level category in a hierarchical structure or be ignored in a flattened structure.

Table 3. The accuracy of the flattened and the ECI-F catalog integration from Yahoo! to Google

Category Test Doc. No.	Autos 451	Movies 1374	Outdoors 226	Photo 222	Software 723	Avg. Accuracy
Flattened (λ=0)	87.58% (395)	83.48% (1147)	82.30% (186)	78.83% (175)	88.52% (640)	84.14%
λ=0.05	97.56% (440)	96.58% (1327)	95.58% (216)	91.89% (204)	91.29% (660)	94.58%
λ=0.1	98.23% (443)	97.67% (1342)	97.35% (220)	94.59% (210)	92.95% (672)	96.16%
λ=0.15	98.00% (442)	97.89% (1345)	97.35% (220)	94.59% (210)	92.95% (672)	96.16%
λ=0.2	97.78% (441)	97.96% (1346)	96.90% (219)	93.69% (208)	92.95% (672)	95.86%
λ=0.3	97.34% (439)	97.96% (1346)	96.90% (219)	91.89% (204)	92.25% (667)	95.27%
λ=0.4	97.12% (438)	97.96% (1346)	96.90% (219)	91.44% (203)	91.84% (664)	95.05%
λ=0.5	96.45% (435)	97.96% (1346)	96.02% (217)	90.99% (202)	91.15% (659)	94.51%
λ=0.6	96.01% (433)	97.96% (1346)	96.02% (217)	90.54% (201)	91.15% (659)	94.34%
λ=0.7	96.01% (433)	97.89% (1345)	93.81% (212)	90.99% (202)	90.73% (656)	93.89%
λ=0.8	95.79% (432)	97.82% (1344)	93.81% (212)	90.99% (202)	90.73% (656)	93.83%
λ=0.9	95.79% (432)	97.82% (1344)	93.81% (212)	90.99% (202)	90.73% (656)	93.83%
λ=1	95.57% (431)	97.82% (1344)	93.81% (212)	90.99% (202)	90.73% (656)	93.78%

Table 4. The accuracy of the flattened and the ECI-F catalog integration from Google to Yahoo!

Category Test Doc. No.	Autos 412	Movies 1415	Outdoors 194	Photo 212	Software 691	Avg. Accuracy
Flattened (λ=0)	88.11% (363)	80.14% (1134)	80.41% (156)	73.58% (156)	92.19% (637)	82.89%
λ=0.05	92.72% (382)	92.44% (1308)	96.39% (187)	87.74% (186)	95.08% (657)	92.87%
λ=0.1	93.20% (384)	94.20% (1333)	98.45% (191)	92.45% (196)	94.65% (654)	94.59%
λ=0.15	92.96% (383)	94.56% (1338)	98.45% (191)	92.92% (197)	94.65% (654)	94.71%
λ=0.2	92.72% (382)	94.56% (1338)	97.94% (190)	92.45% (196)	94.36% (652)	94.40%
λ=0.3	92.72% (382)	94.49% (1337)	97.94% (190)	91.04% (193)	94.21% (651)	94.08%
λ=0.4	91.99% (379)	94.49% (1337)	96.91% (188)	89.62% (190)	93.92% (649)	93.39%
λ=0.5	91.26% (376)	94.35% (1335)	96.91% (188)	88.21% (187)	93.49% (646)	92.84%
λ=0.6	90.29% (372)	94.35% (1335)	95.88% (186)	86.79% (184)	93.34% (645)	92.13%
λ=0.7	90.29% (372)	94.35% (1335)	95.36% (185)	86.79% (184)	93.05% (643)	91.97%
λ=0.8	89.81% (370)	94.35% (1335)	95.36% (185)	86.32% (183)	93.05% (643)	91.78%
λ=0.9	89.81% (370)	94.28% (1334)	95.36% (185)	85.85% (182)	93.05% (643)	91.67%
λ=1	89.81% (370)	94.28% (1334)	95.36% (185)	85.85% (182)	93.05% (643)	91.67%

Since some documents may appear in more than one category of the same catalog, the number of test documents may slightly vary in Yahoo! and Google. In the experiment, we measured the accuracy by the following equation.

$$\frac{\text{Number of docs correctly classified into } D_i}{\text{Total number of docs in the test dataset}}$$

In the preprocessing, we used the stopword list in [5] to remove the stopwords.

5.2 Experimental Results and Discussion

In the experiments, a threshold λ is set to accommodate the weights of the source thesauri to fit in with the destination catalog. In order to optimize the value of λ, different λ values are tested to gain the best accuracy improvement. Table 3 and Table 4 show the integration results of a flattened scheme and an ECI-F with different λ values, in which the ECI-F achieves the best averaged accuracy when $\lambda = 0.15$ in both Yahoo!-to-Google and Google-to-Yahoo! cases. Table 5 and Table 6 show the integration results of a hierarchical scheme and an ECI-H with different λ values, in which the ECI-H achieves the best averaged accuracy when $\lambda = 0.20$ in both Yahoo!-to-Google and Google-to-Yahoo! cases.

Table 5. The accuracy of the hierarchical and the ECI-H catalog integration from Yahoo! to Google

Category	Autos	Movies	Outdoors	Photo	Software	Avg. Accuracy
Test Doc. No.	451	1374	226	222	723	
Hierarchical (λ=0)	80.04% (361)	90.47% (1243)	81.42% (184)	63.96% (142)	89.63% (648)	81.10%
λ=0.05	94.24% (425)	93.45% (1284)	92.48% (209)	78.83% (175)	89.49% (647)	89.70%
λ=0.1	95.34% (430)	96.80% (1330)	96.46% (218)	88.29% (196)	91.98% (665)	93.77%
λ=0.15	95.79% (432)	97.53% (1340)	96.46% (218)	88.74% (197)	92.12% (666)	94.13%
λ=0.2	95.79% (432)	97.74% (1343)	95.58% (216)	89.19% (198)	92.39% (668)	94.14%
λ=0.3	95.57% (431)	97.74% (1343)	95.58% (216)	87.39% (194)	91.70% (663)	93.59%
λ=0.4	95.12% (429)	97.67% (1342)	96.46% (218)	86.04% (191)	90.87% (657)	93.23%
λ=0.5	94.46% (426)	97.53% (1340)	95.58% (216)	85.59% (190)	90.32% (653)	92.69%
λ=0.6	94.24% (425)	97.45% (1339)	94.69% (214)	85.14% (189)	90.04% (651)	92.31%
λ=0.7	94.24% (425)	97.31% (1337)	92.04% (208)	84.68% (188)	89.76% (649)	91.61%
λ=0.8	93.79% (423)	97.16% (1335)	91.15% (206)	85.14% (189)	89.63% (648)	91.37%
λ=0.9	93.79% (423)	97.09% (1334)	88.94% (201)	86.49% (192)	89.63% (648)	91.19%
λ=1	93.57% (422)	97.09% (1334)	88.94% (201)	85.14% (189)	89.63% (648)	90.87%

Table 6. The accuracy of the hierarchical and the ECI-H catalog integration from Google to Yahoo!

Category	Autos	Movies	Outdoors	Photo	Software	Avg. Accuracy
Test Doc. No.	412	1415	194	212	691	
Hierarchical (λ=0)	82.04% (338)	72.37% (1024)	78.35% (152)	61.32% (130)	77.57% (536)	74.33%
λ=0.05	93.20% (384)	90.04% (1274)	94.85% (184)	82.08% (174)	91.90% (635)	90.41%
λ=0.1	93.93% (387)	93.36% (1321)	98.45% (191)	87.26% (185)	94.36% (652)	93.47%
λ=0.15	94.17% (388)	93.85% (1328)	98.45% (191)	88.68% (188)	94.21% (651)	93.87%
λ=0.2	93.93% (387)	93.99% (1330)	98.45% (191)	89.62% (190)	94.07% (650)	94.01%
λ=0.3	93.69% (386)	94.06% (1331)	98.45% (191)	90.09% (191)	93.92% (649)	94.04%
λ=0.4	92.23% (380)	93.92% (1329)	97.42% (189)	88.68% (188)	93.63% (647)	93.18%
λ=0.5	91.02% (375)	94.06% (1331)	97.42% (189)	87.74% (186)	93.49% (646)	92.75%
λ=0.6	90.53% (373)	94.06% (1331)	96.91% (188)	87.26% (185)	93.34% (645)	92.42%
λ=0.7	89.81% (370)	94.06% (1331)	95.88% (186)	86.32% (183)	93.20% (644)	91.85%
λ=0.8	89.56% (369)	94.06% (1331)	95.36% (185)	85.38% (181)	93.05% (643)	91.48%
λ=0.9	89.56% (369)	94.13% (1332)	95.36% (185)	84.43% (179)	93.05% (643)	91.31%
λ=1	89.56% (369)	94.13% (1332)	95.36% (185)	84.43% (179)	93.05% (643)	91.31%

Fig. 6 and Fig. 7 separately depict the best averaged performance of the flattened, ECI-F, hierarchical, and ECI-H schemes for the catalog integration from Yahoo! to Google and the catalog integration from Google to Yahoo!. In both Fig. 6 and Fig. 7, the best averaged accuracy of ECI-F is achieved when $\lambda = 0.15$, and the best averaged accuracy of ECI-H is achieved when $\lambda = 0.20$. According to the results, the best averaged accuracy of the ECI-F and ECI-H is both over 94% for the catalog integration from Yahoo! to Google and the catalog integration from Google to Yahoo!. Moreover, the accuracy improvement of a flattened structure is boosted over 11% on average, and the accuracy improvement of a hierarchical structure is boosted over 16% on average.

The results show that the integration accuracy with the ECI scheme is much better than the original integration scheme. Even though the test documents are only with the label information, that is, the value of $\lambda = 1$, the integration accuracy is still better than the original catalog integration schemes. The result indicates that using the ECI scheme to extract the source label information and the conceptual relationships from the hierarchical thesaurus strongly affects the integration accuracy because most of the labels are mainly the positive training features in the category classifiers. Besides that, the

Catalog Integration from Yahoo! to Google

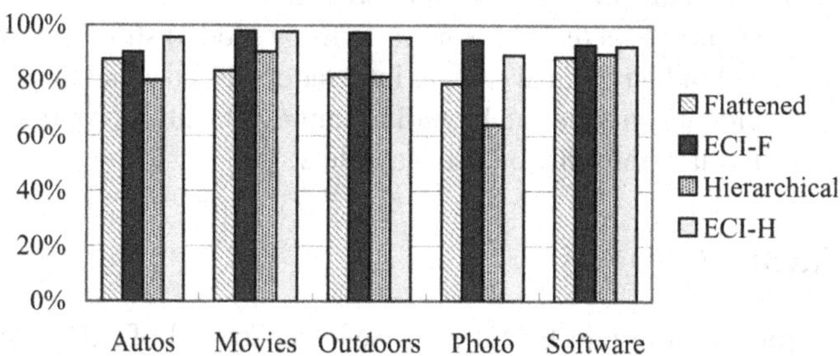

Fig. 6. The accuracy performance from Yahoo! to Google

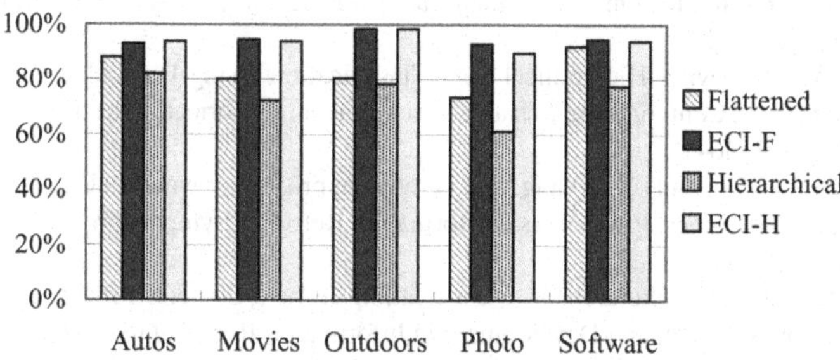

Fig. 7. The accuracy performance from Google to Yahoo!

classifiers are further enhanced with the native category label information of $\lambda = 0.05$, and thus the ECI scheme can greatly boosts the accuracy of Web catalog integration.

6 Conclusions

Catalog integration is an important issue in current digital content management and e-commerce applications. In this paper, we have addressed the problem of integrating documents from a source catalog into a destination catalog. This paper also reports our studies on the effects of an enhanced catalog integration to boost the integration accuracy. By exploiting the hierarchical relationships between categories and subcategories, the improvement of integration accuracy is very promising. The ECI scheme has been employed in both a flattened structure and a hierarchical structure, and the accuracy performance is greatly improved. It shows that our ECI approach consistently boosts the SVM classifiers in catalog integration, and the improvement of a hierarchical structure is more obvious than a flattened one.

This study is still on the first stage to apply an enhanced catalog integration to both a flattened structure and a hierarchical structure. There are still some issues left for further discussion. For example, a systematical mechanism to finding a better kernel function is a more difficult problem but can investigate the power of SVM. Besides that, other advanced classifiers (e.g. Maximum Entropy model) still need to be further studied and employed in hierarchical catalog integration. To conclude, we believe that the accuracy of catalog integration can be further improved with appropriate assistance of more effective auxiliary information and classifiers.

Acknowledgement

This work was supported in part by National Science Council of R.O.C. under grant NSC 94-2213-E-155-050. The authors would also like to express their sincere thanks to anonymous reviewers for their precious comments.

References

1. Agrawal, R., Srikant., R.: On Integrating Catalogs. Proc. the 10th WWW Conf. (WWW10), (May 2001) 603–612
2. Boyapati, V.: Improving Hierarchical Text Classification Using Unlabeled Data. Proc. the 25th Annual ACM Conf. on Research and Development in Information Retrieval (SIGIR'02), (Aug. 2002) 363–364
3. I.-X. Chen, J.-C. Ho, and C.-Z. Yang.: An iterative approach for web catalog integration with support vector machines. Proc. of Asia Information Retrieval Symposium 2005 (AIRS2005), (Oct. 2005) 703–708
4. Dumais, S., Chen, H.: Hierarchical Classification of Web Content. Proc. the 23rd Annual ACM Conf. on Research and Development in Information Retrieval (SIGIR'00), (Jul. 2000) 256–263
5. Frakes, W., Baeza-Yates, R.: Information Retrieval: Data Structures and Algorithms. Prentice Hall, PTR. (1992)
6. Joachims, T.: Text Categorization with Support Vector Machines: Learning with Many Relevant Features. Proc. the 10th European Conf. on Machine Learning (ECML'98), (1998) 137–142
7. Joachims, T.: Making Large-Scale SVM Learning Practical. In Scholkopf, B., Burges, C., Smola, A. (eds): Advances in Kernel Methods: Support Vector Learning. MIT Press. (1999)
8. Keller,A. M.: Smart Catalogs and Virtual Catalogs. In Ravi Kalakota and Andrew Whinston, editors, Readings in Electronic Commerce. Addison-Wesley. (1997)
9. Kim, D., Kim, J., and Lee, S.: Catalog Integration for Electronic Commerce through Category-Hierarchy Merging Technique. Proc. the 12th Int'l Workshop on Research Issues in Data Engineering: Engineering e-Commerce/e-Business Systems (RIDE'02), (Feb. 2002) 28–33
10. Marron, P. J., Lausen, G., Weber, M.: Catalog Integration Made Easy. Proc. the 19th Int'l Conf. on Data Engineering (ICDE'03), (Mar. 2003) 677–679
11. Rennie, J. D. M., Rifkin, R.: Improving Multiclass Text Classification with the Support Vector Machine. Tech. Report AI Memo AIM-2001-026 and CCL Memo 210, MIT (Oct. 2001)
12. Sarawagi, S., Chakrabarti S., Godbole., S.: Cross-Training: Learning Probabilistic Mappings between Topics. Proc. the 9th ACM SIGKDD Int'l Conf. on Knowledge Discovery and Data Mining, (Aug. 2003) 177–186
13. Stonebraker, M. and Hellerstein, J. M.: Content Integration for e-Commerce. Proc. of the 2001 ACM SIGMOD Int'l Conf. on Management of Data, (May 2001) 552–560

14. Sun, A. ,Lim, E.-P., and Ng., W.-K. :Performance Measurement Framework for Hierarchical Text Classification. Journal of the American Society for Information Science and Technology (JASIST), Vol. 54, No. 11, (June 2003) 1014–1028
15. Tsay, J.-J., Chen, H.-Y., Chang, C.-F., Lin, C.-H.: Enhancing Techniques for Efficient Topic Hierarchy Integration. Proc. the 3rd Int'l Conf. on Data Mining (ICDM'03), (Nov. 2003) (657–660)
16. Wu, C.-W., Tsai, T.-H., and Hsu, W.-L.: Learning to Integrate Web Taxonomies with Fine-Grained Relations: A Case Study Using Maximum Entropy Model. Proc. of Asia Information Retrieval Symposium 2005 (AIRS2005), (Oct. 2005) 190–205
17. Yang, Y., Liu, X.: A Re-examination of Text Categorization Methods. Proc. the 22nd Annual ACM Conference on Research and Development in Information Retrieval, (Aug. 1999) 42–49
18. Zadrozny., B.: Reducing Multiclass to Binary by Coupling Probability Estimates. In: Dietterich, T. G., Becker, S., Ghahramani, Z. (eds): Advances in Neural Information Processing Systems 14 (NIPS 2001). MIT Press. (2002)
19. Zhang, D., Lee W. S.: Web Taxonomy Integration using Support Vector Machines. Proc. WWW2004, (May 2004) 472–481
20. Zhang, D., Lee W. S.: Web Taxonomy Integration through Co-Bootstrapping. Proc. SIGIR'04, (July 2004) 410–417

Discovering Authoritative News Sources and Top News Stories

Yang Hu[1,*], Mingjing Li[2], Zhiwei Li[2], and Wei-ying Ma[2]

[1] University of Science and Technology of China, Hefei 230027, China
yanghu@ustc.edu
[2] Microsoft Research Asia, No 49, Zhichun Road, Beijing 100080, China
{mjli, zli, wyma}@microsoft.com

Abstract. With the popularity of reading news online, the idea of assembling news articles from multiple news sources and digging out the most important stories has become very appealing. In this paper we present a novel algorithm to rank assembled news articles as well as news sources according to their importance and authority respectively. We employ the visual layout information of news homepages and exploit the mutual reinforcement relationship between news articles and news sources. Specifically, we propose to use a label propagation based semi-supervised learning algorithm to improve the structure of the relation graph between sources and new articles. The integration of the label propagation algorithm with the HITS like mutual reinforcing algorithm produces a quite effective ranking algorithm. We implement a system TOPSTORY which could automatically generate homepages for users to browse important news. The result of ranking a set of news collected from multiple sources over a period of half a month illustrates the effectiveness of our algorithm.

1 Introduction

According to a recent survey conducted by Nielsen/NetRatings [12], online newspapers have enjoyed double-digit year-over-year growth last year, reaching one out of four Internet users. This should not be astonishing considering the advantages of online news as said by Peter Steyn of Nielsen/NetRatings, "...it provides a different perspective and greater depth of information — statistics, pictures, interactive maps, streaming video, and analyst comments". This kind of growth spurt urges the necessity for efficient organization of large amount of news articles available online. As the traditional process of reading one newspaper after another and selecting relevant stories has become inefficient or even infeasible in this environment, the idea of assembling news articles from multiple sources and digging out the most important stories seems to be very appealing. In this paper, we tackle this challenging problem of automatically ranking news stories assembled from multiple news sources according to their importance.

* This work was performed when the first author was a visiting student at Microsoft Research Asia.

H.T. Ng et al. (Eds.): AIRS 2006, LNCS 4182, pp. 230–243, 2006.

It is not easy to identify important news, as each person may have his/her own interest on news events, and each news source may also have its own preference when reporting news. Measuring the importance of news is inherently a subjective problem. However, one of the major advantages of assembling news from multiple sources is that it allows for the integration of different opinions and could provide unbiased perspective on the most important events currently occurring or happened during a specified period of time.

Generally speaking, important news event is usually covered by multiple news sources. Besides, important news always occupies a visually significant area on the homepage. The visual significance of a piece of news on the homepage can be regarded as the recommendation strength of the news presented by the source. However, the authority of different sources is not the same. News sources with high reputation generally recommend news with proper strength, while others may have certain kinds of local preference. News presented by authoritative sources with highlighted visual representation is more likely to be important. And a news source which issues a lot of important news is expected to be authoritative. Thus the authority of the news sources and the importance of news exhibit a mutual reinforcement relationship, which is quite similar with the relationship between hub pages and authoritative pages in a hyperlinked environment [1].

A primary difference between our problem and the structure of a hyperlinked environment is that news pages are always only pointed to by the homepages they belong to, which will crash the HITS algorithm. Fortunately, different news pages are not absolutely independent. They may cover the same or related news events. Based on this implicit relationship between news articles, we propose to use a label propagation based semi-supervised learning algorithm [2] to predict a news site's recommendation strength on the news articles that are issued by other sites. The local and global consistency property of this label propagation algorithm guarantees a quite smooth and precise prediction. Integration of the label propagation algorithm with the HITS like mutual reinforcing algorithm produces a reasonable ranking of assembled news articles as well as news sources.

The rest of this paper is organized as follows. In Sect.2, we introduce some related works both from commercial and academic communities. In Sect.3, we explain in detail our algorithm for ranking news articles and news sources. We describe the system TOPSTORY which is built to validate our algorithm in Sect.4. Experiment results are presented in Sect.5. Finally, we conclude in Sect.6.

2 Related Works

Many commercial news search engines are already available for indexing online news. Google news [13] gathers and indexes news information from more than 4,500 sources worldwide. In addition to keyword search, it also provides the ability to browse categories of news where headlines are assembled and ranked automatically. Besides, related news articles are grouped to better present reports on the same story from different organizations. Yahoo news [14] performs similar service on more than 5,000 sources. Unlike Google news, the articles and sources are hand-assembled by Yahoo editors and no clustering technique is applied. MSN has also issued its news search service known as Newsbot [15]. What set Newsbot apart from other news aggregators are the history and personalization features designed to help users easily find news relevant to their own interests. However, the information on how commercial news search engines rank news articles is very limited. We only learn that many different metrics have been employed for determining the prominence of news articles such as the importance of the source, timeliness of the article and its relation to other stories in the news currently. We take in account these factors in designing our news ranking algorithm.

The problem of organizing news articles in a meaningful manner to facilitate user navigation has also been exploited in the academic scenario. A research program called Topic Detection and Tracking (TDT) [3] investigates a variety of automatic techniques for discovering and threading together topically related material in streams of data such as newswire and broadcast news. However, it doesn't address the problem in the current web environment where multiple streams of data come across. And ranking is not involved in its formally defined research tasks. There are also many other works which focus on organizing news articles such as summarizing clusters of related news articles [10], providing personalized news [11] and etc.

The first academic discussion on the news ranking problem is addressed by Gianna M. Del Corso et al. in [4]. They proposed a framework to rank a stream of news articles and a set of news sources. Quite similar to our work, they also utilize the mutual reinforcement between news articles and news sources as well as the clustering character of important news. However, they did not take in account the different visual significance of news items on the homepages, which is an extremely valuable metric for evaluating importance of news. In [5], the relationship between homepages, news and latent news events was modeled by a tripartite graph. A hybrid model was presented to identify important news, which combined the mutual reinforcement relationships between homepages and news articles and between news articles and news events. However, the combination process can be regarded as using a very naïve method to predict a news site's recommendation strength on news articles that are not issued by it. The imprecision of the prediction would degrade the overall ranking algorithm.

3 Ranking News Articles and News Sources

3.1 A Graph Model for News Articles and News Sources

Generally speaking, news sources usually maintain a set of homepages which serve as portals for users to access the news articles. Accordingly, news articles are categorized and their titles are listed on the corresponding homepages. The relationship between news articles belonging to a certain category (World, Business, Sports, etc) and their sources can be represented by a graph $G = (V, E)$ where $V = S \cup N$ and $E = E_e \cup E_i$ (Fig. 1). $S = \{s_1, \cdots, s_m\}$ is the set of vertices corresponding to news sources and $N = \{n_1, \cdots, n_n\}$ represents news articles. E is the set of edges which indicate the relations between vertices.

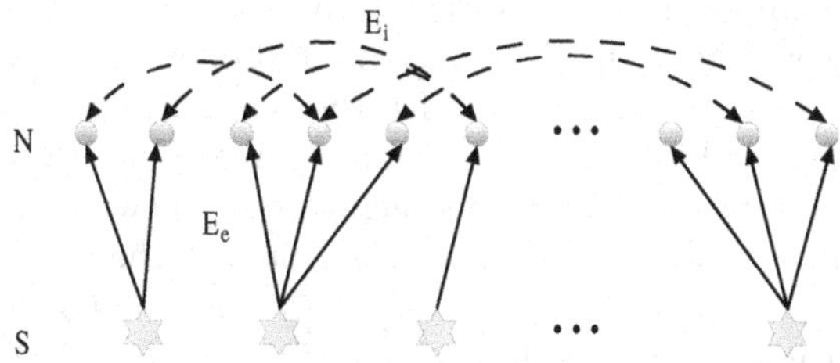

Fig. 1. A graph model for news articles and news sources

The set of edges E_e describe the explicit linkage relation between news sources and news articles. An simple and intuitive method for expressing this relationship is to use a binary function where $e_{ij} = 1$ if the homepage of s_i points to the news article n_j, else let $e_{ij} = 0$. This definition only captures the linkage relationship and treats every link equally. However, news articles are not randomly listed on each homepage. Instead, they are usually carefully arranged by human editors who would assign different visual strength to the titles based on their evaluation of the importance of the news. The title of the most important news is usually put on the top of the page, accompanied by an image and a paragraph of abstract text, while the less important ones are only represented by the title (see Fig. 4 for example). This kind of difference in representation reflects editor's recommendation and ranking of important news, which should be very helpful when considering our ranking problem. Therefore, a more meaningful manner is to use a real-valued function to characterize more accurately the relation between news sources and news articles. Accordingly, we define a matrix $Q_{m \times n}$ where q_{ij} represents the recommendation strength of n_j by s_i.

Besides the explicit linkage, a kind of implicit relation lies between news articles. News is generally triggered by events happened in the world. Different articles from different sources may cover the same or related events. However, as news event detection in news corpus is not a simple task and has not been well resolved yet, a more reasonable method is to use similarity relation to describe this fact. The measure of similarity between text documents has been extensively studied in IR community. The more similar the news articles, the more probably they are reporting the same event. The set of edges E_i represent this relationship accompanied by a matrix $A_{n \times n}$ where a_{ij} indicates the similarity between news articles n_i and n_j. We use the popular Vector-Space-Model (VSM) to measure the similarity in this paper.

3.2 Homepage Model

While our initial focus is primarily on ranking news articles, the reputation of news sources and homepages also varies a lot. Homepages of some sources are more authoritative than others. An interesting fact is, the authority of news sources and the importance of news exhibit a mutually reinforcing relationship. News posted by more authoritative sources with prominent visual representation is more likely to be important. And authoritative sources are expected to recommend important news more reliably. This kind of property between news sources and news articles is extremely similar with the relationship between Web hubs and authorities identified by HITS algorithm [1]. We associate each source with a nonnegative authority weight w_i^s and each piece of news with a nonnegative importance weight w_i^n. We maintain the invariant that the weights of each type are normalized so their squares sum to 1: $\sum_i (w_i^s)^2 = 1$, and $\sum_i (w_i^n)^2 = 1$. Numerically, it is natural to express the mutually reinforcing relationship between news and sources as follows:

$$\boldsymbol{w}^n \leftarrow Q^T \times \boldsymbol{w}^s \ , \tag{1}$$

$$\boldsymbol{w}^s \leftarrow Q \times \boldsymbol{w}^n \ . \tag{2}$$

Unlike the model in [5], we don't normalize Q in (2) here, as we find out that the number of news a source presents should be an important factor when ranking homepages. Besides, according to our experiment, the ranking result of the news sources does not completely depend on the number of news articles they present even without normalization.

These two operations are the basic means by which $\boldsymbol{w}^s$ and $\boldsymbol{w}^n$ reinforce each other. The desired equilibrium values for the weights can be reached by applying (1) and (2) in an alternating fashion while maintaining the normalization conditions. It is expected that $\boldsymbol{w}^s$ converges to the principal eigenvector of $Q \times Q^T$ and $\boldsymbol{w}^n$ converges to the principal eigenvector of $Q^T \times Q$.

Let $B = Q^T \times Q$ with its entries $b_{ij} = \sum_{l=1}^{m} q_{li} \times q_{lj}$. We note that $b_{ij} \neq 0$ if and only if the news articles n_i and n_j are issued by the same source, which is not the case for most pairs of news articles under this multi-source environment. If the news articles are arranged according to the sources they belong to, then B should be a block diagonal matrix. The eigenvectors of block diagonal matrices are the eigenvectors of the blocks, padded with zeros. As a result, only a set of news articles from a certain source have non-zero values in the principal eigenvector of B. So is the case for $\boldsymbol{w}^s$. It fails to achieve our original goal of ranking news articles from multiple sources as well as ranking the sources.

3.3 Label Propagation Based Recommendation Strength Prediction

The homepage model does not work out due to the sparseness of the matrix Q. It is quite similar with the data sparse problem in collaborative filtering if we regard each news source as a user who rates the importance of the news. The problem is, each piece of news is only rated by the site it belongs to and the intersection of the sets of news rated by different sites is extremely small. Fortunately, the news articles are not absolutely independent. They associate with others through the latent news events because different articles may cover the same or related affairs. We have used similarity relationship to model this property in Sect.3.1. Given a site's ratings on a set of news which are the visual significance of the news on the homepage, it is possible to predict its ratings on the news articles that are issued by other sites.

Given a set of news articles $N = \{n_1, \cdots, n_n\}$, the recommendation strength of the first l pieces of news are labeled by a site, let $y(i) = y_l(i), i = 1, \ldots, l$. We intend to predict the unknown recommendation strength on the rest pieces of news, i.e. $y(i) = y_u(i), i = l+1, \ldots, n$ by using the known label information $\boldsymbol{y}_l$ and the similarity information among news articles.

Recently, Zhou et al. [2] proposed a label propagation based semi-supervised learning algorithm, which works by representing labeled and unlabeled examples as vertices in a graph, then iteratively propagating label information from any vertex to nearby vertices through weighted edges, and finally inferring the labels of unlabeled examples after the propagation process converges. This label propagation algorithm is motivated by a local and global consistency assumption. And the learned labels are sufficiently smooth with respect to the intrinsic structure collectively revealed by labeled and unlabeled vertices. It has been applied to many problems such as digit recognition, document classification [2] and image retrieval [6] and has been proven to work quite well. We investigate this label propagation algorithm for our recommendation strength prediction problem.

Let $y_u^0(i) = 0, i = l+1, \ldots, n$. The algorithm is as follows:

1. Two news articles n_i and n_j are connected by an edge if n_i is among n_j's k nearest neighbors or if n_j is among n_i's k nearest neighbors;
2. Form the affinity matrix W defined by $w_{ij} = \exp\left[-d^2(n_i, n_j)/\sigma^2\right]$ if there is an edge linking n_i and n_j. $d(n_i, n_j)$ is the distance between n_i and n_j and is defined by $d(n_i, n_j) = 1 - a_{ij}$. Let $w_{ii} = 0$;
3. Normalize W symmetrically by $L = D^{-1/2} W D^{-1/2}$ where D is a diagonal matrix with entries $d_{ii} = \sum_{j=1}^{n} w_{ij}$;
4. Iterate (3) until it converges where t is iteration index and $\alpha \in [0, 1]$;

$$y_u^{t+1}(i) = \sum_{j=1}^{l} L_{ij} y_l(j) + \alpha \sum_{j=l+1}^{n} L_{ij} y_u^t(j) \ . \tag{3}$$

5. Let $\boldsymbol{y}_u^*$ denote the limit of the sequence $\{\boldsymbol{y}_u^t\}$. Assign $\boldsymbol{y}_u^*$ as the recommendation strength of the unlabeled news, i.e. $y(i) = y_u^*(i), i = l+1, \ldots, n$.

Here, the scaling parameter σ^2 controls how rapidly the affinity w_{ij} falls off with the distance between n_i and n_j. The matrix W fully specifies the data manifold structure. In the update scheme of (3), each piece of unlabeled news receives recommendation information from its neighbors, including both of the labeled news and other unlabeled news. The parameter $\alpha \in [0, 1]$ is used to control the recommendation strength received from unlabeled neighbors.

This algorithm has been proven to converge to a unique solution [2]:

$$\boldsymbol{y}_u^* = \lim_{t \to \infty} \boldsymbol{y}_u^t = (I - \alpha L_{uu})^{-1} L_{ul} \boldsymbol{y}_l \ , \tag{4}$$

where I is a $(n-l) \times (n-l)$ identity matrix. L_{uu} and L_{ul} are acquired by splitting matrix L into 4 blocks after the lth row and column:

$$L = \begin{bmatrix} L_{ll} & L_{lu} \\ L_{ul} & L_{uu} \end{bmatrix} . \tag{5}$$

According to (4), the initialization of $\boldsymbol{y}_u^0$ does not affect the solution $\boldsymbol{y}_u^*$.

Although $\boldsymbol{y}_u^*$ can be expressed in a closed form, for large scale problems, the iteration algorithm is preferable due to computational efficiency. Using Taylor expansion, we have

$$\begin{aligned} \boldsymbol{y}_u^* &= (I - \alpha L_{uu})^{-1} L_{ul}\boldsymbol{y}_l \\ &= \left(I + \alpha L_{uu} + \alpha^2 L_{uu}^2 + \cdots\right) L_{ul}\boldsymbol{y}_l \\ &= L_{ul}\boldsymbol{y}_l + \alpha L_{uu}\left(L_{ul}\boldsymbol{y}_l\right) + \alpha L_{uu}\left(\alpha L_{uu} L_{ul}\boldsymbol{y}_l\right) + \cdots . \end{aligned} \tag{6}$$

According to (6), $\boldsymbol{y}_u^*$ can be regarded as the sum of a serious of infinite terms. The first term spreads the recommendation strength of the labeled news to their nearby vertices, the second term spreads the strength further, and so does the third term etc.

3.4 Comparing with Previous Work

In [5], the relationship between homepages, news and latent events was modeled by a tripartite graph, and a hybrid model was proposed which combined homepage voting model and cross-site similarity model to identify important news as shown in (7) and (8):

$$\boldsymbol{w}^n \leftarrow A \times \boldsymbol{w}^n \leftarrow A \times Q^T \times \boldsymbol{w}^s , \tag{7}$$

$$\boldsymbol{w}^s \leftarrow K \times Q \times \boldsymbol{w}^n , \tag{8}$$

where A is the similarity matrix as defined previously. $K = diag(k_i)$ is a normalization matrix with entries $k_i = 1/\sum_j q_{ij}^2$. The major difference between the hybrid model ((7) and (8)) and the homepage model ((1) and (2)) lies in that in (7) a new matrix $P = A \times Q^T$ is used to substitute the original matrix Q^T in (1). Let $\{\boldsymbol{v}_1, \boldsymbol{v}_2, \cdots, \boldsymbol{v}_m\}$ be the column vectors of Q^T and $\{\boldsymbol{v}'_1, \boldsymbol{v}'_2, \cdots, \boldsymbol{v}'_m\}$ be the column vectors of P. $\boldsymbol{v}_i$ is the recommendation strength of the ith site on news articles with nonzero values on the news it presents and zeros on other news. The nonzero values of $\boldsymbol{v}_i$ also constitute the initial labels $\boldsymbol{y}_l$ that we use in our label propagation algorithm. We can regard the operation $\boldsymbol{v}'_i = A \times \boldsymbol{v}_i$ as another method of predicting unknown recommendation strength. The components of $\boldsymbol{v}'_i$ are defined as the weighted sums of the known strengths where the weights are the similarities between news articles. Besides, the known strengths are also updated based on the same criteria. Comparing with the label propagation algorithm, it looks quite like that this criteria stops the iteration process of (3) at $t = 1$ with $\alpha = 1$. It is obviously not an equilibrium state and it does not possess the local and global consistency property of label propagation algorithm.

We summarize our integrated ranking algorithm in Fig. 2.

Iterate (Q, A, k)

$Q_{m \times n}$: a source vs news matrix

$A_{n \times n}$: a news vs news matrix

k : a natural number

For each row of Q, use the label propagation algorithm to predict the site's recommendation strength on the news that are not issued by it. Let $R_{m \times n}$ be the new source vs news matrix.

Let $\boldsymbol{z}$ denote the vector $(1, 1, \cdots, 1) \in \mathbb{R}^n$.

Set $\boldsymbol{w}_0^n = \boldsymbol{z}$.

Set $\boldsymbol{w}_0^s = \boldsymbol{z}$.

For $i = 1, 2, \ldots, k$

$\boldsymbol{w}_i^n \leftarrow R^T \times \boldsymbol{w}_{i-1}^s$,

$\boldsymbol{w}_i^s \leftarrow Q \times \boldsymbol{w}_{i-1}^n$,

Normalize $\boldsymbol{w}_i^n$, $\boldsymbol{w}_i^s$.

End

Return $(\boldsymbol{w}_k^n, \boldsymbol{w}_k^s)$.

Fig. 2. Integrated ranking algorithm

4 TOPSTORY System

We have implemented a system named TOPSTORY to verify our algorithm (see Fig. 3). It monitors a set of news sources and crawls their homepages in a certain frequency. News pages that are linked by these homepages are also crawled when they are detected for the first time. Useful information in both homepages and news pages is extracted by parsing the crawled pages and then saved into a database.

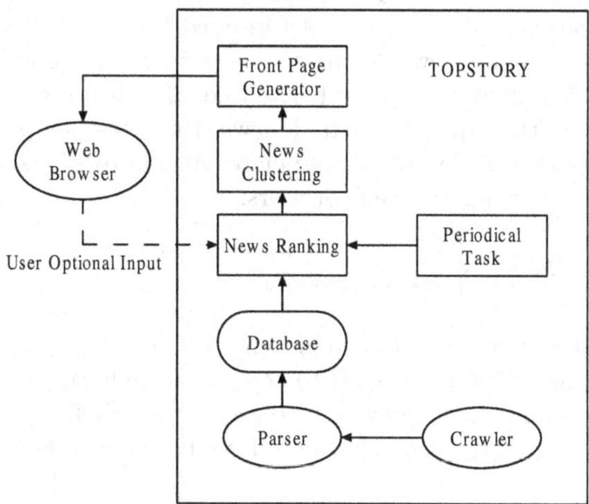

Fig. 3. An overview of TOPSTORY system

The system could interact with users in two ways. Firstly, it periodically detects latest important news and automatically generates homepages for users to browse. Secondly, it can also been driven by users' query. It could detect the most import news during any time period specified by the users. We also implement a simple clustering algorithm to group related news into events. These events are ranked according to the most important piece of news within them.

In the following, we describe how we extract and use the information in homepages to compute matrix Q which indicates the visual recommendation strength.

4.1 Recommendation Strength from Homepage

Each homepage is tracked by a set of snapshot pages $\{S_{t_1}, S_{t_2}, \cdots\}$ with S_{t_i} denotes the homepage at a specific time t_i. Each snapshot presents a set of news with different visual strength. We use a vision-based page segmentation algorithm (VIPS) [9] to analyze snapshot's content structure.

Each snapshot is divided into a set of blocks such that each block is dominated by a piece of news (see Fig. 4 for example). The visual strength of a block is mainly determined by its size, position in the page and whether it contains an image. We use a simple rule to estimate it:

$$\begin{aligned} q\left(S_{t_i}, n_j\right) = & \; BlockSize/MaxBlockSize + (1 - Top/PageHeight) \\ & \; 0.5 \times (ContainImage?1:0) \;\;, \end{aligned} \tag{9}$$

where $q\left(S_{t_i}, n_j\right)$ is the visual strength of news n_j in snapshot S_{t_i}. $BlockSize$ is the area of the block. $MaxBlockSize$ denotes the max area of all blocks in S_{t_i}. Top is the position of the top side of the block. $PageHeight$ is the height of the snapshot page. And $ContainImage$ indicates whether the block contains an image. Here 0.5 is an empirically chosen weight.

The visual strength of a piece of news may evolve over time. We need to summarize these snapshots to have a global view of how the homepage recommends it. The summarization rule is actually determined by users' intention. If a user wants to browse important news during a time period, for example a week, then all snapshots in this week should be treated equally. In another case, when a user wants to know the latest important news, the latest snapshots should be more important than older ones. We associate weights to snapshots in order to meet the different information needs of users:

$$w(S_{t_i}) = \begin{cases} 1 & \text{for the first case} \\ \frac{1}{1+e^{-a(t_i - t_0)}} & \text{for the second case} \end{cases} \;. \tag{10}$$

Here a sigmoid function is used to represent the decaying character of users' interest. a and t_0 are parameters used to control time effect.

For each piece of news, its recommendation strength from a source is the weighted combination of the visual strength from the homepage's snapshots that contain it:

$$q(s_i, n_j) = \frac{\sum_{S_{t_i}} w(S_{t_i}) \times q(S_{t_i}, n_j)}{\sum_{S_{t_i}} w(S_{t_i})} \;. \tag{11}$$

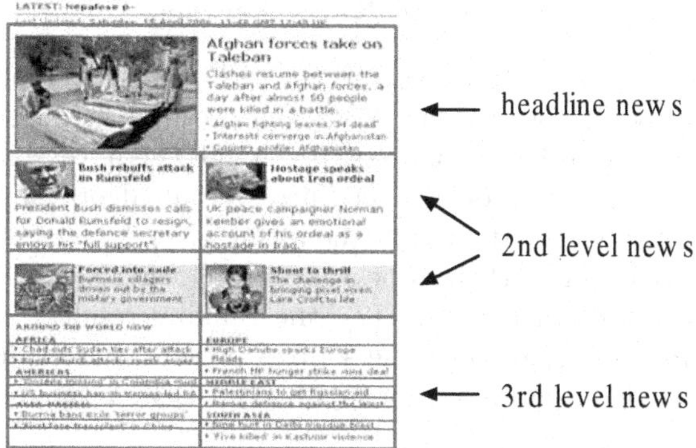

Fig. 4. A snapshot of a news homepage

$q(s_i, n_j)$ is further normalized so that the maximum recommendation strength of each source equals 1.

5 Experimental Results

In this section, we first describe the data set we collected for our evaluation and the parameter setting in our experiment. Then we present the ranking result obtained using our algorithm as well as its comparison with other methods.

5.1 Data Set and Parameter Setting

We monitored 9 continuous updated online news sites (see Table 2) for a period of half a month (from 3/16/2006 to 3/31/2006) and collected about 35,000 pieces of news. These news articles are classified in 7 different categories (see Table 1). And news articles are ranked within each category, as the importance of news from different category is generally incomparable. For space reason, we will only report the result of ranking the news belonging to the "World" category in the following, considering that "World" news is generally most popular.

There are three parameters in the label propagation algorithm: k, σ and α. The algorithm is not very sensitive to the number of neighbors. We set $k = 10$ in our experiment. We choose $\sigma = 0.25$ experimentally. And α is simply fixed at 0.95, which has been proven to work well in [2].

Table 1. News categories and the corresponding number of news crawled

Category	*#News*	*Category*	*#News*
World	6301	Entertainment	4115
Local	5983	Sports	4865
Business	4990	Health	2459
Sci/Tech	4093		

5.2 Ranking News Sources and News Articles

Table 2 shows the ranking of the news sources with respect to reporting "World" news over the period of observation. ABC News results the most authoritative source, followed by REUTERS, CBS and YAHOO. An very interesting fact is the top two authoritative sources returned by our algorithm, i.e. ABC and REUTERS, are exactly consistent with the rating given by Newsknife [16] which rates news sites according to their appearances at Google News [13]. We observe that even without normalizing Q in (2), the ranking result is not completely determined by the number of news the sources present. For example, note that REUTERS is considered more authoritative than BBC although it posts relatively less news.

Table 2. Ranking result of news sources

News source	*#News*
ABC	1451
REUTERS	927
CBS	744
YAHOO	802
BBC	955
NEWSDAY	325
CBC	382
CNN	325
MSNBC	390

We compare the ranking result of our algorithm with two other algorithms. The first one is the hybrid model proposed in [5]. We omit the normalization matrix K in (8) when applying this model so as to better focus on comparing how different prediction methods would influence the overall ranking algorithm. Another simple and intuitive method for predicting a site's recommendation strength on the news that are not posted by it is to find the one-nearest neighbor of each unlabeled piece of news in the labeled set and take the product of the nearest neighbor's recommendation strength and their similarity value as the recommendation strength of the unlabeled news.

We asked a group of five people to manually assess the importance of top articles returned by these algorithms. They mainly take in account the relative importance of the news events these articles report. We define three importance levels and their corresponding weight values as shown in Table 3. For each article,

Table 3. Importance Level

Importance Level	*Weight*
Very important	3
Important	2
Normal	1

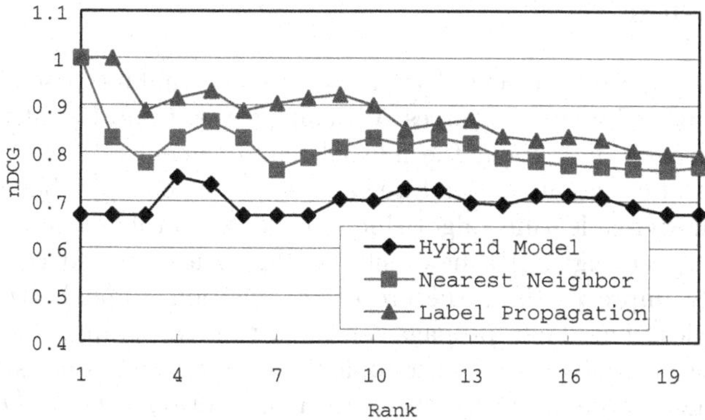

Fig. 5. Ranking quality comparison

Table 4. Top ten events during the observation period for the category "World"

Date	*News Source*	*News Title*
3/23	YAHOO	Basque ETA Announces Permanent Cease-Fire
3/24	CBS	India Offers 'Treaty of Peace' to Pakistan
3/25	CBS	Freed British Hostage Returns Home
3/21	ABC	At Least 28 Dead as Insurgents Storm Iraqi Jail
3/19	REUTERS	Hamas completes formation of Palestinian cabinet
3/25	NEWSDAY	Pressure Grows to Free Afghan Convert
3/17	YAHOO	Bolton: U.N. Will Send Iran Strong Signal
3/31	ABC	Strong Quakes Kill at Least 38 in Iran
3/21	YAHOO	U.S. Calls for New Vote in Belarus
3/17	ABC	Violence erupts in French student protests

the average of the five labels is taken as its importance value. The ranking quality is measured using normalized discounted cumulated gain (NDCG) measure [7]. The results are shown in Fig. 5.

According to Fig. 5, the performance of the hybrid model is worst. As the frameworks of the three algorithms are quite similar, the major reason for its degradation is because its predictions on the recommendation strength of unlabeled news are not accurate. The one step weighted average of labeled values does not reflect properly how a site would recommend the news issued by other sites. Besides, the original known visual significance is also altered by this model, which should not be the case. Our label propagation algorithm performs better than the nearest neighbor method. As analyzed by Zhu et al. in [8], the semi-supervised learning is quite efficient for news data probably because "the common use of quotations within a topic thread: article n_2 quotes part of article n_1, article n_3 quotes part of article n_2, and so on. Thus, although articles far apart in the thread may be quite different, they are linked by edges in the graphical representation of the data, and these links are well exploited by the learning algorithm." The top ten events returned by our algorithm are listed in Table 4.

6 Conclusion

In this paper, we have presented an algorithm for ranking assembled online news articles as well as news sources. We employ the visual layout information of news homepages and exploit the mutual reinforcement relationship between news articles and news sources. And we have proposed to use a label propagation based semi-supervised learning algorithm to tackle the problem of predicting the recommendation strength of a news site on the articles issued by other sites, which actually improves the structure of the relation graph between sources and new articles. This label propagation algorithm works quite effectively on news data as it naturally exploits the topic thread structure of news articles. By integration these two algorithms, the performance of our TOPSTORY system has been improved significantly.

In our future work, we intend to make our ranking algorithm work in an incremental manner, which is more consistent with the stream character of online news. And we will increase the scale of our system and make it fit better with the real web environment.

References

1. Kleinberg, J.: Authoritative sources in a hyperlinked environment. Journal of the ACM, Vol. 46, No. 5, pp. 604-622, 1999.
2. Zhou, D.Y., Weston, J., Gretton, A., Bousquet, O. and Schölkopf, B.: Ranking on Data Manifolds. MPI Technical Report (113), Max Planck Institute for Biological Cybernetics, Tübingen, Germany, 2003.
3. Wayne, C.L.: Multilingual Topic Detection and Tracking: Successful Research Enabled by Corpora and Evaluation. In Proceedings of the Language Resources and Evaluation Conference (LREC), 2000.
4. Corso, G.M., Gulli, A., and Romani, F.: Ranking a stream of news. In Proceedings of the 14th International Conference on World Wide Web, 2005.
5. Yao, J.Y., Wang, J., Li, Z.W., Li, M.J., and Ma, W.Y.: Ranking Web News via Homepage Visual Layout and Cross-site Voting. In Proceedings of the 28th European Conference on IR Research, 2006.
6. He, J.R., Li, M.J., Zhang, H.J., Tong, H.H., and Zhang, C.S.: Manifold-ranking based image retrieval. In Proceedings of the 12th annual ACM International Conference on Multimedia, 2004.
7. Jarvelin, K., and Kekalainen, J.: Cumulated Gain-based Evaluation of IR Techniques. ACM Transactions on Information Systems (ACM TOIS), 20(4), 422-446, 2002.
8. Zhu, X.J., Ghahramani, Z. and Lafferty, J.: Semi-supervised learning using Gaussian fields and harmonic functitons. In Proceedings of the 20th International Conference on Machine Learning, 2003.
9. Cai, D., Yu, S.P., Wen, J.R. and Ma, W.Y.: VIPS: a vision-based page segmentation algorithm. Microsoft Technical Report, MSR-TR-2003-79, 2003.
10. Radev, D.R., Blair-Goldensohn, S., Zhang, Z., and Raghavan, R.S.: Newsinessence: A system for domain-independent, real-time news clustering and multi-document summarization. In Proceedings of the Human Language Technology Conference, 2001.

11. Gabrilovich, E., Dumais, S., and Horvitz, E.: Newsjunkie: Providing personalized newsfeeds via analysis of information novelty. In Proceedings of the 13th International Conference on World Wide Web, 2004.
12. http://www.nielsen-netratings.com/
13. http://news.google.com/
14. http://news.yahoo.com/
15. http://newsbot.msnbc.msn.com/
16. http://www.newsknife.com/

Chinese Question-Answering: Comparing Monolingual with English-Chinese Cross-Lingual Results

Kui-Lam Kwok and Peter Deng

Computer Science Department, Queens College, City University of New York, New York 11367, U.S.A.
kwok@ir.cs.qc.edu, peterqc@yahoo.com

Abstract. A minimal approach to Chinese factoid QA is described. It employs entity extraction software, template matching, and statistical candidate answer ranking via five evidence types, and does not use explicit word segmentation or Chinese syntactic analysis. This simple approach is more portable to other Asian languages, and may serve as a base on which more precise techniques can be used to improve results. Applying to the NTCIR-5 monolingual environment, it delivers medium top-1 accuracy and MRR of .295, .3381 (supported answers) and .41, .4998 (including unsupported) respectively. When applied to English-Chinese cross language QA with three different forms of English-Chinese question translation, it attains top-1 accuracy and MRR of .155, .2094 (supported) and .215, .2932 (unsupported), about ~52% to ~62% of monolingual effectiveness. CLQA improvements via successively different forms of question translation are also demonstrated.

1 Introduction

While IR attempts to retrieve all documents that are relevant to an information need expressed as a query, QA has the goal to provide exact answers in order to satisfy a question. These are different modes of information access for different needs. English QA has been under active investigation since TREC 8 [1]. Chinese QA has been less studied and it is, like Chinese IR, considered more difficult because of difficulties in word segmentation and syntactic analysis during answer extraction or retrieval. Recently, both CLEF and NTCIR introduced mono/cross-lingual QA tasks for their evaluations. CLQA involves posing a question in language X and extracting answers from documents in language Y. Like CLIR, CLQA needs to handle the mismatch between the question and collection languages, and a common flexible approach is to translate questions, do retrieval and answering in language Y.

The complexity of QA requires stronger NLP processing than for IR, such as segmentation, POS tagging, parsing, knowledge sources for semantic operations. As discussed in [2], off-the-shelf QA systems are rare unlike IR systems. There is advantage to a simpler system that is more portable and easier to manage. For example, there are many languages in the world; for a web service offering QA/CLQA in multiple languages, supporting and upgrading the NLP software is expensive. A simple system might be preferred depending on how much effectiveness it can deliver.

H.T. Ng et al. (Eds.): AIRS 2006, LNCS 4182, pp. 244 – 257, 2006.

Researchers have studied CLQA among European languages in CLEF [3], and between Hindu and English [4]. This may be one of the first attempts at large-scale English-Chinese CLQA because evaluated data is not previously available. NTCIR-5 provides a large collection (~3GB) of BIG-5 encoded newspaper published in 2000 and 2001, 200 evaluated factoid questions in Chinese and their corresponding English counterparts, and their answers [5].

The paper is organized as follows: Section 2 presents our monolingual QA approach. Section 3 discusses results of Chinese QA. Section 4 introduces our methods of translating English question to Chinese for E-C CLQA. Section 5 discusses results of CLQA and comparison with monolingual. Section 6 has our conclusion.

2 Question-Answering in Chinese

Our objective is to experiment with a minimal approach to Chinese QA to see how much baseline result it can attain, apply it to English-Chinese CLQA, and compare their effectiveness within the NTCIR-5 environment. Fig.1 shows a common flowchart for QA and CLQA. The top two steps involve preprocessing, translation and are different or active for CLQA to produce a query in Chinese. Monolingual QA assumes questions are in Chinese, and just need preprocessing to produce a query q^C. We divide our QA procedures into several broad steps discussed below.

2.1 Question Classification

An important step in QA is question analysis which tries to understand what a question wants. In NTCIR-5, answers are limited to nine categories [5]: person, location, organization, date, time, money, percentage, 'numex' (numbers) and artifact. 'numex' is any numeric entity other than money and percent; 'artifact' represents other objects like book titles, product names, etc. We denote a question's class, C_Q, by one of these answer categories. Naturally, if one assumes a perfect question classification algorithm, and a question needs a person as answer, then returning a candidate that belongs to any other categories would be erroneous.

In our experiments, two simplifications were made. First, 'artifact' category (a) was not implemented; an 'unknown' class (x) was employed to tag all questions that failed to be classified into the other eight types. 'artifact'-related questions may involve ontology assistance and are left for future studies. Second, results of an English question classification algorithm (Fig.1-i, Sec.4) are assigned to the corresponding Chinese questions. This is done so that comparison with CLQA results can be made with one less confounding factor. We have a simple Chinese question classifier that is based on usage templates of characters/words such as: '谁', '(哪|哪一)(位|个|...) (人|地方|公司|...)', and performs moderately like our English question classifier (Sec. 4.1).

2.2 Question Preprocessing, Indexing and Retrieval

The next step, IR processing, is to obtain samples of text that have high probability of containing an answer to a Chinese question Q^C. This has correlation with the probability of retrieval units being relevant to the query q^C, and our PIRCS probabilistic retrieval engine [6] was employed to rank such units. The following three considerations guide our IR processing: a) questions are pre-processed to include double weighting of extracted entities to form q^C (Fig.1-i): the idea is to improve the focus of the query in order to have better retrieval (extraction is described in Sec.2.3); b) bigram and unigram indexing for retrieval (Fig.1-iii): this simple indexing strategy has been shown to be as effective as word indexing [6]. Moreover, factoid QA questions often contain names. These are difficult to segment accurately, and bigram indexing has the advantage of covering such an entity completely for retrieval; c) sentence as retrieval unit (Fig.1-iv): it has been shown [7] that the restricted context of a single sentence for answer candidates lead to good QA results in English.

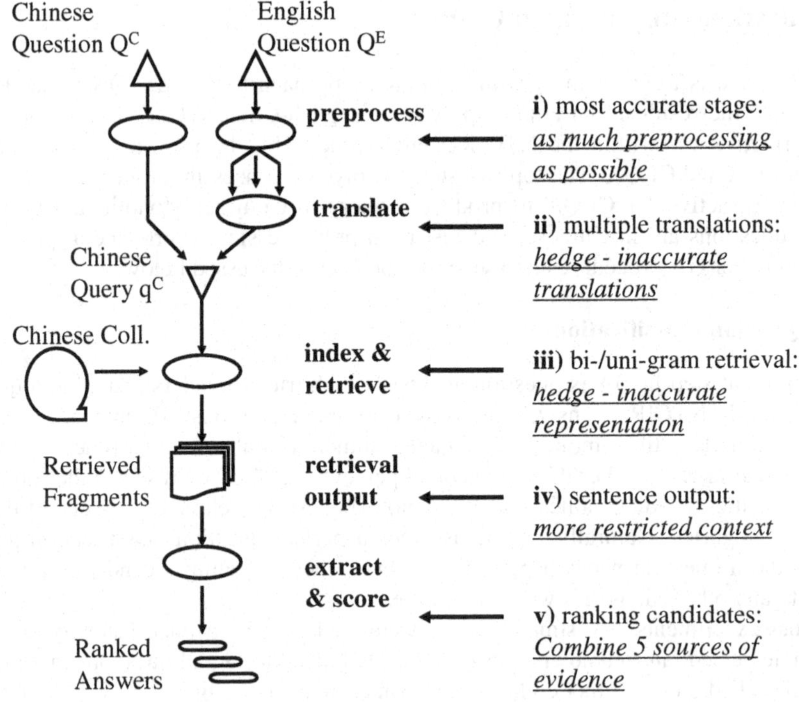

Fig. 1. Flowchart for C-C QA and E-C CLQA

2.3 Answer Candidate Extraction

From the retrieved sentences, one needs to identify substrings as candidate answers to a question. Sophisticated linguistic techniques can be employed to analyze the syntactic and semantic structure of each sentence, and match it with the question in order to decide where an answer may be. Here, we follow the tradition of extracting all possible answer candidates first [e.g. 8], and rank them later statistically. For this purpose, we employ BBN's IdentiFinder [9], a COTS software. It is based on prior training and HMM decoding, can handle both English and Chinese texts separately, brackets substrings as entities, and tags them into seven of the NTCIR-5 categories except 'numex' and artifact. We augment this with our own numeric extraction module. Thus, given retrieved sentences, IdentiFinder extracts many entities whose unique entries we denote as a_i with tag(a_i). All entities extracted by IdentiFinder form a pool with their tags, source sentence and other properties (Fig.1-v). These are used in the next section for their ranking and answer identification.

2.4 Answer Candidate Ranking

The last step (Fig.1-v) in QA is to identify which one in the pool of candidates is the correct answer. One may view a candidate a_i in sentence S_j as having probability $P(a_iS_j|Q)$ of being an answer and support to question Q. The most likely answer would be: $a = \text{argmax}_i \ \Sigma_j P(a_iS_j|Q) = \text{argmax}_i \ \Sigma_j P(a_i|S_jQ)*P(S_j|Q)$, where Σ_j sums over all sentences having a_i. This can be cast into a training problem by using Bayes' inversion if one has sufficient known QA samples. Absent such data, we employ intuitive estimation of factors that may proportionately reflect the value of these probabilities. $P(a_i|S_jQ)$ captures the probability of an answer in a sentence for Q; the influencing factors include category agreement (V_c), existence-in-question filtering (V_w), proximity to question substrings in S_j (V_p). $P(S_j|Q)$ is related to sentence retrieval; the influencing factors include similarity between Q and S_j (V_s), and retrieval depth (d). Finally, the sum is related to a candidate's frequency (V_f) in the retrieval list. These factors are discussed below and in Sec.3.

(a) Categorical Evidence

A most important evidence source is the agreement of a question class C_Q with a candidate's tag(a_i). Since the determination of these classes may have uncertainty, we employ a graded category score V_c for their measure:

```
Vc = 1;          //default value
if (CQ equals tag(ai) )
        Vc = Vc1 ;
else if (CQ ε {p,l,o} and tag(ai) ε {p,l,o})
        Vc = Vc2;
else  if (CQ ε {x} and tag(ai) ε {p,l,o})
        Vc = Vc3;
```

with $V_{c1} > V_{c2} > V_{c3} > 1$. When the question and candidate's class agree, the highest score V_{c1} is given. When they disagree but both are named entities {p,l,o}, a medium matching score (V_{c2}) is assigned since classification has tendency to mix up the named entities, such as identifying a an organization as location. If C_Q equals 'x' (analysis failed), we still assign a small score V_{c3} to V_c if tag(a_i) ε {p,l,o}. We assume there is a prior predominance of the {p,l,o} categories for the questions which may also be reflected within the failed class 'x'. Otherwise, $V_c = 1$, but other factors are still active for ranking.

It is found that $V_{c1} = 200$, $V_{c2} = 50$, $V_{c3} = 2$ can give good results for our monolingual QA. These values (and those below) are estimated by using 3-way cross validation on our 200 question set with 1/3 as held-out data. The search is rough and most likely not optimal.

(b) In-Question/In-Web Evidence

It is usual that a QA question seldom contains the answer explicitly. One can imagine questions like "What is the surname of President Bush's father?", but these circumstances are rare. We assume no such question exists and employ a score V_w to reflect this fact:

```
Vw =1;       //default value
if (ai matches substring(qC) )  Vw =0;
```

This binary evidence score can help eliminate some erroneous candidates. As the subheading implies, V_w is meant to capture existence among web expansion terms as well. This is done for CLQA (Sec.4.3). For monolingual, web expansion has not been found helpful and is not used.

(c) Proximity Evidence

A sentence is retrieved because it has bigrams and/or unigrams covering some keywords or substrings of the query. If an answer candidate exists in this sentence, we assume that the closer this candidate is to these coverings the more probable it is an answer. Suppose multiple coverings are found in a sentence containing one or more candidates. For each candidate, a preceding proximity score $V_{p\text{-}pre}$ and a succeeding score $V_{p\text{-}suc}$ are accumulated. The pseudo-code for evaluating $V_{p\text{-}pre}$ score follows:

```
Let c ε {a Chinese character, a numeric sequence, or an English word};
for (each c preceding a candidate) {
      x=c;
      while (x matches substring(qC)) {
         Vp-pre += f(match-length) / g(distance-from-candidate);
         x = x || next_x;
      }
      c = next_c (no match) or character_after_match;
}
```

A long sequence (<10) of character/word match will be given higher weight because of its length and subsequences. This weight is also a function of the distance of the sequence from a candidate. We experimentally found both f(), g() as log(1+match-length), log(1+distance-from-candidate) are effective for monolingual QA. A similar procedure for evaluating $V_{p\text{-}suc}$ is done for coverings appearing after a candidate. The final score for proximity is:

$$V_p = 1 + \alpha_p * (V_{p\text{-}pre} + V_{p\text{-}suc}), \text{ with } \alpha_p = 0.5.$$

As discussed before in Sec.2, we keep our approach simple by matching substrings rather than segmented words, and to work with returned sentences only without need for collection information.

(d) Sentence Similarity Evidence

If a sentence has high probability of relevance to q^C, its candidates may also more likely be answers. The proximity score in (c) has part of this accounted through scoring the coverings between sentence and query substrings. Our retrieval system also provides a retrieval status value (RSV) for each sentence and reflects the probability of relevance through collection and sentence statistics of the bi-/unigrams. For

Chinese QA, our experiments show that the returned retrieval rank is helpful. The following score V_s is used:

$V_s = 1/h(\text{rank})$, with h() being the identity function.

This not only keeps calculations simple; in certain retrieval situations such as web searching, RSV's may not be available.

(e) Candidate Frequency Evidence
Each answer candidate may appear in different retrieved sentences with total occurrence frequency f. We assume that the more often a candidate occurs, the more likely it is a correct answer based on repeated confirmation. This is independent of proximity or similarity. We employ the following V_f score to capture this information:

$V_f = 1 + 0.1 * \log(f)$

There are uncertainties in every step of our procedure. For example, we do not have 'artifact' as an answer class. The IR processing is itself fuzzy. Entity extraction and question analysis are uncertain. These determine the candidate pool. The functions and assumptions used in the evidence factors for candidate ranking may be unreliable. Combination of the five evidence sources by multiplication: $V=V_c*V_w*V_p*V_s*V_f$ is used for final ranking. This gives a more effective score for determining a candidate's answer-hood than any other subsets. All the factors do not involve segmentation or syntactic analysis for estimation.

3 Monolingual Chinese QA Results

In this section, we present monolingual QA results based on our simple approach and compare to other NTCIR-5 results. We also study in greater details the influence of retrieval depth and question classification accuracy.

3.1 Retrieval Depth

Retrieval depth (Fig.1-iv) refers to the number d of sentences returned from retrieval for QA processing. If too few sentences are used, correct answer(s) might not be recalled, while too many may cause erroneous candidates (fitting the evidence scoring) to rank high. Fig.2 plots retrieval depth vs. QA effectiveness. All five evidence factors are used. Here, top-n (n=1,2,5) denote accuracy evaluation based on 'Correct and Document-Supported' answers for the *first n* suggestions, while top-nU are based on 'Correct including Unsupported', i.e. the sentence is irrelevant to the question but happens to contain the correct answer. Although the graphs are rather flat, both groups show some bimodal behavior (except for top-5U) with peaks at around d=4 and also around d=13-50.

For this monolingual environment, the questions are given in original Chinese and retrieval is relatively more accurate (compared to CLQA). When only a few (4) sentences are returned, some easy questions have correct answers within them, and our system provides accuracy values of .295 (59 questions correctly answered, one suggestion each), and .36 (72 correct, two suggestions each) for top-1,-2 respectively. Mean reciprocal rank MRR=.3381 means that one may expect the first correct answer to appear on average at about the third suggestion. As additional sentences are returned, noisy wrong answers get ranked high and lead to accuracy drop. However, as more variety of sentences (13-50) are returned, more instances of correct answers fitting our evidence factors are encountered, and this leads to accuracy rise, sometimes approximating previous peak values. Thus retrieval relevance, which controls the order of sentences returned, may not necessarily be compatible with the existence of correct answers in returned sentences. The performance at the two approximate peak retrieval depths is shown in Table 1a for answers with sentence support. In general, retrieval depth d=4 is better for precision-oriented (i.e. top-1,-2), while d=25 can be better for recall-oriented (e.g. top-5 accuracy .43) operations. All 13 'artifact' questions failed except #97. This has 'artifact' answer '錢德拉X光' which IdentiFinder tags as 'person' and our evidence criteria ranks it best.

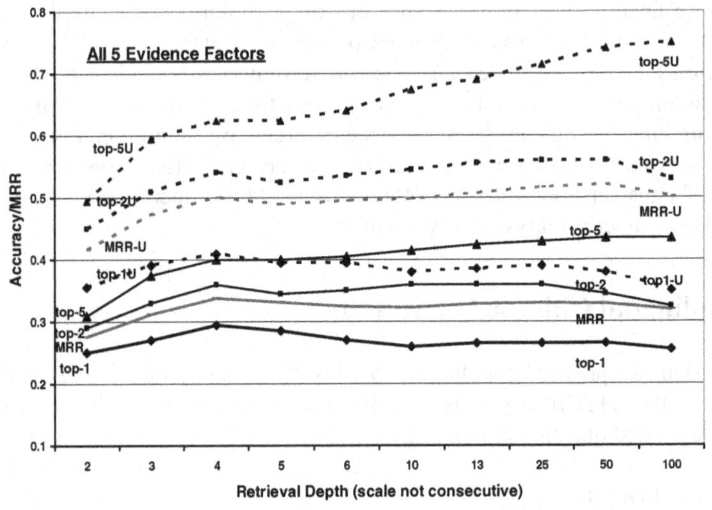

Fig. 2. Monolingual Chinese QA vs. Retrieval Depth

If 'Correct including Unsupported' answers are also included as right (Fig.2 dashed-lines), the peaks at d=25 surpass those at d=4 except at top-1U. top-1U accuracy is .41 at d=4 vs. .39 at d=25. The highest accuracy value is .75 (at d=100) for top-5U, and leads to 126 of the 200 questions having at least one correct answer in top 5. Table 1b shows the evaluation for the two peaks in greater detail. MRR-U values of .4998/.5163 mean a correct answer at position 2 on average approximately.

As comparison, the NTCIR-5 thirteen (including unofficial) *blind* Chinese QA runs attain top-1 accuracy ranging between .1-.375, with median=.29, and top-1U accuracy in the range .105-.445 with median .315 [5]. Our d=4 results are above median but worse than the best top-1 (.375) and top-1U (.445) results by 21% and 8% respectively. The best result from ASQA system was achieved with use of NLP tools including segmentation and POS tagging, Hownet sense and machine learning[10].

Table 1a,b. Top-1 to Top-5 Monolingual QA Results at Two Retrieval Depths (a) Supported, (b) including Unsupported

(1a)	top-1	top-2	top-3	top-4	top-5	MRR
d=4	.295	.36	.37	.395	.4	.3381
d=25	.265	.36	.37	.405	.43	.3296
NTCIR-5 best	.375	not available				
NTCIR-5 median	.290					

(1b)	top-1U	top-2U	top-3U	top-4U	top-5U	MRR-U
d=4	.41	.54	.585	.62	.625	.4998
d=25	.39	.56	.615	.675	.715	.5163
NTCIR-5 best	.445	not available				
NTCIR-5 median	.315					

3.2 Question Classification Accuracy

Our question classification has accuracy of about 80% as discussed in Sec.4.1. Here, we study effects of better classification. We simulate perfect classification by manually assigning the correct class tag to each question. Tables 2a,b show these results compared to Tables 1a,b. As can be seen, having perfect question classification only buys our system about 5-6% improvements at top-1 and top-2 (d=4) for supported answers, and about 7% including 'Unsupported'. The improvement rises to >=9% for top-1, d=25. Perfect question classification is only part of the story; better selection procedures could promote more good answers to top-1 position.

Table 2a,b. Monolingual QA Results with Perfect Question Classification a) Supported, b) including Unsupported

(2a)	top-1	top-2	top-3	top-4	top-5	MRR
d=4	.31	.38	.39	.405	.405	.3521
%imprv	*5.1*	*5.6*	*5.4*	*2.5*	*1.3*	*4.1*
d=25	.29	.375	.39	.425	.44	.3493
%imprv	*9.4*	*4.2*	*5.4*	*4.9*	*2.3*	*6.0*

(2b)	top-1U	top-2U	top-3U	top-4U	top-5U	MRR-U
d=4	.44	.58	.61	.635	.635	.5263
%imprv	*7.3*	*7.4*	*4.3*	*2.4*	*1.6*	*5.3*
d=25	.425	.595	.645	.7	.73	.5329
%imprv	*9.0*	*6.3*	*4.9*	*3.7*	*2.1*	*3.2*

4 English-Chinese CLQA

Here, one starts with an English question and attempts to obtain a translated Chinese query of similar meaning. The English questions are in their most accurate form. We capitalize on this with as much pre-processing as possible (Fig.1-i,ii). This includes question classification, entity extraction, and multiple translations. In addition, the weighting of the evidence factors are also modified.

4.1 Question Classification

An important issue in QA is to discover what a question wants. Since IdentiFinder with our own numeric extraction can provide eight of nine required answer categories (Sec.2.1) from sentences, we used an unknown class (x) as catch-all for questions that failed. Our basic procedure is to detect cue words (e.g.: 'Who', 'When', 'Where', 'What', 'How'..) in a question, and adjacent meta-keywords (e.g.: how 'many', what 'city', how 'long'). The meta-keywords are obtained from the Cognitive Computation Group (University of Illinois, Urbana-Champaign http://l2r.cs.uiuc.edu/~cogcomp/Data/). If this pattern failed, the question is POS tagged by MXPOST (http://www.cogsci.ed.ac.uk/~jamesc/taggers/MXPOST.html), followed with Collins' parser (http://people.csail.mit.edu/mcollins/code.html). This identifies a nearest noun phrase. Based on the words in the noun phrase, meta-keyword list and simple heuristics, we assign a possible answer class to the English question.

We tested this classification procedure on a set of training questions (T0001-T0200), and attains >80% precision. For the test set, it is about 78%. This should be more accurate than other QA processes. In contrast, some participants of NTCIR-5 achieve 92% [10] and 86.5% [11] accuracy.

4.2 English Question Translation

For CLQA, one needs to preserve in a translated query the fidelity of the intent in the original question. Translation technology from English to Chinese is an approximation at best. We attempt to overcome this issue by using multiple translations and web-based question expansion. One single translation is often brittle because an important concept may fail to get translated correctly. Systran MT is employed as a general English translation tool together with our web-based entity/terminology-oriented translation (CHINET [12]). The latter mines translation pairs from patterns in bilingual snippets returned by using an English name/terminology as keyword in a web search engine. It is always current and complements MT nicely.

For each question, three forms are derived: the raw English question, a named-entity list extracted via IdentiFinder, and the original statement plus 20 related terms using our English query expansion based on Google searching on the web [13]. The first form is translated using Systran, producing q^{C1}, the second via our web-based translation producing q^{C2}. The third form is passed through both translation paths, producing q^{C3}. They are concatenated into $q^{C} = q^{C1} \cup q^{C2} \cup q^{C3}$. This arrangement gives extra weight to entity names that occur in the original question, while terms in the question are given higher weight than web-expansion terms.

An interesting observation is that sometimes the web-expansion terms may contain a correct answer (in English) already. Previously, investigators performed *indirect* QA [14] by finding answers from the web, then locate supporting documents from the collection. This however may not be as successful in CLQA because the chance of obtaining an answer in English pages could be small if the question content is locally-oriented Chinese information. Moreover, a translation of the answer needs to be performed correctly. Instead we translate all expanded terms (including potential answers) to enhance retrieval results. Later, expansion terms are also used to confer evidence of answer-hood for candidates (Sec.4.3).

Once a Chinese query q^C is composed, indexing, retrieval and answer candidate pool formation is done similarly as in Sec.2.2-2.3.

4.3 Answer Candidate Ranking

The answer ranking procedures need some changes for CLQA. Because retrieval is much noisier, most of the parameter values are modified to reflect more uncertainty. For Category Evidence (a), V_{c1},V_{c2},V_{c3} were set to 70, 30 and 5. For In-Question/In-Web Evidence (b), V_w=0,1,2 depending on whether a candidate is present, or absent in q^{C1} U q^{C2}, or absent in q^{C1} U q^{C2} but present among web-expansion terms. For Proximity Evidence (c), sharper drop-off with distance appears more useful than logarithm and g() was chosen as $1/(\text{distance})^{2.5}$. For Similarity Evidence (d), it is found that a different similarity score is helpful:

$$V_s = 1 + \alpha_s * (\textstyle\sum_{i=1..5} m_i * \log(1+i)) / \log(1+L_s) / h(\text{rank})$$

Here, L_s is the sentence length, m_i is the number of overlaps between a sentence and q^{C1} U q^{C2} of i characters, $1 < i < 5$; $\alpha_s = 20$. This attempts to calculate the content similarity between a question and a sentence by accumulating the character overlaps between them, and does not need word segmentation, corpus statistics or retrieval RSV. For CLQA, retrieval is less accurate; many more sentences need to be returned to find answers. This leads to the use of a slower decreasing function h(rank) = $\text{rank}^{0.25}$. Moreover, too many sentences cause some common entities like "台湾" having too high a frequency to adversely influence candidate selection, and the frequency evidence is not used.

5 English-Chinese CLQA Results

In this section, we present CLQA results and compare to our monolingual QA and other NTCIR-5 results. We also study in greater details the influence of retrieval depth, question classification accuracy, and translation effects.

5.1 Retrieval Depth

In CLQA, the translated queries lead to less accurate retrieval. Low retrieval depth is not as effective – one needs more sentences to acquire answer candidates with good fit to our evidence formulae. Fig.3 shows CLQA results versus retrieval depths using four evidence factors (without frequency evidence). Most of the curves are uni-modal and attain a maximum at about d~100 (except for top-2U,-5U) to give the best English-Chinese CLQA results (Table 3a,b).

At accuracies of 0.155 (top-1, supported) to .455 (top-5, including 'Unsupported'), our cross-lingual approach attains 53% and 73% of our respective monolingual QA values (Table 1a,b). For top-1, this means getting 31 correct Chinese answers out of

200 English questions compared to getting 59 correct if one starts with original Chinese questions instead. Cross-lingual MRR .2094 (supported) vs. monolingual MRR .3381 means on average correct answers appear at about the 5th position for CLIR vs. at the 3rd position on average for monolingual QA. Thus the loss due to translation is substantial. As comparison, seven NTCIR-5 E-C CLQA runs attain top-1 accuracies ranging from .03 to .125 with median .075. Our value of .155 improves over the best value [15] by 24%. When 'Unsupported' are also counted, the top-1 accuracies from NTCIR-5 runs range from .04 to .165 with median .095. Our results of top-1U .215 and top-5U .455 are substantially better.

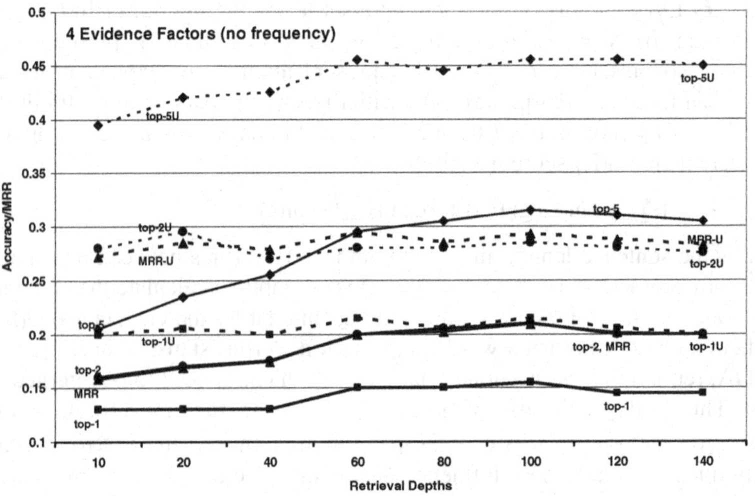

Fig. 3. E-C CLQA vs. Retrieval Depth

Table 3a,b. Top-1 to Top-5 E-C CLIR Results (a) supported; (b) incl. unsupported

(3a)	top-1	top-2	top-3	top-4	top-5	MRR
d=100	.155	.21	.245	.265	.315	.2094
%mono	*53*	*58*	*66*	*67*	*79*	*62*
NTCIR-5 best	.125	not available				
NTCIR-5 median	.075					

(3b)	top-1U	top-2U	top-3U	top-4U	top-5U	MRR-U
d=100	.215	.285	.335	.385	.455	.2932
%mono	*52*	*55*	*57*	*62*	*73*	*59*
NTCIR-5 best	.165	not available				
NTCIR-5 median	.095					

In [8], investigators reported French-English CLQA MRR of 78% compared to English QA using 'lenient' evaluation. Our corresponding performance is 52%. French is more similar to English than Chinese. For English-Hindu CLQA studied in [4], the MRR achieved was .25 for 56 questions compared to our result of .2094.

5.2 Question Classification Accuracy

The above experiments were repeated by assuming perfect question classification. Tables 4a,b show such results compared to Table 3a,b. Except for top-4, improvements are <10%, and reflect similar situations in monolingual where other considerations may be just as important as classification accuracy improvement for this simple approach.

Table 4a,b. CLQA Results with Perfect Question Classification (a) Supported; (b) including Unsupported

(3a)	**top-1**	**top-2**	**top-3**	**top-4**	**top-5**	**MRR**
d=100	.165	.22	.265	.29	.325	.2208
%imprv	*6.5*	*4.8*	*8.2*	*11.3*	*3.2*	*5.4*

(3b)	**top-1U**	**top-2U**	**top-3U**	**top-4U**	**top-5U**	**MRR-U**
d=100	.225	.295	.36	.41	.47	.3062
%imprv	*4.7*	*3.5*	*7.5*	*6.5*	*3.3*	*4.4*

5.3 Effects of Translation Processes

For CLQA, translation accuracy has the greatest effect on effectiveness. As discussed in Sec.4.2, we used multiple translation and redundancy processes in order to hedge wrong results from one single translation. Here, we analyze the contribution of these processes to the final CLQA results.

Fig.4 shows the accuracy, MRR results (same parameters as in Table 3) vs. different translation processes. "1:Systran x2" on the x-axis denotes results using Systran MT alone for translation (q^{C1} U q^{C1}). Here, many entity names are not translated correctly and results are poor. Top-1 accuracy is only 0.08. When entity-oriented web-based translation was added "2:1+webtran", top-1 improved to .09. When entities in questions were emphasized by extraction and translated "3:2+entity", top-1 improved to .11. Best results are obtained by including related terms via web-expansion "3+webxpan": ~40% improvement at top-1 to .155. The last step "4a:2+webexpan" shows that question entity extraction and translation (q^{C2}) is useful in the presence of web-expansion to anchor the retrieved sentences to the focus of the original question. Skipping it depresses results somewhat by ~3% (top-1=.15). These results show the usefulness of exploiting the source language questions.

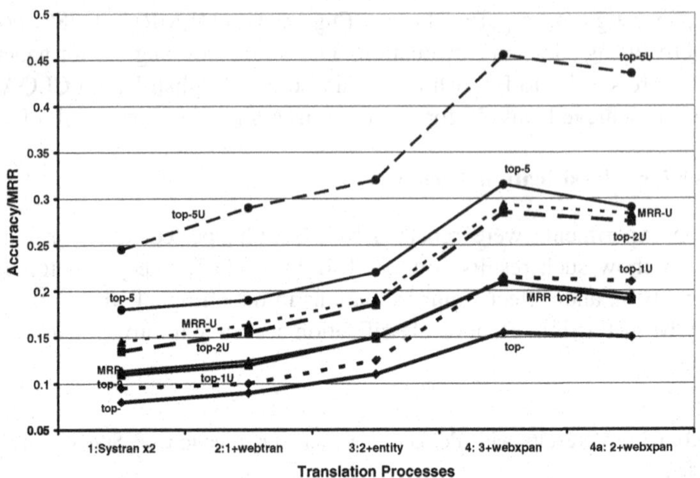

Fig. 4. E-C CLQA Effectiveness vs. Translation Processes

6 Conclusion

We present a simple Chinese factoid QA system that does not use explicit word segmentation, Chinese syntactic or semantic analysis. It employs a COTS entity extraction software and our retrieval engine, and emphasizes on ranking candidates extracted from retrieved sentences via category, in-question/in-web, proximity, similarity, frequency evidence factors together. It attains a medium QA top-1 accuracy of .295, top-5 of .43 and MRR ~.33, all are exact answers with sentence support. Accuracy has a bimodal behavior with retrieval depth d for monolingual QA. d=4 is best for top-1, while d=25 is better for top-5.

When applied to English-Chinese CLQA it attains top-1 accuracy of .155, .315 for top-5, and MRR ~.21, using retrieval depth d=100. These are 53%, 73% and 64% of monolingual values, due to question translation inaccuracies and less effective retrievals. Three separate translation paths were employed for CLQA to hedge translation failures: original question with MT, extracted entities with web-translation, and original question plus web-assisted expansion terms with both translations. Accuracy improves successively with each of these translation processes.

Like IR systems, a simple QA approach is easier to port to other languages. It can serve as a basis for better answer selection techniques to improve and push correct answers within top-5 to top-1 position. Future studies include better question classification, and the ability to extract artifacts. Additional translation paths or disambiguation techniques are helpful for improving CLQA. Reverse translation of answers back to source language is also a necessary future investigation.

Acknowledgments. We like to thank the anonymous reviewers for their constructive comments.

References

1. Voorhees, E.M & Tice, D.M.: The TREC-8 Question Answering Track Evaluation. In: Information Technology: The Eighth Text REtrieval Conference (TREC-8). NIST Special Publication 500-246. (2000) 83-105
2. Ramakrishnan, G., Chakrabarti, S., Paranjpe, D. & Bhattacharyya, P.: Is Question Answering an Acquired Skill? In: Proc. of 13th International WWW Conference (2004) 111-120.
3. Magnini, B, Romagnoli, S, Vallin, A, Herrera, J, Peñas, A, Peinado, V, Verdejo, F & de Rijke, M.: The Multiple Language Question Answering Track at CLEF 2003 (http://clef.iei.pi.cnr.it/)
4. Sekine, S and Grishman, R.: Hindi-English Cross-lingual Question-Answering System. ACM TALIP 2 (2004) 181-192
5. Sasaki, Y, Chen, H-H, Chen, K-H, Lin, C-J.: Overview of the NTCIR-5 Cross-lingual Question Answering Task. In: Proc. Fifth Workshop Meeting on Evaluation of Information Access Technologies: IR, QA and CLIR. NII, Tokyo (2005) 175-185
6. Kwok, K.L.: Improving English & Chinese Ad-Hoc Retrieval: A Tipster Text Phase 3 Project Report. Information Retrieval, 3 (2000) 313-338
7. Chang, Y, Xu, Hongbo & Bai, S.: A Re-examination of IR Techniques in QA Systems. Proc. IJCNLP (2004) 443-450.
8. Plamondon, L & Foster, G.: Quantum, a French/English Cross-Language Question Answering System. In: Comparative Evaluation of Multilingual Information Access Systems, 4th Workshop of the Cross-Language Evaluation Forum, CLEF 2003. LNCS 3237 (2004) 549-558
9. Bikel, D.M, Miller, S, Schwartz, R & Weischedel, R.: A High-Performance Learning Name-Finder. In: Proc. Conference of Applied Natural Language Processing, 1997
10. Lee, C-W, Shih, C-W, Day, M-Y, Tsai, T-H, Jiang, T-J, Wu, C-W, Sung, C-L, Chen, Y-R, Wu, S-H & Hsu, W-L: ASQA: Academia Sinica Question Answering System for NTCIR-5 CLQA. In: Proc. Fifth Workshop Meeting on Evaluation of Information Access Technologies: IR, QA and CLIR. NII, Tokyo (2005) 202-208
11. Lin,,F, Shima, H, Wang, M & Mitamura, T.: CMU JAVELIN System for NTCIR5 CLQA1. In: Proc. Fifth Workshop Meeting on Evaluation of Information Access Technologies: IR, QA and CLIR. NII, Tokyo (2005) 194-201
12. Kwok, K.L, Deng, P, Sun, H.L, Xu, W, Dinstl, N, Peng, P & Doyon, J.: CHINET – a Chinese Name Finder for Docucment Triage. In: Proc. 2005 Intl. Conf. on Intelligence Analysis. (Sasaki, Y, Chen, H-H, Chen, K-H, Lin, C-J.: Overview of the NTCIR-5 Cross-lingual Question Answering Task. In: Proc. Fifth Workshop Meeting on Evaluation of Information Access. (https://analysis.mitre.org/ proceedings_agenda.htm#papers)
13. Kwok, K.L, Grunfeld, L, Sun, H.L & Deng, P.: TREC2004 Robust Track Experiments using PIRCS. In: Information Technology: The Fourteen Text REtrieval Conference (TREC-2004). NIST Special Publication 500-261. (2005)
14. Brill, E, Lin, J, Banko, M, Dumais, S & Ng, A.: Data-Intensive Question Answering. In: Information Technology: The Tenth Text REtrieval Conference, TREC 2001. NIST Special Publication 500-250. (2002) 393-400
15. Kwok, K.L., Deng, P, Dinstl, N. & Choi, S.: NTCIR-5 English-Chinese Cross Language Question-Answering Experiments using PIRCS. In: Proc. Fifth Workshop Meeting on Evaluation of Information Access Technologies: IR, QA and CLIR. NII, Tokyo (2005) 209-214

Translation of Unknown Terms Via Web Mining for Information Retrieval

Qing Li[1], Sung Hyon Myaeng[1], Yun Jin[2], and Bo-Yeong Kang[3]

[1] Information and Communications University, Korea
{liqing, myaeng}@icu.ac.kr
[2] Chungnam National University, Korea
wkim@cnu.ac.kr
[3] Seoul National University, Korea
kby@chord.snu.ac.kr

Abstract. Many English words appear in Asian language texts, especially in the news reports and technical documents. Although a foreign term and its counterpart in English refer to the same concept, they are erroneously treated as independent index units in traditional monolingual IR. For CLIR, one of the major hindrances to achieving retrieval performance at the level of monolingual information retrieval is the translation of terms in queries, which are not found in a bilingual dictionary. This paper describes the degree to which these problems arise in Korean Information Retrieval and suggests a novel approach to solve it. Experimental results based on NTCIR and KT-Set test collections show that the high translation precision of our approach greatly improves the IR performance.

Keywords: cross-language information retrieval, machine translation, indexing.

1 Introduction

For cross-language information retrieval (CLIR), one of the major hindrances to achieving retrieval performance at the level of monolingual information retrieval is the translation of terms in queries, which are not found in a bilingual dictionary. Such terms are also called "out of vocabulary" words (OOV) or unknown words [12]. In traditional CLIR approaches, bilingual dictionaries are used to translate queries. New words and phrases are constantly introduced and it would be hard to keep updating a dictionary to include them all. Approaches such as what we propose in this paper can be used to acquire translation knowledge about these terms.

Mixed use of English and native languages presents a classical problem of vocabulary mismatch in monolingual information retrieval (MIR). The problem is significant especially in Asian language because words in the native language are often mixed with English words. Although English terms and their equivalences in a local language refer to the same concept, they are erroneously treated as independent units in traditional MIR. Such separation of semantically identical words in different languages may limit retrieval performance. For instance, Google search engine indexed 2,220,000

H.T. Ng et al. (Eds.): AIRS 2006, LNCS 4182, pp. 258–269, 2006.

Web pages that contain "삼성" but not its English counterpart "Samsung", about 46,000,000 pages that contain "Samsung" but not "삼성" and about 183,000 for both of them. A user would expect that a query with either "Samsung" or "삼성" would retrieve information from all of these three groups of Korean Web pages. Otherwise some potentially useful information will be ignored. There should be a semantic equivalence relation between them. Furthermore, one English term may have several corresponding terms in a different language. For instance, words "디지탈", "디지틀", and "디지털" are found in Korean texts, which all correspond to the English word "digital" but are in different forms because of different phonetic interpretations. Establishing an equivalence class among the three Korean words and the English counterpart is indispensable. By doing so, although the query is "디지탈", the Web pages containing "디지틀", "디지털" or "digital" can be all retrieved.

Observing the fact that many specific phrases sometimes appear together with their English translation in Web texts, for example, "…노무현 (Roh, MooHyun)….", it is possible to mine the bilingual search-result pages obtained from Web search engines to extract proper translations for specific terms which are treated as queries.

Nagata [10] firstly suggested using the search-result pages for translation of specific Japanese terms which had not been registered in the bilingual lexicon yet. He queried a search engine with Japanese terms to be translated, and downloaded the top 100 Web documents returned from the search engine to find the proper English translations. Cheng [1] developed Nagata's idea to dig out the right translations for specific terms from the search-result pages returned by a search engine. Instead of downloading the documents listed in the search-result pages, Cheng only employed the snippets (including titles and page descriptions) of those documents in the search-result pages. Therefore, it greatly reduced the complexity of mining process with a satisfied result.

While mining the Web to translate words has been studied extensively in the existing literatures (e.g. [1] [2] [12]), almost all existing work solely resort to the statistical methods on Chinese-English pair. In this paper, we are interested in translating Korean specific phrases into English. More specifically, due to the Altai language nature of Korean, we integrate the phoneme and semanteme instead of statistical information alone to pick out the right translation from the search-result pages. Although the technique we developed has values in their own right and can be applied for other language engineering fields, we intend to understand to what extent information retrieval effectiveness can be increased, especially in Korean monolingual IR case when relevant terms in different language are treated as one unit in index and the contribution of specific term translation for Korean-English CLIR.

2 Translation of Specific Phrases

The procedure to translate OOV (out-of-vocabulary)-containing phrases by mining the search-result pages can be generally divided into three main steps. Firstly, query the search engine with an OOV-containing phrase to retrieve the search-result pages, a list of snippets of Web pages containing the phrase. Then, extract candidates as the potential English translations from the search-result pages. Finally, select one or more candidates as right translations for the OOV by statistical, phonetic, and semantic models. While the main thrust of the procedure can be applied to other languages like

Chinese and Japanese, we are only focusing on Korean language in this paper and deal with Chinese in [8] where similar techniques are applied to resolve concept unification for IR.

3 Querying the Search Engine

We query the search engine, Google, with an OOV term to retrieve the snippets of Web pages containing both the term and its translation. However, not all the snippets of these Web pages contain both the term and its translation as in the example in Introduction Section; only 183,000 Web pages contain both “삼성” and its counterpart “Samsung” whereas 2,220,000 pages contain “삼성” only. In order to obtain Web pages with both the OOV query term and its English translation at high ranks, we need to query the search engine with not only the OOV term but also its hint words that would co-occur with the target translation. Hint words can be obtained by searching the engine with the OOV query term first, selecting the top-n topic-relevant words using simple statistics like $tf \times idf$, and translating them into English.

Our method differs from Zhang’s method [12] that expands the query with one hint word at one time and collects all the snippets from several query sessions, in that we query the search engine with all the hint words included with an OR operator. This strategy not only saves the searching time, but also helps avoiding translation ambiguity of individual hint words because the Web pages containing multiple hint words are likely to be semantically more relevant than those containing only one. For example, an OOV term “파우스트” (Faust) can be used as the query to the search engine to retrieve the top-n topic-relevant words. The top-three candidates are “괴테” (Goethe), “소개” (introduction), and “문학” (literature). Thus the query with the hint words to collect the snippets including translation of “파우스트” becomes: {“파우스트”+ <“Goethe” or “introduction” or “literature”>}, instead of three different queries in Zhang’s approach.

In this example, “소개” can be translated into “introduction”, “recommendation”, “presentation”, “dispersal”, “dispersion”, “dissemination” by Korean-English WordNet. While we can choose one based on a sense disambiguation method using the other candidate words, we chose to use them all as hint words, expecting that a pseudo-disambiguation would occur with the other hint words while searching is carried out. The pages retrieved by “Goethe” and “literature” are likely to contain “introduction” rather than “dispersion”. In other words, multiple hint words in a query would mutually disambiguate themselves and hence the retrieved pages are likely to contain the OOV term’s translation.

4 Translation

4.1 Extraction of Candidates for Selection

After querying the search engine with the OOV term and its hint words, the next step is to extract English candidates within a window of a limited size, which includes the

OOV term, in the snippets of Web texts in the returned search-result pages. Because the alignment types of translation pairs are diverse, (e.g. a Korean word "광우병" is aligned with three English words "mad cow disease"), we should combine English words into proper units as candidates, instead of making each single word as candidates. There are two typical ways, one is to group the words based on the co-occurrence information of the words in the corpus [1], and the other is to employ all sequential combination of the words as the candidates [13]. Although the first reduces the number of candidates, it risks losing the right combination of words as candidates. Selecting sequential combination of words actually refuses considering the discontinuous word sequences as candidates. For example, in the phrase "President George W Bush", the best 2-word entity is "President Bush", or perhaps "George Bush", but certainly none of the adjacent combinations ("President George", "George W", "W Bush"). Therefore, we use not only all adjacent combinations but also the discontinuous sequential combinations with skipped words. The maximum number of skipped words is 2. The final candidates for the previous example are shown in Table1.

Table 1. Candidates for translation

String : "*President George W Bush*"					
Adjacent sequential combinations				Discontinuous sequential combinations	
				Skip 1 word	Skip 2 words
President George W Bush *President George W* *President George* *President*	*George W Bush* *George W* *George*	*W Bush* *W*	*Bush*	*President W Bush* *President W* *George Bush*	*President Bush*

4.2 Selection of Candidates

The final step is to select the proper English candidate(s) as the translation(s) of the OOV term. We present a method that considers the statistical, phonetic and semantic features of the English candidates for selection.

4.3 Statistical Model

There are several statistical models to rank the candidates [10][12][13]. In our statistical model, we consider the frequency, length and location of candidates together. The intuition is that if the candidate is the right translation, it tends to co-occur with the query term frequently; its location tends to be close to the specific term; and the longer the candidates' length, the higher the chance to be the right translation. The formula to calculate the ranking score is

$$w_{FL}(q,c_i) = \alpha \times \frac{len(c_i)}{\max_{len}} + (1-\alpha) \times \frac{\sum_k \frac{1}{d_k(q,c_i)}}{\max_{Freq-len}}$$

where $d_k(q,c_i)$ is the word distance between the English phrase q and the candidate c_i in the k-th occurrence of candidate in the search-result pages. If q is adjacent to c_i, the word distance is one. If there is one word between them, it is counted as two and so forth. α is the coefficient constant, and $\max_{Freq-len}$ is the max reciprocal of $d_k(q,c_i)$ among all the candidates.

4.4 Phonetic and Semantic Model

4.4.1 Phonetic and Semantic Match

There has been some related work on extracting term translation based on the transliteration model [3][5]. Different from Korean-English transliteration that attempts to generate English transliteration given a Korean foreign word, our approach is a kind a match problem since we already have English candidates and aim at selecting the right candidates.

While the transliteration method is partially successful, it suffers from the problem that transliteration rules are not applied consistently. Korean phrase for which we are looking for the translation sometimes contains several words that may appear in a dictionary as an independent unit. Therefore, it can only be partially matched based on the phonetic similarity, and the rest part may be matched by the semantic similarity in such situation.

Returning to the above example, "클론" is matched with "clone" by phonetic similarity. "의" and "습격" are matched with "of" and "attack" respectively by semantic similarity. The objective is to find a set of mappings between the English word(s) in candidates and the Korean word(s) in a phrase, which maximize the sum of the semantic and phonetic mapping weights. We call the sum as SSP (Score of semanteme and phoneme). The higher SSP value is, the higher the probability of the candidate to be the right translation.

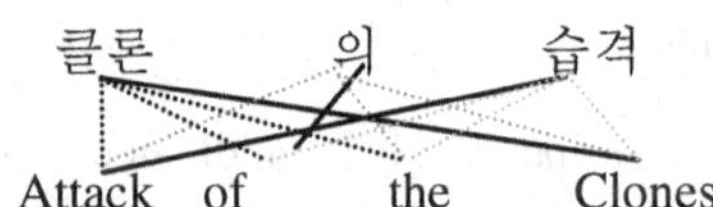

Fig. 1. Matching based on the semanteme and phoneme

The solution for a maximization problem can be found using an exhaustive search method. If the data we process get really big, however, performance degrades to an astonishing extent. As shown in Figure 1, the problem can be represented as a bipartite weighted graph matching problem. Let the English phrase, E, be represented as a sequence of tokens $< ew_1,...,ew_m >$, and the Korean phrase, K, be represented as a sequence of tokens $< kw_1,...,kw_n >$. Each English and Korean token is represented as a graph vertex. An edge (ew_i,kw_j) is formed with the weight $\omega(ew_i,kw_j)$ calculated as the average of normalized semantic and phonetic values, whose calculation details are explained in the following sections. In order to balance the number of vertices on both sides, we add the virtual vertex (vertices) with zero weight on the side with less number of vertices. The SSP is calculated:

$$\text{SSP}=\text{argmax} \sum_{i=1}^{n} \omega(kw_i, ew_{\pi(i)})$$

where π is a permutation of {1, 2, 3, …, n}. It can be solved by the Kuhn-Munkres algorithm (also known as Hungarian algorithm) with polynomial time complexity [9].

4.4.2 Phonetic and Semantic Weights

Phonetic weight is the transliteration probability between Korean and English candidates. We adopt our previous method in [3] with some adjustments. In essence, we compute the probabilities of particular Korean specific phrase, KW, given a candidate in English, EW.

$$\omega_{phoneme}(EW, KW) = \omega_{phoneme}(e_1,\ldots,e_m,k_1,\ldots,k_k)$$
$$= \omega_{phoneme}(g_1,\ldots,g_n,k_1,\ldots,k_k) = \frac{1}{n}\sum_j \log P(g_j g_{j+1} \mid g_{j-1} g_j) P(k_j k_{j+1} \mid g_j g_{j+1})$$

where $P(k_j k_{j+1} \mid g_j g_{j+1})$ is computed from the training corpus as the ratio between the frequency of $k_j k_{j+1}$ in the candidates, which were originated from $g_j g_{j+1}$ in English words, to the frequency of $g_j g_{j+1}$. If $j=1$ or $j=n$, g_{j-1} or g_{j+1}, k_{j+1} is substituted with a space mark.

The semantic weight is calculated from the bilingual dictionary. The current dictionary we employed for the local languages are Korean-English WordNet and LDC Chinese-English dictionary with additional entries inserted manually. The weight relies on the degree of overlaps between an English translation and the candidate

$$w_{semanteme}(E,K) = argmax \frac{No.of\ overlapping\ units}{total\ No.of\ \ units}$$

For example, given the Korean phrase "인하대" (Inha University) and its candidate "Inha University", "University" is translated into "대학교", therefore, the semantic weight between "University" and "대" is about 0.33 because only one third of the full translation is available in the candidate.

Due to the range difference between phonetic and semantic weights, we normalized them by dividing the maximum phonetic and semantic weights in each pair of the Korean phrase and a candidate if the maximum is larger than zero.

The strategy for us to pick up the final translation(s) is distinct on two different aspects from the others. If the SSP values of all candidates are less than the threshold, the top one obtained by statistical model is selected as the final translation. Otherwise, we re-rank the candidates according to the SSP value. Then we look down through the new rank list and draw a "virtual" line if there is a big jump of SSP value. If there is no big jump, the "virtual" line is drawn at the bottom of the new rank list. Instead of the top-1 candidate, the candidates above the "virtual" line are all selected as the final translations. It is because that a Korean phrase may have more than one correct translation. For instance, the Korean phrase "마카오" corresponds to two English translations "Macau" and "Macou". The candidate list based on the statistical information is "Macau, china, macou …". We then calculate the SSP value of these candidates and re-rank the candidates whose SSP values are larger than the threshold which we set to 0.3. Since the SSP value of "macau (0.892)" and "macou (0.875)" are both larger than the threshold and there is no big jump, both of them are selected as the final translation.

5 Experimental Evaluation

We conducted extensive experiments for phrase translation, query translation in CLIR, and indexing for Monolingual IR. We first compared the phrase translation approach against other systems. We then examined the impact of our approach in CLIR. Finally, we wanted to know the impact of phrase translation on monolingual IR, with which terms in different languages are treated as one unit in the index.

We selected 200 phrases from NTCIR-3 Korean corpus and manually found the translations for these phrases as the evaluation data. Table 2 shows the results in terms of the top 1, 3, 5 and 10 inclusion rates. "LiveTrans" and "Google" represent the systems against which we compared the translation ability. Google provides a machine translation function to translate text such as Web pages. Although it works pretty well to translate sentences, it is ineligible for short terms where only a little contextual information is available for translation. LiveTrans [1] provided by the WKD lab in Academia Sinica is the first unknown word translation system based on Web mining. There are two ways in this system to translate words: the fast one with lower precision is based on the "chi-square" method (χ^2) and the smart one with higher precision is based on "context-vector" method (CV) and "chi-square" method (χ^2) together. "ST" and "ST+PS" represent our approaches based on statistic model and statistic model plus phonetic and semantic model, respectively.

Even though the overall performance of LiveTrans' combined method (χ^2 +CV) is better than the simple method (χ^2) in table 2, the same does not hold for each individual. For instance, "Jordan" is the English translation of Korean term "요르단", which ranks 2nd and 5th in (χ^2) and (χ^2 +CV), respectively. The context-vector sometimes misguides the selection. However, in our two-step selection approach, the final selection would not be diverted by the false statistic information. In addition, in order to examine the contribution of distance information in the statistical method, we ran our experiments based on statistical method (ST) with two different conditions. In the first case, we set $d_k(q,c_i)$ to 1 for all candidates, that is, the location information of all candidates is ignored. In the second case, $d_k(q,c_i)$ is calculated based on the real textual distance of the candidates. As in table 2, the later case shows better performance. It also can be observed that "ST+PS" shows the best performance, then followed by "LiveTrans (smart)", "ST", "LiveTrans (fast)", and "Google". The statistical methods seem to give a rough estimate for potential translations without giving high precision. Considering the contextual words surrounding the candidates and the query phrase can further improve the precision but still less than the improvement by phonetic and semantic information in our approach.

The Korean-English CLIR experiments are based on the NTCIR-3 English corpus associated with 32 training topics. One topic has four main parts: title, description, narrative and keywords relevant to the topic. In our experiment, only the title part was used as queries to retrieve the documents from the English document collection because the average length of the titles, about 3 terms, is closer to the real situation of

Table 2. Overall performance

		Top -1	Top-3	Top -5	Top-10
Google		46%	NA	NA	NA
Live Trans	"Fast" χ^2	27%	35%	41.5%	46.5%
	"Smart" χ^2 +CV	21.5%	42%	48%	54%
Our Approach	ST (d_k=1)	25.5%	35.5%	38%	43%
	ST	28.5%	40%	46%	50.5%
	ST+PS	87%	88.5%	89%	89%

user queries. The bilingual dictionary we used is Korean-English WordNet with additional entries inserted manually. Using this dictionary, 9 out of 32 Korean queries can not find English translation for at least one phrase. Since the Korean terms in the query may have many senses, some of the translated terms are not related to the meaning of the original query. We used the mutual information [4] to alleviate this translation ambiguity problem. Finally, we searched for the documents relevant to a translated Korean query as well as an original English query using the Okapi BM25 [11]. The main metric is interpolated 11-point average precision and mean average precision (MAP).

Four runs are compared to investigate the contribution of lexeme translation in CLIR. Since not all of these 32 queries contain phrase(s) containing OOV, we report our experimental results in two different cases. In one case, all of the 32 queries are used. In the other case, only 9 queries containing phrase(s) with OOV are used.

1. RUN1 (Google): to provide a baseline for our test, we used Google Translation module to translate each Korean queries into English.
2. RUN2 (Without OOV): Korean queries were translated using a dictionary look-up ignoring the phrase(s) containing OOV.
3. RUN3 (With OOV): Korean queries were translated using a dictionary look-up while applying our method to phrase(s) containing OOV.
4. RUN4 (Monolingual): to provide a comparison with the "ideal" case, we retrieve the English documents with the corresponding English queries (titles) provided for those Korean titles by NTCIR.

NTCIR has two different criteria to evaluate the retrieval performance. The first is called "Rigid Relevance" where "highly relevant" and "relevant" documents are regarded as relevant documents. The second is called "Relaxed Relevant" where "highly relevant", "relevant" and "partially relevant" documents are all treated as relevant. The results of this series of experiments are shown in Figure 2~7 based on these two criteria. The precision-recall performance on all queries is shown in Figure 2 and 3. Regardless of which criteria are used between "Relaxed Relevance" or "Rigid Relevance", RUN3 (With OOV) shows better performance than RUN1 (Google) and RUN2 (Without OOV), but still performs less than RUN4 (Monolingual). We also observed that commercial translation product provided by Google performed almost the same as the RUN2 (Without OOV), where the lexemes containing OOV are ignored during the word-to-word translation.

Since not all of these 32 queries contain the specific lexemes, we further carried our experiments only with 9 queries containing OOV lexemes. As shown in Figure 4 and 5, translating the OOV lexemes using our approach in CLIR (RUN3) shows quite similar performance to Monolingual case (RUN4). To our surprise, our approach outperformed the monolingual case, especially for high-ranked documents, when we used the "Rigid Relevance" criterion.

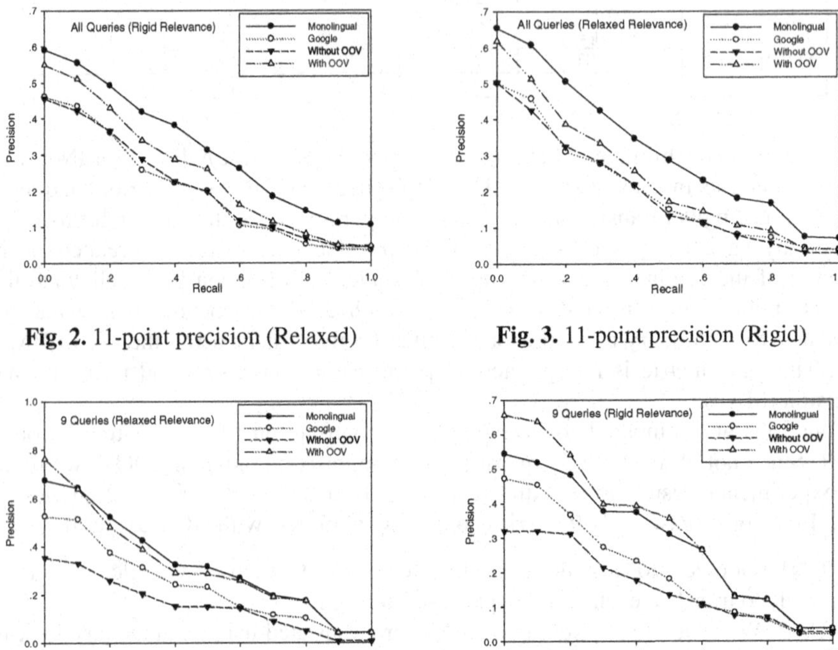

Fig. 2. 11-point precision (Relaxed)

Fig. 3. 11-point precision (Rigid)

Fig. 4. 11-point precision (Relaxed)

Fig. 5. 11-point precision (Rigid)

In order to understand this situation, we calculated the mean average precision (MAP) per query. As shown in Figure 6 and 7, most queries except query 13, 20, and 27 gave similar retrieval performance between RUN3(With OOV) and RUN4 (Monolingual). Query 13 (종말론, "Doomsday thought") was translated into "Eschatology" by our approach. Although it is different from the correct translation "doomsday thought" provided by NTCIR, it is still an acceptable translation. Unfortunately, only a few documents contain "doomsday thought" and no documents contain "eschatology", resulting in retrieval failure. For query 20 (닛산 자동차 회사와 르노사의 합병 "The capital tie-up of the Nissan Motor Company and Renault"), although it performed more poorly in RUN3 (With OOV) based on "Relaxed Relevance" standard in Figure 6, it shows better retrieval accuracy in RUN3 (With OOV) than RUN4 (Monolingual) based on the "Rigid Relevance" standard in Figure 7. Since the "Rigid Relevance" has a more strict rule to select "relevant" documents for test judgment, our approach shows a good ability for accurate information retrieval. In addition, the performance enhancement before 5-point is much higher than the 6~10-point area, that is, our approach performs well, especially for high-ranked documents. For query 27 (마카오의 반환 "Macau returns"), it shows better performance in RUN3 (With OOV) than RUN4 (Monolingual) on both two relevance standards. The corresponding English query of query 27 (마카오의 반환) provided by NTCIR is "Macau returns", but the translation returned by our approach is "macau macou return". Since "macau" and "macou" are both the right translation of the "마카오", this query gave a better result. By applying our approach, the MAP of these 9 queries increased about 115% from 0.1484 in RUN2 to 0.3186 with "Rigid Relevance" judgment criterion and about 105% from 0.1586 to 0.3228 with "Relaxed Relevance" judgment criterion.

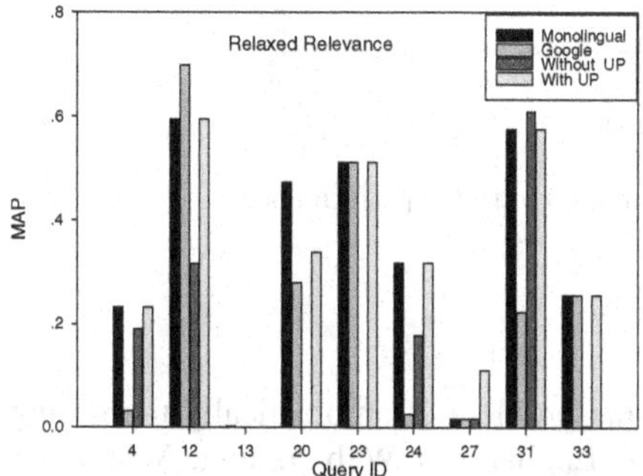

Fig. 6. MAP of 9 queries (Relaxed)

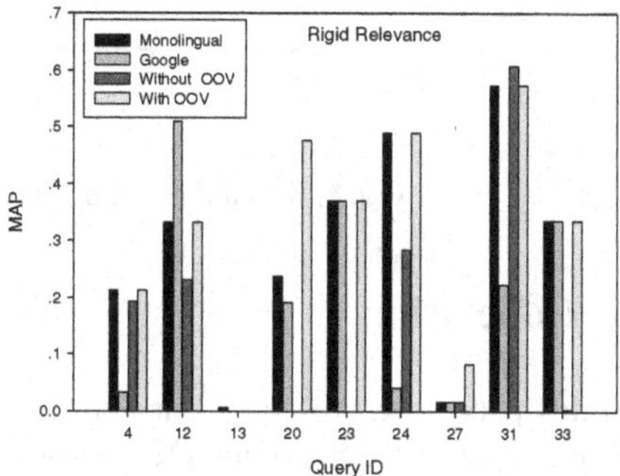

Fig. 7. MAP of 9 queries (Rigid)

We ran experiments to examine the impact of our OOV handling method on IR. The retrieval system is based on the vector space model with our own indexing scheme to which the OOV handling part was added. We employed the standard $tf \times idf$ scheme for index term weighting and *idf* for query term weighting. Our experiment is based on KT-SET test collection [6], which has 30 queries together with relevance judgments for them.

We tested two indexing methods. One was to extract the phrase with OOV word(s), which were recognized as index units. The other one was using our approach to translate these phrases into their English equivalents if possible, so the phrases and their corresponding counterparts were treated as the same index units. The baseline against which we compared our approach applied a relatively simple indexing technique. It uses a bilingual dictionary to identify index terms. The effectiveness of the baseline scheme is comparable with other indexing methods [7]. While there is a possibility that an indexing method with a full morphological analysis may perform better than our rather simple method, it would also suffer from the same problem associated with OOV words, which can be alleviated by our OOV translation approach.

We obtained 9.4% improvement based on mean average 11-pt precision when we simply separated phrases with OOV word(s) and used them as index units. Furthermore, when we applied our translation approach to the entire phrases with OOV words so that a matching occurred between English counterparts, the improvement was 14.9% based on mean average 11-pt precision.

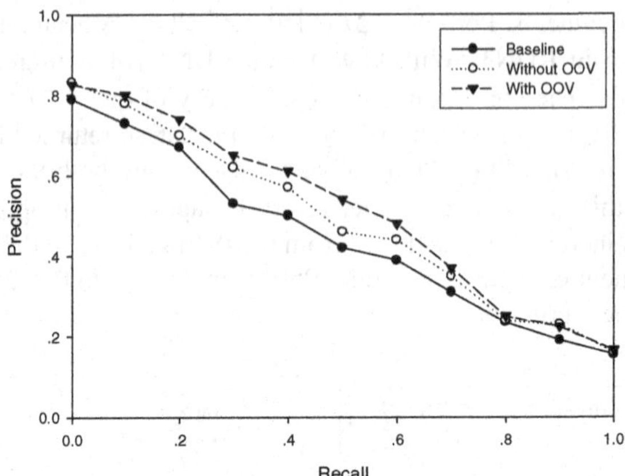

Fig. 8. Korean monolingual information retrieval performance

6 Conclusion

In this paper, we presented our approach to the problem of automatically translating Korean phrases containing OOV words into English via Web mining. While our current work focused on Korean text, the main thrust of the method is applicable to other languages like Chinese and Japanese. We also studied its contribution to monolingual and cross-lingual information retrieval. As witnessed by previous research as well as in our experiments, translating the OOV words in the query is necessary for an effective cross-lingual information retrieval. Due to the wide use of English terms in Korean texts, it is useful to treat these English words and their corresponding Korean words as a single unit for monolingual IR. Treating terms in different languages as one index unit is not only meaningful to English-Korean pairs but also to others like English-Chinese or even triplets English-Chinese-Korean. This is along the line of work where researchers attempt to index documents with concepts rather than words. We would extend our work along this road in the future.

References

1. Cheng, P., Teng, J., Chen, R., Wang, J., Lu,W., Chien, L.. Translating Specific Queries with Web Corpora for Cross-language Information Retrieval. *In Proc. of ACM SIGIR*, (2004).
2. Fung, P. and Yee., L.Y.. An IR Approach for Translating New Words from Nonparallel, Comparable Texts. *In Proc. of ACL98*, (1998).
3. Jeong, K. S., Myaeng, S. H., Lee, J. S., Choi, K. S.. Automatic identification and back-transliteration of foreign words for information retrieval. *Information Processing & Management. 35(4): 523-540*, (1999).
4. Jang, M. G., Myaeng, S. H. and Park, S. Y.. Using Mutual Information to Resolve Query Translation Ambiguities and Query Term Weighting. *In Proc. of ACL99*, (1999).
5. Kang, B. J., and Choi, K. S. Effective Foreign Word Extraction for Korean Information Retrieval. *Information Processing & Management, 38(1)*, (2002).

6. Kim, S.-H. et al.. Development of the Test Set for Testing Automatic Indexing. *In Proc. of the 22nd KISS Spring Conference*. (in Korean), (1994).
7. Lee, J, H. and Ahn, J. S.. Using N-grams for Korean Test Retrieval. *In Proc. of SIGIR*96, (1996).
8. Li, Q., Myaeng, S.H., Jin, Y. and Kang, B.Y.. Concept Unification of Terms in Different Languages for IR, *In Proc. Of ACL06*, (2006).
9. Munkres, J. Algorithms for the Assignment and Transportation Problems. *J. Soc. Indust. Appl. Math., 5* (1957).
10. Nagata, M., Saito, T., and Suzuki, K.. Using the Web as a Bilingual Dictionary. *In Proc. of ACL 2001 DD-MT Workshop*, (2001).
11. Robertson, S. E. and Walker, S.. Okapi/Keenbow at TREC-8. *In Proc. of TREC-8*, (1999).
12. Zhang,Y., Huang,F. and Vogel, S. Mining Translations of OOV Terms from the Web through Cross-lingual Query Expansion, *In Proc. of SIGIR05*, (2005).
13. Zhang, Y. and Vines, P.. Using the Web for Automated Translation Extraction in Cross-Language Information Retrieval. *In Proc. of ACM SIGIR04*, (2004).

A Cross-Lingual Framework for Web News Taxonomy Integration

Cheng-Zen Yang, Che-Min Chen, and Ing-Xiang Chen

Department of Computer Science and Engineering
Yuan Ze University, Taiwan, R.O.C.
{czyang, cmchen, sean}@syslab.cse.yzu.edu.tw

Abstract. There are currently many news sites providing online news articles, and many Web news portals arise to provide clustered news categories for users to browse more related news reports and realize the news events in depth. However, to the best of our knowledge, most Web news portals only provide monolingual news clustering services. In this paper, we study the cross-lingual Web news taxonomy integration problem in which news articles of the same news event reported in different languages are to be integrated into one category. Our study is based on cross-lingual classification research results and the cross-training concept to construct SVM-based classifiers for cross-lingual Web news taxonomy integration. We have conducted several experiments with the news articles from Google News as the experimental data sets. From the experimental results, we find that the proposed cross-training classifiers outperforms the traditional SVM classifiers in an all-round manner. We believe that the proposed framework can be applied to different bilingual environments.

1 Introduction

There are currently many Web news sites providing online news articles. Therefore, many Web news portals, such as AltaVista News [1] and Google News [2], arise to provide clustered news categories for users to browse more related news articles with a consistent interface. With the news clustering services provided by the Web news portals, users can rapidly click the related news links for more comprehensive information. However, to the best of our knowledge, most Web news portals only provide monolingual news clustering services. Users cannot easily find the related news articles of the same news event in different languages.

Consider the following example. Fig. 1 shows two news articles of the same story [3] in English and Chinese separately. The title of the English news focuses on "first impressions". However, the meaning of the Chinese title is somehow changed to "fall in love at first sight". Although these two titles have similar information semantically and a reader, say Alice, can speak English and Chinese fluently, she may hardly find the corresponding Chinese news article from the title of the English news. Furthermore, Fig. 1 shows another problem existing in most translated news articles. The amount of information of a news article is likely not equal to that of the correspondingly translated news article. In Fig. 1, the length of the English article is much longer than that of the

H.T. Ng et al. (Eds.): AIRS 2006, LNCS 4182, pp. 270–283, 2006.

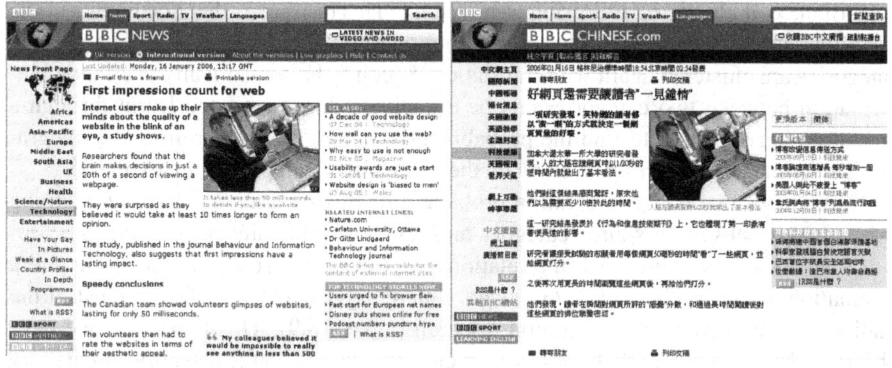

Fig. 1. Two news articles of the same story in two different languages

Chinese article. This example reveals that finding other corresponding news articles in different languages may help readers get more information.

The above example shows that the integration of two Web news taxonomies in different languages can provide Web surfers more abundant information. Such an automatic integration framework has two advantages as follows. First, it can save much news searching time resulted from the cumbersome searching procedure in which readers need to query different monolingual news sources in a trial-and-error manner. Second, the relevance accuracy can be still maintained no matter from what kind of language barriers the readers may suffer.

Recently, the Web taxonomy integration problem [4] has been obtained much attention to integrate the documents in a source taxonomy into a destination taxonomy by utilizing the implicit source information. In the past few years, many studies have proposed excellent approaches with which the integration accuracy can be highly improved [4,5,6,7,8,9]. Several classification techniques have been comprehensively studied in the recent research work, such as enhanced Naive Bayes (e.g. [4]), SVM-based approaches (e.g. [5,6,9]), and the maximum entropy model (e.g. [8]). Their experimental results show that improvements can be effectively achieved.

For some researchers as stated in [10], this may be a general *cross-lingual information retrieval* (CLIR) problem for news documents and can be easily solved with the assistance of some high-quality training resources and high-performance machine translation tools. However, there are still two major challenges in constructing a framework to perform cross-lingual Web news taxonomy integration. First, since the content size of the source taxonomy may be very different with that of the destination taxonomy as shown in Fig. 1, traditional integration techniques may have poor performance. Second, the integration problem is more complicated because the viewpoints of the source news taxonomy may be very different from the viewpoints of the destination taxonomy. The perceptional differences cannot be dealt with at the translation level.

In this paper, we propose a framework to automatically find the taxonomy mapping relationships from bilingual news categories with the aids of the semantically overlapped information. The proposed framework employs two major past research observations. First, the performance of clustering prior to translation is better than the performance of clustering following translation [11]. This means that documents in the same language are preferably clustered into monolingual categories first. Then all the categories are clustered again with translation. Second, the semantically overlapped information between two categories can be effectively discovered and learnt through a cross-training approach to find the probabilistic mappings [5]. Therefore, the integration performance can be improved when the probabilistic mappings are explored.

To the news taxonomy integration problem, our framework considers the mapping relationships between a source category and a destination category (the *category-to-category* mappings) rather than the relationships between a source news article and a destination category (the *article-to-category* mappings), usually studied in most of the traditional Web taxonomy integration research [4,5,6,7,8,9]. This is because we have observed that even if a less related news article is integrated on the category-to-category basis, most readers will not notice this less accurate integration. Besides, if the integration is performed on an article-to-category basis, some highly important news article may be miss-classified into other categories due to the performance limitation of current article-to-category integration approaches [4,5,6,7,8,9]. In this case, the readers will be confused when they read some totally unrelated but miss-integrated news articles. Although the integration on the category-to-category may also incorrectly find associations, this can be easily controlled with thresholds. Therefore, the cross-lingual Web news integration problem considered in this paper is to find the mapping relationship of two corresponding categories in different languages.

We have constructed the framework with SVM-based classifiers called SVM-BCT (Bilingual Cross-Training). We have also conducted several experiments to compare the accuracy performance of SVM-BCT with that of SVM by 92 manually identified mappings in a 14-day period. The experimental results show that SVM-BCT can be tuned to outperform SVM in all cases. If SVM-BCT is not properly tuned, it still has the same performance as SVM in the worst case.

The rest of the paper is organized as follows. In Section 2, we present the problem definitions, and briefly review previous related research on Web taxonomy integration. Section 3 elaborates the proposed cross-training framework. Section 4 describes our experiments in which English news and Chinese news articles from Google News were used as the data sets. Section 5 concludes the paper and discusses future directions.

2 Problem Statement and Related Research

Like the definitions in [4], we assume that the integration process deals with two Web news taxonomies which are in different languages. One taxonomy is the source $\mathcal{S}$ in which the news articles have been classified into m event groups (i.e. clusters) S_1, S_2, ..., S_m. Another is the destination taxonomy $\mathcal{D}$ in which the news articles have been also classified into n clusters D_1, D_2, ..., D_n. The integration is to discover the category-to-category mapping relationships between $\mathcal{S}$ and $\mathcal{D}$, i.e. $S_i \rightarrow D_j$. Optimistically, if S_i and D_j are of the same news event, their feature sets should be completely semantically overlapped. However, their mapping relationship may not be accurately identified because of the following two factors: the unbalanced amount of information between $\mathcal{S}$ and $\mathcal{D}$, and the translation mismatches resulted from the perceptional differences.

There are many approaches proposed for the general taxonomy integration problem. For example, an enhanced Naive Bayes (ENB) approach is first presented in [4] and then becomes the baseline for the following research. In [6], a Cluster Shrinkage (CS) approach is applied to SVM to improve the performance by shrinking the documents with semantic overlapping. In [7], a co-bootstrapping approach shows its enhancement by utilizing many weak hypotheses in a co-training procedure. In [9], an iterative SVM-based (SVM-IA) approach is proposed to improve the classification accuracy by iterative pseudo-relevance feedbacks and exploring auxiliary source information. In [8], a maximum entropy model is employed to enhance the performance. However, since these approaches perform the integration process on an article-to-category basis, they are not suitable for the news taxonomy integration problem.

Our framework uses a bilingual cross-training (BCT) approach adapted from the cross-training (CT) approach proposed in [5] to solve the taxonomy integration problem. The CT approach is a semi-supervised learning strategy. The idea behind CT is that a better classifier can be built with the assistance of another catalog that has semantic overlap relationships. The overlapped document set is fully-labelled and partitioned into a tune-set and a test set where the tune-set is used to tune the system performance and the test set is used to evaluate the system. Through the cross-training process, the implicit information in the source taxonomy is learnt, and more source documents can be accurately integrated into the destination taxonomy.

3 The Cross-Training Framework

To improve the bilingual Web news taxonomy integration performance, we propose a bilingual cross-training (BCT) framework with the aids of the source taxonomy information. The BCT framework performs the integration process on a category-to-category basis and is implemented with SVM classifiers.

3.1 The Processing Flow

Fig. 2 depicts the processing flow in the BCT framework. Without the loss of the generosity, here we use English and Chinese as two language representatives for our bilingual processing discussion in this paper.

In the framework, the classification system first retrieves English and Chinese news articles from news portals, say Google News or Yahoo! News [12]. Since these news articles usually have been well-clustered by the news portals, SVM-BCT does not recluster these articles. Because the performance of clustering prior to translation is better than the performance of clustering following translation [11], SVM-BCT can be benefitted form this observation to process these monolingual clusters.

The preprocessor then parses the Web news, and eliminates unnecessary words, such as stop-words. After the preprocessing, the source categories are translated into the target language. To study the effectiveness of SVM-BCT, we further interchange the source/destination roles. Suppose the Chinese news categories are originally the source categories. In the role-interchanged experiments, the English news articles will be the source news articles and translated into Chinese. After the translation process, all the

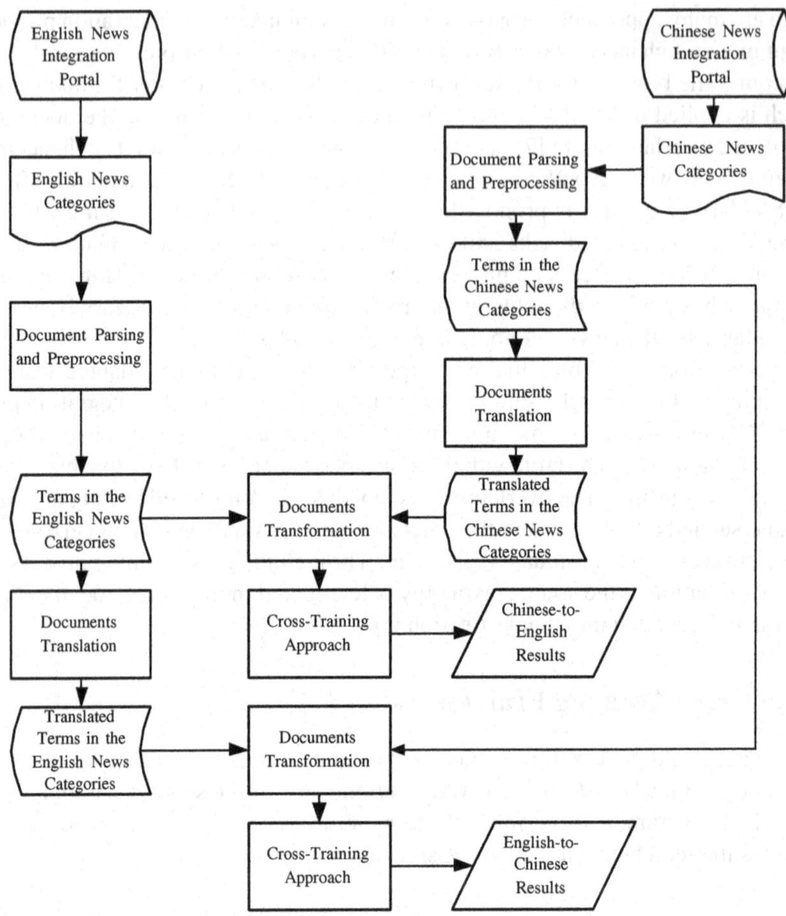

Fig. 2. The process flow of bilingual news taxonomy integration in SVM-BCT

source and target news articles are transformed into SVM vectors for further cross-training classification.

3.2 Parsing and Preprocessing

Since each Web news article is composed of plain text and HTML tags, it needs to be parsed first to extract useful information. Although several past studies [13,14] show that considering these HTML tags helps the classification accuracy, the parsing procedure is currently designed in a conservative manner by considering only the plain text and ignoring the HTML tags. This issue is left for future study.

Both Chinese and English news articles are preprocessed before they are classified with SVM-BCT. There are four steps for English news articles: (1) tokenization, (2) stop-word removing, (3) stemming, and (4) generation of term-frequency vectors. Because there is no word boundary in Chinese sentences, the Chinese articles need to be segmented first [15,16,17]. We use a hybrid approach proposed in [18] which can achieve a high precision rate and a considerably good recall rate by considering unknown words. The hybrid approach combines a longest match dictionary-based segmentation method and a fast keyword/key-phrase extraction algorithm. With this hybrid approach, each sentence is scanned sequentially and the longest matched words based on the dictionary entries are extracted. This process is repeated until all characters are scanned.

When the fast keyword extraction algorithm is executed, the preprocessor collects the occurrence frequencies of the terms to determine whether they are keywords. It first converts the text strings to an ordered list in which each element is either an unsegmented Chinese character or a segmented word. If there is an English word in the article, it is treated as a segmented word and preprocessed as if it is in an English article. Then the preprocessor will combine two adjacent elements into one longer element if their occurrence frequencies exceed a predefined threshold. If the frequencies of these two terms are not higher enough simultaneously, the term which cannot cross the threshold will be dropped. Therefore, a term is identified as a keyword due to its high-frequency and the low-frequency neighbors. This process is repeated to find all keywords.

3.3 Translation and Transformation

After the preprocessing, the Chinese and English news articles in each category are tokenized. Then the Chinese news documents are translated. The translation can be based on a bilingual dictionary or a well-trained machine translation system. In the translation, each Chinese word is translated to a set of English terms which are listed in the bilingual dictionary or derived from the machine translation system. The same procedure is also applied to the English news articles.

Because our framework is constructed based on discriminative SVM classifiers, each news article is converted to a feature vector. For each index term in the feature vector, a weight is associated with the term to express the importance, which could be the term frequency or its influential factor.

3.4 The Cross-Training Process

From previous studies on the general Web taxonomy integration problem, we find that if a source article can be integrated into a destination category, there must be a sufficiently large semantic overlap between them. This observation is also common for the Web news taxonomy integration problem. If an English/Chinese news category can be associated with a Chinese/English news category, they must be related to the same news event and there is a sufficiently large semantic overlap. Fig. 3 elaborates this relation. The semantic overlap can then be iteratively explored and used in SVM-BCT to enhance the SVM classifiers. This is the basic concept behind the proposed cross-training approach as shown in Fig. 4. The target news categories (destination taxonomy) are used as the training data sets, and the source news categories (source taxonomy) are used as the testing data sets. The SVM classifiers then calculate the positively mapped ratios as the *mapping score* (MS_i) to predict the semantical overlapping. The mapping score MS_i of $S_i \rightarrow D_j$ is defined as follows.

$$MS_i = \frac{\text{Number of the source news articles classified into } D_j}{\text{Total number of the source news articles in } S_i}.$$

For each destination category, the source category with the highest mapping score is predicted as the associative source news category, and vice versa in the cross-training process.

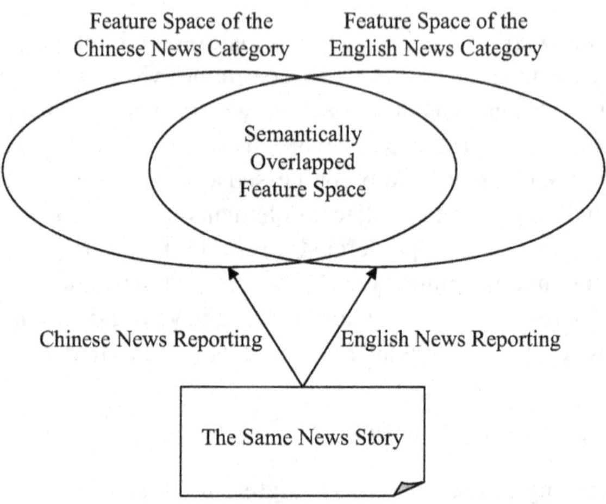

Fig. 3. The relation of the correspondent English/Chinese news categories and the news event

To get more accurate mapping scores, the classifiers are fine-tuned with cross-training. In the proposed cross-training process, the source news articles are translated into the target language first. All the news articles in the destination taxonomy are then used as the training set to build the first-stage SVM-BCT classifiers. Fig. 5 depicts the detailed process of the cross-training approach. For each destination news category, a source news category is the associative category only if it has the highest mapping score. This means that they should be related to the same news event and have semantic-overlapped label features which appear in the titles.

In SVM-BCT, the semantic-overlapped information is explored by first feeding the SVM-BCT classifiers with source news articles as test documents to find the preliminary possible mappings. Then we integrate the label information of the associative target category into the feature vectors of each source document to train SVM classifiers for next stage. Each vector comprises three parts: S_F, the test output, and S_L, where S_F is the term features of the source news article, S_L is the label features of the source news event, and the test output contains the label features of the associative target category. With the predicted mapping information, the discriminative power of the classifiers of the next stage can be enhanced for the source taxonomy.

At the second stage, the augmented feature vectors are used to train source SVM classifiers to fine-tune the exploration of the semantically overlapped information. For controlling the discriminative power of the added semantically overlapped label information as in [5], the score of the each feature term in the augmented feature vectors can be scaled by a factor of f, and the score of each label feature by a factor of $1 - f$.

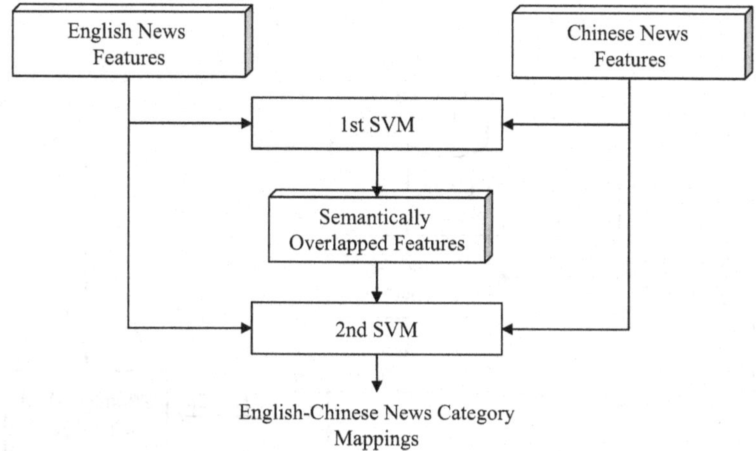

Fig. 4. The basic concept of the cross-training approach

The parameter f is used to decides the relative weights of the label and term features, and can be tuned for different application environments. In our current prototype, the influence of the parameter f is left for future discussion. The augmented feature vectors are currently calculated for SVM on a TF-IDF basis.

At the third stage, the target term features T_F, the target label features T_L, and the semantically overlapped source label features are used to train the third-stage SVM classifiers. We can then get the fine-tuned classifiers for target categories to obtain the predicted mapping scores. Generally, this cross-training process can be interchangeably continued for several rounds to further fine-tune the SVM classifiers.

4 Experimental Analysis

We have conducted several experiments to compare the performance of SVM-BCT with that of traditional SVM for bilingual Web news taxonomy integration. In the experiments, English news taxonomy and Chinese news taxonomy were used as the representatives to demonstrate the classification performance of the proposed SVM-BCT framework. The experimental results show that the classification accuracy of SVM-BCT is higher than that of traditional SVM comprehensively. We believe that the proposed framework can be applied to different bilingual environments.

4.1 The Experimental Environment

The SVM-BCT framework is currently implemented in Java 5.0 with Eclipse. The segmentation corpus is based on the Standard Segmentation Corpus published by the Association for Computational Linguistics and Chinese Language Processing (ACLCLP) [19]. We use SVM^{light} [20] as the SVM tool with a linear kernel.

The bilingual word lists published by Linguistic Data Consortium (LDC) [21] are the bilingual dictionaries. The Chinese-to-English dictionary ver. 2 (ldc2ce) has about

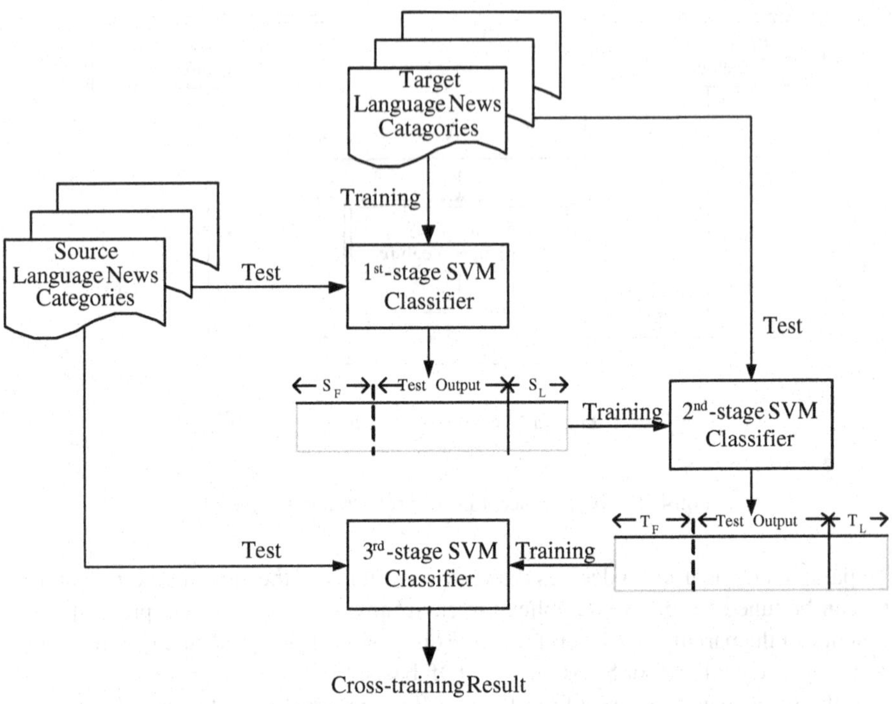

Fig. 5. The process of cross-training

120,000 records, and the English-to-Chinese dictionary (ldc2ec) has about 110,000 records. In ldc2ce and ldc2ce, each entry is composed of one single word and several translated words separated by slashes without any indication of the importance. Therefore, the translated words are treated equally in our experiments.

4.2 The Data Sets

In our experiments, two news portals were chosen as the bilingual news sources: Google News U.S. version for English news and Google News Taiwan version for Chinese news. Both the Chinese and English news articles were retrieved from the *world news* category from May 10 to May 23, 2005. The experiments were performed on the data set of each day. Twenty news categories were collected per day. All the English news articles were translated into Chinese with bilingual dictionaries. The size of the English-to-Chinese data set is 80.2 Mbytes. All the Chinese news articles were also translated, and the size of the Chinese-to-English data set is 46.9 Mbytes. Both data sets contain 29,182 news articles in total.

In the experiments, the mapping relations between the Chinese and English news reports were first identified manually by three graduate students independently. An English news category can be associated with a Chinese news category if at least two students have the same mapping identification. These manually-identified mapping relations were used to evaluate the performance of the bilingual integration systems.

The experiments were conducted in two ways: Chinese-to-English and English-to-Chinese. In the Chinese-to-English experiments, the English news taxonomy was first treated as the training set. To find a correspondent Chinese category (S_i) of an English destination category (D_j), the news articles in D_j were all used as the positive training examples, and the news articles in the other English news categories were used as the negative training examples. Then all mapping scores between English categories and Chinese categories were measured. For each destination English category, the Chinese category with the highest mapping score was considered the associative category. We measured the accuracy performance for each day using the correct mappings as the evaluation data sets, and calculated it by the following equation:

$$Accuracy = \frac{\text{Number of the mappings correctly discovered}}{\text{Total number of the correct mappings}},$$

which is similar to [4]. Accuracy rather than precision or recall is used because the integration is performed on a category-to-category basis. Apparently, the error rate is the complement of the accuracy. In the English-to-Chinese experiments, the roles of two taxonomies were switched.

4.3 The Results

Table 1 lists the experimental results of Chinese-to-English integration and Table 2 lists the experimental results of English-to-Chinese integration. In Table 1, the average accuracies are from 46.74% to 60.87%. SVM-BCT achieves the best accuracy of 66.67%, and SVM achieves the best accuracy of 50.00%. From Table 1, we can find that SVM-BCT outperforms SVM in 13 days. In Table 2, the average accuracies are from 52.17% to 67.39%. SVM-BCT achieves the best accuracy of 80.00%, and SVM achieves the best accuracy of 62.50%. From Table 1 and Table 2, we can find that SVM-BCT outperforms SVM very comprehensively.

In SVM-BCT, label features of the category with the highest mapping score are used to improve the accuracy performance. However, the predicted mapping score may not be correct due to classification errors. Therefore, we use a threshold θ to control the accuracy. Only if the mapping score is greater than θ, the category mapping is regarded as a correct mapping. In the experiments, we examined θ of different values from 0 to 1.0. If $\theta = 0$, the framework behaves like the original SVM-BCT. If $\theta = 1$, the framework behaves like the original SVM. In other cases, the framework filters out the noise information.

Table 3 depicts the results of the Chinese-to-English threshold control and only shows 5 cases. From the results, we can find that the classifiers achieve the best performance when $\theta = 0.5$. The experiments of the English-to-Chinese threshold control have the similar results in which the classifiers also achieve the best performance when $\theta = 0.5$.

4.4 Discussion

According to the experimental results with SVM-BCT and SVM, the accuracy can be further improved by threshold tuning. We further explored the reasons that hinder the

Table 1. The experimental results of Chinese-to-English news taxonomy integration

Data Set	Correct Mappings	SVM	SVM-BCT
510	8	4 (50.00%)	5 (62.50%)
511	6	3 (50.00%)	4 (66.67%)
512	8	4 (50.00%)	5 (62.50%)
513	5	2 (40.00%)	3 (60.00%)
514	7	3 (42.86%)	4 (57.14%)
515	7	3 (42.86%)	3 (42.86%)
516	6	3 (50.00%)	4 (66.67%)
517	7	3 (42.86%)	4 (57.14%)
518	6	3 (50.00%)	4 (66.67%)
519	5	2 (40.00%)	3 (60.00%)
520	6	3 (50.00%)	4 (66.67%)
521	8	4 (50.00%)	5 (62.50%)
522	6	3 (50.00%)	4 (66.67%)
523	7	3 (42.86%)	4 (57.14%)
Total	92	43	56
Average		46.74%	60.87%

Table 2. The experimental results of English-to-Chinese news taxonomy integration

Data Set	Correct Mappings	SVM	SVM-BCT
510	8	4 (50.00%)	5 (62.50%)
511	6	3 (50.00%)	4 (66.67%)
512	8	5 (62.50%)	6 (75.00%)
513	5	3 (60.00%)	4 (80.00%)
514	7	4 (57.14%)	5 (71.43%)
515	7	4 (57.14%)	5 (71.43%)
516	6	3 (50.00%)	4 (66.67%)
517	7	4 (57.14%)	5 (71.43%)
518	6	3 (50.00%)	4 (66.67%)
519	5	2 (40.00%)	3 (60.00%)
520	6	3 (50.00%)	4 (66.67%)
521	8	4 (50.00%)	5 (62.50%)
522	6	3 (50.00%)	3 (50.00%)
523	7	3 (42.86%)	5 (71.43%)
Total	92	48	62
Average		52.17%	67.39%

accuracy improvement. From the experiments, we found that the following two issues are very important to the accuracy performance and need to be future studied. First, name entity recognition (NER) is a serious issue because the bilingual dictionaries lack some name entities, such as "Yasukuni Shrine", which are common in news articles.

Table 3. The experimental results of Chinese-to-English news taxonomy integration in SVM-BCT ($0.0 \leq \theta \leq 1.0$)

Data set	Correct Mappings	$\theta = 0.0$	$\theta = 0.2$	$\theta = 0.5$	$\theta = 0.7$	$\theta = 1.0$
510	8	5 (62.50%)	5 (62.50%)	5 (62.50%)	5 (62.50%)	4 (62.50%)
511	6	4 (66.67%)	4 (66.67%)	4 (66.67%)	3 (50.00%)	3 (50.00%)
512	8	5 (62.50%)	5 (62.50%)	6 (75.00%)	5 (62.50%)	4 (62.50%)
513	5	3 (60.00%)	3 (60.00%)	3 (60.00%)	3 (60.00%)	2 (40.00%)
514	7	4 (57.14%)	4 (57.14%)	5 (71.43%)	4 (57.14%)	3 (42.86%)
515	7	3 (42.86%)	3 (42.86%)	4 (57.14%)	3 (42.86%)	3 (42.86%)
516	6	4 (66.67%)	4 (66.67%)	4 (66.67%)	4 (66.67%)	3 (50.00%)
517	7	4 (57.14%)	4 (57.14%)	4 (57.14%)	4 (57.14%)	3 (42.86%)
518	6	4 (66.67%)	4 (66.67%)	4 (66.67%)	4 (66.67%)	3 (50.00%)
519	5	3 (60.00%)	3 (60.00%)	3 (60.00%)	3 (60.00%)	2 (40.00%)
520	6	4 (66.67%)	4 (66.67%)	4 (66.67%)	4 (66.67%)	3 (50.00%)
521	8	5 (62.50%)	5 (62.50%)	6 (75.00%)	5 (62.50%)	4 (50.00%)
522	6	4 (66.67%)	4 (66.67%)	4 (66.67%)	3 (50.00%)	3 (50.00%)
523	7	4 (57.14%)	4 (57.14%)	5 (71.43%)	4 (57.14%)	3 (42.86%)
Total	92	56	56	61	54	43
Average		60.87%	60.87	66.30%	58.70%	46.74%

Because these name entities are excluded from the dictionary, their translations are completely ignored, and the accuracy is thus lowered. Second, in English and Chinese language processing, one word may have different transliterations, especially for the names of people and places. These issues are also left for future study.

5 Conclusions

As the amount of news information explosively grows over the Internet, on-line news services have played an important role to deliver news information to people. Although these Web news portals have provided users with integrated monolingual news services, cross-lingual search still has no effective solutions. If a user wants to find some related English news articles from a Chinese news article, she needs to surf English news portals and input keywords to query the related news article.

In this paper, we propose a cross-lingual framework with a cross-training approach called SVM-BCT to get high accuracy performance in finding the mapping relationships between two news categories in different languages. From the experimental results, we can find that the proposed cross-training approach outperforms the traditional SVM very comprehensively. We believe that the proposed framework can be applied to other bilingual environments.

There are still many research issues left for our future work. For example, feature weighting plays an important role to the system performance. Meaningful features should be explored and employed for integration. In addition, we only consider the accuracy rate of correct mappings in current experiments. The correct rejection rate needs to be further studied for non-overlapped source/destination categories. One of the most challenging issues is how to translate new words which are created daily due to the fast changing Web. A better automatic bilingual translation system is needed to fulfill the requirements of effective term translation for the NER problem and the transliteration problem.

Acknowledgement

This work was supported in part by National Science Council of R.O.C. under grant NSC 94-2213-E-155-050. The authors would also like to express their sincere thanks to anonymous reviewers for their precious comments.

References

1. Altavista News: `http://www.altavista.com/news/default` (2006)
2. Google News: `http://news.google.com/` (2006)
3. BBC News: First impressions count for web. English version available at `http://bbc.co.uk/2/hi/technology/4616700.stm`, Chinese version available at `http://news.bbc.co.uk/chinese/trad/hi/newsid_4610000/newsid_4618500/4618552.stm` (2006)
4. Agrawal, R., Srikant, R.: On Integrating Catalogs. In: Proceedings of the 10th International Conference on World Wide Web. (2001) 603–612
5. Sarawagi, S., Chakrabarti, S., Godbole, S.: Cross-training: Learning Probabilistic Mappings between Topics. In: Proceedings of the 9th ACM SIGKDD International Conference on Knowledge Discovery and Data Mining. (2003) 177–186
6. Zhang, D., Lee, W.S.: Web Taxonomy Integration using Support Vector Machines. In: Proceedings of the 13th international conference on World Wide Web. (2004) 472–481
7. Zhang, D., Lee, W.S.: Web Taxonomy Integration Through Co-Bootstrapping. In: Proceedings of the 27th annual international ACM SIGIR Conference on Research and development in information retrieval. (2004) 410–417
8. Wu, C.W., Tsai, T.H., Hsu, W.L.: Learning to Integrate Web Taxonomies with Fine-Grained Relations: A Case Study Using Maximum Entropy Model. In: Proceedings of 2nd Asia Information Retrieval Symposium, (AIRS 2005). (2005) 190–205
9. Chen, I.X., Ho, J.C., Yang, C.Z.: An Iterative Approach for Web Catalog Integration with Support Vector Machines. In: Proceedings of 2nd Asia Information Retrieval Symposium, (AIRS 2005). (2005) 703–708
10. Rogati, M., Yang, Y.: Resrouce Selection for Domain-Specific ross-Lingual IR. In: Proceedings of the 27th annual international ACM SIGIR Conference on Research and development in information retrieval. (2004) 154–161
11. Chen, H.H., Kuo, J.J., Su, T.C.: Clustering and Visualization in a Multi-lingual Multi-document Summarization System. In: Proceedings of 25th European Conference on Information Retrieval Research. (2003) 266–280
12. Yahoo! News. `http://news.yahoo.com/` (2006)
13. Jenkins, C., Inman, D.: Adaptive Automatic Classification on the Web. In: Proc. of the 11th International Workshop on Database and Expert Systems Applications, Greenwich, London, U.K. (2000) 504–511

14. Chen, I.X., Shih, C.H., Yang, C.Z.: Web Catalog Integration using Support Vector Machines. In: Proceedings of the 1st Workshop on Intelligent Web Technology (IWT 2004), Taipei, Taiwan (2004) 7–13
15. Nie, J.Y., Ren, F.: Chinese Information Retrieval: Using Characters or Words. Information Processing and Management **35**(4) (1999) 443–162
16. Nie, J.Y., Gao, J., Zhang, J., Zhou, M.: On the Use of Words and N-grams for Chinese Information Retrieval. In: Proceedings of the 5th International Workshop on on Information Retrieval with Asian Languages. (2000) 141–148
17. Foo, S., Li, H.: Chinese Word Segmentation and Its Effect on Information Retrieval. Information Processing and Management **40**(1) (2004) 161–190
18. Tseng, Y.H.: Automatic Thesaurus Generation for Chinese Documents. Journal of the American Society for Information Science and Technology **53**(13) (2002) 1130–1138
19. The Association for Computational Linguistics and Chinese Language Processing. `http://www.aclclp.org.tw/use_ssc.php` (2006)
20. Thorsten Joachims: SVM^{light}. `http://svmlight.joachims.org/` (2006)
21. Linguistic Data Consortium. `http://projects.ldc.upenn.edu/Chinese/LDC_ch.htm`(2006)

Learning Question Focus and Semantically Related Features from Web Search Results for Chinese Question Classification

Shu-Jung Lin and Wen-Hsiang Lu

Department of Computer Science and Information Engineering
National Cheng Kung University, Taiwan, R.O.C.
shu-jung@dns.csie.ncku.edu.tw, whlu@mail.ncku.edu.tw

Abstract. Recently, some machine learning techniques like support vector machines are employed for question classification. However, these techniques heavily depend on the availability of large amounts of training data, and may suffer many difficulties while facing various new questions from the real users on the Web. To mitigate the problem of lacking sufficient training data, in this paper, we present a simple learning method that explores Web search results to collect more training data automatically by a few seed terms (question answers). In addition, we propose a novel semantically related feature model (SRFM), which takes advantage of question focuses and their semantically related features learned from the larger number of collected training data to support the determination of question type. Our experimental results show that the proposed new learning method can obtain better classification performance than the bigram language modeling (LM) approach for the questions with untrained question focuses.

Keywords: Question Answering, Question Classification, Web Search Results, Language Model, Question Focus, Semantically Related Feature.

1 Introduction

Open-domain Question Answering (QA) systems are expected to response exact answers to a variety of free-formed users' questions in natural languages. Typically, most existing QA systems consist of the following components: question analysis, information retrieval, and answer extraction. Question analysis accounts for the processes of analyzing a given question, extracting keywords, and determining the question type. The process of determining the type of a given question is usually called question classification. According to error analysis about QA systems [1], 36.4% of the errors were generated by the question classification. Thus, in this paper, we focus on dealing with the following problems to improve the performance of question classification for a Chinese QA system.

Traditional regular expression methods using a set of handcrafted patterns are time consuming and ineffective to handle classification of questions with unseen

H.T. Ng et al. (Eds.): AIRS 2006, LNCS 4182, pp. 284–296, 2006.

(untrained) patterns. To improve such drawbacks, Li [2] propose a statistical language model (LM) to enhance the performance of question classification. However, the LM approach only using a small number of questions as training data (about 700 questions in Li's report) suffers from the problems of insufficient training data. Recently, a few approaches to classifying factoid questions perform well by using effective machine learning techniques [3] like support vector machines (SVM) [4], [5], [6], [7]. However, these approaches using machine learning techniques heavily depend on the availability of large amounts of training questions, and still suffer many difficulties while facing various real questions from the Web users. In fact, it is still difficult to collect sufficient training questions because most questions currently available is generated manually.

Brill et al. (2002) take advantage of the Web as a tremendous data resource and employ simple information extraction techniques for question answering [8]. Ravichandran and Hovy (2002) tried to utilize Web search results to learn surface patterns for finding correct question answers [9]. Solorio et al. and others improve the performance of question classification based on the numbers of the returned search results by submitting keywords in questions in combination with all the possible question types [5].

Different from the aforementioned works, in this paper, we present a simple learning method that explores Web search results to collect more training data (sentences) automatically by a few seed terms (question answers) of each question type. In addition, we propose a novel semantically related feature model (SRFM), which takes advantage of question focus (QF) [10] and semantically related features learned from the collected training sentences to determine question type. Our experimental results show that for the test questions with untrained QFs, the proposed SRFM can obtain better performance than the bigram LM approach while mitigating the problem of insufficient training data.

2 Chinese Question Classification

2.1 Problems

According to our analyses on the 400 Chinese questions from the task of NTCIR-5[1] Cross-Language Question Answering, some interrogative (question) words can be used to determine question type directly. For examples, a Chinese question with a interrogative word "誰" (who) can be classified as question type PERSON, and "哪裡" (where) can be used to determine question type LOCATION. However, some interrogative words can not be used to determine question type, such as "哪/哪個/哪一個" (which), "什麼" (what), because these interrogative words always result in ambiguities of question types. Two examples are shown in Table 1. Both questions contain the interrogative word "哪" but have different question types: the first one is type PERSON, and the other type LOCATION.

---

[1] http://research.nii.ac.jp/ntcir/

The above problem can be solved easily due to the fact that in general, there is a indicative keyword near the interrogative word in the question, called question focus (QF), that can be used to disambiguate the sense of the interrogative words and determine the type of a question. For the examples in Table 1, the QF of the question in the first example is "公司" (company), and the QF of the question in the second example is "機場" (airport). Another challenging problem is resulted from the case that if we didn't know the QF in a new question, the type of the question may not be determined.

Table 1. An example showing ambiguity of type of a question with the interrogative word "哪" (which)

	Example 1	Example 2
Chinese Question	DB2資料庫管理軟體是哪一個公司的產品？	請問獲選為2001年全球最佳機場的是哪個機場？
English Translation	Which company issued DB2 Database Management Software?	Which airport was voted as the world's best airport in 2001?
Interrogative Word	"哪" (which)	"哪" (which)
Question Type	ORGANIZATION	LOCATION

Table 2. An example showing the question focus and its semantically related features in a question

Chinese Question	請問獲得2001年諾貝爾文學獎的是哪一位作家?
English Translation	Which writer won the Nobel Prize in Literature in 2001?
Question Focus (QF)	作家(writer)
Related Verb (RV)	獲得(won)
Related Quantifier (RQ)	位(a Chinese quantifier, no English equivalent)
Related Noun (RN)	文學獎(Prize in Literature)

2.2 Ideas

In this paper, we intend to solve the problem of classifying questions without QF information. Our idea is that first, some words which have semantically related information to QFs in questions may support the lacking QF information like the verbs, quantifiers, and nouns in a question. We call these words related verbs (RV), related quantifiers (RQ), and related nouns (RN), and think that these semantically related features may be easier to be learned using our proposed SRFM based on the abundant training data from Web search results (described in Section 3.3). In Table 2, if we didn't learn the word "作家" (writer) which belongs to type PERSON, then we can use the three semantically related features: the RV "獲得" (won), the RN "文學獎" (Prize in Literature), and the RQ "位" (a Chinese quantifier unit for type PERSON, no English equivalent) to easily obtain the information of type PERSON for determining the type of this question. It is quite possible that they can be trained in advance if these verbs, nouns, and quantifiers frequently co-occurred with those name entities labeled by type PERSON in Chinese texts.

3 The Proposed Approach

Fig. 1 illustrates the architecture of our question classification approach. In the first stage, we use a few basic classification rules to handle the simple questions without the ambiguous interrogative words. Second, the LM or our proposed SRFM is used to deal with the complicated questions that can't be processed in the first stage. In the following, we describe the basic classification rules, the LM and the SRFM in details.

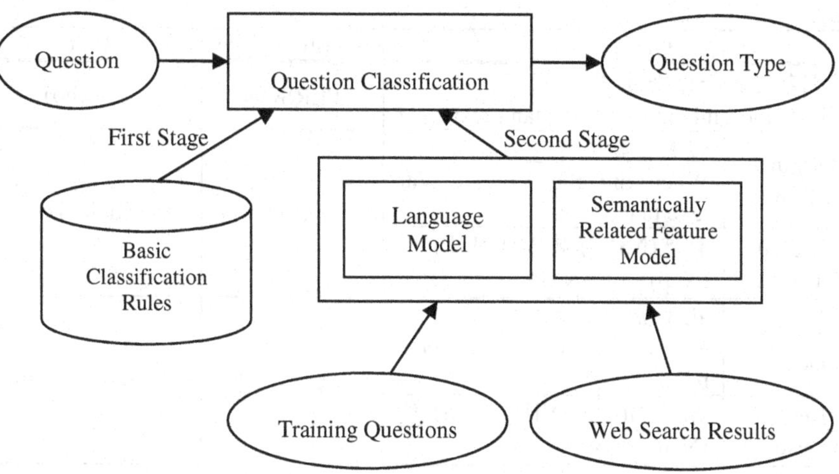

Fig. 1. Architecture of our question classification approach

3.1 Basic Classification Rules

To quickly classify some simple questions, we particularly generate 17 basic classification rules for four types of questions. These basic rules are shown in Table 3. Actually, these rules are very efficient and effective. Table 4 shows three examples. The first question "請問芬蘭第一位女總統為<u>誰</u>? " (Who is the Finland's first woman president?) can be easily determined as the type PERSON according to its interrogative word "誰" (who). However, these basic rules are too simple to classify many complicated questions correctly as mentioned in Section 2, whereas the complicated questions are able to be handled by the language model or our proposed semantically related feature model.

Table 3. Seventeen basic classification rules, where Term_{Nd} indicates a word with tag "Nd" (time noun), Term_{Nf} indicates a word with tag "Nf" (quantifier), and Term_{VH} indicates a word with tag "Vh" (stative intransitive verb). We use the CKIP[2] tagger (a representative Chinese tagger) to label the POS tag of each word in a question.

Question Type	Rule
PERSON	誰(who), 哪位(who)
LOCATION	哪裡(where), 何處(where)
DATE	何時(when), 幾年(which year), 幾月(which month), 幾日(which date),什麼(what)/何(which)/哪(which)·Term_{Nd}
NUMBER	多少(how many/much),約為(about),幾個(how many/much), 第幾(ordinal number), 幾(how many/much)·Term_{Nf}, 多(how)·Term_{Vh}

Table 4. Effective examples of question classification using the basic classification rules

Question		Question Type	Rule
請問芬蘭第一位女總統為誰? Who is the Finland's first woman president?		PERSON	誰(who)
Original Quesiton	韓戰爆發於哪一年? When did the Korean War take place?	DATE	哪(which)·Term_{Nd}
Tagged Question	韓戰(Na) 爆發(VJ) 於(P) 哪(Nep) 一年(Nd) ?		
Original Quesiton	孔子活了幾歲? How old was Confucius when he died?	NUMBER	幾(how)·Term_{Nf}
Tagged Question	孔子(Nb) 活(VH) 了(Di) 幾(Neu) 歲(Nf) ?		

3.2 Language Model

To handle the problems of classifying some complicated questions, we initially refer to the language modeling approach proposed by Li (2002). The major advantage of language model over the basic classification rules is its flexibility, but its current drawback is the lack of sufficient training data as mentioned before. When given a question $Q = q_1q_{2...}q_n$, where q_i represents the word in Q. we calculate the probability $P(Q \mid C)$ by the following language models for each question type (category) C and select the best possible C with the highest probability. Here we construct the unigram and bigram language models for each question type C of training questions:

$$P(Q \mid C) = P(q_1 \mid C) \times P(q_2 \mid C) \times \cdots \times P(q_n \mid C), \quad (1)$$

[2] http://ckipsvr.iis.sinica.edu.tw/

and

$$P(Q \mid C) = P(q_1 \mid C) \times P(q_2 \mid q_1, C) \times \cdots \times P(q_n \mid q_{n-1}, C)\,. \tag{2}$$

3.3 Learning Question Focus and Semantically Related Features

We try to use a different training method of only using a certain type of few seed terms and explore Web search results to obtain the same performance.

3.3.1 Collecting Training Data from Web Search Results

We describe below the process of collecting question focus and semantically related features from Web search results.

1. Collect a few seed terms (e.g. question answers) for each question type. (For example, for the type PERSON, we can use some famous person names, such as "李安" (Ang Lee), "拿破崙" (Napoleon) and so on, and for type LOCATION, we can use some location names, such as "新加坡" (Singapore), "紐約" (New York), etc.)
2. Submit each seed term to search engines (e.g., Google), and then retrieve a number of (e.g., 500) search results.
3. Extract the sentences containing the seed term, and then use the CKIP tagger to label the POS tags for each sentence.
4. Extract the terms labeled with the POS tag: verbs, quantifiers, and nouns, and then take them as the question focuses and semantically related features.

3.3.2 Question Focus Identification

The QF of a question provides important information to determine the type of a question. We design a simple algorithm to identify QF based on the POS tags and a certain of interrogative words, e.g., 什麼(what), 何(what), 哪(which), 那(which), 為何 (what), 為(is), 是(is). Our algorithm of question focus identification is described below:

Algorithm. Question_Focus_Identification;

Input: Question
Output: Question focus

1. Use the CKIP tagger to obtain the POS tags of all words in a question.
2. Determine whether a question contains an interrogative word, if it does, then go to step 3, else go to step 4.
3. Seek a word with the tag "Na" or "Nc" following the interrogative word, and take it as the question focus ("Na" is a common noun and "Nc" is a place noun). Otherwise, seek a word with the tag "Na" or "Nc" preceding the interrogative word, and take it as the question focus.
4. Seek a word with the tag "Na" or "Nc" preceding the end of the question until the question focus is found.

Please note that during the identification process, if two consecutive words with the same tag "Na" or "Nc", we identify the latter word as the QF. If the first one is "Na", and the second one is "Nc", then we identify the word with "Na" to be the QF. If the first word is "Nc", and the second one is "Na", then the word with "Na" is considered the QF.

Table 5. The first effective example for question focus identification

Chinese Question	請問九一一恐怖攻擊事件的主謀是哪個人？
EnglishTranslation	Who was the prime suspect behind the 911 attack?
Tagged Question	請問(VE) 九一一(Neu) 恐怖(VH) 攻擊(VC) 事件(Na) 的(DE) 主謀(Na) 是(SHI) <u>哪</u>(Nep) 個(Nf) <u>人</u>(Na)？
Interrogative Word	哪(Which)
Question Focus	人(person)

Table 6. The second effective example for question focus identification

Chinese Question	世界上第二大的沙漠是哪一個?
EnglishTranslation	What is the name of the world's second largest desert?
Tagged Question	世界(Nc) 上(Ncd) 第二(Neu) 大(VH) 的(DE) <u>沙漠</u>(Na) 是(SHI) <u>哪</u>(Nep) 一(Neu) 個(Nf)？
Interrogative Word	哪(Which)
Question Focus	沙漠(desert)

In Table 5, we can find the term "人" (person) tagged with "Na" following the interrogative "哪" (which),, thus it can be identified as the QF. Table 6 shows that the term "沙漠" (desert) with the tag "Na" can be identified as the QF by seeking a noun preceding the interrogative word "哪" (which).

3.3.3 Semantically Related Feature Model

To mitigate the problems of lacking sufficient training questions in LM approach, we intend to exploit the abundant Web search results to learn more QFs frequently occurring in questions and their semantically related features, and then train an effective model for question classification. Fig. 2 illustrates our SRFM. We assume that some QFs and semantically related features in question Q possibly occur in Web search results frequently and thus is easier to be trained.

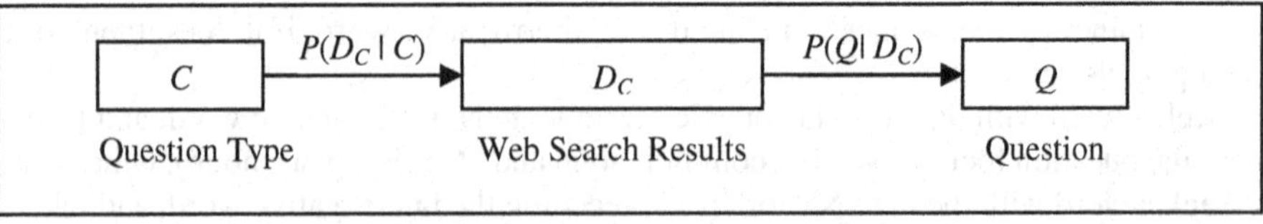

Fig. 2. Illustration of the SRFM

We consider that the question generation under question type C is as follows. At first, each retrieved training search results D_c can be generated under the constraint of question type C according to the distribution $P(D_c \mid C)$. Then, the question is generated according to the distribution $P(Q \mid D_c, C)$.

The semantically related feature model generates the probability of the query $P(Q \mid C)$ over all training texts related to question type C. The model is formulated as:

$$P(Q \mid C) = \sum_{D_C} P(D_C \mid C) P(Q \mid D_C, C). \tag{3}$$

To make the model of Equation (3) tractable, we assume that the probability $P(D_C \mid C)$ is uniform for each D_C under C. Additionally, assume that given D_C, Q and C are independent of each other, thus the probability $P(Q \mid D_C, C)$ can be simplified to the following:

$$P(Q \mid D_C, C) = P(Q \mid D_C). \tag{4}$$

We assume that C is determined based on the two kinds of important features in Q: the question focus Q_F and their semantically related features $Q_R = \{R_V, R_Q, R_N\}$, where R_V, R_Q, and R_N represent the related verbs, related quantifiers, and related nouns, respectively. For convenience of training parameters, we assume these features are mutually independent in this initial work. Therefore, Equation (4) can be decomposed as follows:

$$\begin{aligned} P(Q \mid D_C, C) &= P(Q_F, Q_R \mid D_C) \\ &= P(Q_F \mid D_C) P(Q_R \mid D_C) \\ &= P(Q_F \mid D_C) P(R_V, R_Q, R_N \mid D_C) \\ &= P(Q_F \mid D_C) P(R_V \mid D_C) P(R_Q \mid D_C) P(R_N \mid D_C) \\ &= P(Q_F \mid D_C) \prod_{r_v \in R_V} P(r_v \mid D_C) \prod_{r_q \in R_Q} P(r_q \mid D_C) \prod_{r_n \in R_N} P(r_n \mid D_C). \end{aligned} \tag{5}$$

We substitute Equation (5) into Equation (3), and then get the final form:

$$P(Q \mid C) = \sum_{D_C} \left[P(D_C \mid C) P(Q_F \mid D_C) \prod_{r_v \in R_V} P(r_v \mid D_C) \prod_{r_q \in R_Q} P(r_q \mid D_C) \prod_{r_n \in R_N} P(r_n \mid D_C) \right]. \tag{6}$$

Finally, we consider that Q_F and the semantically related features in R_V, R_Q, and R_N have different importance for supporting the determination of question types. Therefore, we may give QFs and each kind of semantically related features different weights in the parameter estimation.

4 Experiments

4.1 Data Sets

The data set of 680 Chinese questions used in this work are collected from two sources: 400 Chinese questions from the NTCIR-5 CLQA Task and 280 translated Chinese questions for a few rare question types from TREC[3] 2005, 2004, 2002, and 2001 English questions. They are classified into six different question types: PERSON, LOCATION, ORGANIZATION, NUMBER, DATE, and ARTIFACT. We use the

[3] TREC: http://trec.nist.gov/

evaluation technique of four-fold cross-validation which divides the whole question set into four equally-sized subsets and perform four different experiments. For evaluation of the LM, one subset is chosen to be the testing set, and the other three subsets are considered the training set. Table 7 shows the distribution of six question types in the training and testing sets. As for evaluation of the proposed SRFM, the training and testing set are the same with the LM, but in the training process, we only take the answers of questions in the training set as the seed terms for retrieving Web search results and learning QFs and semantically related features (see Section 3.3.1).

Table 7. Distribution of six question types in the training and testing sets

Question Type	Training Set	Testing Set
PERSON	90	30
LOCATION	90	30
ORGANIZATION	60	20
NUMBER	90	30
DATE	90	30
ARTIFACT	90	30

4.2 Results

We conduct the following experiments to realize the effectiveness of our proposed SRFM and compare classification performance with the LM. Basically, these two models have different training process. The LM using questions as training data can easily learn reliable probabilities for QFs of questions. However, for the unseen (untrained) QFs, the SRFM using the seed terms (question answers) to collect large number of Web search results can utilize the semantically related features (related verbs, related quantifiers, and related nouns) to support the lack of QFs. To observe the effectiveness of semantically related features to support the lack of QFs, we thus divided the testing data into three testing classes: the first class, Unneeded_Question_Focus, is that test questions can be directly classified according the basic classification rules, such as the questions containing interrogative words "誰" (who) or "哪裡" (where) (see Section 3.1). The second class, Trained_Question_Focus, is that the QFs of test questions have been trained by the LM. The third class, Untrained_Question_Focus, is that the QFs of testing questions have not been trained by the LM. The number of questions in the three classes is listed in Table 8. Please note that the classification accuracy for the testing class Unneeded_Question_Focus is about 98%. Then, the second and third testing classes are particularly used to compare classification performance for these two models.

Some preliminary results are shown in Table 9. For the testing class Trained_Question_Focus, the LM (word_unigram) (71%) performs better than the SRFM (QF, RV, RQ, RN) (59%). The major reason might be that the training questions are suitable to train a good bigram LM, but currently our collected training data may contain too much noises to train a good SRFM. For the testing class Untrained_Question_Focus, we can see that the SRFM (QF, RV, RQ, RN) (54%)

performs better than the LM (word_unigram) (49%). It shows that our hypothesis of supporting untrained QFs with the related verbs, related quantifiers, and related nouns should be reasonable. Some detailed results on all six question types are shown in Table 10. According to our analyses, for the testing class Untrained_Question_Focus, our SRFM performs better than the LM for types LOCATION and ORGANIZATION, but is worse than the LM for types PERSON and ARTIFACT.

Table 8. The total number of questions in the three testing classes

	Testing Class		
	Unneeded Question Focus	Trained Question Focus	Untrained Question Focus
Number of Questions	327	216	137

Table 9. Classification accuracy for the LM and the SRFM, where QF, RV, RQ, RN indicated question focus, related verb, related quantifier, and related noun

Model	Accuracy	
	Trained Question Focus	Untrained Question Focus
LM (word_bigram)	**153/216=71%**	67/137=49%
SRFM (only QF)	137/216=63%	60/137=44%
SRFM (only RV)	74/216=34%	37/137=27%
SRFM (only RQ)	40/216=19%	35/137=26%
SRFM (only RN)	46/216=21%	31/137=23%
SRFM (QF, RV, RQ, RN)	127/216=59%	**74/137=54%**

Table 10. The detailed analyses about classification accuracy for the bigram LM and the SRFM with uniform weighting scheme (QF, RV, RQ, RN)

Question Type	Accuracy			
	Trained Question Focus		Untrained Question Focus	
	SRFM	LM	SRFM	LM
PERSON	11/19=58%	**12/19=63%**	7/13=54%	**8/13=62%**
LOCATION	53/85=62%	**68/85=80%**	**10/14=71%**	6/14=43%
ORGANIZATION	**37/53=70%**	33/53=62%	**17/27=63%**	12/27=44%
NUMBER	**2/2=100%**	0/2=0%	2/4=50%	2/4=50%
DATE	5/16=31%	**13/16=81%**	0/0=0%	0/0=0%
ARTIFACT	19/41=46%	**27/41=66%**	38/79=48%	**39/79=49%**
Total	127/216=59%	153/216=71%	**74/137=54%**	67/137=49%

Table 11 shows some examples of question classification in the testing class Trained_Question_Focus to explain why some classifications are effective based on the LM or SRFM. For the first test question, the LM is effective since the QF "國家"

(country) has been trained on the training questions, but the SRFM is ineffective because the QF "國家" (country) was trained among different question types, and there are not enough semantically related features in this short question to support the QF. For the second question, the SRFM is still ineffective. Although the QF "時候" (time) has been trained by the SRFM, the semantically related feature "槍擊案" (shooting), which has not been trained in type DATE but trained in other question types, causes noise to make the classification failed. For the third question, the SRFM model is effective due to that the QF "公司" (company) and the semantically related feature the RV "推出" (provide) is closely related to type ORGANIZATION, and has been trained by the SRFM. For the last question (with long length), the QF "金額" (amount of money) and the RV "賠償" (pay out) have been trained by the SRFM and thus make the classification succeed, whereas the LM suffers from a difficult problem that is caused by too much noise on long questions.

In Table 12, some results of question classification are demonstrated for the testing class Untrained_Question_Focus. The first question contains the QF "河"(river), the RV "整治" (renovate) and the RQ "條" (no English equivalent), which are related to type LOCATION, and trained by the SRFM, therefore it is classified correctly. On the contrary, the LM is not effective since all the features were not trained in training questions. For the second question, the QF "企業" (company) has been trained by the SRFM, but the QF was also not trained by the LM. However, the third question with the

Table 11. Some results of question classification in the testing class Trained_Question_Focus, where 'R' indicates the right classification, and 'W' the wrong classification

Question Type	Question	SRFM	LM
LOCATION	世界上最小的國家是什麼? Which country is the world's smallest country?	W	R
DATE	金凱爾校園槍擊案發生在什麼時候? When did the Kip Kinkel school shooting occur?	W	R
ORGANIZATION	率先推出台灣手機保單的公司是哪個? Which company provides the first cell phone insurance program in Taiwan?	R	W
NUMBER	請問2000年7月佛羅里達法院判決包括飛利浦莫里斯在內的美國五大煙草公司應賠償該州共同利害關係人的賠償金額? How much money has the Miami jury ordered America's big cigarette manufacturers including Philips Morris to pay out for knowingly causing smoking related illnesses in 2000?	R	W

QF "醫師"(doctor) trained by the SRFM is classified incorrectly because the QF and some semantically related features have also been trained in other question types and cause incorrect noise. Oppositely, the QF has not been trained by the LM, but because some of the semantically related features in this question have been trained, thus the question is classified correctly using the LM.

Table 12. Some results of question classification in the testing class Untrained_Question_Focus, where 'R' indicates the right classification, and 'W' the wrong classification

Question Type	Question	SRFM	LM
LOCATION	請問高雄哪一條河原本為臭水，現已整治成功? Which river in Kaohsiung used to be stinking, but has been renovated successfully?	R	W
ORGANIZATION	請問美國史上最大宗的企業破產事件為哪一家企業? What is largest company bankruptcy case in the US history?	R	W
PERSON	請問全球首位進行心臟移植手術的南非醫師是? Who is the first doctor, a South African, to perform heart transplant?	W	R

4.3 Discussions

According to the above analyses on our experimental results, we can see that the LM is effective to determine question type for the questions with trained QFs from the training questions. However, the LM is not appropriate to handle the classification of questions with untrained QFs, but our proposed SRFM performs better than the LM by employing the QFs and their semantically related features trained from Web search results. In fact, we think that these two models are complementary, and it is worthy to investigate an integrated technique to improve both models in the future.

5 Conclusion

We have presented a novel approach to learning QFs and semantically related features from Web search results for Chinese question classification. We have also demonstrated a simple technique to collect training data from Web search results by starting with a few seed question answers, and to extract QFs and semantically related features from the collected training data. Our contribution in this work is to mitigate the problems of lacking sufficient training data while training a classifier by using machine learning techniques. Our experimental results show that for the testing questions with untrained QFs using our proposed SRFM, the classification accuracy can be increased from 49% (LM) to 54%.

In fact, the techniques we propose in this paper are more broadly applicable. Our collected Web search results might be exploited to train better LMs and SVM classifiers, or even semantically related features would be beneficial to the SVM classifiers which always require good features for training. We are trying to propose a good bootstrapping technique to effectively collect more and more training data from the Web for various question types, and some new methods to improve our SRFM.

Acknowledgments. We would like to thank Yi-Che Chan for his support to collect training data and test data. We are also grateful to Dr. Zhao-Ming Gao for helpful remarks and suggestions.

References

1. Dan Moldovan, Marius Pasca, Sanda Harabagiu, Mihai Surdeanu: Performance Issues and Error Analysis in an Open-Domain Question Answering System, ACM Transactions on Information systems (2003)
2. Wei Li: Question Classification Using Language Modeling, CIIR Technical Report (2002)
3. Xin Li, Dan Roth: Learning Question Classifiers, COLING 2002 (2002)
4. Min-Yuh Day, Cheng-Wei Lee, Shih-Hung Wu, Chorng-Shyong Ong, Wen-Lian Hsu: An Integrated Knowledge-based and Machine Learning Approach for Chinese Question Classification, IEEE NLPKE2005 (2005)
5. Thamar Solorio, Manuel Perez-Coutino, Manuel Montes-y-Gomez, Luis Villasenor-Pineda, Aurelio Lopez-Lopez: A Language Independent Method for Question Classification, CLING 2004 (2004)
6. Jun Suzuki, Hirotoshi Taira, Yutaka Sasaki, and Eisaku Maeda: *Question Classification using HDAG Kernel*, ACL 2003 Workshop on Multilingual Summarization and Question Answering (2003)
7. Dell Zhang, Wee Sun Lee: Question Classification using Support Vector Machines, ACM SIGIR2003 (2003)
8. Eric Brill, Susan Dumais and Michele Banko: An analysis of the Ask MSR question-answering system, Proceedings of 2002 Conference on Empirical Methods in Natural Language Processing (2002)
9. Deepak Ravichandran and Eduard Hovy: Learning surface text patterns for a question answering system, Association for Computational Linguistics Conference (ACL) (2002)
10. Dan Moldovan, Sanda Harabagiu, Marius Pasca, Rada Mihalcea, Richard Goodrum, Roxana Gîrju, and Vasile Rus: Lasso: A Tool for Surfing the Answer Net, Proceedings of the 8th TExt Retrieval Conference (TREC-8) (1999) 175-183

Improving the Robustness to Recognition Errors in Speech Input Question Answering

Hideki Tsutsui, Toshihiko Manabe, Mika Fukui, Tetsuya Sakai, Hiroko Fujii, and Koji Urata

Knowledge Media Laboratory, Corporate R&D Center, TOSHIBA Corp.
Kawasaki 212-8582, Japan
{hideki.tsutsui, toshihiko.manabe, mika.fukui, tetsuya.sakai, hiroko.fujii, kouji.urata}@toshiba.co.jp

Abstract. In our previous work, we developed a prototype of a speech-input help system for home appliances such as digital cameras and microwave ovens. Given a factoid question, the system performs textual question answering using the manuals as the knowledge source. Whereas, given a HOW question, it retrieves and plays a demonstration video. However, our first prototype suffered from speech recognition errors, especially when the Japanese interrogative phrases in factoid questions were misrecognized. We therefore propose a method for solving this problem, which complements a speech query transcript with an interrogative phrase selected from a pre-determined list. The selection process first narrows down candidate phrases based on co-occurrences within the manual text, and then computes the similarity between each candidate and the query transcript in terms of pronunciation. Our method improves the Mean Reciprocal Rank of top three answers from 0.429 to 0.597 for factoid questions.

1 Introduction

In recent years, home appliances are becoming increasingly difficult for the user to handle, due to advanced features such as network connectivity. Given the versatility of the appliances, it is almost impossible for the user to take advantage of every available feature. Manuals that accompany a product are supposed to contain all the information necessary for the user. However, in reality, customer inquiries at a call center are often to do with a specific passage within a manual. That is, reading the manuals does not always solve the user's problem at hand. These problems will probably become serious with the advent of more and more complex digital products.

In order to tackle the above problems, we previously reported on a prototype help system for home appliances that combines multimodal knowledge search and textual question answering [17]. The multimodal knowledge search subsystem retrieves text-annotated video contents for dealing with the user's HOW questions, while the question answering subsystem returns a short, exact answer in response to factoid questions (Fig. 1). Our prototype help system allows speech input so that the user can consult the system in any situation, e.g., when his hands are occupied.

H.T. Ng et al. (Eds.): AIRS 2006, LNCS 4182, pp. 297 – 312, 2006.

However, this meant that the retrieval effectiveness of the system depended heavily on speech recognition accuracy. Our initial system seriously suffered from recognition errors especially for factoid questions, because recognition errors of interrogative phrases such as “How many grams” and “How many meters” were fatal for question classification.

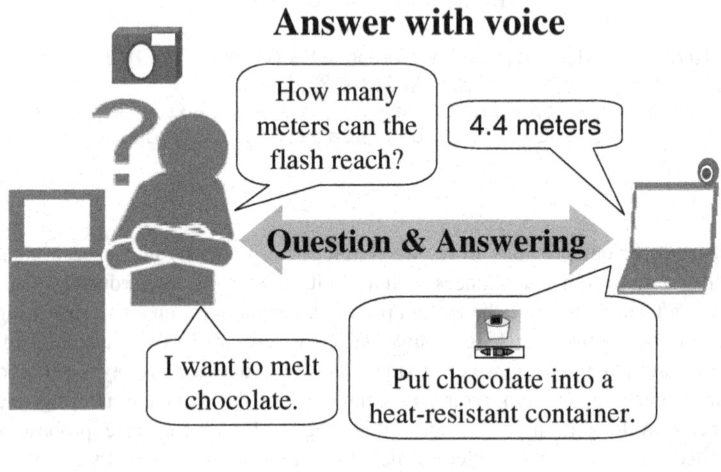

Fig. 1. A multimodal help system based on question answering technology

The objective of this paper is to make the above question answering subsystem more robust to speech recognition errors. To this end, we complement the speech recognition transcript with interrogative phrases that help the system classify questions accurately. These interrogative phrases are selected from a pre-determined list by computing the edit-distance between each list entry and each substring of the transcript, and also by examining within-document co-occurrences between terms and interrogative phrases.

The remainder of this paper is organized as follows. Section 2 discusses related works. Section 3 provides an overview of our multimodal help system, and Section 4 analyses the effect of speech recognition errors on the system. Section 5 proposes a method to overcome the above problem and Section 6 reports on an evaluation experiment. Finally, Section 7 concludes this paper.

2 Related Work

Question Answering (QA) technology is attracting worldwide attention, as exemplified by the English TREC QA Track [18], the European QA@CLEF Track [10], and the Japanese NTCIR QAC Track [4]. These efforts not only tackle the problem of providing an accurate answer to one-off, factoid questions, but also that of handling question series/context as well as definitional questions. A typical QA system consists of at least three modules: Question classification (or answer type identification), named entity recognition for tagging candidate answer strings within the knowledge source documents, and document/passage retrieval for selecting documents that are likely to contain a good answer to the given question.

While the aforementioned efforts focus on textual output in response to a textual input, some researches have pursued speech input information access techniques. Barnett et al. [1] performed comparative experiments related to search-driven retrieval, where an existing speech recognition system was used as an input interface for the INQUERY text retrieval system. They used as test inputs 35 queries collected from the TREC 101-135 topics, dictated by a single male speaker. Crestani et al. [2] also used the above 35 queries and showed that conventional relevance feedback techniques marginally improved the accuracy for speech-driven text retrieval. These studies focused solely on improving text retrieval methods and did not address problems of improving speech recognition accuracy, and speech recognition and text retrieval modules were fundamentally independent and were simply connected by way of an input/output protocol. However, when speech input is used, it is known that retrieval performance seriously suffers from speech recognition errors.

Nishizaki et al. [13] developed a speech input retrieval system for spoken documents of news, for which they alleviated the effect of recognition errors by considering alternative keyword candidates in addition to the best candidate obtained through speech recognition, and the relationships between keywords. Hori et al. [5] reported on a voice-activated question answering system for large-scale corpora, such as a newspaper articles. In this research, they performed syntactic analysis of the question and generated dialogue sentences for disambiguation it. Fujii et al. [3] exploited the vocabulary in the target documents for handling unknown words in a spoken query for speech input document retrieval. Our approach differs from this in that it automatically complements an interrogative expression to the query transcript for improving the question classification accuracy in question answering.

Kiyota et al. [7] developed an interactive, speech input online help system for Microsoft Windows, which can answer questions in natural language and disambiguate questions through a dialogue. Their main target is "asking back to disambiguate a question". When a user's query contains ambiguity, the system generates a query statement to disambiguate it and asks back to the user. Also in order to eliminate misunderstandings caused by speech recognition errors, asking back is applied. The system asks back to the user the part which carried out erroneous recognition. And the system finds erroneous recognition by calculating relevance score based on perplexity. Moreover, the system calculates significance for retrieval using N-best candidates of speech recognition. When search results differ greatly by each candidate, the system asks the user to select the correct candidate. However, this system focuses on What/How/Symptom questions. In contrast, while our system currently does not have a dialogue/interactive feature, it can handle both HOW questions and factoid questions. Moreover, our system is for home appliance.

3 System Overview

The configuration of our prototype help system is shown in Fig. 2. As shown, the system consists of a help interface on a client PC, a help manager, a video search engine, a question answering engine and a media server on the server side. The system's outward appearance is shown in Fig. 3. The help interface recognizes a speech-input query and either outputs an exact answer through speech synthesis, or retrieves and displays a video content. The help manager is responsible for classifying questions and for calling either the video search engine or the question answering engine (Table 1). The help manager extracts interrogative phrases of question by a rule base. When a query contains an interrogative phrase which asks a number (time, quantity, etc), a name or place of a control unit, the query is classified as a factoid question, and the other queries are classified as HOW question. For a factoid question, if the interrogative phrase is misrecognized, the query is classified as HOW question and the system cannot reply to the query correctly.

Table 1. Question type

question type	search engine	Example
Factoid	question answering search engine	ask a number(time, quantity, etc.), a name or place of a control unit when, how many, how long, how many quantity, where, which, what e.g., which is the key to choose the automation menu?
HOW	video search engine	operating instructions how, operating instructions, want to, what should I do, the way to, I cannot e.g., I would like to photograph a night view.

The video search engine searches a database comprising text-annotated virtual streams that have been created by extracting and concatenating relevant segments from video, audio and image files using MPEG7 [12]. It adopts the vector space model for information retrieval, and retrieves three items in response to a HOW question [16]. If the top-ranked candidate is a video stream, it is displayed on the help interface as shown in Fig. 4, so that the user can start playing it. For the second and the third candidates, only their titles are shown at the bottom of the window [8].

The question answering engine extracts answer candidates such as quantity, the names of the switches on the control panel of the appliance, time expressions and so on off-line. It stores these candidates together with the source documents (i.e., supporting documents for the answer strings) [15] [6]. The help interface outputs three answer candidates in decreasing order of answer confidence values, and utters only the first candidate via speech synthesis (Fig. 5). In our prototype, we adopted LaLaVoice2001 [9] for continuous speech recognition for query input and speech synthesis for answer output. LaLaVoice2001 is a large vocabulary continuous speech recognition and speech synthesis system developed by Toshiba. It recognizes input voice based on the statistical language model which estimates the linguistic validity of a sentence and the acoustic model expressing the sound feature of the voice [11] [14]. It is for speaker independent speech recognition and voice enrollment is not required.

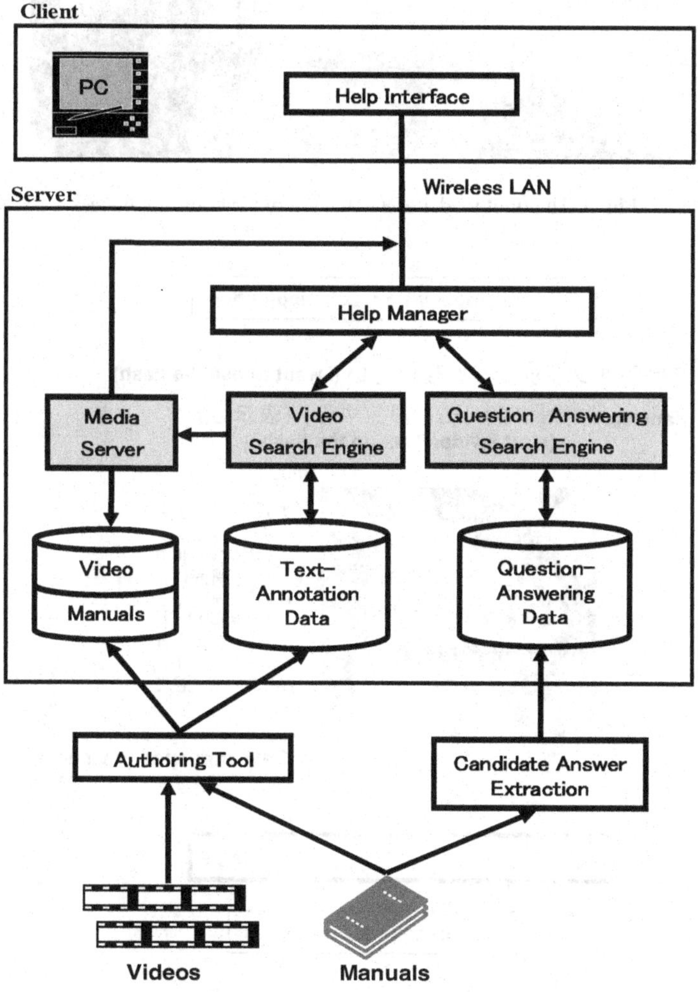

Fig. 2. Configuration of our prototype help system

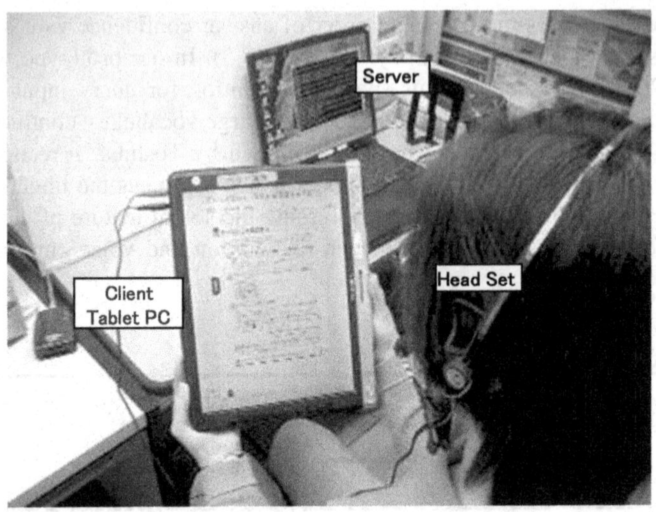

Fig. 3. The pilot model of a help system for home appliances

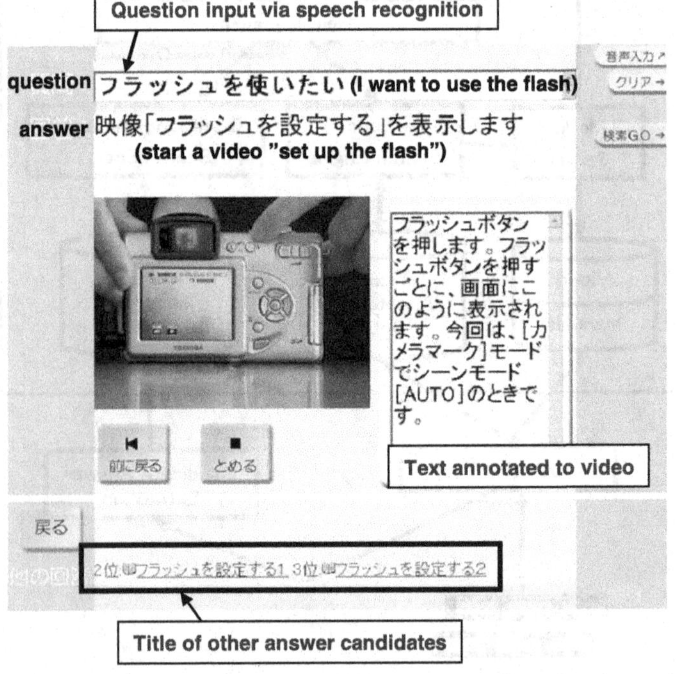

Fig. 4. The screen for reproducing a video stream

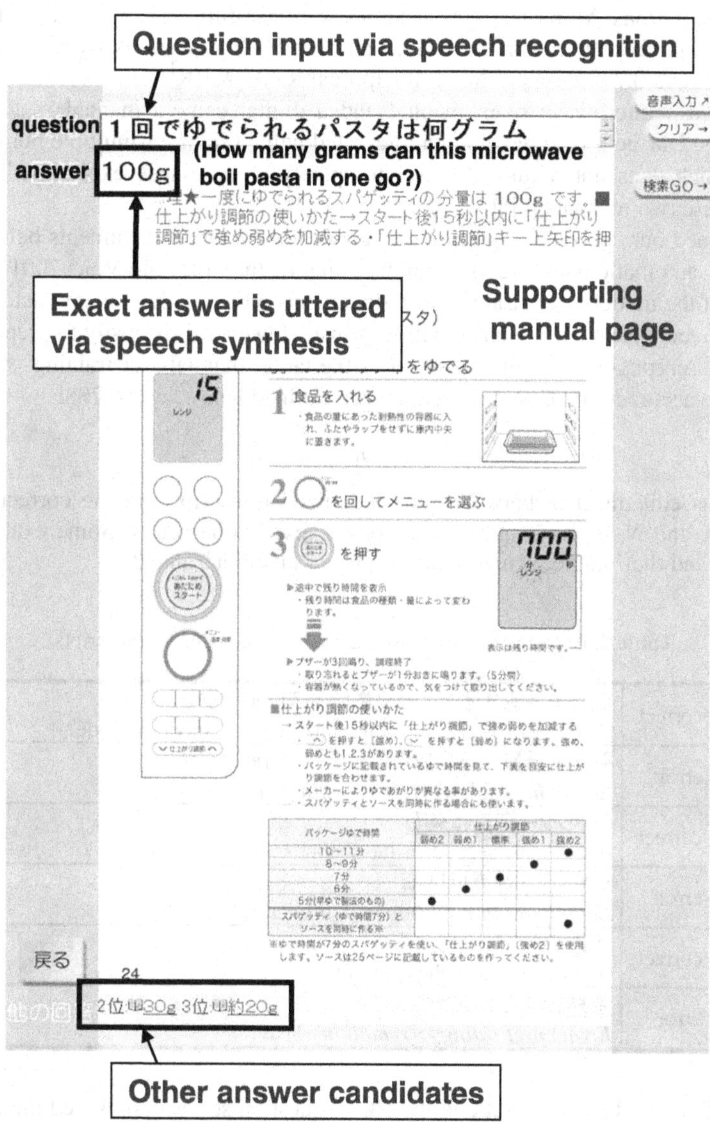

Fig. 5. Help interface outputs search results

4 Effect of Speech Recognition Errors

We first conducted an experiment to investigate the effect on speech recognition errors on the performance of our initial prototype system. The experiment involved ten subjects and 96 real questions collected from a Toshiba call center, yielding 960 trials altogether. We tried to make the number of factoid questions and that of HOW questions roughly equal, and as a result the question set actually contained 40 factoid and

56 HOW questions. Moreover, the questions were carefully selected so that the average system performances for the factoid and the HOW questions were comparable, namely, around 0.7 in terms of Mean Reciprocal Rank (MRR[1]).

We used a microwave oven manual and a digital camera manual as the target documents. Our goal is to develop an integrated help system for multiple home appliances, which does not require the user to specify an appliance explicitly. Thus we treated these two manuals jointly.

We turned our speech recognition dictionary for the target documents beforehand, using the automatic word registration/learning feature of LaLaVoice2001. In this training, if the models were trained with documents which include many factoid questions, the recognition performance will improve. However, it is costly to prepare such an external corpus. We therefore used only the target manuals for training. After this tuning, the accuracy of our speech recognizer, defined below, was 0.788:

$$1 - D_l / N_l \quad (1)$$

Here, D_l is edit distance between the recognized transcript and the correct text in characters, and N_l is the length of the correct text in characters. Some examples of questions and their misrecognized transcripts are given in Table 2.

Table 2. Example of questions and their misrecognized transcripts

Factoid	correct	動画を撮影するときの画像サイズは**何ピクセル** (How many pixels does this camera record in video mode?)
	error	動画を撮影するときの画像サイズ**阿南ピクセル** ***Error: how many pixels -> south pixels***
Factoid	correct	シャッター速度は**何秒** (How many seconds is the shutter speed of this camera?)
	error	シャッター速度は**難病** ***Error: how many seconds -> intractable disease***
HOW	correct	**予熱**の設定はどうやるの (How can I set preheating of this oven?)
	error	**余熱**の設定はどうやるの ***Error: preheating -> remaining heat***

For both factoid question set and the HOW question set, we compared the MRR of speech input to that of keyboard input (i.e. when the system is error-free), based on the top three candidate answers returned by the system for each question. Table 3 summarizes the results.

For a factoid question, the correct answer is an exact answer string, while for a HOW question, the correct answer is either a relevant video content or a page from a manual. The returned answer was counted as correct only if question classification was successful and the most appropriate answer medium was selected. For example,

[1] MRR (Mean Reciprocal Rank): The results were evaluated using the score called "mean reciprocal rank" (MRR) defined as the average of "reciprocal ranks" for all queries. The reciprocal rank of a query is calculated as 1/r where r is the rank of first correct answer [15].

Table 3. Summarizes the results

	(1)	(2)	(3) = (2) / (1)
Factoid question	0.742	0.429	57.8%
HOW question	0.732	0.625	85.4%

(1) keyboard input
(2) speech input
(3) rate of retrieval performance degradation

when a factoid question was misclassified as a HOW question, and a retrieved video content or a manual page actually contained a correct exact answer string somewhere in it, we regarded this system output as a failure.

From the table, we can observe that the performance degradation for the factoid question set is much more serious than that for the HOW question set. The HOW question set does not suffer very much from speech recognition errors because the video search engine is a simple bag-of-words system based on the vector space model. In general, there are several terms in a bag-of words query, so misrecognizing one or two terms is not necessarily fatal. On the other hand, the question answering engine, which we call ASKMi [15], is less robust to speech recognition errors because of its intricate mechanisms. The system consists of question classification, document retrieval and named entity recognition modules, and the question classification module identifies the question type/answer type based on interrogative phrases within a given question. Thus, if interrogative phrases are misrecognized, question type/answer type classification can fail completely.

For both keyboard input and speech input, we investigated the performance of each subcomponent of ASKMi, and the results are shown in Table 4.

Table 4. Accuracy in each step

	(1)	(2)	(3)
keyboard input	1.000	1.000	95.3%
speech input	0.700	0.964	67.1%

Column (1) shows the accuracy of question type identification: that is, whether a factoid question was correctly identified as a factoid question.

Column (2) shows that of answer type identification: that is, given a question correctly identified as a factoid one, whether its answer type (e.g., weight, length etc.) was correctly identified also.

Column (3) shows the final MRR based on the ranked answer list divided by the MRR based on the retrieved list of supporting documents (i.e., documents containing the correct answer strings).

The table shows that the failure of question type identification has a substantial negative effect on the overall system performance. As mentioned earlier, this is because the interrogative phrases within the questions, which are key to the success of question type identification, are often misrecognized by speech recognition. It can also be observed that, once the question type has been determined successfully, answer type identification is highly successful for both keyboard input and speech input, because this process also relies on interrogative phrases.

On the other hand, Column (3) of Table 4 shows that the "MRR ratio" for the speech input case is much lower than that for the keyboard input case. This is because, while answer ranking is directly affected by answer type identification, document retrieval is fairly robust to speech recognition errors in the input question since it relies on query terms rather than interrogative phrases.

Speech recognition errors occur frequently for interrogative phrases than for query terms. This is probably because, while the initial system tuning improves the recognition accuracy for terms that occur in the manuals, it does not improve recognition accuracy for interrogative phrases, since interrogative phrases themselves seldom occur in the manual texts. That is, while expressions such as "100 grams" may occur in the texts, "how many grams" probably never occurs, so speech recognition never "learns" the interrogative phrases.

5 Proposed Method

Our analysis in the previous section showed that the speech recognition errors of interrogative phrases are the main cause of the final performance degradation. We therefore devised a method to alleviate this problem under the assumption that misrecognized text and the correct text are phonetically similar. More specifically, we tried to restore the interrogative phrases lost in speech recognition as follows:

Step1: Make a list of possible interrogative phrases beforehand;
Step2: For every question that was not classified as a factoid question, narrow down the above list of phrases based on co-occurrences of the phrases and the query terms in the target documents;
Step3: For every phrase in the filtered list, compute the edit distance between its pronunciation and that of query transcript substrings. Select the phrase with minimum edit distance that is below a pre-determined threshold, and concatenate it to the original query transcript.

An example of this procedure is shown in Fig. 6.

Here, the correct interrogative phrase in the question "何ピクセル(how many pixels)" has been misrecognized as "南ピクセル(south pixels)", and as a result the question is not identified as a factoid question. In this case, our proposed method first narrows down the list of interrogative phrases to those which co-occur with query terms "video", "image and "size", based on a co-occurrence table of interrogative phrases and terms. Then, it computes the edit distance between each candidate interrogative phrase and every substring of the question transcript and restores the correct interrogative phrase "how many pixels". In this particular case, the edit distance is zero since "how many pixels" and "south pixels" are phonetically identical in Japanese.

Table 5 shows a nonexhaustive list of our interrogative phrases which we used Step 1. We obtained this list by automatically extractive phrases from ASKMi's pattern matching rules for question type / answer type identification. The list actually contains 113 interrogative phrases.

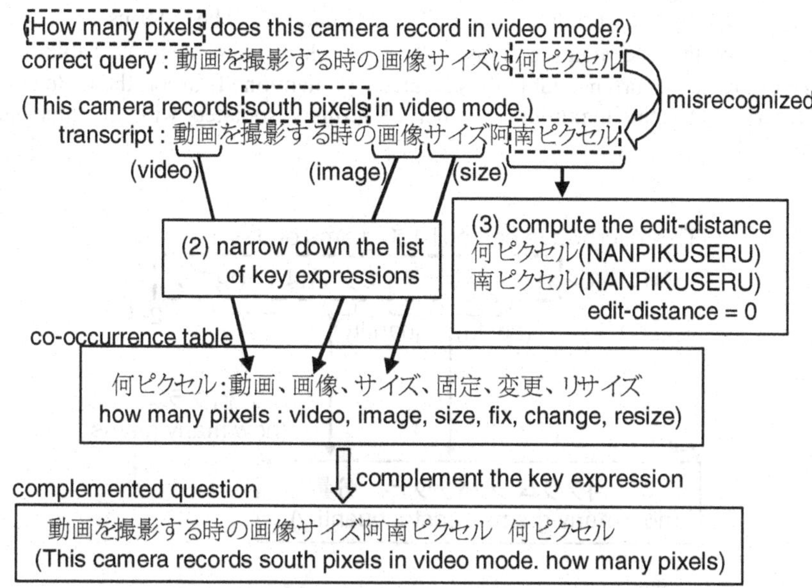

Fig. 6. Example of a complement of key expression

Table 5. Example of key expressions

何回(how many times)	何グラム(how many grams)	何枚(how many sheets)
何倍(what times)	何秒(how many seconds)	何分(how many minutes)
何時間(how many hours)	何週間(how many weeks)	何年(how many years)
何ミリ (how many millimeters)	何センチ (how many centimeters)	何ピクセル (how many pixels)

The above mentioned co-occurrence table was created in advance as follows:

Step 2-1 From the target documents, extract all answer candidates that may be used in response to a factoid question;
Step 2-2 For every answer candidate thus obtained, do Steps 2-3 and 2-4;
Step 2-3 Select an interrogative phrase that may be used in a question to which the answer candidate may be an appropriate answer;
Step 2-4 From a fixed-size text snippet that surrounds the answer candidate, extract all terms (e.g., morphemes that are nouns), and register them with the aforementioned interrogative phrase onto the co-occurrence table.

An example of the above procedure is shown in Fig. 7.

For this example, Step 2-1 extracts the string “100 g” as an answer candidate, which consists of a number and a unit. For this answer, Step 2-3 selects “何グラム (how many grams)” from the list of possible interrogative phrases. Then, Step 2-4 obtains a text snippet "れるスパゲッティの分量は100gです。(the quantity of pasta... is 100 g.)" and extracts “pasta” and “quantity”. These terms are registered together with the interrogative phrase "何グラム(how many grams)". Table 6 shows a part of our co-occurrence table thus created. As mentioned earlier, the table contains 113 entries, and the maximum number of terms associated with an interrogative phrase is 49.

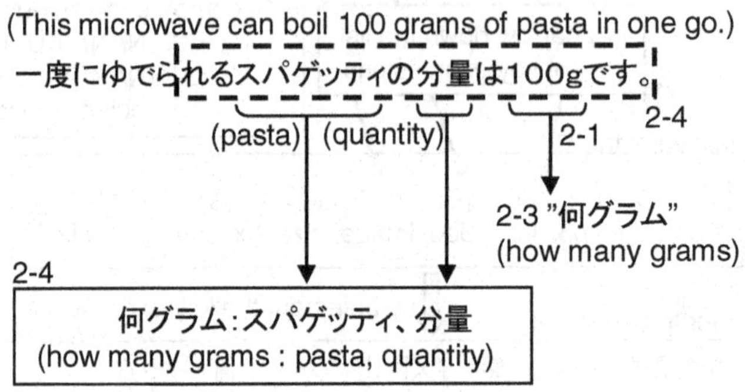

Fig. 7. Example of collocation table creation procedure

Table 6. Example of collocation table

何回 (how many times)	連┌ ,動き,被┌ 体,赤目,┌ 減,フラッシュ,┌ 光,... (continuance,move,target,red eyes,reduce,flash,emit)
何倍 (what times)	ズ┌ム,被┌ 体,光┌ ,デジタル,距離,最大,┌ 大,┌ 素,... (zoom,target,optical,digital,distance,max,expand,pixel)
何枚 (how many sheets)	ソフトウェア,ケ┌ブル,撮影,間隔,枚┌ ,再生,モ┌ド,... (software,cable,photo,interval,sheets,playback,mode)
何杯 (how many cups)	一度,分量,解凍,キ┌,中華,バタ┌,ロ┌ル,調理,パン,... (once,quantity,thaw,key,chinese,butter,roll,cook,bread)
何秒 (how many seconds)	撮影,間隔,枚┌ ,ビュ┌,┌ 像,表示,プレビュ┌,┌ 光,... (photo,interval,sheets,view,picture,display,preview,emit)

As mentioned earlier, the edit distance computation is computed after narrowing down the list of interrogative phrases based on co-occurrences with query terms. This is beneficial from two viewpoints. Firstly, this enhances the efficiency of our algorithm, interrogative phrases are often very short, e.g., "何度 (NANDO: how many times)" so the edit distance computation between question substrings the phrases are generally time-consuming. Secondly, our method has a positive effect on the accuracy of phrase recovery, as it can filter out homonyms (e.g., " 何℃ (NANDO: what degrees)" as opposed to "何度 (NANDO: how many times)") from the list based on the vocabulary in the co-occurrence table. As for interrogative expressions that do not contain specific units (e.g., "長さは (how long)", "どのくらい (how many)", we carry out the edit distance computation for any speech input question, regardless of the question terms.

As mentioned earlier, Step 3 computes the edit distance between the pronunciation (represented by a romaji string) of each candidate interrogative phrase and that of a substring of the query transcript. Since raw edit distances would favor short phrases, we normalize them as follows:

$$1 - D_r / N_r \qquad (2)$$

Here, D_r is the edit distance between a candidate interrogative phrase and a query transcript substring (in terms of romaji), and N_r is the length of the candidate in romaji characters.

6 Evaluation

Using the experimental environment we described in Section 3, we compared the final MRR performance with and without the question complementation using our proposed algorithm. Table 7 compares the final MRR performances of the error-free, keyboard input system, the speech input system without complementation, and that with complementation. The results from Tables 3 and 7 are visualized in Fig. 8. It can be observed that our question complementation method successfully improved the final MRR performance from 0.429 to 0.597.

Table 7. Retrieval performance by complemented query

	(1)	(2)	(3)
Factoid question	0.742	0.429	0.597

(1) keyboard input
(2) speech input
(3) speech input with complementation

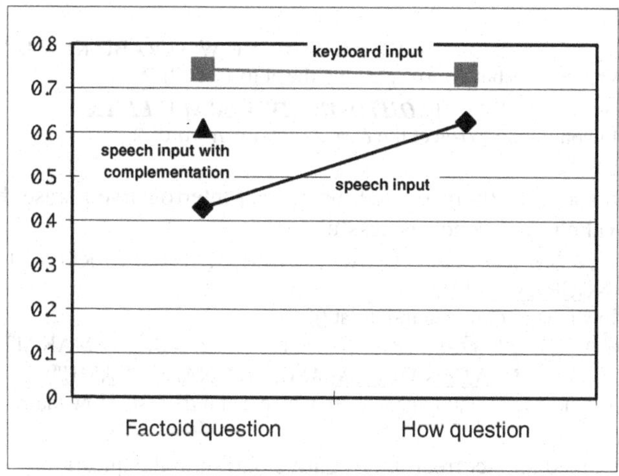

Fig. 8. Result of Experiments

Table 8. Performance of each processing stop

	(1)	(2)	(3)
keyboard input	1.000	1.000	95.3%
speech input	0.700	0.964	67.1%
speech input with complementation	0.923	0.959	92.6%

We also investigated the performance of each question answering component, and the results are shown in Table 8, which subsumes the results we already discussed in Table 4.

This table also shows that the accuracy of question classification has improved significantly by question complementation. Thus we successfully raised the performance level for factoid questions so that it is comparable to that for HOW questions.

We finally conducted a failure analysis of the proposed question complementation method. Of 152 questions which originally suffered from speech recognition errors, as many as 80 questions were not improved by our method. We have classified the failures as follows.

- For 35 queries,
  Our system did not complement an interrogative phrase because the interrogative phrase of question was recognized correctly and the question was classified as a factoid question correctly. Nevertheless, the system could not search a correct answer because other query terms were misrecognized and supporting documents were searched incorrectly.
  e.g. "カラ┌ きするときの時間は (KARAYAKI SURU TOKI NO JIKAN WA) "
  (How long should I run it empty?)
  → "空息するときの時間は (***KARAIKI*** SURU TOKI NO JIKAN WA")
  (How long should I breathe it empty?)
- For 28 queries,
  Our method failed to complement an interrogative phrase due to a misrecognized query term.
  e.g. "オ┌トメニュ┌を選ぶキ┌は (OHTOMENYU WO ERABU KI WA) ") "
  (Which is the button for selecting the auto-menu?)
  → "オ┌トメには選ぶ┌ は (***OHTOME NIWA*** ERABU ***KI WA***")
  (Do you intend to select the *otome* [unknown word]?)
- For 8 queries,
  Our method successfully recovered the correct interrogative phrase, but the document retrieval phase was not successful.
  e.g. "生解凍できる分量は何グラムまで (NAMAKAITOU DEKIRU BUNRYOU WA NANGURAMU MADE")
  (How many grams can it defrost?)
  → "生解凍できる分量判┌ グラムまでに 何グラム (NAMAKAITOU DEKIRU BUNRYOU ***HANDAN*** GURAM ***MADENI*** NANGURAMU")
  (By the quantity judgment gram which can it defrost? How many grams)
- For 9 queries,
  Our method added an incorrect interrogative phrase to the question.

e.g. " ·回でゆでられるパスタは何グラム (IKKAI DE YUDERARERU PASUTA WA NANGURAMU")
(How many grams of pasta can it boil in one go?)
→ " ·回で┌ られるパスタは玩具ラグ 何分 (IKKAI DE ***YUZURARERU*** PASUTA WA ***GANGURAGU*** NANPUN")
(The pasta yielded at once is a toy lug. How many minutes)

Here, misrecognised query terms are in bold italics, original interrogative phrases are underlined with a straight solid line, and the complemented interrogative phrases are underlined with a wavy dotted line.

7 Conclusions

This paper proposed and evaluated a method for complementing a speech input query for improving the robustness of our help system for home appliances to speech recognition errors. To enhance the chance of success at the question type identification step for factoid questions, our method recovers an interrogative phrase based on co-occurrences between answer candidates and terms in the target documents, and adds it to the query transcript. Prior to introducing this method, the accuracy of question type identification via speech input was 70.0% of that for the keyboard input case, but the method raised this number to 92.3%.

Through our experiments and analyses, it became clear to us that a simple combination of question answering and speech recognition does not achieve an expected level of performance, namely, the product of the performances of individual modules. Methods such as the one we proposed are necessary for the system to enjoy a synergy effect. Our future work includes feeding back the results of our question complementation to the speech recognition module, and incorporating spoken dialogues for overcoming the negative effects of speech recognition errors. We would like to combine several different approaches in order to achieve a satisfactory level of the overall system performance.

References

1. J. Barnett, S. Anderson, J. Broglio, M. Singh, R. Hudson, and S. W. Kuo, Experiments in Spoken Queries for Document Retrieval, In Proceedings of Eurospeech97, (1997) 1323-1326
2. F. Crestani, Word recognition errors and relevance feedback in spoken query processing, Proceedings of the Fourth International Conference on Flexible Query Answering Systems, (2000) 267-281.
3. A. Fujii, K. Itou, and T. Ishikawa, Speech-Drive Text Retrieval: Using Target IR Collections for Statistical Language Model Adaptation in Speech Recognition, ACM SIGIR 2001 Workshop on Information Retrieval Techniques for Speech Application, (2001)
4. J. Fukumoto, T. Kato, and F. Masui, Question Answering Challenge (QAC-1): An Evaluation of QA Tasks at the NTCIR Workshop 3, Proceedings of AAAI Spring Symposium: New Directions in Question Answering, (2003) 122-133,
5. C. Hori, T. Hori, H. Isozaki, E. Maeda, S. Katagiri, and S. Furui, Deriving Disambiguous Queries in a Spoken Interactive ODQA System, In Proc. the 2003 IEEE International Conference on Acoustics, Speech, and Signal Processing (ICASSP), (2003) 624-627.

6. Y. Ichimura, Y. Yoshimi, T. Sakai, T. Kokubu, and M. Koyama, The Effect of Japanese Named Entity Extraction and Answer Type Taxonomy on the Performance of a Question Answering System, IEICE Journal vol.J88-D2 No.6, (2005) 1067-1080
7. Y. Kiyota, S. Kurohashi, T. Misu, K. Komatani, T. Kawahara, Dialog Navigator: A Spoken Dialog Q-A System based on Large Text Knowledge Base, Proceedings of 41st Annual Meeting of the Association for Computer Linguistics, (2003) 149-152
8. T. Kokubu, T. Sakai, Y. Saito, H. Tsutsui, T. Manabe, M. Koyama, and H. Fujii, The Relationship between Answer Ranking and User Satisfaction in a Question Answering System, Proceedings of NTCIR-5 Workshop Meeting, (2005) 537-544.
9. LaLaVoice2001, http://www3.toshiba.co.jp/pc/lalavoice/index_j.htm
10. B. Magnini, A. Vallin, C. Ayache, G. Erbach, A. Pe as, M. de Rijke, P. Rocha, K. Simov, and R. Sutcliffe, Overview of the CLEF 2004 Multilingual Question Answering Track, Working Notes of the Workshop of CLEF 2004, (2004) 15-17.
11. Y. Masai, S. Tanaka, and T. Nitta, Speaker-independent keyword recognition based on SMQ/HMM, In Proceedings of International Conference on Spoken Language Processing, (1992) 619-622
12. MPEG7, http://www.itscj.ipsj.or.jp/mpeg7/
13. H. Nishizaki, S. Nakagawa, A System for Retrieving Broadcast News Speech Documents Using Voice Input Keywords and Similarity between Words, Proceedings of ICSLP2000 Vol.3, (2000) 1073-1076
14. T. Nitta and A. Kawamura, Designing a reduced feature-vector set for speech recognition by using KL/GPD competitive training, In Proceedings of the 7th European Conference on Speech Communication and Technology, (1997) 2107-2110
15. T. Sakai, Y. Saito, Y. Ichimura, M. Koyama, T. Kokubu, T. Manabe, ASKMi: A Japanese question answering system based on semantic role analysis, RIAO 2004 Proceedings, (2004) 215-231.
16. M. Suzuki, T. Manabe, K. Sumita, and Y. Nakayama, Customer Support Operation with a Knowledge Sharing System KIDS: An Approach based on Information Extraction and Text Structurization, SCI 2001 Proceedings Vol.7, (2001) 89-96
17. K. Urata, M. Fukui, H. Fujii, M. Suzuki, T. Sakai, Y. Saito, Y. Ichimura and H. Sasaki, A multimodal help system based on question answering technology, IPSJ SIG Technical Reports FI-74-4, (2004) 23-29.
18. EM. Voorhees, Overview of the TREC 2004 Question Answering Track, In Proceedings of the Thirteenth Text REtreival Conference (TREC 2004), 2005

An Adjacency Model for Sentence Ordering in Multi-document Summarization

Yu Nie, Donghong Ji, and Lingpeng Yang

Institute for Infocomm Research, 21 Heng Mui Keng Terrace, Singapore 119613
{ynie, dhji, lpyang}@i2r.a-star.edu.sg

Abstract. In this paper, we proposed a new method named adjacency based ordering to order sentences for summarization tasks. Given a group of sentences to be organized into the summary, connectivity of each pair of sentences is learned from source documents. Then a top-first strategy is implemented to define the sentence ordering. It provides a solution of ordering texts while other information except the source documents is not available. We compared this method with other existing sentence ordering methods. Experiments and evaluations are made on data collection of DUC04. The results show that this method distinctly outperforms other existing sentence ordering methods. Its low input requirement also makes it capable to most summarization and text generation tasks.

1 Introduction

Information overload has created an acute need for multi-document summarization. There have been a lot of works on multi-document summarization [3][5][7][10][11][12]. So far the issue of how to extract information from source documents is the main topic of summarization area. Being the last step of multi-document summarization tasks, sentence ordering attracts less attention up to now. But since a good summary must be fluent and readable to human being, sentence ordering which organizes texts into the final summary can not be ignored.

Sentence ordering is much harder for multi-document summarization than for single-document summarization. The main reason is that unlike single document, multi-documents don't provide a natural order of texts to be the basis of sentence ordering judgment. This is more obvious for sentence extraction based summarization systems.

Chronological ordering [1][2][4] is a generally used method for sentence ordering in multi-document summarization. It orders sentences by published date of source documents or time information within texts. This method is suitable for event based documents. Sometimes it gets good results [4]. But on the one hand, it doesn't work for all kinds of summarization tasks. On the other hand, the chronological information is not always available. Because of these limitations, chronological ordering can only solve part of sentence ordering problems.

Majority ordering is another way of sentence ordering. This method groups sentences to be ordered into different themes or topics of texts in source documents, and the order of sentences is given based on the order of themes. The idea of this

H.T. Ng et al. (Eds.): AIRS 2006, LNCS 4182, pp. 313 – 322, 2006.

method is reasonable since the summary of multi-documents usually covers several topics in source documents to achieve representative, and the theme ordering can suggest sentence ordering somehow. However, there are two challenges for this method. One is how to cluster sentences into topics, and the other is how to order sentences belonging to the same topic. Barzilay et al. [2] combined topic relatedness and chronological ordering together to order sentences. Besides chronological ordering, sentences were also grouped into different themes and ordered by the order of themes learned from source documents. The experiment results show that chronological ordering is not sufficient for sentence ordering, and the results could be improved combining with topic relatedness.

Probabilistic model was also used to order sentences. Lapata [8] ordered sentences based on conditional probabilities of sentence pairs. The condition probabilities of sentence pairs were learned from a training corpus. With condition probability of each sentence pairs, the approximate optimal global ordering was achieved with a simple greedy algorithm. The condition probability of a pair of sentences was calculated by condition probability of feature pairs occurring in the two sentences. The experiment results show that it gets significant improvement compared with randomly sentence ranking. This suggests that without any linguistic knowledge, statistical information does help to improve performance of sentence ordering.

Bollegala et al. [4] combined chronological ordering, probabilistic ordering and topic relatedness ordering together. He used a machine learning approach to learn the way of combination of the three ordering methods. The combined system got better results than any of the three methods.

In this paper, we propose a new sentence ordering method named adjacency-based ordering. It orders sentences with sentence adjacency or connectivity. Sentence adjacency or connectivity between two sentences means how closely they should be put together in a set of summary sentences. Although there is no ordering information in sentence adjacency, an optimal ordering of summary sentences can be derived by use of such information of all sentence pairs with help of the first sentence selection.

We also implemented majority ordering and probability based sentence ordering to make a comparison with our method. The probabilistic model we implemented in this paper is similar to the work of [8], but probabilities of feature pairs are learned from source documents instead of external corpus.

We test our methods on DUC2004 data collection and evaluated our results against manually produced summaries provided by DUC as "ideal" summaries. We also present manually evaluation results.

The remainder of this paper is organized as follows. Section 2 introduces our adjacency-based ordering model, with a comparison with probabilistic model. Chapter 3 introduces our experiment results as well as evaluation metrics and results. In Chapter 4 we present the discussion and conclusion.

2 Methodology

Before we describe our newly proposed adjacency based ordering, the method of majority ordering and probabilistic ordering will be briefly introduced. All thiese three methods can order sentences without any external sources but with the source documents to be summarized. This makes them capable to most summarization tasks.

2.1 Majority Ordering

Majority ordering assumes that sentences in the summary belong to different themes or topics, and the ordering of sentences in the summary can be determined by the occurring sequence of themes in source documents. To define the order of themes, Barzilay et al. [2] presented themes and their relations as a directed graph first. The nodes of the graph are themes and each edge from one node to another denotes the occurring of one theme before another theme in source documents. The weight of each edge is set to be the frequency of the pair occurring in the texts. Each theme is given a weight that equals to the difference between outgoing weights and incoming weights. By finding and removing a theme with the biggest weight in the graph recursively, an ordering of themes is defined.

2.2 Probabilistic Model

The probabilistic model treats the ordering as a task of finding the sentence sequence with the biggest probability [8]. For a sentence sequence T= $S_1\ S_2 \ldots S_n$, with an assumption that the probability of any given sentence is determined only by its previous sentence, the probability of a sentence sequence can be generated given the condition probabilities $P(S_i|S_{i-1})$ of all adjacent sentence pairs of the given sentence sequence. The condition probability $P(S_i|S_{i-1})$ can be further resolved as the product of condition probabilities of feature pairs $P(f_l|f_m)$, where f_l is the feature of S_i, f_m is the feature of S_{i-1}.

By finding the sentence with the biggest condition probability with the previous sentence recursively, an ordering of sentences is defined. A null sentence can be introduced as the beginning of the ordering to get the first sentence found. For more details please refer to [8].

2.3 Adjacency Model

Using conditional probabilities of sentence pairs to order sentences is a natural idea. However, we notice that the learned probability might lose important information presented by the data. Consider examples below:

Example 1:
Source Document = ……ABA……
Example 2:
Source Document 1 = ……AB……
Source Document 2 = ……BA……

Here A and B denote two sentences. Let's assume that A and B are both selected as the summary sentences.

With probabilistic model described in the previous section, since the times of A preceding B equal to the times of B preceding A in both examples, the system will learn that p(A|B) equals to p(B|A). Thus it will get confused when ordering sentences A and B. In other words, text structures in the two examples do not contribute to ordering in probabilistic model. But in fact we can understand intuitively that sentence A and B shall be put adjacently, despite the sequence between them.

Understanding that the conditional probabilities learned from sequences of sentences or features in source documents only focus on feature sequence information but ignore feature adjacent information, we propose the adjacency model which focuses on sentence adjacency instead of feature sequence. Given a group of sentences $\{S_1,\ldots,S_n\}$, for each pair of sentences S_i and S_j, the adjacency of S_i and S_j can be captured by their connectivity defined as below:

$$C_{i,j} = \frac{1}{K * L} \sum_{k,l} C_f(f_{ik}, f_{jl}) \tag{1}$$

Here f_{ik} are the features of sentence S_i, f_{jl} are the features of sentence S_j. C_f is the connectivity function for feature pairs. $C_f(f_{ik} f_{jl})$ is the connectivity of features f_{ik} and f_{jl} in source documents. K and L are numbers of features in sentence S_i and S_j. In general the connectivity of a pair of sentence can be understood as the average connectivity of all feature pairs derived from the sentence pair.

The connectivity of feature pair $C_f(f_i, f_j)$ is defined as below:

$$C_f(f_i, f_j) = \frac{f(f_i, f_j)^2}{f(f_i) f(f_j)} \tag{2}$$

Here *f()* is the frequency function, $f(f_i)$ and $f(f_j)$ respectively denote the frequency of feature f_i and f_j in the source documents, $f(f_i, f_j)$ denotes the frequency of f_i and f_j cooccurring in the source documents within a limited range (say, one or several sentences).

Intuitively, $C_{i,j}$ denotes how close the two sentences S_i and S_j are in the source documents. It can be reasonably inferred that sentences that are close to each other in source documents tend to remain close to each other in the summary produced from the source documents. Notice that with equation (1) and (2) we can get connectivity learned from source documents between any two sentences, not only for sentences occurring in source documents. This helps to implement this model to sentence ordering of any summarization task, not only sentence-extraction based summarization tasks.

We consider that those sentence pairs with bigger $C_{i,j}$ are more likely to be put adjacently in the summary. Notice that connectivity doesn't directly suggest the ordering between two sentences but only the information of how close they are. Given a sentence S_i, the sentence S_j which makes $C_{i,j}$ biggest is the most likely sentence to be put right before or after S_i. To avoid the selection between the two possible positions while ordering, we could always find the next sentence after sentences at previous positions are all found already. Then the next selected sentence with the biggest $C_{i,j}$ with the last sentence of the confirmed sentence serial could only be added right behind the last sentence, because the position right before that sentence has been taken already.

Given an already ordered sentence serial $S_1S_2\ldots S_i$ which is a subset R of the whole sentence set T, the task of finding the (i+1)th sentence can be described as:

$$s_{i+1} = \arg\max_{s_j \in T-R}(C_{i,j}) \tag{3}$$

Now the sentence sequence become $S_1 S_2 \ldots S_i S_{i+1}$. By repeating this step the whole sentence sequence could be derived, if given the first sentence. The method of finding the first sentence S_1 can be defined in the same way as above:

$$s_1 = \arg\max_{s_j \in T}(C_{0,j}) \tag{4}$$

Here $C_{0,j}$ denotes how close the sentence j and a null sentence are. By adding a null sentence to the beginning of each source document and assuming it contains one null feature, $C_{0,j}$ can be calculated with equation (2).

Noises Elimination

In experiments, we found that the majority amount of feature pairs were assigned with quite low connectivity values, and they produce noises for the models, especially for long sentences. Assume that there are K features in one sentence and L features in another sentence, there are K*L possible feature pairs. We observed that the relationship of the two sentences is mainly determined by distinguishable feature pairs, the feature pairs with bigger connectivity values. The other feature pairs are producing noises. Thus we don't need to consider all K*L feature pairs.

For this sake we modified equation (1) as:

$$C_{i,j} = \frac{1}{biggest\_top\_n} \sum_{biggest\_top\_n} C_f(f_{ik}, f_{jl}) \tag{5}$$

The biggest_top_n is number of feature pairs which will be considered while calculating the connectivity between a pair of sentences. Given a pair of sentences S_i and S_j, all $C_f(f_{ik}, f_{jl})$ of feature pairs derived from the sentence pair are sorted, and only the biggest_top_n feature pairs with the biggest connectivity values are used to produce the connectivity of sentence pairs. Notice that if the biggest_top_n is set to K*L, where K is number of features in sentence S_i and L is number of features in sentence S_j., the modified equation is the same with original equation (1).

3 Experiments

In this section we describe the experiments made with majority ordering, probability-based ordering and adjacency-based ordering. Lapata [8] tested probabilistic model with an external training corpus, he also extracted sentence features such as nouns, verbs and dependencies. In this paper we focus on the methodology itself with raw input data, ie with source input documents only without external corpus and supporting semantics or grammar knowledge. We use single words except stop words as features to represent sentences.

3.1 Test Set and Evaluation Metrics

DUC 04 provided 50 document sets and four manual summaries of each set for its task 2. Each document set consist of 10 documents. After excluding summaries containing more than 8 sentences for the sake of running efficiency, we chose 157 summaries. The average number of sentences in each summary is 5.92. Sentences of each summary are token as input to the ordering algorithm. Original sequential information of sentences is abandoned. This is to simulate the circumstance of sentence ordering step in summarization tasks. The ordering algorithm then produces orderings which are compared with those orderings in manually generated summaries.

Here we assume that each manually generated summary has an "ideal" ordering. With this assumption, given any produced ordering, we can compare it with the "ideal" ordering by automatic evaluation. This assumption is reasonable. But at the same time, given a group of sentences, there might be not only one "ideal" ordering. In experiments of Barzilay et al. [2], 10 human participants often get 10 unique readable orderings for the same sentence group with an average size of 8.8. While number of human participants increased to 50, number of unique orderings raised to 21. This suggests that automatic evaluation for sentence ordering is hard to represent the actual performance, and certainly will get worse evaluating result than human evaluation. But we believe that with a statistically big enough test set, automatic evaluation is not meaningless. Considering the shortcoming of automatic evaluation and the difficulty and subjectivity of human evaluation for large amount of experiment results, we present automatic evaluation result on whole test set (157 summaries) and human evaluation on 10 randomly selected summaries.

A number of metrics can be used to measure the difference between two orderings. In this paper we use Kendall's τ [9], which is defined as:

$$\tau = 1 - \frac{2(number\_of\_inversions)}{N(N-1)/2} \tag{6}$$

Here N is number of objects to be ordered (ie sentences). Number_of_inversions is the minimal number of interchanges of adjacent objects to transfer an ordering into another. τ can be understood as how easy an ordering can be transfer to another. The value of τ changes from -1 to 1, where 1 denotes the best situation ---- the two orderings are the same, and -1 denotes the worst situation. Given an ordering, randomly produced orderings of same objects get an average τ of 0. In examples in Table 1, the τ values with natural sequences (123…n) are 0.67, 0.4, 0.33 respectively.

Table 1. Ordering Examples

Example 1	1 2 4 3
Example 2	1 5 2 3 4
Example 3	2 1 3

3.2 Results

We use single words as features to calculate condition probabilities and connectivity values of sentence pairs. We did experiments with majority ordering (run_Mo), probabilistic model (run_Pr) and connectivity model (run_Cn). A run producing random sentence orderings (run_Rd) was also presented to make the comparison more illustrative.

Automatic Evaluation

For automatic evaluation, the Kendall's τ is calculated between each output ordering and the ordering of the manual summary. This evaluation measures how similar the output orderings and orderings of manual summaries are. Since we believe that the orderings of manual summaries are ideal, hence the bigger Kendall's τ got, the better is the ordering performance of our ordering applications. The evaluation was made on all 157 summaries. The value of τ varies from -1 to 1.

Table 2 describes automatic evaluation results of our experiments. We can see that the majority ordering got a τ of 0.143, which is very close to probabilistic model. Meanwhile the adjacency model got a τ of 0.276, with an improvement of 91.7%.

Table 2. Automatic evaluation results on 157 summaries

	τ
Run_Rd	-0.007
Run_Mo	0.143
Run_Pr	0.144
Run_Cn	**0.276**

Table 3 gives out a further comparison between the four methods. The first data column describes how many output orderings put the 1st sentence of the 157 manual standard summaries at the "correct" position (ie the first position in output orderings). The adjacency model run significantly outperforms the other three runs. 94 out of 157 output orderings got the same 1st sentences with manual summaries. The reason why adjacency model performs better than other two models for the first sentence selection may be that when determining the first sentence, not only the occurrences of features after the null feature but also their occurrences elsewhere are both considered, while other models only consider their occurrence after the null features.

Table 3. Correction ratio of 1st sentence ranking

	Number of correctly found 1st sentences	Positive orderings	Negative orderings	Median Orderings
Run_Rd	22 (14.0%)	76 (48.4%)	70 (44.6%)	11 (7.0%)
Run_Mo	41 (26.1%)	97 (61.8%)	50 (31.8%)	10 (6.4%)
Run_Pr	64 (40.8%)	96 (61.1%)	46 (29.3%)	15 (9.6%)
Run_Cn	**94 (59.9%)**	**117 (74.5%)**	**28 (17.8%)**	12 (7.6%)

In Table 3, we use "positive ordering" to denote the output ordering that gets a positive τ, which means it can be considered as a better ordering than a random ordering. Similarly, "negative ordering" means the output ordering with a negative τ, which means it can be considered as a worse ordering than a random ordering. "median ordering" is the ordering get a τ of 0. We can find from the last 3 columns of Table 3 that the adjacency based ordering significantly outperforms the other 3 runs on finding positive orderings and avoiding negative orderings. 82.1% (positive plus median ordrings) of its output orderings are not worse than random produced orderings.

Table 4. Comparison of correct sentence inferrence

Correct Inferrence	1st sentence → 2nd sentence	2nd sentence → 3rd sentence
Run_Mo	10 (24.4%)	1 (10%)
Run_Pr	16 (25%)	4 (25%)
Run_Cn	**30 (31.9%)**	**15 (50%)**

We compared the correction ratio of sentence inferrence from correctly ordered previous sentences in Table 4. The first data column describes the number and percentage of runs that the second sentence is correctly ordered compared with the manual summary, given the first sentence being correctly ordered. The second data column discribes that of correctly ordered third sentence, given the first two sentences are correctly ordered. Because the number of runs with first 3 sentences correctly orderred is too few, we didn't make further comparison on inferrence of later sentences. Table 4 shows that given correctly ordered previous sentences, probabilistic model outperforms majority ordering and adjacency model outperforms both of them in inferring the next correct sentence.

We also find that though majority ordering got close Kendall's τ with probabilistic ordering, but it's weaker than probabilistic ordering to correctly order the first sentence and infer the next correct sentence. This may be because that majority ordering select the next sentence based on relations among unordered sentences, ie. it's dependent with ordered sentences. Meanwhile the probabilistic model and adjacency model depend on the previously ordered sentence to select the next sentence. This makes majority ordering weaker when a good previous ordering acquired, but stronger when a bad previous ordering acquired.

Table 5. Results with varing parameters

top_n	τ (top-n=1)	τ (top-n=2)	τ (top-n=3)	τ (top-n=4)	τ (top-n=5)	τ (top-n=10)
Run_Cn_range=2	0.184	0.213	0.253	0.262	0.261	0.224
Run_Cn_range=3	0.251	0.252	**0.273**	0.268	0.257	0.213
Run_Cn_range=4	0.201	0.253	0.268	**0.276**	0.272	0.248

Table 5 describes the experiment results of adjacency model with varing connectivity window ranges and noise elimination parameters. Cooccurrences of feature pairs are counted within the given window range in source documents. For example, range being set to 3 means that feature pairs co-occurring within adjacent 3 sentences are counted. Top-n is the noise elimilation parameter introduced in section 2.3. Table 5 shows that noise elimination does affect the performance, and the best performance was acquired with top-n being set to 3 or 4. The connectivity window size slightly affect the performance. The proper setting of connectivity window size might depend on the length of source documents. The longer the source documents, the bigger connectivity window size may be expected.

Manual Evaluation

As mentioned in previous sections, there are usually more than one acceptable orderings for a given group of sentences. This means that the automatic evaluation with Kendall's τ, which takes only one given ordering as the standard answer, will certainly underestimate the performance of ordering applications. To more accurately measure how efficient our ordering applications are, we randomly selected 10 manual summaries from the test set to manually evaluate output orderings.

Table 6. Manual evaluation results on 10 summaries

	τ
Run_Rd	0.299
Run_Mo	0.437
Run_Pr	0.445
Run_Cn	**0.584**

In manual evaluation, the number of inversions is defined as the minimal number of interchanges of adjacent objects to transfer the output ordering to an acceptable ordering judged by human. The Kendall's τ of all 3 runs are listed in table 6.

We can see from table 6 that all runs get bigger Kendall's τ than in automatic evaluation. The result of connectivity model is quite good.

4 Discussion and Conclusion

In this paper we proposed an adjacency model for sentence ordering in multi-document summarization. It learns the adjacency information of sentences from the source documents and orders sentences accordingly. Unlike the conditional probability between two sentences which helps to decide the sequence of them, the adjacency between two sentences denotes only the information about how close they should be in sumaries, despite of their relative sequence. But with the first sentence determined, an ordering can be acquired from adjacency information of sentence pairs. The experiment results show that this method significantly outperforms two other existing sentence ordering methods on dataset of DUC04.

With the adjacency based sentence ordering, sentences are ordered based on only the input source documents without any extra sources. This means that it is capable to almost any text ordering task for document summarization. While multi-document summarization becomes more and more neccessary in varies domains and occasions, supporting knowledge is getting harder to be predicted and prepared. This makes the adjacency based sentence ordering even more meaningful.

From the experiment results we found that adjacency based model outperforms probabilistic model in finding both the first sentence and the next sentence given previous ones. The reason might be that probabilistic model focuses on feature sequences, however, they might mutually contradict during learning, which will make the conditional probability useless. In contrast, adjacency model focuses on feature proximity, which will not contradict with each other.

There are some aspects for future improvement. First, noise reduction is a key step in sentence ordering, so we may use some feature weighting to filter noisy features. Second, currently the connectivity between two features is based on counts of their superficial co-occurrence, how to disclose the latent co-occurrence is an interesting problem. Third, the features in current experiments are all single words, and we can try some multi-word units as features in future.

References

1. Regina Barzilay, Noemie Elhadad, and Kathleen R. McKeown. 2001. Sentence ordering in multidocument summarization. Proceedings of the First International Conference on Human Language Technology Research (HLT-01), San Diego, CA, 2001, pp. 149–156..
2. Regina Barzilay, Noemie Elhadad, and Kathleen R. McKeown. 2002. Inferring strategies for sentence ordering in multidocument news summarization. Journal of Artificial Intelligence Research 17:35–55.
3. Sasha Blair-Goldensohn, David Evans. Columbia University at DUC 2004. In Proceedings of the 4th Document Understanding Conference (DUC 2004). May, 2004.
4. Danushka Bollegala, Naoaki Okazaki, Mitsuru Ishizuka. 2005. A machine learning approach to sentence ordering for multidocument summarization and it's evaluation. IJCNLP 2005, LNAI 3651, pages 624-635, 2005.
5. Endre Boros, Paul B. Kantor and David J. Neu. A Clustering Based Approach to Creating Multi-Document Summaries. DUC 2001 workshop.
6. R. C. Dubes and A. K. Jain. Algorithms for Clustering Data. Prentice Hall, 1988.
7. Hilda Hardy, Nobuyuki Shimizu. Cross-Document Summarization by Concept Classification. In SIGIR 2002, page 121-128.
8. Mirella Lapata. Probabilistic text structuring: Experiments with sentence ordering. Proceedings of the annual meeting of ACL, 2003., pages 545–552, 2003.
9. Guy Lebanon and John Lafferty. 2002. Combining rankings using conditional probability models on permutations. In C. Sammut and A. Hoffmann, editors, In Proceedings of the 19th International Conference on Machine Learning. Morgan Kaufmann Publishers, San Francisco, CA.
10. Dragomir Radev, Timothy Allison, Sasha Blair-Goldensohn, John Blitzer, Arda Çelebi, Stanko Dimitrov, Elliott Drabek, Ali Hakim, Wai Lam, Danyu Liu, Jahna Otterbacher, Hong Qi, Horacio Saggion, Simone Teufel, Michael Topper, Adam Winkel, and Zhang Zhu. MEAD - a platform for multidocument multilingual text summarization. In Proceedings of LREC 2004, Lisbon, Portugal, May 2004.
11. Advaith Siddharthan, Ani Nenkova and Kathleen McKeown. Syntactic Simplication for Improving Content Selection in Multi-Document Summarization. In Proceeding of COLING 2004, Geneva, Switzerland.
12. Gees C. Stein, Amit Bagga, G. Bowden Wise. Multi-Document Summarization: Methodologies and Evaluations. In Conference TALN 2000, Lausanne, 16-18 October 2000.

Poor Man's Stemming: Unsupervised Recognition of Same-Stem Words

Harald Hammarström

Chalmers University, 412 96 Gothenburg, Sweden

Abstract. We present a new fully unsupervised human-intervention-free algorithm for stemming for an open class of languages. Since it does not rely on existing large data collections or other linguistic resources than raw text it is especially attractive for low-density languages. The stemming problem is formulated as a decision whether two given words are variants of the same stem and requires that, if so, there is a concatenative relation between the two. The underlying theory makes no assumptions on whether the language uses a lot of morphology or not, whether it is prefixing or suffixing, or whether affixes are long or short. It does however make the assumption that 1. salient affixes have to be frequent, 2. words essentially are variable length sequences of random characters, and furthermore 3. that a heuristic on what constitutes a systematic affix alteration is valid. Tested on four typologically distant languages, the stemmer shows very promising results in an evaluation against a human-made gold standard.

1 Introduction

The problem at hand can be described as follows:

Input : An unlabeled corpus of an arbitrary natural language and two arbitrary words w_1, w_2 from that language

Output : A YES/NO answer as to whether w_1 and w_2 are morphological variants of one and the same stem (according to traditional linguistic analysis).

Restrictions : We consider only concatenative morphology and assume that the corpus comes already segmented on the word level.

The relevance of the problem is that of stemming as applied in Information Retrieval (IR). The issues of stemming in IR has been discussed at length elsewhere and need not be repeated here. It suffices to say that, though not uncontroversial, stemming continues to be a feature of modern IR systems for languages like English (e.g Google[1]), and is likely to be of crucial importance for languages which make more use of morphology (cf. [1]).

The reasons for attacking the problem in an unsupervised manner include advantages in elegance, economy of time and money (no annotated resources required), and the fact that the same technology may be used on new languages.

[1] According to http://www.google.com/help/basics.html accessed 20 March 2006.

H.T. Ng et al. (Eds.): AIRS 2006, LNCS 4182, pp. 323–337, 2006.

The latter two reasons are especially important in the context of resource-scarce languages.

Our proposed unsupervised same-stem decision algorithm proceeds in two phases. In the first phase, a ranked list of salient affixes are extracted from an unlabeled text corpus of a language. In the second phase, an input word pair is aligned to shortlist affixes that could potentially be added to a common stem to alternate between the two. Crucially, this shortlist of affix alternations is analyzed to check whether they form a *systematic* alternation in the language as a whole (i.e not just in the pair at hand). This analysis depends strongly on the ranked affix list from the first phase.

An outline of the paper is as follows: we start with some notation and basic definitions, with which we describe the theory that is intended to model the assumed behaviour of affixation in natural languages. Then we describe in detail and with examples the thinking behind the affix extraction phase, which actually requires only a few lines to define mathematically. Following that, we present our ideas on how to distinguish a systematic morphological alternation from a spurious one. This part is the more experimental one but at least it requires no guiding, tuning or annotation whatsoever. The algorithm is evaluated against a human gold standard on four languages chosen to span the full width of morphological typology. Finally, we briefly discuss related work, draw some tentative conclusions and hint at future directions.

2 Affix Extraction

We have chosen to illustrate using suffixes but the method readily generalizes to prefixes as well (and even prefixes and suffixes at the same time).

2.1 A Naive Theory of Affixation

Notation and definitions:

- $w, s, b, x, y, \ldots \in \Sigma^*$: lowercase-letter variables range over strings of some alphabet Σ and are variously called words, segments, strings, etc.
- $s \lhd w$: s is a terminal segment of the word w i.e there exists a (possibly empty) string x such that $w = xs$
- $W, S, \ldots \subseteq \Sigma^*$: capital-letter variables range over sets of words/strings/segments
- $f_W(s) = |\{w \in W | s \lhd w\}|$: the (suffix) frequency, i.e the number of words in W with terminal segment s
- $S_W = \{s | s \lhd w \in W\}$: all terminal segments of the words in W
- $uf_W(u) = |\{(x, y) | xuy = w \in W\}|$: the substring frequency of u, i.e the number times u occurs as a substring in the set of words W (x and y may be empty).
- $nf_W(u) = uf_W(u) - f_W(u)$: the non-final frequency of u, i.e. the substring frequency minus those in which it occurs as a suffix.
- $|\cdot|$: is overloaded to denote both the length of a string and the cardinality of a set

Assume we have two sets of random strings over some alphabet Σ:

- Bases $B = \{b_1, b_2, \ldots, b_m\}$
- Suffixes $S = \{s_1, s_2, \ldots, s_n\}$

Such that:

Arbitrary Character Assumption (ACA): Each character $c \in \Sigma$ should be equally likely in any word-position for any member of B or S.

Note that B and S need not be of the same cardinality and that any string, including the empty string, could end up belonging to both B and S. They need neither to be sampled from the same distribution; pace the requirement, the distributions from which B and S are drawn may differ in how much probability mass is given to strings of different lengths. For instance, it would not be violation if B were drawn from a a distribution favouring strings of length, say, 42 and S from a distribution with a strong bias for short strings.

Next, build a set of affixed words $W \subseteq \{bs | b \in B, s \in S\}$, that is, a large set whose members are concatenations of the form bs for $b \in B, s \in S$, such that:

Frequent Flyer Assumption (FFA): The members of S are frequent. Formally: Given any $s \in S$: $f_W(s) >> f_W(x)$ for all x such that 1. $|x| = |s|$; and 2. not $x \lhd s'$ for all $s' \in S$).

In other words, if we call $s \in S$ a *true suffix* and we call x an *arbitrary segment* if it neither a true suffix nor the terminal segment of a true suffix, then any true suffix should have much higher frequency than an arbitrary segment of the same length.

2.2 An Algorithm for Affix Extraction

The key question is, if words in natural languages are constructed as W explained above, can we recover the segmentation? That is, can we find B and S, given only W? The answer is yes, we can partially decide this. To be more specific, we can compute a score Z_W such that $Z_W(x) > Z_W(y)$ if $x \in S$ and $y \notin S$. In general, the converse need not hold, i.e if both $x, y \in S$, or both $x, y \notin S$, then it may still be that $Z_W(x) > Z_W(y)$. This is equivalent to constructing a ranked list of all possible segments, where the true members of S appear at the top, and somewhere down the list the junk, i.e non-members of S, start appearing and fill up the rest of the list. Thus, it is not said *where* on the list the true-affixes/junk border begins, just that there is a consistent such border. We shall now define three properties that we argue will be enough to put the S-belonging affixes at the top of the list. For a terminal segment s, define:

Frequency. The frequency $f_W(s)$ of s (as a terminal segment).

Curve Drop. The Curve Drop of s is the minimal percentage drop in freqency if s is extended to the left with one character, normalized to the best possible such precentage drop.

$$\overline{C}(s) = \frac{1 - \max_c \frac{f_W(cs)}{f_W(s)}}{1 - \frac{1}{|\Sigma|}} \tag{1}$$

Random Adjustment. First, for s, define its probability as:

$$P_W(s) = \frac{f_W(s)}{\sum_{s' \in S_W} f_W(s')} \tag{2}$$

Second, equally straighfowardly, for an arbitrary segment u, define its non-final probability as:

$$nP_W(u) = \frac{nf_W(u)}{\sum_{u'} nf_W(u')} \tag{3}$$

Finally, for a terminal segment s, define its *random adjustment* $RA(s)$ to be the ratio between the two:

$$RA(s) = \begin{cases} \frac{P_W(s)}{nP_W(s)} & \text{if } nP_W(s) > 0 \\ 1.0 & \text{otherwise} \end{cases} \tag{4}$$

It is appropriate now to show the intuition behind the definitions. There isn't much to comment on frequency, so we'll go to curve drop and random adjustment. All examples in this section come from the Brown corpus [2] of one million tokens ($|W| = 47178$ and $|S_W| = 154407$).

The curve drop measure is meant to predict when a suffix is well-segmented to the left. Consider a suffix s, in all the words on which it appears, there is a preceding character c. For example, *-ing* occurs 3258 times, of which it is preceded by t 640 times, of l 329 times, r 317 times, d 258 times, n 249 times and so forth. This contrasts with *-ng* which occurs 3352 times, of which it is preceded by i 3258 times, by o 35 times, by a 26 times and so on. The reasoning is thus as follows. If s is a true suffix and is well-segmented to the left, then its curve-drop value should be high. Frequent true suffixes that attach to bases whose last character is random should have a close to uniform curve. On the other hand, if the curve drop value is low it means there is a character that suspiciously often precedes s. However, if s weren't a true suffix to begin with, perhaps just a frequent but random character, then we expect it's curve drop value to be high too! To exemplify this, we have $\overline{C}(ing) \approx 0.833$, $\overline{C}(ng) \approx 0.029$ and $\overline{C}(a) \approx 0.851$.

The random adjustment measure it precisely to distinguish what a "frequent but random segment" is, that is, discriminate e.g *-a* versus *-ing* as well as *-a* versus *-ng*. Now, how does one know whether something is random or not? One approach would be to say the shorter the segment the more random. Although it's possible to get this to work reasonably well in practice, it has some drawbacks. First, it treats all segments of the same length the same, which may be too brutal, e.g should *-s* be penalized as much as *-a*? Second, it might be considered too vulnerable to orthography. For example if a language has an odd trigraph for some phoneme, we are clearly going to introduce an error source. Instead we propose that a segment is random iff it has similar probability in any position of the word. Instead we propose that a segment is random iff it has similar probability in any position of the word. This avoids the "flat length"-problems but has others, which we think are less harmful. First, we might get sparse data which can either be back-off smoothed or, like here, effectively ignored (where we lack occurence we set the RA to 1.0). Second, phonotactic or orthographic constraints may cause curiousities, e.g. English y is often spelled i when medial as in *fly* vs. *flies*.

To put it all together, we propose the characterization of suffixes in terms of the three properties as shown in table 1. The terms high and low are of course idealized, as they are really gradient properties.

Table 1. The logically possible condigurations of the three suffix properties, accompanied by an appropriate linguistically inspired label and an example from English

f_W	$\overline{C}$	RA	Example	Label
high	high	high	*-ing*	True suffix
high	high	low	*-a*	Frequent random segment
high	low	high	*-ng*	Tail of true suffix
high	low	low	N/A	Second part of a digraph
low	high	high	*-oholic*	Infrequent true suffix
low	high	low	*-we*	Happenstance low RA-segment?
low	low	high	*-icz*	Tail of foreign personal name ending
low	low	low	*-ebukadnessar*	Infrequent segment

As seen from the table, we hold that true suffixes (and only true suffixes) are those which have a high value for all three properties. Therefore, we define our final ranking score, the $Z_W : S_W \rightarrow \mathbf{Q}$:

$$Z_W(s) = \overline{C}(s) \cdot RA(s) \cdot f_W(s) \tag{5}$$

The final Z_W-score in equation 5 is the one that purports to have the property that $Z_W(x) > Z_W(y)$ if $x \in S_W$ and $y \notin S_W$ – at least if purged (see below). We cannot give a formal proof that languages satisfying ACA and FFA should get a faultless ranking list because this is true only in a heuristic sense. To set bounds on the probability for it to hold is also depends on a lot of factors that are hard, or at least inelegant, to characterize. We hope, however, to have sketched the how the ACA and FFA assumptions are used.

2.3 Affix Extraction Sample Results

On the affix extraction part as such, we will only give some impressionistic results rather than a full-scale evaluation. The reason for this is that, although undoubtedly the list has some valid meaning, it is at present unclear to the author what a gold standard should be in every detail in every language. Furthermore, different applications, such as the final objective in this paper, may not require that a context-less choice between two related affixes, e.g *-ation* and *-tion*, be asserted.

For an English bible corpus [3] we get the top 30 plus bottom 3 suffixes as shown in table 2.

Table 2. Top 30 and bottom 3 extracted suffixes for an English bible corpus. The high placement of English *-eth* and *-iah* are due to the fact that the bible version used has drinketh, sitteth etc and a lot of personal names in *-iah*.

-ed	15448.4	*-ity*	6917.6	*-ts*	3783.1	*-y*	2239.2	*-ded*	1582.2
-eth	12797.1	*-edst*	6844.7	*-ah*	3766.9	*-leth*	2166.3	*-neth*	1540.0
-ted	11899.4	*-ites*	5370.2	*-ness*	3679.3	*-nts*	2122.6	...	...
-iah	11587.5	*-seth*	5081.6	*-s*	3407.3	*-ied*	1941.7	...	...
-ly	10571.2	*-ned*	4826.7	*-ions*	2684.5	*-ened*	1834.9	*-io*	0.0
-ings	8038.9	*-s'*	4305.2	*-est*	2452.6	*-ers*	1819.5	*-ti*	0.0
-ing	7292.8	*-nded*	3833.8	*-sed*	2313.7	*-ered*	1796.7	*-ig*	0.0

The results largely speak for themselves but some comments are in order. A good sign is that the list and its order seems to be largely independent of corpus size (as long as the corpus is not extremely small) but we do get some significant differences between bible English and newspaper English. As is easily seen from the lists, some suffixes are suffixes of each other so one could *purge* the list in some way to get only the most "competitive" suffixes. For a fuller discussion of purging, other languages and all other matters pertaining to the affix extraction algorithm, the reader is referred to the longer exposition in [4].

3 Affix Alternation Analysis

Having a list of salient affixes is not sufficient to parse a given word into stem and affix(es). For example, *sing* happens to end in the most salient suffix yet it is not composed of *s* and *ing* because crucially, there is no **s*, **sed* etc. Thus to parse a given word we have to look at additional evidence beyond the word itself, such as the existence of other inflections of potentially the same stem as the given word, or further, look at inflections of other stems which potentially share an affix with the given word. This line of thought will be pursued below.

The problem at hand, namely, to decide if two given words w_1, w_2 share a common stem (in the linguistic sense) is easier than parsing one word. Essentially, there are four interesting kinds of situations the same-stem-decider must face:

1. w_1 and w_2 do share the same stem and have a salient affix each, e.g *played* vs. *playing*.
2. w_1 and w_2 do share the same stem but one of them has the "zero" affix, e.g *play* vs. *playing*.
3. w_1 and w_2 do not share the same stem (linguistically) but do share some initial segment, e.g *playing* vs. *plough*.
4. w_1 and w_2 do not share the same stem (linguistically) and do not share any initial segment, e.g *playing* vs. *song*.

Number 4 is trivial to decide in the negative. Number 1 is also easy to affirm using a list of salient affixes, whereas the special case of number 2 requires some care. The real difficulty lies in predicting a negative answer for case number 3 (while, of course, at the same time predicting a positive for cases 1 and 2). We will go for an extended discussion of this matter below.

Consider two words $w_1 = xs_1$ and $w_2 = xs_2$ that share some non-empty initial segment x. Except for chance resemblances, which by definition are rare, we would like to say that w_1 and w_2 belong to the same stem iff:

1. s_1 and s_2 are well-segmented salient suffixes in the language, i.e *-w* and *-lt* for *saw* and *salt* are **not**; and
2. s_2 and s_2 must systematically contrast in the language, that is, there must be a large set of stems which can take both s_1 and s_2. For example, the word pair *sting* and *station* align to *-ing* and *-ation* which are both salient suffixes but they do **not** systematically contrast.

The key difficulty is to decide, in an unsupervised manner, when something is systematic and when it isn't. In order to tackle this, we will propose a heuristic for measuring how much two suffixes contrast. This will give a score between 0 and 1 where it is not clear at which value "systematic" begins. We could say that, at this point, the user has to supply a threshold value. However, instead, we devise another heuristic that obviates the need for a threshold at all. The resulting system thus supplies a YES/NO answer to the same-stem decising problem without any human interaction.

3.1 Formalizing Same Stem Co-occurence

From the word distributions characteristic of natural language corpora, it is surprisingly difficult to come up with a measure of how much a set of suffixes show up on the "same stems" that is not such that it favours the inclusion of any simply frequent, rather than truly contrasting, terminal segment. For example, the author has not had much success with standard vector similarity measures. Instead, we propose the following usage of co-occurence statistics. The measure presented is valid for an arbitrary set of suffixes (called P for "paradigm") even though the relevance in this paper is for the case where $|P| = 2$.

First, for each suffix x, define its quotient function $H_x(y) : S_W \rightarrow [0, 1]$ as:

$$H_x(y) = \frac{|Stems(x) \cap Stems(y)\}|}{|Stems(x)|} \tag{6}$$

where $Stems(x) = \{z|zx \in W\}$. The formula is conveying the following: We are given a suffix x, and we want to construct a quotient function which is a function from any other suffix to a score between 0 and 1. The score is calculated as: look at all the stems of x, other suffixes y will undoubtedly also occur on some of these stems. For each other suffix y, find the proportion of x:s stems on which y also appears. This proportion will be the quotient associated with y. Two examples of quotient functions (sorted on highest value) are given in table 3.

Now, given a set of affixes P, construct a rank by summing the quotient functions of the members of P:

$$V_P(y) = \sum_{x \neq y \in P} H_x(y) \tag{7}$$

Table 3. Sample quotient functions/lists for *ing* and *ed* on the Brown Corpus. H_{ing} and H_{ed} have 68337 and 75853 nonzero values respectively.

y	$H_{ing}(y)$	y	$H_{ed}(y)$
ing	1.00	*ed*	1.00
ed	0.59	*ing*	0.42
''	0.41	*''*	0.33
s	0.25	*e*	0.21
e	0.24	*s*	0.20
es	0.19	*es*	0.17
er	0.12	*er*	0.08
ers	0.10	*ion*	0.07
ion	0.07	*ers*	0.05
y	0.05	*y*	0.04
ings	0.05	*ions*	0.03
ions	0.03	*ation*	0.03
in	0.03	*able*	0.02
ation	0.03	*ings*	0.02
's	0.03	*'s*	0.02
ingly	0.03	*or*	0.02
or	0.02	*in*	0.01
able	0.02	*ly*	0.01
ive	0.02	*ive*	0.01
ors	0.02	*ingly*	0.01
ations	0.01	*al*	0.01
er's	0.01	*ment*	0.01
ment	0.01	*ors*	0.01
ly	0.01	*ations*	0.01
...	...	...	...

Table 4. Example ranks for $P = \{a, an, as, ans, or, orna, ors, ornas\}$ (left) and $P = \{ungen, ig, ar, ts, s, de, ende, er\}$ (right)

y	$VI_P(y)$	y	$VI_P(y)$
a	3.93	"	3.32
an	2.82	t	1.48
or	2.71	a	1.19
"	1.91	r	1.18
orna	1.76	s	1.15
ar	1.13	en	1.14
as	1.06	iga	0.86
ade	1.05	d	0.80
ans	0.94	igt	0.73
at	0.89	as	0.66
en	0.82	**de**	0.59
s	0.76	des	0.57
t	0.73	ade	0.55
e	0.71	ung	0.49
er	0.66	**er**	0.49
ad	0.61	at	0.48
ande	0.52	n	0.46
ades	0.47	**ar**	0.45
ats	0.40	an	0.44
i	0.36	e	0.42
...	...	...	...
ors	0.35		
...	...		
ornas	0.27		
...	...		

The $x \neq y$ is just there so that the y:s that are also in P do not get an "extra" 1.0, since $H_x(x) = 1.0$ regardless of the data. The rank is just y sorted on highest $V_P(y)$.

As an example, take W from the Swedish PAROLE-Corpus [5]. We can compare in table 4 the very common paradigm $\{a, an, as, ans, or, orna, ors, ornas\}$ with the nonsense paradigm $\{ungen, ig, ar, ts, s, de, ende, er\}$ consisting only of individually frequent suffixes. In table 4, the ranks of the member of P to the left are [0, 1, 2, 4, 6, 8, 22, 31], and for P to the right the ranks are [115044, 127, 17, 28, 4, 10, 100236, 14].

Now, if we can generalize from these cases it seems that we can rank different hypotheses of paradigms (of the same size) by looking at their quotient ranks. If the members of P "turn up high in" the quotient rank then the members of P tend to turn up on the same stems. There are several issues in formalizing the notion of "turn up high in". The places in the ranked list alone? Also incorporate the scores? Average place or total sum of places? For now we will just do a simple sum of places in the ranked list, divide by the optimum sum (which depends on

$|P|$ and is $0 + \ldots + |P| - 1$), and take the inverse. This gives a score between 0 and 1 where a high score means the members of P tend to appear on the same stems:

$$VI(P) = \frac{|P|(|P| - 1)}{2 \sum_{x \in P} place(x, V_P)} \tag{8}$$

According to the desiderata 1 and 2 in section 3 (p. 329) we finally define an affix-systematicity likelihood score as:

$$A(P) = VI(P) \sum_{s \in P} Z_W(s) \tag{9}$$

As a convention we set $Z_W('') = 0$.

3.2 Escaping Thresholds

The VI-score from the last section may be used for a greedy hill-climbing search through the affix set space. For example, we may start with an affix, a one member set, and see whether we can improve the affix score by including another member, and perhaps another after that until we can't improve the score anymore. In this process, we may also entertain the possibility of kicking some member out if that improves the score – as long as there is no backtracking the search remains polynomial. Formally, define the growing function of a set P of affixes as:

$$G(P) = argmax_{p \in \{P\} \cup \{P \text{ xor } s | s \in S_W\}} VI(p) \tag{10}$$

$$G^*(P) = \begin{cases} P & \text{if } G(P) = P \\ G^*(G(P)) & \text{if } G(P) \neq P \end{cases} \tag{11}$$

Two growth-examples are shown in table 5, one which attains a perfect 1.0 score and one in which the original member is expelled in a later iteration.

Table 5. Example iterations of $G^*('ation')$ and $G^*('xt')$

P	$VI(P)$	P	$VI(P)$
{'ation'}	0.00	{'xt'}	0.00
{'ated', 'ation'}	0.14	{'xt', 'n'}	0.04
{'ate', 'ated', 'ation'}	0.40	{'xt', 'n', 'ns'}	0.12
{'ate', 'ated', 'ating', 'ation'}	0.75	{'n', 'ns'}	0.55
{'ate', 'ated', 'ating', 'ation', 'ations'}	1.00	...	...

Now, how does this help us work around a threshold for deciding how systematically a pair of suffixes have to co-occur to conflate their stems? Recall the writing convention $w_1 = xs_1$ and $w_2 = xs_2$. Instead of having a threshold we may conjecture that:

w_1, w_2 have the same stem iff $s_1 \in G^*(s_2)$ and $s_2 \in G^*(s_1)$

For example, this predicts that *sting* and *station* are not the same stem because neither $G^*(ing) = \{'', e, ed, er, es, ing, s\}$ contains 'ation' nor does $G^*(ation) = \{ate, ated, ating, ation, ations\}$ contain 'ing'. From our experience this test is quite powerful. However, there are of course cases where it predicts wrongly, due to the greedy nature of the G^*-calculation, e.g $G^*(ing)$ does not contain 'ers'. Moreover, if one of the affixes is the empty affix, we need a special fix (see below).

3.3 Same-Stem Decision Algorithm

We can now put all pieces together to define the full algorithm as shown in table 6.

Table 6. Summary of same-stem decision algorithm

Input: A text corpus C and two words w_1, w_2
Step 1. Calculate Z_W as in equation 5
Step 2. Form the set of candidate alignment pairs as:

$$C(w_1, w_2) = \{(s1, s2) | xs_1 = w_1 \text{ and } xs_2 = w_2\} \tag{12}$$

Step 3. If $C(w_1, w_2)$ is empty then answer NO, otherwise pick the best candidate pair as:

$$argmax_{(s1,s2) \in C(w_1,w_2)} A(\{s_1, s_2\}) \tag{13}$$

Step 4. For the winning pair, answer YES/NO acccordingly as $s_1 \in G^*(s_2)$ and $s_2 \in G^*(s_1)$

If one of s_1, s_2 is the empty string then step 3 and 4 should be restated as follows (using s to denote the non-empty one of the two). The maximization value in step 3 should be modified to: $\frac{Z_W(s)}{1+place('',H_s)}$. Step 4 should be modified to: answer YES/NO acccordingly as $'' \in G^*(s)$.

The bad news is that the computation of the G^*:s tends to be slow due to the summing and sorting of typically very long (50 000-ish items) lists. On my standard PC with a Python implementation it typically takes 30 seconds to decide whether two words share the same stem.

4 Evaluation

Several authors, e.g [6,7], have evaluated their stemming algorithms on Information Retrieval performance. While IR is the undoubtedly the major application area we feel that evaluating on retrieval performance does not answer all relevant questions of stemming performance. For instance, a stemmer may make conflations and miss conflations that simply did not affect the test queries. In fact, one may get different best stemmers depending on the test collection. There is also difference as to whether the whole document collection, an abstract of each document or just the query is stemmed.

We find it more instructive to test stemming separately against a stemming gold standard and assess the relevance of stemming for IR by testing the stemming gold standard on IR performance. If stemming turns out to be relevant for IR, then researchers should continue to develop stemming algorithms towards the gold standard. In the other case, one wonder whether IR-improving term conflation methods should be called stemmers.

In order to assess the cross-linguistic applicability of our stemming algorithm we have chosen languages spanning spectrum of morphological typology – from isolating to highly suffixing – Maori, English, Swedish and Kuku Yalanji [8]. As training data we used only the set of words from a bible translation to emphasize the applicability to resource-scarce languages.

For these four languages we devised a stemming gold standard using [12,13] for Maori and [14,15] for Kuku Yalanji, languages not generally known to the author. So as not to let the test set be dominated by too many simple test cases, the selection of test set cases was done as follows:

1. Select a random word w_1 from W for the corresponding language
2. Select a random number i in $0 \leq i \leq |w_1| - 1$
3. Select a random word w_2 from the subset of words from $W \setminus \{w_1\}$ sharing i initial characters with w_1
4. Mark the pair w_1, w_2 to be of the same stem or not, according to traditional linguistic analysis

This was repeated until 200 pairs of words for each language had been selected, 100 same-stem and 100 not same-stem. Except for Maori where we could only really find 13 same-stem cases this way, all involving active-passive alternating verbs (described in detail in [16]).

Table 7. Evaluation results

	Same-stem		Diff.-stem				
Language	Correct	Total	Correct	Total	Language Type	Corpus	Size
Maori	10	13	100	100	Isolating	[9]	NT & OT
English	97	100	100	100	Mildly Suffixing	[3]	NT & OT
Swedish	96	100	100	100	Suffixing	[10]	NT & OT
Kuku Yalanji	94	100	100	100	Strongly Suffixing	[11]	NT & OT Parts

The evaluation results are shown in table 7. Errors fall into just one major type, in which the algorithm is too cautious to conflate; it is when two words do share the stem but where one of the suffixes is rather uncommon (possibly because it is really composite) and therefore it is not in the grow-set of the other suffix; for example Swedish *skap-ade-s* (past passive) and *skap-are-n-s* (agent-noun definite genitive). We also expected false positives in the form of random resemblances involing short words and short affixes; e.g *as* versus *a* but no such cases seem to have occurred in the test set in any of the languages.

We have done attempted a comparison with other existing stemmers, mainly because they tend not be aimed at an open set of languages and those which are,

are really not fully supervised and we fear we might not do justice to them in setting parametres (see Related Work section). The widely known Porter stemmer [17] for English scores exactly the same result for English as our stemmer, which suggests than an unsupervised approach may come very close to explicitly human-informed stemmers. Many other stemmers, however, are superior to ours in the sense that they can stem a single word correctly whereas ours requires a pair of words to make a decision. This is especially relevant when large bodies of data needs to be stem-indexed as it would take quadratic time (in the number of words) in our setting.

5 Related Work

A full survey of stemming algorithms for specific languages or languages like English has more or less fully been done elsewhere (the technology becoming relatively mature cf. [18,19,6,7,20,21,22,23] and references therein). We will focus instead on unsupervised approaches for a wider class of languages.

Melucci and Orio [7] present a very elegant unsupervised stemming model. While training does not require any manually annotated data, some architectural choices depending on the language still has to be supplied by a human. If this can be overcome in an easy way, it would be very interesting to test their Baum-Welch training approach versus the explicit heuristics in this paper, especially on a wider scope of languages than given in their paper. The unsupervised stemmer outlined in [6] actually requires a lot of parametres to be tweaked humanly and mainly targets languages with one-slot morphology.

Other systems for unsupervised learning of morphology which do not explicitly do stemming could easily be transformed into stemmers. Work includes [24,25,26,27,28,29,30,31,32,33,34,35,36,37] and other articles by the same authors. All of these systems, however, require some parametre tweaking as it is and perhaps one more if transformed to stemmers, so there is still work wanting before they can be compared on equal grounds to the stemmer described here. Given that they use essentially the same kind of evidence, it is likely that some of them, especially [38], will reach just as competitive results on the same task.

Of course, we also wish to acknowledge that traditional stemmers output the actual stem, which is one (significant) step further than deciding the same-stem problem for word pairs.

6 Conclusion

We have presented a fully unsupervised human-intervention-free algorithm for stemming for an open class of languages showing very promising accuracy results. Since it does not rely on existing large data collections or other linguistic resources than raw text it is especially attractive for low-density languages. Although polynomial in time, it appears rather slow in practice and may not be suitable for stemming huge text collections. Future directions include investigating whether there is a speedier shortcut and a better, more systematic, approach to layered morphology i.e for languages which allow affixes to be stacked.

Acknowledgements

The author has benefited much from discussions with Bengt Nordström. We also wish to extend special thanks to ASEDA for granting access to electronic versions of the Kuku Yalanji bible texts.

References

1. Pirkola, A.: Morphological typology of languages for IR. Journal of Documentation **57**(3) (2001) 330–348
2. Francis, N.W., Kucera, H.: Brown corpus. Department of Linguistics, Brown University, Providence, Rhode Island (1964) 1 million words.
3. King James: The Holy Bible, containing the Old and New Testaments and the Apocrypha in the authorized King James version. Thomas Nelson, Nashville, New York (1977)
4. Hammarström, H.: A naive theory of morphology and an algorithm for extraction. In Wicentowski, R., Kondrak, G., eds.: SIGPHON 2006: Eighth Meeting of the Proceedings of the ACL Special Interest Group on Computational Phonology, 8 June 2006, New York City, USA, Association for Computational Linguistics (2006) 79–88
5. Borin, L.: Parole-korpusen vid språkbanken, göteborgs universitet. `http://spraakbanken.gu.se` accessed the 11th of Febuary 2004. (1997) 20 million words.
6. Goldsmith, J., Higgins, D., Soglasnova, S.: Automatic language-specific stemming in information retrieval. In Peters, C., ed.: Cross-Language Information Retrieval and Evaluation: Proceedings of the CLEF 2000 Workshop. Lecture Notes in Computer Science. Springer-Verlag, Berlin (2001) 273–283
7. Melucci, M., Orio, N.: A novel method for stemmer generation based on hidden markov models. In: CIKM '03: Proceedings of the twelfth international conference on Information and knowledge management, New York, NY, USA, ACM Press (2003) 131–138
8. Dryer, M.S.: Prefixing versus suffixing in inflectional morphology. In Comrie, B., Dryer, M.S., Gil, D., Haspelmath, M., eds.: World Atlas of Language Structures. Oxford University Press (2005) 110–113
9. The British & Foreign Bible Society: Maori Bible. The British & Foreign Bible Society, London, England (1996)
10. Svenska Bibelsällskapet: Gamla och Nya testamentet: de kanoniska böckerna. Norstedt, Stockgholm (1917)
11. Summer Institute of Linguistics: Bible: New testament and old testament selctions in kuku-yalanji (1985)
12. Bauer, W., Parker, W., Evans, T.K.: Maori. Descriptive Grammars. Routledge, London & New York (1993)
13. Williams, H.W.: A dictionary of the Maori language. 7 edn. GP Books, Wellington (1971)
14. Patz, E.: A Grammar of the Kuku Yalanji Language of North Queensland. Volume 257 of Pacific Linguistics. Research School of Pacific and Asian Studies, Australian National University, Canberra (2002)

15. Hershberger, H.D., Hershberger, R.: Kuku-Yalanji dictionary. Volume 7 of Work Papers of SIL - AAB. Series B. Summer Institute of Linguistics, Darwin (1982)
16. Sanders, G.: On the analysis and implications of maori verb alternations. Lingua **80** (1990) 149–196
17. Porter, M.F.: An algorithm for suffix stripping. Program **14**(3) (1980) 130–137
18. Erjavec, T., Džeroski, S.: Machine learning of morphosyntactic structure: Lemmatizing slovene words. Applied Artificial Intelligence **18** (2004) 17–41
19. Frakes, W.B., Fox, C.J.: Strength and similarity of affix removal stemming algorithms. SIGIR Forum **37**(1) (2003) 26–30
20. Rogati, M., McCarley, S., Yang, Y.: Unsupervised learning of arabic stemming using a parallel corpus. In: ACL '03: Proceedings of the 41st Annual Meeting on Association for Computational Linguistics, Morristown, NJ, USA, Association for Computational Linguistics (2003) 391–398
21. Hull, D.A.: Stemming algorithms: A case study for detailed evaluation. Journal of the American Soicety for Information Science **47**(1) (1996) 70–84
22. Galambos, L.: Multilingual Stemmer in Web Environment. PhD thesis, Faculty of Mathematics and Physics, Charles University in Prague (2004)
23. Flenner, G.: Ein quantitatives morphsegmentierungssystem für spanische wortformen. In Klenk, U., ed.: Computatio Linguae II: Aufsätze zur algorithmischen und Quantitativen Analyse der Sprache. Volume 83 of Zeitschrift für Dialektologie und Linguistik: Beihefte. Franz Steiner, Stuttgart (1994) 31–62
24. Jacquemin, C.: Guessing morphology from terms and corpora. In: Proceedings, 20th Annual International ACM SIGIR Conference on Research and Development in Information Retrieval (SIGIR '97), Philadelphia, PA. (1997)
25. Yarowsky, D., Wicentowski, R.: Minimally supervised morphological analysis by multimodal alignment. In: Proceedings of the 38th Annual Meeting of the Association for Computational Linguistics (ACL-2000). (2000) 207–216
26. Baroni, M., Matiasek, J., Trost, H.: Unsupervised discovery of morphologically related words based on orthographic and semantic similarity. In: Proceedings of the Workshop on Morphological and Phonological Learning of ACL/SIGPHON-2002. (2002) 48–57
27. Clark, A.: Learning morphology with pair hidden markov models. In: ACL (Companion Volume). (2001) 55–60
28. Ćavar, D., Herring, J., Ikuta, T., Rodrigues, P., Schrementi, G.: On induction of morphology grammars and its role in bootstrapping. In Jäger, G., Monachesi, P., Penn, G., Wintner, S., eds.: Proceedings of Formal Grammar 2004. (2004) 47–62
29. Brent, M.R., Murthy, S., Lundberg, A.: Discovering morphemic suffixes: A case study in minimum description length induction. In: Fifth International Workshop on Artificial Intelligence and Statistics, Ft. Lauderdale, Florida. (1995)
30. Déjean, H.: Concepts et algorithmes pour la découverte des structures formelles des langues. PhD thesis, Université de Caen Basse Normandie (1998)
31. Snover, M.G., Jarosz, G.E., Brent, M.R.: Unsupervised learning of morphology using a novel directed search algorithm: Taking the first step. In: Workshop on Morphological and Phonological Learning at Association for Computational Linguistics 40th Anniversary Meeting (ACL-02), July 6-12, ACL Publications (2002)
32. Argamon, S., Akiva, N., Amit, A., Kapah, O.: Efficient unsupervised recursive word segmentation using minimum description length. In: COLING-04, 22-29 August 2004, Geneva, Switzerland. (2004)
33. Goldsmith, J.: Unsupervised learning of the morphology of natural language. Computational Linguistics **27**(2) (2001) 153–198

34. Neuvel, S., Fulop, S.A.: Unsupervised learning of morphology without morphemes. In: Workshop on Morphological and Phonological Learning at Association for Computational Linguistics 40th Anniversary Meeting (ACL-02), July 6-12. ACL Publications (2002) 9–15
35. Gaussier, É.: Unsupervised learning of derivational morphology from inflectional lexicons. In: Proceedings of the 37th Annual Meeting of the Association for Computational Linguistics (ACL-1999), Association for Computational Linguistics, Philadephia (1999)
36. Sharma, U., Kalita, J., Das, R.: Unsupervised learning of morphology for building lexicon for a highly inflectional language. In: Proceedings of the 6th Workshop of the ACL Special Interest Group in Computational Phonology (SIGPHON), Philadelphia, July 2002, Association for Computational Linguistics (2002) 1–10
37. Oliver, A.: Adquisició d'informació lèxica i morfosintàctica a partir de corpus sense anotar: aplicació al rus i al croat. PhD thesis, Universitat de Barcelona (2004)
38. Creutz, M., Lagus, K.: Unsupervised models for morpheme segmentation and morphology learning. ACM Transactions on Speech and Language Processing (2006) 1–33

NAYOSE: A System for Reference Disambiguation of Proper Nouns Appearing on Web Pages

Shingo Ono[1], Minoru Yoshida[2], and Hiroshi Nakagawa[2]

[1] Graduate School of Information Science and Technology,
The University of Tokyo,
7-3-1 Hongou, Bunkyo, Tokyo, 113-0033, Japan
ono@r.dl.itc.u-tokyo.ac.jp
[2] Information Technology Center, The University of Tokyo
mino@r.dl.itc.u-tokyo.ac.jp, nakagawa@dl.itc.u-tokyo.ac.jp

Abstract. We are developing a reference disambiguation system called *NAYOSE System*. In order to cope with the case the same person name or place name appears over two or more Web pages, we propose a system classifying each page into a cluster which corresponds to the same entity in the real world. For this purpose, we propose two new methods involving algorithms to classify these pages. In our evaluation, the combination of local text matching and named entities matching outperformed the previous baseline algorithm used in simple document classification method by 0.22 in the overall F-measure.

1 Introduction

Have you ever had trouble when you have used a common person/place name as a query in a search engine? For example, when you want to know about a person "George Bush" who is not the president but an ordinary person, many pages describing about the president must bring you trouble. According to the circumstances, we have to look for once more which Web page has the information about the target person/place among too many search results. This problem frequently occurs when the different people (or places, and organization) have the same name. We can find the target Web page, but this often forces us to do hard and time consuming work.

Let us now describe this problem in more detail. Each string appearing as a name on a Web page has a reference to a certain entity in the real world, i.e., each name refers to an entity. A problem occurs when the same name appears on many Web pages, where we do not know whether the same name refers to the same entity or different entities. This problem is the subtask of the co-referencing task [5] in the field of natural language processing. Although it has been recognized in the past, research on it has actively been done in recent years. However, it is not clear whether the previous research is suitable for practical use since most of the experiments were done with artificial data sets.

H.T. Ng et al. (Eds.): AIRS 2006, LNCS 4182, pp. 338–349, 2006.

To rectify this undesirable situation, we considered the NAYOSE System on the Web. NAYOSE means reference disambiguation of names in Japanese. For simplicity, we have described reference disambiguation as NAYOSE in the rest of the paper. Reference disambiguation results in page clusters, which makes an target information accessible. We also propose two algorithms to improve system performance. The system is aimed at practical use. To achieve this, we used real-world data sets that were composed of the top 100 - 200 results from search engines. The queries input to search engines were not restricted to person names, as we used some place names in the experiments. As a result of an experiment, we found our proposed method significantly outperformed the baseline by 0.22 in the overall F-measure.

The remainder of this paper is organized as follows. We first discuss related works in Section 2. In Section 3, we define the task and discuss our methodology for solving it. In Section 4, we propose two new methods for disambiguating references, which are *Local Context Matching* and *Named Entities Matching*. We describe our system in Section 5, and we show the results of our evaluation in Section 6. We conclude the paper in Section 7.

2 Related Works

There have been several important works that tried to solve a reference disambiguation problem. Bagga and Baldwin [1] applied the vector space model to calculating similarity between names using only co-occurring words. Based on this, Niu et al. [6] presented an algorithm using information extraction results in addition to co-occurring words. However, these methods had only been tested on artificial small test data. Therefore, it is doubtful that these methods are suitable for practical use. Mann and Yarowsky [4] employed a clustering algorithm to generate person clusters based on extracted biographic data. However, this method was also only tested on artificial test data. Wan et al. [7] proposed a system that rebuilt search results for person names. Their system called WebHawk was aimed at practical use like ours, but their task was somewhat different from ours. Their system was designed for actual frequent queries. Their algorithm for the system was specialized for English person-name query, consisting of three words: family name, first name, and middle name. They mainly assumed queries as "<*first name*>" or "<*first name*> <*family name*>", and took middle names into consideration, which may have improved accuracy. However, it would not be suitable for other types of names such as those in Japanese (consisting of two words) or place names.

As another approach to this task, Bekkerman and McCallum [2] proposed two methods of finding Web pages referring to a particular person. Their work consists of two distinct mechanisms: the first based on link structure and the second using agglomerative/conglomerative double clustering. However, they focused on disambiguating an existing social network of people, which is not the case for searching for people in reality. In addition, based on our experience, as the number of direct links between pages that contain the same name were fewer than we expect, information on link structures would be difficult to use to resolve our task. Although there may be indirect links (i.e., one page can be found from another page via other pages), it is too time consuming to find them.

3 Bag of Words Model

3.1 Task Definition

Our task, which is the reference disambiguation of names appearing on Web pages, is formalized as follows. The query (target name) is referred to as q. The set of Web pages obtained by inputting query q to a search engine is denoted by $\mathcal{P} = \{p_1, p_2, \cdots, p_k\}$. Each Web page p_i has at least one string q. Then, the jth appearance of string q on Web page p_i is assumed to be s_{ij}. Each s_{ij} only indicates one entity from the set $\mathcal{E} = \{e_1, e_2, \cdots, e_n\}$ of entities in the real world having the name q. Now, the set of s_{ij} is assumed to be $\mathcal{S}$. We define function $\Phi : \mathcal{S} \to \mathcal{E}$. Function Φ means mapping from the name appearing in the document to entities in the real world. In other words, Φ means mapping from a string to an entity. Our purpose is to find function $\breve{\Phi}$ that will approximate function Φ.

The modeling above permits the same string q appearing in the same document to refer to different entities. Web pages with such properties actually exist. However, these pages are quite rare, and dealing with them makes the system more complicated. We therefore decided to ignore these pages. Therefore, We assumed that all of the same string q on a certain Web page d_i would refer to the same entity, i.e., for each i, there exists $e_m \in \mathcal{E}$, such that $\forall j$, $\Phi(S_{ij}) = e_m$. This supposition means that the same name appeared on one page only refers to one entity. Therefore, this results in a simpler model i.e., $\Phi' : \mathcal{P} \to \mathcal{E}$. In this research, our aim was to estimate Φ'. The problem here is n (that appears in the definition of $\mathcal{E}$) is not known previously. In other words, we do not know how many distinct entities have the string q. We actually estimated Φ' by clustering Web pages.

Our system works as follows. Given query q, the system retrieves Web pages that have string q using a search engine and does the reference disambiguation task. Finally, the output of the system is a set of page clusters, each of which refers to the same entity. Although queries have been limited to people's names in previous methods, our aim was to accept not only people's names but all general proper nouns such as place names and organization names. To achieve this goal, the system does not require any knowledge about the query.

3.2 Baseline System

We first implemented a simple system as our baseline. Note that all we need to do is to cluster Web pages. We adopted Agglomerative Hierarchical Clustering (AHC) with the *Bag of Words Model* to calculate similarity between pages.

The string sequence for the page in *Bag of Words Model* (so-called *Vector Space Model*) is represented as a vector whose elements are the weights of words, and word i on page p_j is weighted as

$$weight(i, j) = \mathrm{tf}(i, j) * \log \frac{D}{\mathrm{df}(i)},$$

where $\mathrm{tf}(i, j)$ is the frequency of word i on page p_j; D is the total number of pages; and $\mathrm{df}(i)$ is the number of the pages containing word i. The cosine of the angle between the two resulting vectors is then used as the context similarity measure.

We evaluated this baseline system using the test set described in Section 6.1 and found the F-measure was lower than 0.5. The main problem is shortcoming with the *Bag of Words Model*, which only focuses on the frequency of words. Other significant information in the words is ignored, such as word positions or word meanings. We therefore propose two methods to overcome this kind of shortcoming.

4 Local Context Matching and Named Entities Matching

Now, we propose two methods for calculating similarity. One method is *Local Context Matching* using location information of words on Web pages, and the other method is *Named Entities Matching* using the meanings of surrounding words on the pages.

4.1 Local Context Matching

As discussed in the previous section, the *Bag of Words model* cannot utilize word positions. *Local Context Matching* copes with this shortcoming by giving *nearby words* (i.e., words close to the query) higher scores than others.

Procedure: Local Context Matching

1. For all documents d_j $(1 \leq j \leq k)$ involving query q, do the following.

1.1. Find all appearance positions p_i $(1 \leq i \leq n_j)$ of query q in each document d_j.

1.2. For all p_i, put words whose positions are from $p_i - \alpha$ to $p_i + \alpha$ into S_j, where S_j is a set of nearby words in document d_j.

1.3. Weight each word t in set S_j as

$$w_{tj} = \begin{cases} 0 & \text{if } t \text{ is involved in a stop word list which} \\ & \text{is described later} \\ 1 & \text{otherwise} \end{cases}$$

2. For all document pairs d_x, d_y $(1 \leq x < y \leq k)$, do the following.

2.1. Calculate *local context similarity* $\mathrm{sim}_{\mathrm{LC}}(d_x, d_y)$ as

$$\mathrm{sim}_{\mathrm{LC}}(d_x, d_y) = \sum_{t \in S_x \cap S_y} w_{tx} w_{ty}$$

2.2. If $\mathrm{sim}_{\mathrm{LC}}(d_x, d_y) \geq \theta_{\mathrm{LC}}$, then $\Phi'(d_x) = \Phi'(d_y)$, where θ_{LC} is the threshold.

Here, α is the window size and is currently set to three.

Note that $\Phi'(d_x) = \Phi'(d_y)$ means two pages, d_x and d_y, are to be in the same cluster (see Section 3.1) and clustering is done as follows. Let G be an undirected graph with vertex set V and edge set E. Each vertex $v_i \in V$ corresponds to page d_i. The result of *Local Context Matching* gives edge set E. Each edge $e_{ij} \in E$ exists if and only if constraint $\Phi'(d_i) = \Phi'(d_j)$ was added in Step 2.2 of the algorithm. Then, graph $G = \langle V, E \rangle$ has some connected components. Each connected components means one cluster of Web pages all of which refer to the same entity.

In the case of a person's name, personal data such as his/her age, affiliation, and position will appear near the target name. Therefore, we expect to extract important features by *Local Context Matching* for disambiguating references and checking whether the pages refer to the same entity or not without having to use information extraction methods. All we need to do is focus on the words near the query string.

The stop word list in Step 1.3 prevents meaningless words from exerting an influence. The list, which was made by the hand, consists of stop words in Japanese, such as: *shi*, and *san* (mean "Mr." and "Ms." in Japanese), *namae* (meaning "name"), and *nen*, *tuki*, and *hi* (meaning "year", "month", and "day").

4.2 Named Entities Matching

Named Entities (NEs) are generally more discriminating (i.e. document specific) than general words. We attempted to determine the same person by using NEs. For example, person names can be disambiguated by using heuristics such as:

- If the target person's name and another person's name co-occur on many pages, we can determine the target person on the pages refers to the same entity.
- The organization name that appears near the position of the person's name is thought to be that of the organization the person belongs to. Therefore, if the same organization name is near the position of the target name on many pages, they all refer to the same entity.
- The same person names co-occuring with the same date refer to the same entity.

Place names can be disambiguated in the same way.

Although these heuristics are not necessarily correct, they indicate that the probability of referring to the same entity is very high. The method of determining whether the same name appearing on many pages refers to the same entity using such criteria is called *Named Entities Matching.* NEs characterize documents very strongly, and typically represent their main topics. Therefore, if NEs can be correctly used, they can be expected to become powerful clues for determining whether the same names refer to the same entity or not.

To find out NEs, we used *Sen*[1] as not only a Japanese morphological analyzer but the NE tagger which tags each proper noun in obedience to context, such

[1] http://ultimania.org/sen/

as "person-name (family name)" "person-name (first name)", "place-name", or "organization-name". *Sen* works faster than information extraction systems, and provides more general information. According to our manual benchmarking[2], *Sen* achieved 0.8371 F-measure as the NE tagger for person names and 0.7969 F-measure for place names.

The algorithm for *Named Entities Matching* we used in our system is presented below.

Procedure: Named Entities Matching

1. From all documents d_j $(1 \leq j \leq k)$, extract person names (full name) and place names with a NE tagger.
2. Calculate NE similarity $\text{sim}_{\text{NE}}(d_x, d_y)$ as,

$$\begin{aligned}&\text{sim}_{\text{NE}}(d_x, d_y)\\&= \beta * (\text{number of person names appearing in both } d_x \text{ and } d_y)\\&+\gamma * (\text{number of place names appearing in both } d_x \text{ and } d_y)\end{aligned}$$

3. If $\text{sim}_{\text{NE}}(d_x, d_y) \geq \theta_{\text{NE}}$, then $\Phi'(d_x) = \Phi'(d_y)$, where θ_{NE} is the threshold.

This is where β and γ are parameters for weighting. Having constrains $\Phi'(d_x) = \Phi'(d_y)$, clustering is done in the same way as *Local Context Matching.*

5 NAYOSE System

We describe our *NAYOSE System* in this section. One problem we found in dealing with Web pages is that there were several noisy pages that were inappropriate for our task, such as those page with no meaning and those which refers to multiple entities. We defined these as *junk pages.* We also defined a rule to filter these out. The *NAYOSE System* is outlined at the end of this section.

5.1 Filtering Junk Pages

When the system tries to automatically disambiguate references, some Web pages have a negative influence that decreases efficiency. As previously mentioned, these are called *junk pages.*

Wan et al. [7] have pointed out that it is important to filter junk pages out. Though their task was different from ours, they coped with pages in which the target name referred to non-person entities as junk pages. Since our system was designed to receive proper nouns as general as possible, we determined junk pages as follows. If a Web page conforms to more than one of following properties, the page is determined to be a junk page:

J1. The page has disappeared from the Web (i.e., if "`404 not found`" appears),
J2. The page does not contain the query string,

[2] We checked manually randomly selected 250 NEs (170 person names and 80 place names).

J3. Most of the page is occupied by the enumerations of names or numbers, and
J4. The same name on the page refers to multiple entities.

J1 and J2 are for changes in Web pages. Our system collects Web pages by using results from a search engine. Because there is a time lag between pages cached by the search engine and original page, we need to check if the page we collect actually contains the query string. If the page does not have the query string, it is regarded as a junk page.

J3 is a page that is hard to use. In many cases, it does not have enough information for each person on the page. However, a few pages do have useful information. It is not easy to distinguish these useful pages from junk pages. Therefore, none of the pages extracted under J3 should be dumped. But they are marked as "J3" and treated in a special manner as later mentioned.

J4 is a page that is beyond the scope of our task where we have assumed one page would refer to only one entity.

We defined the following rules to filter out junk pages:

F1. The URL contains Japanese characters,
F2. The title contains the string "search result" (*kensaku kekka*, in Japanese),
F3. Named entities appear too frequently, and
F4. There is no string corresponding to the query.

Each rule copes with certain junk pages. The F1 and F2 rules are for junk pages J4, the F3 rule is for J3, and the F4 rule is for J1 and J2. The F1 and F2 rules are based on our experience. We found the pages on which the same name refers to multiple entities were mostly the output automatically generated by some systems. Therefore, these page may have the title contains the string "search result" or have the URL contains Japanese characters such as:
`http://*****.jp/keyword/%E4%B8%AD%E5%B7%9D%E8%A3%95%E5%BF%97/`or
`http://*****.jp/search?q=%E4%B8%AD%E5%B7%9D%E8%A3%95%E5%BF%97/`,
where `%E4%B8%AD%E5%B7%9D%E8%A3%95%E5%BF%97` is the Japanese string "Hiroshi Nakagawa" encoded for URL. We regarded the encoding string `%80-%FF` for URL were regarded as a Japanese character.

According to junk page properties, the F1, F2, and F4 rules dump junk pages, and the F3 rule marks junk pages J3 with the label "J3". It is feared that J3 pages have meaningless and noisy NEs which lead to errors in Named Entities Matching. Therefore, NEs in these "J3" labeled pages are ignored in Named Entities Matching.

We have preliminarily examined our filtering rules. We checked 1362 Web pages and found 20 pages (1.5%) conformed J4. In addition, 78 pages in the rest of 1342 pages (5.8%; because the F1 and F2 rules dump J4 pages and the F3 rule does not dump.) conformed J3. Although junk pages do not appear so frequently, these pages often lead to critical clustering errors.

We used precision (P), recall (R), and F-measure (F) for the metrics. The metrics were calculated as follows: let J be a set of junk pages and let S be a set of pages which were regarded as junk pages by filtering rules,

$$P = \frac{|S \cap J|}{|S|},\ R = \frac{|S \cap J|}{|J|},\ F = \frac{2PR}{P+R}.$$

Table 1. Results of preliminary experimentation about filtering rules

Rules	F-measure	Precision	Recall
F1 and F2	0.7143	0.5556 (20/36)	1.0000 (20/20)
F3	0.5379	0.3679 (78/212)	1.0000 (78/78)

Table 1 summarizes results of our preliminary experimentation. Note that the F4 rule which was for unavailable pages was excepted from this experimentation since it is no doubt that the rule works clearly. Since the F1 and F2 rules have the same target J4, we grouped the rules together. We see filtering rules work effectively. Recall is 1.0 as shown in Table 1. That means that the rules filtered all junk pages out perfectly at least in this preliminary experiment. Some pages were dumped by errors. However, the number of pages dumped by errors is very small in all pages because junk pages are minority of retrieved pages.

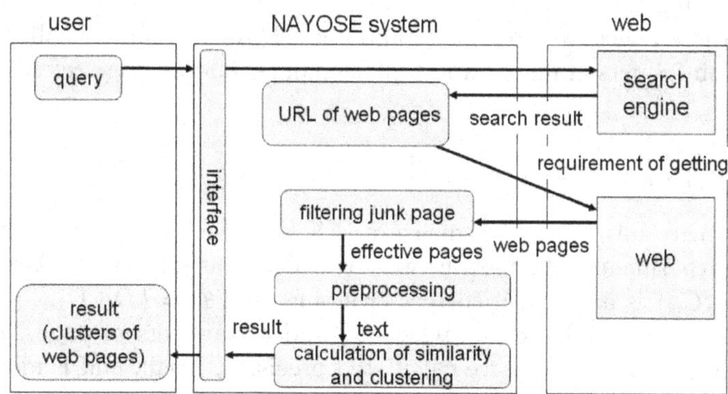

Fig. 1. Overview of *NAYOSE System*

5.2 Outline of Nayose System

There is an overview of the *NAYOSE System* in Fig. 1. The system works as follows. It:

1. Receives a query from an user,
2. Retrieves Web pages' URLs with a search engine[3] using the query and obtains the top k search results,
3. Downloads all top k pages, and executes preprocessing (junk page filtering, morphological analysis and NE tagging),

[3] We use Yahoo! JAPAN Web Service. (`http://developer.yahoo.co.jp/start/`)

4. Calculates the similarity between Web pages and does clustering,
5. Outputs the results.

6 Experimentation

6.1 Data Set

As far as we know, no gold standard for the task has yet been proposed. We originally developed the test set for this task for this reason.

We first input Japanese person-name queries and Japanese place-name queries into a search engine. Part of the person queries and all the place queries were chosen from ambiguous popular names. For example, the person name "Taro Kimura" is a very common name in Japan: we found there were many people called "Taro Kimura" such as a famous commentator, a member of the Diet, a translator, and a schoolmaster. As another example, the place name "Fujimicho" is found throughout Japan. Some other person queries were selected from persons in our laboratory, and other person name queries were generated automatically.

Second, we tried to extract Web pages containing these names. We retrieved these pages with a search engine. If the query hit many pages, we collected the top 100-200 Web pages. All junk pages were labeled as "J{1-4}", and treated with the filtering rules.

Finally, these pages were manually annotated[4]. As a result, we collected 3859 Web pages on 28 person names and 9 place names, and all page references were clarified.

6.2 Evaluation

Precision (P), recall (R), and F-measure (F) were used as the evaluation metrics in our experiments. All metrics were calculated as follows [3]. Assume $\mathcal{C} = \{C_1, C_2, \cdots, C_n\}$ is a set with correct grouping, and $\mathcal{D} = \{D_1, D_2, \cdots, D_m\}$ is a set for the result of clustering, where C_i and D_j are sets of pages. For each correct cluster $C_i (1 \le i \le n)$, we calculated precision, recall, and F-measure for all clusters $D_j (1 \le j \le m)$ as

$$\mathrm{P}_{ij} = \frac{|C_i \cap D_j|}{|D_j|}, \ \mathrm{R}_{ij} = \frac{|C_i \cap D_j|}{|C_i|}, \ \mathrm{F}_{ij} = \frac{2\mathrm{P}_{ij}\mathrm{R}_{ij}}{\mathrm{P}_{ij} + \mathrm{R}_{ij}}.$$

The F-measure of C_i (F_i) was calculated by $\mathrm{F}_i = \max_j \mathrm{F}_{ij}$. Using $j' = \mathrm{argmax}_j \mathrm{F}_{ij}$, P_i and R_i were calculated as $\mathrm{P}_i = \mathrm{P}_{ij'}, \mathrm{R}_i = \mathrm{R}_{ij'}$.

The entire evaluation was conducted by calculating the weighted average where weights were in proportion to the number of elements of clusters, calculated as

[4] Note that annotators were unable to determine pages perfectly. There were a few pages that were too ambiguous to determine. To standardize them, if annotators encountered these ambiguous pages, each of them was regarded as referring to another independent entity, i.e., each of them composed a cluster by itself in correct grouping.

$$F = \sum_{i=1}^{n} \frac{|C_i| F_i}{|\mathcal{C}|},$$

where $|\mathcal{C}| = \sum_{i=1}^{n} |C_i|$. The weighted average precision and recall were also calculated in the same way for the F-measure.

We investigated which of the *Bag of Word model* (**BoW**), *Local Context Matching* (**LC**), *Named Entities Matching* (**NE**) or their combinations were the best. We compared the following methods: "BoW (Baseline)", "LC", "LC, BoW", "NE", "NE, BoW", "NE, LC", "NE, LC, BoW". Combinations of two or three methods means using different methods together. More precisely, the result of the combination of LC and NE is given by considering graph $G = \langle V, E_{NE} \cup E_{LC} \rangle$ where $G_{LC} = \langle V, E_{LC} \rangle$ is the result for LC and $G_{NE} = \langle V, E_{NE} \rangle$ is the result for NE. When BoW was combined with NE/LC methods, NE/LC methods were applied first, and BoW was then applied to the NE/LC results. Table 2 lists the results of the pre-experimentation that revealed that applying NE/LC methods before BoW were more effective than applying NE/LC methods after it.

Table 2. Results of pre-experimentation about the order between NE/LC and BoW

methods	LC, BoW	NE, BoW	NE, LC, BoW
NE/LC → BoW (F-measure)	0.5549	0.6149	0.6452
BoW → NE/LC (F-measure)	0.5158	0.5726	0.5149

Table 3 lists the results of an average of 37 queries. According to the results, all the methods that involve at least one proposed methods (NE/LC) showed higher performance than the baseline. As we intended, *Named Entities Matching* and *Local Context Matching* achieved high performance by using significant information about words in the pages, such as word meanings and word positions. The combination of *Named Entities Matching* and *Local Context Matching* outperformed the baseline by a significant 0.22 in the overall F-measure. In addition, the highest precision of almost 0.95 was obtained by only using *Named Entities Matching*, which contributed to the high F-measure of *NE and LC*.

Table 3. Results: Average of 37 queries

Applied method(s)	F	P	R
BoW(Baseline)	0.4596	0.6821	0.5154
LC	0.5871	0.8302	0.5852
LC, BoW	0.5510	0.6889	0.6597
NE	0.6311	**0.9484**	0.5585
NE, BoW	0.5874	0.7126	0.6579
NE, LC	**0.6834**	0.7991	0.7357
NE, LC, BoW	0.6225	0.6639	**0.7811**

Table 4. Results: Average 26 person queries

Applied method(s)	F	P	R
BoW (Baseline)	0.4883	0.6950	0.5488
LC	0.6384	0.8507	0.6380
LC, BoW	0.5884	0.6966	0.7125
NE	0.7240	**0.9297**	0.6561
NE, BoW	0.6492	0.7347	0.7306
NE, LC	**0.7437**	0.8058	0.8168
NE, LC, BoW	0.6710	0.6640	**0.8563**

Table 5. Results: Average of 9 place queries

Applied method(s)	F	P	R
BoW (Baseline)	0.3798	0.5853	0.4536
LC	0.4793	0.7335	0.4999
LC, BoW	0.4410	0.6048	0.5525
NE	0.3809	**0.9912**	0.3063
NE, BoW	0.3947	0.5921	0.4701
NE, LC	**0.5266**	0.7351	0.5564
NE, LC, BoW	0.4685	0.5963	**0.6064**

Using the *Bag of Words Model* improved the recall, but decreased precision. Since the drop in precision was more than the gain in recall, the overall F-measure decreased. We observed calculating similarity with the *Bag of Words Model* often induced mistaken cluster merging and caused serious drop in precision.

More detailed results are listed in Tables 4 and 5. They reveal similar tendencies: the combination of NE and LC yields the highest F-measure. A precision score of NE for person queries was over 0.92. However, these methods are less efficient for place than for person queries.

7 Conclusion

We introduced the reference disambiguation system NAYOSE that we had developed in this paper. It is aimed at accurately classifying Web pages by entity without assuming knowledge concerning queries. We first proposed two methods for calculating similarity to improve the accuracy of reference disambiguation, i.e., *Local Context Matching* and *Named Entities Matching.* We next defined junk pages as those with a thin meaning on the Web, and we explained the mechanism to remove these. As we intended, we showed two methods were very effective in our evaluation. The combination of *Local Context Matching* and *Named Entities Matching* outperformed the previous algorithm by 0.22 in the overall F-measure.

In our future work, we have to make the system more adaptive for any types of query, such as a place-name or an organization-name. Our experimentation showed our proposed methods were not effective enough to make high quality clusters for a place-name query. Therefore, we will explore new methods for calculating and clustering.

References

1. A. Bagga and B. Baldwin: Entity-Based Cross-Document Coreferencing Using the Vector Space Model. In Proceedings of COLING-ACL 1998, pp. 79–85, 1998.
2. R. Bekkerman and A. MacCallum: Disambiguating Web Appearances of People in a Social Network. In Proceedings of WWW2005, pp. 463–470, 2005.
3. B. Larsen and C. Aone: Fast and effective text mining using linear-time document clustering. In Proceedings of the 5th ACM SIGKDD. pp. 16–22, 1999.
4. G. S. Mann and D. Yarowsky: Unsupervised Personal Name Disambiguation. In Proceedings of CoNLL2003, pp. 33–40, 2003.
5. T. S. Morton: Coreference for NLP Applications. In Proceedings of ACL-2000, pp. 173–180, 2000.
6. C. Niu, W. Li, and R. K. Srihari: Weakly Supervised Learning for Cross-document Person Name Disambiguation Supported by Information Extraction. In Proceedings of ACL-2004, pp. 598–605, 2004.
7. X. Wan, J. Gao, M. Li, and B. Ding: Person Resolution in Person Search Results: WebHawk. In Proceedings of CIKM2005, pp. 163–170, 2005.

Efficient and Robust Phrase Chunking Using Support Vector Machines

Yu-Chieh Wu[1], Jie-Chi Yang[2], Yue-Shi Lee[3], and Show-Jane Yen[3]

[1] Department of Computer Science and Information Engineering, National Central University,
[2] Graduate Institute of Network Learning Technology, National Central University, No.300, Jhong-Da Rd., Jhongli City, Taoyuan County 320, Taiwan, R.O.C.
bcbb@db.csie.ncu.edu.tw, yang@cl.ncu.edu.tw
[3] Department of Computer Science and Information Engineering, Ming Chuan University, No.5, De-Ming Rd, Gweishan District, Taoyuan 333, Taiwan, R.O.C.
{leeys, sjyen}@mcu.edu.tw

Abstract. Automatic text chunking is a task which aims to recognize phrase structures in natural language text. It is the key technology of knowledge-based system where phrase structures provide important syntactic information for knowledge representation. Support Vector Machine (SVM-based) phrase chunking system had been shown to achieve high performance for text chunking. But its inefficiency limits the actual use on large dataset that only handles several thousands tokens per second. In this paper, we firstly show that the state-of-the-art performance (94.25) in the CoNLL-2000 shared task based on conventional SVM learning. However, the off-the-shelf SVM classifiers are inefficient when the number of phrase types scales to high. Therefore, we present two novel methods that make the system substantially faster in terms of training and testing while only results in a slightly decrease of system performance. Experimental result shows that our method achieves 94.09 in F rate, which handles 13000 tokens per second in the CoNLL-2000 chunking task.

1 Introduction

With the exponential growth of textual information available in the World Wide Web, there has been many research efforts addressed on effectly discover important knowledge in text data, i.e. text mining and knowledge discovery in text (KDT). The ultimate goal of text chunking is to identify the non-recursive, non-overlap phrase structures in natural language text which is also the intermediate step between part-of-speech tagging (POS) and full parsing. Text chunking does not only support knowledge-based systems to extract syntactic relations, but also provide an important knowledge representation fundamental. For example, Sagae (Sagae et al., 2005) adopted a machine learning-based evaluating system to measure the numerical score for grammatical complexity to children instead of human judgments. In addition, chunk information is also the key technology in natural language processing areas, e.g., chunking-based full parsing (Tjong Kim Sang, 2002), clause identification (Carreras et al., 2005), and machine translation (Watanabe et al., 2003). To effect acquire

H.T. Ng et al. (Eds.): AIRS 2006, LNCS 4182, pp. 350 – 361, 2006.

and represent knowledge from text, a high-performance phrase chunking system is indispensable. To handle huge text dataset, the efficiency of a chunking system should be taken into account.

Over the past few years, many researches have been addressed on developing a high-performance chunking system and have applied various machine learning methods to it (Ando and Zhang, 2005; Carreras and Marquez, 2003; Kudoh and Matsumoto, 2001; Molina and Pla, 2002). CoNLL-2000 provided a competition on arbitrary phrase chunking where all systems should be evaluated on the same benchmark corpus. Till now, there have been over 30 papers investigated in this issue. Currently, many high performance phrase chunking systems had been proposed, such as SVM-based (Kudoh and Matsumoto, 2001; Wu et al, 2006). SVM-based algorithms showed an excellent performance in terms of accuracy. However, its inefficiency in actual analysis limits practical purpose, such as information retrieval and question answering. For example, a linear kernel SVM-based phrase chunker (Wu et al., 2006) runs at a rate of about 1600 tokens/sec, while achieved 94.12 in F rate. In comparison, the language model-based approaches (Molina and Pla, 2002) can process 40000-50000 tokens per second[1]. Such inefficient execution time is inadequate for real time processing where fast chunking of large quantities of text is indispensable.

In this paper, we propose two methods that make the SVM-based chunking model substantially more efficient in terms of training and testing. These methods are applicable not only to the chunking tasks but also to other similar tasks such as POS tagging and named entity recognition (NER) that do not conflict to the selected kernels. The proposed two fast classification algorithms are designed based on the word-classification model (Kudoh and Matsumoto, 2001; Ando and Zhang, 2005; Zhang et al., 2002). One is C-OVA (Constraint One-Versus-All), which is an extension of traditional one-versus-all classification scheme. The other is HC-OVA (Hierarchical Constraint One-Versus-All) where the classification process can be viewed as visiting a skewed binary tree. By applying HC-OVA, both training and testing times are largely reduced. Both theoretical and experimental results show that HC-OVA and C-OVA outperform conventional methods for testing (247% and 152%) and substantially faster than classic OVA in terms of training. In the CoNLL-2000 shared task, our method can handle 13000 tokens per second while the performance is kept in 94.09 in F rate.

2 SVM-Based Phrase Chunking Model

Ramshaw and Marcus (1995) firstly proposed an inside/outside label style to represent noun phrase chunks. This method involves in three main tags, B, I, and O. I tag indicates the current token which is inside a chunk, B tag is used to represent the beginning of a chunk which immediately follows another chunk, O tag means the current token does not belong to a part of chunk. This method is also called IOB1. Tjong Kim Sang [12] derived the other three alternative versions, IOB2, IOE1, and IOE2.

IOB2: is different from IOB1, which uses the B tag to mark every beginning token of a chunk and the other inside tokens are labeled as I tag.

[1] The HMM-based chunking system only achieves 92.19 in F rate.

IOE1: An E tag is denoted as the ending token of a chunk which is immediately before a chunk.

IOE2: The E tag is given for every token that is the end of a chunk.

Let us illustrate the four representation styles with an example, considering an incomplete sentence "In early trading in busy Hong Kong Monday". The four representation styles of the sentence are listed in Table 1. This example only encodes the noun phrase chunk type. It can be extent to mark the other phrase types with labeling the specific type behinds the I/B/E tags. For example, the B-VP is used to represent the beginning of a verb phrase (VP) in the IOB2 style.

Table 1. An example for IOB1/2 and IOE1/2 chunk representation styles

	IOB1	IOB2	IOE1	IOE2
In	O	O	O	O
early	I	B	I	I
trading	I	I	I	E
in	O	O	O	O
busy	I	B	I	I
Hong	I	I	I	I
Kong	I	I	E	E
Monday	B	B	I	E

2.1 Phrase Chunking Model

In general, the contextual information is often used as the basic feature type; the other features can then be derived based on the surrounding words, such as words and their POS tags. The chunk tag of the current token is mainly determined by the context information. Similar to previous researches (Gimenez and Marquez, 2003; Kudo and Matsumoto, 2001), we employ a classification algorithm (i.e. SVM) learns to classify the chunk class for each token via encoding the context features. In this paper, we adopt the following feature types.

- Lexical information (Unigram/Bigram)
- POS tag information (UnPOS/BiPOS/TriPOS)
- Affix (2~4 suffix and prefix letters)
- Previous chunk information (UniChunk/ BiChunk)
- Orthographic feature type (See (Wu et al., 2006))
- Possible chunk classes (See (Wu et al., 2006))
- Word + POS Bigram (current token + next token's POS tag)

In addition, the chunking directions can be reversed from left to right into right to left. The original left to right chunking process classifies tokens with the original directions, i.e. the class of the current token is determined after chunking all preceding tokens of the current word. In the reverse version, the chunking process begins from the last token of the sentence to the first token. We name the original chunking process as **forward chunking**, while the reverse process as **backward chunking**.

In this paper, we employ SVM^{light} (Joachims, 1998) as the classification algorithm, which has been shown to perform well on classification problems (Gimenez and Marquez, 2003; Kudoh and Matsumoto, 2001). Since the SVM algorithm is a binary classifier, we have to convert it into multiple binary problems. Here we use the One-Versus-One (OVO) type to solve the problem. As discussed in (Gimenez and Marquez, 2003; Wu et al., 2006), working on linear kernel is much more efficient than polynomial kernels. To take the time efficiency into account, we choose the linear kernel type.

2.2 Preliminary Chunking Results

To fairly compare with previous studies, we use the chunking data set from CoNLL-2000 shared task. This corpus is derived from the English treebank Wall Street Journals sections: 15-18 for training and section 20 for testing. The POS tag information for each token is mainly produced by Brill-tagger (Brill, 1995) as consistent with the CoNLL-2000 shared task. In this task, there are 11*2+1 chunk classes (11 phrase types with B/I for IOB1,2 or I/E for IOE1,2 tags and one outside tag O). Table 2 shows the experimental results of the four representation styles with forward (F) and backward (B) chunking directions.

As shown in Table 2, the best system performance is obtained by applying IOE2 with backward chunking (94.25 in F rate) in the CoNLL-2000 chunking task. Compared to previous best results (94.12 reported by (Wu et al., 2006), we further improve the chunking accuracy by combining word+POS bigram feature and IOE2-B method. In terms of efficiency, the training time is about 4.5 hours while it handles 5200 tokens per second. Although the better system performance is reached, the cost of training time increased from 2.8 to 4.5 hours. It is worth to note that we do not solve the inconsistent chunk class problem, instead, we simply use the output result produced by SVM, i.e., deterministic chunking. This simplified version largely increases the testing time efficiency (from 1200 to 5200 tokens per second).

Table 2. Chunking results for different representation styles and directions of the CoNLL-2000 shared and Chinese base-chunking tasks

	CoNLL-2000	Chinese base-chunking
IOB1-F	93.96	92.09
IOB1-B	93.92	92.09
IOB2-F	94.22	**92.30**
IOB2-B	93.77	92.10
IOE1-F	93.80	92.02
IOE1-B	93.77	91.87
IOE2-F	93.81	90.27
IOE2-B	**94.25**	92.10

Similar results were obtained by porting our chunking system to Chinese base-chunking task. The right hand side of Table 2 is the experimental result of different representation styles with combining forward and backward chunking directions. The best performance is obtained by employing the IOB2-F. Testing corpus of the Chinese base-chunking task was derived from the Chinese Treebank, 0.27 million words for training while 0.08 million words for testing. In this task, the training time is 16 hours while the chunking speed is 4700 tokens per second.

3 Efficient Phrase Chunking Schema

3.1 Hierarchical Constraint-One-Versus-All

For OVA, the chunk class of each token is mainly determined by comparing with all existing categories. The linear kernel SVM compares each example once for each category. But when the number of category scales to high, the comparison times also increase. In the CoNLL-2000 chunking dataset, there are 11 phrase types that produce 11*2+1=23 chunk classes, while in the Chinese base-chunking task, 23 phrase types contribute to 23*2+1=47 chunk classes. The situation is even worse for other non-linear kernels as they aim to compare with all the support vectors or applying OVO (one-versus-one) multiclass strategy that involves in |C|*(|C|-1)/2 comparisons where |C| is the number of chunk classes. Although the original chunking system is somewhat fast (by applying linear kernel SVM), when the phrase type tends to be hundreds, the large number of categories considerably decrease the chunking efficiency. For instance, in the Chinese base-chunking task, the chunking speed decreases to 4700 compared to the CoNLL-2000 shared task where the number of chunk classes increase from 23 to 47.

The main reason is given rise to the more number of chunk classes that increase several times larger for SVM classifying. In chunking task, the chunk tag is encoded as IOB styles as described in Section 2.1. About half of the categories are unnecessarily compared. These comparisons are useless to determine the chunk class of current token. For example, when previous token was chunked as B-NP, the next token is impossible to be the I-VP, I-PP,…etc.

To find the relationships between chunk classes, a consistent matrix is used, which marks the validity of previous and current chunk classes. For example, Table 3 lists the consistent matrix of noun phrase (NP) and verb phrase (VP) types using IOB2. The first column indicates the chunk class of previous token while the first row gives possible chunk classes of current token. Intuitively, if previous chunk tag was classified as B-NP, the current chunk class should belong to one of the four chunk classes, B-NP/B-VP/I-NP/O, whereas I-VP is invalid. We then generalize this relation to the other three IOB styles, IOB1, IOE1, and IOE2. Here, we define all chunk classes should belong to one of the following two groups.

Begin-group: start of a chunk, for example, the B-tag for IOB2, I-tag for IOE2.
Interior-group: inside of a chunk, for example, I-tag for IOB2, and E-tag for IOE2.

Based on the relationship between the two groups (see Fig. 1.), we can easily build the consistent matrix by connecting the validity between the two groups. That is, whether previous chunk class is Begin or Interior-group, the chunk class of current token potentially belongs to any member of the Begin-group or a specific chunk class of the Interior-group in which the phrase type is equivalent to previous. In other words, from Begin/Interior-groups to Begin-group is valid, while from Begin/Interior-groups to Interior-group is valid only when they share the same chunk.

Table 3. Consistent matrix

	B-NP	B-VP	I-NP	I-VP	O
B-NP	1	1	1	0	1
B-VP	1	1	0	1	1
I-NP	1	1	1	0	1
I-VP	1	1	0	1	1
O	1	1	0	0	1

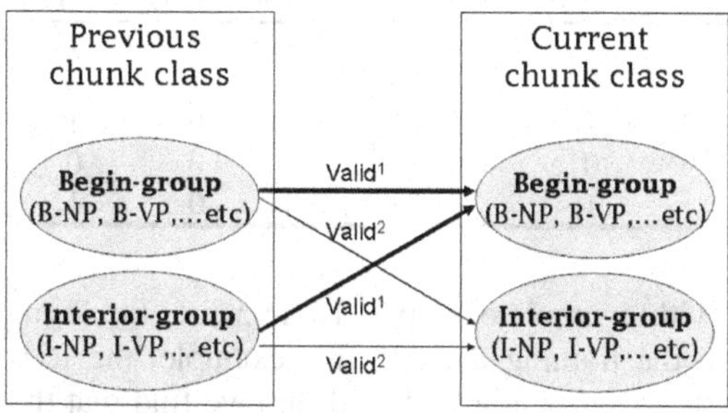

Fig. 1. The validity relationship between begin and interior groups

Obviously, for IOB2 representation style, it is not difficult to construct the consistent matrix manually. However, for the other three representation styles with forward and backward chunking directions, manual-constructing this matrix requires lots of human effort especially to the numerous chunk types, like Chinese. To overcome this, we propose an algorithm (see algorithm 1) to build the consistent matrix automatically.

Algorithm 1: Auto-construct consistent matrix
Notation: IOB-tag = {I, B, E} (Obviously, O tag should be the Begin-group)
Chunk type = {NP, VP, PP,...}
Chunk class is the combinations of each chunk type and IOB-tag

1. Initialize the consistent matrix (CM) by verifying the valid chunk class pairs of previous and current tokens in the training data
2. Get IOB-tag and set them as Interior-group initially.
   (for example, **B**-NP => **B**, **I**-NP => **I**)
3. For each IOB-tag,
   For each chunk class pair (Ch_j and Ch_k) with the same IOB-tag where $j \neq k$
   Check whether there exists a valid item between Ch_j and Ch_k in CM
   If the validity exists, then this IOB-tag is assigned to Begin-group
4. Re-organizes CM with the relations between Begin-group and Interior-group.

As outlined in algorithm 1, the first two steps aim to initialize the consistent matrix by scanning the previous-current chunk class pairs and extracting the IOB tag from training data. In the third step, we check the consistency of a chunk class pairs of the same IOB-tag in the initial consistent matrix. If the two chunk classes are valid, then the IOB tag belongs to Begin-group. Since for Interior-group, the validity only exists when previous and current chunk share the same phrase type. Once the Begin and Interior groups were determined, the final consistent matrix could be generated by connecting the relationships between the two types. It is worth to note that we ignore the O-tag here since it should be the Begin-group clearly.

Table 4. Initial consistent matrix

	B-NP	B-VP	I-NP	I-VP	O
B-NP	1	1	1	0	1
B-VP	0	0	0	1	0
I-NP	1	0	1	0	1
I-VP	0	0	0	0	0
O	1	0	0	0	1

Let us illustrated with a simple example, Table 5 lists the initial consistent matrix that is gathered from the training data. In this example, the IOB tags are B and I which are set to Interior-group initially. For B tag, we find that the validity occurs in the chunk class pair, B-NP and B-VP in which they share the same IOB tag (i.e. B tag) in the initial consistent matrix. Thus, B tag is assigned as Begin-group. For I tag, we conclude that the validity only exists when previous and current chunk classes belong to the same phrase type, for example, NP and VP. Therefore I-tag is still in the Interior-group. After classifying Begin/Interior-groups for each IOB tag, the final consistent matrix is reconstructed by bridging the relationship between the two types. That is, we set the column of each Begin-group as valid, and the validity of the Interior-group is valid only when its previous chunk is the same as current chunk. After this step, we can automatically construct the consistent matrix as Table 3.

Even the consistent matrix method can reduce half of the comparison times for SVM. However when the number of chunk types is large, comparing with all the categories is time-consuming. Analytically, in the chunking task, we observe that about 50% of the tokens belong to NP, in particular the three phrase types, NP, VP, and PP cover 90% of the training data. In Chinese base-chunking, the proportion of NP in the training set is 57% and the three phrase types NP, ADVP and VP denominate 83%. In most cases, the chunk class of each token should be one of those high-frequent phrase types. Therefore, we design a new multi-class strategy for SVM to restrict the comparison of high-frequent chunk classes at higher priority. The main spirit of this method is to reduce unnecessary comparisons through comparing the high-frequent chunk class first. This schema can be illustrated with a skewed binary tree (see Fig. 2.).

Every internal node in Figure 2 represents a decision of one class and the remaining non-visited classes. Each leaf node denotes as a chunk class, the higher frequent the chunk class observed, the higher level it is in the tree. Determining the chunk class of a token is equivalent to visit the skewed binary tree. Once the leaf node is visited in the higher level, the remaining nodes will no longer be compared. Therefore, most rare chunk classes are not compared in testing phase. Combining with the consistent matrix, some of the internal nodes can be ignored by checking the validity. The overall training and testing algorithm is described as follows.

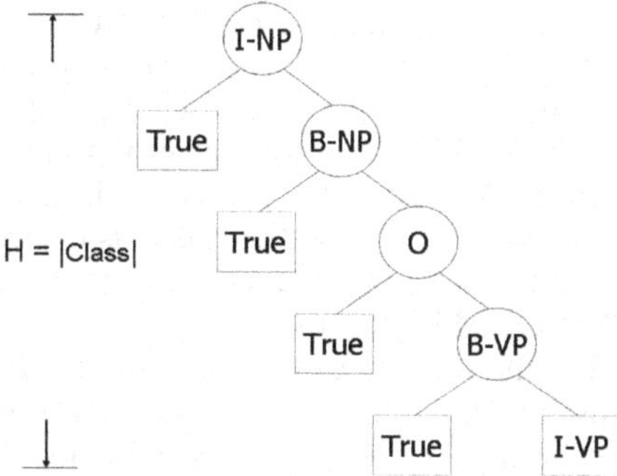

Fig. 2. Skewed tree structures for multi-class chunking strategy

```
Algorithm 2: Training algorithm
Notation: |C| is the number of chunk classes.
Algorithm:
 1. Construct the consistent matrix using Algorithm 1.
 2. Build the ordering list: Or[i] based on counting the frequency of each chunk class in the
    training set. The Or list stores all chunk classes.
 3. For i :=1 to |C|-1
           Collecting the positive examples from chunk class Or[i].
           For j := i+1 to |C|
                Collecting negative examples from chunk class Or[j].
           Training SVM for chunk class Or[i].
```

```
Algorithm 3: Testing algorithm
Notation: t_i: token i. Ch_i: is the chunk class of t_i.
Algorithm:
    For each token t_i,
         For j := 1 to |C|
              If (consistent (Ch_{i-1}, Or[j])) // check the validity between previous and current
                                                                                  chunk classes
                   Ch_i := Or[j].
                   Classify t_i with SVM_{Or[j]}
                   If t_i is classified as positive, then
                        Stop comparing.
```

As outlined in algorithm 2, we construct an ordering list via estimating the frequency in the training data. For example, in Figure 1, the item of Or[0] is I-NP. These chunk classes are trained by following this order (the third step). Note that iteratively, we discard a subset of training example that belongs to previous chunk class. On the other hand, the testing algorithm (algorithm 3) is to compare each support vector machine according to the ordering list. In addition, the consistent matrix limits the comparisons only when previous-current chunk class pair is valid in the matrix.

3.2 Constraint-One-Versus-All

The main idea of constraint OVA method is to reduce half of the comparison times using the consistent matrix. During testing, we only put emphasis on picking up the whole Begin-group and one from Interior-group. As described in Section 3.1, the Interior-group should follow the Begin-group with the same phrase type. For the previous example, three chunk classes belongs to Begin-group, B-NP, B-VP, and O, while the I-NP and I-VP was assigned to Interior-group via algorithm 1. In testing phase, for the first token, we only focus on classifying with all of the chunk classes in the Begin-group, i.e. B-NP, B-VP, O. If the first token is classified as B-NP, then the second token should not be the other chunk classes in Interior-group other than I-NP. Therefore the B-NP, B-VP, O and I-NP is used.

Based on this constraint, the training examples of the Interior-group could be further reduced. It is impossible to compare two chunk classes of the Interior-group simultaneously. On the contrary, the Begin-group is invited to be compared with a specified chunk class of Interior-group. Considerably, we can not reduce any training example for Begin-group. In overall, training the chunk class of the Begin-group is equivalent to classic one-versus-all method, while training the Interior-group, only one chunk class is used to against to the whole Begin-group.

3.3 Time Complexity Analysis

The training time complexity of single SVM is ranged from $O(m^2)$ to $O(m^3)$ (assume $O(m^2)$) where m is the number of training examples [10]. Based on this assumption, we can easily derive the training time complexity for classic OVA is $O(|C|m^2)$ in which we train $|C|$ single SVM for each category. Testing time for OVA involves in comparing with all categories, i.e. $|C|$ times denoted by $O(|C|)$.

Different from OVA, the OVO (One-Versus-One) constructs $\binom{|C|}{2}$ SVMs for arbitrary two class pairs. When training data is balanced, i.e. for each category there are $m/|C|$ examples, the training time complexity is

$$O(C_2^{|C|}(\frac{2m}{|C|})^2) = O(\frac{2(|C|-1)}{|C|}m^2) = O((\frac{2|C|}{|C|} - \frac{2}{|C|})m^2) \cong O(2m^2)$$

In the bias case, all training examples centralize to a specific class, the training time complexity is $O((|C|-1)m^2)$.

In 2000, Plat [10] had proved that the DAG (Directed acyclic graph)-based multi-class SVM had the same training time complexity as OVO method, while testing time complexity was equivalent to OVA by assuming training single SVM is $O(m^2)$. The main idea of DAG is based on the results of the OVO where testing phase is similar to visit the acyclic graph. But the selecting of the classification order is heuristically.

On the other hand, we analyze the time complexity of the proposed two methods, COVA, and HC-OVA. For HC-OVA, the training time complexity is only $O(m^2)$ when data is familiar to one class. In the balanced case, each category contains $m/|C|$ examples, we have,

$$(\frac{|C|}{|C|}m)^2 + (\frac{|C|-1}{|C|}m)^2 + \ldots + (\frac{3}{|C|}m)^2 + (\frac{2}{|C|}m)^2$$

As described in Section 3.1, at each time, we use single SVM learns to classify one class. At the next step, the training data is reduced by removing previous positive examples. In the balanced case, we can almost remove $m/|C|$ examples at each step from the whole set. The above equation can be simplified as follows.

$$=(\frac{2^2+3^2+...+|C|^2}{|C|^2})m^2=(\frac{|C|(|C|+1)(2|C|+1)}{6|C|^2})m^2=(\frac{(|C|+1)(2|C|+1)}{6|C|})m^2\cong(\frac{|C|}{3})m^2$$

For testing time complexity, in the best case HC-OVA only compares once for testing, i.e., O(1), while in the worst case, it should compare all of the chunk classes that belong to Begin-group and a class from Interior-group, i.e. O(B+1) where B is the number of chunk classes in the Begin-group. On the other hand, the training time complexity of C-OVA in the balanced case is:

Begin-group: $O(Bm^2)$
Interior-group: $O((|C|-B)\frac{(B+1)^2}{|C|^2}m^2)$

Overall: $O((B+\frac{(|C|-B)(B+1)^2}{|C|^2})m^2)$

In the unbalanced case, when dataset biased to the Begin-group, the training time complexity is $O(Bm^2)$. On the contrary, if the dataset biased to a chunk class of Interior-group, the overall training time is $O(Bm^2+m^2)$. For the testing time, C-OVA compares the whole Begin-group and one chunk class from Interior-group, thus, the time complexity is precisely O(B+1) which is the same as the worst case of HC-OVA.

In this paper, we focus on the chunking task which is not usually a balanced case. In most situations, only a few phrase types cover 90% of the data, such as NP, and VP. In the biased situation, the proposed HC-OVA and C-OVA is more efficient than the other methods since $|C| > B$ (usually $B=|C|/2$). Although the training times of our methods are worse than DAG and OVO in the balanced case, our method provide more efficient theoretical testing speed for fast text analysis.

4 Experimental Results

In the unbalanced case, we had shown the proposed methods more efficient than the other approaches. In this section we concern the actual chunking performance of the two methods. We combine the two methods with the original chunking model (as discussed in Section 2) and applied them to the same chunking tasks. Table 5 lists the actual results of the CoNLL-2000 shared task, and the experimental results of the Chinese base-chunking task is given in Table 6. In the CoNLL-2000 shared task, the actual training time of HC-OVA/C-OVA are 1.85 and 4 hours which outperforms the conventional OVA method (4.5 hours). Besides, the testing speeds of HC-OVA/C-OVA are 12902 and 7915 tokens per second that are 247% and 152% substantially faster than OVA method. Both training and testing times are largely reduced. In the Chinese base-chunking task, the testing times were 219% and 126% faster than OVA. In average, for each token, the HC-OVA gives rise to compare 2.34 times to determine the chunk class, while C-OVA and OVA are 10.92, and 22 respectively in the CoNLL-2000 shared task. In Chinese base-chunking task, the three methods cost 2.80, 14.20 and 40 times for comparing in average.

As shown in Table 5, although the HC-OVA is very efficient in terms of testing, it decreases the chunking accuracy. However, the decreasing rate is somewhat marginal (from 94.25 to 94.09). Compared to previous studies, our method is not only efficient but also accurate. In this comparison, we only focus on pure training with the

CoNLL-2000 provided dataset, while the use of external knowledge, like parsers, or additional corpus is not compared. It is no conflict to employ these methods to further improve the performance. Zhang [16] had shown that the use of external parsers could enhance their chunker from 93.57 to 94.17. In his later research [1], he also incorporated with unlabeled data to boost the learner which increased the accuracy from 93.60 to 94.39.

Table 5. Comparison of chunking performance for CoNLL-2000 shared task

Chunking system	Recall	Precision	$F_{(\beta)}$
SVM-COVA	**94.18**	**94.31**	**94.25**
SVM-OVA	**94.19**	**94.30**	**94.25**
SVM [15]	94.12	94.13	94.12
SVM-HCOVA	**94.03**	**94.16**	**94.09**
Voted-SVMs [7]	93.89	93.92	93.91
Voted-perceptrons [4]	93.38	94.20	93.79
Structural Learning [1]	93.83	93.37	93.60

Table 6. Comparison of chunking performance for Chinese Base-chunking task

Chunking system	Recall	Precision	$F_{(\beta)}$
SVM-COVA	91.50	93.14	**92.31**
SVM-OVA	91.49	93.13	**92.30**
SVM [16]	91.30	93.12	92.20
SVM-HOVA	91.23	92.77	92.00

We should truly report a very efficient method based on hidden markov model [8] which can handle 45000~50000 tokens in one second and only spend few seconds for training. But this method is not accurate. Molina made use of one million words training data which is four times larger than us to obtain the improved performance where the F rate is enhanced from 92.19 to 93.25.

5 Conclusion

In this paper, we present two methods to speed up training and testing for phrase chunking. Experimental results show that our method can handle 13000 tokens per second which kept the accuracy in 94.09 in the CoNLL-2000 shared task. The online demonstration of our phrase chunking systems can be found at (http://dblab87.csie.ncu.edu.tw/bcbb/fast_chunking.htm).

References

1. Rie Kubota Ando, and Tong Zhang. 2005. A high-performance semi-supervised learning method for text chunking. *In Proceedings of 43rd Annual Meetings of the Association for Computational Linguistics*, pages 1-9.
2. Eric Brill. 1995. Transformation-based error-driven learning and natural language processing: a case study in part of speech tagging, Computational Linguistics, 21(4):543-565.

3. Xavier Carreras and Lluis Marquez. 2003. Phrase recognition by filtering and ranking with perceptrons. *In Proceedings of the International Conference on Recent Advances in Natural Language Processing (RANLP).*
4. Xavier Carreras, Lluis Marquez, and J. Castro. 2005. Filtering-ranking perceptron learning for partial parsing. Machine Learning Journal, 59: 1-31.
5. Jesús Giménez and Lluís Márquez. 2003. Fast and accurate Part-of-Speech tagging: the SVM approach revisited. *In Proceedings of the International Conference on Recent Advances in Natural Language Processing (RANLP)*, pages 158-165.
6. Thorsten Joachims. 1998. Text categorization with support vector machines: learning with many relevant features. *In Proceedings of the European Conference on Machine Learning*, pages 137-142.
7. Taku Kudoh and Yuji Matsumoto. 2001. Chunking with support vector machines. *In the Proceedings of the 2nd Meetings of the North American Chapter and the Association for the Computational Linguistics.* Pages 192-199.
8. Antonio Molina and Ferran Pla. 2002. Shallow Parsing using Specialized HMMs. Journal of Machine Learning Research, pages 595-613.
9. John C. Platt, Nello Cristianini, John Shawe-Taylor. 2000. Large margin dags for multiclass classification. Advanced in Neural Information Processing Systems, 12: 547-553.
10. Lance A. Ramshaw and Mitchell P. Marcus. 1995. Text chunking using transformation-based learning. *In Proceedings of the 3rd Workshop on Very Large Corpora*, pages 82-94.
11. Kenji Sagae, Alon Lavie, and Brian MacWhinney. 2005. Automatic Measurement of Syntactic Development in Child Language. *In Proceedings of 43rd Annual Meetings of the Association for Computational Linguistics*, pages 197-204.
12. Eric F. Tjong Kim Sang and Sabine Buchholz. 2000. Introduction to the CoNLL-2000 shared task: chunking. *In Proceedings of Conference on Natural Language Learning (CoNLL)*, pages 127-132.
13. Eric F. Tjong Kim Sang. 2002. Memory-based shallow parsing. Journal of Machine Learning Research, pages 559-594.
14. Taro Watanabe, Eiichiro Sumita, and Hiroshi G. Okuno. 2003. Chunk-based statistical translation. *In Proceedings of 41st Annual Meetings of the Association for Computational Linguistics*, pages 303-310.
15. Yu-Chieh Wu, Chia-Hui Chang, and Yue-Shi Lee. 2006. A general and multi-lingual phrase chunking model based on masking method. *In Proceedings of 7th International Conference on Intelligent Text Processing and Computational Linguistics,* pages 144-155.
16. Tong Zhang, Fred Damerau, and David Johnson. 2002. Text Chunking based on a Generalization Winnow. Journal of Machine Learning Research, 2: 615-637.

Statistical and Comparative Evaluation of Various Indexing and Search Models

Samir Abdou and Jacques Savoy

Computer Science Department, University of Neuchatel,
rue Emile Argand 11, 2009 Neuchatel, Switzerland
{Samir.Abdou, Jacques.Savoy}@unine.ch

Abstract. This paper first describes various strategies (character, bigram, automatic segmentation) used to index the Chinese (ZH), Japanese (JA) and Korean (KR) languages. Second, based on the NTCIR-5 test-collections, it evaluates various retrieval models, varying from classical vector-space models to more recent developments in probabilistic and language models. While no clear conclusion was reached for the Japanese language, the bigram-based indexing strategy seems to be the best choice for Korean, and the combined "unigram & bigram" indexing strategy is best for traditional Chinese. On the other hand, *Divergence from Randomness* (DFR) probabilistic model usually results in the best mean average precision. Finally, upon an evaluation of the four different statistical tests, we find that their conclusions correlate, even more when comparing the non-parametric bootstrap with the t-test.

1 Introduction

In order to promote IR activities involving Asian languages and also to facilitate technological transfers into products, the latest NTCIR evaluation campaign [1] created test-collections for the traditional Chinese, Japanese and Korean languages. Given that English is an important language for Asia and that we also wanted to verify that the various approaches suggested might also work well with European languages, a fourth collection of newspaper articles written in English was used.

Even with all participants working with the same newspapers corpora and queries, it is not always instructive to directly compare IR performance results achieved by two search systems. In fact, given that their performance is usually based on different indexing and search strategies involving a large number of underlying variables (size and type of stopword lists, stemming strategies, token segmentation, n-grams generation procedures, indexing restrictions or adaptations and term weighting approaches).

Based on the NTCIR-5 test-collections [1], this paper empirically compares various indexing and search strategies involving East Asian languages. In order to obtain more solid conclusions, this paper also considers various IR schemes, and all comparisons are analyzed statistically. The rest of this paper is organized as follows: Section 2 describes the main features of the test-collections.

H.T. Ng et al. (Eds.): AIRS 2006, LNCS 4182, pp. 362–373, 2006.

Section 3 contains an overview of the various search models, from vector-space approaches to recent developments in both probabilistic and language models. Section 4 portrays the different indexing strategies used to process East Asian languages, and Section 5 contains various evaluations and analyzes of the resultant retrieval performance. Finally, Section 6 compares decisions that might result from using other statistical tests and Section 7 presents the main findings of our investigation.

2 Overview of NTCIR-5 Test-Collections

The test-collections used in our experiments include various newspapers covering the years 2000-2001 [1]. The Chinese and Japanese corpora were larger in size (1,100 MB) but the Chinese collection contained a slightly larger number of documents (901,446) than did the Japanese (858,400). The Korean and English corpora were smaller, both in terms of size (438 MB for the English and 312 MB for the Korean) and number of newspaper articles (259,050 for the English and 220,374 for the Korean).

When analyzing the number of pertinent documents per topic, only rigid assessments were considered, meaning that only "highly relevant" and "relevant" items were viewed as being relevant, under the assumption that only highly or relevant items would be useful for all topics. A comparison of the number of relevant documents per topic indicates that for the English collection the median number of relevant items per topic is 33, while for the Asian languages corpora it is around 25 (ZH: 26, JA: 24, KR: 25.5). The number of relevant articles is also greater for the English (3,073) corpus, when compared to the Japanese (2,112), Chinese (1,885) or Korean (1,829) corpora.

The 50 available topics covered various subjects (e.g., "Kim Dae-Jun, Kim Jong Il, Inter-Korea Summit," or "Harry Potter, circulation"), including both regional/national events ("Mori Cabinet, support percentage, Ehime-maru") or topics having a more international coverage ("G8 Okinawa Summit"). The same set of queries was available for the four languages, namely Chinese, Japanese, Korean and English. According to the TREC model, the structure of each topic consisted of four logical sections: brief title (<TITLE>), one-sentence description (<DESC>), narrative (<NARR>) specifying both the background context (<BACK>) and a relevance assessment criterion (<REL>) for the topic. Finally a concept section (<CONC>) provides some related terms. In our experiments, we only use the title field of the topic description.

3 Search Models

In order to obtain a broader view of the relative merit of the various retrieval models, we examined six vector-space schemes and three probabilistic models. First we adopted the classical *tf idf* model, in which the weight (denoted w_{ij}) attached to each indexing term t_j in document D_i was the product of its term occurrence frequency (or tf_{ij}) and its inverse document frequency (or

$idf_j = ln(n/df_j)$, where n indicates the number of documents in the corpus, and df_j the number of documents in which the term t_j appears). To measure similarities between documents and requests, we computed the inner product after normalizing indexing weights (model denoted "document=ntc, query=ntc" or "ntc-ntc").

Other variants might also be created, especially in cases when the occurrence of a particular term in a document is considered as a rare event. Thus, the proper practice may be to give more importance to the first occurrence of a term, as compared to any successive occurrences. Therefore, the tf component might be computed as the $ln(tf) + 1$ (denoted "ltc", "lnc", or "ltn") or as $0.5 + 0.5 \cdot [tf \ / \ max \ tf \ in \ D_i]$ ("atn"). We might also consider that a term's presence in a shorter document would be stronger evidence than its occurrence in a longer document. More complex IR models have been suggested to account for document length, including the "Lnu" [2], or the "dtu" IR models [3] (more details are given in the Appendix).

In addition to vector-space approaches, we also considered probabilistic IR models, such as the Okapi probabilistic model (or BM25) [4]. As a second probabilistic approach, we implemented the PB2 taken from the *Divergence from Randomness* (DFR) framework [5], based on combining the two information measures formulated below:

$w_{ij} = Inf^1_{ij}(tf) \cdot Inf^2_{ij}(tf) = -log_2 \left[Prob^1_{ij}(tf)\right] \cdot (1 - Prob^2_{ij}(tf))$

where w_{ij} indicates the indexing weight attached to term t_j in document D_i, $Prob^1_{ij}(tf)$ is the pure chance probability of finding tf_{ij} occurrences of the indexing unit t_j in the document D_i. On the other hand, $Prob^2_{ij}(tf)$ is the probability of encountering a new occurrence of t_j in the document given that we have already found tf_{ij} occurrences of this indexing unit. Within this framework, the PB2 model is based on the following formulae:

$$Prob^1_{ij}(tf) = \left[e^{\lambda_j} \cdot \lambda_j^{tfn_{ij}}\right] / tf_{ij}! \quad \text{with } \lambda_j = tc_j/n \tag{1}$$

$$Prob^2_{ij}(tf) = 1 - \left[\frac{tc_j + 1}{df_j \cdot (tfn_{ij} + 1)}\right] \quad \text{with} \tag{2}$$

$$tfn_{ij} = tf_{ij} \cdot \log_2 [1 + ((c \cdot mean \ dl)/l_i)] \tag{3}$$

where tc_j indicates the number of occurrences of t_j in the collection, *mean dl* the mean length of a document and l_i the length of document D_i.

Finally, we also considered an approach based on a language model (LM) [6], known as a non-parametric probabilistic model (the Okapi and PB2 are viewed as parametric models). Probability estimates would thus not be based on any known distribution (as in Equation 1) but rather be estimated directly, based on occurrence frequencies in document D or corpus C. Within this language model paradigm, various implementations and smoothing methods might also be considered, and in this study we adopted a model proposed by Hiemstra [6], as described in Equation 4, which combines an estimate based on document ($P[t_j \mid D_i]$) and corpus ($P[t_j \mid C]$).

$$P[Di|\ Q] \ = \ P[D_i] \ \cdot \ \prod_{t_j \in Q} [\lambda_j \cdot P[t_j \mid D_i] + (1-\lambda_j) \cdot P[t_j \mid C]] \qquad (4)$$

with $P[t_j \mid D_i] = tf_{ij}/l_i$, $P[t_j \mid C] = df_j/lc$, $lc = \sum_k df_k$, and where λ_j is a smoothing factor (fixed at 0.3 for all indexing terms t_j) and lc an estimate of the corpus size.

4 Indexing Strategies

In the previous section, we described how each indexing unit was weighted to reflect its relative importance in describing the semantic content of a document or a request. This section will explain how such indexing units are extracted from documents and topic formulations.

For the English collection, we used words as indexing units and we based the indexing process on the SMART stopword list (571 terms) and stemmer. For European languages, it seems natural to consider words as indexing units, and this assumption has been generally confirmed by previous CLEF evaluation campaigns [7].

For documents written in the Chinese and Japanese languages, words are not clearly delimited. We therefore indexed East Asian languages using an overlapping bigram approach, an indexing scheme found to be effective for various Chinese collections [8], [9]. In this case, the "ABCD EFG" sequence would generate the follow-ing bigrams "AB," "BC," "CD," "EF," and "FG". Our choice of an indexing tool also involves other factors. As an example for Korean, Lee *et al.* [10] found more than 80% of nouns were composed of one or two Hangul characters, while for Chinese Sproat [11] reported a similar finding. An analysis of the Japanese corpus reveals that the mean length of continuous Kanji characters to be 2.3, with more than 70% of continuous Kanji sequences being composed of one or two characters (for Hiragana: mean=2.1, for Katakana: mean=3.96).

In order to stop bigram generation in our work, we generated overlapping bigrams for Asian characters only, using spaces and other punctuation marks (as collected for each language from its respective encoding). Moreover, in our experiments, we did not split any words written in ASCII characters, and the most frequent bigrams were removed before indexing. As an example, for the Chinese language we defined and removed a list of 90 most frequent unigrams, 49 most frequent bigrams and 91 most frequent words. For the Japanese language, we defined a stopword list of 30 words and another of 20 bigrams, and for Korean our stoplist was composed of 91 bigrams and 85 words. Finally, as suggested by Fujii & Croft [12], before generating bigrams for the Japanese documents we removed all Hiragana characters, given that these characters are mainly used to express grammatical words (e.g., *doing*, *do*, *in*, *of*), and the inflectional endings of verbs, adjectives and nouns. Such removal is not error-free because Hiragana could also be used to write Japanese nouns.

For Asian languages, there are of course other indexing strategies that might be used. In this vein, various authors have suggested that words generated by a segmentation procedure could be used to index Chinese documents. Nie &

Ren [13] however indicated that retrieval performance based on word indexing does not really depend on an accurate word segmentation procedure and this was confirmed by Foo & Li [14]. They also stated that segmenting a Chinese sentence does affect retrieval performance and that recognizing a greater number of 2-characters words usually contributes to retrieval enhancement. These authors did not however find a direct relationship between segmentation accuracy and retrieval effectiveness. Moreover, manual segmentation does not always result in better performance when compared to character-based segmentation.

To analyze these questions, we also considered automatic segmentation tools, namely Mandarin Tools (MTool, `www.mandarintools.com`) for the traditional Chinese language and the Chasen (`chasen.aist-nara.ac.jp`) morphological analyzer for Japanese. For Korean, the presence of compound construction could harm retrieval performance. Thus, in order to automatically decompose them, we applied the Hangul Analyser Module (HAM, `nlp.kookmin.ac.kr`) tool. With this linguistic approach, Murata *et al.* [15] obtained effective retrieval results while Lee *et al.* [9] showed that n-gram indexing could result in similar and sometimes better retrieval effectiveness, compared to word-based indexing applied in conjunction with a decompounding scheme.

5 Evaluation of Various IR Models

To measure retrieval performance, we adopted mean average precision (MAP) as computed by `TREC_EVAL`. To determine whether or not a search strategy might be better than another, we applied a statistical test. More precisely, we stated the null hypothesis (denoted H_0) specifying that both retrieval schemes achieved similar performance levels (MAP), and this hypothesis would be rejected at the significance level fixed at $\alpha = 5\%$ (two-tailed test). As a statistical test, we chose the non-parametric bootstrap test [16]. All evaluations in this paper were based on the title-only query formulation.

The MAP achieved by the six vector-space schemes, two probabilistic approaches and the language model (LM) are shown in Table 1 for the English and Chinese collections. The best performance in any given column is shown in bold and this value served as baseline for our first set of statistical tests. In this case, we wanted to verify whether this highest performance was statistically better than other performances depicted in the same column. When performance differences were detected as significant, we placed an asterisk (*) next to a given search engine performance. In the English corpus for example, the PB2 model achieved the highest MAP (0.3728). The difference in performance between this model and the "Lnu-ltc" approach (0.3562) was statistically significant while the difference between it and the Okapi model (0.3692) was not significant.

For the Chinese corpus, the PB2 probabilistic model also resulted in the best performance, except for the unigram-based indexing scheme where the best performance was obtained by the language model LM (0.2965). With these various indexing schemes, the difference between either the PB2, the LM, the Okapi or the "Lnu-ltc" models were not statistically significant. PB2 was the

Table 1. MAP for English and Chinese corpora (T queries)

	Mean average precision (MAP)				
	English	Chinese			
Model	word	unigram	bigram (base)	MTool	uni+bigram
PB2-nnn	**0.3728**	0.2774	**0.3042**	**0.3246**	**0.3433**
LM	0.3428*	**0.2965**	0.2594*	0.2800*	0.2943*
Okapi-npn	0.3692	0.2879	0.2995	0.3231	0.3321
Lnu-ltc	0.3562*	0.2883	0.2999	0.3227	0.3356
dtu-dtn	0.3577	0.2743	0.2866	0.2894*	0.3094*
atn-ntc	0.3423*	0.2329*	0.2527*	0.2578*	0.2729*
ltn-ntc	0.3275*	0.2348*	0.2886	0.2833*	0.3068*
ltc-ltc	0.2509*	0.1464	0.1933*	0.1772*	0.2202*
ntc-ntc	0.2345*	0.1162*	0.2130*	0.1645*	0.2201*
Improvement (7 best mod.)		-5.0%	0%	+4.5%	+10.2%

preferred model but by slightly changing the topic set, other models might perform better.

Based on an analysis of the four different indexing schemes used with the Chinese corpus, the data in Table 1 indicates that the combined "uni+bigram" indexing scheme tends to result in the best performance levels. As shown in the last row of this table, we computed mean improvements over the bigram indexing strategy, considering only the 7-best performing IR models (rows ending with the "ltn-ntc" model). From this overall measure we can see for example that the character-based indexing strategy results in lower performance level than does the bigram scheme (-5.0%). Using the bigram indexing strategy as a baseline, we verified whether performance differences between the various indexing schemes were statistically significant, and then underlined those that were statistically significant. Table 1 illustrates that the differences between the bigram and word-based indexing strategies (row labeled "MTool") are usually not significant. The differences between the bigram approach and the combined indexing strategy (last column) are usually significant and in favor of the combined approach.

Table 2. MAP for Japanese corpus (T queries)

	Mean average precision (MAP)			
Model	unigram	bigram (base)	Chasen	uni+bigram
PB2-nnn	**0.2240**	**0.2816**	**0.3063**	**0.3026**
LM	0.1369*	0.1791*	0.1968*	0.1944*
Okapi-npn	0.2208	0.2660*	0.2655*	0.2802
Lnu-ltc	0.2239	0.2579*	0.2743*	0.2736
dtu-dtn	0.2126	0.2461*	0.2735*	0.2735
atn-ntc	0.1372*	0.1799*	0.2109*	0.1901*
ltn-ntc	0.1518*	0.2651	0.2723	0.2726*
ltc-ltc	0.0580*	0.0992*	0.0945*	0.1154*
ntc-ntc	0.0706*	0.1292*	0.1227*	0.1295*
Improvement	-22.0%	0%	+7.4%	+6.6%

Evaluations done on the Japanese corpus are given in Table 2. With this language, the best performing search model was always PB2, often showing significant improvement over others (indicated by "*"). Comparing the differences between the four indexing strategies shows that both Chasen (automatic segmentation) and the combined indexing approaches ("uni+bigram") tend to result in the best performance levels. Using the bigram indexing strategy as baseline, the differences between the word (Chasen) or the combined ("uni+bigram") indexing strategies are however usually not significant. Moreover, performances that result from applying the bigram scheme are always better than with the unigram approach.

Table 3. MAP for Korean corpus (T queries)

	Mean average precision (MAP)		
Model	word	bigram (base)	HAM
PB2-nnn	0.2378	0.3729	**0.3659**
LM	0.2120*	0.3310*	0.3135*
Okapi-npn	0.2245*	0.3630*	0.3549
Lnu-ltc	0.2296	**0.3973***	0.3560
dtu-dtn	**0.2411**	0.3673*	0.3339*
atn-ntc	0.2242*	0.3270*	0.2983*
ltn-ntc	0.2370	0.3708	0.3383*
ltc-ltc	0.1606*	0.2260*	0.2299*
ntc-ntc	0.1548*	0.2506*	0.2324*
Improvement	-36.5%	0%	-6.6%

Our evaluations on the Korean collection are reported in Table 3. In this case, the best performing search model varies according to the indexing strategy. The performance differences between the best performing models ("dtu-dtn", "Lnu-ltc", PB2) are usually not significant. Using the bigram scheme as baseline, the performance differences with the word-based indexing approach were always detected as significant and in favor of the bigram approach. Comparing bigrams with the automatic decompounding strategy (under the label "HAM" in Table 3), the bigram indexing strategy tends to present a better performance, but the differences are usually not significant.

General measurements such as MAP always hide irregularities found among queries. It is interesting to note for example that for some queries, retrieval performance was poor for all search models. For example, for Topic #4 entitled "the US Secretary of Defense, William Sebastian Cohen, Beijing", the first relevant item appears in rank 37 with the PB2 model (English corpus). When inspecting top-ranked articles for this query, we found that these articles more or less contained all words included in the topic description. Moreover, their length was relatively short and these two aspects were taken into account when ranking these documents high in the response list. From a semantic point of view, these short and non-pertinent articles do not specify the reason or purpose of the visit made by the US Secretary of Defense, with content being limited to facts such as "the US Secretary of Defense will arrive next week" or "William Sebastian Cohen will leave China tomorrow".

Topic #45 "population issue, hunger" was another difficult query. After stemming, the query is composed by the stem "hung" present in 3,036 documents, the indexing term "populat" (that occurs in 7,995 articles), and "issu" (appearing in 44,209 documents). Given this document frequency information, it would seem natural to assign more importance to the stem "hung", compared to the two other indexing terms. The term "hunger" however does not appear in any relevant document, resulting in poor retrieval performance for this query. The inclusion of the term "food" (appearing in the descriptive part of the topic) resulted in some pertinent articles being found by the search system.

6 Statistical Variations

In the previous section, we based our statistical validation on the bootstrap approach [16] in order to determine whether or not the difference between two given retrieval schemes was really significant. The null hypothesis (denoted H_0) stated that both IR systems produce the same performance level and the observed difference was simply due to random variations. To verify this assumption statistically, other statistical tests could be considered.

The first might be the Sign test [17, , pp. 157–164], in which only the direction of the difference (denoted by a "+" or "-" sign) is taken into account. This non-parametric test does not take the amount of difference into account, but only the fact that a given system performs better than the other for any given query. For example, for a set of 50 queries, System A produced better MAP for 32 queries (or 32 "+"), System B was better for 16 (or 16 "-"), and for the two remaining requests both systems showed the same performance. If the null hypothesis were true, we would expect to obtain roughly the same number of "+" or "-" signs. In the current case involving 48 experiments (the two ties results are ignored), we had 32 "+" and only 16 "-" signs. Assuming that the null hypothesis is true, the probability of observing a "+" is equal to the probability of observing a "-" (= 0.5). Thus for 48 trials the probability of observing 16 or fewer occurrences of the same sign ("+" or "-", for a two-tailed test) is only 0.0293. This value is rather small (but not null) and, in this case, when the limit was fixed at $\alpha = 5\%$, we must reject the H_0 and accept the alternative hypothesis that there were truly retrieval performance differences between System A and B.

Instead of observing only the direction of the difference between two systems, we might also consider the magnitude of the difference, not directly but by sorting them from the smallest to the largest difference. Then we could apply the Wilcoxon signed ranking test [17, pp. 352-360]. Finally, we might apply the paired t-test, a parametric test assuming that the difference between two systems follows a normal distribution. Even if the distribution of the observations was not normally shaped but the empirical distribution found to be roughly symmetric, the t-test would still be useful, given that it is a relatively robust test, in the sense that the significance level indicated is not far from the true level. However, previous studies have shown that IR data do not always follow a normal distribution [16].

Based on 264 comparative evaluations (most of them are shown in Section 5), we applied the four statistical tests to the resultant differences. Among them for all four tests, 143 comparisons were found to be significant and 88 non-significant. Thus, for 231 (143+88) comparisons out of 264 (or 87.5%), the four tests resulted in the same decision. These four statistical tests thus are clearly in agreement, even though they use different kinds of information (e.g., for the Sign test, only the difference direction).

For the other 33 (264-231) comparisons, there was some disagreement and these cases can be subdivided into three categories. First, in 11 cases, three tests were detected to have a significant difference while the other one did not. Following inspection, we found that in 10 (out of 11) observations only the Sign test did not detect a significant difference by obtaining a *p*-value greater than 0.05 (see Example A in the second row of Table 4). Second, for 16 cases, two tests indicated a significant difference while the other two did not. After inspecting this sample, we found 8 observations for which both the t-test and the bootstrap detected a significant difference (see for example Case C in Table 4). In 7 other cases, both the Sign and Wilcoxon tests detected significant retrieval performance differences (see Case D in Table 4). Finally, in 6 only one test detected a significant difference while for the three others the performance difference could be due to random variations (see, for example, Case E in Table 4).

Table 4. Description and *p*-value for some comparisons

	Comparison	MAP	Sign test	Wilcoxon	Bootstrap	t-test
A.	ZH unigram	0.2965	**0.0595**	0.0122	0.0085	0.0084
	LM vs. ltn-ntc	0.2348	(31+ vs. 17-)			
B.	JA bigr. vs unigr.	0.1799	0.0186	0.0073	0.0430	**0.0528**
	atn-ntc vs. atn-ntc	0.1372	(32+ vs. 15-)			
C.	ZH MTools	0.3246	0.3916	0.0574	**0.0260**	**0.0299**
	PB2 vs. dtu-dtn	0.2894	(28+ vs. 21-)			
D.	JA uni+bigram	0.3026	**0.0011**	**0.0040**	0.1555	0.1740
	PB2 vs. Okapi	0.2802	(35+ vs. 12-)			
E.	KR HAM	0.3659	0.3916	**0.0297**	0.1215	0.1354
	PB2 vs. Okapi	0.3549	(28+ vs. 21-)			

To provide a more general overview of the relationship between two tests, in Figure 1 we plotted the *p*-values for performance comparisons from the two tests. We also computed the Pearson correlation coefficient and drew a line representing the corresponding slope. The first plot in the top left corner of Figure 1 indicates a strong correlation (r=0.9996) between the bootstrap *p*-values and those obtained by the t-test. Clearly, the bootstrap test agrees with the t-test results, without having to assume a Gaussian distribution.

We also tested to find out whether or not the differences distribution follows a normal distribution. In 228 (out of 264) observations, the underlying distribution of performance difference did not follow a Gaussian distribution (Shapiro-Wilk test, significance level $\alpha = 5\%$ [18]). In both cases, the Pearson correlation coefficient between the bootstrap and t-test p-values is very high.

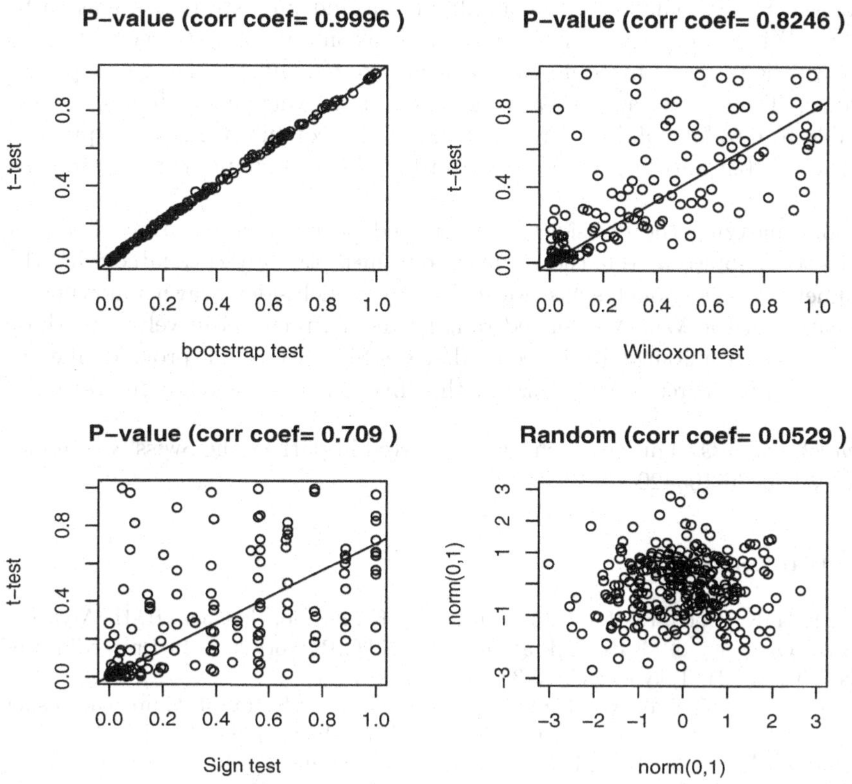

Fig. 1. Three plots of two related tests (p-values) and a random example

The relationship between the t-test and the Wilcoxon test is not as strong (top right) but still relatively high (Pearson coefficient correlation of 0.8246). When comparing p-values obtained from the t-test and the Sign test, the correlation coefficient is lower (0.709) but statistically different from 0. Finally, we plotted the same number of points obtained by generating values randomly according to the normal distribution. In this case, the true correlation coefficient is a null value, even though the depicted value is not (0.0529). The latter picture is an example of no correlation between two variables.

7 Conclusion

The experiments conducted with the NTCIR-5 test-collections show that the PB2 probabilistic model derived within the *Divergence from Randomness* framework usually produces the best mean average precision, according to different indexing strategies and languages. For the Chinese language (Table 1), the best indexing strategy seems to be a combined approach (unigram & bigram) but when compared with a word-based approach (obtained with an automatic segmentation system), the difference is not always statistically significant.

For the Korean language, the simple bigram indexing strategy seems to be the best. When compared with the automatic decompounding strategy (HAM in Table 3), the performance difference is usually not-significant. For the Japanese language (Table 2), we may discard the unigram indexing approach, but we were not able to develop solid arguments in favor of a combined indexing approach (unigram + bigram), compared to a word-based or a simple bigram indexing scheme.

Upon analyzing the decisions that resulted from our application of a non-parametric bootstrap test, the evidence obtained strongly correlated with the (parametric) t-test conclusions. Moreover, the conclusions drawn following an application of the Wilcoxon signed ranking test correlate positively with those of the t-test. From our data, it seems that the Sign test might provide different results than the three other tests, but this divergence is not really important.

Acknowledgments. This research was supported in part by the Swiss NSF under Grant #200020-103420.

References

1. Kishida, K., Chen, K.-H., Lee, S., Kuriyama, K., Kando, N., Chen, H.-H., Myaeng, S.H.: Overview of CLIR Task at the Fifth NTCIR Workshop. In Proceedings of NTCIR-5. NII, Tokyo (2005) 1–38
2. Buckley, C., Singhal, A., Mitra, M., Salton, G.: New Retrieval Approaches using SMART. In Proceedings TREC-4. NIST, Gaithersburg (1996) 25–48
3. Singhal, A., Choi, J., Hindle, D., Lewis, D.D., Pereira, F.: AT&T at TREC-7. In Proceedings TREC-7. NIST, Gaithersburg (1999) 239–251
4. Robertson, S.E., Walker, S., Beaulieu, M.: Experimentation as a Way of Life: Okapi at TREC. Information Processing & Management **36**, (2000) 95–108
5. Amati, G., van Rijsbergen, C.J.: Probabilistic Models of Information Retrieval Based on Measuring the Divergence from Randomness. ACM Transactions on Information Systems **20** (2002) 357–389
6. Hiemstra, D.: Using Language Models for Information Retrieval. CTIT Ph.D. Thesis (2000)
7. Peters, C., Clough, P.D., Gonzalo, J., Jones, G.J.F., Kluck, M., Magnini, B. (Eds.): Multilingual Information Access for Text, Speech and Images: Results of the Fifth CLEF Evaluation Campaign. Lecture Notes in Computer Science, Vol. 3491. Springer, Berlin (2005)
8. Kwok, K.L. Employing Multiple Representations for Chinese Information Retrieval. Journal of the American Society for Information Science **50** (1999) 709–723
9. Luk, R.W.P., Kwok, K.L.: A Comparison of Chinese Document Indexing Strategies and Retrieval Models, ACM Transactions on Asian Languages Information Processing **1** (2002), 225–268
10. Lee, J.J., Cho, H.Y., Park, H.R.: N-gram-based Indexing for Korean Text Retrieval. Information Processing & Management **35** (1999) 427–441

11. Sproat, R.: Morphology and Computation. The MIT Press, Cambridge (1992)
12. Fujii, H., Croft, W.B.: A Comparison of Indexing Techniques for Japanese Text Retrieval. In Proceedings ACM-SIGIR. The ACM Press, New York (1993) 237–246
13. Nie, J.Y., Ren, F. Chinese Information Retrieval: using Characters or Words? Information Processing & Management **35** (1999) 443–462
14. Foo, S., Li, H.: Chinese Word Segmentation and its Effect on Information Retrieval. Information Processing & Management **40** (2004) 161–190
15. Murata, M., Ma, Q., Isahara, H.: Applying Multiple Characteristics and Techniques to Obtain High Levels of Performance in Information Retrieval. In Proceedings of NTCIR-3. NII, Tokyo (2003)
16. Savoy, J.: Statistical Inference in Retrieval Effectiveness Evaluation. Information Processing & Management **33** (1997) 495–512
17. Conover, W.J.: Practical Nonparametric Statistics. 3rd edn. John Wiley & Sons, New York (1999)
18. Maindonald, J., Braun, J.: Data Analysis and Graphics Using R. Cambridge University Press, Cambridge (2003)

Appendix: Term Weighting Formulae

In Table 5, n indicates the number of documents in the collection, t the number of indexing terms, df_j the number of documents in which the term t_j appears, the document length of D_i (the number of indexing terms) is denoted by nt_i. We assigned the value of 0.55 to the constant b, 0.1 to *slope*, while we fixed the constant k_1 at 1.2 for the English, Korean and Japanese collection and 1.0 for the Chinese corpus. For the PB2 model, we assigned $c = 3$ for the English and Korean corpus, $c = 6$ for the Japanese and $c = 1$ for the Chinese collection. These values were chosen because they usually result in improved levels of retrieval performance. Finally, the value *mean dl*, *slope* or *avdl* were fixed according to the corresponding statistics (e.g., for bigram-based indexing, 321 for ZH, 133 for JA, and 233 for KR).

Table 5. Various Weighting Schemes

ntc	$w_{ij} = \frac{tf_{ij} \cdot idf_j}{\sqrt{\sum_{k=1}^{t}(tf_{ik} \cdot idf_k)^2}}$	atn	$w_{ij} = idf_j \cdot \left[\frac{0.5 + 0.5 \cdot tf_{ij}}{max\ tf_{i.}}\right]$
dtn	$w_{ij} = [ln(ln(tf_{ij}) + 1) + 1] \cdot idf_j$	ltn	$w_{ij} = [ln(tf_{ij}) + 1] \cdot idf_j$
ltc	$w_{ij} = \frac{[ln(tf_{ij})+1] \cdot idf_j}{\sqrt{\sum_{k=1}^{t}([ln(tf_{ik})+1] \cdot idf_k)^2}}$	npn	$w_{ij} = tf_{ij} \cdot ln\left(\frac{n-df_j}{df_j}\right)$
dtu	$w_{ij} = \frac{[ln(ln(tf_{ij})+1)+1] \cdot idf_j}{(1-slope) \cdot pivot + (slope \cdot nt_i)}$	lnc	$w_{ij} = \frac{ln(tf_{ij})+1}{\sqrt{\sum_{k=1}^{t}(ln(tf_{ik})+1)^2}}$
Lnu	$w_{ij} = \frac{\frac{ln(tf_{ij})+1}{ln\left(\frac{l_i}{nt_i}\right)+1}}{(1-slope) \cdot pivot + (slope \cdot nt_i)}$	nnn	$w_{ij} = tf_{ij}$
Okapi	$w_{ij} = \frac{(k_1+1) \cdot tf_{ij}}{K + tf_{ij}}$ with $K = k_1 \cdot [(1 - b) + b \cdot \frac{l_i}{avdl}]$		

Bootstrap-Based Comparisons of IR Metrics for Finding One Relevant Document

Tetsuya Sakai

Toshiba Corporate R&D Center, Kawasaki 212-8582, Japan
tetsuya.sakai@toshiba.co.jp

Abstract. This paper compares the sensitivity of IR metrics designed for the task of finding one relevant document, using a method recently proposed at SIGIR 2006. The metrics are: P$^+$-measure, P-measure, O-measure, Normalised Weighted Reciprocal Rank (NWRR) and Reciprocal Rank (RR). All of them except for RR can handle graded relevance. Unlike the ad hoc (but nevertheless useful) "swap" method proposed by Voorhees and Buckley, the new method derives the sensitivity and the performance difference required to guarantee a given significance level directly from Bootstrap Hypothesis Tests. We use four data sets from NTCIR to show that, according to this method, "P($^+$)-measure $\geq$ O-measure $\geq$ NWRR $\geq$ RR" generally holds, where "$\geq$" means "is at least as sensitive as". These results generalise and reinforce previously reported ones based on the swap method. Therefore, we recommend the use of P($^+$)-measure and O-measure for practical tasks such as known-item search where recall is either unimportant or immeasurable.

1 Introduction

Different Information Retrieval (IR) tasks require different evaluation metrics. For example, a patent survey task may require a *recall-oriented* metric, while a *known-item search* task [16] may require a *precision-oriented* metric. When we search the Web, for example, we often stop going through the ranked list after finding *one* good Web page even though the list may contain some more relevant pages, either knowing or assuming that the rest of the retrieved pages lack *novelty*, or additional information that may be of use to him. Thus, finding exactly *one* relevant document with high precision is an important IR task.

Reciprocal Rank (RR) [16] is commonly used for the task of finding one relevant document: $RR = 0$ if the ranked output does not contain a relevant document; otherwise, $RR = 1/r_1$, where r_1 is the rank of the retrieved relevant document that is nearest to the top of the list. However, RR is based on binary relevance and therefore cannot distinguish between a retrieved *highly* relevant document and a retrieved *partially* relevant document. Thus, as long as RR is used for evaluation, it is difficult for researchers to develop a system that can rank a highly relevant document above partially relevant ones. In light of this, Sakai [10] proposed a metric called *O-measure* for the task of finding one *highly* relevant document. O-measure is a variant of *Q-measure* which is very highly

H.T. Ng et al. (Eds.): AIRS 2006, LNCS 4182, pp. 374–389, 2006.

	Binary relevance	Graded relevance
Finding as many relevant documents as possible	AveP PDoc	Q-measure R-measure n(D)CG
Finding one relevant document only	RR	P+-measure P-measure O-measure NWRR

Fig. 1. Four categories of IR metrics, with some examples

correlated with Average Precision (AveP) but can handle graded relevance [8,14]. O-measure can also be regarded as a generalisation of RR (See Section 3).

Eguchi *et al.* [4], the organisers of the NTCIR Web track, have also proposed a metric for the task of finding one highly relevant document, namely, *Weighted Reciprocal Rank* (WRR). WRR assumes that ranking a *partially* relevant document at (say) Rank 1 is more important than ranking a *highly* relevant document at Rank 2. It has never actually been used for ranking the systems at NTCIR (See Section 3) and its reliability is unknown. We point out in Section 3 that, if WRR must be used, then it should be normalised before averaging across topics: We call the normalised version *Normalised Weighted Reciprocal Rank* (NWRR).

Just like RR, both O-measure and NWRR rely on r_1, the rank of the first relevant document in the list. This means that all of these metrics assume that *the user stops examining the ranked list as soon as he finds one relevant document, even if it is only partially relevant.* This assumption may be valid in some retrieval situations, but not always, as we shall discuss in Section 3. In contrast, *P-measure*, recently proposed at the SIGIR 2006 poster session [12], assumes that *the user looks for a highly relevant document even if it is ranked below partially relevant documents.* We shall also discuss its variant called P^+(pee-plus)-measure in Section 3.

Thus we have at least five metrics for the task of finding one relevant document: P^+-measure, P-measure, O-measure, NWRR and RR. All of them except for RR can handle graded relevance, as illustrated in Figure 1. (Section 2 will touch upon all of the metrics shown in this figure.) This paper compares the sensitivity of these five metrics using a method recently proposed at SIGIR 2006 [11]. This method derives the sensitivity and the performance difference required to guarantee a given significance level directly from Bootstrap Hypothesis Tests, and therefore has a theoretical foundation unlike the ad hoc (but nevertheless useful) "swap" method proposed by Voorhees and Buckley [17]. We use four data sets (i.e., test collections and submitted runs) from NTCIR [6] and conduct extensive experiments. Our results show that "$\mathrm{P}(^+)$-measure $\geq$ O-measure $\geq$ NWRR $\geq$ RR" generally holds, where "$\geq$" means "is at least as sensitive as". These results generalise and reinforce previously reported ones which were based only on ad hoc methods such as the swap method. We therefore recommend

Table 1. Representative work on evaluating the sensitivity of IR metrics

publication (year)	principal metrics		methods	collections
	binary	graded		
Buckley and Voorhees SIGIR (2000) [1]	AveP R-Prec PDoc Recall@1000	-	stability	TREC1 (TREC8 query)
Voorhees and Buckley SIGIR (2002) [17]	AveP PDoc	-	swap	TREC3-10
Buckley and Voorhees SIGIR (2004) [2]	AveP R-Prec PDoc bpref		swap	TREC8,10,12
Soboroff SIGIR (2004) [16]	AveP R-Prec PDoc **RR**	-	swap	TREC11,12 Web
Voorhees TREC 2004 (2005) [18]	AveP PDoc area G_AveP	-	swap, stability	TREC13
Sakai AIRS (2005) [8,14]	AveP R-Prec PDoc	Q-measure R-measure n(D)CG	swap, stability	NTCIR-3 Chinese NTCIR-3 Japanese
Sakai IPSJ TOD* (2006) [10]	**RR**	**O-measure**	swap, stability	NTCIR-3 Chinese NTCIR-3 Japanese
Sakai SIGIR (2006) [11]	AveP PDoc G_AveP	Q-measure n(D)CG G_Q-measure	swap, stability Bootstrap	NTCIR-3 Chinese NTCIR-3 Japanese
Sakai SIGIR (2006) [12]	**RR**	**P-measure** **O-measure**	swap, stability	NTCIR-5 Chinese NTCIR-5 Japanese
This paper	**RR**	**P($^+$)-measure** **O-measure** **NWRR**	Bootstrap	NTCIR-3 Chinese NTCIR-3 Japanese NTCIR-5 Chinese NTCIR-5 Japanese

* Information Processing Society of Japan Transactions on Databases.

the use of P($^+$)-measure and O-measure for practical tasks such as known-item search where recall is either unimportant or immeasurable.

The remainder of this paper is organised as follows. Section 2 discusses previous work on evaluating the sensitivity of IR metrics to clarify the contribution of this study. Section 3 formally defines and characterises the metrics we examine. Section 4 describes the Bootstrap-based method for assessing the sensitivity of metrics [11]. Section 5 describes our experiments for comparing the sensitivity of P($^+$)-measure, O-measure, NWRR and RR. Section 6 concludes this paper.

2 Previous Work

Table 1 provides an overview of representative studies on the stability and sensitivity of IR evaluation metrics or evaluation environments. Metrics for the task of finding one relevant document (as opposed to finding as many relevant documents as possible) are shown in bold, as we are focussing on these metrics. Below, we shall briefly review these studies and clarify the contribution of the present one.

At SIGIR 2000, Buckley and Voorhees [1] proposed the *stability* method which can be used for assessing the stability of IR metrics with respect to change in the topic set. The primary input to the stability method are a test collection, a set of runs submitted to the task defined by the test collection and an IR metric. The essence of the method is to compare systems X and Y in terms of an IR metric using different topic sets and count how often X outperforms Y, how often Y outperforms X and how often the two are regarded as equivalent. To this end, the stability method generates B new topic sets (where typically $B = 1000$) by sampling without replacement from the test collection topics. Using this method, Buckley and Voorhees examined binary IR metrics such as Average Precision (AveP), R-Precision (R-Prec), Precision at a given document cut-off (PDoc) and Recall at document cut-off 1000 (Recall@1000).

At SIGIR 2002, Voorhees and Buckley [17] proposed the *swap* method for assessing the sensitivity of IR metrics. The essence of the swap method is to estimate the *swap rate*, which represents the probability of the event that two experiments (using two different topic sets) are contradictory given an overall performance difference. Just like the stability method, the swap method samples topics without replacement, but generates B *disjoint pairs* of new topic sets to establish a relationship between the swap rate and the overall performance difference between a system pair.

Several studies followed that used the stability and swap methods for assessing different IR metrics: Buckley and Voorhees [2] assessed their *bpref* (binary preference) metric to deal with incomplete relevance judgments; Voorhees [18] assessed the *area* measure and *Geometric Mean* AveP (G_AveP) to emphasise the effect of worst-performing topics. Soboroff [16] assessed RR for the TREC Web known-item search task. Moreover, Sanderson and Zobel [15] and Sakai [9] explored some variants of the swap method, including one using topic sampling *with* replacement which allows duplicate topics within each set.

All of the aforementioned studies (except for Sakai [9]) used the TREC data and considered *binary* IR metrics only. Among them, only Soboroff [16] considered the task of finding one relevant document, by examining RR.

At AIRS 2005, Sakai [8,14] reported on stability and swap experiments that included IR metrics based on *graded* relevance, namely, *Q-measure*, *R-measure* and *normalised (Discounted) Cumulative Gain* (n(D)CG) [7]. But this study was limited to IR metrics for the task of finding as many relevant documents as possible. As for the task of finding one highly relevant document, two previous studies based on the stability and the swap methods reported that P-measure [12] and O-measure [10] may be more stable and sensitive than RR. As Table 1 shows, all of these studies involving graded-relevance metrics [8,10,12,14] used either the NTCIR-3 or the NTCIR-5 data, but not both.

All of the aforementioned studies used the stability and the swap methods, which have proven to be very useful, but are rather ad hoc. In light of this, Sakai [11] proposed a new method for assessing the sensitivity of IR metrics at SIGIR 2006, which relies on nonparametric *Bootstrap Hypothesis Tests* [3]. This method obtains B *bootstrap samples* by sampling *with* replacement from the original topic set, conducts a Bootstrap Hypothesis Test for every system pair, and estimates an absolute difference required to guarantee a given significance level based on Achieved Significance Levels. However, the above SIGIR paper only dealt with IR metrics for finding as many relevant documents as possible, including AveP, Q-measure and *Geometric Mean Q-measure* (G_Q-measure).

As it is clear from the bottom of Table 1, this paper *complements* a SIGIR 2006 poster [12] by assessing IR metrics for the task of finding one relevant document based on the aforementioned Bootstrap-based method instead of the ad hoc stability and swap methods. This paper also *extends* the poster by using four data sets, namely, NTCIR-3 and NTCIR-5 Chinese/Japanese data, and also by examining two additional metrics (P^+-measure and NWRR). In addition, it studies how changing *gain values* (See Section 3) affects the sensitivity of $P(^+)$-measure and O-measure. This paper can also be regarded as an extension of the Bootstrap paper [11], where the IR tasks being considered are different: finding as many relevant documents as possible versus finding one relevant document.

3 Metrics

This section formally defines and characterises $P(^+)$-measure, O-measure and NWRR. (We have already defined RR in Section 1.) Prior to this, we also define AveP and Q-measure since we include them in our experiments just for comparison. (The AveP and Q-measure results have been copied from [11], and are not part of our contribution.)

3.1 AveP and Q-measure

Let R denote the number of relevant documents for a topic, and $count(r)$ denote the number of relevant documents within top r of a system output of size L (≤ 1000). Clearly, Precision at Rank r can be expressed as $P(r) = count(r)/r$. Let $isrel(r)$ be 1 if the document at Rank r is relevant and 0 otherwise. Then, AveP can be defined as:

$$AveP = \frac{1}{R} \sum_{1 \leq r \leq L} isrel(r) P(r) . \qquad (1)$$

Next, we define Q-measure [8,14], which is very highly correlated with AveP but can handle graded relevance. Let $R(\mathcal{L})$ denote the number of $\mathcal{L}$-relevant documents so that $\sum_{\mathcal{L}} R(\mathcal{L}) = R$, and let $gain(\mathcal{L})$ denote the *gain value* for retrieving an $\mathcal{L}$-relevant document. In the case of NTCIR, $\mathcal{L} = S$ (highly relevant), $\mathcal{L} = A$ (relevant) or $\mathcal{L} = B$ (partially relevant), and we use $gain(S) = 3, gain(A) = 2, gain(B) = 1$ by default. Let $cg(r) = \sum_{1 \leq i \leq r} g(i)$ denote the *cumulative gain* at Rank r for a system output [7], where $g(i) = gain(\mathcal{L})$ if the document at Rank i is $\mathcal{L}$-relevant and $g(i) = 0$ otherwise. Similarly, let $cg_I(r)$ denote the cumulative gain at Rank r for an *ideal* ranked output: For NTCIR, an ideal ranked output lists up all S-, A- and B-relevant documents in this order.

Then, Q-measure is defined as:

$$\textit{Q-measure} = \frac{1}{R} \sum_{1 \leq r \leq L} isrel(r) BR(r)$$

where

$$BR(r) = \frac{cg(r) + count(r)}{cg_I(r) + r} . \tag{2}$$

$BR(r)$ is called the *blended ratio*, which measures how a system output deviates from the ideal ranked output *and* penalises "late arrival" of relevant documents. (Unlike the blended ratio, it is known that *weighted precision* $WP(r) = cg(r)/cg_I(r)$ cannot properly penalise late arrival of relevant documents and is therefore not suitable for IR evaluation [8,10,14].)

3.2 O-measure and NWRR

Traditional IR assumes that *recall* is important: Systems are expected to return as many relevant documents as possible. AveP and Q-measure, both of which are recall-oriented, are suitable for such tasks. (Note that the number of relevant documents R appear in their definitions.) However, as was discussed in Section 1, some IR situations do not necessarily require recall. More specifically, some IR situations require *one* relevant document only. Although RR is commonly used in such a case, it cannot reflect the fact that users prefer highly relevant documents to partially relevant ones. Below, we describe O-measure and Normalised Weighted Reciprocal Rank (NWRR), both of which can be regarded as graded-relevance versions of RR.

O-measure [10] is defined to be zero if the ranked output does not contain a relevant document. Otherwise:

$$\textit{O-measure} = BR(r_1) = \frac{g(r_1) + 1}{cg_I(r_1) + r_1} . \tag{3}$$

That is, O-measure is the blended ratio at Rank r_1. (Since the document at r_1 is the *first* relevant one, $cg(r_1) = g(r_1)$ and $count(r_1) = 1$ hold.) In a binary relevance environment, $\textit{O-measure} = RR$ holds iff $r_1 \leq R$, and $\textit{O-measure} > RR$ holds otherwise. Moreover, if small gain values are used with O-measure, then it behaves like RR [10].

Next, we define Weighted Reciprocal Rank (WRR) proposed by Eguchi *et al.* [4]. Our definition looks different from their original one, but it is easy to show that the two are equivalent [13]. In contrast to cumulative-gain-based metrics (including Q-measure and O-measure) which require the gain values ($gain(\mathcal{L})$) as parameters, WRR requires "penalty" values $\beta(\mathcal{L})$ (> 1) for each relevance level $\mathcal{L}$. We let $\beta(S) = 2, \beta(A) = 3, \beta(B) = 4$ throughout this paper: note that the smallest penalty value must be assigned to highly relevant documents. WRR is defined to be zero if the ranked output does not contain a relevant document. Otherwise:

$$WRR = \frac{1}{r_1 - 1/\beta(\mathcal{L}_1)} \tag{4}$$

where $\mathcal{L}_1$ denotes the relevance level of the relevant document at Rank r_1.

WRR was designed for the NTCIR Web track, but the track organisers always used $\beta(\mathcal{L}) = \infty$ for all $\mathcal{L}$, so that WRR is reduced to binary RR. That is, the graded relevance capability of WRR has never actually been used.

WRR is not bounded by one: If the highest relevance level for a given topic is denoted by $\mathcal{M}$, WRR is bounded above by $1/(1 - 1/\beta(\mathcal{M}))$. This is undesirable for two reasons: Firstly, a different set of penalty values yields a different range of WRR values, which is inconvenient for comparisons; Secondly, the highest relevance level $\mathcal{M}$ may not necessarily be the same across topics, so the upperbound of WRR may differ across topics. This means that WRR is not suitable for averaging across topics if $\mathcal{M}$ differs across the topic set of the test collection.

This paper therefore considers Normalised WRR (NWRR) instead. NWRR is defined to be zero if the ranked output does not contain a relevant document. Otherwise:

$$NWRR = \frac{1 - 1/\beta(\mathcal{M})}{r_1 - 1/\beta(\mathcal{L}_1)} \,. \tag{5}$$

The upperbound of NWRR is one for any topic and is therefore averageable.

There are two important differences between NWRR and O-measure.

(a) *Just like RR, NWRR disregards whether there are many relevant documents or not. In contrast, O-measure takes the number of relevant documents into account by comparing the system output with an ideal output.*

(b) *NWRR assumes that the rank of the first retrieved document is more important than the relevance levels.* Whereas, O-measure is free from this assumption.

We first discuss (a). From Eq. (5), it is clear that NWRR depends only on the rank and the relevance level of the first retrieved relevant document. For example, consider a system output shown in the middle of Figure 2, which has an S-relevant document at Rank 3. The NWRR for this system is $(1-1/\beta(S))/(3-1/\beta(S)) = (1-1/2)/(3-1/2) = 1/5$ for *any* topic. Whereas, the value of O-measure for this system depends on how many $\mathcal{L}$-relevant documents there are. For example, if the system output was produced in response to Topic 1 which has only one S-relevant document (and no other relevant documents), then, as shown on the left hand side of Figure 2, $O\text{-}measure = (g(3)+1)/(cg_I(3)+3) = (3+1)/(3+3) = 2/3$. On the other hand, if the system output was produced in response to Topic 3 which has at least three S-relevant documents, then, as shown in the right hand side of the figure, $O\text{-}measure = (3+1)/(9+3) = 1/3$. Thus, *O-measure assumes that it is relatively easy to retrieve an $\mathcal{L}$-relevant document if there are many $\mathcal{L}$-relevant documents in the database.* If the user has no idea as to whether a document relevant to his request exists or not, then one could argue that NWRR may be a better model. On the other hand, if the user has some idea about the number of relevant documents he might find, then O-measure may be more suitable. Put another way, O-measure is more *system-oriented* than NWRR.

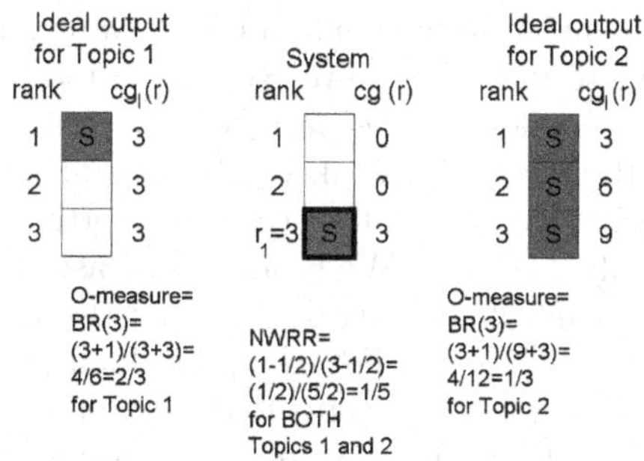

Fig. 2. O-measure vs NWRR: Topics 1 and 2

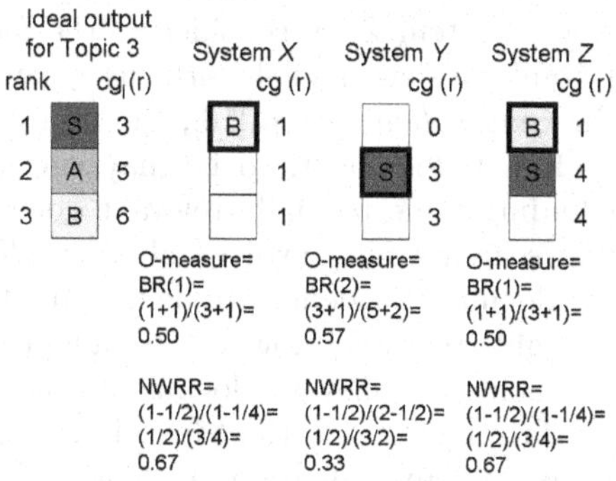

Fig. 3. O-measure vs NWRR: Topic 3

Next, we discuss (b) using Topic 3 shown in Figure 3, which has one S-relevant, one A-relevant and one B-relevant document. System X has a B-relevant document at Rank 1, while System Y has an S-relevant document at Rank 2. Regardless of the choice of penalty values ($\beta(\mathcal{L})$), X always outperforms Y according to NWRR. Thus, *NWRR is unsuitable for IR situations in which retrieving a highly relevant document is more important than retrieving any relevant document in the top ranks.* In contrast, O-measure is free from the assumption underlying NWRR: Figure 3 shows that, with default gain values, Y outperforms X. But if X should be preferred, then a different gain value assignment (e.g. $gain(S) = 2, gain(A) = 1.5, gain(B) = 1$) can be used. In this respect, O-measure is more flexible than NWRR.

3.3 P-measure and P$^+$-measure

Despite the abovementioned differences, both NWRR and O-measure rely on r_1, the rank of the first retrieved relevant document. Thus, both *NWRR and O-measure assume that the user stops examining the ranked list as soon as he finds one relevant document, even if it is only a partially relevant one.* This assumption may be counterintuitive in some cases: Consider System Z in Figure 3, which has a B-relevant document at Rank 1 *and* an S-relevant document at Rank 2. According to both NWRR and O-measure, System Z and System X are always equal in performance regardless of the parameter values, because only the B-relevant document at Rank $r_1 = 1$ is taken into account for Z. In short, both NWRR and O-measure ignore the fact that there is a better document at Rank 2.

This is not necessarily a flaw. NWRR and O-measure may be acceptable models for IR situations in which it is difficult for the user to spot a highly relevant document in the ranked list. For example, the user may be looking at a plain list of document IDs, or a list of vague titles and poor-quality text snippets of the retrieved documents. Or perhaps, he may be examining the content of each document one-by-one without ever looking at a ranked list, so that he has no idea what the next document will be like. However, if the system can show a high-quality ranked output that contain informative titles and abstracts, then perhaps it is fair to assess System Z by considering the fact that it has an S-relevant document at Rank 2, since a real-world user can probably spot this document. Similarly, in *known-item search* [16], the user probably knows that there exists a highly relevant document, so he may continue to examine the ranked list even after finding some partially relevant documents. (A "narrow" definition of known-item search would involve only one relevant document per topic. That is, the target document is defined to be the one that the user has seen before, and it is always highly relevant. However, we adopt a broader definition: In addition to the known highly relevant document, there may be unvisited documents which are in fact relevant to the topic. It is possible to treat these documents as partially relevant in evaluation. Moreover, there is a related task called *suspected-item* search [5], which does not require that the user has actually *seen* a relevant document. It is clear that more than one relevant document may exist in such cases too, possibly with different relevance levels.)

We now define P-measure [12] for the task of finding one highly relevant document, under the assumption that *the user continues to examine the ranked list until he finds a document with a satisfactory relevance level.* P-measure is defined to be zero if the system output does not contain a relevant document. Otherwise, let the *preferred rank* r_p be the rank of the first record obtained by sorting the system output, using the relevance level as the primary sort key (preferring higher relevance levels) and the rank as the secondary sort key (preferring the top ranks). Then:

$$P\text{-}measure = BR(r_p) = \frac{cg(r_p) + count(r_p)}{cg_I(r_p) + r_p} \, . \tag{6}$$

That is, P-measure is simply the blended ratio at Rank r_p. For System Z in Figure 3, $r_p = 2$. Therefore, $P\text{-}measure = BR(2) = (cg(2) + 2)/(cg_I(2) + 2) = (4 + 2)/(5 + 2) = 0.86$. Whereas, since $r_p = r_1$ holds for systems X and Y, $P\text{-}measure = O\text{-}measure = 0.50$ for X and $P\text{-}measure = O\text{-}measure = 0.57$ for Y. Thus, only Z is handsomely rewarded, for retrieving both B- and S-relevant documents.

Because P-measure looks for a most highly relevant document in the ranked output and then evaluates by considering all (partially) relevant documents ranked above it, it is possible that P-measure may be more stable and sensitive than O-measure, as we shall see later. Moreover, it is clear that P-measure inherits some properties of O-measure: It is a system-oriented metric, and is free from the assumption underlying NWRR.

However, just like R-measure [8,14], P-measure is "forgiving", in that it can be one for a suboptimal ranked output. For example, in Figure 3, supppose that there is a fourth system output, which is a *perfect inverse* of the ideal output. For this system output, $r_p = 3$ and therefore $P\text{-}measure = BR(3) = (6+3)/(6+3) = 1$. One could argue that this is counterintuitive. We therefore examine P$^+$(pee-plus)-measure in addition, which does not have this problem:

$$P^{+}\text{-}measure = \frac{1}{count(r_p)} \sum_{1 \leq r \leq r_p} isrel(r) BR(r) \; . \tag{7}$$

For example, for the above perfect inverse output, $BR(1) = (1+1)/(3+1)$, $BR(2) = (3+2)/(5+2)$ and $BR(3) = P\text{-}measure = 1$. Thus $P^{+}\text{-}measure = (2/4 + 5/7 + 1)/3 = 0.74$. Note also that $P^{+}\text{-}measure = P\text{-}measure = O\text{-}measure$ holds if there is no relevant document above Rank r_p, i.e., if $r_p = r_1$.

In practice, a document cut-off may be used with P($^+$)-measure, since these metrics assume that the user is willing to examine an "unlimited" number of documents. That is, in theory, r_p can be arbitrarily large. However, a small cut-off makes IR evaluation unstable, and requires a larger topic set [1,8,14].

4 Bootstrap-Based Method for Evaluating IR Metrics

This section briefly describes Sakai's Bootstrap-based method for assessing the sensitivity of IR metrics [11].

First, we describe the paired Bootstrap Hypothesis Test, which, unlike traditional significance tests, is free from the normality and symmetry assumptions and yet has high power [3]. The strength of the Bootstrap lies in its reliance on the computer for directly estimating any data distribution through *resampling* from observed data. Let Q be the set of topics provided in the test collection, and let $|Q| = n$. Let $\mathbf{x} = (x_1, \ldots, x_n)$ and $\mathbf{y} = (y_1, \ldots, y_n)$ denote the per-topic performance values of systems X and Y as measured by some performance metric M. A standard method for comparing X and Y is to measure the difference between *sample means* $\bar{x} = \sum_i x_i/n$ and $\bar{y} = \sum_i y_i/n$ such as *Mean* Average Precision values. But what we really want to know is whether the *population means* for X and Y (μ_X and μ_Y), computed based on the population P of topics, are any different. Since we can regard $\mathbf{x}$ and $\mathbf{y}$ as *paired* data, we let $\mathbf{z} = (z_1, \ldots, z_n)$ where $z_i = x_i - y_i$, let $\mu = \mu_X - \mu_Y$ and set up the following hypotheses for a two-tailed test:

$$H_0: \quad \mu = 0 \quad vs \quad H_1: \mu \neq 0 \; .$$

Thus the problem has been reduced to a *one-sample problem* [3]. As with standard significance tests, we assume that $\mathbf{z}$ is an independent and identically distributed sample drawn from an unknown distribution.

```
for b = 1 to B
    create topic set Q*b of size n = |Q| by randomly
    sampling with replacement from Q;
    for i = 1 to n
        q = i-th topic from Q*b;
        w_i*b = observed value in w for topic q;
```

Fig. 4. Algorithm for creating Bootstrap samples Q^{*b} and $\mathbf{w}^{*b} = (w_1^{*b}, \ldots, w_n^{*b})$

```
count = 0;
for b = 1 to B
    t(w*b) = w̄*b/(σ̄*b/√n);
    if( |t(w*b)| ≥ |t(z)| ) then count++;
ASL = count/B;
```

Fig. 5. Algorithm for estimating the Achieved Significance Level

In order to conduct a Hypothesis Test, we need a *test statistic* t and a *null hypothesis distribution*. Here, let us consider a Studentised statistic:

$$t(\mathbf{z}) = \frac{\bar{z}}{\bar{\sigma}/\sqrt{n}}$$

where $\bar{\sigma}$ is the standard deviation of $\mathbf{z}$, given by

$$\bar{\sigma} = (\sum_i (z_i - \bar{z})^2/(n-1))^{\frac{1}{2}} \ .$$

Moreover, let $\mathbf{w} = (w_1, \ldots, w_n)$ where $w_i = z_i - \bar{z}$, in order to create *bootstrap samples* of per-topic performance differences $\mathbf{w}^{*b}$ that obey H_0. Figure 4 shows the algorithm for obtaining B bootstrap samples of topics (Q^{*b}) and the corresponding values of $\mathbf{w}^{*b}$. (We let $B = 1000$ throughout this paper.) For example, let us assume that we only have five topics $Q = (001, 002, 003, 004, 005)$ and that $\mathbf{w} = (0.2, 0.0, 0.1, 0.4, 0.0)$. Suppose that, for trial b, sampling with replacement from Q yields $Q^{*b} = (001, 003, 001, 002, 005)$. Then, $\mathbf{w}^{*b} = (0.2, 0.1, 0.2, 0.0, 0.0)$.

For each b, let $\bar{w}^{*b}$ and $\bar{\sigma}^{*b}$ denote the mean and the standard deviation of $\mathbf{w}^{*b}$. Figure 5 shows how to compute the Achieved Significance Level (ASL) using $\mathbf{w}^{*b}$. In essence, we examine how *rare* the observed difference would be under H_0. If $ASL < \alpha$, where typically $\alpha = 0.01$ (very strong evidence against H_0) or $\alpha = 0.05$ (reasonably strong evidence against H_0), then we reject H_0. That is, we have enough evidence to state that μ_X and μ_Y are probably different.

We now describe Sakai's method for assessing the sensitivity of IR metrics. Let C denote the set of all possible combinations of two systems. First, perform a Bootstrap Hypothesis Test for every system pair in C and count how many of the pairs satisfy $ASL < \alpha$: The result represents the sensitivity of a given IR metric. We thereby obtain the values of $\bar{w}^{*b}$ and $t(\mathbf{w}^{*b})$ for each

$DIFF = \phi$;
for each system pair $(X, Y) \in C$
  sort $|t(\mathbf{w}^{*1}_{X,Y})|, \ldots, |t(\mathbf{w}^{*B}_{X,Y})|$;
  if $|t(\mathbf{w}^{*b'}_{X,Y})|$ is the $B\alpha$-th largest value
  then add $|\bar{w}^{*b'}_{X,Y}|$ to $DIFF$;
$estimated\_diff = \max\{diff \in DIFF\}$ (rounded to two significant figures);

Fig. 6. Algorithm for estimating the minimum absolute performance difference that correspond to a statistically significant difference

Table 2. Statistics of the NTCIR CLIR data (including monolingual runs)

	$\|Q\|$	R	$R(S)$	$R(A)$	$R(B)$	#runs used
		per topic				
NTCIR-3 Chinese	42	78.2	21.0	24.9	32.3	30
NTCIR-3 Japanese	42	60.4	7.9	31.5	21.0	30
NTCIR-5 Chinese	50	61.0	7.0	30.7	23.3	30
NTCIR-5 Japanese	47	89.1	3.2	41.8	44.2	30

system pair (X, Y), which we shall denote explicitly by $\bar{w}^{*b}_{X,Y}$ and $t(\mathbf{w}^{*b}_{X,Y})$. Since each $\bar{w}^{*b}_{X,Y}$ is a performance difference computed based on $|\mathbf{w}^{*b}_{X,Y}| = |Q| = n$ topics, we can use the algorithm shown in Figure 6 to obtain a natural estimate of the minimum performance difference required for guaranteeing $ASL < \alpha$, given the topic set size n. For example, if $\alpha = 0.05$ is chosen, the algorithm looks for the $B\alpha = 1000 * 0.05 = 50$-th largest value among $|t(\mathbf{w}^{*b}_{X,Y})|$ and takes the corresponding value of $|\bar{w}^{*b}_{X,Y}|$ for each (X, Y). Among the $|C|$ values thus obtained, the algorithm takes the maximum value just to be conservative.

Note that the estimated differences themselves are not necessarily suitable for comparing metrics, since some metrics tend to take small values while others tend to take large values. The sensitivity of metrics should primarily be compared in terms of how many system pairs satisfy $ASL < \alpha$, that is, how many pairs show a statistically significant difference.

5 Experiments

5.1 Data

Table 2 shows some statistics of the NTCIR CLIR data (i.e., test collections and submitted runs) we used, which were kindly provided by National Institute of Informatics (NII), Japan. Currently, the NTCIR-3 and NTCIR-5 CLIR data are the only data publicly available from NII. For each data set, we selected the top 30 runs as measured by "Relaxed" Mean AveP [8,14]. Thus we conducted four sets of experiments, each with $|C| = 30 * 29/2 = 435$ system pairs.

Table 3. Sensitivity and the estimated difference required based on Sakai's method ($\alpha = 0.05$)

data	metric	sensitivity (ASL $< \alpha$)	estimated diff.
NTCIR-3 Chinese (42 topics)	Q-measure	242/435=56%	0.10
	AveP	240/435=55%	0.11
	P-measure	170/435=39%	0.18
	P$^+$-measure	167/435=38%	0.18
	O-measure	165/435=38%	0.19
	NWRR	136/435=31%	0.20
	RR	126/435=29%	0.22
NTCIR-3 Japanese (42 topics)	Q-measure	305/435=70%	0.13
	AveP	296/435=68%	0.11
	P$^+$-measure	272/435=63%	0.18
	P-measure	271/435=62%	0.20
	O-measure	255/435=59%	0.22
	NWRR	247/435=57%	0.19
	RR	246/435=57%	0.23
NTCIR-5 Chinese (50 topics)	Q-measure	174/435=40%	0.11
	AveP	159/435=37%	0.11
	P$^+$-measure	134/435=31%	0.15
	O-measure	125/435=29%	0.15
	P-measure	123/435=28%	0.16
	NWRR	114/435=26%	0.16
	RR	94/435=22%	0.16
NTCIR-5 Japanese (47 topics)	Q-measure	136/435=31%	0.09
	AveP	113/435=26%	0.10
	P$^+$-measure	77/435=18%	0.14
	P-measure	73/435=17%	0.15
	O-measure	63/435=14%	0.16
	NWRR	63/435=14%	0.16
	RR	54/435=12%	0.17

5.2 Results and Discussions

Table 3 shows the results of our Bootstrap-based experiments. It shows, for example, that if P-measure is used for assessing the "top" 30 systems (as measured by AveP) that were submitted to the NTCIR-3 Chinese document retrieval subtask, it can detect a statistically significant difference at $\alpha = 0.05$ for 39% of the system pairs; The estimated overall performance difference required for showing a statistical significance is 0.18.

The *absolute* sensitivity values depend heavily on the set of runs: It can be observed, for example, that P-measure can detect a significant difference for 62% of the system pairs for the NTCIR-3 Japanese data, but for only 17% for the NTCIR-5 Japanese data. That is, the NTCIR-5 Japanese runs are much harder to distinguish from each other because a larger number of teams performed well at NTCIR-5 than at NTCIR-3. (For both NTCIR-3 and NTCIR-5, the top 30 Japanese runs we used came from 10 different teams.)

On the other hand, the *relative* sensitivity, which is the focus of this study, is quite consistent across the four data sets: We can observe that "P($^+$)-measure $\geq$ O-measure $\geq$ NWRR $\geq$ RR" generally holds, where "$\geq$" means "is at least as sensitive as". (O-measure outperforms P-measure by two system pairs for the NTCIR-5 Chinese data, but this difference is probably not substantial.) The difference in sensitivity between P($^+$)-measure and O-measure arises from the fact that P($^+$)-measure consider all relevant documents ranked above r_p; That

Table 4. The effect of changing gain values on sensitivity and the estimated difference required ($\alpha = 0.05$). The default results have been copied from Table 3.

data	metric	sensitivity (ASL < α)	estimated diff.
NTCIR-5 Chinese (50 topics)	P^+3:2:1 (default)	134/435=31%	0.15
	P^+0.3:0.2:0.1	132/435=30%	0.15
	P^+30:20:10	128/435=29%	0.16
	P^+10:5:1	125/435=29%	0.14
	P^+1:1:1	124/435=29%	0.15
	P10:5:1	125/435=29%	0.14
	P3:2:1 (default)	123/435=28%	0.16
	P30:20:10	121/435=28%	0.15
	P0.3:0.2:0.1	120/435=28%	0.14
	P1:1:1	113/435=26%	0.15
	O3:2:1 (default)	125/435=29%	0.15
	O30:20:10	123/435=28%	0.15
	O10:5:1	118/435=27%	0.16
	O0.3:0.2:0.1	107/435=25%	0.17
	O1:1:1	94/435=22%	0.16
NTCIR-5 Japanese (47 topics)	P^+3:2:1 (default)	77/435=18%	0.14
	P^+30:20:10	73/435=17%	0.15
	P^+0.3:0.2:0.1	69/435=16%	0.17
	P^+1:1:1	67/435=15%	0.14
	P^+10:5:1	67/435=15%	0.15
	P10:5:1	85/435=20%	0.14
	P30:20:10	81/435=19%	0.15
	P3:2:1 (default)	73/435=17%	0.15
	P0.3:0.2:0.1	64/435=15%	0.16
	P1:1:1	59/435=14%	0.15
	O30:20:10	65/435=15%	0.16
	O3:2:1 (default)	63/435=14%	0.16
	O10:5:1	62/435=14%	0.17
	O0.3:0.2:0.1	59/435=14%	0.16
	O1:1:1	54/435=12%	0.18

between O-measure and NWRR arises from the fact that the former considers an ideal ranked ouput (i.e., the number of relevant documents); That between NWRR and RR arises from the use of graded relevance. Our findings generalise and reinforce previously-reported results which suggested the superiority of P-measure and O-measure over RR [10,12]. However, these previous studies were based only on the ad hoc stability and swap methods, and examined neither P^+-measure nor NWRR.

It is also clear that the five metrics for the task of finding one relevant documents are not as sensitive as Q-measure and AveP. This is because, while Q-measure and AveP considers *all* relevant documents, the four metrics do not: They generally examine the very top of a ranked output only. That is, because they are based only on a small number of observations, they are inherently less stable than Q-measure and AveP. Therefore, one should prepare a larger set of topics if any of the four metrics is to be used instead of more stable metrics such as Q-measure and AveP.

As for the estimated difference required for a given significance level, the NTCIR-5 results suggest that, with around 50 topics, we need an absolute difference of around 0.14-0.16 in terms of P($^+$)-measure or O-measure in order to detect a significant difference at $\alpha = 0.05$.

5.3 Changing the Gain Values for P(+)-measure and O-measure

We finally focus on P(+)-measure and O-measure which we have shown to be the three most sensitive metrics for the task of finding one relevant document, and study the effect of chaning *gain values* (See Section 3) using the NTCIR-5 data.

Table 4 summarises the results, where, for example, "P10:5:1" represents P-measure with $gain(S) = 10, gain(A) = 5, gain(B) = 1$. (Hence the default gain value results, labelled with "3:2:1", have been copied from Table 3.) Recall that: (a) Using small gain values makes O-measure resemble RR (See Eq. (3)); (b) Given that the gain values are all one, $O\text{-}measure = RR$ holds iff $r_1 \leq R$; (c) $P(^+)\text{-}measure = O\text{-}measure$ holds if there is no relevant document above Rank r_p. Thus, we can expect "flat" and small gain values to reduce the sensitivity of these metrics. Indeed, Table 4 shows that the gain value assignments $gain(S) = 1, gain(A) = 1, gain(B) = 1$ and $gain(S) = 0.3, gain(A) = 0.2, gain(B) = 0.1$ tend to hurt sensitivity, especially the former. On the other hand, using large gain values (which implies less penalty on late arrival of relevant documents) and using "steeper" gain values (which emphasises the relevance levels) generally do not seem to have a substantial impact on sensitivity. In summary, P(+)-measure and O-measure are fairly robust to the choice of gain values as long as graded relevance is properly utilised.

6 Conclusions

This paper compared the sensitivity of five evaluation metrics designed for the task of finding one relevant document, using Sakai's method based on Bootstrap Hypothesis Tests. Using four data sets from NTCIR, we showed that "P(+)-measure $\geq$ O-measure $\geq$ NWRR $\geq$ RR" generally holds, where "$\geq$" means "is at least as sensitive as". These results generalise and reinforce previously reported ones [10,12] based on the ad hoc stability and swap methods, which suggested the superiority of P-measure and O-measure over RR. Moreover, in terms of sensitivity, P(+)-measure and O-measure are fairly robust to the choice of gain values. Thus we recommend the use of P(+)-measure and O-measure for practical tasks such as known-item search where recall is either unimportant or immeasurable. But it should be remembered that while O-measure represents a user who is satisfied with finding one partially relevant document, P(+)-measure approximate one who looks for one "good" relevant document.

As for how the above metrics resemble each other in terms of system ranking, we already have results that generalise Sakai's previous analysis [10] which did not include NWRR and P$^+$-measure: Metrics for finding one relevant document (P(+)-measure, O-measure, NWRR and RR) produce system rankings that are substantially different from those produced by Q-measure and AveP; Rankings by P(+)-measure, O-measure and NWRR generally resemble each other. Moreover, system rankings by P(+)-measure are fairly robust to the choice of gain values. But due to lack of space, we will discuss these additional results elsewhere.

References

1. Buckley, C. and Voorhees, E. M.: Evaluating Evaluation Measure Stability. ACM SIGIR 2000 Proceedings (2000) 33–40
2. Buckley, C. and Voorhees, E. M.: Retrieval Evaluation with Incomplete Information. ACM SIGIR 2004 Proceedings (2004) 25–32
3. Efron, B. and Tibshirani, R. J.: An Introduction to the Bootstrap. Chapman & Hall/CRC (1993)
4. Eguchi, K. *et al.*: Overview of the Web Retrieval Task at the Third NTCIR Workshop. National Institute of Informatics Technical Report NII-2003-002E (2003)
5. Hawking, D. and Craswell, N.: The Very Large Collection and Web Tracks. In TREC: Experiment and Evaluation in Information Retrieval, MIT Press (2005) 199-231
6. Kando, N: Overview of the Fifth NTCIR Workshop. NTCIR-5 Proceedings (2005)
7. Kekäläinen, J.: Binary and Graded Relevance in IR Evaluations – Comparison of the Effects on Ranking of IR Systems. Information Processing and Management **41** (2005) 1019–1033.
8. Sakai, T.: The Reliability of Metrics based on Graded Relevance. AIRS 2005 Proceedings, LNCS **3689**, Springer-Verlag (2005) 1–16
9. Sakai, T.: The Effect of Topic Sampling on Sensitivity Comparisons of Information Retrieval Metrics. NTCIR-5 Proceedings (2005) 505–512
10. Sakai, T.: On the Task of Finding One Highly Relevant Document with High Precision. Information Processing of Japan Transactions on Databases **TOD-29** (2006)
11. Sakai, T.: Evaluating Evaluation Metrics based on the Bootstrap. ACM SIGIR 2006 Proceedings, to appear (2006)
12. Sakai, T.: Give Me Just One Highly Relevant Document: P-measure. ACM SIGIR 2006 Proceedings, to appear (2006)
13. Sakai, T.: A Further Note on Evaluation Metrics for the Task of Finding One Highly Relevant Document. Information Processing of Japan SIG Technical Reports **FI-82** (2006) 69–76
14. Sakai, T.: On the Reliability of Information Retrieval Metrics based on Graded Relevance, Information Processing and Management, to appear (2006)
15. Sanderson, M. and Zobel, J.: Information Retrieval System Evaluation: Effort, Sensitivity, and Reliability. ACM SIGIR 2005 Proceedings (2005) 162–169
16. Soboroff, I.: On Evaluating Web Search with Very Few Relevant Documents. ACM SIGIR 2004 Proceedings (2004) 530–531
17. Voorhees, E. M. and Buckley, C.: The Effect of Topic Set Size on Retrieval Experiment Error. ACM SIGIR 2002 Proceedings (2002) 316–323
18. Voorhees, E. M.: Overview of the TREC 2004 Robust Retrieval Track. TREC 2004 Proceedings (2005)

Evaluating Topic Difficulties from the Viewpoint of Query Term Expansion

Masaharu Yoshioka

Graduate School of Information Science and Technology
Hokkaido University
N-14 W-9, Kita-ku, Sapporo 060-0814, Japan
National Institute of Informatics
2-1-2 Hitotsubashi, Chiyoda-ku, Tokyo 101-8430, Japan
yoshioka@ist.hokudai.ac.jp

Abstract. Query term expansion is an important technique for achieving higher retrieval performance. However, since many factors affects the quality of this technique, it is difficult to evaluate this technique in isolation. Feature quantities that characterize the quality of the initial query are defined in this study for evaluating topic difficulties from the viewpoint of query term expansion. I also briefly review the result of the NTCIR-5 query term expansion subtask that uses these quantities for evaluating the effectiveness of the query term expansion techniques. I also describe detailed analysis results on the effect of query term expansion based on topic-by-topic analysis.

1 Introduction

It is very difficult for many users of Information Retrieval (IR) system to select appropriate query terms to represent their information need. This difficulty cause the mismatch between query terms and information need; e.g., the query terms are ambiguous and it is difficult to focus on relevant documents only; a number of relevant documents are difficult to retrieve because they have only a small part of initial query terms.

To reduce this mismatch, many IR systems use query term expansion techniques to find better query terms[1]. However, the effectiveness of this technique depends on the quality of initial query terms. Cronen-Townsend et al. [2] used a query clarity score based on a language model to decide if the query terms contain relevant information for the query term expansion; this approach was shown to be effective.

The Reliable Information Access (RIA) Workshop [3] conducted a failure analysis [4] for a set of topics using seven different popular IR systems and proposed a topic categorization based on the types of failures they encountered. They also conducted a relevance feedback experiment using a different IR systems [5]. This study, however, did not examine the relationship between topic difficulty based on the mismatch and the effect of relevance feedback. Because the relevance feedback technique is used for reducing the mismatch between the

H.T. Ng et al. (Eds.): AIRS 2006, LNCS 4182, pp. 390–403, 2006.

initial query and information need, it is important to determine the effectiveness of relevance feedback when used with query expansion.

For characterizing topic difficulty based on the mismatch, query term expansion subtask was designed for evaluating topic difficulties from the viewpoint of query term expansion in the NTCIR-5 Web task [6]. In this subtask, several feature quantities that characterize this mismatch were investigated for analyzing the topic difficulty, but single quantity is not enough to characterize this topic difficulty.

Therefore, in this paper, I analyze the effect of query term expansion technique of IR experiment data in topic-by-topic manner and discuss how these quantities are affect the quality of this technique.

The remainder of this paper is divided into four sections. Section 2 briefly review various statistical features for defining the topic difficulty and the effectiveness of the query expansion term. In Section 3, the result of query term expansion subtask in the NTCIR-5 web task is briefly summarized. Section 4 analyzes the experimental results, and Section 5 gives the conclusions of the paper.

2 Statistical Features for Evaluation of the Query Term Expansion Technique

Buckley et al. [7] hypothesized a possible reason why query expansion improves the query performance as follows.

1. one or two good alternative words to original query terms (synonyms)
2. one or two good related words
3. a large number of related words that establish that some aspect of the topic is present (context)
4. specific examples of general query terms
5. better weighting to original query terms

The first four reasons relate to query term expansion. Reasons 1, 2, and 4 can be evaluated using a thesaurus. However, since Voorhees [8] confirmed simple automatic query term expansion based on a general thesaurus did not improve query performance, it may be inappropriate to use a general thesaurus for this evaluation.

Therefore, in the query term expansion subtask in the NTCIR-5 Web task, a mismatch between different information-need expressions (query terms and relevant documents) were used for this evaluation.

2.1 Feature Quantities for Characterizing Mismatch Between Initial Query and Relevant Documents

When a user carefully selects good query terms, query term expansion may not improve the retrieval performance. Therefore, it is crucial for this subtask to evaluate the quality of the initial query based on the mismatch between the initial query and the relevant documents.

In the NTCIR-4 web test collection, each query has a Boolean operator information. In order to estimate the information need precisely, it is better to use this Boolean information[1].

When the query is represented with a Boolean operator, this mismatch is characterized as a mismatch between the documents that satisfy this query and the relevant documents. When the initial query is precise enough, documents that satisfy the Boolean query (Boolean satisfied documents) and relevant documents are equivalent ((1) and (3) in Figure 1 are an empty set).

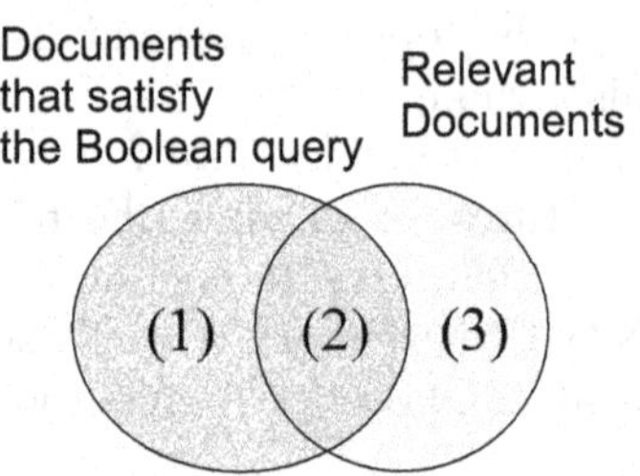

Fig. 1. Mismatch between Initial Query and Relevant Documents

However, because it is difficult to construct good queries, (1) and (3) are empty sets in almost no query. The sizes of (1) and (3) characterizes the quality of the query from the viewpoint of query expansion. For example, when there are many documents in (1), the initial query is too general and requires new query terms that define the context of the query. Conversely, when (3) has many documents, the initial query is too strict and it is necessary to determine alternative words to relax the query.

Since the number of documents in (1) and (3) are affected by the number of relevant documents, the following two feature quantities were used for this evaluation.

$R\&B/R$ The ratio between the size of relevant documents that satisfy the Boolean query ((2)) and the size of relevant documents ((2)+(3)).

$R\&B/B$ The ratio between the size of relevant documents that satisfy the Boolean query ((2)) and the size of the Boolean satisfied documents ((1)+(2)).

Since a set of B is a result of the Boolean IR system without term weighting, $R\&B/R$ and $R\&B/B$ represents the appropriateness of initial query quality from the view point of recall and precision respectively.

These two feature quantities are the values that represent the characteristics of the set of initial query terms. Therefore, this value is also effective for the different IR models (probabilistic, vector-space, and etc.) evaluation even though it is based on the Boolean IR model.

[1] For the test collection without Boolean operator, all terms with "AND" operator or Boolean query generation system proposed in [9] or [10] can be used as a starting point.

2.2 Feature Quantities for Evaluating Effectiveness of the Query Term

Feature quantities proposed in 2.1 can also be used to evaluate a query term when the Boolean query is constructed using only this term.

In addition to these features, the following three criteria were proposed for selecting feature quantities.

1. Appropriateness of the alternative term for each initial query term.
2. Appropriateness of the context definition term for the query.
3. Appropriateness of the term that characterizes the relevant documents.

A good alternative term should exist for relevant documents that do not contain the initial query term. Therefore, the number of documents that have a query expansion term and do not have an initial query term are useful for evaluation.

A good term for context definition is a distinct term that exists in relevant documents. Therefore, the number of documents that have a query expansion term in the relevant documents, the Boolean satisfied documents, and total documents are useful for evaluation.

The following feature quantities are defined for each query expansion term.

total : *Rel* The number of relevant documents that have a query expansion term.

total : *Bool* The number of Boolean satisfied documents that have a query expansion term.

total : *R&B* The number of Boolean satisfied relevant documents that have a query expansion term.

total : *All* The number of documents that have a query expansion term in the document database.

Feature quantities that are based on mutual information content were used for evaluating the distinctiveness of each term [9]. These quantities are the mutual information content between relevant documents r and the term w. $p(w)$ is the probability of the term w in the document database and $p(w|r)$ is the probability of the relevant documents.

$$MI(w) = p(w|r) log_2 \frac{p(w|r)}{p(w)}$$

When term w exists explicitly in the relevant documents, $MI(w)$ increases.

3 Results of the NTCIR-5 Query Term Expansion Subtask

The NTCIR-5 query term expansion subtask was designed for evaluating the effect of query term expansion technique by using NTCIR-4 web test collection[6]. In order to evaluate the query term expansion technique by itself, several feature quantities discussed in previous section were introduced for analyzing the topic difficulty.

3.1 The NTCIR-4 Web Test Collection

The NTCIR-4 Web test collection [11] is a set of 100 gigabytes of html document data and 80 topics for retrieval experiments. 35 out of 80 topics are for the survey retrieval topics and the other 45 are for the target retrieval topics. The survey retrieval topics are designed for finding most relevant documents and the target retrieval topics are for finding just one, or only a few relevant documents of the highly ranked documents. Since the target retrieval topics may miss relevant candidate documents, only the survey retrieval topics (topic numbers 1, 3, 4, 6, 19, 21, 22, 23, 28, 29, 34, 44, 45, 55, 58, 61, 62, 63, 65, 68, 70, 71, 73, 74, 76, 80, 82, 84, 86, 88, 91, 95, 97, 98, and 99) were used for this query term expansion subtask.

Figure 2 shows a sample topic in this test collection. <TITLE> includes 1-3 terms with Boolean expressions. The attribute "CASE" in <TITLE>, <ALT0>, <ALT1>, <ALT2>, <ALT3> means:

(a) All the terms are related to one another by the OR operator.
(b) All the terms are related to one another by the AND operator.
(c) Only two terms can be related using the OR operator; the rest are specified by the attribute "RELAT."

For the sample topic described in Figure 2, the Boolean query (オフサイド(offside) and (サッカー(soccer) or ルール(rule))) from TITLE and (オフサイド(offside) and サッカー(soccer) and ルール(rule)) is formulated from ALT3.

3.2 Feature Quantities of Topics in the NTCIR-4 Web Survey Retrieval Topics

The graph in Figure 3 shows a characteristic of the topics in the Survey Retrieval Topics by using Boolean formula defined in title field. The X axis of the graph corresponds to $R\&B/R$ and the Y axis corresponds to $R\&B/B$. The radius of each circle indicates the number of the relevant documents.

All statistical values were calculated using an organizer reference IR system named "Appropriate Boolean Query Reformulation for Information Retrieval" (ABRIR) [9]. Such values may differ according to the method of extracting index keywords from the documents.

From this graph, all initial queries were not sufficiently appropriate to distinguish all relevant documents from the other documents. For the topics that have higher $R\&B/R$ and lower $R\&B/B$, such as topics 1, 4, 6, 55, and 98, the term for context definition may be good query expansion terms. For the topics that have lower $R\&B/R$ and higher $R\&B/B$, such as 65, 76, and 82, alternative terms may be good query expansion terms. The topics that have lower $R\&B/R$ and lower $R\&B/B$, such as 45, 62, 63, 80, and 84, may require various types of query expansion terms.

3.3 Retrieval Experiments in the Subtask

Four participants, listed below in alphabetical order of RUN-id, and one organizer reference system submitted their completed run results.

<TOPIC> <NUM>0001</NUM>
<TITLE CASE=“c” RELAT=“2-3”> オフサイド, サッカー, ルール </TITLE>
<DESC> サッカーのオフサイドというルールについて説明されている文書を探したい </DESC>
<NARR><BACK> サッカーでオフサイドとはどういうルールなのかを知りたい。</BACK><TERM> オフサイドはオフェンス側の反則である。オフサイドが適用される状況にはいくつかのパターンがあり、サッカーのルールの中で最もわかりにくいものである。</TERM><RELE> 適合文書はオフサイドが適用される状況を説明しているもの </RELE></NARR>
<ALT0 CASE=“b”> オフサイド </ALT0>
<ALT1 CASE=“b”> オフサイド, 選手, 位置 </ALT1>
<ALT2 CASE=“b”> オフサイド, サッカー </ALT2>
<ALT3 CASE=“b”> サッカー, オフサイド, ルール </ALT3>
<USER> 大学 2 年, 男性, 検索歴 4 年, 熟練度 3, 精通度 5</USER>
</TOPIC>

(a) An original sample topic

<TOPIC> <NUM>0001</NUM>
<TITLE CASE=“c” RELAT=“2-3”> offside, soccer, rule </TITLE>
<DESC> I want to find documents that explain the offside rule in soccer. </DESC>
<NARR> <BACK> I want to know about the offside rule in soccer. </BACK> <TERM> Offside is a foul committed by a member of the offense side. There are several patterns for situations in which the offside rule can be applied, and it is the most difficult soccer rule to understand. </TERM> <RELE> Relevant documents must explain situations where the offside rule applies</RELE> </NARR>
<ALT0 CASE=“b”> offside </ALT0>
<ALT1 CASE=“b”> offside, player, position </ALT1>
<ALT2 CASE=“b”> offside, soccer </ALT2>
<ALT3 CASE=“b”> soccer, offside, rule </ALT3>
<USER> 2nd year undergraduate student, male, 4 years of search experience, skill level 3, familiarity level 5 </USER>
</TOPIC>

(b) An English translation of the sample topic

Fig. 2. A sample topic from the NTCIR-4 Web test collection [11]

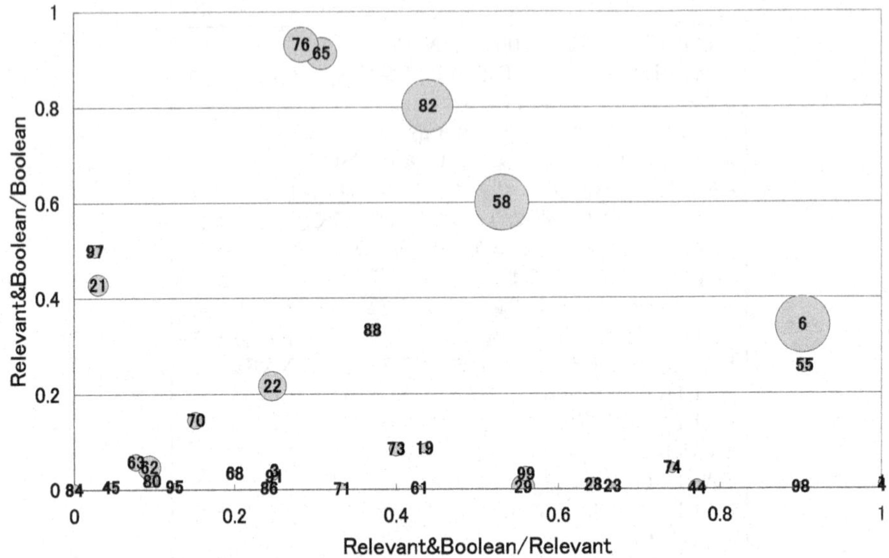

Fig. 3. Characteristics of the Topics

Several query term expansion techniques and information retrieval models were used in these runs for which results were submitted.

JSWEB[12]. Experimented with relevant document vectors that were generated based on the existence of the keyword in the relevant documents. They also proposed combining relevant document vectors (one from the user selected relevant documents and the other from the pseudo-relevant documents). The retrieval method was based on a vector space IR model.

NCSSI[13]. Experimented with a clustering technique for the initial retrieval results and a named entity recognition technique for selecting query expansion terms from the appropriate cluster (user selected cluster or pseudo-relevant cluster). They used an organizer reference model, ABRIR, based on a probabilistic model as an IR system.

R2D2[14]. Experimented with Robertson's Selection Value (RSV) for selecting query expansion terms using pseudo relevant documents. The retrieval method was based on the modified Okapi. They also used link information for scoring the retrieved documents.

ZKN [15]. Experimented with Larvenko's relevance model for selecting query expansion terms using pseudo relevant documents. The retrieval method was based on the inference network and language model.

ABRIR: Organizer Reference System [9]. Experimented with mutual information between terms and relevant documents for selecting query expansion terms. The retrieval method was based on the Okapi.

In order to evaluate the effect of the query term expansion technique, all participants submitted a set of run with and without query term expansion. In addition, following two different (pseudo-)relevant document selection methods were used for evaluating the effect of the quality of the feedback documents.

- Automatic selection: `auto`
  Each system selects pseudo-relevant document automatically (e.g., use the top-N ranked documents).
- Simulation of the user relevance document selection: user
  Each system uses relevant document information defined in the test collection and compare system output (e.g., initial retrieval results) to select relevant documents.

Each participant submitted sets of retrieval results, expansion term candidates lists, and document lists that were used for query term expansion.

3.4 Summary of Overall Evaluation Results

Table 1 shows the overall evaluation of the submitted runs. In most of the runs, the query term expansion technique improved the retrieval performance on average, but there was no run that improved the query performance for all topics.

Table 2 shows the number of Run IDs where query performance improved by using query term expansion techniques. There is no direct correlation between the effectiveness of the query term expansion technique and the mismatch between the initial query and the relevant documents showed in Figure 3.

It is interesting that there are several topics whose number of Run IDs for a user is lower than that for the automatic (e.g., 21, 58, 76, 82). Common char-

Table 1. Evaluation Results for Average

Run ID	type of relevant document selection	No. of topics where performance improve			10 query terms expansion (average)			No query term expansion (average)		
		AP	RP	RR	AP	RP	RR	AP	RP	RR
JSWEB-auto-01	auto	2	1	0	0.011	0.0236	212	0.0743	0.0992	1512
JSWEB-auto-02	auto	3	2	0	0.0197	0.0344	516	0.0743	0.0992	1512
JSWEB-auto-03	auto	17	15	11	0.0714	0.1094	1101	0.0743	0.0992	1512
NCSSI-auto-01	auto	22	17	23	0.1708	0.2107	2432	0.1511	0.1991	2256
NCSSI-auto-02	auto	21	12	17	0.1536	0.1962	2322	0.1511	0.1991	2256
R2D2-auto-01	auto	19	15	21	0.1747	0.2239	2257	0.162	0.2066	2155
R2D2-auto-02	auto	19	19	21	0.181	0.2236	2257	0.162	0.2066	2155
ZKN-auto-01	auto	25	17	11	0.1523	0.2011	2139	0.1405	0.1839	2152
ZKN-auto-02	auto	23	18	14	0.1537	0.1968	2153	0.1405	0.1839	2152
ABRIR-auto	auto	28	20	25	0.2198	0.2506	2591	0.169	0.2085	2422
JSWEB-user-B-02	user	7	5	5	0.0235	0.049	755	0.0743	0.0992	1512
JSWEB-user-B-03	user	18	18	13	0.0976	0.1466	1453	0.0743	0.0992	1512
NCSSI-user-01	user	27	18	17	0.2434	0.2705	2508	0.173	0.2258	2353
NCSSI-user-02	user	28	15	16	0.2196	0.2487	2415	0.173	0.2258	2353
ABRIR-user	user	32	18	20	0.2569	0.2834	2689	0.1801	0.2268	2469

AP: Average Precision, RP: R-Precision, RR: Relevant Retrieved

Table 2. Effectiveness of Query TermExpansion Technique for Each Topic

Topic	No. of Run IDs where perform-ace improve						Topic	No. of Run IDs where perform-ace improve					
	auto			user				auto			user		
	AP	RP	RR	AP	RP	RR		AP	RP	RR	AP	RP	RR
1	6	3	2	4	3	0	65	5	5	4	4	4	4
3	4	3	2	4	2	1	68	5	2	1	3	1	0
4	9	7	0	3	1	0	70	5	6	7	4	3	4
6	5	3	5	3	2	3	71	2	0	5	1	1	2
19	2	1	1	2	1	0	73	6	7	0	4	4	1
21	5	4	5	1	1	1	74	5	4	1	4	4	1
22	8	7	6	5	4	5	76	2	4	7	0	0	0
23	4	3	1	2	0	1	80	5	2	3	3	1	2
28	5	4	4	5	5	5	82	6	4	6	2	1	1
29	4	3	8	5	5	5	84	0	1	0	4	1	1
34	7	5	6	3	1	1	86	3	2	8	2	2	2
44	1	1	1	2	0	0	88	8	6	5	4	2	2
45	7	6	8	3	3	3	91	10	2	0	5	1	2
55	4	5	4	4	4	2	95	5	1	3	3	0	3
58	7	5	7	1	0	1	97	7	6	4	5	5	5
61	0	0	3	2	0	1	98	6	4	6	4	3	4
62	6	6	6	2	1	1	99	5	3	0	3	3	2
63	3	4	6	2	1	2							

AP: Average Precision, RP: R-Precision, RR: Relevant Retrieved

acteristics of those topics is that they have higher $R\&B/B$ values compared with other topics. It means that good query expansion terms for these topics are related terms to find out terms that are alternative to the initial query terms.

This results show that it is not necessary to use real relevant documents to find out these expansion terms.

4 Detailed Analysis on the Effect of Query Term Expansion

4.1 Topic-by-Topic Analysis of the Organizer's Reference System Run

In NTCIR-5 query term expansion subtask, the correlation between topic difficulties based on Figure 3 and effect of query term expansion is not so clear. So topic-by-topic analysis is conducted by using organizer's reference system (ABRIR) run.

Table 3 shows the comparison results between query expansion run (300 terms expansion and 10 terms expansion) and no expansion run of the organizer's reference system. In most of the topics, retrieval performance improves with use of query term expansion. It is also difficult to find a direct correlation between

Table 3. Difference between Query Term Expansion Run and No Expansion Run

		300 Terms	10 Terms
Retrieved	+	28 topics	24 topics
	=	3 topics(1,23,73)	9 topics(1,4,19, 61,68,73,80,84,99)
	-	4 topics(19,68,82, 99)	2 topics(71,74)
MAP	+	25 topics	27 topics
	=	0 topics	1 topics(84)
	-	10 topics(1,19,23, 29,34,61,68,71, 76,82)	7 topics (3,29,61, 68,71,74,76) 68,71,74,76)
Prec@5	+	16 topics	13 topics
	=	14 topics(1,4,23, 34,44,55,58,63, 68,73,74,76,82, 84)	18 topics (1,3,4, 23,34,44,55,58, 63,68,71,74,76, 82,84,86,91,95)
	- -	5 topics(6,19,29, 61,71)	4 topics(6,19,29 61)

Table 4. Quality of Pseudo-Relevant Documents

1	S(1),C(4)	65	S(1),A(4)
3	S(1),A(1),B(3)	68	C(5)
4	A(1),C(4)	70	S(2),A(1),B(2)
6	S(2),A(2),C(1)	71	A(2),C(3)
19	A(1),C(4)	73	A(1),B(2),C(2)
21	C(5)	74	S(1),B(1),C(3)
22	A(2),B(1),C(2)	76	S(1),A(4)
23	C(5)	80	A(1),B(1),C(3)
28	A(1),C(4)	82	S(1),A(4)
29	C(5)	84	B(1),C(4)
34	S(1),C(4)	86	A(1),B(2),C(2)
44	A(2),B(3)	88	S(1),A(3),C(1)
45	A(1),B(1),C(3)	91	C(5)
55	S(2),A(3)	95	S(1),A(2),C(2)
58	S(1),A(4)	97	A(2),C(3)
61	C(5)	98	S(1),B(1),C(3)
62	A(2),C(3)	99	A(2),C(3)
63	A(1),C(4)		

the effectiveness of the query term and the mismatch between the initial query and the relevant documents shown in Figure 3.

The correlation between the pseudo-relevant document quality and improvement of retrieval performance using query term expansion is investigated.

Table 4 shows the quality of pseudo-relevant documents used in the automatic feedback("S" is highly relevant, "A" is relevant, "B" is partially relevant, and

"C" is non-relevant). From this table, I confirmed that there are topics that improve query performance without real relevant documents (e.g., Topic 21 and 91). On the contrary, there are topics that degrade query performance with real relevant documents (e.g., Topic 76 and 82).

The mismatch between the initial query terms and the terms that are distinctive terms in relevant document sets is also checked. In 33 out of 35 topics, initial query terms are included in the top 5 distinctive terms. Two topics (topic 29: ((バイオ"Vaio" or ソニー"Sony") and パソコン"Personal Computer") and topic 98: (世界遺産"World Heritage" and 日本"Japan")) do not have initial query terms in this list. Considering retrieval performance of these two topics, topic 29 degrades retrieval performance with query term expansion but topic 98 improves. I assume this difference comes from the difference of the pseudo-relevant documents; e.g., there are no real relevant documents in a pseudo-relevant document set of topic 28 and there are two real relevant documents (one "S" and one "B") in one of topic 98 (Table 3). From this result, I assume that topics with inappropriate initial query terms are easily affected by the quality of the pseudo-relevant documents.

Before discussing the effect of the pseudo-relevant documents quality, I would like to discuss the characteristics of the query expansion terms generated from pseudo-relevant documents. Since pseudo-relevant documents contains initial query terms, good related terms to find out document without initial query terms may found in the documents. However, in order to find out context terms that improves the precision of the retrieval performance, it is necessary to use real relevant documents.

Based on these understandings, retrieval performance difference for the topics that do not use real relevant documents in a pseudo-relevant document set is checked. There are six topics which use no real relevant documents in a pseudo-relevant document set. In this topic sets, two topics (topic 21 (作家の値打ち"Value of Writer" and 福田和也"FUKUDA Kazuya") and topic 91 (TOEIC and 高得点"high score" and 方法"Method")) improve all retrieval performance measure. Since these topics have smaller $B\&R/R$ (Figure 3), addition of related terms extracted from non-relevant documents may improve the query performance.

In contrast, there are two topics (topic 76 (ヴィトゲンシュタイン "Wittgenstein" and 思想"Thought") and topic 82 (社会主義市場経済"Socialist market economy" and 中国"China")) which degrade performance by adding query expansion terms from relevant documents. These topics have smaller $B\&B/B$ (Figure 3). In such cases, addition of related terms works well compared to the addition of context terms.

4.2 Discussion

Since there are varieties of mismatches between the initial query and relevant documents list, most of the query term expansion techniques are effective on average. In addition, there is no topic where all query expansion technique degrade retrieval performance. From this results, I conclude that it is difficult to define simple topic difficulties measure from the viewpoint of query term expansion.

Table 5. Appropriate Query Term Expansion Strategy for the Topic Difficulty Type

Characteristic feature for topic difficulty	Type of appropriate expansion terms	Appropriate strategy for query term expansion
$B\&R/R$ is small	alternative words to original query terms	Thesaurus
	one or two good related words	Pseudo-Relevance feedback Pseudo-Relevance feedback
$B\&R/B$ is small	many related terms that define context	Relevance feedback with appropriate relevant documents
	specific examples of general query terms	Named entity extraction by using relevance feedback

However, topic-by-topic analysis tells that several factors (a type of the mismatch between query and relevant documents, quality of pseudo-relevant documents, etc.) affects the retrieval performance.

Table 5 summarize the type of appropriate expansion terms and candidate strategies for query term expansion according to the type of the mismatch based on this experiment.

This table suggests that the analysis on the effect of simple query term expansion technique (e.g., query term expansion by using thesaurus only) in average may be biased the number of each type of topics stored in the test collection.

Therefore, it is necessary to consider the characteristics of the test collection for simple query term expansion technique. It is also important to estimate the type of the mismatch for integrating these different strategies to establish good query term expansion technique.

5 Conclusion

In this paper, I briefly review previous research on feature quantities that characterize the quality of the initial query that are defined in this study for evaluating topic difficulties from the viewpoint of query term expansion, and summarize the results of the NTCIR-5 query term expansion subtask. I also describe detailed analysis concerning how these feature quantities affect the query term expansion technique. From this analysis, I found that the effect of the simple query term expansion technique may vary according to the characteristics of the test collection and it is necessary to select appropriate test collection for evaluating the simple query term expansion technique.

Further analysis is necessary to establish a framework to evaluate the query term expansion technique in isolation. For example, proposal of feature quantities other than $R\&B/R$ and $R\&B/B$ may be useful. In addition, in order to integrating several query term expansion strategies, it is also important to make a framework to estimate the type of the mismatch.

Acknowledgments

This research was partially supported by a Grant-in-Aid for Scientific Research on Priority, 16016201 from the Ministry of Education, Culture, Sports, Science, and Technology, Japan.

I would also like to thank all participants of the NTCIR-5 query expansion subtask for their fruitful contribution.

References

1. Baeza-Yates, R., Ribeiro-Neto, B.: 5 Query Operations. In: Modern Information Retrieval. Addison-Wesley (1999) 19–71
2. Cronen-Townsend, S., Zhou, Y., Croft, W.B.: Predicting query performance. In: Proceedings of the 25th Annual International ACM SIGIR Conference on Research and Development in Information Retrieval. (2002) 299–306
3. Harman, D., Buckley, C.: SIGIR 2004 workshop: RIA and "where can IR go from here?". SIGIR Forum **38** (2004) 45–49
4. Buckley, C.: Why current IR engines fail. In: SIGIR '04: Proceedings of the 27th annual international conference on Research and development in information retrieval, New York, NY, USA, ACM Press (2004) 584–585
5. Warren, R.H., Liu, T.: A review of relevance feedback experiments at the 2003 Reliable Information Access (RIA) workshop. In: SIGIR '04: Proceedings of the 27th annual international conference on Research and development in information retrieval, New York, NY, USA, ACM Press (2004) 570–571
6. Yoshioka, M.: Overview of the NTCIR-5 web query term expansion subtask. In: Proceedings of the Fifth NTCIR Workshop Meeting on Evaluation of Information Access Technologies: Information Retrieval, Question Answering and Cross-Lingual Information Access. (2005) 443–454 http://research.nii.ac.jp/ntcir/workshop/OnlineProceedings5/data/WEB/NTCIR5-OV-WEB-YoshiokaM.pdf.
7. Buckley, C., Harman, D.: Reliable information access final workshop report. Technical report, Northeast Regional Research Center, MITRE (2004) http://nrrc.mitre.org/NRRC/Docs_Data/RIA_2003/ria_final.pdf.
8. Voorhees, E.: Query expansion using lexical-semantic relations. In: Proceedings of the 17th Annual International ACM SIGIR Conference on Research and Development in Information Retrieval. (1994) 61–69
9. Yoshioka, M., Haraguchi, M.: On a combination of probabilistic and Boolean IR models for WWW document retrieval. ACM Transactions on Asian Language Information Processing (TALIP) **4** (2005) 340–356
10. Yoshioka, M., Haraguchi, M.: An appropriate Boolean query reformulation interface for information retrieval based on adaptive generalization. In: International Workshop on Challenges in Web Information Retrieval and Integration. (2005) 145–150
11. Eguchi, K., Oyama, K., Aizawa, A., Ishikawa, H.: Overview of the informational retrieval task at NTCIR-4 web. In: Proceedings of the Fourth NTCIR Workshop on Research in Information Access Technologies Information Retrieval, Question Answering and Summarization. (2004) http://research.nii.ac.jp/ntcir/workshop/OnlineProceedings4/WEB/NTCIR4-OV-WEB-A-EguchiK.pdf.

12. Tanioka, H., Yamamoto, K., Nakagawa, T.: A distributed retrieval system for NTCIR-5 web task. In: Proceedings of the Fifth NTCIR Workshop Meeting on Evaluation of Information Access Technologies: Information Retrieval, Question Answering and Cross-Lingual Information Access. (2005) 472–480 http://research.nii.ac.jp/ntcir/workshop/OnlineProceedings5/data/WEB/NTCIR5-WEB-TaniokaH.pdf.
13. Toda, H., Kataoka, R.: Search result clustering method at NTCIR-5 web query expansion subtask. In: Proceedings of the Fifth NTCIR Workshop Meeting on Evaluation of Information Access Technologies: Information Retrieval, Question Answering and Cross-Lingual Information Access. (2005) 481–485 http://research.nii.ac.jp/ntcir/workshop/OnlineProceedings5/data/WEB/NTCIR5-WEB-TodaH.pdf.
14. Masada, T., Kanazawa, T., Takasu, A., Adachi, J.: Improving web search by query expansion with a small number of terms. In: Proceedings of the Fifth NTCIR Workshop Meeting on Evaluation of Information Access Technologies: Information Retrieval, Question Answering and Cross-Lingual Information Access. (2005) 486–493 http://research.nii.ac.jp/ntcir/workshop/OnlineProceedings5/data/WEB/NTCIR5-WEB-MasadaT.pdf.
15. Eguchi, K.: NTCIR-5 query expansion experiments using term dependence models. In: Proceedings of the Fifth NTCIR Workshop Meeting on Evaluation of Information Access Technologies: Information Retrieval, Question Answering and Cross-Lingual Information Access. (2005) 494–501 http://research.nii.ac.jp/ntcir/workshop/OnlineProceedings5/data/WEB/NTCIR5-WEB-EguchiK.pdf.

Incorporating Prior Knowledge into Multi-label Boosting for Cross-Modal Image Annotation and Retrieval

Wei Li and Maosong Sun

State Key Lab of Intelligent Technology and Systems
Department of Computer Science and Technology, Tsinghua University
Beijing 100084, P.R. China
wei.lee04@gmail.com, sms@mail.tsinghua.edu.cn

Abstract. Automatic image annotation (AIA) has proved to be an effective and promising solution to automatically deduce the high-level semantics from low-level visual features. In this paper, we formulate the task of image annotation as a multi-label, multi class semantic image classification problem and propose a simple yet effective joint classification framework in which probabilistic multi-label boosting and contextual semantic constraints are integrated seamlessly. We conducted experiments on a medium-sized image collection including about 5000 images from Corel Stock Photo CDs. The experimental results demonstrated that the annotation performance of our proposed method is comparable to state-of-the-art approaches, showing the effectiveness and feasibility of the proposed unified framework.

1 Introduction

With the growing amount of digital information such as image and video archives, multimedia information retrieval has become increasingly important in the last few years and draws much more attention from computer vision and object recognition community. Through the sustained efforts of many researchers in image retrieval domain, two successful retrieval architectures have been proposed: one is query-by-example (QBE), a mono-media retrieval paradigm based on visual similarity and the other one is query-by-keyword (QBK), a cross-media retrieval paradigm based on associated text. More specifically, query-by-keyword is a more intuitive and desirable choice for common users to conduct image search, since both users' information needs and natural image semantics can be described more accurately by keywords or captions. However, for other applications, especially face recognition and texture image retrieval, QBE is more preferred, because it is hard to represent texture information in natural languages in most cases. However, the key issue for QBK is the semantic gap, in other words, how to deduce the high-level semantic concepts from low-level perceptual features. To bridge this gap, two effective methods base on machine learning techniques have been explored: automatic image annotation and query concept learning. In this paper, we focus on the branch of automatic image annotation. Automatic image annotation is the task of automatically generating multiple semantic labels to describe image semantics based on the image appearance, which is a crucial step for object recognition and semantic scene interpretation. In recent years,

H.T. Ng et al. (Eds.): AIRS 2006, LNCS 4182, pp. 404–415, 2006.

many generative models and discriminative approaches have been proposed to automatically annotate images with descriptive textual words to support multi-modal image retrieval using different keywords at different semantic levels. Many of them have achieved state-of-the-art performance. However, most of the approaches may face two key challenges. First, high-quality training set is required to improve the label prediction accuracy, but, in most cases, associated text are assigned to images instead of image regions, large amount of hand-segmented, hand-labeled images with explicit correspondence is hard to obtain or create in large quantities. Second, the fundamental problem for image annotation is the ambiguity or incompatibility of label assignment, contextual constraints are little examined in the annotation process, that is to say, each image segment or block is identified independently without considering the word-to-word correlations, which may degrade annotation accuracy due to the ambiguities inherent to visual properties. For example, like "*sky*" and "*ocean*" region, even for human beings, it is often difficult to identify these two regions accurately without context. In literature, relevance models, coherent language models and context-dependent classification models have been explored to solve these above problems. In this paper, we propose a simple yet effective annotation model in which multi-label boosting algorithm and contextual semantic constraints are integrated seamlessly.

The main contribution of this paper is two-fold: First, we formulate the task of image annotation as a multi-label, multi class semantic image classification problem under ensemble classification framework. Second, as a prior knowledge, contextual constraints between the semantic labels are taken into consideration to avoid label ambiguity or label incompatibility problem. To our best knowledge, multi-label boosting combined with contextual semantic constraints has not been carefully investigated in the domain of automatic image annotation.

This paper is organized as follows: Section 2 discusses related work. Section 3 first reviews the underlying theory of boosting, and then describes image annotation model based on the multi-label boosting and contextual constraints. Section 4 demonstrates our experimental results and some theoretical analysis. Conclusions and future work are discussed in Section 5.

2 Related Work

Recently, many models using machine learning techniques have been proposed for automatic and semi-automatic image annotation and retrieval, Including hierarchical aspect model [1][2], translation model [3][19], relevance model [4][9][17][22], classification methods [8][10][12-15][30], latent space approaches [20][21] and random fields models [23][24]. Mori et al [5] is the earliest to performance image annotation, they collects statistical co-occurrence information between keywords and image grids and uses it to predict annotated keywords to unseen images. Dyugulu et al [3] views annotating images as similar to a process of translation from "visual information" to "textual information" by the estimation of the alignment probability between visual blob-tokens and textual keywords based on IBM model 2. K. Fang [19] improved IBM model 1 by regularizing the imbalance of keywords distribution in the training set. Barnard et al [1][2] proposed a hierarchical aspect model to capture the joint distribution of words and image regions using EM algorithm. Jeon et al [4] presented a cross-media relevance model (CMRM) similar to the cross-lingual retrieval techniques to perform the image annotation and ranked image retrieval. Lavrenko et al [8] proposed continuous relevance model (CRM) to extend the cross-media relevance model using actual continuous-valued features extracted from image regions. This method avoids the clustering and constructing the discrete visual vocabulary stage. S. L. Feng et al [17] improved the CRM model by assuming a multiple Bernoulli distribution to generate the keyword annotations instead of multinomial distribution to model the Blei et al [11] proposed a correspondence LDA and assumes that a Dirichlet distribution can be used to generate a mixture of latent factors that can further relates words and image regions. Wang and Li [8] introduced a 2-D multiresolution HMM model to automate linguistic indexing of images. Clusters of fixed-size blocks at multiple resolution and the relationships between these clusters is summarized both across and within the resolutions. E. Chang et al [6] proposed content-based soft annotation (CBSA) for providing images with semantic labels using (BPM) Bayesian Point Machine. Cusano C et al [10] proposed using Multi-class SVM to classify each square image region into one of seven pre-defined concepts of interest and then combine the partial decision of each classifier to produce the overall description for the unseen image. E. Chang, B. Li and K. Goh [12-15] introduced a multi-level confidence-based ensemble scheme to assist the discovery of new semantics and useful low-level perceptual features. F. Monay [20][21] presented to use the latent variables to link image features and words based on the latent semantic analysis (LSA) and probabilistic latent semantic analysis (pLSA). Instead of predicting the annotation probability of a single word given an image, R. Jin et al[18] estimated a coherent language model for each image to infer a set of words with the word-to-word correlation to be considered. J. Fan et al [16] presented a concept hierarchy model and adaptive EM algorithm to deduce multi-level image semantics. More recently, R. Zhang et al [22] introduced a latent variable model to connect image features and textural words, X. He et al [23] and Kumar. S et al [24] have used context-dependent classifiers such as random fields to perform image annotation.

3 The Framework of Image Annotation Model

3.1 Formulation of Automatic Image Annotation

Given a training set of annotated images, where each image is associated with a number of semantic labels. We make an assumption that each image can be considered as a multi-modal document containing both the visual component and semantic component. Visual component provides the image representation in visual feature space using low-level perceptual features including color and texture, etc. While, semantic component captures the image semantics in semantic feature space based on textual annotations derived from a generic vocabulary, such as "*sky*", "*ocean*", etc. Automatic image annotation is the task of discovering the association model between visual and semantic component from such a labeled image database and then applying the association model to generate annotations for unlabeled images. More formally, let *ID* denote the training set of annotated images:

- $ID = \{I_1, I_2, \ldots, I_N\}$
- each image I_j in ID can be represented by the combination of visual features and semantic labels in a multi-modal feature space, i.e., $I_j = \{L_j; V_j\}$
- semantic component L_j is a bag of words described by a binary vector $L_j = \{l_{j,1}, l_{j,2}, \cdots, l_{j,m}\}$, where m is the size of generic vocabulary, $l_{j,i}$ is a binary variable indicating whether or not the i-th label l_i appears in I_j
- visual component V_j may be more complex due to a large variety of methods for visual representation, in general, it can also have the vector form $V_j = \{v_{j,1}, v_{j,2}, \ldots, v_{j,n}\}$, for patch-based image representation, i.e., image I_j is composed of a number of image segments or fixed-size blocks, each of them is described by a feature vector $v_{j,i}$, and n is the number of image components; for global image representation, $v_{j,i}$ only denotes a feature component and n is the dimension of selected feature space

3.2 Underlying Theory of Boosting

Boosting is a general framework for improving the label prediction accuracy of any given learning algorithm under the PAC (probably approximately correct) style model. The basic idea behind boosting is to combine many inaccurate or moderately accurate classification rules into a single, highly accurate classification rule in an iterative scheme. More specifically, a distribution or set of weights is associated with each training instance. These weighted training instances are used to train the simple rules, on each round of iterations, the weights of incorrectly classified examples are increased so that the simple rule is forced to concentrate on hard examples which are most difficult to classify by the preceding rules [26].

3.3 Multi-label Boosting with Semantic Knowledge for Annotation Model

In traditional classification problems, to reduce the model complexity, class labels are assumed to be mutually exclusive and each instance to be classified belongs to only one class. However, in the context of image annotation or semantic image classification, it is natural that one image belongs to multiple classes simultaneously due to the richness of image content, causing the actual classes overlap in the feature space. For example, as shown in Fig. 4.1, it is quite hard and insufficient to describe the image content using only a keyword because image semantics is represented by both basic semantic entities contained in that image and the relationships between them. In literature, only a few papers are concerned with multi-label classification problems, R. Schapire et al [25] extended AdaBoost to address multi-label text categorization which motivated us to attack the problem of multi-label image annotation. More formally, given a training set of labeled images, let D be the collection of images to be classified and L be the finite set of semantic labels, each image $I_j \in D$ is associated with a binary label vector $L_j = \{l_{j,1}, l_{j,2}, \cdots, l_{j,m}\}$, then we convert the multi-label images into several single-label image pairs (I_j, l_i) using the criterion that each image serves as an observation for each of the classes to which it belongs, finally these converted single-label instances are taken as input to train a multi-label annotation model based on the boosting framework. The detailed boosting algorithm for multi-label, multi class semantic image classification is shown as follows.

Multi-label Boosting for Image Annotation Algorithm

Input: $I = \{(I_1, L_1), (I_2, L_2), \ldots, (I_n, L_n)\}$, a sequence of multi-labeled images
T, number of iterations

Output: $f(I, l)$ final accurate annotation model

Algorithm:

1. initialize $D_1(j, l) = 1/(M)$ (M is the total number of single-label image pairs)
2. for $t = 1 : T$
3. pass distribution D_t to weak learner
4. get weak annotator $h_t : ID * L \rightarrow \Re$
5. choose $\alpha_t \in \Re$
6. update

$$D_{t+1}(j, l) = \frac{D_t(j, l)\exp(-\alpha_t l_j h_t(I_j, l))}{Z_t}$$

where Z_t is a normalization factor to ensure that D_{t+1} is a distribution

$$Z_t = \sum_{i=1}^{n} \sum_{l \in L} D_t(j, l)\exp(-\alpha_t l_j h_t(I_j, l))$$

7. output the final annotation model

$$f(I, l) = \sum_{i=1}^{T} \alpha_t h_t(I, l)$$

Distribution D_t denotes the importance weights over single-label image pairs. Initially, the distribution is set to uniform. For each iteration, the sequence of multi-labeled images together with the D_t are taken as input to compute a weak annotator $h(I_j, l): ID * L \rightarrow \Re$, then a parameter α_t is chosen and D_t is updated based on the actual labels and the predicted labels of the weak annotator. Through iterations, the next weak annotator will pay more attention to these example pairs with higher weights that are most difficult to classify by the preceding weak annotator. Finally, a highly accurate annotation model is constructed by combining all the weak models through weighted voting. To annotate an unlabeled image, the final annotation model can output a weight vector. In order to produce a reasonable ranking of the labels to be used as annotations, we map the associated weights in the weight vector via the following logistic function to get confidence scores or annotation probabilities for each label.

$$P(y = 1|f) = \frac{1}{1 + \exp(Af + B)} \tag{1}$$

where f is the output of the final annotation model for a given unseen image, A and B are real-valued parameters estimated by Maximum Likelihood criterion. To solve the ambiguity or incompatibility of label assignment, label co-occurrence statistics and pair-wise semantic similarity defined by WordNet [27] [28] are incorporated respectively. Finally the confidence scores associated with each label can be re-weighted and sorted again by using the label-to-label correlations.

4 Experimental Results

Our experiments are carried out using a mid-sized image collection, comprising about 200 images with the size of 180*116 pixels from Corel Stock Photo CDs. In this collection, each image is manually labeled with 2-5 keywords to describe the image content. In our experiments, 130 images are randomly selected for training and the remaining 70 images for test.

water trees sky

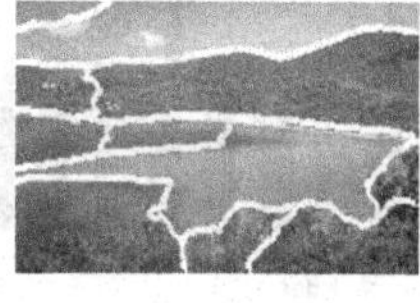

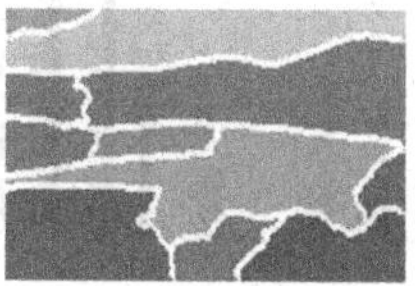

Fig. 4.1. Images, segmentation results and blob-token representation

Fig. 4.1 shows the image example used in our experiments, left-most one is the original color image together with the associated text, middle is the segmentation results by Ncuts [7] and the right-most is blob-token representation. In general, the performance of such patch-based image annotation may be affected by two factors: one is the quality of segmentation, since perfect image segmentation is still an open issue, semantic objects don't correspond to segmented regions in most cases. Second, blob-tokens are generated by clustering algorithm usually based on visual features. However, in some cases, two image patches with different semantics can share the same visual appearances which may lead to poor clustering results. For example, in the above images, part of "*sky*" and "*water*" patch is represented by the same blob-token. To avoid the influence of segmentation error and poor clustering quality, only global visual features including color and texture is considered for the weakly labeled images (no correspondence between labels and image patches is provided) used in this paper.

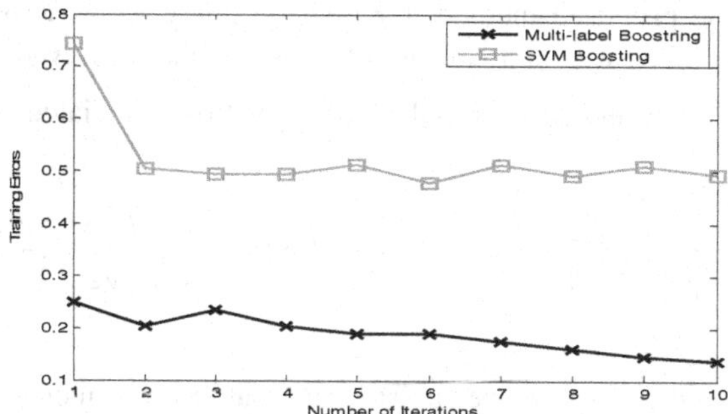

Fig. 4.2. Comparison of training error between Multi-label boosting and SVM-boosting

Fig.4.2 demonstrates comparison of training error between Multi-label boosting and SVM-boosting which illustrates that SVMs [29] is not suitable for multi-label classification problems although it is powerful in single-label multi class situations and boosting scheme fail to improve the performance of SVMs. In our experiments, the weak classifier used for multi-label boosting is one-level decision tree. For each round of iterations, a threshold is generated from the corresponding feature value, decision is determined based on whether the observation value is above or below the given threshold and α_t is set to 1.

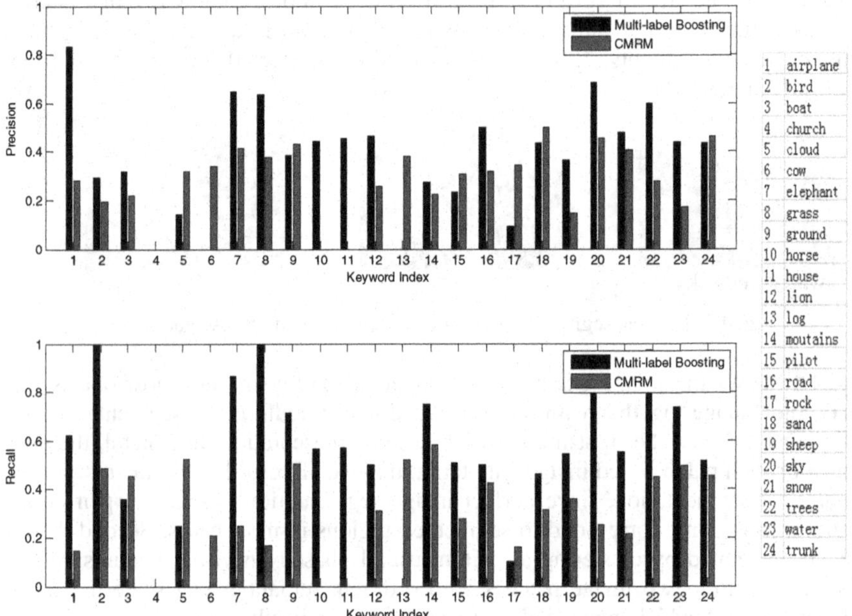

Fig. 4.3. Precision and recall curve compared to CMRM

Fig. 4.3 shows the performance of our method compared to Cross-Media Relevance Model (CMRM). In this paper, we also use the precision and recall to evaluate the performance of the proposed method, for a single query word *w*, precision and recall are defined as follows, and we use I_j denotes the retrieved *j*-th image, t_j and a_j represent the true annotations and auto annotation associated with the *j*-th image.

$$precision(w)=\frac{\left|\{I_j \middle| w\in t_j \wedge w\in a_j\}\right|}{\left|\{I_j \middle| w\in a_j\}\right|} \qquad recall(w)=\frac{\left|\{I_j \middle| w\in t_j \wedge w\in a_j\}\right|}{\left|\{I_j \middle| w\in t_j\}\right|} \tag{2}$$

Fig. 4.4 illustrates the co-occurrence statistics and pair-wise semantic similarity in the finite set of semantic labels. By mapping the associated weights with each label via

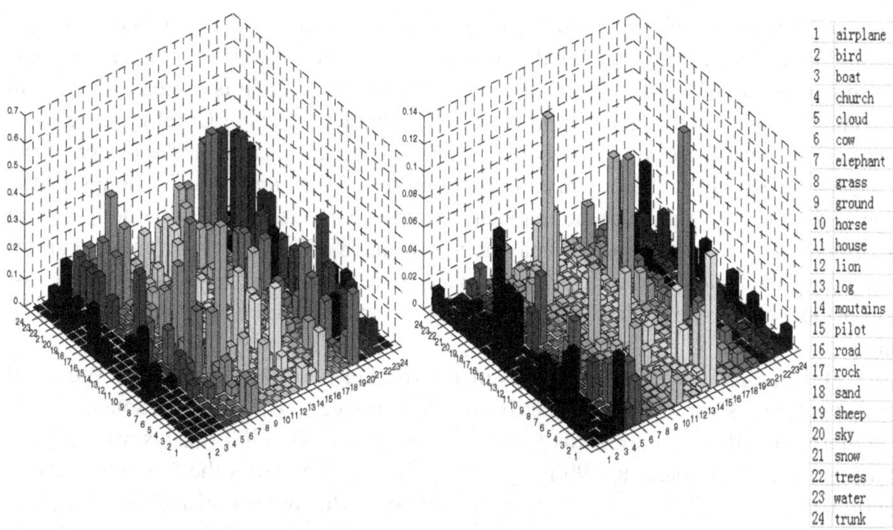

Fig. 4.4. Co-occurrence Statistics and pair-wise semantic similarity between labels

logistic function, confidence scores for each predicted label can be produced. Then we can re-weight the confidence scores by considering the label co-occurrence frequency and pair-wise semantic similarity using the following re-weighting formula.

$$P_i^k = \frac{P_i^k + \prod_j R_{label(i,\, j)} P_j^k}{P^k} \tag{3}$$

$$P^k = P_i^k + \prod_j R_{label(i,\, j)} P_j^k \tag{4}$$

$$R_{label}(i,\, j) = \begin{cases} P_{co}(i,\, j) \\ wns(i,\, j) \end{cases} \tag{5}$$

$$P_{co}(i,\, j) = IF(l_i \cap l_j) / IF(l_i \cup l_j) \tag{6}$$

where P_i^k denotes the confidence score of i-th label to be used as annotation for k-th image, $R_{label}(i,\, j)$ measures the correlation between i-th and j-th label, P^k is a normalization factor, $p_{co}(i,\, j)$ represents the co-occurrence frequency between i-th label and j-th label, $IF(l_i \cap / \cup\, l_j)$ is the number of images that contain both i-th label and (or) j-th label, $wns(i,\, j)$ is the pair-wise similarity of these two labels defined by WordNet. When considering label co-occurrence statistics, we use $p_{co}(i,\, j)$ to replace $p_{label}(i,\, j)$ in formula (2) and for the pair-wise similarity, $wns(i,\, j)$ is utilized.

However, experimental results show that although label ambiguity and incompatibility can be solved in some way, nevertheless the precision and recall has not improved. Furthermore, empirical studies show that co-occurrence statistics can give a better result than pair-wise similarity concerning label incompatibility. For example, by using label co-occurrence statistics, all the co-occurrence frequency between "*elephant*" and "*lion*", "*airplane*" and "*church*", "*boat*" and "*cow*", etc is zero so that-these two labels will not co-occur in image annotations, which is more reasonable in command sense and more compatible with the characteristics of Corel image collections, since each Corel image only has a core object and located in the center of the image in most cases. To contrast, since both "*elephant*" and "*lion*" belongs to the "*animal*" category, they may be relevant to some degree in terms of the definition in WordNet, so pair-wise similarity may be not so much effective in solving label ambiguity and label incompatibility in our image collection. After re-weighting process, top 5 labels with the highest confidence scores are selected to annotate the unseen images.

To further verify the effectiveness of multi-label boosting algorithm, we carried out experiments on a large collection of 4999 images, in which the semantic vocabulary contains 374 keywords (mainly English nouns), 4500 images are used for training, and the remaining 499 images for test. Fig. 4.5 shows the frequency distribution of keywords in the semantic vocabulary. The precision and recall curves using the top 20 keywords with highest frequency as single word queries are shown in Fig. 4.6.

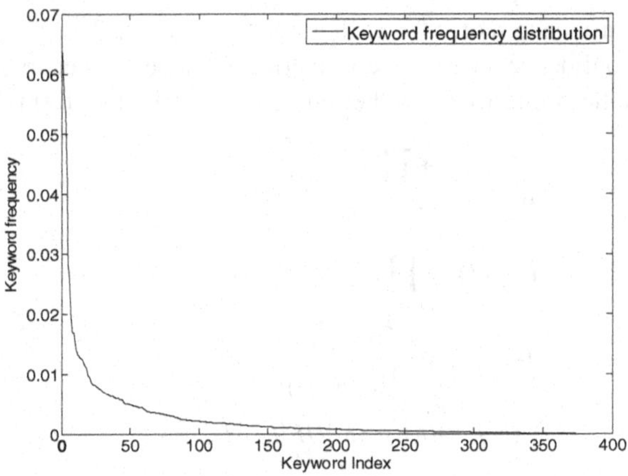

Fig. 4.5. Frequency distribution of keywords in the semantic vocabulary

An inverted-file can be constructed for each semantic label in the pre-defined vocabulary, text-based information retrieval technique is then used to retrieval images. Fig. 4.7 shows the some of the retrieval results using the keyword "*elephant*" as a single word query.

5 Conclusion and Future Work

In this paper, we propose a general framework for automatic image annotation and retrieval based on multi-label boosting and semantic prior knowledge. Experimental

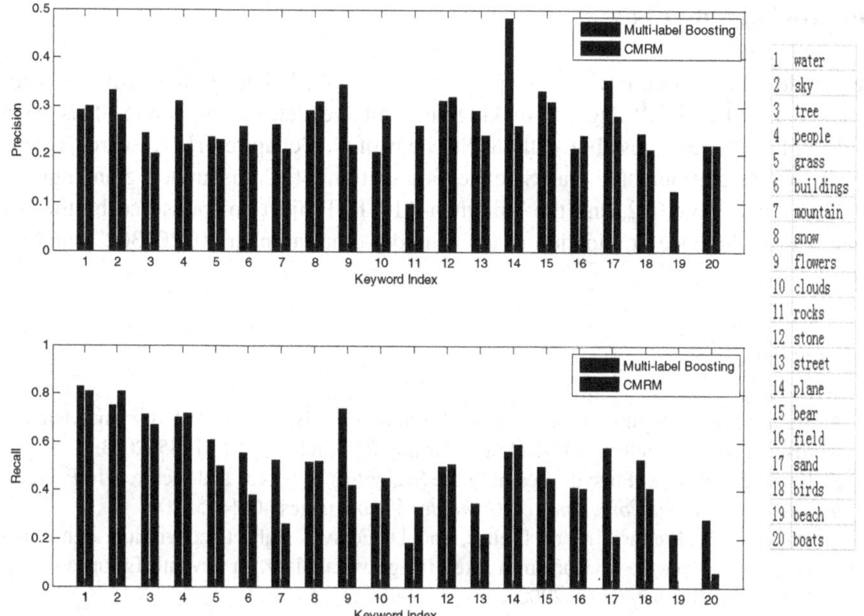

Fig. 4.6. Precision and recall compared to CMRM using high frequency keywords

Fig. 4.7. Some of the retrieved images using "*elephant*" as a single word query

results and theoretical analysis shows that multi-label boosting learning is a simple yet effective method for multi-label, multi class classification problems, especially forimage annotation and semantic scene interpretation. By mapping the output of multi-label boosting via logistic function, we can obtain confidence scores for each label, then we can re-weight the confidence scores by incorporating the prior knowledge such as co-occurrence statistics and pair-wise semantic similarity to avoid the label ambiguity, label incompatibility problem to some degree and get a relatively reasonable annotation results. Nevertheless, empirical studies show that this post-processing step based on language models has not too much effect on the annotation accuracy in some cases.

In the future, more work should be done to provide more expressive image content representation, efficient algorithms and precise models for semantic image classification. In addition, more sophisticated language models will be taken into consideration as a post-processing step to achieve a more scalable and accurate image annotation model.

Acknowledgements

We would like to express our deepest gratitude to Michael Ortega-Binderberger, Kobus Barnard and J.Wang for making their image datasets available. This work would not have been possible without the help of R. Schapire. The research is supported by the National Natural Science Foundation of China under grant number 60573187 and 60321002, and the Tsinghua-ALVIS Project co-sponsored by the National Natural Science Foundation of China under grant number 60520130299 and EU FP6.

References

1. K. Barnard, P. Dyugulu, N. de Freitas, D. Forsyth, D. Blei, and M. I. Jordan. Matching words and pictures. Journal of Machine Learning Research, 3: 1107-1135, 2003.
2. K. Barnard and D. A. Forsyth. Learning the Semantics of Words and Pictures. In *Proceedings of International Conference on Computer Vision,* pages 408{415, 2001.
3. P. Duygulu, K. Barnard, N. de Freitas, and D. Forsyth. Ojbect recognition as machine translation: Learning a lexicon fro a fixed image vocabulary. In Seventh European Conf. on Computer Vision, 97-112, 2002.
4. J. Jeon, V. Lavrenko and R. Manmatha. Automatic image annotation and retrieval using cross-media relevance models. In Proceedings of the 26th intl. SIGIR Conf, 119-126, 2003.
5. Y. Mori, H. Takahashi, and R. Oka, Image-to-word transformation based on dividing and vector quantizing images with words. First International Workshop on Multimedia Intelligent Storage and Retrieval Management, 1999.
6. Edward Chang, Kingshy Goh, Gerard Sychay and Gang Wu. CBSA: Content-based soft annotation for multimodal image retrieval using bayes point machines. IEEE Transactions on Circuts and Systems for Video Technology Special Issue on Conceptual and Dynamical Aspects of Multimedia Content Descriptions, 13(1): 26-38, 2003.
7. J. shi and J. Malik. Normalized cuts and image segmentation. IEEE Transactions On Pattern Analysis and Machine Intelligence, 22(8): 888-905, 2000.
8. J. Li and J. A. Wang. Automatic linguistic indexing of pictures by a statistical modeling approach. IEEE Transactions on PAMI, 25(10): 175-1088, 2003.
9. V. Lavrenko, R. Manmatha and J. Jeon. A model for learning the semantics of pictures. In Proc of the 16th Annual Conference on Neural Information Processing Systems, 2004.
10. Cusano C, Ciocca G, Schettini R, Image Annotation using SVM. Proceedings of SPIE-IS&T Electronic Imaging, 330-338, SPIE Vol. 5304, 2004.
11. D. Blei and M. I. Jordan. Modeling annotated data. In Proceedings of the 26th intl. SIGIR Conf, 127–134, 2003.
12. K.-S. Goh, E. Chang and K.-T. Cheng, SVM binary classifier ensembles for image classification, in Proceedings of the tenth international conference on Information and knowledge management, ACM Press, 2001,pp. 395-402.
13. B. Li and K. Goh, Confidence-based dynamic ensemble for image annotation and semantics discovery, in Proceedings of the eleventh ACM international conference on Multimedia, ACM Press, 2003, pp. 195-206.
14. K.Goh, B. Li and E. Chang, Semantics and feature discovery via confidence-based ensemble, ACM Transactions on Multimedia Computing, Communications, and Applications, 1(2), 168-189, 2005.

15. K.Goh, E. Chang and B. Li, Using on-class and two-class SVMs for multiclass image annotation, IEEE Trans. on Knowledge and Data Engineering, 17(10), 1333-1346, 2005.
16. J. Fan, Y. Gao, and H. Luo, Multi-level annotation of natural scenes using dominant image components and semantic concepts," in *Proc. of ACM MM*, 540-547, 2004.
17. S. L. Feng, V. Lavrenko and R. Manmatha. Multiple Bernoulli Relevance Models for Image and Video Annotation. In Proceedings of CVPR04, 2004.
18. R. Jin, J. Y. Chai, and L. Si. Effective Automatic image annotation via a coherent language model and active learning. In *Proceedings of ACM MM'--*, 2004.
19. F. Kang, R. Jin, and J. Y. Chai. Regularizing Translation Models for Better Automatic Image Annotation. In *Proceedings of ACM MM'04,* 2004.
20. F. Monay and D. Gatica-Perez. On image auto-annotation with latent space models. In Proc. of ACM Int. Conf. on Multimedia, Berkeley, Nov. 2003.
21. F. Monay and D. Gatica-Perez. PLSA-based image auto-annotation: Constraining the latent space. In Proc. ACM Int. Conf. on Multimedia, New York, Oct. 2004.
22. R. Zhang, Z. Zhang, M. Li, WY. M and HJ. Zhang. A probabilistic semantic model for image annotation and multi-modal image retrieval. IEEE Int'l ICCV'05, 2005.
23. X. He, R. Zemel, and M. Carreira-Perpinan. Multiscale conditional random fields for image labeling. In *IEEE Conf. CVPR'04*, 695–702, 2004.
24. Kumar, S., & Hebert, M. (2003). Discriminative fields for modeling spatial dependencies in natural images. *NIPS'03*.
25. R. Schapire, Y. Singer, Boostexter: A boosting-based system for text categorization, *Machine Learning* 39, 135-168, 2000.
26. R.E. Schapire. The boosting approach to machine learning: An overview. In *Workshop on Nonlinear Estimation and Classification*. MSRI, 2002.
27. C. Fellbaum. WordNet: An electronic lexical database, MIT Press, 1998.
28. Pedersen T., Patwardhan S., and Michelizzi J. WpordNet::Similarity - measuring the relatedness of concepts. In Proceedings of the Nineteenth National Conference on Artificial Intelligence (AAAI-04), 2004.
29. Chih-Chung Chang and Chih-Jen Lin. LIBSVM: a library for support vector machines, 2001. Software available at http://www.csie.ntu.edu.tw/~cjlin/libsvm.
30. M. Boutell, X. Shen, J. Luo, and C. Brown. Multi-label semantic scene classification. Technical report, Dept. Comp. Sci. U. Rochester, 2003.

A Venation-Based Leaf Image Classification Scheme

Jin-Kyu Park[1], EenJun Hwang[1,*], and Yunyoung Nam[2]

[1] School of Electrical Engineering, Korea University
Anam-dong, Seongbuk-Gu, Seoul, Korea
{saanin, ehwang04}@korea.ac.kr
[2] Graduate School of Information and Communication,Ajou University
Wonchon-dong, Youngtong-Gu, Suwon, Kyunggi-Do, Korea
youngman@ajou.ac.kr

Abstract. Most content-based image retrieval systems use image features such as textures, colors, and shapes. However, in the case of leaf image, it is not appropriate to rely on color or texture features only because such features are similar in most leaves. In this paper, we propose a novel leaf image retrieval scheme which first analyzes leaf venation for leaf categorization and then extracts and utilizes shape feature to find similar ones from the categorized group in the database. The venation of a leaf corresponds to the blood vessel of organisms. Leaf venations are represented using points selected by the curvature scale scope corner detection method on the venation image, and categorized by calculating the density of feature points using non-parametric estimation density. We show its effectiveness by performing several experiments on the prototype system.

1 Introduction

Recently, due to the development of computer and network technologies, generating, processing and sharing digital contents have become popular. And this leads to a huge amount of images being available. As the number of digital images has increased, the need for sophisticated image retrieval has been emphasized. Traditional image retrievals relied on the textual information such as file names or keywords describing the image. However, as digital images are so populated these days, attaching and memorizing such text information is not manageable any more by human beings. To solve this problem, the content- based image retrieval (CBIR) has been researched. For example, CBIR can detect main objects from the image, and generate automatically some useful information describing those objects including shapes, textures and colors.

CBIR techniques have huge diverse applications. Especially, due to the popularity of hand-held devices, this technique will lead a strong trend in the recent ubiquitous information retrieval. For example, during some field trip or visit to a botanical garden, people may encounter some unfamiliar plant. In this case, instead of looking up a botanical book for detailed information, we may inquire it on the spot by

* Corresponding author.

H.T. Ng et al. (Eds.): AIRS 2006, LNCS 4182, pp. 416 – 428, 2006.

drawing or taking a picture of it and query to the database via wireless connection [1]. As another example, if someone is on a fishing trip and he wants to know about some fish that he just caught, but do not know its name, then he may rely on some CBIR technique by describing its features on PDA and the system will provide him with information about the fish [2].

For the effective content based image retrieval, the system needs to figure out and represent most effective feature points from the image and based on those features, match any similar ones from the database. Content based image retrieval typically uses images features such as textures, colors, or shapes.

In the case of leaf image, it may not be appropriate to rely on color or texture features because such features are similar in most leaves. Instead, leaf shapes can provide some clue to finding similar ones. In addition, if we examine the leaf venation, we can easily figure out that they have distinct venation patterns. The leaf venation corresponds to the blood vessel of organisms. In this paper, we propose a novel leaf image retrieval scheme which first analyzes the leaf venation for the leaf categorization and then extracts and utilizes shape features from the image to find similar ones from its corresponding categorized group in the database. Leaf venations are represented using points which are selected by the curvature scale scope corner detection method on the venation image. Categorization is performed by calculating the density of feature points using non-parametric estimation density.

The rest of this paper is organized as follows. Section 2 introduces related work. Section 3 describes a venation-based leaf image categorization scheme which collects leaf venation's feature points and calculates their distribution using the Parzen Window [3]. Based on this distribution, the type of leaf venation is identified. Section 4 describes several experiments and the last section concludes the paper and discusses future work.

2 Related Work

Well-designed image retrieval tools enable people to make an efficient use of digital image collections. Typical image retrieval systems in the late 1970's were mainly based on keyword annotation. This approach suffered from many difficulties including vast amount of human labors required and challenging problem of maintaining annotation consistency among images in large databases. In order to overcome these difficulties, Content-Based Image Retrieval (CBIR) has been researched in the last decade. Examples of some of the prominent systems using this approach are VIRAGE, QBIC, Photobook, and VisualSEEk.

In many CBIR systems, an image is represented by low level features such as color, texture, shape, and structure. Relevant images are retrieved based on the similarity of their image features. In particular, shape-based image retrieval is regarded as an efficient and interesting approach. For example, shape recognition methods have been proposed and implemented into face recognition, iris recognition, and fingerprint recognition. In case images show similar color or texture, shape-based retrieval can be more effective than other approaches using color or texture. For instance, leaves of most plants are green or brown; but the leaf shapes are distinctive and can thus be used for identification.

Previous studies on leaf image retrieval were mostly based on the shape of leaf. One of the main issues in these studies is how to represent the shape of a leaf image. One approach is to get an approximated polygon of leaf shape via Minimum Perimeter Polygon (MPP) algorithm [4]. A leaf shape can be represented by positions of vertexes [5] or a set of line segments connecting two adjacent vertexes and angles at vertexes [6]. These representations may contain different vertexes between same species images. Because of this mismatch of vertexes, these representations are less effective.

On the other hand, there is a representation scheme called Center-Contour Distance Curve (CCD) which calculates distances between the center and external points of leaf [7]. But this scheme may generate two distances for an angle. So this is not suitable for the general leaf image retrieval.

3 Venation-Based Categorization

In this section, we describe how to identify leaf venations. Leaf venation corresponds to the blood vessel of organisms. Before we describe details, we first explain the types and characteristics of leaves that are generally observed from leaves. And then, we will show how to select feature points for the representation of venation. Finally, we will describe how to identify the type of leaf venation (categorization) based on the distribution of those selected feature points.

3.1 Venation Types of Leaves

Fig. 1 shows four typical venation types [8] which need to be distinguished for the categorization.

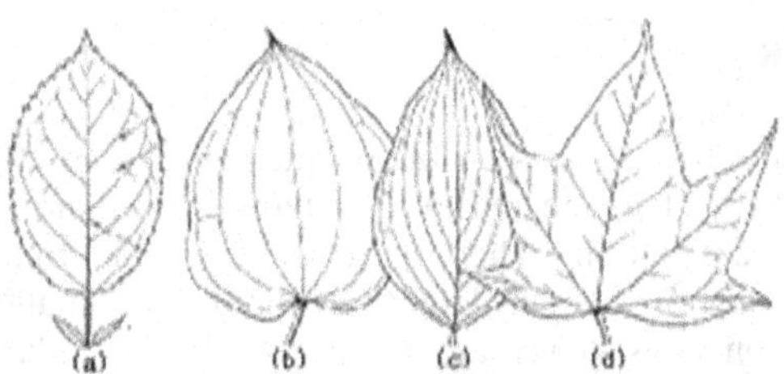

Fig. 1. Four Types of venation

From Fig. 1, type (a) is called a pinnate venation. It has a large primary vein and several secondary veins, which look like a bird's feather and are split from the primary vein. The type (b) and (c) are called parallel venation. In type (b), we can observe a lot of primary veins, which are parallel up to the end of the leaf, split from petiole and secondary veins are not clear or ladder-like. Type (c) would have a primary vein with several secondary veins parallel up to the end of the leaf. For the last, type (d) is palmate venation, in which three or more primary veins are split from the petiole and form palm-like shape.

The key characteristic of pinnate venation is that there exists one primary vein and secondary veins are split from the primary vein. Therefore, for this type of categorization, we need to find out the points where the secondary veins get split, and check the distribution of those points. If the observed distribution is in the line-type, then this line is considered as a primary vein and the points along this line are where the secondary veins get split.

The key characteristic of parallel venation is the fact that all the veins merge at the end of the leaf. In this case, we check the distribution of the points where the vein gets ended. If there are heavy densities at the top of the leaf, then the leaf is considered to have parallel venation. The most efficient way to separate (b) and (c) would be to check locations with heavy density, where vein gets split. Another method to separate them would be to check whether it has line-type distribution or not.

In the case of palmate venation, we can observe the vein gets split at the bottom of the image. So palmate venation can be distinguished by checking the distribution of points where the vein gets split.

3.2 Leaf Feature Extraction

As mentioned earlier, feature points of leaves are where the venation gets branched and ended. By checking the distribution of such feature points, we can categorize a given leaf into one of the four different venation types. If the leaves in the database were also categorized by the venation type, then we can reduce the number of images to consider for matching greatly. This section explains the Curvature Scale Space Corner Detection algorithm [9] in order to extract those feature points. This algorithm calculates curvature from the points on curves, and extracts some points as corners which have the maximum value of curvature. In the case of venation, feature points are the maximum curvature points.

3.2.1 Edge Detection

Before we apply the CSS algorithm to image, we first perform the Canny Edge Detection [10] to detect the shape of the venation. In Fig. 2, the original venation image is on the left side, and on the right side is the image which results from applying the Canny Edge Detection algorithm to it.

Fig. 2. Venation (left) and edge detection (right)

During this step, if the leaf venation is somehow broken or the vein is too thin, the leaf image is recognized to have several curves. Ideal case would be when it is detected as a single closed loop. If we have several curves due to the reasons already mentioned, the order of feature points extracted will be mixed up and cause a problem, which we will explain later. This problem can be solved by adding image pre-processing step. That is, before the Edge Detection process, making the venation thicker and increasing the contrast of the image would solve the broken vein problem.

3.2.2 Feature Points Detection

Feature points can be obtained from the curves in the previous step. The CSS algorithm will be used at this point. Applying the algorithm to the image on the right side of Fig. 2 will give the venation feature points_where the curvature gets maximum values. At these points, the venation gets branched and ended.

However, as you can see from Fig. 3, two points are detected where the venation gets branched, because they both have the maximum curvature value. In order to solve this problem, we need to do the following process.

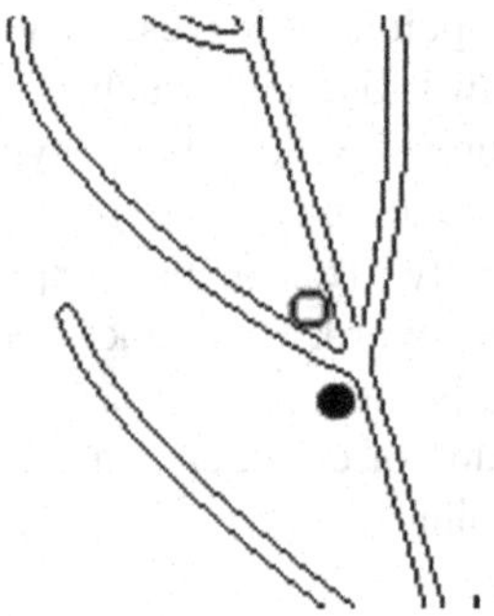

Fig. 3. Maximum curvature positions

Two feature points are not necessarily representing one position. Therefore, we first calculate the angles of the maximum value points to the curve for these two points, and then detect a point which have less than 90 degree angle value. In the case of Fig. 3, the black point, which is located below, will be ignored, and the white point will be selected as a feature point. The feature points after applying the CSS algorithm to the image of Fig. 2 are showed in Fig. 4.

3.2.3 Branching Points / Ending Points Distinction

As you might notice already, extracted feature points can be classified either as Branching Points (BP) or as Ending Points (EP). For example, for the ending point ① in Fig. 5(b), the direction changes from the ending point to the left side, and changes to the right side at the branching point ②. As in Fig. 5(c), when there are three connected points (C_1, C_2, C_3) along the proceeding direction, making a decision of BP/EP for the middle point C_2 is based on previous proceeding direction of $\overrightarrow{C_1C_2}$,

Fig. 4. Feature points by CSS algorithm

which means checking the location of later point C_3, (whether C_3 is located at upper position or bottom position) will be the decision maker. If C_3 is located at upper position, then the proceeding direction will be changed to the left and C_2 will be the EP based on previous proceeding direction. If it is the other way, then C_2 will be BP. If the angle between the line $\overrightarrow{C_1C_2}$ to X-axis is θ, then rotating C_3, which has C_2 as reference point and checking the y-coordinate of C_3 which is rotated up to $-\theta$ will be enough. As in Fig. 5(a), the proceeding direction is counter-clockwise and if the y-coordinate of C_3 has positive value, then it is Ending Point; if the coordinate has negative value, then it is Branching Point. Detailed algorithm is showed in Table 1.

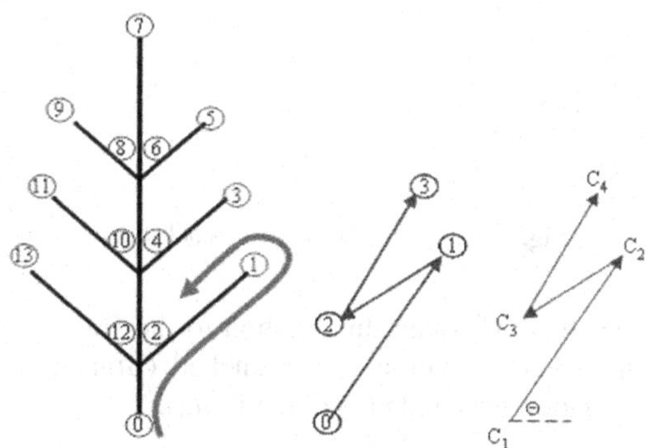

Fig. 5. (a) Feature points (b) Direction (c) Feature point distinction

To judge the proceeding direction, we set the bottom venation starting point as base point and check whether the direction is clockwise or counter-clockwise compared to the x-coordinates of previous point and later point.

Table 1. Algorithm for distinguishing feature points

```
function CornerDistinct( C_1, C_2, C_3, direction)
{
    θ ← an angle between vector C_1C_2 (→) and x-axis
    C'_3 ← rotation of C_3 around C_2 at −θ
    if C'_3.y > 0
        state ← Ending Point
    else
        state ← Branching Point
    end if
    if direction is counter-clockwise
        return state
    else
        return !state
    end if
}
```

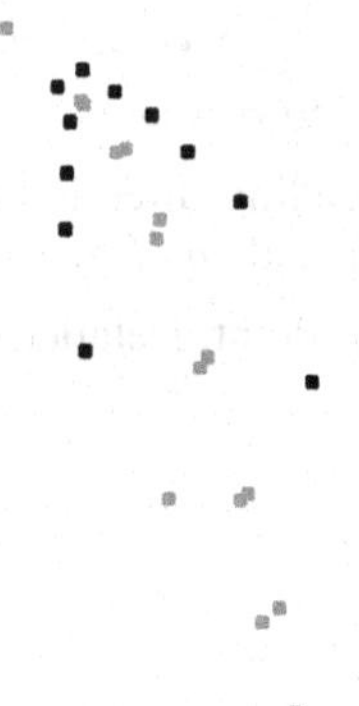

Fig. 6. BP(gray-dot) and EP(black-dot)

Fig. 6 shows the result of applying this method to the feature points in Fig. 4. In Fig 6, grey points are where venation gets branched (Branching Point) and black points are where the venation gets ended (Ending Point).

3.3 Density Distribution of Feature Points

To classify the venation, we need to decide whether feature points are distributed along a line or around one point. The density of Branching Points and Ending Points can be calculated by the Parzen Window method [5] which is non-parametric density. This method estimates the density function from a limited data. We first explain how to get the standard line that is necessary to calculate the density of feature points, and then how to decide the distribution type by calculating the density of distances from this standard line.

3.3.1 Pseudo Primary Vein

To check whether the feature points have line-type distribution or point-type distribution, we need to obtain the density of feature points by calculating the distance between some line and feature points. The lines to be used here are the pseudo primary vein and the pseudo normal line. The pseudo primary vein will be the line that connects from the top point of the venation to the bottom point. And the calculated perpendicular line is used as the pseudo normal line. We can check the row distribution by calculating the density between BP and imaginary venation line and find where density gets maximum value. Similarly to this, by calculating the density between BP or other feature points and the pseudo normal line, we can possibly check the column distribution.

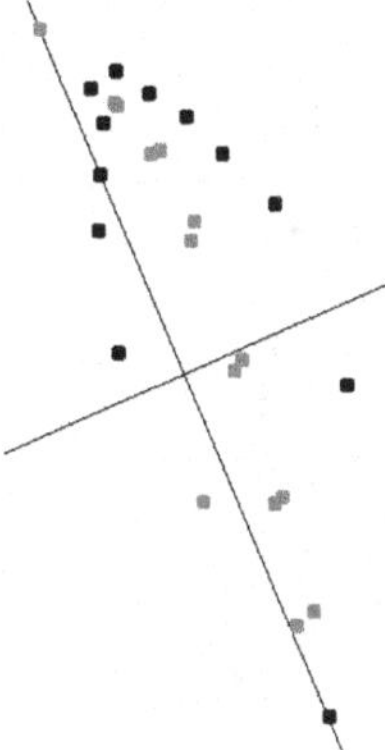

Fig. 7. Pseudo primary vein and pseudo normal line

3.3.2 Parallel Venation Verification

As mentioned before, the characteristic of parallel venation is that there are points densely dispersed at end of the leaf. To check this type of distribution, we have to see the column distribution of feature points. Table 2 shows the algorithm for calculating the density of distance between a line and feature points.

In order to get the distance where the density gets maximum value, we need to check the number of maximum values. If the number of maximum values is one and the average distance of feature points that is close to this maximum value is located above the venation, then we may consider it parallel venation. And if the number of maximum values is two and the average distances of feature points on each maximum value are distributed above and below the venation, then we may consider it parallel venation too. Fig. 8 shows a sample graph from calculating density of feature points in the parallel venation.

Table 2. Algorithm for calculating density

```
function Density (distances, w_size)
{// distances is an array of distances of corners from a line
    minDist ← min(distances)
    maxDist ← max(distances)
    foreach r such that minDist <= r <= maxDist
        sum ← 0
        foreach d in distances
          if | r –d | / w_size < 0.5
             sum++
          endif
        endforeach
        kde[r] ← sum
    endforeach
  return kde
  }
```

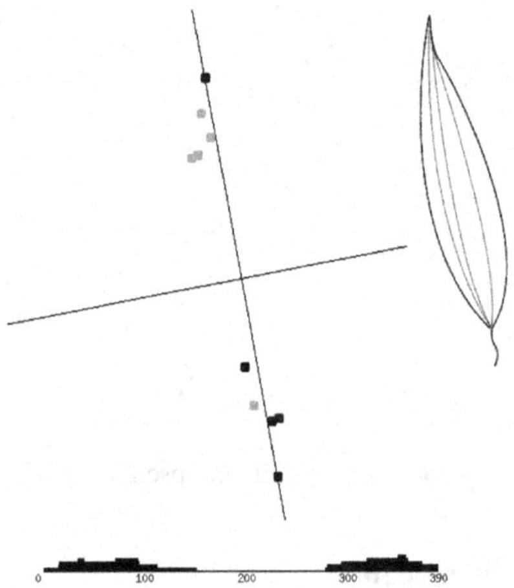

Fig. 8. Calculating the density

3.3.3 Relationships of Branching Points

In this step, we calculate the density of BPs to check whether the primary vein does exist or not and to decide whether it is palmate venation or not.

To check whether the distribution of BPs is line-shaped or densely dispersed around one point, we need to calculate both the row density and the column density. We find out points where the density gets maximum value by calculating the row and column density. And then we need to check BPs whose distance is around this maximum value and decide whether they are related or not. Namely, we can get the vertically related BPs which has the pseudo primary vein as standard line and the horizontally related BPs from the pseudo normal line.

If the leaf image contains a real primary vein, vertically related BPs form a line parallel to the pseudo primary vein. Like a palmate venation, if some BPs are gathered around one point, the BPs are related in both directions. Fig. 9 shows the relationships among BPs using this method. In the figure, small circles indicate BPs with no relationships. They are considered as secondary branching point or noise and will be ignored in the vein classification process. In the figure, a black box indicates a horizontally related BP and a cross indicates a vertically related BP. Finally, a small black point is an EP.

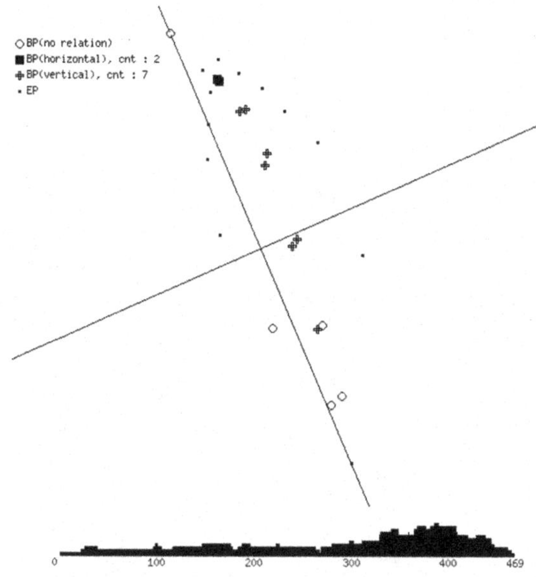

Fig. 9. Distribution of branching points

3.3.4 Venation Classification

Parallel venation is identified by a scheme described in section 3.3.2. If a vertically related BP is dominant, a real primary vein is found and the BPs are on it. In this case, the leaf image can be classified as pinnate venation or parallel venation. If maximum density value is found at the top of the venation, this leaf will be classified as parallel venation. Otherwise, it is pinnate venation. Dominant BPs of palmate venation have relations in both directions.

4 Experiment

In order to measure the effects of the venation-based categorization in the leaf image retrieval, we have used the CLOVER system as our test-bed. CLOVER is a shape-based image retrieval system that we have built for retrieving domestic aqua-plants in Korea. The algorithms were tested on the PC with Pentium 4 3.0GHz CPU, 1GB RAM. In the test, we extracted leaf images from Illustrated flora of Korea [8], which contains native plants in Korea. MATLAB was used to calculate the feature points from the images. Also, we used PHP to categorize leaf venations from the feature points. Fig. 10 and Table 3 show the result.

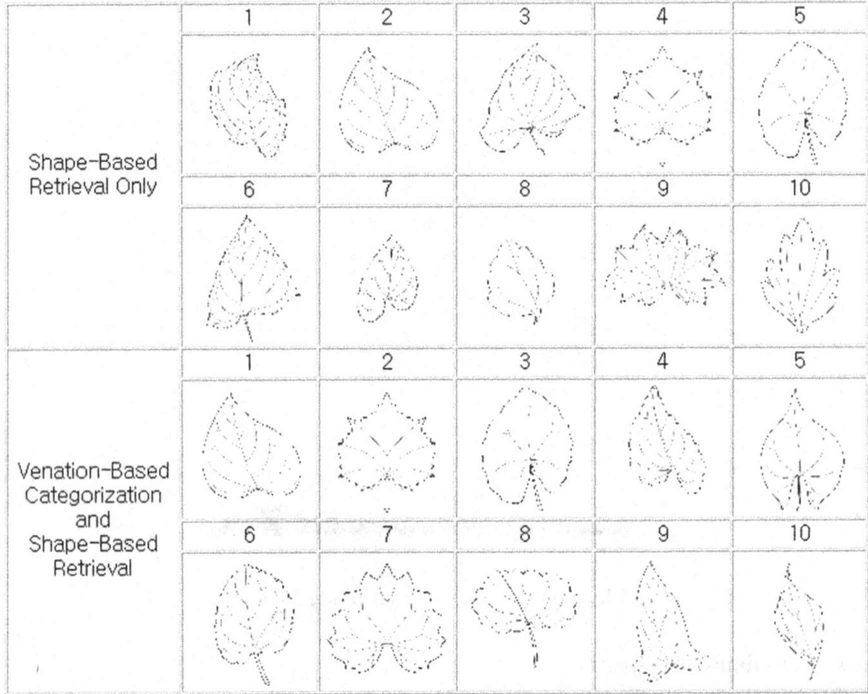

Fig. 10. Image retrieval result

Table 3. Average retrieval success rate

Method	**r = 1%**	**r = 2%**	**r = 3%**	**r = 4%**	**r = 5%**
Without Categorization	25%	58.33%	58.33%	83.33%	83.33%
With Categorization	50%	60.67%	75%	100%	100%

In Fig. 10, a query image is drawn by a user. For the comparison, we have performed two types of retrievals: shape-based retrieval only and shape-based retrieval with venation categorization which was described in this paper. Each retrieval scheme calculates image similarities between the query image and images in the database. In the shape-based retrieval with venation categorization, we categorize the query image first, and calculate the similarity against database images that belong to the same category as the query image. Based on it, Fig. 10 shows ten most similar images in the database.

Table 3 shows average rates of success retrievals which the querying images are found in top 1%, 2%, 3%, 4% and 5% of retrieved images. From Table 3, we can observe that our categorization scheme enhances the retrieval effectiveness a lot.

Average time taken to extract the feature points by MATLAB was 0.55 second, and average time to categorize the leaf venation using this result was 0.08 second.

By incorporating this scheme, we can reduce the time for retrieving similar images from the database. Ideally, if there are N types of venations and M images in the database, then by this scheme, we could reduce the search space by M/N using the proposed categorization scheme.

5 Discussion

In this paper, we have presented a novel leaf image categorization scheme by using its venation feature. Typically, there are several common venation types among leaves. By extracting and representing these venation types, we can improve the retrieval performance by reducing the number of images to be searched for. Through the experiment, we showed that extracting and categorizing the venation features took very little time (less than 1 sec) whereas it enhanced the retrieval time and accuracy a lot. The proposed categorization scheme needs further studies. A classification of the feature points depends on the extraction order and direction of such points. In the section 3.2.1, a faint venation or broken vein influenced the order of feature points. As a result, some BPs and EPs are not classified properly. This may be corrected by thickening the vein and increasing its contrast. But, if the position of a fault point was around the petiole, the extraction direction of feature points might be inversed. So, the whole BP/EP classification would be mixed up. We need more studies to solve these problems.

Acknowledgments. This research was supported by the MIC(Ministry of Information and Communication), Korea, under the ITRC support program supervised by the IITA and a grant(no.BDM0100211) to JRL from the Strategic National R&D Program through the Generic Resources and Information Network Center funded by the Korean Ministry of Science and Technology.

References

1. S. Kim, Y. Tak, Y. Nam, and E. Hwang, "mClover: mobile Content-based Leaf Image Retrieval System", ACM Multimedia 2005.
2. H. Sonobe, S. Takagi, and F. Yoshimoto, "Mobile Computing System for Fish Image Retrieval," in Proc. of International Workshop on Advanced Image Technology (IWAIT) 2004 (poster session), pp. 33-37, Singapore, January 2004.
3. Parzen, E. 1962. On Estimation of a Probability Density Function and Mode. Ann. Math. Statist. 33, 1065-10763.
4. Sklansky, Chazin et al.: Minimum perimeter polygons of digitized silhouetts. (1972).
5. Y. Nam, E. Hwang, A Shape-Based Retrieval Scheme for Leaf Image. Lecture Notes in Computer Science, Springer-Verlog, Vol. 3767, pp.876-887, Nov. 2005.

6. C Im, H Nishida, TL Kunii, Recognizing Plant Species by Normalized Leaf Shapes, Vision Interface '99, Trois-Rivières, Canada, 19-21 May, 397-404.
7. Zhiyong Wang, Zheru Chi, Dagan Feng, Qing Wang, Leaf Image Retrieval with Shape Features. Lecture Note in Computer Science, Springer-Verlog, Vol.1929.(2000) 477-487.
8. Lee, C.B.: Illustrated flora of Korea. ISBN-8971871954, Hangmoonsa, (1999).
9. Mokhtarian, F. and R. Suomela, Curvature Scale Scope Based Image Corner Detection, Proc. European Signal Processing Conference, pp. 2549-2552, Greece, 1998.
10. J. Canny, "A Computational Approach to Edge Detection," IEEE Trans. Pattern Analysis and Machine Intelligence, vol. 8, no. 6, pp. 679-698, Nov. 1986.
11. Alt, H., Behrends, B. and Blomer, J.: Approximate matching of polygonal shapes. Ann. Math. Artif. Intell. Vol.13. (1995) 251-266
12. Wang, Z., Chi, Z., Feng, D., Wang, Q.: Leaf Image Retrieval with Shape Features. Lecture Notes in Computer Science, Vol.1929. (2000) 477 – 487
13. Mokhtarian, F., and S. Abbasi.: Matching Shapes with Self-Intersections: Application to Leaf Classification. IEEE Transactions on Image Processing, Vol.13. No.5. (2004) 653-661
14. The MathWorks - MATLAB and Simulink for Technical Computing http://www.mathworks.com

Pic-A-Topic: Gathering Information Efficiently from Recorded TV Shows on Travel

Tetsuya Sakai, Tatsuya Uehara, Kazuo Sumita, and Taishi Shimomori

Toshiba Corporate R&D Center, Kawasaki 212-8582, Japan
tetsuya.sakai@toshiba.co.jp

Abstract. We introduce a system called Pic-A-Topic, which analyses closed captions of Japanese TV shows on travel to perform topic segmentation and topic sentence selection. Our objective is to provide a table-of-contents interface that enables efficient viewing of desired topical segments within recorded TV shows to users of appliances such as hard disk recorders and digital TVs. According to our experiments using 14.5 hours of recorded travel TV shows, Pic-A-Topic's F1-measure for the topic segmentation task is 82% of manual performance on average. Moreover, a preliminary user evaluation experiment suggests that this level of performance may be indistinguishable from manual performance.

1 Introduction

Nowadays, hard disk recorders that can record more than *one thousand hours* of TV shows are on the market, so that people can watch TV shows at the time of their convenience. But there is a problem: Nobody can spend one thousand hours just watching TV! Thus, unless there are ways to let the user handle recorded TV contents efficiently, the recorded contents will eventually be deleted or forgotten before put to use in any way.

Many researchers have tackled the problem of efficient information access for *broadcast news* [3,4,5,6,8,13,14,20,22], by means of *news story segmentation*, *segment/shot retrieval*, *topic labelling* (i.e., assigning a closed-class category) and so on. Broadcast news is clearly an important type of video contents, especially for professionals and for organisations such as companies and governments. Both timely access to incoming news and retrospective access to news archives are required for such applications.

However, at the personal level, there are many other TV genres that need to be considered: soap, drama, comedy, quiz, talkshow, sport, music, wildlife, cookery, education, and so on. In fact, one could argue that these types of contents are more important than news for general consumers, as these are the kind of contents that tend to accumulate in hard disks, waiting to be accessed by the user some day, often in vain.

Among the aforementioned "entertaining" kinds of TV genres, we are currently interested in *separable* contents. By "separable", we casually mean that a TV show can be broken down into several segments, where each segment is independent enough to provide the user with a useful piece of information. Thus,

H.T. Ng et al. (Eds.): AIRS 2006, LNCS 4182, pp. 429–444, 2006.

according to our definition, most factual TV shows are separable, while most soaps and dramas are not. That is to say, a short segment drawn from a drama is less informative on its own than one drawn from a factual TV show.

For separable TV contents, we believe that *topic segmentation* is useful for solving the aforementioned "hard disk information overload" problem. For example, suppose that there is a recorded TV show that is two-hour long, which contains several distinct topics. If it is possible to segment the TV show according to topics and provide the user with a clickable "table-of-contents" interface from which he can select an interesting topic or two, then the user may be able to obtain useful information by viewing the selected segments only, which may only last for several minutes. In short, we believe that efficient, selective viewing of separable TV contents is important.

As a first step to tackling separable contents, we decided to handle TV shows on *travel*, since they are very popular in Japan and are representative (i.e., general approaches that work for travel TV shows should also work for those on *cookery*, *wildlife*, and so on). A typical Japanese travel TV show involves a group of non-professional reporters (typically an actor and his family), and may run as follows:

- They visit a place in the country by train or by car;
- They visit a couple of sightseeing spots or a cafe;
- They check in at a hotel in a hot springs resort area;
- They have a good open-air bath;
- They enjoy their dinner;
- They go to sleep;
- Next morning, they have breakfast and check out;
- They go to a souvenir shop and head for home.

In fact, one TV show may contain two or three such sequences. Each of the above list items could viewed as a *topic*, that may correspond to a video *segment*, that typically runs for several minutes. In contrast to *news story segmentation* [4,20], our definition of a "topic" is loose: if a video segment is useful on its own for the user for obtaining some information, we regard it as an acceptable topical segment. Some users might choose to view segments that are to do with food; others might choose to view segments that provides information on the hotel facilities.

We currently provide the user with a clickable table-of-contents interface like the one shown in Figure 1, although this is only a prototype mainly for debugging and testing purposes and not intended for the end user. The interface relies on two functionalities which we have developed:

Topic Segmentation. Our current algorithm analyses *closed captions* (or *transcripts*) for topic segmentation. Figure 1 shows our automatically detected topical segments (together with start times and *confidence values*): Topic 1 discusses a dinner featuring Koshu beef; Topic 2 discusses a sightseeing spot called Shosenkyo; Topic 3 discusses a visit to another place in the country called Shiotadaira; Topic 4 discusses a temple in Shiotadaira. Our algorithm relies on both *cue phrase detection* and *vocabulary shift detection*, which will be described in Section 2.2.

Topic Sentence Selection. We currently select, from the closed-caption text, two topic sentences per segment based on a *relevance feedback* algorithm which will be described in Section 2.3. (The choice of the number of sentences per segment is arbitrary: We simply found that two is much more informative than one in our preliminary investigations.) The user is expected to select the segments he would like to view by reading these sentences. Sample Japanese topic sentences are shown in Figure 1, and their rough English translations are provided in Table 1.

Fig. 1. A prototype interface of Pic-A-Topic

In Figure 1, two thumbnails per segment are shown to the user. These key frames are selected based on the timestamps of the aforementioned topic sentences. The thumbnail on the left corresponds to the first topic sentence, and

Table 1. Rough English translations of the topic sentences shown in Figure 1

Segment 1	
Sentence 1	Koshu beef is the best Japanese beef available in Yamanashi prefecture.
Sentence 2	You just cannot stop even when you are full.
Segment 2	
Sentence 1	From Tenjin Forest, which is the entrance to Shosenkyo, along the valley...
Sentence 2	We have now reached the goal of our trip.
Segment 3	
Sentence 1	Next, we will visit Shiotadaira, south of Ueda.
Sentence 2	From Shiotadaira Station to Bessho Hot Springs, there is a hiking trail...
Segment 4	
Sentence 1	It is called the "completed pagoda that is incomplete" because...
Sentence 2	There is a statue of Yakushi Nyorai inside, which the hall has protected...

the one on the right corresponds to the second topic sentence. (Our current user interface is admittedly not user-friendly, but we stress that it is not intended for the end user.) By clicking a segment, the user can start playing it.

We call our prototype system *Pic-A-Topic*, because it enables the user to "pick a topic" from a TV show for selective viewing, and also because our clickable table-of-contents contains not only topic sentences but also key frames (i.e., *Pic*tures). The objective of this paper is to describe our current topic segmentation and topic sentence selection algorithms and to report on some experimental results, primarily on topic segmentation, for the travel domain.

The remainder of this paper is organised as follows. Section 2 describes the Pic-A-Topic system. Section 3 reports on two sets of experiments for evaluating our topic segmentation algorithm. Section 4 discusses previous work, with an emphasis on those that deal with TV genres other than broadcast news. Finally, Section 5 concludes this paper.

2 Pic-A-Topic

2.1 Overview

Figure 2 provides an overview of the Pic-A-Topic system. As can be seen, it consists of three main components: *topic segmentation*, *topic sentence selection* and *table-of-contents creation*.

The input to the topic segmentation component are Japanese closed captions and Electronic Program Guide (EPG) data. In our initial experiments, we also used *shot boundaries* as input, for treating them as candidate topical boundaries. However, this approach was not successful because there simply were too many shot boundaries, and most of them were not *topical* boundaries.

Currently, the EPG data (if available) is used only for the purpose of obtaining names of celebrities, in order to automatically augment our *named entity recognition* dictionaries used in *vocabulary shift detection*. This is because celebrities

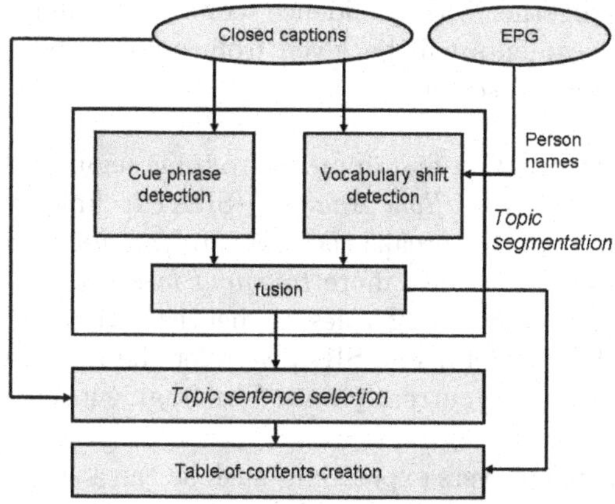

Fig. 2. A system overview of Pic-A-Topic

such as comedians tend to have "anomalous" names, which named entity recognition tend to overlook (e.g., "Beat Takeshi" and "Papaya Suzuki").

The topic segmentation component performs both *cue phrase detection* and *vocabulary shift detection*, and then finally "fuses" the two results. The contribution of each approach can be controlled by a parameter. Details will follow in Section 2.2.

The input to the topic sentence selection component are the result of topic segmentation as well as the raw closed-caption text. For each topical segment, this component first selects topic words based on a relevance feedback algorithm used widely in information retrieval. Subsequently, it selects topic sentences based on the weights of the aforementioned topic words. Note that this functionality is different from "topic labelling" [5] which assigns a *closed-class category* to (say) a news story. Topic Sentence Selection can assign *any* text extracted from closed captions, and are more flexible. Details will follow in Section 2.3.

The input to the table-of-contents creation component are the results of topic segmentation and topic sentence selection. This component performs several postprocessing functions, including:

Keyframe selection/thumbnail creation. One keyframe is selected for each topic sentence, based on its timestamp.

Start/End time adjustment. Because there may be a time lag between the closed-caption timestamps and the actual audio/video, the timestamps output by topic segmentation are heuristically (but automatically) adjusted. An example heuristic would be to avoid playing the video from the middle of an utterance, since this would be neither informative nor user-friendly.

2.2 Topic Segmentation

The task of topic segmentation is to take the timestamps in the closed captions as candidates and output a list of selected timestamps that are likely to represent topical boundaries, together with confidence scores. We assume that the number of required topical segments will be given from outside, based on constraints such as the size of the TV screen.

Cue Phrase Detection. Our first approach to topic segmentation is cue phrase detection using the *Semantic Role Analysis* (SRA) techniques. SRA first performs morphological analysis, breaks the text into *fragments* (which in our case are sentences), and assigns one or more *fragment labels* to each fragment based on hand-written pattern-matching rules. A heuristically-determined weight is assigned to each rule. For details on SRA, we refer the reader to [18,19].

For the present study, we currently have four fragment labels:

CONNECTIVE. This covers expressions such as "as a starter", "at last" and "furthermore". (Hereafter, all examples of Japanese words and phrases will be given in English translations.)
MOVEMENT. This covers verbs such as "head for" and "visit".
TIME_ELAPSED. This covers expressions that refer to the passage of time, such as "next morning" and "lunchtime".
OTHER. Anything else that are useful as cues.

Note that the above labels are not necessarily domain-specific. We currently have 45 regular expression pattern-matching rules in total.

For each candidate boundary (i.e., timestamp), cue phrase detection calculates the raw confidence score by summing up the weights of all rules that matched the corresponding sentence. Finally, it obtains the *normalised* confidence scores c by dividing the raw confidence scores with the maximum one among all candidates.

The result of cue phrase detection may be used on its own for defining topical segments. In this case, we *sieve* the topical boundary timestamps before handing them to topic sentence selection: For each timestamp s (in milliseconds) obtained, we examine all its "neighbours" (i.e., timestamps that lie within $[s - 30000, s + 30000]$), and overwrite its confidence score c with zero if any of the neighbours has a higher confidence score than s. This is for obtaining "local optimum" timestamps which are at least 30 seconds apart from one another.

Vocabulary Shift Detection. Our second approach to topic segmentation is vocabulary shift detection, which is similar in spirit to standard topic segmentation algorithms such as *TextTiling* [7]. Although these algorithms originally designed for "written" text are often directly applied to closed captions (e.g., [11]), our preliminary experiments suggested that they are not satisfactory for analysing closed-captions which mainly consist of dialogues. Since closed captions contain timestamps, our algorithm uses timestamps explicitly and extensively. Moreover, as our preliminary experiments showed that *domain specific* knowledge is effective for our topic segmentation task, we use *named entity recognition* tuned specifically for the travel domain. Our algorithm is described below.

We first analyse the closed-caption text and extract *morphemes* and *named entities*, which we collectively refer to as *terms*. We have over one hundred *generic* named entity classes covering person names, place names, organization names, numbers and so on, originally developed for *open-domain question answering* [18]. In addition, we devised four *domain-specific* classes for the travel domain:

TRAVEL_ACTIVITY. This class covers typical activities of a tourist, such as "dinner", "walk" and "rest".

TRAVEL_ATTRACTION_CLASS. This class covers concepts that represent tourist attractions and events such as "sightseeing spot", "park", "show" and "festival". Note that this is not for detecting specific *instances* such as "Tokyo Disneyland".

TRAVEL_HOTEL_CLASS. This class covers words such as "hotel" and "inn".

TRAVEL_BATH_CLASS. This class covers words such as "hot spring" and "bath".

It is clear that, in order to deal with other TV genres such as *cookery*, a different set of domain-specific classes will be required. We feel optimistic about this issue since our named entity recogniser is fairly easy to customise.

Let t denote a term in a given closed-caption text, and s denote a candidate topical boundary (represented by a timestamp in milliseconds). For a fixed widow size S, let WL denote the set of terms whose timestamps (by which we actually mean *start* times) lie within $[s - S, s)$, and WR denote the set of terms whose timestamps lie within $[s, s + S)$. For each $t \in WL \cup WR - WL \cap WR$ (i.e., term included in either WL or WR but not both), we define a downweighting factor $dw(t)$ as follows:

$$dw(t) = \max\{dw_{domain}(t), dw_{generic}(t), dw_{morph}(t)\} \tag{1}$$

where

$dw_{domain}(t) = DW_{domain}$ if t is a domain-specific named entity; Otherwise 0;
$dw_{generic}(t) = DW_{generic}$ if t is a generic named entity; Otherwise 0;
$dw_{morph}(t) = DW_{morph}$ if t is a single morpheme; Otherwise 0.

Here, DW_{domain}, $DW_{generic}$ and DW_{morph} are tuning constants between 0 and 1.

Next, for each $t \in WL \cup WR - WL \cap WR$ whose timestamp is $s(t)$, we compute:

$$f(t) = 0.5 - 0.5 \cos \pi (\frac{s(t) - s}{S} + 1) . \tag{2}$$

$f(t)$ takes the maximum value of 1 when $s(t) = s$ and the minumum value of 0 when $|s(t) - s| = S$. That is, $f(t)$ gets smaller as the term moves away (along the timestamp) from the candidate boundary.

Meanwhile, for each $t \in WL \cap WR$, let $s_{WL}(t)$ and $s_{WR}(t)$ denote the timestamps of t that correspond to WL and WR. (If there are multiple occurrences within the interval covered by WL or WR, then we take the timestamp that is closest to s.) Then we compute:

$$g(t) = 0.5 - 0.5 \cos \pi (\frac{s_{WR}(t) - s_{WL}(t)}{2S} + 1) \; . \tag{3}$$

If $s_{WL}(t)$ and $s_{WR}(t)$ are close (i.e., the term occurs just before the candidate boundary *and* just after it), then $g(t)$ is close to 1. If $s_{WR}(t) - s_{WL}(t)$ is close to $2S$ (i.e., the two occurrences of the same term are far apart), then $g(t)$ is close to 0.

The term weighting functions $f(t)$ and $g(t)$ have been designed in order to make the algorithm robust to the choice of window size S, which is currently fixed at 30 seconds. We currently do not to use global statistics such as *idf* (e.g., [21]).

Based on $dw(t)$, $f(t)$ and $g(t)$ as well as two positive parameters α and β, we compute the *novelty* of each candidate boundary s as follows:

$$novelty = \sum_{t \in WR-WL} dw(t) * f(t) + \alpha * \sum_{t \in WL-WR} dw(t) * f(t) - \beta * \sum_{t \in WL \cap WR} g(t) \tag{4}$$

Thus, a candidate boundary receives a high novelty score if WR has many terms that are not in WL (and vice versa) *and* if WL and WR have few terms in common. Using $\alpha < 1$ implies that WR and WL are not treated symmetrically, unlike the cosine-based segmentation methods (e.g., [7,8]). This corresponds to the intuition that terms that occur *after* the candidate boundary may be more important than those that occur before it. However, preliminary experiments suggested that this asymmetrical treatment may not be beneficial for our data set: We let $\alpha = 1$ and $\beta = 0.5$ hereafter.

The final confidence score v based on vocabulary shift is given by:

$$v = \frac{novelty - minnovelty}{maxnovelty - minnovelty} \tag{5}$$

where *maxnovelty* and *minnovelty* are the maximum and minimum values among all the novelty values computed for the closed-caption text.

The result of vocabulary shift detection may be used on its own for defining topical segments. Again, *sieving* is performed in such a case.

Fusion. The confidence scores based on cue phrase detection and vocabulary shift detection can be fused as follows:

$$confidence = \gamma * v + (1 - \gamma) * c \tag{6}$$

In fact, we fix γ to 0.5. Sieving is performed *after* the above fusion.

2.3 Topic Sentence Selection

Our topic sentence selection relies on a standard relevance feedback algorithm [15]. For each topical segment, we select 10 topic words from the closed-caption text by regarding the segment as a set of relevant documents and the other segments as nonrelevant ones. (A "document" is usually a sentence or two, defined by a start time and an end time.)

Let N be the total number of "documents" in the closed-caption text, and R be the number of "documents" within the segment in question. Let w denote a candidate keyword (morpheme, actually). Moreover, let $n(w)$ denote the number of "documents" containing w, and let $r(w)$ denote the number of "documents" within the segment containing t. Then the term selection value for w is the *offer weight* $ow(w)$:

$$ow(w) = r(w) * rw(w) \tag{7}$$

where

$$rw(w) = \log \frac{(r(w)+0.5)(N-n(w)-R+r(w)+0.5)}{(n(w)-r(w)+0.5)(R-r(w)+0.5)} . \tag{8}$$

Next, we select a given number of sentences as follows:

1. Let L be the list of 10 topic words obtained above;
2. For each sentence in the segment in question, compute the sentence score by summing $rw(w)$ for all words included in the sentence.
3. Take the sentence with the highest score as the topic sentence, and remove from L all topic words included in the selected sentence;
4. Repeat from 2, until a desired number of topic sentences (which is our case is 2) is obtained.

The above algorithm tries to obtain *unique* sentences that contain different topic words. This is because our preliminary experiments showed that selecting sentences *independently* based on the keyword weights (as was done for questionnaire analysis in [17]) generally yield two topic sentences that are too similar to each other. This meant that the corresponding two key frames were also similar, and the entire table-of-contents did not look so informative.

3 Experiments

This section reports on our experiments primarily designed for validating our topic segmentation algorithms. Section 3.1 describes our video test collection. Section 3.2 discusses the segmentation accuracy of Pic-A-Topic for the travel domain, by comparing its performance to those of humans. Section 3.3 discusses what the accuracy values would actually mean to real users by conducting some subjective evaluations.

3.1 Topic Segmentation Test Collection for Travel TV Shows

Unfortunately, there is no standard test collection available for evaluating topic segmentation for TV genres such as travel TV shows. We therefore had no choice but to create our own collection by recording real Japanese broadcast TV shows on travel, extracting the closed captions and preparing the "right answers" for ourselves. Our test collection consists of ten clips from four different travel TV show series, totalling approximately 14.5 hours. Note that, with only ten clips, it is difficult for us to discuss statistical significance of experimental results.

Recall that, unlike the news story segmentation task, our definition of "topic" segmentation is rather ill-defined: We are happy as long as Pic-A-Topic is useful to the user for quickly obtaining desired information from a long TV show. Because of this subjective nature of the task, we let *four* assessors manually segment each clip, after giving them a common set of instructions. (We used a different set of assessors for each clip.) Each assessor was asked to view a clip and provide the timestamps of topical boundaries, using a simple graphical user interface with buttons such as "play", "pause", "fast forward" and "record this timestamp as a topical boundary". Based on some pilot studies, we encouraged the assessors to find about 20 boundaries per hour. (An extremely short segment would not be informative; whereas, an extremely long segment would prevent the user from viewing the TV show efficiently.)

For each clip, the timestamp files produced by the *first three* assessors were merged, to create a single "ground truth" file. Each ground truth file is a set of timestamp *intervals*, reflecting the individual assessment. The philosophy is that, if the system agrees with *any* of the three assessors about a topical boundary, then that boundary is acceptable.

The *fourth* assessor's timestamps were used for providing the "best-possible" performance by comparing it with the ground truth. (Note that the Fourth Assessor is not a single person.) The output of Pic-A-Topic was compared with the ground truth in exactly the same way, and we examined the *relative* performance of Pic-A-Topic, by dividing the system performance by the best-possible one. The method of comparison with the ground truth will be described below.

3.2 Segmentation Accuracy

For both Pic-A-Topic and the Fourth Assessor, the topic segmentation accuracy was computed in terms of Precision, Recall and F1-measure (i.e., harmonic mean of Precision and Recall). Suppose that the ground truth file contains N timestamp intervals, and that the Fourth Assessor file contains M timestamps. In general, $N > M$ holds, because the ground truth file covers the timestamps of three assessors. Since Pic-A-Topic requires the target number of topical boundaries as a parameter, M was given to Pic-A-Topic to compare its performance with the Fourth Assessor. (Note that giving N to Pic-A-Topic would not be a fair comparison: the recall of the Fourth Assessor would suffer very much since $N > M$, so the relative performance of Pic-A-Topic would be overestimated.)

A topical boundary detected by Pic-A-Topic (or the Fourth Assessor) is counted as correct if it lies within the window $[start - 10000, end + 10000]$, where *start* and *end* are the start/end times in millisecs of one of the ground-truth interval. (Of course, at most one boundary can be counted as correct for each ground-truth interval: If there are plural boundaries that lie within a single ground-truth interval, then only the one that is closest to the center of the interval is counted as correct.) The ten-second margins are for handling the time

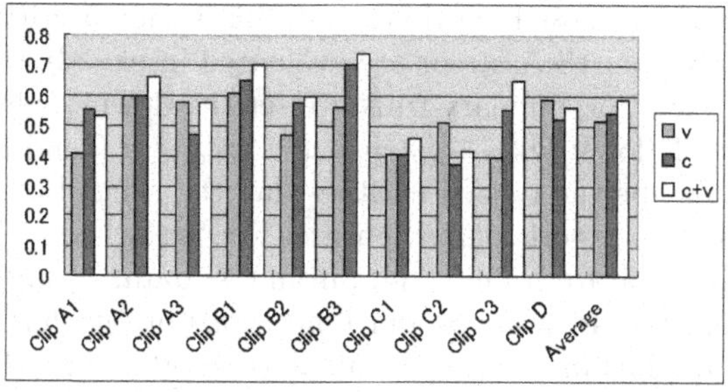

Fig. 3. The effect of fusion on Absolute F1-measure

Table 2. Relative segmentation accuracy after fusion

		Precision	Recall	F1-measure
Clip A1 (2 hours)	Fourth Assessor	34/37 = 0.92	34/61 = 0.56	0.69
	Pic-A-Topic	26/37 = 0.70	26/61 = 0.43	0.53
	relative	76%	77%	77%
Clip A2 (2 hours)	Fourth Assessor	43/56 = 0.77	43/50 = 0.86	0.81
	Pic-A-Topic	35/56 = 0.62	35/50 = 0.70	0.66
	relative	81%	81%	81%
Clip A3 (2 hours)	Fourth Assessor	24/26 = 0.92	24/46 = 0.52	0.67
	Pic-A-Topic	21/26 = 0.81	21/46 = 0.46	0.58
	relative	88%	88%	87%
Clip B1 (1 hour)	Fourth Assessor	17/19 = 0.89	17/27 = 0.63	0.74
	Pic-A-Topic	16/19 = 0.84	16/27 = 0.59	0.70
	relative	94%	94%	95%
Clip B2 (2 hours)	Fourth Assessor	29/36 = 0.81	29/54 = 0.54	0.64
	Pic-A-Topic	27/36 = 0.75	27/54 = 0.50	0.60
	relative	93%	93%	94%
Clip B3 (1 hour)	Fourth Assessor	18/20 = 0.90	18/23 = 0.78	0.84
	Pic-A-Topic	16/20 = 0.80	16/23 = 0.70	0.74
	relative	89%	90%	88%
Clip C1 (1 hour)	Fourth Assessor	13/19 = 0.68	13/20 = 0.65	0.67
	Pic-A-Topic	9/19 = 0.47	9/20 = 0.45	0.46
	relative	69%	69%	69%
Clip C2 (1 hour)	Fourth Assessor	18/21 = 0.86	18/22 = 0.82	0.84
	Pic-A-Topic	9/21 = 0.43	9/22 = 0.41	0.42
	relative	50%	50%	50%
Clip C3 (1 hour)	Fourth Assessor	14/17 = 0.82	14/23 = 0.61	0.70
	Pic-A-Topic	13/17 = 0.76	13/23 = 0.57	0.65
	relative	93%	93%	93%
Clip D (1.5 hours)	Fourth Assessor	19/25 = 0.76	19/36 = 0.53	0.62
	Pic-A-Topic	17/25 = 0.68	17/36 = 0.47	0.56
	relative	89%	89%	90%
Average over 10 clips	Fourth Assessor	83%	65%	72%
	Pic-A-Topic	69%	53%	59%
	relative	**82%**	**82%**	**82%**

lags between video and closed captions, but this is not a critical choice since Pic-A-Topic and the Fourth Assessor are evaluated in exactly the same way.

Based on a couple of preliminary runs, we set the downweighting parameters (See Section 2.2) as follows: $DW_{domain} = 1$, $DW_{generic} = 0$ and $DW_{morph} = 0.1$. That is, the domain-specific named entities play the central role in vocabulary shift detection, but the generic named entities are "switched off". Although our experiments do not separate training data from test data, we argue that they are useful for exploring the advantages and limitations of our current approaches. "Open data" evaluations may have to wait until good standard test collections become available.

Table 2 summarises the results of our topic segmentation experiments, in which the four TV series on travel are represented by A, B, C and D. For example, for Clip A1, the ground-truth file contained 61 intervals but the Fourth Assessor (and therefore Pic-A-Topic) produced only 37 topical boundaries. As a result, relative Precision, relative Recall and relative F1-measure are 76%, 77% and 77%, respectively. The results were obtained by fusing the cue phrase detection ouput and the vocabulary shift detection output: Figure 3 shows the absolute F1-measure values of the individual approaches as well, in which "c" and "v" represent cue phrase detection and vocabulary shift detection, respectively. It can be observed that fusion ("c+v") is generally beneficial, with a few exceptions.

The best relative F1-measure performance is for Clip B1 (95%), while the worst one is for Clip C2 (50%). We found that the TV series C is generally challenging, because unlike the other series, it has a complex program structure. For example, in the middle of a footage showing reporters having a walk in the countryside, a studio scene is inserted several times, in which the presenters of the show and the same reporters make some comments over a coffee. Thus this probably shows a limitation of our purely linguistic approach: Some travel TV shows may require video feature analysis (e.g., indoor/outdoor detection and face recognition) for accurate topic segmentation.

More generally, our failure analysis found that Pic-A-Topic tends to break up a single dinner sequence into several "subtopics". This is because, in a typical Japanese travel TV show, a dinner sequence is rather long, and it does contain some vocabulary shifts, involving words that are to do with the starter, the main dish, the dessert and so on. That is, it appears that Pic-A-Topic is currently too sensitive to change in "food" vocabularies! There are ways to remedy this problem: Since we already use named entity recognition, we could ignore food-related named entities and morphemes in vocabulary shift detection. A possibly more robust method would be to incorporate (lack of) scene changes, by analysing the video. We have also encountered some cases in which we felt that music/sound effect detection would be useful for detecting the beginning of a topical segment.

Another limitation of our approach is that closed captions do not contain all textual information. Some important textual information, such as the title of a travel episode and the names of hotels and restaurants, are shown as *overlay* text, and are *not* included in closed captions. That is, closed captions and overlays are complementary. This suggests that overlay text recognition/detection may be useful for topic segmentation of travel TV shows. Moreover, when a travel TV show consists of several episodes, tiny banners that represent each episode are often overlaid in the corner of each frame. This would be useful for obtaining the overall structure of the show. To sum up, image analysis techniques known to be useful for broadcast news may transfer well to our domain in some cases.

In spite of all these limitations, however, our overall relative performances (each obtained by averaging the ten corresponding relative values in the table) are all 82%, which is quite impressive even if they do not necessarily represent performances for "open" data. (Dividing the average absolute performance of Pic-A-Topic by that of the Fourth Person yields similar results.) The question is, "What does a relative F1-measure of 82% mean in terms of practical usefulness?"

3.3 Blind Tests: Preliminary Subjective Evaluation

We conducted a preliminary experiment to answer the above question. From Table 2, we first selected Clips A2, B1 and C2, representing the relative F1-measure of 81%, 95% and 50%, respectively. Thus, the three clips represent our "average", "best" and "worst" performances. Then, for each clip, we generated two table-of-contents interfaces similar to the one shown in Figure 1, one based on the Fourth Assessor's segmentation and the other based on Pic-A-Topic's segmentation. The idea was to conduct "blind" tests, and see which output is actually preferred by the user for an information seeking task.

For each of the three clips, five subjects were employed. Each subject was given a clip with a table-of-contents interface based on either the Fourth Assessor's output or Pic-A-Topic's, and was given two questions per interface in random order. The subjects were blind as to how each interface was created, and were asked to locate the answers to the above questions within the given clip using the given interface. Each subject was finally asked which of the two interfaces he preferred for efficient information access. An example from our question set is: "What was the name of the restaurant that Actress X and Comedian Y visited?". Thus this experiment is similar in spirit to the evaluation of summarisation using reading comprehension questions [10].

Initially, we tried to measure the time the subjects took to find the answer through the interface. However, we quickly gave this up because the variance across subjects was far greater than the efficiency differences between the two interfaces. Hence we decided to compare the two interfaces based on the subjects' preferences only. Note that, since Pic-A-Topic and the Fourth Assessor interfaces have different topical boundaries, the topic sentences presented to the subjects are also different.

Table 3 summarises the results of our subjective evaluation experiments. It can be observed that, for Clip B1 for which the relative F1-measure is 95%, Pic-A-Topic was actually a little more popular than the Fourth Assessor. Moreover, even for Clip A2, which represents a typical performance of Pic-A-Topic, only one out of the five subjects preferred the Fourth Assessor interface, and the other four could not tell the difference. Thus, although this set of experiments may be preliminary, it is possible that 80% relative F1-measure is a practically

Table 3. Subjective evaluation results

	#subjects who preferred the Fourth Assessor	#subjects who preferred Pic-A-Topic	#subjects who did not feel any difference
Clip B1 (95%)	1	2	2
Clip A2 (81%)	1	0	4
Clip C2 (50%)	2	1	1
Total	5	3	6

acceptable level of performance. Note also that, even at 50% relative F1-measure, one subject said that Pic-A-Topic was better than the Fourth Assessor. This reflects the subjective nature of our topic segmentation task.

In summary, our results are encouraging: On average, Pic-A-Topic achieves a relative F1-measure of 82% at least for a known data set, and it is possible that this level of performance is practically acceptable to the end user.

4 Related Work

We have already mentioned in Section 1 that many researchers focus on the problem of efficient information access from *broadcast news* video. Below, we briefly mention some researches that handle TV genres other than news, and point out how their approaches differ from ours.

Extracting highlights from *sports* TV programs is a popular research topic, for which *audio* features [16] or manually transcribed utterances [23] are often utilised. Aoki, Shimotsuji and Hori [1] used *colour and layout analysis* for selecting unique keyframes from *movies*. More recently, Aoki [2] reported on a system that can structuralise *variety shows* based on *shot interactivity*. While these approaches are very interesting, we believe that audio and image features alone are not sufficient for identifying *topics* within a separable and informative TV content. Lack of language analysis also implies that providing topic words or topic sentences to the user is difficult with these approaches. Zhang *et al.* [24] handled non-news video contents such as *travelogue* material to perform *video parsing*, but their method is based on *shot boundary detection*, not topic segmentation. As we mentioned in Section 2.1, we feel that shot boundaries are not suitable for the purpose of viewing a particular topical segment.

There exist, of course, approaches that effectively combine audio, image and textual evidence. Jasinschi *et al.* [9] report on a combination-of-evidence system that can deal with *talk shows*. However, what they refer to as "topic segmentation" appears to be to segment closed captions based on speaker change markers for the purpose of labelling each "closed caption unit" with either *financial news* or *talk show*. Nitta and Babaguchi [12] structuralise *sports* programs by analysing both closed captions and video. Smith and Kanade [21] also combine image and textual evidence to handle news and non-news contents: They first select *keyphrases* from closed captions based on *tf-idf* values, and use them as the basis of a *video skim*. While their keyphrase extraction involves detection of breaks between utterances, it is clear that this does not necessarily corresond to *topical* boundaries. Thus, even though their Video Skimming interface may be very useful for viewing the entire "summary" of a TV program, whether it is also useful for selecting and viewing a particular topical segment or two is arguably an open question.

5 Conclusions

We introduced a system called Pic-A-Topic, which analyses closed captions of Japanese TV shows on travel to perform topic segmentation and topic sentence selection. According to our experiments using 14.5 hours of recorded travel TV shows, Pic-A-Topic's F1-measure for the topic segmentation task is 82% of manual performance on average. Moreover, a preliminary user evaluation experiment suggested that this level of performance may be indistinguishable from manual performance.

Some of our approaches are domain-specific, but we believe that domain-specific knowledge is a necessity for practical, high-quality topical segmentation of TV shows. Since genre information can be obtained from EPGs, we would like to let Pic-A-Topic select the optimal segmentation strategy for each genre. Our future work includes the following:

- Incorporating video and audio information in our topic segmentation algorithms;
- Expanding our TV genres;
- Developing a user-friendly table-of-contents interface;
- Building other applications, such as selective downloading of TV content segments for mobile phones.

References

1. Aoki, H., Shimotsuji, S. and Hori, O.: A Shot Classification Method of Selecting Key-Frames for Video Browsing. ACM Multimedia '96 Proceedings (1996)
2. Aoki, H.: High-Speed Topic Organizer of TV Shows Using Video Dialog Detection (in Japanese). IEICE Transactions on Information and Systems **J88-D-II-1**, pp. 17-27 (2005)
3. Boykin, S. and Merlino, A.: Machine Learning of Event Segmentation for News on Demand. Communications of the ACM **43-2**, pp. 35-41 (2000)
4. Chua, T.-S. *et al.*: Story Boundary Detection in Large Broadcast News Video Archives - Techniques, Experience and Trends. ACM Multimedia 2004 Proceedings (2004)
5. Hauptmann, A. G. and Lee, D.: Topic Labeling of Broadcast News Stories in the Informedia Digital Video Library. ACM Digital Libraries '98 Proceedings (1998)
6. Hauptmann, A. G. and Witbrock, M. J.: Story Segmentation and Detection of Commercials in Broadcast News Video. Advances in Digital Libraries '98 (1998)
7. Hearst, M. A.: Multi-Paragraph Segmentation of Expository Text. ACL '94 Proceedings, pp. 9-16 (1994)

8. Ide, I. *et al.*: Threading News Video Topics. ACM SIGMM Workshop on Multimedia Information Retrieval (MIR 2003), pp. 239-246 (2003)
9. Jasinschi, R. S. *et al.*: Integrated Multimedia Processing for Topic Segmentation and Classification. IEEE ICIP Proceedings (2001)
10. Mani, I. *et al.*: The TIPSTER SUMMAC Text Summarization Evaluation. EACL '99 Proceedings, pp. 77-85 (1999)
11. Miyamori, H. and Tanaka, K.: Webified Video: Media Conversion from TV Program to Web Content and their Integrated Viewing Method. ACM WWW 2005 Proceedings (2005)
12. Nitta, N. and Babaguchi, N.: Story Segmentation of Broadcasted Sports Videos for Semantic Content Acquisition (in Japanese). IEICE Transactions on Information and Systems **J86-D-II-8**, pp. 1222-1233 (2003)
13. Over, P., Kraaij, W. and Smeaton, A. F.: TRECVID 2005 - An Introduction. TREC 2005 Proceedings (2005)
14. Pickering, M., Wong, L. and Rüger, S. M.: ANSES: Summarisation of News Video. CIVR 2003 Proceedings (2003)
15. Robertson, S. E. and Sparck Jones, K.: Simple, Proven Approaches to Text Retrieval. University of Cambridge Computer Laboratory, TR356 (1997)
16. Rui, Y., Gupta, A. and Acero, A.: Automatically Extracting Highlights for TV Baseball Programs. ACM Multimedia 2000 Proceedings (2000)
17. Sakai, T. *et al.*: Efficient Analysis of Student Questionnaires using Information Retrieval Techniques (in Japanese). Proceedings of the National Conference 2003/Spring of the Japan Society for Management Information, pp. 182-185 (2003)
18. Sakai, T. *et al.*: ASKMi: A Japanese Question Answering System based on Semantic Role Analysis. RIAO 2004 Proceedings, pp. 215-231 (2004)
19. Sakai, T.: Advanced Technologies for Information Access. International Journal of Computer Processing of Oriental Languages **18-2**, pp. 95-113 (2005)
20. Smeaton, A. F. *et al.*: The Físchlár-News-Stories System: Personalised Access to an Archive of TV News. RIAO 2004 Proceedings (2004)
21. Smith, M. A. and Kanade, T.: Video Skimming and Characterization through the Combination of Image and Language Understanding. IEEE ICCV '98 Proceedings (1998)
22. Uehara, T., Horikawa, M. and Sumita, K.: Navigation System for News Programs Featuring Direct Access to Desired Scenes (in Japanese). Toshiba Review **55-10** (2000)
23. Yamada, I. *et al.*: Meta-Data Generation for Football Games using Announcer's Commentary (in Japanese). Forum on Information Technology 2004 Proceedings, pp. 177-178 (2004)
24. Zhang. H.-J. *et al.*: Video Parsing, Retrieval and Browsing: An Integrated and Content-Based Solution. ACM Multimedia '95, pp. 15-24 (1995)

A Music Retrieval System Based on Query-by-Singing for Karaoke Jukebox

Hung-Ming Yu[1], Wei-Ho Tsai[2], and Hsin-Min Wang[1]

[1] Institute of Information Science, Academia Sinica, Taipei, Taiwan
[2] Dept. of Electronic Engineering, National Taipei Univ. of Technology, Taipei, Taiwan
donny@iis.sinica.edu.tw, whtsai@en.ntut.edu.tw, whm@iis.sinica.edu.tw

Abstract. This paper investigates the problem of retrieving Karaoke music by singing. The Karaoke music encompasses two audio channels in each track: one is a mix of vocal and background accompaniment, and the other is composed of accompaniment only. The accompaniments in the two channels often resemble each other, but are not identical. This characteristic is exploited to infer the vocal's background music from the accompaniment-only channel, so that the main melody underlying the vocal signals can be extracted more effectively. To enable an efficient and accurate search for a large music database, we propose a phrase onset detection method based on Bayesian Information Criterion (BIC) for predicting the most likely beginning of a sung query, and adopt a multiple-level multiple-pass Dynamic Time Warping (DTW) for melody similarity comparison. The experiments conducted on a Karaoke database consisting of 1,071 popular songs show the promising results of query-by-singing retrieval for Karaoke music.

Keywords: music information retrieval, Karaoke, query-by-singing.

1 Introduction

In recent years, out of the burgeoning amount of digital music circulating on the Internet, there has been an increasing interest in the research for music information retrieval (MIR). Instead of retrieving music with metadata, such as title, performer, and composer, it is desirable to locate music by simply humming or singing a piece of tune to the system. This concept has been extensively studied in various content-based music retrieval research [1-7], collectively called *query-by-humming* or *query-by-singing*.

Depending on the applications, the design of a music retrieval system varies with the type of music data. In general, digital music can be divided into two categories. One is the symbolic music represented by musical scores, e.g., MIDI and Humdrum. The second category relates to those containing acoustic signals recorded from real performances, e.g., CD music and MP3. This type of music is often polyphonic, in which many notes may be played simultaneously, in contrast to monophonic music, in which at most one note is played at any give time. Consequently, extracting the main melody directly from a polyphonic music proves to be a very challenging task [6-10],

H.T. Ng et al. (Eds.): AIRS 2006, LNCS 4182, pp. 445 – 459, 2006.

compared to dealing with the MIDI music, which is easy to acquire the main melody by selecting one of the symbolic tracks.

On the other hand, the development of a query-by-singing MIR system relies on an effective melody similarity comparison. Since most users are not professional singers, a sung query may contain inevitable tempo errors, note dropout errors, note insertion errors, etc. To handle these errors, various approximate matching methods, such as dynamic time warping (DTW) [5][11-12], hidden Markov model [13], and N-gram model [8][10], have been studied, with DTW being the most popular. However, due to the considerable time consumption for DTW, another key issue on designing a query-by-singing MIR system is how to speed up the similarity comparison, so that a large scale music database can be searched efficiently [5][14-15].

In this study, we focus on a sort of music data called *Karaoke*. It stems from Japanese popular entertainment, which provides prerecorded accompaniments to popular songs so that any users can sing live as a professional singer. Karaoke is gaining popularity in East Asia. Nowadays, there is a plenty of Karaoke music in either VCD or DVD format. Each piece of Karaoke track comes with two audio channels: one is a mix of vocal and accompaniment, and the other is composed of accompaniment only. The music in the accompaniment-only channel is usually very similar but not identical to that in the accompanied vocal channel. In this work, methods are proposed to extract vocal's melody from accompanied Karaoke tracks by reducing the interference from the background accompaniments. In parallel, we apply Bayesian Information Criterion (BIC) [16] to detect the onset time of each phrase in the accompanied vocal channel, which enables the subsequent DTW-based similarity comparison to be performed more efficiently. The proposed system further uses multiple-level multiple-pass DTW to improve the retrieval efficiency and accuracy. We evaluate our approaches on a Karaoke database consisting of 1,017 songs. The experimental results indicate the feasibility of retrieving Karaoke music by singing.

The remainder of this paper is organized as follows. Section 2 introduces the configuration of our Karaoke music retrieval system. Section 3 presents the methods for background music reduction and main melody extraction. Section 4 describes the phrase onset detection. In Section 5, we discuss the similarity comparison module and the schemes to improve the retrieval accuracy and efficiency. Finally, Section 6 presents our experimental results, and Section 7 concludes this study.

2 Overview

Our Karaoke music retrieval system is designed to take as input an audio query sung by a user, and to produce as output the song containing the most similar melody to the sung query. As shown in Fig. 1, it operates in two phases: indexing and searching.

2.1 Indexing

The indexing phase consists of two components: main melody extraction and phrase onset detection. The main melody extraction is concerned with the symbolic description of the melody related to the vocal sung in each song in the collection. Since the channel containing vocal signals also encompasses accompaniments, the melody extracted from raw audio data may not be the tune performed by a singer, but the instruments instead. To reduce the interference from the background accompaniments to main melody extraction, we propose exploiting the signal of accompaniment-only channel to approximate the vocal's background music. The desired vocal signal can thus be distilled by subtracting its background music. Then, the fundamental frequencies of the vocal signals are estimated, whereby converting the waveform representation into a sequence of musical note symbols.

The phrase onset detection aims to locate the expected beginning of a query that users would like to sing to the system. In view of the fact that the length of a popular song is normally several minutes, it is virtually impossible that a user sings a whole song as a query to the system. Further, a user's singing tends to begin with the initial of a sentence of lyrics. For instance, a user may query the system by singing a piece of *The Beatle's* "Yesterday" like this, "Suddenly, I'm not half to man I used to be. There's a shadow hanging over me." In contrast, a sung query like "I used to be. There's a shadow" or "half to man I used to be." is believed almost impossible. Therefore, pre-locating the phrase onsets could not only match users' queries better, but also improve the efficiency of the system in the searching phase.

After indexing, the database is composed of the note-based sequences and the labels of phrase onset times for each individual song.

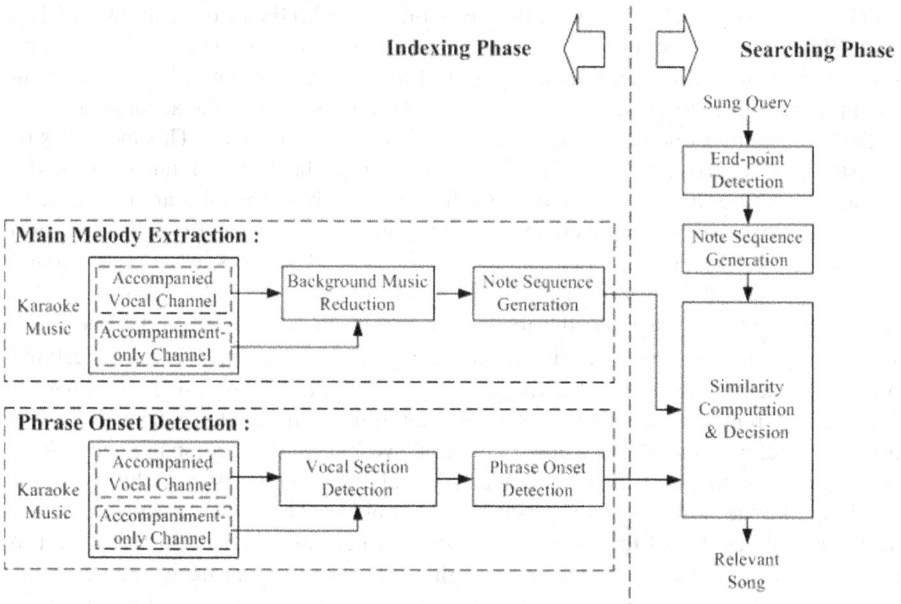

Fig. 1. The proposed Karaoke music retrieval system

2.2 Searching

In the searching phase, the system determines the song that a user looks for based on what he/she is singing. It is assumed that a user's sung query can be either a complete phrase or a partial phrase but starting from the beginning of a phrase. The system commences with the end-point detection that records the singing voice and marks the salient pauses within the singing waveform. Next, the singing waveform is converted into a sequence of note symbols via the note sequence generation module as used in the indexing phase. Accordingly, the retrieval task is narrowed down to a problem of comparing the similarity between the query's note sequence and each of the documents' note sequences. The song associated with the note sequence most similar to the query's note sequence is regarded as relevant and presented to the user.

3 Main Melody Extraction

3.1 Background Music Reduction

Main melody extraction plays a crucial role in music information retrieval. In contrast to the retrieval of MIDI music, which is easy to acquire the main melody by selecting one of the symbolic tracks, retrieving a polyphonic object in CD or Karaoke format requires to extract the main melody directly from the accompanied singing signals, which proves difficult to handle well simply using the conventional pitch estimation. For this reason, a special effort in this module is put on reducing the background music from the accompanied singing signals.

The indexing phase begins with the extraction of audio data from Karaoke VCD's or DVD's. The data consists of two-channel signals: one is a mix of vocal and accompaniment, and the other is composed of accompaniment only. Fig. 2 shows an example waveform for a Karaoke music piece. It is observed that the accompaniments in the two channels often resemble each other, but are not identical. This motivates us to estimate the pristine vocal signal by inferring its background music from the accompaniment-only channel. To do this, the signals in both channels are first divided into frames by using a non-overlapping sliding window with length of W waveform samples. Let $\mathbf{c}_t=\{c_{t,1}, c_{t,2},\ldots, c_{t,W}\}$ denote the t-th frame of samples in the accompanied vocal channel and $\mathbf{m}_t=\{m_{t,1}, m_{t,2},\ldots, m_{t,W}\}$ denote the t-th frame of the accompaniment-only channel. It can be assumed that $\mathbf{c}_t = \mathbf{s}_t + a\mathbf{m}'_t$, where $\mathbf{s}_t=\{s_{t,1}, s_{t,2},\ldots, s_{t,W}\}$ is the pristine vocal signal and $\mathbf{m}'_t=\{m'_{t,1}, m'_{t,2},\ldots, m'_{t,W}\}$ is the underlying background music. Usually, $\mathbf{m}'_t \neq \mathbf{m}_t$, since the accompaniment signals in one channel may be different from another in terms of amplitude, phase, etc., where the phase difference reflects the asynchronism between two channels' accompaniments. As a result, direct subtraction of one channel's signal from another's is of little use for distilling the desired vocal. To handle this problem better, we assume that the accompanied vocal is of the form $\mathbf{c}_t = \mathbf{s}_t + a\mathbf{m}_{t+b}$, where $\mathbf{m}_{t+b}$ is the b-th frame next to $\mathbf{m}_t$, which is most likely corresponding to $\mathbf{m}'_t$, and a is a scaling factor reflecting the amplitude difference between $\mathbf{m}_t$ and $\mathbf{m}'_t$. The optimal b can be found by choosing one of the possible values within a pre-set range, B, that results in the smallest estimation error, i.e.,

$$b^* = \arg\min_{-B\le b\le B} |\, \mathbf{c}_t - a_b^* \mathbf{m}_{t+b} \,|, \tag{1}$$

where a_b^* is the optimal amplitude scaling factor given $\mathbf{m}_{t+b}$. Letting $\partial|\mathbf{c}_t - a_b\mathbf{m}_{t+b}|^2/\partial a_b = 0$, we have a minimum mean-square-error solution of a_b as

$$a_b^* = \frac{\mathbf{c}_t' \mathbf{m}_{t+b}}{\| \mathbf{m}_{t+b} \|^2}. \tag{2}$$

Accordingly, the underlying vocal signal in frame t can be estimated by $\mathbf{s}_t = \mathbf{c}_t - a_b^* \mathbf{m}_{t+b^*}$. Fig. 2(c) shows the resulting waveform of the accompanied vocal channel after background music reduction. We can see that the accompaniment in the accompanied vocal channel is largely reduced.

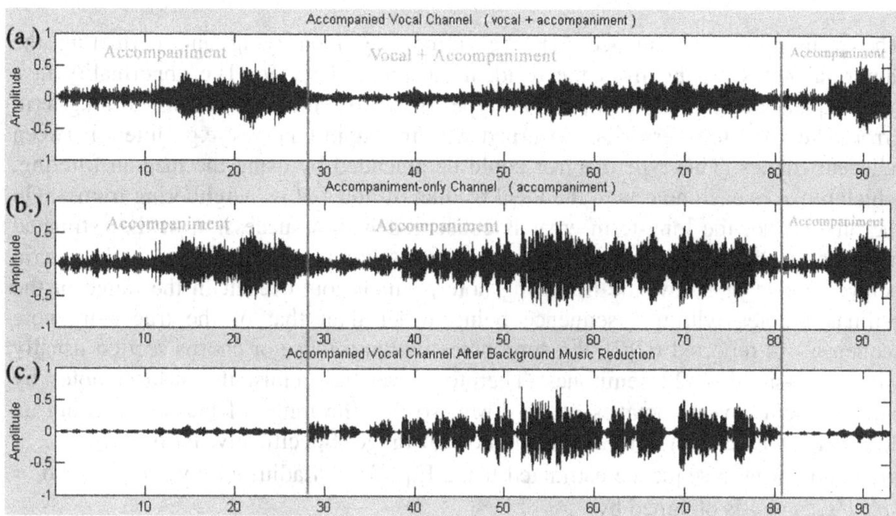

Fig. 2. (a) An accompanied vocal channel. (b) An accompaniment-only channel. (c) The accompanied vocal channel after background music reduction.

3.2 Note Sequence Generation

After reducing the undesired background accompaniments, the next step is to convert each recording from its waveform representation into a sequence of musical notes. Following [7], the converting method begins by computing the short-term Fast Fourier Transform (FFT) of a signal. Let $e_1, e_2, \ldots, e_N$ be the inventory of possible notes performed by a singer, and $x_{t,j}$ denote the signal's energy with respect to FFT index j in frame t, where $1 \leq j \leq J$. The sung note of a recording in frame t is determined by

$$o_t = \arg\max_{1 \leq n \leq N} \left(\sum_{c=0}^{C} h^c \, y_{t,n+12c} \right), \tag{3}$$

where C is a pre-set number of harmonics concerned, h is a positive value less than 1 for discounting higher harmonics, and $y_{t,n}$ is the signal's energy on note e_n in frame t, estimated by

$$y_{t,n} = \arg\max_{\forall j, U(j)=e_n} x_{t,j}, \tag{4}$$

and

$$U(j) = \left\lfloor 12 \cdot \log_2\left(\frac{F(j)}{440}\right) + 69.5 \right\rfloor, \tag{5}$$

where $\lfloor\ \rfloor$ is a floor operator, $F(j)$ is the corresponding frequency of FFT index j, and $U(\cdot)$ represents a conversion between the FFT indices and the MIDI note numbers.

3.3 Note Sequence Smoothing

The resulting note sequence may be refined by identifying and correcting the abnormal notes arising from the residual background music. The abnormality in a note sequence can be divided into two types of errors: short-term error and long-term error. The short-term error is concerned with the rapid changes, e.g., jitters between adjacent frames. This type of error could be amended by using the median filtering, which replaces each note with the local median of notes of its neighboring frames. On the other hand, the long-term error is concerned with a succession of the estimated notes not produced by a singer. These successive wrong notes are very likely several octaves above or below the true sung notes, which could result in the range of the estimated notes within a sequence being wider than that of the true sung note sequence. As reported in [7], the sung notes within a verse or chorus section usually vary no more than 22 semitones. Therefore, we may adjust the suspect notes by shifting them several octaves up or down, so that the range of the notes within an adjusted sequence can conform to the normal range. Specifically, let $\mathbf{o} = \{o_1, o_2,\ldots, o_T\}$ denote a note sequence estimated using Eq. (3). An adjusted note sequence $\mathbf{o}' = \{o'_1, o'_2,\ldots, o'_T\}$ is obtained by

$$o'_t = \begin{cases} o_t & , \text{ if } |o_t - \bar{o}| \leq (R/2) \\ o_t - 12 \times \left\lfloor \dfrac{o_t - \bar{o} + R/2}{12} \right\rfloor, & \text{if } o_t - \bar{o} > (R/2) \\ o_t - 12 \times \left\lfloor \dfrac{o_t - \bar{o} - R/2}{12} \right\rfloor, & \text{if } o_t - \bar{o} < (-R/2) \end{cases}, \tag{6}$$

where R is the normal varying range of the sung notes in a sequence, say 22, and $\bar{o}$ is the mean note computed by averaging all the notes in $\mathbf{o}$. In Eq. (6), a note o_t is considered as a wrong note and needs to be adjusted if it is too far away from $\bar{o}$, i.e., $|o_t - \bar{o}| > R/2$. The adjustment is done by shifting the wrong note $\lfloor (o_t - \bar{o} + R/2)/12 \rfloor$ or $\lfloor (o_t - \bar{o} - R/2)/12 \rfloor$ octaves.

4 Phrase Onset Detection

In general, the structure of a popular song involves five sections: *intro*, *verse*, *chorus*, *bridge*, and *outro*. The verse and chorus contain the vocals sung by the lead singer, while the intro, bridge, and outro are largely accompaniments. This makes it natural that a verse or chorus is the favorite that people go away humming when they hear a good song, and hence is often the query that a user may hum or sing to a music retrieval system.

Since a user's singing query tends to begin with the initial of a sentence in lyrics, we can consider a song's lyrics as a collection of phrases. The beginning of each phrase is likely the starting point of a user's query. Therefore, if the onset time of each phrase can be detected before the DTW comparison is performed, it is expected that both the search efficiency and the retrieval accuracy can be improved.

Our strategy for detecting the phrase onsets is to locate the boundaries that the signal in the accompanied vocal channel is changed from accompaniment-only to a mix of vocal and accompaniment. As mentioned earlier, the accompaniment of one channel in a Karaoke track often resembles that of the other channel. Thus, if no vocal is performed in a certain passage, the difference of signal spectrum between the two channels is tiny. In contrast, if a certain passage contains vocal signals, there must be a significant difference between the two channels during this passage. We therefore could examine the difference of signal spectrum between the two channels, thereby locating the phrase onsets. In our system, the Bayesian Information Criterion (BIC) [16] is applied to characterize the level of spectrum difference.

4.1 The Bayesian Information Criterion (BIC)

The BIC is a model selection criterion which assigns a value to a stochastic model based on how well the model fits a data set, and how simple the model is. Given a data set $\mathbf{X} = \{\mathbf{x}_1, \mathbf{x}_2,\ldots, \mathbf{x}_N\} \subset R^d$ and a model set $\mathbf{H} = \{H_1, H_2,\ldots, H_K\}$, the BIC value for model H_k is defined as:

$$BIC(H_k) = \log p(\mathbf{X}|H_k) - 0.5\,\lambda\,\#(H_k)\log N, \tag{7}$$

where λ is a penalty factor, $p(\mathbf{X}|H_k)$ is the likelihood that H_k fits $\mathbf{X}$, and $\#(H_k)$ is the number of free parameters in H_k. The selection criterion favors the model having the largest value of BIC.

Assume that we have two audio segments represented by feature vectors, $\mathbf{X} = \{\mathbf{x}_1, \mathbf{x}_2,\ldots, \mathbf{x}_N\}$ and $\mathbf{Y} = \{\mathbf{y}_1, \mathbf{y}_2,\ldots, \mathbf{y}_N\}$, respectively. If it is desired to determine whether $\mathbf{X}$ and $\mathbf{Y}$ belong to the same acoustic class, then we have two hypotheses to consider: one is "yes", and the other is "no". Provided that hypotheses "yes" and "no" are characterized by a certain stochastic models H_1 and H_2, respectively, our aim will be to judge which among H_1 and H_2 is better. For this purpose, we represent H_1 by a single Gaussian distribution $\mathcal{N}(\boldsymbol{\mu},\boldsymbol{\Sigma})$, where $\boldsymbol{\mu}$ and $\boldsymbol{\Sigma}$ are the sample mean and covariance estimated using vectors $\{\mathbf{x}_1, \mathbf{x}_2,\ldots, \mathbf{x}_N, \mathbf{y}_1, \mathbf{y}_2,\ldots, \mathbf{y}_N\}$, and represent H_2 by two Gaussian distributions $\mathcal{N}(\boldsymbol{\mu}_x,\boldsymbol{\Sigma}_x)$ and $\mathcal{N}(\boldsymbol{\mu}_y,\boldsymbol{\Sigma}_y)$, where the sample mean $\boldsymbol{\mu}_x$ and covariance $\boldsymbol{\Sigma}_x$ are estimated using vectors $\{\mathbf{x}_1, \mathbf{x}_2,\ldots, \mathbf{x}_N\}$, and the sample mean $\boldsymbol{\mu}_y$ and covariance $\boldsymbol{\Sigma}_y$ are estimated using vectors $\{\mathbf{y}_1, \mathbf{y}_2,\ldots, \mathbf{y}_N\}$. Then, the problem of judging which model is better can be solved by computing a difference value of BIC between $BIC(H_1)$ and $BIC(H_2)$, i.e.,

$$\Delta BIC = BIC\,(H_2) - BIC(H_1). \tag{8}$$

Obviously, the larger the value of ΔBIC, the more likely segments $\mathbf{X}$ and $\mathbf{Y}$ are from different acoustic classes, and vice versa. We can therefore set a threshold of ΔBIC to determine if two audio segments belong to the same acoustic class.

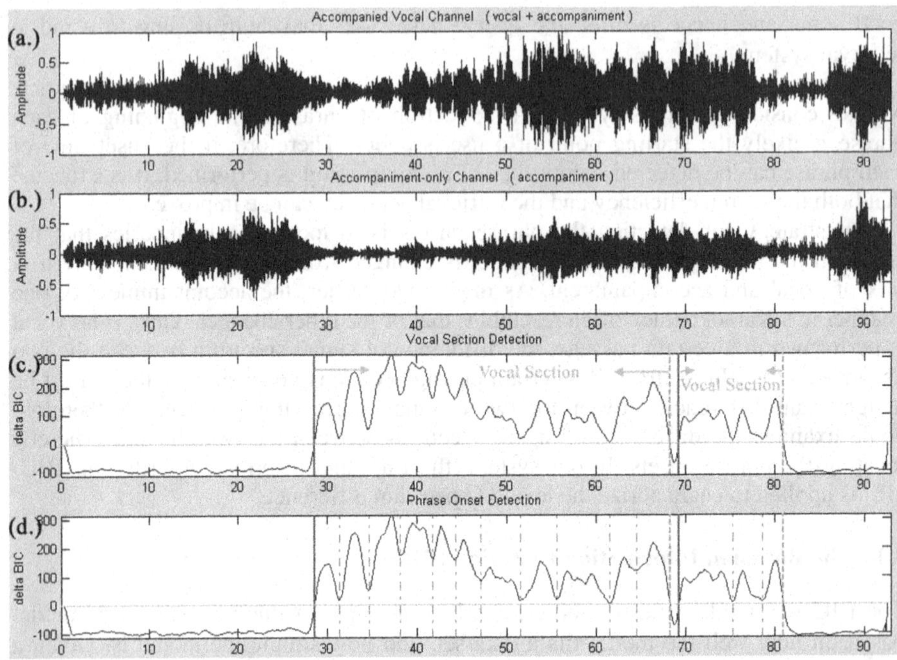

Fig. 3. An example of phrase onset detection. (a) The waveform of an accompanied vocal channel. (b) The waveform of an accompaniment-only channel. (c) The ΔBIC curve and the detected vocal sections. (d) The results of phrase onset detection.

4.2 Phrase Onset Detection Via BIC

In applying the concept of BIC to the phrase onset detection problem, our goal is to judge whether the signals in the two channels belong to the same acoustic class during a certain time interval, where one class represents accompaniment only, and the other represents vocal over accompaniment. Thus, by considering **X** and **Y** as two concurrent channels' signals, ΔBIC can be computed along the entire recording, and then plotted as a curve. Fig. 3 shows an example Karaoke music clip underwent our phrase onset detection. Figs. 3(a) and 3(b) are the waveforms in the accompanied vocal channel and the accompaniment-only channel, respectively. The phrase onset detection begins by chopping the waveform in each of the channels into non-overlapping frames of 20ms. Each frame is represented as 12 Mel-scale Frequency Cepstral Coefficients (MFCCs). Then, the ΔBIC value between each pair of one-second segments in the two channels is computed frame by frame, thereby forming a ΔBIC curve over time, as shown in Fig. 3(c). The positive value of ΔBIC indicates that the frame of the accompanied vocal channel contains vocals but the concurrent frame in the accompaniment-only channel does not. In contrast, the negative value of ΔBIC indicates that both frames contain accompaniments only. Therefore, the intervals where positive values of ΔBIC appear are identified as vocal sections.

In the example shown in Fig. 3(c), two vocal sections are identified, each surrounded by a solid line and a dashed line. Within each vocal section, the local minimums on the ΔBIC curve can be further located as phrase onsets because the frame corresponds to rest or breathing between phrases usually yields a small ΔBIC value. Fig. 3 (d) depicts the detected phrase onsets in this way. Note that the trade-off between the retrieval accuracy and retrieval efficiency are highly dependent on the number of detected phrase onsets. In general, the larger the number of the likely phrase onsets is detected, the higher the retrieval accuracy can be achieved. However, increasing the number of the candidate phrase onsets often decreases the retrieval efficiency drastically.

5 Melody Similarity Comparison

Given a user's sung query and a set of music documents, each of which is represented by a note sequence, the task here is to find a music document whose partial note sequence is most similar to the query's note sequence.

5.1 Dynamic Time Warping Framework

Let $\mathbf{q} = \{q_1, q_2,..., q_T\}$ and $\mathbf{u} = \{u_1, u_2,..., u_L\}$ be the note sequences extracted from a user's query and a particular music segment to be compared, respectively. As the lengths of **q** and **u** are usually different, computing the distance between **q** and **u** directly is infeasible. To handle this problem, the most prevalent way is to find the temporal mapping between **q** and **u** by Dynamic Time Warping (DTW). Mathematically, DTW constructs a $T{\times}L$ distance matrix $\mathbf{D} = [D(t,\ell)]_{T \times L}$, where $D(t,\ell)$ is the distance between note sequences $\{q_1, q_2,..., q_t\}$ and $\{u_1, u_2,..., u_\ell\}$, computed using:

$$D(t,\ell) = \min\begin{cases} D(t-2,\ell-1) + 2\times d(t,\ell) \\ D(t-1,\ell-1) + d(t,\ell) - \varepsilon \\ D(t-1,\ell-2) + d(t,\ell) \end{cases}, \tag{9}$$

and

$$d(t,\ell) = |\, q_t - u_\ell |, \tag{10}$$

where ε is a small constant that favors the mapping between notes q_t and u_ℓ, given the distance between note sequences $\{q_1, q_2,...,q_{t\text{-}1}\}$ and $\{u_1, u_2,..., u_{\ell\text{-}1}\}$.

To compensate for the inevitable errors arising from the phrase onset detection, we assume that the true phrase onset time associated with the automatically detected phrase onset time t_{os} is within $[t_{os}\text{-}r/2,\ t_{os}+r/2]$, where r is a predefined tolerance. Though the DTW recursion in Eq. (9) indicates that the best path exists only when the length of the music note sequence is within half to twice length of the query note sequence (i.e., between $T/2$ and $2T$), we prefer to limit the best mapping length of **u** to **q** to be between $T/2$ and kT, where k is a value between 1 and 2, so that the mapping can be more precisely. In other words, the tempo of the query is allowed to be $1/k$ times to twice the tempo of the target music document. Therefore, in the implementation, we set the new phrase onset time t'_{os} as $t_{os}\text{-}r/2$, clone the subsequence of notes of a music document starting from t'_{os} with a length L of $kT+r$ to **u**, and define the boundary conditions for the DTW recursion as,

$$\begin{cases} D(1,1) = d(1,1) \\ D(t,1) = \infty,\ 2 \le t \le T \\ D(t,2) = \infty,\ 4 \le t \le T \\ D(1,\ell) = \begin{cases} d(1,\ell), & 1 \le \ell \le r \\ \infty, & r < l \le L \end{cases} \\ D(2,\ell) = \begin{cases} d(1,\ell-1) + d(2,\ell), & 2 \le \ell \le r+1 \\ \infty, & r+1 < l \le L \end{cases} \\ D(3,2) = d(1,1) + 2 \times d(3,2) \end{cases} . \tag{11}$$

After the distance matrix **D** is constructed, the similarity between **q** and **u** can be evaluated by

$$S(\mathbf{q},\mathbf{u}) = \min_{T/2 \le \ell \le L} D(T,\ell). \tag{12}$$

5.2 Multiple-Pass DTW to Improve Retrieval Accuracy

Since a query may be sung in a different key or register than the target music document, i.e., the so-called *transposition*, the resulting note sequences of the query and the document could be rather different. This problem can be alleviated by shifting the query's note sequence upward or downward several semitones, so that the mean of the shifted query's note sequence can equal that of the document to be compared. In addition, considering that a user's transposition or key change may occur in a partial sung query, we further perform multiple DTW similarity comparisons by shifting a query sequence upward or downward v semitones. The distance $S(\mathbf{q},\mathbf{u})$ is then defined as,

$$S(\mathbf{q},\mathbf{u}) = \min_{-V \le v \le V} S(\mathbf{q}^{(v)},\mathbf{u}), \tag{13}$$

where $\mathbf{q}^{(v)}$ denotes the query sequence obtained by shifting **q** upward or downward v semitones. As reported in [7], the retrieval performance improves as the value of V increases. However, increasing the value of V substantially increases computational costs, because the similarity comparison requires two extra DTW operations whenever the value of V is increased by one. Thus, an economic value of $V = 1$, i.e., three-pass DTW, is adopted in this work.

In addition to the difference of key and tempo existing between queries and documents, another problem to be addressed is the existence of voiceless regions in a sung query. The voiceless regions, which may arise from the rest, pause, etc., result in some notes being tagged with "0" in the query note sequence. However, the corresponding non-vocal regions in the document are usually not tagged with "0", because there are accompaniments in those regions. Although the voiceless regions in a sung query can be detected by simply using the energy information, the accurate detection of non-vocal regions in a music document remains a very difficult problem. To sidestep this problem, we modify the computation of $d(t,\ell)$ in Eq. (10) to

$$d(t,\ell)=\begin{cases}|q_t-u_\ell|, & q_t\neq 0\\ \varphi, & q_t=0\end{cases}, \tag{14}$$

where φ is a small constant. Implicit in Eq. (14) is equivalent to bypassing the voiceless regions of a query.

5.3 Multiple-Level DTW to Improve Retrieval Efficiency

As shown in Fig. 4(a), the computational complexity in terms of the number of the necessary distance computation $D(\cdot)$ for constructing a $T\times L$ table is

$$\begin{aligned} Complexity &= T\times L-\frac{T\times T/2}{2}-\frac{L\times L/2}{2} \\ &= TL-\frac{(T^2+L^2)}{4}. \end{aligned} \tag{15}$$

As mentioned earlier, since L is usually set to be kT, where $1/2\leq k\leq 2$, the computational complexity can be rewritten as

$$Complexity=\frac{(4k-k^2-1)T^2}{4}. \tag{16}$$

Although the DTW recursion allows a document sequence within half to twice the length of the query sequence, empirical evidence shows that the document length can be simply limited to 1.2 times the length of a query, i.e., $k = 1.2$, without significantly degrading the retrieval performance. Hence, by substituting $k = 1.2$ into Eq. (16), the computational complexity is $0.59T^2$. As shown in Fig. 4(b), the introduction of a phrase onset tolerance in the DTW will increase the computational complexity by rT. If r is small compared to T, the increase in complexity is negligible. If $r = 0.59T$, the computational complexity is twice that of the typical DTW.

Since the complexity is $O(T^2)$, the most promising way to speed up the searching process is to reduce the value of T. Motivated by Keogh and Pazzani's Piecewise Aggregate Approximation (PAA) [14], we propose a dimensionality reduction technique, called Multi-Level Data Abstraction (MLDA). Unlike PAA, which divides a time series into equal-length frames, and then calculates the mean value of the data falling within a frame, MLDA aims to prune the less likely music clips in a step-by-step manner. In MLDA, the compression rate of data is power of two at each level. For example, if we go through the note sequence and pick one note every two notes, the note sequence is reduced to half length, while the computation is reduced to quarter complexity. We can first use the reduced note sequences to prune less likely music documents. If the original computational complexity is CC and the pruning rate is R_{pr}, then the total computational complexity of the two-level DTW is $[1/4+(1-R_{pr})]CC$. If a three-level DTW is applied, the complexity is reduced to $[(1/4)^2+(1/4)(1-R_{pr})+(1-R_{pr})^2]CC$. In this way, when the pruning rate R_{pr} is set at 0.75, the complexity of the two-level DTW and three-level DTW is 1/2 and 3/16 that of the single-level DTW, respectively.

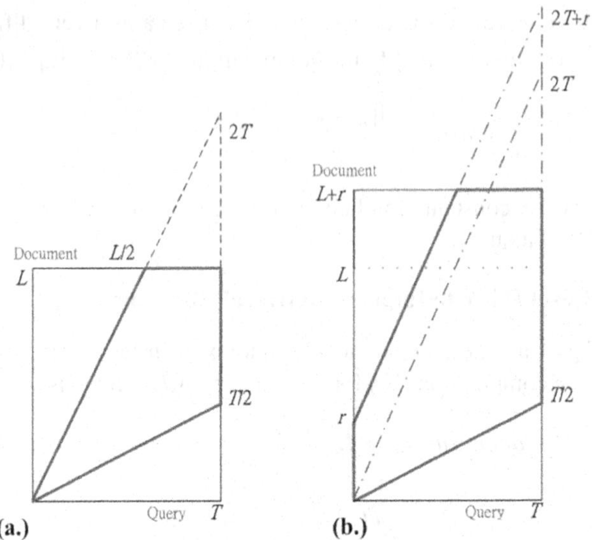

Fig. 4. (a) The search space for the typical DTW, in which the starting frame of a query and that of the target phrase are aligned; (b) The search space for the DTW we apply, in which the starting frame of a query can be mapped to any one of the first *r* frames of the target phrase

6 Experiments

6.1 Music Database

The music database used in this study consisted of 1,071 songs extracted from Karaoke VCDs. The extracted waveform signals were down-sampled from the sampling rate of 44.1 kHz to 22.05 kHz. The database was divided into two subsets. The first subset consisted of 95 songs, denoted as DB-1. The second subset consisted of 976 songs, denoted as DB-2. To compare the system performances achieved with automatic and manual detection of phrase onset, we manually labeled the phrase onsets of the songs in DB-1. There were 775 phrase onsets marked.

We collected 90 queries from 9 male and 4 female users. The duration of each query ranged from 15 seconds to 45 seconds, but only the first 8-second portion was input to the system. The performance was measured on the basis of *song accuracy* defined as,

$$Song\ accuracy(\%) = \frac{\#queries\ receiving\ the\ correct\ songs}{\#queries} \times 100\%. \tag{17}$$

In addition, considering a more user-friendly scenario where a list of Top-N ranked documents can be provided for user's choices, we also computed the Top-N accuracy defined as the percentage of the queries whose target songs are among Top-N. The overall system performance was evaluated on both DB-1 and DB-2.

6.2 Experimental Results

The first experiment was conducted to evaluate the effectiveness of the background music reduction using DB-1. Here, the phrase onsets were labeled manually. The retrieval performance is given in Table 1. It is clear that the background music reduction improves the retrieval performance.

Table 1. Retrieval performance obtained with and without background music reduction

	Song accuracy (%)		
	Top 1	Top 3	Top 10
without background music reduction	56.67	67.78	76.67
with background music reduction	72.22	76.67	83.33

The second experiment was conducted to compare the retrieval performance obtained with the manual phrase onset labeling and the automatic phrase onset detection. The automatic phrase onset detection approach marked 2,719 phrase onsets in the 95 songs in DB-1, which is about 3.5 times that of hand-labeled phrase onsets. The retrieval performance is given in Table 2. We observe that the retrieval accuracy only decreases slightly when the way to mark the phrase onsets was changed from manual to automatic.

Table 2. Retrieval performance based on manual and automatic phrase onset detections

	Song accuracy (%)		
	Top 1	Top 3	Top 10
Manual phrase onset labeling	72.22	76.67	83.33
Automatic phrase onset detection	70.00	74.44	77.78

Lastly, we evaluated the performance of the Karaoke retrieval system using both DB-1 and DB-2, with the same 90 queries. The system automatically marked 27,397 phrase onsets in the 1,017 songs in DB-1 and DB-2. The retrieval performance is given in Table 3. We observe that the Top-1 accuracy drops from 70.00% to 51.11% as the database expands from 95 songs to 1,071 songs, while the Top-10 accuracy only slightly drops from 77.78% to 70.00%. To speed up the searching, the four-level DTW, with various pruning rates, R_{pr}, was implemented. We observe from Table 3 that the searching time for the multiple-level DTW can be greatly reduced at a small cost of retrieval accuracy degradation.

Table 3. Retrieval performance evaluated using both DB-1 and DB-2

	Song accuracy (%)		
	Top 1	Top 3	Top 10
Single-level DTW, complexity *CC*	51.11	57.78	70.00
Four-level DTW, $R_{pr} = 0.6$ (complexity reduced to 0.145 *CC*)	51.11	57.78	67.78
Four-level DTW, $R_{pr} = 0.75$ (complexity reduced to 0.063 *CC*)	50.00	55.56	65.56
Four-level DTW, $R_{pr} = 0.9$ (complexity reduced to 0.025 *CC*)	46.67	54.44	63.33

7 Conclusions

We have presented a Karaoke music retrieval system that allows users to locate their desired music by singing to the system. Since the vocals and various concurrent accompaniments are mixed together in an accompanied vocal channel, we proposed a method to reduce the accompaniments in the accompanied vocal channel so that the accuracy of main melody extraction could be improved. In addition, we applied Bayesian Information Criterion (BIC) to detect the onset time of a musical phrase which reflects the most likely beginning of a sung query. The phrase onset detection, in conjunction with multiple-level multiple-pass DTW matching for similarity comparison, enables an efficient and effective search for a large music database. The experiments conducted on a music database consisting of 1,017 songs confirmed the feasibility of our retrieval system.

Acknowledgments. This work was supported in part by the National Science Council, Taiwan, under Grants: NSC94-2422-H-001-007 and NSC95-2422-H-001-008.

References

1. Ghias, A., H. Logan, D. Chamberlin, and B. C. Smith, "Query by Humming: Musical Information Retrieval in an Audio Database," *Proc. ACM International Conference on Multimedia*, 1995.
2. Kosugi, N., Y. Nishihara, T. Sakata, M. Yamamuro, and K. Kushima, "Music Retrieval by Humming," *Proc. IEEE Pacific Rim Conference on Communications, Computers and Signal Processing*, 1999.
3. Kosugi, N., Y. Nishihara, T. Sakata, M. Yamamuro, and K. Kushima, "A Practical Query-By-Humming System for a Large Music Database," *Proc. ACM International Conference on Multimedia*, 2000.
4. Nishimura, T., H. Hashiguchi, J. Takita, J. X. Zhang, M. Goto, and R. Oka, "Music Signal Spotting Retrieval by a Humming Query Using Start Frame Feature Dependent Continuous Dynamic Programming," *Proc. International Symposium on Music Information Retrieval,* 2001.

5. Jang, J. S. Roger, and H. R. Lee, "Hierarchical Filtering Method for Content-based Music Retrieval via Acoustic Input," *Proc. ACM International Conference on Multimedia,* 2001.
6. Song, J., S. Y. Bae, and K. Yoon, "Mid-Level Music Melody Representation of Polyphonic Audio for Query-by-Humming System," *Proc. International Conference on Music Information Retrieval*, 2002.
7. Yu, H. M., W. H. Tsai, and H. M. Wang, "A Query-by-singing Technique for Retrieving Polyphonic Objects of Popular Music," *Proc. Asian Information Retrieval Symposium*, 2005
8. Doraisamy, S. and S. M. Ruger, "An Approach Towards a Polyphonic Music Retrieval System," *Proc. International Symposium on Music Information Retrieval,* 2001.
9. Goto, M., "A Predominant-F0 Estimation Method for CD Recordings: MAP Estimation Using EM Algorithm for Adaptive Tone Models," *Proc. IEEE International Conference on Acoustics, Speech, and Signal Processing,* 2001.
10. Doraisamy, S. and S. Rüger, "Robust Polyphonic Music Retrieval with *N*-grams," *Journal of Intelligent Information Systems*, 21(1), pp. 53–70, 2003.
11. Liu, C. C., A. J. L. Hsu, and A. L. P. Chen, "An Approximate String Matching Algorithm for Content-Based Music Data Retrieval," *Proc. IEEE International Conference on Multimedia Computing and Systems,* 1999.
12. Mo, J. S., C. H. Han, and Y. S. Kim, "A Melody-Based Similarity Computation Algorithm for Musical Information," *Proc. Workshop on Knowledge and Data Engineering Exchange*, 1999.
13. Shifrin, J. and W. Burmingham, "Effectiveness of HMM-based Retrieval on Large Databases," *Proc. International Conference on Music Information Retrieval*, 2003.
14. Keogh, E. and M. Pazzani, "Scaling up Dynamic Time Warping for Datamining Applications," *Proc. ACM SIGKDD*, 2000.
15. Salvador, S. and P. Chan, "FastDTW: Toward Accurate Dynamic Time Warping in Linear Time and Space," *Proc. KDD Workshop on Mining Temporal and Sequential Data*, 2004
16. Schwarz, G., "Estimation the Dimension of a Model," *The Annals of Statistics*, 6, pp. 461-364, 1978.

A Semantic Fusion Approach Between Medical Images and Reports Using UMLS

Daniel Racoceanu[1,2], Caroline Lacoste[1], Roxana Teodorescu[1,3], and Nicolas Vuillemenot[1,4]

[1] IPAL-Image Perception, Access and Language - UMI-CNRS 2955
Institute for Infocomm Research, A*STAR, Singapore
{visdaniel, viscl, sturot}@i2r.a-star.edu.sg
http://www.i2r.a-star.edu.sg/
[2] University of Franche-Comte, Besancon, France
[3] "Politehnica" University from Timisoara, Romania
[4] Ecole Nationale Superieure de Mecaniques et Microtechniques de Besancon, France

Abstract. One of the main challenges in content-based image retrieval still remains to bridge the gap between low-level features and semantic information. In this paper, we present our first results concerning a medical image retrieval approach using a semantic medical image and report indexing within a fusion framework, based on the Unified Medical Language System (*UMLS*) metathesaurus. We propose a structured learning framework based on Support Vector Machines to facilitate modular design and extract medical semantics from images. We developed two complementary visual indexing approaches within this framework: a global indexing to access image modality, and a local indexing to access semantic local features. Visual indexes and textual indexes - extracted from medical reports using *MetaMap* software application - constitute the input of the late fusion module. A weighted vectorial norm fusion algorithm allows the retrieval system to increase its meaningfulness, efficiency and robustness. First results on the CLEF medical database are presented. The important perspectives of this approach in terms of semantic query expansion and data-mining are discussed.

1 Introduction

Many programs and tools have been developed to formulate and execute queries based on the visual content and to help browsing large multimedia repositories. Still, no general breakthrough has been achieved with respect to large varied databases with exogenous documents. Many questions with respect to speed, semantic descriptors or objective image interpretations are still unanswered. In the medical field, digital images are produced in ever-increasing quantities and used for diagnostics and therapy. With digital imaging and communications in medicine (DICOM), a standard for image communication has been set and patient information can be stored with the actual image(s), although still a few problems prevail with respect to the standardization. In several articles, content-based access to medical images for supporting clinical decision-making has been

H.T. Ng et al. (Eds.): AIRS 2006, LNCS 4182, pp. 460–475, 2006.

proposed that would ease the management of clinical data and scenarios for the integration of content-based access methods into picture archiving and communication systems (PACS) have been created [16].

There are several reasons why there is a need for additional, alternative image retrieval methods apart from the steadily growing rate of image production. For the clinical decision making process it can be beneficial or even important to find other images of the same modality, the same anatomic region of the same disease. Although part of this information is normally contained in the DICOM headers and many imaging devices are DICOM-compliant at this time, there are still some problems. DICOM headers contain a high rate of errors: error rates of 16% have been reported by [9] for the field anatomical region. This can hinder the correct retrieval of all wanted images. Clinical decision support techniques such as case-based reasoning [12] or evidence-based medicine [4] can even produce a stronger need to retrieve images that can be valuable for supporting certain diagnoses. It could even be imagined to have Image-Based Reasoning (IBR) as a new discipline for diagnostic aid. Decision support systems in radiology [10] and computer-aided diagnostics for radiological practice as demonstrated at the RSNA (Radiological Society of North America) are on the rise and create a need for powerful data and meta-data management and retrieval [1].

It needs to be stated that the purely visual image queries as they are executed in the computer vision domain will most likely not be able to ever replace text-based methods as there will always be queries for all images of a certain patient, but they have the potential to be a very good complement to text-based search based on their characteristics. Still, the problems and advantages of the technology have to be stressed to obtain acceptance and use of visual and text-based access methods up to their full potential.

Besides diagnostics, teaching and research especially are expected to improve through the use of visual access methods as visually interesting images can be chosen and can actually be found in the existing large repositories. The inclusion of visual features into medical studies is another interesting point for several medical research domains. Visual features do not only allow the retrieval of cases with patients having similar diagnoses but also cases with visual similarity but different diagnoses. In teaching it can help lecturers as well as students to browse educational image repositories and visually inspect the results found.

According to those remarks, this study introduces a semantic indexing approach of the medical report and the medical image, according to the concepts associated to an unified medical modeling system. This approach give a complementary description of the textual and visual characteristic of a medical case, using a balanced weighted vectorial norm fusion method at the conceptual level, by taking into account the confidence, the localization and the frequency of the associated concepts.

The remainder of this paper is organized as follows. Section 2 provides a brief state of the art in the content-based image retrieval, focusing on text and image fusion. In section 3, we introduce the semantic indexing approach, by presenting the main ideas used to close the semantic gap and to develop a complementary text-image fusion approach. Finally, some operational approaches and results are synthesized in the section 4 in CLEF [1] benchmark context, by extracting the main points for the conclusions and the future developments.

2 Brief Overview of the Content-Based Image Retrieval and Fusion Methods

Although access methods in image databases already existed in the beginning of the 1980s [5], the content-based image retrieval (CBIR) started in the 1990s [17],[8] and has been an extremely active research area over the last 10 years [16]. There is growing interest in CBIR because of the limitations inherent in metadata-based systems. Textual information about images can be easily searched using existing technology, but requires humans to personally describe every image in the database. Moreover, it is possible to miss images that use different synonyms in their descriptions.

Content-based image retrieval (CBIR) is the application of computer vision to the image retrieval problem, that is, the problem of searching for digital images in large databases. "Content-based" means that the search makes use of the contents of the images themselves, rather than relying on human-imputed metadata such as captions or keywords.

The visual features, used for indexing and retrieval, are classified in [11] into three classes:

- primitive features that are low-level features such as color, shape, and texture;
- logical features that are medium-level features describing the image by a collection of objects and their spatial relationships;
- abstract feature that are semantic/contextual features.

Current CBIR systems generally make use of primitive features [17], [18]. However, some general semantic layers are insufficient to model medical knowledge. Consequently results are poor when common algorithms are general system offers interpretation of images or even medium level concepts as they can easily be captured with text. This loss of information from image data to a representation by features is called the semantic gap [21]. To reduce this semantic gap, specialized retrieval systems have been proposed in literature [11],[20],[6]. Indeed, the more a retrieval application is specialized for a limited domain, the smaller the gap can be made by using domain knowledge.

Some interesting initiative like Image Retrieval in Medical Applications (IRMA) [13] and medGIFT [16] propose a general content-based medical image retrieval system. Even if the results are considerably improved with those systems in the general content-based retrieval framework, still remain the challenge to bridge the semantic gap and the fusion between heterogeneous medical multimedia sources.

[1] Cross Evaluation Forum Language: `http://www.clef-campaign.org/`

Some recent studies [3], [2] associate explicitly the image and the text. Statistic methods are used for modeling the occurrence of document keywords and visual characteristics. This system is very sensitive to the quality of the segmentation of the images.

Some published works [19] propose to find a semantic of the text using Latent Semantic Analysis. Visual information are extracted using color histograms and edges,and next clustered using the Principal Component Analysis method. In this approach, the two modalities are not combined using the LSA.

Other more recent initiatives study the use of LSA techniques. [22] apply the LSA method to color and texture features extracted from the two media. The conclusion of this study is that combining the image and the text through the LSA method is not always efficient. [23] use the LSA method on the textual and visual combined information. The tests are not really consistent, like the database contain only few documents.

Interesting recent initiatives using LSA for combining text and image are presented in [7] for general images retrieval by the combination of different sources (text and image) at the features level, after clustering. This approach, even promising, is still subject to some empiric thresholds and parameters like the number of visual clusters associated to the image indexes. The database - more consistent that in [23] -, contains nevertheless only 4500 documents.

In our approach, we initiate a semantic level indexing and fusion, according to a well known medical metathesaurus: the Unified Medical Language System. Even if this ontology is still perfectible (some incoherences and inconsistencies still remain), this approach allows us to fuse the medical report and image at the homogeneous medical conceptual level and the tests give us some reliable indicators about its efficiency (the tests are operated on Clef Medical Image Database containing 50 000 medical images). Moreover, the conceptual level indexing and fusion will facilitate all future works concerning the semantic query and case expansion, the context-aware navigation and query and the data-mining.

3 General Framework: A Unified Medical Indexing

The main idea of our approach is to take deeply into account existing structured medical ontology/knowledge to reduce the indexing semantic gap and improve the efficiency of the current general CBIR systems for medical images. The semantic indexing framework proposed consists of three main modules:

1. Medical image analysis and conceptualization
2. Medical report analysis and conceptualization
3. Medical conceptual processing and balanced fusion

This article proposes a semantic approach by focusing on the medical image and report fusion. This approach uses concepts from the Unified Medical Language System (UMLS) to index both images and medical reports (Fig. 1). After

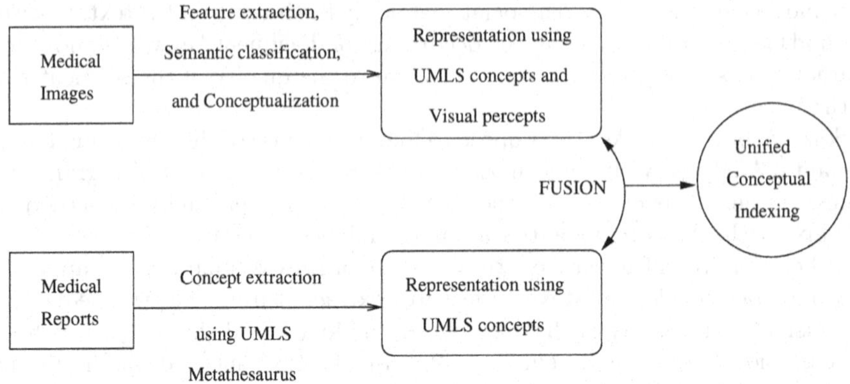

Fig. 1. Unified conceptual indexing

an introduction to *UMLS* in section 3.1, each module of our system is described in detail in the following sections.

3.1 Use of the UMLS Ontology for Improving the Fusion and Retrieval Efficiency

The purpose of NLM's[2] Unified Medical Language System (UMLS)[3] is to facilitate the development of computer systems that behave as if they "understand" the meaning of the language of biomedicine and health.

The UMLS Metathesaurus is a very large, multi-purpose, and multi-lingual vocabulary database that contains information about biomedical and health related concepts, their various names, and the relationships among them. The Metathesaurus is organized by concept or meaning. All concepts in the Metathesaurus are assigned to at least one semantic type from the Semantic Network. This provides consistent categorization of all concepts in the Metathesaurus at the relatively general level represented in the Semantic Network.

In order to filter the UMLS concepts and relationships needed for the fusion, medical case semantic indexing and query expansion, the Metathesaurus UMLS Knowledge Source has been used. This Metathesaurus is composed by medical concepts and is distributed with several tools (programs) that facilitate their use, including the MetamorphoSys install and customization program. MetamorphoSys is the UMLS installation wizard and customization tool included in each UMLS release. It may be used to exclude vocabularies that are not required or licensed for use in local applications and to select from a variety of data output options and filters.

Using MetamorphoSys, we have extracted all the Concept Unique Identifiers (CUI) from the UMLS Metathesaurus UMLS file in order to build the concept layer used for medical cases (image+report) indexing.

[2] US National Library of Medecine - *http://www.nlm.nih.gov/*

[3] Unified medical Language System - *http://www.nlm.nih.gov/research/umls/*

3.2 Medical Image Analysis and Conceptualization

Concerning the image indexing part, our aim is to associate to each image, or to each image region, a semantic label that corresponds to a combination of UMLS concepts and visual percepts. We define three types of UMLS concepts that could be associated to one image or one region:

- modality concepts that belong to the following UMLS semantic type: "Diagnostic Procedure";
- anatomy concepts that belong to the following UMLS semantic types: "Body Part, Organ, or Organ Component", "Body Location or Region", "Body Space or Junction", or "Tissue";
- pathology concepts that belong to the following UMLS semantic types: "Acquired Abnormality", "Disease or Syndrome", or "Injury or Poisoning"

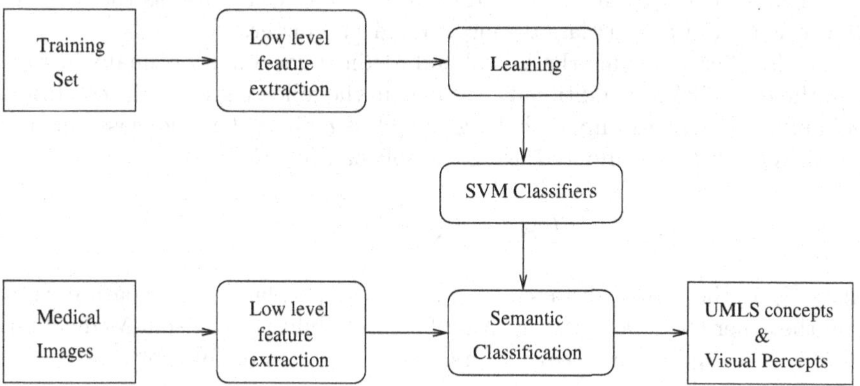

Fig. 2. Conceptual image indexing using SVMs

We propose a structured learning framework based on Support Vector Machines (SVMs) to facilitate modular design and extract medical semantics from images (Fig. 2). We developed two complementary indexing approaches within this statistical learning framework:

- a global indexing to access image modality (chest X-ray, gross photography of an organ, microscopy, etc.);
- a local indexing to access semantic local features that are related to one modality concept, one anatomy concept, and, sometimes, to one pathology concepts.

After presenting our general learning framework, we detail both approaches hereafter.

General Statistical Learning Framework: Firstly, a set of disjoint semantic tokens with visual appearance in medical images is selected to define a Visual and Medical vocabulary. This notion of using a visual and semantic vocabulary to represent and index image has been applied to consumer images in [15,14]. Here, we use UMLS concepts to represent each token in the medical domain.

Secondly, low-level features are extracted from image region instances to represent each token in terms of color, texture, shape, etc.

Thirdly, these low-level features are used as training examples to build a semantic classifier according the Visual and Medical Vocabulary. We use a hierarchical classification based on SVMs. First, a tree whose leaves are the Visual-Medical terms is constructed. The upper levels of the tree consist to auxiliary classes that cluster similar terms with respect to their visual appearance. A learning process is performed at each node in the following way. If a node corresponding to a cluster $\mathcal{C}$ has $N_{\mathcal{C}}$ direct children, $N_{\mathcal{C}}$ SVM classifiers are learned to classify in one class against the $N_{\mathcal{C}} - 1$ other classes. The positive and negative examples for a class $c \in \mathcal{C}$ is respectively given by the instances of the term(s) associated to the class c and the instances of the terms associated to the $N_{\mathcal{C}} - 1$ other classes. This is a conditional learning in a sense as the classifiers are learned given that a class belongs to a given cluster.

The classifier according the Visual and Medical vocabulary is finally designed from the tree of SVM conditional classifiers in the following way. The conditional probability that an example z belong to a class c given that the class belong to the cluster $\mathcal{C}$ is first computed using the softmax function:

$$P(c|z,\mathcal{C}) = \frac{\exp^{\mathcal{D}_c(z)}}{\sum_{j \in \mathcal{C}} \exp^{\mathcal{D}_j(z)}} \quad (1)$$

where $\mathcal{D}_c$ is the signed distance to the SVM hyperplane that separate class c from the other classes of the cluster $\mathcal{C}$. The probability of a Visual-Medical term VMT_i (i. e. a leave of the tree) for an example z is finally given by:

$$P(\text{VMT}_i|z) = P(\text{VMT}_i|z, \mathcal{C}^1(\text{VMT}_i)) \prod_{l=1}^{L-1} P(\mathcal{C}^l(\text{VMT}_i)|z, \mathcal{C}^{l+1}(\text{VMT}_i)) \quad (2)$$

where L is the number of hierarchical levels, $\mathcal{C}^1(\text{VMT}_i))$ denotes the cluster to which VMT_i belongs, and $\mathcal{C}^l(\text{VMT}_i)$ the cluster to which $\mathcal{C}^{l-1}(\text{VMT}_i)$ belongs. $\mathcal{C}^L(\text{VMT}_i)$ is the cluster containing all the defined classes (it corresponds to the tree root).

Global UMLS Indexing According Modality: The global UMLS indexing is based on a two level hierarchical classifier according modality concepts. This modality classifier has been learned from 4000 images separated in 32 classes: 22 grey level modalities, and 10 color modalities. This images come from the CLEF database (2500 examples), from the IRMA database (300 examples), and from the web (1200 examples). The training images from CLEF database was obtained from modality concept extraction using medical report. A manual filtering on this extraction to remove irrelevant examples had to be performed. We plan to automatize this filtering in the near future.

The first level corresponds to a classification according grey level versus color images. Indeed, some ambiguity can appear due to the presence of colored images, or the slightly blue or green appearance of X-ray images. This first classifier uses HSV three first moments computed on the entire image. The second level corresponds to the classification according Modality UMLS concepts given that the image is in the grey or the color cluster. For the grey level cluster, we use grey level histogram (32 bins), texture features (mean and variance of Gabor filtering for 5 scales and 6 orientations), and thumbnails (grey values of 16x16 resized image). For the color cluster, we have adopted HSV histogram (125 bins), Gabor texture features, and thumbnails. Zero-mean normalization is applied to each feature. For each SVM classifier, we adopted a RBF kernel with modified city-block distance between feature vectors y and x that equally takes into account each type of feature:

$$|y - x| = \frac{1}{F} \sum_{f=1}^{N} \frac{|y^f - x^f|}{N_f} \tag{3}$$

where F is the number of type of features ($F = 1$ for the grey versus color classifier, $F = 3$ for the conditional modality classifiers: color, texture, thumbnails), x_f is the feature vector of type f, and $x = \{x_1, ..., x_F\}$.
The probability of a modality MOD_i for an image z is given by equation 2. More precisely, we have:

$$P(\text{MOD}_i|z) = \begin{cases} P(\text{MOD}_i|z, C)P(C|z) & \text{if } \text{MOD}_i \in C \\ P(\text{MOD}_i|z, G)P(G|z) & \text{if } \text{MOD}_i \in G \end{cases} \tag{4}$$

where C and G respectively denote the color and the grey level clusters.

A modality concept can thus be assigned to an image z using the following formula:

$$L(z) = \max_i P(\text{MOD}_i|z) \tag{5}$$

The classifier has been first learned on the half on the training dataset to evaluate its performance in the other half of the training dataset. The error rate on the test set is about 18%, with recall and precision rates larger than 70% for a large majority of the classes. The classification is quite good given the intra-variability of some classes with respect to the class inter-variability. For example, differentiate a brain MRI and a brain CT can be a hard task, even for a human operator.

After learning (using the entire learning set), each database image is indexed according modality given its low-level features. The index values are the probability values given by equation (4).

Local UMLS Indexing: To better capture the medical medical image content, we propose to extend this first modeling for local patch classification in local visual and semantic tokens (LocVisMed terms). Each LocVisMed term is expressed as a combination of Unified Medical Language System (UMLS) concepts from Modality, Anatomy, and Pathology UMLS semantic types. In these experiments, we have adopted color and texture features from patches (i. e. small image blocks) and a non hierarchical classifier based on SVMs and the softmax function given by equation (1). A Semantic Patch Classifier was finally designed to classify a patch according 64 LocVisMed terms. The color features are the three first moments of the Hue, the Saturation, and the Value of the patch. The texture features are the mean and variance of Gabor filtering using 5 scales and 6 orientations. Zero-mean normalization is applied to both the color and texture features. We adopted a RBF kernel with modified city-block distance given by equation (3).

The training set is composed of 3631 patches extracted from images mostly coming from the web (1182 images coming from the web and 158 images from the CLEF collection). The classifier has been first learned on a first half of this training set to be evaluated on a second half. The error rate of this classifier is about 30%.

After learning, the LocVisMed terms are detected during image indexing from image patches without region segmentation to form semantic local histograms. Essentially, an image is tessellated into image overlapping blocks of size 40x40 pixels after area standardization. Each patch is then classified in 64 semantic classes using the Semantic Patch Classifier. An image containing P overlapping patches is then characterized by the set of P LocVisMed histograms and their respective location in the image. An histogram aggregation per block gives the final image index : $M \times N$ LocVisMed histograms (each bin corresponding to the probability of a LocVisMed term presence).

3.3 Medical Report Analysis and Conceptualization

In order to improve conventional text Information Retrieval approaches, we pass form the text syntactic level to the semantic one. In the medical field, this step requires the use of a specialized concepts from available thesaurus and metathesaurus. In this sense, the Unified Medical Language System help us to acquire this pertinent higher level indexing, using a specific $UMLS$ concepts extractor like $MetaMap$, provided by the National Library of Medicine (NLM). A concept is then an abstraction of this synonymous set of terms. Thus, conceptual text indexing consists of associating a set of concepts to a document and uses it as index. This set of concepts should cover the document theme. Conceptual indexing naturally solves the term mismatch and the multilingual problem.

3.4 Medical Conceptual Processing and Balanced Fusion

Even if from the medical point of view, a medical case can have mode than one associated medical image, for the retrieval purpose, we consider one case c composed by one medical report and one medical image.

The main idea of our approach is to use the medical rapport and medical image conceptual and/or visual-concept indexing in order to build an homogeneous high level fusion approach for improve the retrieval system performances.

Such a medical case c will bring thus an indexing from the associated image and respectively medical report:

$$\Lambda_{txt}^{c} = \begin{bmatrix} CUI_{i_1} & \lambda_{txt_{i_1}}^{c} \\ \dots & \dots \\ CUI_{i_n} & \lambda_{txt_{i_n}}^{c} \end{bmatrix}, \; \Lambda_{img}^{c} = \begin{bmatrix} CUI_{j_1} & \lambda_{img_{j_1}}^{c} \\ \dots & \dots \\ CUI_{j_m} & \lambda_{img_{j_m}}^{c} \\ VC_{k_1} & \lambda_{img_{vc\_k_1}}^{c} \\ \dots & \dots \\ VC_{k_p} & \lambda_{img_{vc\_k_p}}^{c} \end{bmatrix} \tag{6}$$

with: $\lambda_{txt_{i_l}}^{c} = \mu_{txt_{i_l}}^{c} \cdot \nu_{txt_{i_l}}^{c} \cdot \omega_{txt_{i_l}}^{c} \cdot \gamma_{txt_{i_l}}^{c},\ l = 1, ..., n$

$\lambda_{img_{j_q}}^{c} = \mu_{img_{j_q}}^{c} \cdot \nu_{img_{j_q}}^{c} \cdot \omega_{img_{j_q}}^{c} \cdot \gamma_{img_{j_q}}^{c},\ q = 1, ..., m$

$\lambda_{img_{vc\_k_s}}^{c} = \mu_{img_{vc\_k_s}}^{c} \cdot \nu_{img_{vc\_k_s}}^{c} \cdot \omega_{img_{vc\_k_s}}^{c} \cdot \gamma_{img_{vc\_k_s}}^{c},\ s = 1, ..., p$

where:

CUI is the notation for an UMLS concepts,
VC are visual concepts, mix between UMLS concepts and perception concepts,
μ means the indexing confidence degree,
ν is associated to the local relative frequency of the concept,
ω corresponds to the spatial localization fuzzy weight,
γ represents the semantic tree (modality, anatomy, biology, pathology and direction) fuzzy weight.

For the medical case c (medical image and associated medical report document) and the corresponding UMLS extracted concept CUI_i, we obtain the next fuzzy confidence indexing vector:

$$\lambda_i^c = \left[\lambda_{txt\ i}^{c},\ \lambda_{img\ i}^{c}\right] = \left[\mu_{txt_i}^{c} \cdot \nu_{txt_i}^{c} \cdot \omega_{txt_i}^{c} \cdot \gamma_{txt_i}^{c},\ \mu_{img_i}^{c} \cdot \nu_{img_i}^{c} \cdot \omega_{img_i}^{c} \cdot \gamma_{img_i}^{c}\right] \tag{7}$$

If the concept is a visual concept VC_s, we have:

$$\lambda_{vc\_s}^{c} = \left[0,\ \mu_{img_{vc\_s}}^{c} \cdot \nu_{img_{vc\_s}}^{c} \cdot \omega_{img_{vc\_s}}^{c} \cdot \gamma_{img_{vc\_s}}^{c}\right] \tag{8}$$

As the semantic medical report and image use the same homogeneous ontology, the fusion takes into account the norm of the so obtained global credibility vector λ_i^c as a projection of the c medical case to each concept CUI_i:

$$pr_{CUI_i}(c) = \|\lambda_i^c\|$$

Obviously, we will find the same kind of projection of the case c to each associated visual concept VC_s of the corresponding medical image:

$$pr_{VC_s}(c) = \|\lambda_{vc\_s}^{c}\|$$

A medical case c will be then characterized by the vector Λ_c of the global credibilities of all associated *UMLS* and *visual* concepts:

$$\Lambda^c = \left[\|\lambda_{i_1}^c\|,\ \|\lambda_{i_2}^c\|,\ \dots,\ \|\lambda_{in}^c\|,\ \|\lambda_{vc1}^c\|,\ \dots,\ \left\|\lambda_{vc_p}^c\right\|\right]^T$$

The parameters used for the projection of a medical case on CUI or VC are obtained by direct image and text pre-computation (indexing) - e.g. fuzzy indexing confidence degree μ and the local relative frequency ν - or by fuzzy-fication - for the spatial localization ω and semantic tree membership γ fuzzy weights.

The indexing confidence degree μ is a fuzzy result directly given by the classifiers (equation (4) used for the medical image indexing. For the text pre-processing, a text indexing software (in our case, $MetaMap$) has been used for calculate the local relative frequency $\nu^c_{txt_i}$ of the concept occurrence in the given medical report. If a patch extraction method is used for the image, the local relative frequency $\nu^c_{img_i}$ will be computed using the relative weight of this concept versus all the patches of the image. The spatial localization ω corresponds - for

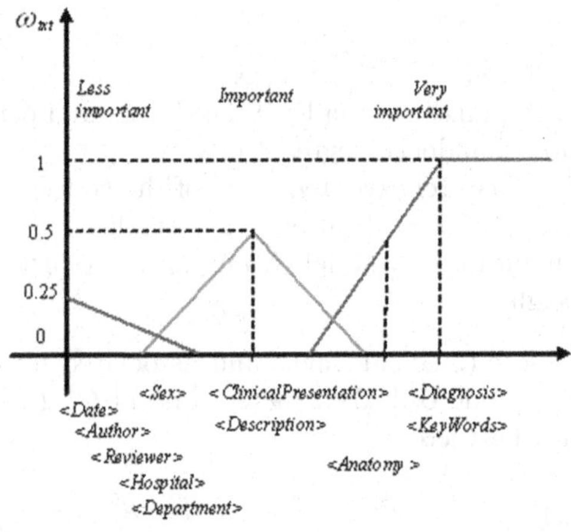

Fig. 3. Spatial localization parameter ω_{txt} fuzzyfication

the text indexing - to the importance of the medical report XML tag or the section from which the concept comes. For example, the $< Diagnosis >$ paragraph of the medical report, the physician synthesized the most important keywords describing the disease (pathology) and the anatomic part. This tag will thus be more important than (par example) the $< Description >$ tag. In order to fuzzy this subjective parameter, we propose the fuzzy membership sets presented in the Fig. 3.

For the image indexing, the spatial localization corresponds to a special weight accorded to a particular place in the image (e.g. the central patches for a CT medical image or a circle sector shape object for a Doppler indexing).

The semantic tree membership γ intent to give a particular weight to a concept belonging to a particular semantic tree (modality, anatomy, pathology, biology, direction) according to the indexing source (image or text) of this concept. Par example, a modality concept coming from the medical image indexing will have more importance that a modality extracted from the medical report. Opposite, a pathology concept will have more "confidence" if extracted by the medical report indexing service (Fig.4), with:

$$\begin{aligned}
&\textbf{if } CUI_i \in MOD \Rightarrow \alpha = \sigma \\
&\textbf{if } CUI_i \in ANA/BIO \Rightarrow \alpha = 2\sigma \\
&\textbf{if } CUI_i \in PAT \Rightarrow \alpha = 3\sigma
\end{aligned}$$

where σ corresponds to the size of the fuzzy membership function associated to the influence zone of each semantic type.

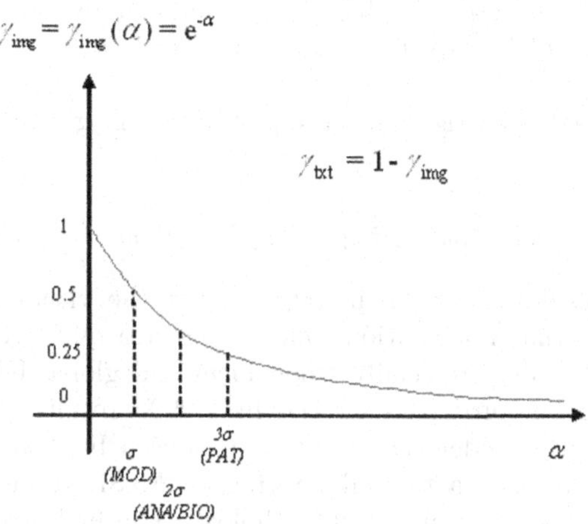

Fig. 4. Semantic tree membership γ fuzzyfication

4 Results on CLEF 2005 Medical Images Benchmark

We apply our approach on the medical image collection of CLEF [4]Cross Language Image Retrieval track. This database consists of four public datasets (CASImage [5], MIR [6], PathoPic [7], PEIR [8]) containing 50000 medical images with the associated medical report in three different languages. In 2005, 134 runs were evaluated on 25 queries containing at least one of the following axes: anatomy (ex: heart), modality (ex: X-ray), pathology or disease (ex: pneumonia), abnormal visual observation (ex: enlarged heart).

We test five approaches on these 25 queries to evaluate the benefit of using a UMLS indexing, especially in a fusion framework. First, three UMLS image indexing were tested on the visual queries:

1. Global UMLS image indexing presented in section 3.2 / Retrieval based on the Manhattan distance between two modality indexes (modality probabilities);

[4] CLEF - Cross Language Evaluation Forum - *www.clef – campaign.org/*
[5] http://www.casimage.com
[6] http://gamma.wustl.edu/home.html
[7] http://alf3.urz.unibas.ch/pathopic/intro.html
[8] http://peir.path.uab.edu

2. Local UMLS image indexing presented in section 3.2 / Retrieval based on the Manhattan distances between LocVisMed histograms.
3. Late fusion of the two visual indexing approaches (1) and (2)

The two last tests respectively concerns textual indexing and retrieval using UMLS and the fusion between this textual indexing and the Global UMLS image indexing (i.e. modality indexing). The integration of local information is under developmen. The text indexing uses $MetaMap$ software and give us the $UMLS$ associated concepts CUI with the corresponding relative frequencies ν. The medical image indexing uses a global image approach giving essentially $UMLS$ modalities concepts CUI_{MOD}, the associated frequency ν and SVM confidence degree μ.

With these partial indexing information, we build the global confidence degree λ such as:

$$\lambda_i^c = \left[\lambda_{txt\ i}^c \ \lambda_{img\ i}^c\right] = \left[1 \cdot \nu_{txt_i}^c \gamma_{txt_i}^c \ \mu_{img_i}^c \nu_{img_i}^c \gamma_{img_i}^c\right]$$

Comparative results are given in table 1. For the visual indexing and retrieval without textual information, our results are quite good with respect to the best 2005 results, especially when local and global UMLS indexes are mixed. Our textual approach is on the first 2005 results (between 9% and 20%) but significantly under the approach proposed by Jean-Pierre Chevallet (IPAL) in 2005 that uses a textual filtering on MeSH terms according three dimensions: Modality, Anatomy, and Pathology. The high average precision is principally due to this textual filtering. We have to notice that the association between MeSH terms and a dimension had to be done manually. With the use to UMLS metha-thesaurus, we have the advantage to have access to these dimension thanks to the semantic type associated to each UMLS concept.

Table 1. Comparative results on the medical task of CLEF 2005

Method	Visual	Textual	MAP
Fusion between UMLS image indexes	X		12.11%
Global UMLS image indexing	X		10.38%
Local UMLS image indexing	X		06.56%
Best automatic visual run in 2005 (GIFT)	X		09.42%
UMLS textual indexing		X	16.41%
Best automatic textual run in 2005 (DFMT) Dimension Filtering on MESH Terms		X	20.84%
UMLS mixed indexing	X	X	24.07 %
Best automatic mixed run in 2005 (DFMT) + local semantic indexing	X	X	28.21 %
Best automatic mixed run in 2005 without dimension filtering	X	X	23.89 %

We can verify the benefit of the fusion between image and text : from 16% for the text only and 10% for the image only, we reach the 24% with a mixed indexing and retrieval. We are slightly superior in average precision than the mixed approaches that do not use dimension filtering.

5 Conclusion

The presented researches constitutes a very promising approach of semantic indexing and late balanced fusion, in the medical cases retrieval framework.

One important contribution of this study is the use of a web-based up to date medical metathesaurus (UMLS) in order to "standardize" the semantic indexing of the text and the medical image. This allows to work on the same level for both medical media and to have thus an homogeneous complementary point of view about the medical case.

The introduction of a late semantic fusion for each common UMLS concept, using multiple criteria weighted norm involving the frequency of a concept, the indexing confidence degree, the spatial localization weight and the semantic tree belonging, constitutes another important point to be underlined. The so obtained vectorial norm represents a balanced projection of the medical case (document and image) to the given UMLS concept, a consistent information enabling a reliable and robust semantic retrieval.

Future developments become very promising using this homogeneous balanced semantic fusion. Appropriate clustering methods should be able to bridge to medical multimedia data-mining, opening the way to the evidence-based medicine and other advanced medical research applications and studies.

In future evolutions of this application, our fusion approach will be improved using the local visual information derived from the proposed local patch classifier. Indeed, this method is complementary to the global medical image analysis and will certainly improve the global retrieval results.

Otherwise, the use of the incremental learning based on the initial database clustering, should be able to facilitate the development of an efficient real-time medical case-based reasoning.

Finally, a semantic query and case expansion policy can be deployed using the symbolic and statistic relation available in $UMLS$ at the first time, and the contextual behavior information extracted from the real use of the retrieval system in the second time.

References

1. H. Abe, H.and MacMahon, R. Engelmann, Q. Li, J. Shiraishi, S. Katsuragawa, M. Aoyama, T. Ishida, K. Ashizawa, C. E. Metz, and K. Doi. Computer-aided diagnosis in chest radiography: Results of large-scale observer tests at the 1996-2001 rsna scientific assemblies. *RadioGraphics*, 23(1):255–265, 2003.
2. K. Barnard, P. Duygulu, D. Forsyth, N. de Freitas, D.M. Blei, and M.I Jordan. Matching words and pictures. *Journal of machine learning research*, 3:1107–1135, 2003.

3. K. Barnard and D. Forsyth. Learning the semantics of words and pictures. In *Proceedings of the International Conference on Computer Vision*, volume 2, pages 408–415, 2001.
4. A. A. T. Bui, R. K. Taira, J. D. N. Dionision, D. R. Aberle, S. El-Saden, and H. Kangarloo. Evidence-based radiology. *Academic Radiology*, 9(6):662–669, 2002.
5. N.-S. Chang and K.-S. Fu. Query-by-pictorial-example. *IEEE Transactions on Software Engineering*, 6(6):519–524, 1980.
6. W.W. Chu, F. C. Alfonso, and K.T. Ricky. Knowledge-based image retrieval with spatial and temporal constructs. *IEEE Transactions on Knowledge and Data Engineering*, 10:872–888, 1998.
7. C. Fleury. Apports reciproques des informations textuelles et visuelles par analyse de la semantique latente pour la recherche d'information. Master's thesis, Intelligence, Interaction and Information, CLIPS-MRIM Grenoble, France, IPAL UMI CNRS 2955, Singapore, June 2006.
8. M. Flickner, H. Sawhney, W. Niblack, J. Ashley, Q. Huang, B. Dom, M. Gorkani, J. Hafner, D. Lee, D. Petkovic, D. Steele, and P. Yanker. Query by image and video content: The qbic system. *IEEE Computer*, 28(9):23–32, 1995.
9. M. O. Guld, M. Kohnen, D. Keysers, H. Schubert, B. B. Wein, J. Bredno, and T. M. Lehmann. Quality of dicom header information for image categorization. In *Proceedings of the International Symposium on Medical Imaging*, volume 4685, pages 280–287, San Diego, CA, USA, 2002. SPIE.
10. C. E. Kahn. Artificial intelligence in radiology: Decision support systems. *RadioGraphics*, 14:849–861, 1994.
11. P. Korn, N. Sidiropoulos, C. Faloutsos, E. Siegel, and Z. Protopapas. Fast and effective retrieval of medical tumor shapes. *IEEE Transactions on Knowledge and Data Engineering*, 10:889–904, 1998.
12. C. LeBozec, M.-C. Jaulent, E. Zapletal, and P. Degoulet. Unified modeling language and design of a case-based retrieval system in medical imaging. In *Proceedings of the Annual Symposium of the American Society for Medical Informatics*, Nashville, TN, USA, 1998.
13. T. M. Lehmann, M.O. Gld, C. Thies, B. Fischer, K. Spitzer, D. Keysers, H. Ney, M. Kohnen, H. Schubert, and B. B. Wein. Content-based image retrieval in medical application. *Methods of Information in Medicine*, 43(4):354–361, 2004.
14. J. Lim and J.P. Chevallet. Vismed: a visual vocabulary approach for medical image indexing and retrieval. In *Proceedings of the Asia Information Retrieval Symposium*, pages 84–96, 2005.
15. J. Lim and J. Jin. A structured learning framework for content-based image indexing and visual query. *Multimedia Systems*, 10:317–331, 2005.
16. H. Muller, N. Michoux, D. Bandon, and A. Geissbuhler. A review ofcontent-based image retrieval systems in medical applications - clinical benefits and future directions. *International Journal of Medical Informatics*, 73:1–23, 2004.
17. W. Niblack, R. Barber, W. Equitz, M. D. Flickner, E. H. Glasman, D. Petkovic, P. Yanker, C. Faloutsos, and G. Taubin. QBICproject: querying images by content, using color, texture, and shape. In W. Niblack, editor, *Storage and Retrieval for Image and Video Databases*, volume 1908, pages 173–187. SPIE, 1993.
18. A. Pentland, R. W. Picard, and S. Sclaroff. Photobook: Tools for content-based manipulation of image databases. *International Journal of Computer Vision*, 18:233–254, 1996.
19. S. Sclaroff, M. la Cascia, and S. Sethi. Unifyng textual and visual cues for content-based image retrieval on the world wide web. *Computer Vision and Image Understanding*, 75(1/2):86–98, 1998.

20. C.-R. Shyu, C. E. Brodley, A. C. Kak, A. Kosaka, A. M. Aisen, and L. S. Broderick. ASSERT: A physician-in-the-loop content-based retrieval system for HRCT image databases. *Computer Vision and Image Understanding*, 75:111–132, 1999. Special issue on content-based access for image and video libraries.
21. A. W. M. Smeulders, M. Worring, S. Santini, A. Gupta, and R. Jain. Content-based image retrieval at the end of the early years. *IEEE Transactions on Pattern Analysis and Machine Intelligence*, 22:1349–1380, 2000.
22. T. Westerveld. Image retrieval : Content versus context. In *Recherche d'Information Assistee par Ordinateur*, 2000.
23. R. Zhao and W. Grosky. Narrowing the semantic gap - improved text-based web document retrieval using visual features. *IEEE Transactions on Multimedia*, 4(2):189–200, 2002.

Automated Object Extraction for Medical Image Retrieval Using the Insight Toolkit (ITK)

Henning Müller, Joris Heuberger,
Adrien Depeursinge, and Antoine Geissbühler

University and Hospitals of Geneva, Medical Informatics, Geneva, Switzerland
henning.mueller@sim.hcuge.ch

Abstract. Visual information retrieval is an emerging domain in the medical field as it has been in computer vision for more than ten years. It has the potential to help better managing the rising amount of visual medical data. One of the most frequent application fields for content–based medical image retrieval (CBIR) is diagnostic aid. By submitting an image showing a certain pathology to a CBIR system, the medical expert can easily find similar cases. A major problem is the background surrounding the object in many medical images. System parameters of the imaging modalities are stored around the images in text as well as patient name or a logo of the institution. With such noisy input data, image retrieval often rather finds images where the object appears in the same area and is surrounded by similar structures. Whereas in specialised application domains, segmentation can focus the research on a particular area, PACS–like (Picture Archiving and Communication System) databases containing a large variety of images need a more general approach. This article describes an algorithm to extract the important object of the image to reduce the amount of data to be analysed for CBIR and focuses analysis to the important object. Most current solutions index the entire image without making a difference between object and background when using varied PACS–like databases or radiology teaching files. Our requirement is to have a fully automatic algorithm for object extraction. Medical images have the advantage to normally have one particular object more or less in the centre of the image. The database used for evaluating this task is taken from a radiology teaching file called *casimage* and the retrieval component is an open source retrieval engine called *medGIFT*.

1 Introduction

Content–based visual information or image retrieval (CBIR) has been an extremely active research area in the computer vision and image processing domains [1,2]. A large number of systems has been developed, mostly research prototypes [3] but also commercial systems such as IBM's QBIC [4] or Virage [5]. The main reason for the development of these systems is the fact that an ever–growing amount of visual data is produced in many fields, for example with the availability of digital consumer cameras at low prices, but also with the possibility to make the visual data accessible on the Internet. The data commonly

H.T. Ng et al. (Eds.): AIRS 2006, LNCS 4182, pp. 476–488, 2006.

analysed includes photographs, graphics, and videos. One typical application of image retrieval is trademark images [6].

The medical field is no exception to this, and a rising amount of visual data is being produced [7]. The radiology department of the University Hospitals of Geneva alone produces currently more than 25,000 images per day, mostly in multi–slice tomographic series. The importance of retrieval of medical images was identified early [8,9,10] and a large number of projects has started to index various kinds of medical images [11]. Not all of the projects are analysing the visual image content, some simply use the accompanying textual information for retrieval [12]. This is often called CBIR but should rather be called context–based retrieval as the text describes the context, in which the image was taken [13] rather than its content. Very few projects are currently used in clinical routine. Most projects have been developed as research prototypes but without a direct link to a need in a clinical application [14,15]. An example for a system that was used as a prototype in a clinical setting is *Assert* that showed a significant improvement of correct diagnosis when using the system [16,17], especially among less experiences radiologists. Another active medical image retrieval project is *IRMA*[1] [18,19], that concentrates on image classification and the segmentation of medical images for retrieval. An annotated database was developed in this project and a multi–axial code for image annotation [20].

medGIFT[2], our CBIR tool, extracts mainly local and global features such as textures (based on Gabor filter responses) and grey level descriptors for the visual similarity retrieval from our teaching file. It is based on the GNU Image Finding Tool (*GIFT*[3]) [21] and includes modifications to use a slightly different feature space. One of the identified problems of retrieval is that a large part of the images does not contain any important information for retrieval but rather noise. This noise can be in the form of black areas around the principal object, but more often in the form of text and logos that occur around a large number of objects in the images. In this case, the area where the main object appears and the kind of noise in the image has a significant influence on the retrieval that is sometimes hindering a good performance. Our goal was to develop a completely automatic algorithm to reduce this background noise and extract the important object in the medical images of our database *casimage*[4], a teaching file containing almost 9000 extremely varied images from several modalities (CT, MRI, PET, ...) as well as photographs and even powerpoint slides. To our knowledge no such algorithm for the extraction of objects from a large variation of medical images exists as of yet. Another goal was to have only an extremely small number of images with too much information being removed from the images, so a rather conservative approach was taken. Some roughly related algorithms have already been used in the analysis of images and also videos to identify text regions in the visual data [22,23] but we do not only have to deal with text but with a large variety of structures that need to be removed.

[1] `http://www.irma-project.org/`
[2] `http://www.sim.hcuge.ch/medgift/`
[3] `http://www.gnu.org/software/gift/`
[4] `http://www.casimage.com/`

2 Insight Toolkit

ITK[5] is an open–source (OS) software system for medical image segmentation and registration. It has been developed since 1999 on initiative of the US National Library of Medicine (NLM) of the National Institutes of Health (NIH). As an OS project, *ITK* is used, debugged, maintained, and upgraded by several developers around the world. It is composed of a large collection of functions and algorithms designed specifically for medical images, and particularly for registration and segmentation tasks. The entire library is implemented in C++ and can be used on most platforms such as Linux, Mac OS and Windows. The decision to use *ITK* was taken because of the quantity of simple manipulation tools and filters it offers and the amount of medical segmentation research done based on it [24]. This allows us to concentrate on integrating tools rather than reprogramming and reinventing them. Many medical images are stored in DICOM [25], a complex standard, and ITK offers to open these images as well as other common standards such as JPG and GIF, which constitute much of our medical teaching file. ITK is the standard open source environment for medical image processing at the moment.

3 Methods Applied

3.1 Functions Used for Background Removal

The algorithm employed assumes that the object to extract has gray levels highly different from the background. Basically, an edge detection method is used to find these fast transitions. Several other steps are needed to handle a maximum of image types and remove some very specific problems that we identified. Much of the fine tuning was performed based on results on a small subset of images. Steps for the object extraction are:

– Removal of specific structures (University logo, typical large structures such as a grey square at the bottom right of images);
– Smoothing;
– Edge detection;
– Removal of small structures;
– Cropping;

The *casimage* collection that we use [26] presents two main image parts that interfere with the planned method. Several images contain the logo of the University hospitals in the upper left corner. Another problem is caused by a gray level square in the lower right corner (Figure 3). These two structures are too large to be removed during the foreseen noise cleaning step that well removes text. For this reason, the logo and the square have to be removed first. The hospital logo can be seen in Figure 1. It always appears in the upper left area in roughly the same size, so we cut 90 pixels horizontally and 30 pixels vertically

[5] `http://www.itk.org/`

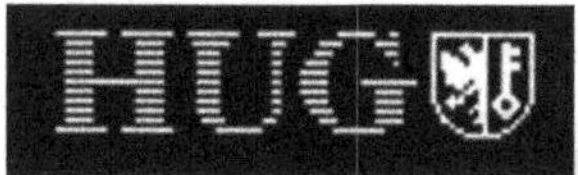

Fig. 1. Logo of the University Hospitals of Geneva

for further analysis containing mainly the large characters of the logo. Part of the remaining text next to the logo is removed automatically in the following steps. To detect whether there is a logo in the upper left corner two criteria were defined. First, the number of white pixels in this area has to be in a certain range (350–400 pixels) and an erosion with a structuring element of size 1 has to eliminate all white pixels. The second criterion is based on a specific aspect of the logo: it is composed of fine horizontal lines. Thus, the erosion with a one pixel radius structuring element is highly destructive. If the two criteria are positive, the logo is removed by filling the region in black.

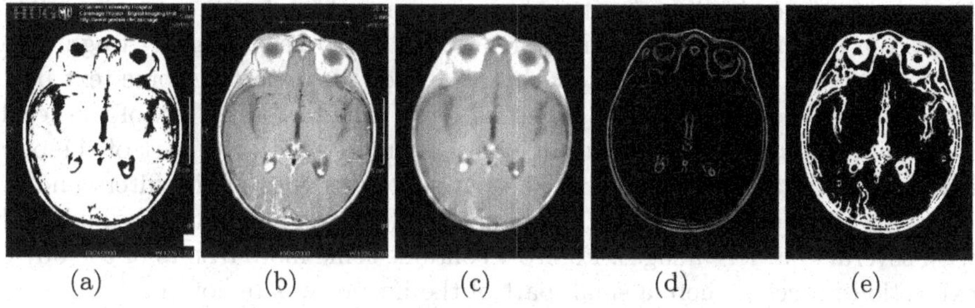

(a) (b) (c) (d) (e)

Fig. 2. The steps of the removal process: (a) thresholding for logo and grey square removal, (b) logo and gray square removed, (c) median filtering for smoothing and removal of small structures, (d) edge detection (e) thresholding

Gray squares (see Figure 3, first image, bottom right) also appear in many images. To remove them, the lower right area is thresholded to select only very light pixels. The resulting binary image is eroded then dilated to eliminate small objects. If a square object is remaining, it is the gray square. This binary image is used as a mask to eliminate the square from the original image.

Then, a median filter of size 4 is applied to smooth the image and already remove many small structures such as part of the text on the black background. Examples for the results of the various treatment steps can be seen in Figure 2.

The edge detection filter has as a consequence a weaker response and only the main structures will remain. A `GradiantMagnitudeImageFilter` edge detection filter from the itk package is used in this step to detect the structures in the image. The aim of this structure removal step is to remove structures not being part of the main object. This can be annotations (patient name, system parameters, ...) but also simple frames around the image or a ruler. A binary image is produced by thresholding the result of the edge detection (threshold 5,

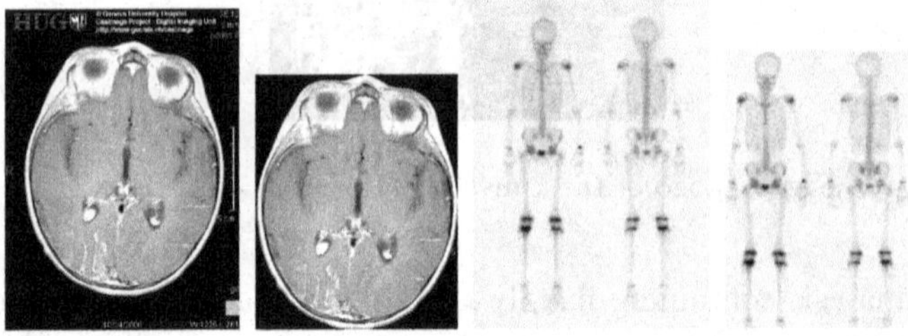

Fig. 3. Result of the algorithm on a CT scan and a scintigraphy image

5-255 are mapped to 255 the rest to 0). Remaining small objects are removed by measuring the size of connected components. The size for removing objects depends on the image size itself. We use a simple cutoff for images having more than 1000×1000 pixels where we remove objects up to a size of 300 pixels and smaller images where we remove objects of a maximum size of 50 pixels. Some small structures can be part of the main object, so the image is dilated (filter size 5) and then eroded (filter size 4) before the removal (a closing operation). This leads to a merging of neighbouring structures, which keeps slightly fragmented structures together. Unfortunately this also leads to connecting some text parts where the characters are fairly close and large. The bounding box of the result is finally computed, and the output image is created with the part of the original image contained in the box. The parameter settings of the filters and for removing connected components were obtained by systematic testing and trials with several "harder" images. Figure 4 contains some more results of the object extraction process, where a small part of the images was removed.

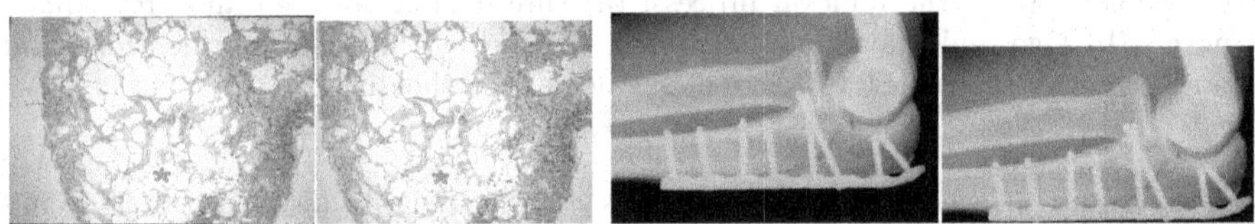

Fig. 4. Results on a colour image and a radiograph with small enhancements

On a Pentium IV computer with 2.8 GHz and 1 GB of RAM, the entire extraction process takes slightly more than 1 second making it feasible for a collection of 9000 images on a simple desktop computer.

3.2 Encountered Problems

Due to the variety of image types and acquisition systems, our algorithm can not handle all specific problems. In particular, the text part in images can be too large or too dense to be considered as noise (see Figure 5). Another problem can occur in CT scans, when the patient support under the patient appears on the image and is considered as part of the main object (Figure 6).

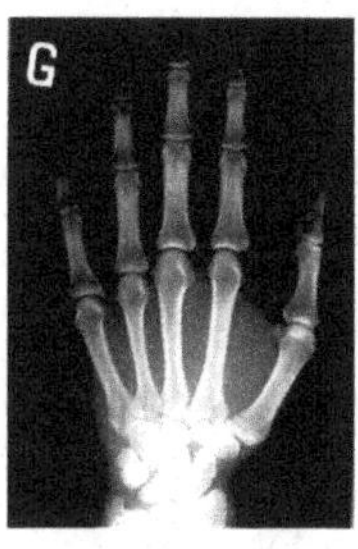
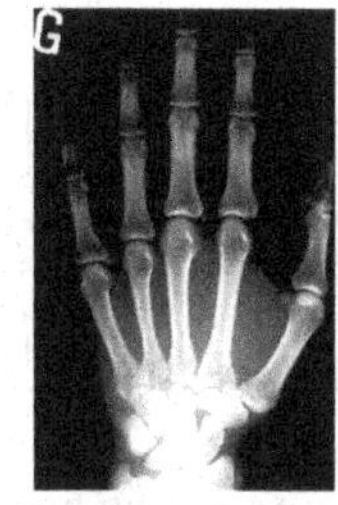
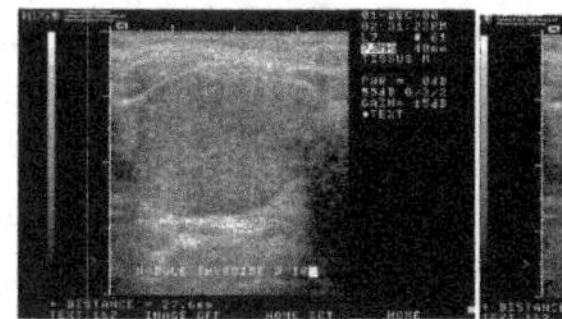
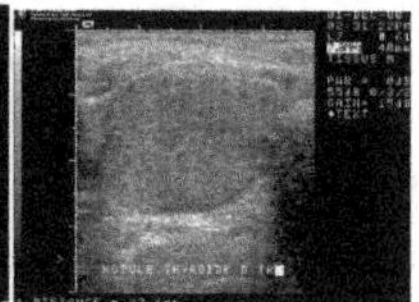

Fig. 5. Images where the text is not recognised and as a consequence not removed

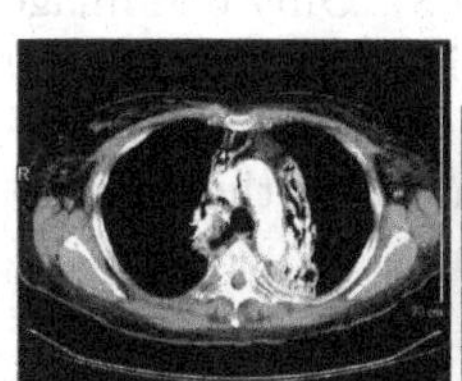
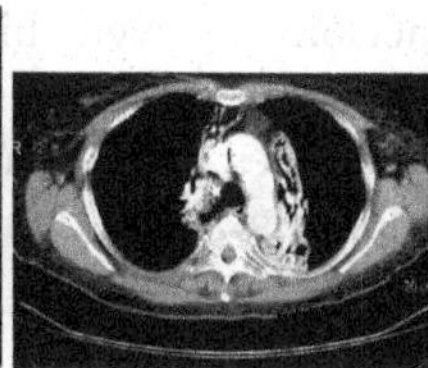
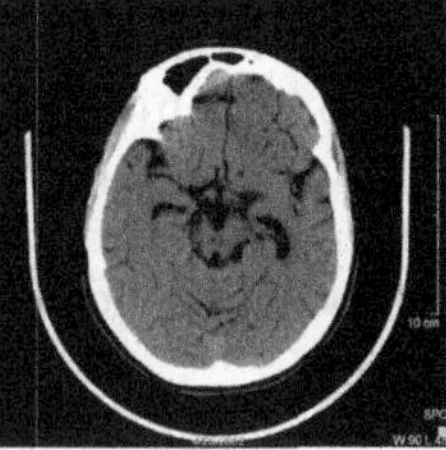
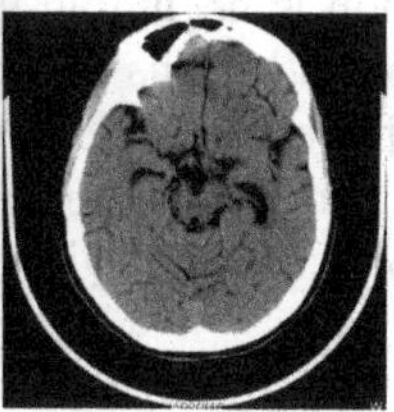

Fig. 6. In these CTs, the structure under the body is too important to be discarded

4 Results

4.1 Extraction Results

To evaluate the accuracy of the extraction algorithm, a subset of 500 medical images from the 9000–image collection was randomly extracted from the *casimage* database and the algorithm was executed on these images. Then, each result was rated with respect to extraction quality. To simplify this task, a visual PHP interface was built that presents each extracted object next to its corresponding original image. It allows the validator to classify the result into one of these four quality classes:

1. Class 1: The object is extracted as wanted.
2. Class 2: The image is fine but no work was needed.
3. Class 3: The result contains the object, but some background remains.
4. Class 4: Parts of the object are lost.

The 500 results were classified into these four categories by one validator familiar with the database. An optimal result was achieved for 389 images (204 in class 1 and 185 in class 2). For 105 images, parts of the text or background could not be removed, with most of the images having at least part of the background removed, so quality is at least better than before, although not perfectly satisfying. Six images have too much of the image being removed in error.

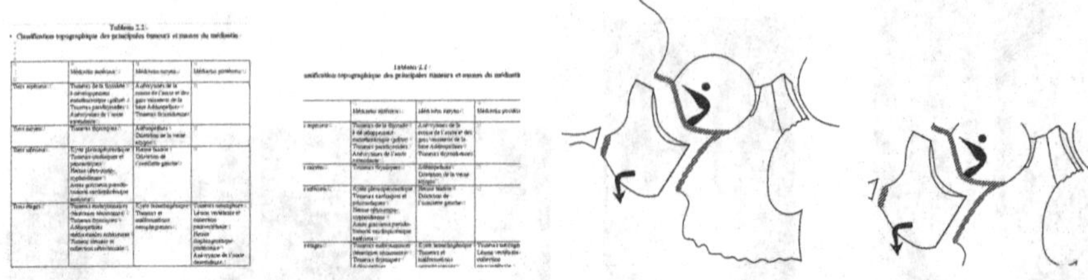

Fig. 7. Structures on these two images are too fine and taken for noise

Of these, two are not medical images but drawings and text (Figure 7). Part of the fine structures was removed in these images. The quantity of lost information is negligible for three of the other four images (Figure 8). Only one image has a serious loss due to the object extraction but even this image contains all diagnostically relevant image information. It can clearly be seen that the algorithm has some problems with very slow changes in the images as the contrast of the main object is not marked enough for edge detection.

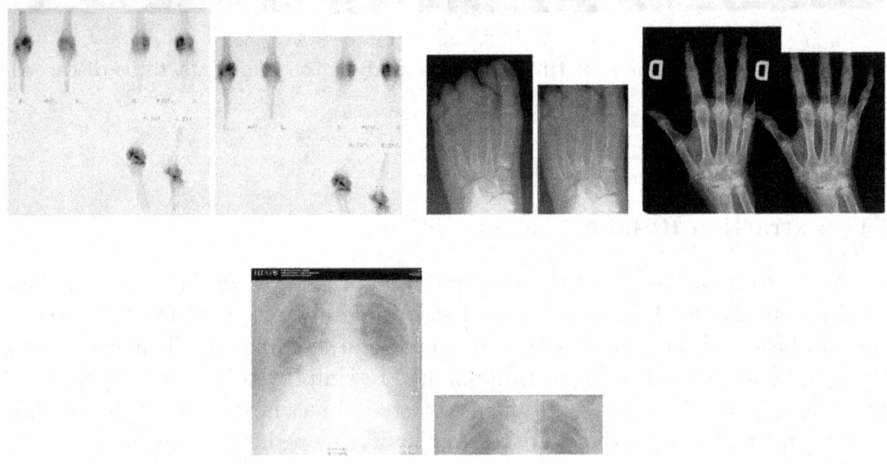

Fig. 8. The four other images where too much was removed

For our goal of CBIR, it is important to eliminate or reduce the amount of useless information but not lose any important parts of the images for retrieval. The results have to be evaluated in terms of improvement and deterioration. Classes 1 and 3 constitute an improvement of the image for content–based image retrieval (about 60% of the images), and class 4 which contains around 1% of the images is a deterioration of the images. For about 39% of the images, the object extraction was not necessary.

An analysis of the number of removed pixels per image on the entire 9000–image dataset shows that 3000 images have more than 10% of the pixels removed,

1600 images more than 20% and 500 images even more than 50% of the pixels. This shows the large amount of data that could be removed in a simple object extraction algorithm.

4.2 A Simple Improvement

To improve the results of our system on the images of class 3, we ran our algorithm a second time on all the already segmented images. We regarded this step as a necessary trial because some images contained a manually introduced border plus the actual image background. The majority of the images stayed unchanged but results were much better for 11 images among the 500 images observed. On the images of the class 1 and 2, no change appeared, but 2 images of class 4 present worse results. These two images are the text fragment and the thoracic radiograph shown beforehand. An analysis of the second step of segmentation shows that in 90% of the images less than 10% of the pixels were removed in this second step. Still, it also shows that more parts of the background could be removed by simply applying the same algorithm twice.

4.3 Retrieval Results

The entire *casimage* database in its original form as well as after a simple segmentation and after running the segmentation twice were indexed using *medGIFT*. *medGIFT* first scales the images to 256 × 256 pixels and then indexes them with the following feature groups:

- a global colour and gray level histogram;
- local gray level information by partitioning the image successively four times into four subregions and taking the mode colour of each region as feature;
- a global histogram of Gabor filter responses (4 directions, 3 scales and 10 strengths for quantisation);
- local Gabor filter responses within the entire images in blocks of size 16.

To compare the features, a frequency–based weighting similar to the text retrieval tf/idf weighting is used (see [21] for more details):

$$\textit{feature weight}_{jq} = \frac{1}{N}\sum_{i=1}^{N}\left(\textit{tf}_{ij}\cdot R_i\right)\cdot \textit{log}^2\left(\frac{1}{\textit{cf}_j}\right), \tag{1}$$

where *tf* is the *term frequency* of a feature, *cf* the *collection frequency*, j a *feature number*, q corresponds to a query with $i = 1..N$ input images, and R_i is the *relevance* of an input image i within the range $[-1; 1]$.

Subjective impressions when using the system show an important improvement in retrieval quality. A few queries deliver worse results in a first step, but much better results once feedback is applied. Figure 9 shows a retrieval result without the use of feedback using a single image as query, and Figure 10 shows a result after one step of feedback for the same image once on the segmented and once on the non–segmented database.

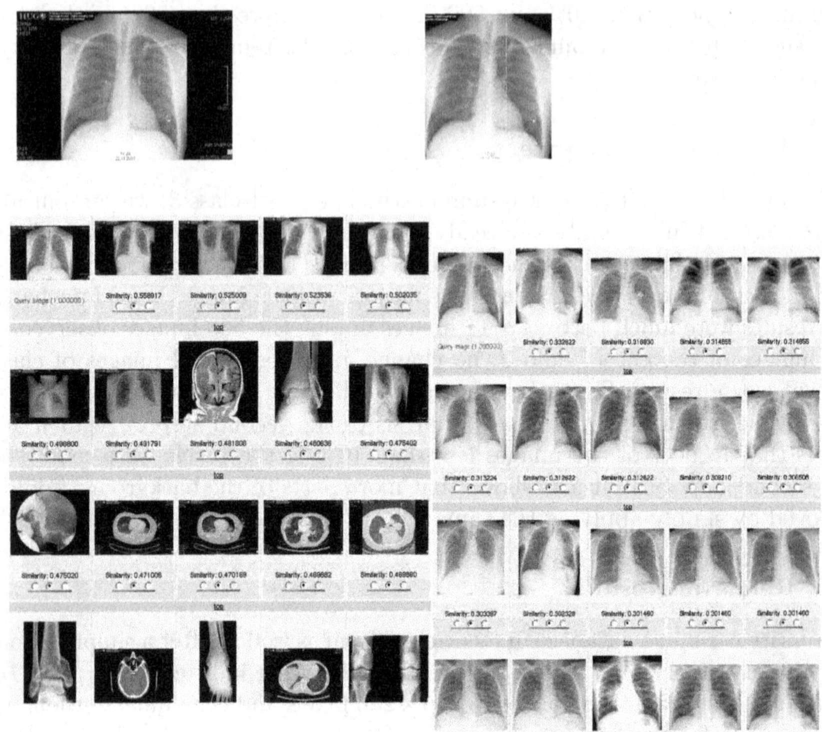

Fig. 9. Comparison of retrieval results without user feedback

Besides the subjective evaluation of the retrieval results, we also used the query topics and relevance judgements that were created in the *ImageCLEF*[6] competition [27,28] for retrieval and compared them with the *medGIFT* base system. This competition created 26 query topics that contain one example image per topic, only, and no text. The lead measure for the competition is the mean average precision (MAP) that is used in most text retrieval benchmarking events such as TREC (Text REtrieval Conference) and CLEF (Cross Language Evaluation Forum). This measure is averaged over all 26 query topics. The *medGIFT* system has a MAP of 0.3757 and was among the three best visual systems in the competition. The results for the segmented database are slightly surprising as they do not appear to be better but rather slightly worse (MAP 0.3562). We also tried out two more configurations of grey levels and Gabor filters. Using 8 grey levels instead of 4 and 8 directions for the Gabor filters leads to even worth results (MAP 0.3350). When using 8 grey levels but the same Gabor filters, the results are even slightly worth than the latter (MAP 0.3322). The second run of the segmentation lead to very similar results (MAP 0.3515). This surprising result can partly be explained with several effects that are due to the way the groundtruth is being produced after the CLEF submissions. As only part of

[6] `http://ir.shef.ac.uk/imageclef/`

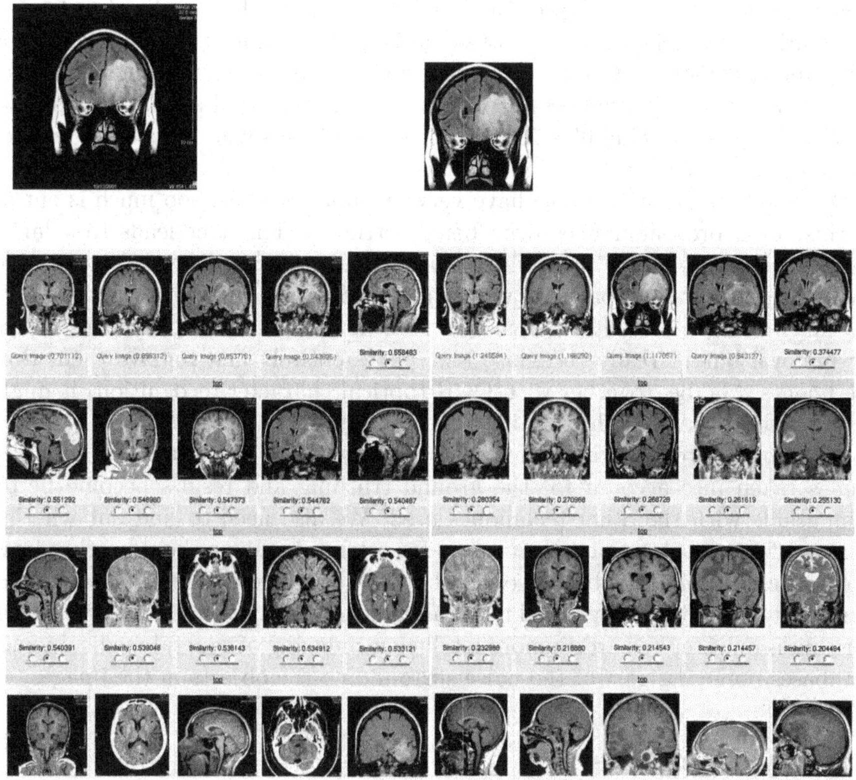

Fig. 10. Comparison of retrieval results with one step of user feedback

the database (a ground truth pool) is controlled for the ground truthing, systems with very differing techniques have a disadvantage [29], even more so if not included into the pool before the ground truthing [30], which was the case as this technique did not participate at the actual competition. We discovered that some of the images found to be relevant with the algorithm do not appear in the relevance set as in the competition no other system retrieved these images at a high enough position to be included into the pool.

Another problem is the loss of shape information when cutting off directly next to the object. Leaving a few background pixels around the object might improve retrieval as artifacts of the Gabor filters are reduced and more shape information is available in this case.

5 Conclusions

In image retrieval systems for specialised medical domains, image segmentation can focus the search very precisely. Retrieval in broad, PACS–like databases needs different algorithms to extract the most important parts of the image for indexing and retrieval. We present such an algorithm that works completely+ automatic and quickly, and as a consequence can be applied to very large databases such as teaching files or even entire PACS systems. Some of the recognised problems might be particular for our setting but they will appear in a similar fashion in other teaching files. The particular problems will need to be detected for any other collection.

Our solution is optimised to have very few images where too much is cut off as this could prevent images from being retrieved. This fact leads to a larger number of images where part of the background remains. We need to work on removing these missed parts as well while keeping the number of images with too much being removed low. One idea is to not only use the properties of text for removal but to really recognise text boxes entirely and remove them from the images. Maybe, together with OCR (optical character recognition) it might be possible to even use the obtained textual data to improve retrieval quality.

We also plan to participate in the 2005 *ImageCLEF* competition so our system is taken into account for the ground truthing and results become better comparable with the other techniques used. We also need to find out whether we cut off too much around the objects for retrieval and we should rather leave a few pixels around the objects so our shape detectors work better on the object form. Data reduction for general medical image retrieval is necessary and our algorithm is one step towards a more intelligent indexing of general medical image databases removing part of the noise surrounding the objects in the images.

Acknowledgements

Part of this research was supported by the Swiss National Science Foundation with grant 632-066041.

References

1. Smeulders, A.W.M., Worring, M., Santini, S., Gupta, A., Jain, R.: Content–based image retrieval at the end of the early years. IEEE Transactions on Pattern Analysis and Machine Intelligence **22 No 12** (2000) 1349–1380
2. Rui, Y., Huang, T.S., Chang, S.F.: Image retrieval: Past, present and future. In Liao, M., ed.: Proceedings of the International Symposium on Multimedia Information Processing, Taipei, Taiwan (1997)
3. Carson, C., Thomas, M., Belongie, S., Hellerstein, J.M., Malik, J.: Blobworld: A system for region–based image indexing and retrieval. In Huijsmans, D.P., Smeulders, A.W.M., eds.: Third International Conference on Visual Information Systems (VISUAL'99). Number 1614 in Lecture Notes in Computer Science, Amsterdam, The Netherlands, Springer (1999) 509–516
4. Flickner, M., Sawhney, H., Niblack, W., Ashley, J., Huang, Q., Dom, B., Gorkani, M., Hafner, J., Lee, D., Petkovic, D., Steele, D., Yanker, P.: Query by Image and Video Content: The QBIC system. IEEE Computer **28** (1995) 23–32
5. Bach, J.R., Fuller, C., Gupta, A., Hampapur, A., Horowitz, B., Humphrey, R., Jain, R., Shu, C.F.: The Virage image search engine: An open framework for

image management. In Sethi, I.K., Jain, R.C., eds.: Storage & Retrieval for Image and Video Databases IV. Volume 2670 of IS&T/SPIE Proceedings., San Jose, CA, USA (1996) 76–87

6. Graham, M.E., Eakins, J.P.: Artisan: A prototype retrieval system for trademark images. Vine **107** (1998) 73–80
7. Müller, H., Rosset, A., Vallée, J.P., Geissbuhler, A.: Integrating content–based visual access methods into a medical case database. In: Proceedings of the Medical Informatics Europe Conference (MIE 2003), St. Malo, France (2003)
8. Tagare, H.D., Jaffe, C., Duncan, J.: Medical image databases: A content–based retrieval approach. Journal of the American Medical Informatics Association **4** (1997) 184–198
9. Lowe, H.J., Antipov, I., Hersh, W., Arnott Smith, C.: Towards knowledge–based retrieval of medical images. The role of semantic indexing, image content representation and knowledge–based retrieval. In: Proceedings of the Annual Symposium of the American Society for Medical Informatics (AMIA), Nashville, TN, USA (1998) 882–886
10. Tang, L.H.Y., Hanka, R., Ip, H.H.S.: A review of intelligent content–based indexing and browsing of medical images. Health Informatics Journal **5** (1999) 40–49
11. Müller, H., Michoux, N., Bandon, D., Geissbuhler, A.: A review of content–based image retrieval systems in medicine – clinical benefits and future directions. International Journal of Medical Informatics **73** (2004) 1–23
12. Le Bozec, C., Zapletal, E., Jaulent, M.C., Heudes, D., Degoulet, P.: Towards content–based image retrieval in HIS–integrated PACS. In: Proceedings of the Annual Symposium of the American Society for Medical Informatics (AMIA), Los Angeles, CA, USA (2000) 477–481
13. Westerveld, T.: Image retrieval: Content versus context. In: Recherche d'Informations Assistée par Ordinateur (RIAO'2000) Computer–Assisted Information Retrieval. Volume 1., Paris, France (2000) 276–284
14. Orphanoudakis, S.C., Chronaki, C.E., Vamvaka, D.: I^2Cnet: Content–based similarity search in geographically distributed repositories of medical images. Computerized Medical Imaging and Graphics **20** (1996) 193–207
15. Liu, C.T., Tai, P.L., Chen, A.Y.J., Peng, C.H., Lee, T., Wang, J.S.: A content–based CT lung retrieval system for assisting differential diagnosis images collection. In: Proceedings of the second International Conference on Multimedia and Exposition (ICME'2001), Tokyo, Japan, IEEE Computer Society, IEEE Computer Society (2001) 241–244
16. Aisen, A.M., Broderick, L.S., Winer-Muram, H., Brodley, C.E., Kak, A.C., Pavlopoulou, C., Dy, J., Shyu, C.R., Marchiori, A.: Automated storage and retrieval of thin–section CT images to assist diagnosis: System description and preliminary assessment. Radiology **228** (2003) 265–270
17. Shyu, C.R., Brodley, C.E., Kak, A.C., Kosaka, A., Aisen, A.M., Broderick, L.S.: ASSERT: A physician–in–the–loop content–based retrieval system for HRCT image databases. Computer Vision and Image Understanding (special issue on content–based access for image and video libraries) **75** (1999) 111–132
18. Lehmann, T., Güld, M.O., Thies, C., Spitzer, K., Keysers, D., Ney, H., Kohnen, M., Schubert, H., Wein, B.B.: Content–based image retrieval in medical applications. Methods of Information in Medicine **43** (2004) 354–361
19. Güld, M.O., Kohnen, M., Keysers, D., Schubert, H., Wein, B.B., Bredno, J., Lehmann, T.M.: Quality of DICOM header information for image categorization. In: International Symposium on Medical Imaging. Volume 4685 of SPIE Proceedings., San Diego, CA, USA (2002) 280–287

20. Lehmann, T.M., Schubert, H., Keysers, D., Kohnen, M., Wein, B.B.: The IRMA code for unique classification of medical images. In: Medical Imaging. Volume 5033 of SPIE Proceedings., San Diego, California, USA (2003)
21. Squire, D.M., Müller, W., Müller, H., Pun, T.: Content–based query of image databases: inspirations from text retrieval. Pattern Recognition Letters (Selected Papers from The 11th Scandinavian Conference on Image Analysis SCIA '99) **21** (2000) 1193–1198 B.K. Ersboll, P. Johansen, Eds.
22. Chen, D., Odobez, J.M., Thiran, J.P.: A localization/verification scheme for finding text in images and video frames based on contrats independent features and machine learning methods. Signal Processing: Image Communication **19** (2004) 205–217
23. Agnithotri, L., Dimitrova, N.: Text detection for video analysis. In: IEEE Workshop on Content–based Access of Image and Video Libraries (CBAIVL'99), Fort Collins, Colorado, USA (1999) 109–113
24. Cates, J., Lefohn, A., Whitaker, R.: GIST: An interactive GPU-based level-set segmentation tool for 3d medical images. Journal on Medical Image Analysis **8** (2004 – to appear)
25. Revet, B.: DICOM Cook Book for Implementations in Modalities. Philips Medical Systems, Eindhoven, Netherlands (1997)
26. Rosset, A., Müller, H., Martins, M., Dfouni, N., Vallée, J.P., Ratib, O.: Casimage project – a digital teaching files authoring environment. Journal of Thoracic Imaging **19** (2004) 1–6
27. Clough, P., Sanderson, M., Müller, H.: A proposal for the CLEF cross language image retrieval track (imageCLEF) 2004. In: The Challenge of Image and Video Retrieval (CIVR 2004), Dublin, Ireland, Springer LNCS (2004)
28. Müller, H., Rosset, A., Geissbuhler, A., Terrier, F.: A reference data set for the evaluation of medical image retrieval systems. Computerized Medical Imaging and Graphics (2004 (to appear))
29. Zobel, J.: How reliable are the results of large–scale information retrieval experiments? In Croft, W.B., Moffat, A., van Rijsbergen, C.J., Wilkinson, R., Zobel, J., eds.: Proceedings of the 21st Annual International ACM SIGIR Conference on Research and Development in Information Retrieval, Melbourne, Australia, ACM Press, New York (1998) 307–314
30. Sparck Jones, K., van Rijsbergen, C.: Report on the need for and provision of an ideal information retrieval test collection. British Library Research and Development Report 5266, Computer Laboratory, University of Cambridge (1975)

Stripe: Image Feature Based on a New Grid Method and Its Application in ImageCLEF

Bo Qiu, Daniel Racoceanu, Chang Sheng Xu, and Qi Tian

Institute for Infocomm Research, A-star,
21, Heng Mui Keng Terrace,
119613, Singapore
{qiubo, visdaniel, xucs, tian}@i2r.a-star.edu.sg

Abstract. There have been many features developed for images, like Blob, image patches, Gabor filters, etc. But generally the calculation cost is too high. When facing a large image database, their responding speed can hardly satisfy users' demand in real time, especially for online users. So we developed a new image feature based on a new region division method of images, and named it as 'stripe'. As proved by the applications in ImageCLEF's medical subtasks, stripe is much faster at the calculation speed compared with other features. And its influence to the system performance is also interesting: a little higher than the best result in ImageCLEF 2004 medical retrieval task (Mean Average Precision — MAP: 44.95% vs. 44.69%), which uses Gabor filters; and much better than Blob and low-resolution map in ImageCLEF 2006 medical annotation task (classification correctness rate: 75.5% vs. 58.5% & 75.1%).

Keywords: Stripe, image feature, image retrieval, image annotation.

1 Introduction

In the medical field, with the increasingly important needs from clinicians, researchers, and patients, interactive or automatic image retrieval and annotation based on content are being paid more and more attention. An open medical image collection has been published by ImageCLEF since 2003 and the realistic scenarios are used to test performance of different retrieval and annotation systems [1]. ImageCLEF[1] is a part of Cross language Evaluation Forum (CLEF) and includes two parts: Adhoc (for historic photographs) and ImageCLEFmed (for medical image retrieval and annotation). In 2005, 13 groups from the world joined its medical image retrieval task, and 12 groups joined the annotation task.

Whatever the methods they used, the common basic problem is how to choose and calculate image features. Some features (like Blob, image patches, etc.) are powerful but the computational cost is too high, so as to be difficult to be used for online purpose when facing an open dataset. Some others are a little weak for all kinds of medical images. To explore more efficient features will be a permanent objective for the researchers in this field.

[1] http://ir.shef.ac.uk/imageclef/

H.T. Ng et al. (Eds.): AIRS 2006, LNCS 4182, pp. 489 – 496, 2006.

In this section firstly we will review the features reported in the past literature in ImageCLEF. Then we will give our solution for a fast and efficient image feature.

1.1 Image Features Used in ImageCLEF

In [2], based on some small regions/blocks like 5x5 or 3x3, HSV histogram (16x4x4), Canny edges (8 directions) and texture (first 5 DCT coefficients, each quantized into 3 values) are used. Then the three components are connected to a feature vector.

In [3], all the images are resized into 256x256 at first. Then they are divided into smaller blocks with the size 8x8. Average gray values of the blocks are taken to form the image feature vector.

In [4] and [5], RGB color quantization is processed and the image pixels are classified as border or interior according to 4-neighbor connection in the 4x4x4=64 color space. Then color histogram is computed based on border pixels or interior pixels and used as the feature vector. Besides of this, global texture histogram is computed from local edge pattern descriptors, using a Sobel filtering, and edge map of a whole image is divided into 4 equal squares and projected to vertical and horizontal axes to form the feature vector.

In [6], 32x32 down-scaled low resolution map (DLSM) of each image is used as layout feature. Further more, color histogram, global texture descriptors, and invariant feature histogram are also used, especially for Tamura features, which include coarseness, contrast, directionality, line-likeness, regularity, and roughness.

In [7], Gabor features and Tamura features are used.

In [8], there are following features: global texture features based on co-occurrence matrix, histogram of edge direction based on Canny edge detection, global shape features based on 7 invariant moments.

In [9], Tamura features, in which only 3 features are chosen: coarseness, contrast, directionality, are quantized into 6x8x8=384 bins based on the normalized image 256x256, ignoring the aspect ratio.

In [10], 4 features are used: 8x8 DLSM, global gray histogram by dividing the image into 9 sections, coherent moment based on pixel clustering and 8-neighbor connection, color histogram.

In [11], after partitioning the images successively into four equally sized regions (4 times), color histogram and Gabor filters are used in the different scales, and global and local features are both considered.

In [12], 16x16 DLSM is used, with regional feature Blob, and 3 texture features: contrast, anisotropy, and polarity.

Image patches [13] and random sub-windows [14] are also used in ImageCLEF and they had very good results in the annotation subtask. Both of them directly use pixel gray values of the small blocks. For image patches, salient points are extracted at first in the work of [13].

According to the last report of ImageCLEF's committee, in visual-only medical image retrieval task, Blob features reached the highest position in 2005; in annotation task, image patches got the No.1 position in 2005. But whether Blob or image patches, their computation cost is very high. This forces researchers to keep on searching more efficient features. We put forward the 'stripe'.

1.2 Stripe: An Efficient Feature Based on a New Grid Method of Images

By dividing an image into different kinds of grids, we can extract features based on the cells of grids. To make it easily handled, histogram of each cell is calculated and we call the cells 'stripes', as shown in Fig.1. For one image, all its stripes' histogram queues in order and forms a feature vector.

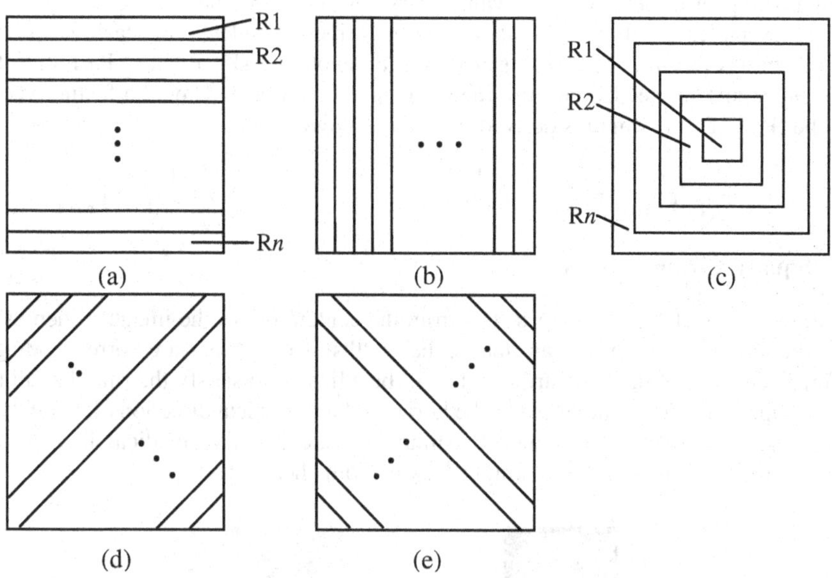

Fig. 1. Stripes' kinds: (a) horizontal stripes (R1, R2, … , R*n*); (b) vertical stripes; (c) square-cirque stripes (R1, R2, … , R*n*); (d) left-tilt stripes; (e) right-tilt stripes

Compared with other image features, stripes have following characteristics:

- Faster at the calculation speed;
- More adaptive to different sizes of images without resizing them;
- More suitable for 'vague' matching where for medical images, they are so different even in a same class, and precise shape matching is impossible but statistical features can match well
- Without losing the information of spatial positions

2 Stripe Generation

Stripe is a kind of local feature of images. To generate it, firstly we divide an image into regions; secondly we choose suitable features to represent the regions.

Considering the division of an image as shown in Fig.1, we can define different 'stripes': Horizontal, vertical, square-cirque, left-tilt, right-tilt.

2.1 Horizontal and Vertical Stripes

In this case, we use straight lines equally to divide an image into some horizontal and vertical regions. For each division, the widths of all the regions are probably equal while sometimes except the last region, because of different image sizes in an image database. But for all the regions, the bins of histogram are the same, starting from 2 at least. The number of bins will be decided by experiments and training dataset. It should be bigger if the image gray values vary a lot in the regions.

For example, for Fig.1 (a), there are n stripes: R1~Rn. For each stripe the histogram has m bins. When using percentage denotation (P_1, P_2, … , Pm), only the first (m-1) bins are needed because the total sum should be 1. Then the feature vector will be ($P_{1,1}$…$P_{n,m-1}$) which is derived from the following queue:

$$\begin{array}{cccc} (\mathrm{R1} & \mathrm{R2} & \ldots & \mathrm{R}n) \\ P_{1,1}\ P_{1,2} \ldots P_{1,m-1} & P_{2,1}\ P_{2,2} \ldots P_{2,m-1} & \ldots & P_{n,1}\ P_{n,2} \ldots P_{n,m-1} \end{array}$$

2.2 Square-Cirque Stripes

As shown in Fig.1 (c), the stripes start from the center/core of the image. When users give the number of stripes in an image, the widths of the stripes are correspondingly defined. The size of the core stripe (R1) will be adjusted to satisfy the rule that all the other stripes have the same width in the horizontal and vertical directions separately.

This kind of stripe is designed to simulate some disc-like medical images like skull, heart, etc. It is useful to distinguish skull from chest x-ray.

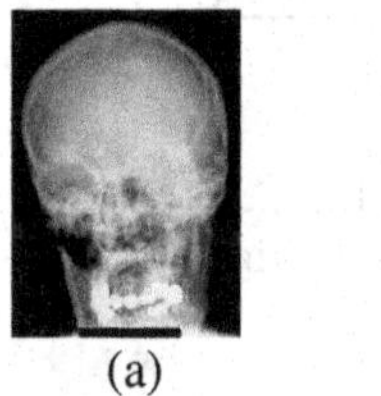
(a)

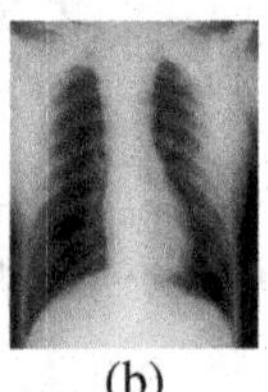
(b)

Fig. 2. Medical images: (a) skull; (b) chest

On this sense, square-cirque stripes can provide some shape information. How to generate the square-cirque feature vector is similar to section 2.1.

2.3 Left- and Right-Tilt Stripes

Tilt stripes are designed to track the rotation of an image. Fig.3 gives two rotation examples. Statistics following the direction of tilt angles will make the feature vector more meaningful. And it can decrease the influence of the white border regions.

Though there are different rotation angles for different images, only ±45° lines are chosen to form the tilt stripes in our program. Surely for more precisely calculation, more angles should be considered. But this will increase the computation cost. Why choosing ±45° lines? Besides the advantage of geometrical calculation without the need to judge by pixels, another one is that if rotate the image in a 90° angle, the left-tilt program can be reused to calculate right-tilt stripes.

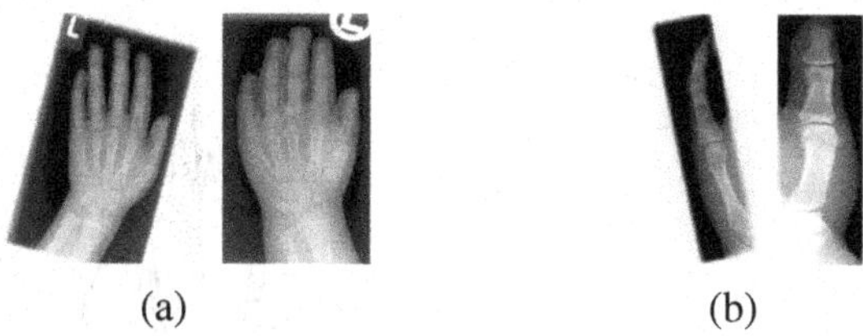

(a) (b)

Fig. 3. Rotation vs. normal: (a) hand; (b) finger

How to generate the tilt feature vector is similar to section 2.1 when the stripes queue is up-to-bottom.

2.4 Stripe Feature Vector for an Image

For an image, its feature vector is formed when connected all the stripe vectors together. For example, given an image *w* x *h* (*w*: width, *h*: height), in our program users just need to define two parameters: N_s—the number of stripes for each kind, and N_b—the number of histogram bins. Then the length of the last feature vector of the image will be N_s x (N_b-1) x 4.5. N_s should be even because the square-cirque stripes are symmetric to axes. Then it is clear that the image sizes won't influence the feature vector's length, so that it is no need to resize all the images to a fixed width and height.

3 Application of Stripes in ImageCLEF Tasks

We apply the stripe feature method into some tasks of ImageCLEF. The first experiment is based on the medical image retrieval task in 2004; the second experiment is based on the medical annotation task in 2006. The effect of stripes is compared with other features and the system performance is evaluated correspondingly.

3.1 ImageCLEF 2004: Medical Image Retrieval Task

According to [15], near to 9000 images are available in this task, and 26 topics are included. In [16], it says in all published results based on the same frame, the best result is from GIFT, with its MAP 44.69%. By Blob features and PCA as described in [17], we reached a higher MAP 45.35%.

- **Precision**

Using the same dataset and PCA method in [17], when STRIPE features take place of Blob features, the best MAP is 44.95%, which is 0.58% higher than GIFT, but 0.88% less than Blob.

- **Time**

Though Blob's result is slightly better, its computation cost is very high. In our case to process 8717 images the total time is longer than one week, with 13 computers working simultaneously. The worst PC has a 1.1G CPU and 128M RAM.

However in the case of STRIPES, when using the 'worst' PC (1.1G CPU + 128M RAM), the longest process costs 35 minutes with the parameter N_s=8, N_b=30.

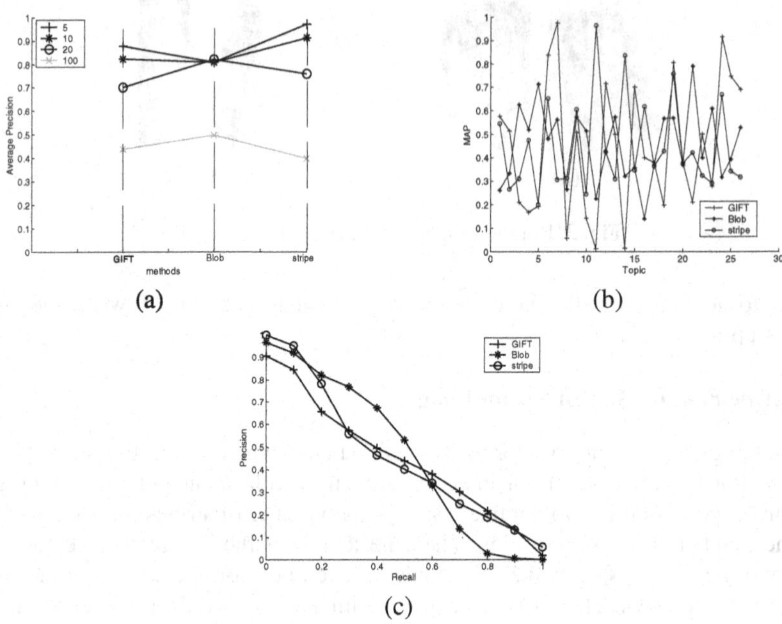

Fig. 4. Comparison: (a) MAPs at recall points (5, 10, 20, 100) for 3 methods; (b) MAPs of 26 topics; (c) PR-curve

Obviously for future online application, STRIPE has great advantage at the speed compared with Blob. In Fig.4 the comparison in MAP and PR-curve is shown.

As we can see, in Fig.4(a), at the first 5 and 10 recall points, MAP of STRIPE is higher than the others'; in Fig.4(b) and (c), STRIPE performs better than GIFT and close to Blob.

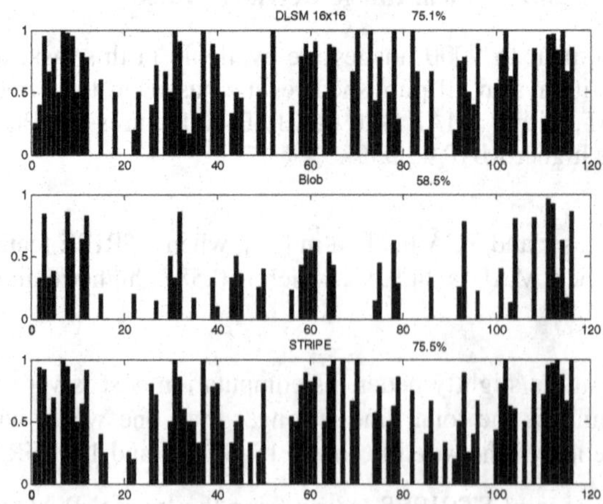

Fig.5. Correctness rates of 3 methods: DLSM, Blob, STRIPE, for all 116 classes

3.2 ImageCLEF 2006: Medical Image Annotation Task[2]

Slightly different from the task in 2005, which is described in [12], in this year the annotation task includes 116 classes, and 1000 images are provided as development set, whose ground truth is known. The training set's volume is the same as before, including 9000 images, as well as for testing set, including 1000 images.

Our experiment is based on the training set and development set. The STRIPE features, Blob features, and DLSM features (16x16) are used separately. And we use SVM as the classifier. At last the correctness rates are 75.5%, 58.5%, and 75.1%. In Fig.5 the correctness for each class is shown.

4 Conclusions and Future Work

Stripe features have great advantage at computation speed, which is very interesting in online applications. With the comparison in two tasks from ImageCLEF, the reliability of the stripe features is also proved. It can enhance the system's performance to a higher degree. For future work, more kinds of stripes should be mined, like more rotation angles for tilt stripes. And more features can be considered besides histogram.

References

1. P. Clough, H. Müller, T. Deselaers , M. Grubinger, T. Lehmann, J. Jensen, W. Hersh, The CLEF 2005 Cross-Language Image Retrieval Track, *Cross Language Evaluation Forum 2005 Workshop*, Vienna, Austria, Sep. 2005
2. Gareth J. F. Jones, K. McDonald, Dublin City University at CLEF 2005: Experiments with the ImageCLEF St Andrew's Collection, *Cross Language Evaluation Forum 2005 Workshop*, Vienna, Austria, Sep. 2005
3. Yih-Chen Chang, Wen-Cheng Lin, Hsin-Hsi Chen, Combining Text and Image Queries at ImageCLEF2005, *Cross Language Evaluation Forum 2005 Workshop*, Vienna, Austria, Sep. 2005
4. R. Besançon, C. Millet, Merging Results from Different Media: Lic2m Experiments at ImageCLEF 2005, *Cross Language Evaluation Forum 2005 Workshop*, Vienna, Austria, Sep. 2005
5. R. O. Stehling, M. A. Nascimento, and A. X. Falcao, A compact and efficient image retrieval approach based on border/interior pixel classification, *CIKM '02: Proceedings of the eleventh international conference on information and knowledge management*, McLean, Virginia, USA, 2002
6. T. Deselaers, T. Weyand, D. Keysers, W. Macherey, H. Ney, FIRE in ImageCLEF 2005: Combining Content-based Image Retrieval with Textual Information Retrieval, *Cross Language Evaluation Forum 2005 Workshop*, Vienna, Austria, Sep. 2005
7. D. Petkova, L. Ballesteros, Categorizing and Annotating Medical Images by Retrieving Terms Relevant to Visual Features, *Cross Language Evaluation Forum 2005 Workshop*, Vienna, Austria, Sep. 2005
8. Md. M. Rahman, B. C. Desai, P. Bhattacharya, Supervised Machine Learning based Medical Image Annotation and Retrieval, *Cross Language Evaluation Forum 2005 Workshop*, Vienna, Austria, Sep. 2005

[2] http://www-i6.informatik.rwth-aachen.de/~deselaers/imageclef06/medicalaat.html

9. M. O Güld, C. Thies, B. Fischer, T. M. Lehmann, Combining Global features for Content-based Retrieval of Medical Images, *Cross Language Evaluation Forum 2005 Workshop*, Vienna, Austria, Sep. 2005
10. Pei-Cheng Cheng, Been-Chian Chien, Hao-Ren Ke, Wei-Pang Yang, NCTU_DBLAB@ ImageCLEFmed 2005: Medical Image Retrieval Task, *Cross Language Evaluation Forum 2005 Workshop*, Vienna, Austria, Sep. 2005
11. H. Müller, A. Geissbühler, J. Marty, C. Lovis, P. Ruch, Using medGIFT and easyIR for the ImageCLEF 2005 Evaluation Tasks, *Cross Language Evaluation Forum 2005 Workshop*, Vienna, Austria, Sep. 2005
12. Bo Qiu, Qi Tian, Chang Sheng Xu, Report on the Annotation Task in ImageCLEFmed 2005, *Cross Language Evaluation Forum 2005 Workshop*, Vienna, Austria, Sep. 2005
13. T. Deselaers, D. Keysers, H. Ney, Improving a Discriminative Approach to Object Recognition using Image Patches, *Proc. DAGM 2005*, LNCS 3663, pp. 326-333, Vienna, Austria, Springer
14. R. Maree, P. Geurts, J. Piater, L. Wehenkel, Biomedical Image Classification with Random Subwindows and Decision Trees, *ICCV2005 workshop-CVBIA*
15. W. X., Bo Qiu, Qi Tian, Chang Sheng Xu, et al, Content-based Medical Retrieval Using Dynamically Optimized Regional Features, *ICIP2005*, Genova, Italy, Sep.11-14, 2005
16. H. Müller, A. Geissbühler and P. Ruch, "Report on the CLEF Experiment: Combining Image and Multilingual Search for Medical Image Retrieval", *2005 imageCLEF proceedings*, Springer Lecture Notes, to appear.
17. P. Clough, M. Sanderson and H. Müller, "The CLEF Cross Language Image Retrieval Track (ImageCLEF) 2004", *the 8th in the series of European Digital Library Conferences*, ECDL 2004, September, Bath, UK, 2004

An Academic Information Retrieval System Based on Multiagent Framework

Toru Abe, Yuu Chiba, Suoya, Baoning Li, and Tetsuo Kinoshita

Tohoku University, Katahira 2–1–1, Aoba-ku, Sendai, 980–8577 Japan

Abstract. In real-life searches in information, a set of information retrieved by a query influences user's knowledge. Usually this influence inspires the user with new ideas and new conception of the query. As a result, the search in information is iterated while the user's query is continually shifting in part or whole. This sort of search is called an "evolving search," and it performs an important role also in academic information retrieval. To support the utilization of digital academic information, this paper proposes a novel system for academic information retrieval. In the proposed system, which is based on a multiagent framework, each piece of academic information is structured as an agent and provided with autonomy. Consequently, since a search is iterated by academic information itself, part of an evolving search is entrusted to the system, and the user's load to retrieve academic information can be reduced effectively.

1 Introduction

To support the utilization of digital academic information (e.g., scholarly monographs, valuable books, historical materials, etc.), various academic information retrieval systems have been proposed. However, even if users use those systems, many tasks (e.g., evaluating the retrieved information, generating new queries based on the evaluation results, etc.) are left for the users to find the needed academic information.

To overcome this problem, we propose a novel system for academic information retrieval. The proposed system is based on a multiagent framework, and designed especially for supporting part of an "evolving search," which performs an important role in information retrieval. In this system, each piece of academic information is structured as an agent and provided with autonomy. Consequently, since a search is iterated by academic information itself, part of an evolving search is entrusted to the system, and the user's load to retrieve academic information can be reduced effectively.

2 Retrieval of Academic Information

This section explains an "evolving search," and describes the existing methods for information retrieving and their limitations on the evolving search support.

H.T. Ng et al. (Eds.): AIRS 2006, LNCS 4182, pp. 497–507, 2006.

2.1 Evolving Search

In real-life searches in information, a set of information retrieved by a query influences user's knowledge. Usually this influence inspires the user with new ideas and new conception of the query. Consequently, as shown in Fig. 1, the search in information is iterated while the user's query is continually shifting in part or whole. This sort of search is called an "evolving search" [1]. Since the change of user's knowledge is fairly active in searching academic information, the evolving search performs an important role also in this process. Therefore, it is expected that a system for supporting the evolving search is effective in academic information retrieval.

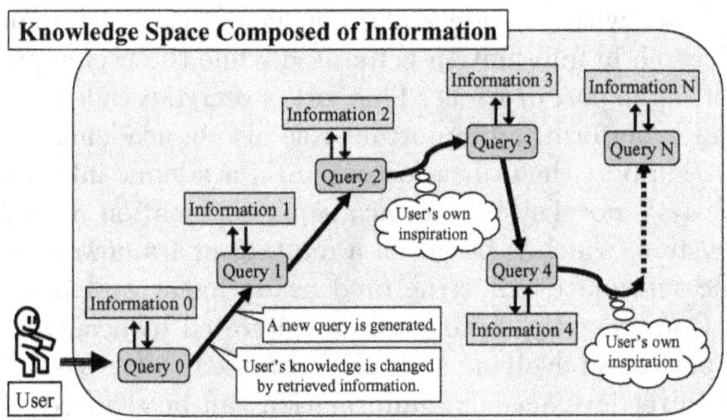

Fig. 1. Evolving search

Query shifts in the evolving search are of two types; we refer to the one as an "unexpected shift" and the other as an "expected shift." Since the unexpected shift is chiefly due to the user's own inspiration, it is difficult to predict on a system. Compared to this, the expected shift is mainly caused by the data obtained from the information that is retrieved with the preceding query. This type shift corresponds to routine procedures such as retrieve one set of information after another according to the data obtained from the previously retrieved information. Digital academic information is generally given several metadata (e.g., subject, references, etc.). Therefore, by extracting metadata from the retrieved information and generating new queries from them, it is possible to produce the expected shifts on the system and release the user from the routine procedures. Hence, in this paper, to support the utilization of academic information, we propose an information retrieval system that produces the expected shifts and carries out part of the evolving search.

2.2 Existing Methods for Information Retrieval

To support the utilization of academic information, various systems have been proposed. They assist the users mainly with information searches based on metadata such as Dublin Core [2], and use the following methods for improving the accuracy in information searching.

- Statistics are computed from the user's search records, and they are introduced into information search process as the user's interests [3].
- With the aid of thesaurus or ontology, a query given by the user is converted or expanded to more appropriate ones [4,5].
- Retrieval results evaluated by the user are fed back to the system, and used for the succeeding information search process [6,7].

Although those methods can improve the accuracy in information searching, they have the following limitations on the evolving search support:

- Since it takes time to detect the change of user's interests from his/her search records, the user's interests change cannot be introduced into information search process immediately.
- Even if a new knowledge or viewpoint has been discovered and published, that knowledge or viewpoint cannot be employed for information search process until it is registered on the thesaurus or ontology.
- Retrieved information is analyzed by the user, and the iteration of information search process based on the analysis is entrusted to the user.

In the process of academic information retrieval, various metadata can be obtained from the retrieved information. However, the existing methods cannot employ them for information search process at an opportune moment in an appropriate way, and thus they cannot produce the expected shifts in the evolving search effectively.

3 Evolving Search Support

This section proposes a method for supporting part of the evolving search, and then explains the concept of active information resource, which is a key technique to actualize the proposed method.

3.1 Evolving Search Support Cycle

For supporting part of the evolving search in academic information, the proposed method iterates the cycle of the following three steps until the number of iterations reaches a given limit or new queries cannot be generated any further. This evolving search support cycle aims to produce the expected shifts on a system effectively.

Step 1. Basic Search
A given query is compared to the metadata in each piece of academic information, and a set of appropriate academic information is acquired.

Step 2. Selection
The pieces of information acquired by the preceding "Basic Search" are classified into groups, and several pieces of information important to the user are selected from each group.

Step 3. Query Generation
Metadata are extracted from the information selected by the preceding "Selection," and new queries for the succeeding "Basic Search" are generated by modifying the current query according to the extracted metadata.

This cycle can be summarized as Fig. 2. Through the iteration of the cycle, the proposed method intends that the set of information retrieved until the t th cycle $c^{(t)}$ should be used by the steps in the $t+1$ th cycle $c^{(t+1)}$ at an opportune moment in an appropriate way.

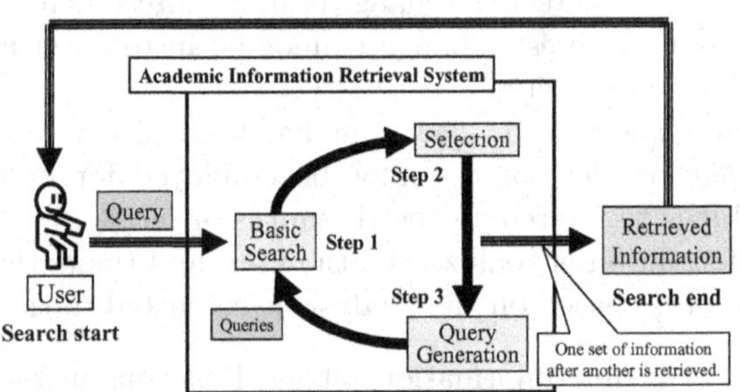

Fig. 2. Evolving search support cycle

3.2 Active Information Resource

In every evolving search support cycle, the proposed method generates new queries from the set of retrieved academic information. For generating highly effective new queries, it is necessary to make full use the metadata obtained from each piece of retrieved academic information individually. To achieve this, by employing the concept of active information resource, each piece of academic information is provided with autonomy.

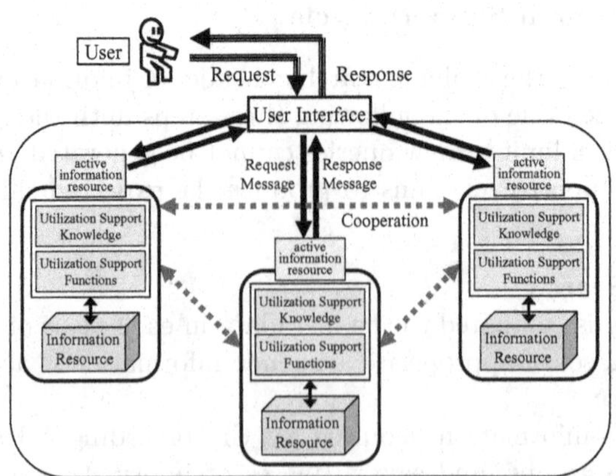

Fig. 3. Active information resource

An active information resource is a distributed information resource (e.g., text, image, sound, etc.) that is structured as an agent to enhance and extend its activity and flexibility. As shown in Fig. 3, an agent maintains not only the contents of information resource themselves, but also the knowledge and functions for supporting the utilization of its contents. When an agent receives a request message from the user or other agents, by using the utilization support knowledge (USK) and invoking the utilization support functions (USFs), the agent carries out tasks for its own contents and responds to the request message actively. Furthermore, through the exchange of messages, each agent can cooperate autonomously with others in a distributed environment, and achieves complicated tasks flexibly.

By applying this concept to academic information, academic information can actively carry out the steps in the evolving search support cycle, and employ the various metadata obtained from the retrieved academic information at an opportune moment in an appropriate way.

4 Academic Information Retrieval System with Active Information Resource

This section describes the structure of the proposed academic information retrieval system, and explains its process for supporting the user's retrieval work.

4.1 Structure of the Proposed System

As shown in Fig. 4, the proposed system consists of two areas; one is a "public area" and the other is a "personal area." In the public area, each piece of academic information is structured as an agent (AI-Ag) based on the concept of active information resource, and these AI-Ags are open to all users of the system. Every user has his/her own personal area, where retrieved AI-Ags are stored and an agent (User-Ag) serving as an interface between the user and the system is activated.

Each AI-Ag maintains the contents of academic information themselves, and preserves the metadata of the contents as part of USK. Part of the metadata (currently used in the system) are listed in Table 1. In addition to these, every AI-Ag has USK and USFs for carrying out the evolving search support cycle actively.

Table 1. Part of the metadata in AI-Ag's utilization support knowledge (USK)

(1) identifier	An unambiguous reference to the resource within a given context.
(2) title	A name given to the resource.
(3) creator	An entity primarily responsible for making the content of the resource.
(4) subject	The topic of the content of the resource.
(5) coverage	The extent or scope of the content of the resource.
	((2)$\sim$(5) are regarded as the sets of keywords.)

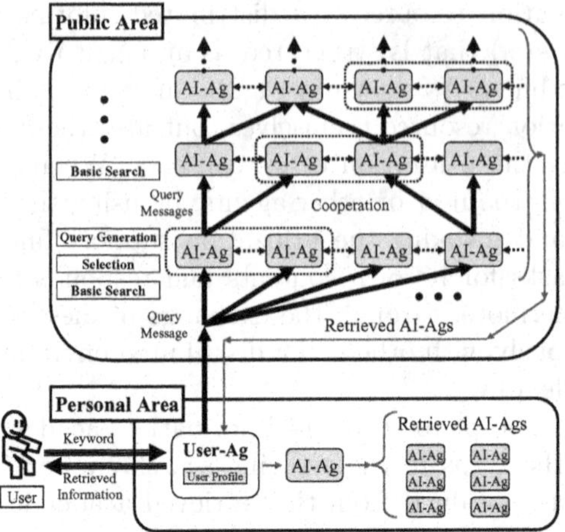

Fig. 4. Structure of the proposed system

A User-Ag maintains a user profile that specifies search conditions and the user's interests (the items specifying the user's interests are represented by the sets of keywords and priorities) as part of USK. Part of the items (currently used in the system) are listed in Table 2. In addition to these, the User-Ag has USK and USFs for serving as the interfaces between the user and the system as well as between the personal area and the public area.

Table 2. Part of the items in User-Ag's utilization support knowledge (USK)

(1) max_{t}	The maximum number of the evolving search support cycle to be iterated.
(2) max_{n}	The maximum number of AI-Ags to be selected in a cycle.
(3) min_{k}	The threshold of accepting a query. (The minimum number of the matched keywords.)
(4) coverage	The extent or scope of the user's interests.
(5) subject	The topic of the user's interests. ((4), (5) are represented by the sets of keywords and priorities.)

4.2 Process in the Proposed System

Step 0. Initial Query Generation

The User-Ag receives a keyword from the user, and generates an initial query by composing the search conditions (Table 2 (1)∼(3)), keywords in the USK (Table 2 (4), (5)), and the received keyword. For every keyword in the query, a priority is given; the received keyword is provided with the top priority, and others are provided with the priorities in the USK. The generated query is sent to AI-Ags in the public area as a message, and the first cycle is started.

Step 1. Basic Search

When AI-Ags in the public area receive a query message, each AI-Ag compares the keywords in its own USK (Table 1 (2)~(5)) with the keywords in the received query. If the number of matched keywords is larger than min_{k} (Table 2 (3)), the AI-Ag accepts the query message and decides that its own academic information is suitable for the query.

Step 2. Selection

To prevent redundant query generation, the following procedures are carried out through the cooperation among the AI-Ags that accept the query message.

1. The AI-Ags that accept the query message classify themselves into several groups. Each group is comprised of AI-Ags that have the same keyword (we refer to it as a "group keyword").
2. In each group, every AI-Ag computes its own importance I by

$$I = \prod_{n=1}^{N} P_n \ , \tag{1}$$

where N is the number of keywords in the query, and P_n is the penalty for mismatched keyword of the n th priority (currently, $P_1 = 0.80, P_2 = 0.85, P_3 = 0.90, P_{4\sim N} = 0.95$ are used for mismatched keywords, and $P_n = 1$ is used for matched keywords).

3. The AI-Ag of the highest I is chosen from every group (let it be AI-Ag*). Each AI-Ag* searches the personal area for the AI-Ags having its own group keyword, compares the keywords in its own USK with the keywords in the found AI-Ags' USK, and determines m, which is the number of mismatched keywords. To reflect the user's interests, each AI-Ag* revises its own importance by

$$I = I \times \alpha^m \ , \tag{2}$$

where α is a penalty (currently, $\alpha = 0.90$ is used).

In this step, not more than max_{n} (Table 2 (2)) AI-Ags are selected from the set of AI-Ag*s in order of revised I.

Step 3. Query Generation

Each selected AI-Ag generates new query by modifying the current query. The mismatched keyword with the top priority in the current query is replaced by the keyword with the top priority in the USK (the replaced keyword is provided with the top priority). Each generated new query is sent to AI-Ags in the public area, and the succeeding Basic Search is started.

The cycle of Step 1~3 is iterated until the number of iterations reaches a given limit max_{t} (Table 2 (1)) or new queries cannot be generated any further.

5 Experiments

To confirm the effectiveness of the proposed system, we performed experiments of academic information retrieval.

5.1 Experimental Environment

A prototype system was implemented on a PC (Celeron 2.0GHz, 768MB memory, RedHat9.0). AI-Ags and User-Ag were actualized as agents in ADIPS/DASH framework [8]. As shown in Fig. 5, each agent in the ADIPS/DASH framework consists of a communication module CM, knowledge module KM, action module AM, and base process BP. To cooperate with other agents, an agent exchanges messages with others by using the CM, and to perform assigned tasks, the agent controls its own BP through the AM. The KM holds rule sets used for exchanging messages and controlling BP. The USK in AI-Ags and User-Ag was installed as the rule set in the KM, and the USFs were implemented as the BPs (Java programs).

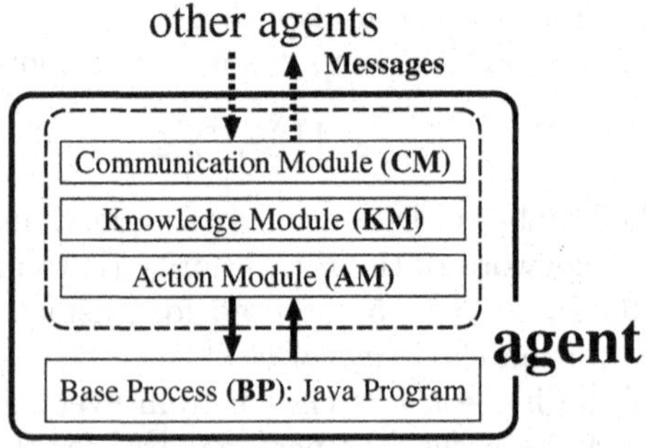

Fig. 5. Structure of an agent in the ADIPS/DASH framework

For the experiments, 150 journal papers (information retrieval, network, knowledge engineering, image processing, and semiconductor) were used as a test set. Each of them was structures as an AI-Ag, and the metadata extracted from a journal paper were installed as part of the USK in the corresponding AI-Ag.

Ten graduates of information science enrolled the experiments as test users. Every user selected the required journal papers from the test set in advance. The journal papers selected by each user were regarded as the correct result of information retrieval for the corresponding user in the experiments. Every user set the USK in the User-Ag severally in advance, and through the experiments, each user used the same setting of his/her own User-Ag ($max_{\mathrm{t}}, max_{\mathrm{n}}, min_{\mathrm{k}}$ were set to 5, 3, 2, respectively).

5.2 Results of the Experiments

Experiment 1.
This experiment was equivalent to the evolving search without the evolving search support cycle.
Step 0. To begin with, each user chose a keyword.
Step 1. The user inputted the keyword into the system.
Step 2. The system carried out the Basic Search only once, and presented the user with the AI-Ags (academic information) selected in the Selection.

Step 3. The user chose one new keyword among the metadata obtained from the presented AI-Ags.

Step 1∼3 were iterated 5 times, and AI-Ags selected through this iteration were regarded as the retrieval result.

Experiment 2.

This experiment was equivalent to the proposed method except that AI-Ags stored in the personal area (i.e., the AI-Ags retrieved by the previous cycles) were not referred.

Step 1. The user inputted the keyword into the system.

Step 2. The system iterated the cycle until the number of iterations reached $max_t = 5$ times or new queries could not be generated any further.

Experiment 3.

This experiment was equivalent to the proposed method. In the experiment, the same procedures as Experiment 2 were performed except that AI-Ags stored in the personal area were referred.

Table 3. Results of the experiments

Experiment 1: without the evolving search support cycle

	avg.	min.	max.
Total number of			
retrieved pieces	**46.7**	24	73
retrieved correct pieces	**1.4**	0	4
Recall [%]	**46.0**	0.0	100.0
Precision [%]	**3.6**	0.0	9.1
F measure [%]	**6.8**	0.0	16.7
Necessary time [s]	**715.7**	444.0	1787.0

Experiment 2: with the evolving search support cycle referring no retrieved information in the personal area

	avg.	min.	max.
Total number of			
retrieved pieces	**23.6**	17	33
retrieved correct pieces	**1.5**	0	3
Recall [%]	**40.5**	0.0	75.0
Precision [%]	**6.5**	0.0	10.5
F measure [%]	**11.1**	0.0	15.8
Necessary time [s]	**94.2**	87.0	101.0

Experiment 3: with the evolving search support cycle referring retrieved information in the personal area

	avg.	min.	max.
Total number of			
retrieved pieces	**25.2**	15	31
retrieved correct pieces	**2.3**	1	4
Recall [%]	**62.8**	25.0	100.0
Precision [%]	**9.2**	3.4	13.3
F measure [%]	**16.0**	6.3	23.0
Necessary time [s]	**93.4**	87.0	108.0

Every test user performed Experiment 1, 2, and 3 one time. In Experiment 2 and 3, each user inputted the same keyword as he/she chose in Step 0 of Experiment 1. Table 3 summarizes the results of the experiments. The evolving search support cycle improved the precision in academic information retrieval (Experiment 2, 3); moreover, by using the pieces of academic information stored in the personal area, it increased both the recall and the precision substantially (Experiment 3). These results show that the proposed system could employ the metadata obtained from the retrieved academic information effectively. Furthermore, the evolving search support cycle reduced the necessary time for academic information retrieval drastically (Experiment 2, 3). The reason for this is that the evolving search support cycle carried out the Query Generation and eliminated the time required for the user to generate new queries. These results mean that the proposed system could release the user from the routine procedures.

6 Conclusions

This paper has proposed a novel system for academic information retrieval, and discussed the details and the effectiveness of the proposed system.

The proposed system is based on a multiagent framework, and designed especially for supporting the evolving search caused by expected query shifts. In the proposed system, a search in information is iterated by academic information itself, and the metadata obtained from the retrieved academic information can be introduced effectively into the search process. The experimental results showed that the proposed system improved the accuracy and the efficiency in the academic information retrieval.

For future research directions, we would like to address the following issues:

- Currently, the proposed system used naive keyword matching algorithms in the steps in the evolving search support cycle. For increasing the accuracy in academic information retrieval, it is necessary to improve the algorithms for these steps, particularly for introducing the change of user's knowledge into these steps effectively.
- To apply the proposed system to a huge set of academic information, some mechanisms are required for reducing the message traffic among AI-Ags. For this purpose, it is expected that methods for clustering AI-Ags and managing them as hierarchically structured groups are effectual.

References

1. Bates, M.J.: The design of browsing and berrypicking techniques for the online search interface. Online Review **13**(5) (1989) 407–424
2. Dublin Core Metadata Initiative: DCMI metadata terms. (http://dublincore.org/documents/dcmi-terms)
3. Liu, F., Yu, C., Meng, W.: Personalized web search for improving retrieval effectiveness. IEEE Trans. Knowledge Data Eng. **16**(1) (2004) 28–40

4. Gonzalez, M., de Lima, V.L.S.: Semantic thesaurus for automatic expanded query in information retrieval. In: Proc. 8th Int. Symposium String Processing and Information Retrieval. (2001) 68–75
5. Sim, K.M., Wong, P.T.: Toward agency and ontology for web-based information retrieval. IEEE Trans. Syst., Man, Cybern. C **34**(3) (2004) 257–269
6. Chakrabarti, K., Ortega-Binderberger, M., Mehrotra, S., Porkaew, K.: Evaluating refined queries in top-k retrieval systems. IEEE Trans. Knowledge Data Eng. **16**(2) (2004) 256–270
7. Jing, F., Li, M., Zhang, H.J., Zhang, B.: Relevance feedback in region-based image retrieval. IEEE Trans. Circuits Syst. Video Technol. **14**(5) (2004) 672–681
8. Dash — Distributed Agent System based on Hybrid architecture —. (http://www.agent-town.com/dash/index.html)

Comparing Web Logs: Sensitivity Analysis and Two Types of Cross-Analysis

Nikolai Buzikashvili

Institute of system analysis, Russian Academy of Science
9 prospect 60 Let Oktyabrya, 117312 Moscow, Russia
buzik@cs.isa.ru

Abstract. Different Web log studies calculate the same metrics using different search engines logs sampled during different observation periods and processed under different values of two controllable variables peculiar to the Web log analysis: a client discriminator used to exclude clients who are agents and a temporal cut-off used to segment logged client transactions into temporal sessions. How much are the results dependent on these variables? We analyze the sensitivity of the results to two controllable variables. The sensitivity analysis shows significant varying of the metrics values depending on these variables. In particular, the metrics varies up to 30-50% on the commonly assigned values. So the differences caused by controllable variables are of the same order of magnitude as the differences between the metrics reported in different studies. Thus, the direct comparison of the reported results is an unreliable approach leading to artifactual conclusions. To overcome the method-dependency of the direct comparison of the reported results we introduce and use a cross-analysis technique of the direct comparison of logs. Besides, we propose an alternative easy-accessible comparison of the reported metrics, which corrects the reported values accordingly to the controllable variables used in the studies.

1 Introduction

Different works describe the results of the user studies centered on the search behavior on the Web which analyze different search engines logs (e.g. [1, 2, 5]), and some metrics of the search behavior reported in these studies have the same name (session length, terms per query, fractions of certain queries, etc.). Few works [3, 5] directly compare these results to evaluate the differences between the user interactions with different engines. However, *no work* is devoted to the techniques of the Web log analysis and to the comparison techniques. As a result, we cannot be sure that a direct comparison of the reported results of the differently conducted studies is well grounded. We can't be sure that the differences discovered in the comparative analysis are not artifacts resulting from the differences in the methods used in the compared studies. This paper tries to fill this gap.

The differences of the results may be caused by the combinations of the features of interfaces, query languages and search methods used by the different search engines, the differences in the contexts, the cultural differences, and *the differences between*

H.T. Ng et al. (Eds.): AIRS 2006, LNCS 4182, pp. 508 – 513, 2006.

the techniques used in different studies. The latter instrumental factor has been out of a rigorous investigation so far. If this factor may explain only minor differences (i.e. those smaller than the differences between the reported results) we may confine ourselves to comparison of the reported results. On the contrary, if the method varying induces varying of the results at lest of the same order of magnitude as the reported differences, then (1) the role of this factor should be taken into account in the comparison of the *reported* results, and (2) this is a sound reason to conduct a cross-analysis of the logs.

In the *Web* log analysis, the sources of the results differences are *controllable variables* assigned by the researchers. These are *a client discriminator* to exclude non-human users (agents) and multi-user clients (local networks), and *a temporal cut-off* to segment a time series of user transactions into temporal sessions. Sensitivity of the calculated metrics to these variables is investigated in Chapter 4. For each log we investigate the effect of the controllable variables on the considered metrics. Since this influence is significant we should use approaches alternative to the direct comparison of the reported results.

In Chapter 5, we consider the cross-analysis *of the logs*, in which the same tool under the same conditions analyses logs of different search engines. However, the direct cross-analysis of the logs is an expensive procedure and its applications are limited by the fact that the explored logs of not all search engines are available.

In Chapter 6, we describe another approach — to take into account the values of controllable variables used in the different log studies and to compare the accordingly corrected reported results rather the reported results themselves. This *indirect cross-analysis of the reported results* is an easy-accessible procedure based on the reported results: if each study reports the values of the controllable variables we can use corresponding values of the correction factors to correct and to compare the cognominal metrics reported in these studies. The sensitivity analysis allows to estimate the interval values of the correction factors.

2 Controllable Variables of the Web Log Analysis

The knowledge about the user search behavior that can be obtained from the Web transaction logs radically differs from the knowledge extracted from the logs of earlier non-HTTP-based information retrieval systems. The limits of the Web log analysis result of: 1) impossibility to reliably detect an individual human user; 2) the Web logs contain transactions rather than queries segmented into search sessions.

To avoid these problems the Web log analysis uses two controllable variables: 1) *a client discriminator* (usually measured in transactions) to exclude local networks and agents conducting more transactions than a client discriminator (*CD*) value and 2) a *temporal cut-off* (*TCO*) to segment a client transactions into temporal sessions. These controllable variables are arbitrary and differently set in different Web log studies. For example the *Excite* project used 15 min *TCO* [5], the *Yandex* study [1] reports about two *TCO* values (30 min and 1 hour). In turn, the *CD* values used in the different log studies seem to be incomparable: *CD* may be measured either in unique queries or in transactions, it may be assigned to the entire observation period, while observation periods may vary from few hours to weeks, so *CD* values assigned for the whole periods are incomparable.

3 Terms, Data and Technique Used

In this paper, we distinguish *unique queries* submitted by a client during some time interval (e.g. during the whole observation period) and *transactions*. While transactions are frequently referred to as queries we avoid doing this. We consider only one of possible session classes, a *temporal session* as any sequence of the single user transactions with the same search engine cut from the previous and successive sessions by *TCO* interval. To analyze logs we developed the *Crossover* mobile programs. This pack is used in both the sensitivity analysis and the direct cross-analysis of logs. The main program of the *Crossover* is adjustable to the specific search engine query languages and log formats.

The logs of two search engines were used: the Russian-language *Yandex* (7 days, 2005, 175,000 users accepted cookies), and the *Excite*: 1999 log fragment (8 hours, 537,639 clients) and 2001 log sample (24 hours, 305,000 clients). The *Excite* data were used to elaborate the results comparable with both (a) the *Yandex* results yielded *by the same analyzer* and (b) results previously yielded *on the same logs*.

The observation periods of different log samples are different. When CD is measured in unique queries, it creates a problem. E.g., what a compatible CD value should be assigned to the *week* sample if 10 unique queries are set for the *8-hour* sample? To overlook this problem we use the *sliding temporal window* technique. We select a sliding window size *T* (not longer than the smallest observation period), assign certain client discriminator *N* (measured in unique queries or transactions) to this window and slide the window over time series of the client transactions comparing a number of client's queries transactions covered by the window with *N*. If in some position of the window this number is bigger than *N* we exclude this client as an agent.

In this paper, *CD* is measured in unique queries covered by 1-hour sliding window.

4 Results Sensitivity to Controllable Variables

The metrics used in the Web log analysis are divided into 2 classes: (1) depending on both *CD* and *TCO*, and (2) depending only on *CD*. The "*per transaction*" metrics (a query length or fractions of some kind of queries, e.g., Boolean queries, queries containing quotations, etc.) don't depend on *TCO*. The same metrics considered "*per unique query*" also don't depend on *TCO* when we consider unique queries *per client* but these metrics depend when "unique queries *per session*" are considered.

All the metrics depend on the *CD* variable. Fig. 1 shows how several *TCO*-independent metrics are changed over unique queries (left) and transactions (right) submitted during the entire observation period: the metric value *m(n)* on the figure corresponds to all clients the number of queries (transactions) submitted by whom is not bigger than *n*. The behavior of the fraction of AND-queries is surprising, especially as a function of the number of transactions.

Table 1 reports the "unit" values of some metrics corresponding to the combination <*TCO*=15 min, *CD*=1 unique query per 1-hour sliding window>. We use these unit values to normalize the values corresponding to combinations of longer *TCO*s and bigger *CD*s. For example, if the 'average session length' metric equals to 1.53 transactions for <*TCO*=15 min, *CD*=1 unique query per 1-hour sliding window> and it equals to 2.11 transactions under <*TCO*=30 min, *CD*=3 u. q. per 1-hour window>

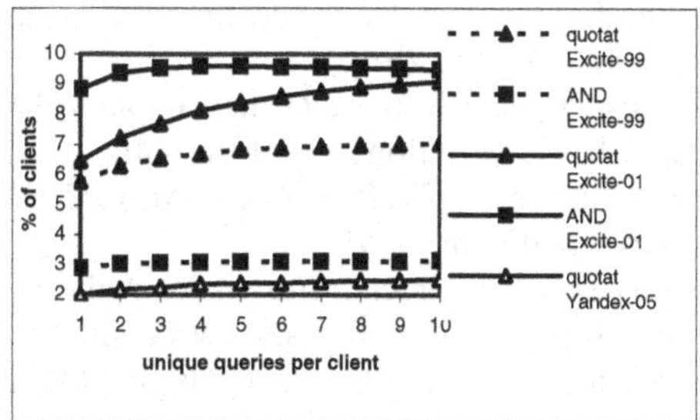

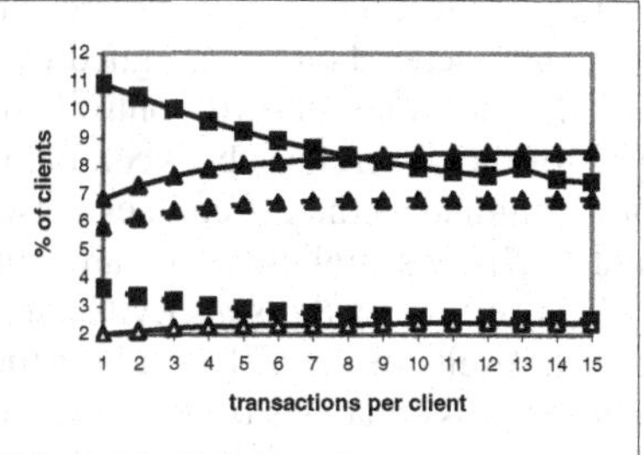

Fig. 1. Cumulative fractions of AND- and quotations- queries in the Excite-99, Excite-01 and the Yandex logs over unique queries and transactions submitted by the client

combination then the normalized value of this metric is 2.11/1.53 = 1.38. Table 2a shows the normalized values of "transactions per temporal session" metric, which depends on both *TCO* and *CD*. Table 2b shows values of *TCO*-independent metrics.

Table 1. Metrics values corresponding to the 'unit' combination of the controllable variables

	transactions per temporal session	*AND*	*OR*	*quotations*	*plus*	*minus*
Excite-1999	1.53	2.78%	0.13%	5.76%	2.07%	0.05%
Excite-2001	1.59	8.28%	0.08%	6.36%	2.08%	0.06%
Yandex-2005	1.66	0.19%	0.03%	1.98%	0.30%	0.11%

Table 2a. Normalized values of the *transactions per temporal session* metric

	Excite 1999			*Excite 2001*			*Yandex 2005*		
TCO (min)	15	30	60	15	30	60	15	30	60
CD (unique queries)									
1	1	1.03	1.05	1	1.03	1.04	1	1.04	1.07
2	1.19	1.25	1.28	1.21	1.26	1.29	1.16	1.23	1.28
3	1.31	1.38	1.44	1.35	1.42	1.47	1.28	1.37	1.45
5	1.44	1.53	1.61	1.52	1.62	1.70	1.43	1.57	1.67

Table 2b. Normalized values of *TCO*-independent metrics corresponding to different *CDs*

	Excite 1999			*Excite 2001*			*Yandex 2005*		
***CD*:**	**2**	**3**	**5**	**2**	**3**	**5**	**2**	**3**	**5**
AND	1.08	1.10	1.12	1.11	1.14	1.16	1.05	1.11	1.21
PLUS	1.13	1.20	1.28	1.20	1.39	1.60	1.17	1.20	1.37
quotations	1.10	1.13	1.19	1.13	1.21	1.39	1.10	1.15	1.21

The influence of *TCO* is predictable: the greater this cut-off value, the longer a temporal session and the greater are corresponding metrics. On the other hand, as seen from Table 2a this influence is small. Thus, we can directly compare the reported results of the studies, which differ only by the *TCO* values. However, the results significantly depend on the assigned *CD* value (Tables 2a and 2b).

The variations of two controllable variables, especially the *CD* may explain 30-50% of the differences between metric values. These variations are of the same order of magnitude as the differences between the results of the *Excite*, *AlataVista*, *Fireball* and *FAST* logs studies, which are considered as significant [3, 5].

The results of the sensitivity analysis show that the direct comparison of the reported results of the different log studies is unreliable. This is a reason for conducting a direct cross-analysis *of the logs*. On the other hand, the sensitive analysis suggests a way of comparison of *the reported results* more reliable than the direct comparison.

5 Direct Cross-Analysis of Logs: *Crossover* Study

Undoubtedly reliable results are produced by a *cross-analysis of logs*, in which the same tool under the same or maximum similar conditions analyzes the logs. To conduct the cross-analysis *Crossover* study we use our *Crossover* tool. Tables 1 and 3 illustrate the results of the cross-analysis of the *Excite*-99, *Excite*-2001 and *Yandex*-2005 logs. The *Yandex* users (Table 1) use Boolean operators very rarely but we should take into account a *usage context*: the *Yandex* help recommends the users not to use Boolean operators. When compared, the results of the *Crossover* study and the results of the *Excite* project produced on the same *Excite* logs (Table 3) appear to be close within the limits induced by controllable variables.

Table 3. *Excite* project and *Crossover* study results on the same logs and the same15 min *TCO*

	Excite-1999 log					***Excite-2001 log***				
	Exc. prj.	***Crossover study***				***Exc. prj.***	***Crossover study***			
CDs used:	**N/A**	**1**	**2**	**3**	**5**	**N/A**	**1**	**2**	**3**	**5**
terms/query	2.4	2.40	2.46	2.49	2.52	2.6	2.51	2.60	2.64	2.69
AND	3%	2.8%	3.0%	3.1%	3.1%	10% all	8.3%	9.2%	9.4%	9.6%
PLUS	2%	2.1%	2.3%	2.5%	3.0%	Boolean	2.1%	2.8%	2.9%	3.3%
quotations	5%	5.8%	6.3%	6.5%	6.9%	9%	6.4%	7.2%	7.7%	8.4%

6 Indirect Cross-Analysis of the Reported Results

As mentioned above, the sensitivity analysis suggests a way to compare the reported results: to re-calculate the results as if they had been calculated under the same combination of controllable variables. Let some metric m be estimated for two combinations of controllable variables (TCO_1,CD_1) and (TCO_2,CD_2). A traditional approach directly compares $m(TCO_1,CD_1)$ and $m(TCO_2,CD_2)$. Alternatively, the indirect cross-analysis of the reported results compares $m(TCO_1,CD_1)/c_m(TCO_1,CD_1)$ and $m(TCO_2,CD_2)/c_m(TCO_2,CD_2)$, where $c_m(TCO,CD)$ is a conversion function analogous to functions tabulated in Tables 2a and 2b. While the stability of these functions estimated on the different logs is an issue, the values estimated on the *8-hour Excite*-99 fragment and the *week Yandex* sample are surprisingly similar (Tables 2a and 2b).

Let's consider an example. Let one study report 5% of queries containing *PLUS*, and the other speaks about 4%. The studies use correspondingly the 5 and 1 unique query per 1-hour sliding window *CD* values. Since *CD*=5 corresponds to [1.28; 1.60] interval value for the *PLUS*-fraction metric (Table 2b) we should compare 4% with the interval value [5/1.60; 5/1.28] ≈ [3.1; 3.9]. Thus, a fraction of the *PLUS* queries is bigger in the study reported the smaller value.

7 Conclusion

To make a step to valid comparison of the different search engines logs we introduced and applied the techniques of the *sensitivity analysis* and two types of the *cross-analysis*. The sensitivity analysis shows significant varying of the metrics values depending on the controllable variables of the Web log analysis. The differences caused by controllable variables are of the same order of magnitude as the differences between the metrics reported in different studies. The direct comparison of the reported results is an unreliable approach. To overcome the method-dependency of the direct comparison of the reported results we introduce and evaluate an expensive cross-analysis technique of the direct comparison *of the logs* and an easy-accessible indirect comparison *of the reported metrics*, which corrects the reported values of the metrics accordingly to the used values of the controllable variables.

Acknowledgements. Finally, the author feels obliged to thank Ilya Segalovich, Ian Ruthven and anonymous reviewers for helpful comments.

References

1. Buzikashvili, N.: The Yandex study: First findings. *Internet-math.* Yandex (2005), 95–120
2. Holscher, C., Strube, G.: Web search behavior of internet experts and newbies. International Journal of Computer and Telecommunications Networking, 33 (1-6) (2000), 337–346
3. Jansen, B.J., Spink, A.: How are we searching the World Wide Web? An analysis of nine search engine transaction logs, *Inf. Processing & Management*, 42(1) (2006), 248–263
4. Silverstein, C., Henzinger, M., Marais, H., Moricz, M.: Analysis of a very large web search engine query log, *SIGIR Forum*, 33 (1) (1999), 6–12
5. Spink A., Ozmutlu H.C., Ozmutlu S., Jansen B.J.: U.S. versus European Web search trends. *SIGIR Forum*, 36 (2) (2002), 32–38

Concept Propagation Based on Visual Similarity
Application to Medical Image Annotation

Jean-Pierre Chevallet, Nicolas Maillot, and Joo-Hwee Lim

IPAL French/Singapore Join Lab, CNRS/A-STAR
Universite Joseph Fourier/Institute for Infocomm Research
21 Heng Mui Keng Terrace, Singapore 119613
viscjp@i2r.a-star.edu.sg,
nmaillot@i2r.a-star.edu.sg,
joohwee@i2r.a-star.edu.sg

Abstract. This paper presents an approach for image annotation propagation to images which have no annotations. In some specific domains, the assumption that visual similarity implies (partial) semantic similarity can be made. For instance, in medical imaging, two images of the same anatomic part in a given modality have a very similar appearance. In the proposed approach, a conceptual indexing phase extracts concepts from texts; a visual similarity between images is computed and then combined with conceptual text indexing. Annotation propagation driven by prior knowledge on the domain is finally performed. Domain knowledge used is a meta-thesaurus for both indexing and annotations propagation. The proposed approach has been applied on the imageCLEF medical image collection.

Keywords: Conceptual Indexing, Annotation Propagation, Visual Similarity.

1 Introduction

Automatic image annotation is still a very difficult task. Achieving reliable automatic annotation is even more difficult when the resulting annotations are to be used for information retrieval purposes. The information retrieval task is highly semantic: it consists of finding a document which content matches the user information need specified most of the time at an abstract level. In the case of medical domain, the task can be to find similar cases of a patient in order to find some similar way to cure a disease. This level of abstraction is difficult to reach because of the so-called "semantic gap" between meaning and signal.

In some cases, the assumption that visual similarity implies (partial) semantic similarity can be made. Assuming we have a collection of images with free text annotations, the goal is to transfer some image annotations to images that are visually similar. Compared to other approaches found in the literature, the proposed approach makes use of a priori knowledge in order to select the annotations which are strongly characterized by the visual appearance. This paper

H.T. Ng et al. (Eds.): AIRS 2006, LNCS 4182, pp. 514–521, 2006.

presents some results on concept propagation using a medical database of images associated with text.

We propose an approach composed of three main phases: *conceptual text indexing*, *visual indexing*, and *concept propagation*. A graphical representation of the processing flow can be found in Fig. 1. The input of the process is a set of images coupled with textual documents. The role of conceptual text indexing is to extract concepts from the annotations (1). Visual indexing (2) computes the similarity between pair-wise images of the collection. Annotation propagation (3) uses the results of (1) and (2) to produce new annotations.

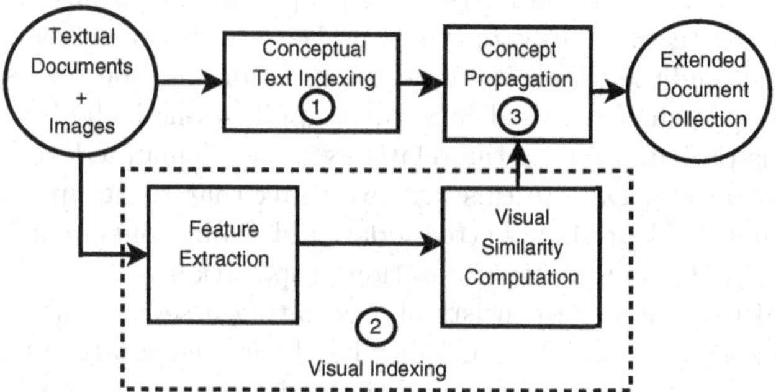

Fig. 1. Overview of the proposed approach

This paper is structured as following. Section 2 gives an overview of the state of the art of annotation propagation. Section 3 explains the conceptual indexing process of textual annotations. Section 4 shows how visual indexing leads to a similarity matrix of the images contained in the collection. Section 5 details the principles of concept propagation. Section 6 details results obtained on ImageCLEFmed [1] collection. Section 7 concludes this paper and presents future research directions.

2 Related Works

As explained in [2], two categories of approaches are usually used for dealing with the automatic image annotation problem. The first one poses the automatic image annotation problem as a *supervised learning* problem. In this case, classes are defined a priori and annotations and images are considered separately. A set of manually annotated samples is then used to train a set of classifiers. Image annotation is finally performed by using the resulting classifiers. This type of approach has been used in [3] and in [4]. The second category of approaches consists of a *joint* modelling of the statistical links between textual annotations and image features. The process of annotation propagation is then modeled as *statistical inference*. An example of such work can be found in [2] and [5].

The novelty of the approach proposed in this paper is that the semantic propagation is performed at a *conceptual level* whereas most other works stay at a text level. This means that propagation is not based on textual annotations but on a conceptual representation derived from the textual annotations. Propagating annotations at a conceptual level enables to solve ambiguity problems when a given image is inconsistently annotated. It is also the only way to apply a semantic control of propagated annotations. In this paper, we will show that this control is crucial to ensure positive benefit of propagation.

3 Conceptual Text Indexing

Conventional text Information Retrieval (IR) approaches extract words or terms from text and use them directly for indexing. Despite staying at the text "signal" level (syntactic) and the simplicity of word extraction, this method is most of the time effective enough to be used in various applications including web search. This success is probably due to the relatively higher semantic level of text compared to other media. Despite this fact, we think that rising up one more step in abstraction should produce better index and should offer more control like semantic filtering for automatic annotation propagation.

Conceptual text indexing consists of associating a set of concepts to a document and using it as index. Conceptual indexing naturally solves the term mismatch problem described in [6], and the multilingual problem.

In this paper, we focus on concept propagation for completing missing information in a restricted domain (medical), and we use a large dedicated concept set: UMLS[1] provided by the National Library of Medicine (NLM). For concept extraction, we uses MetaMap [7] provided by NLM. This tool associates part of speech to input text to detect noun phrases. One of the difficulties in concept finding relies on term variation. Even a large meta-thesaurus like UMLS cannot cover all possible term variations. MetaMap computes a large set of possible variation and try to match the largest best combination to an UMLS entry. After concept extraction, we use the classical vector space model for indexing and matching.

4 Visual Indexing

The visual indexing process is based on patch extraction on all the images of the collection followed by feature extraction on the resulting patches. Let I be the set of the N images in the document collection. First, each image $i \in I$ is split into n_p patches $\{p_i^k\}_{1 \leq k \leq n_p}$. Patch extraction on the whole collection results in $n_p \times N$ patches. Fig. 2 shows the result of patch extraction for one image.

For a patch p_i^k, the numerical feature vector extracted from p_i^k is noted $fe(p_i^k) \in \mathcal{R}^n$. The low-level features extracted on patches are RGB histograms. 32 bins are considered for each color component. This implies that the numerical vector extracted for one patch is of dimension 96.

[1] http://www.nlm.nih.gov/research/umls/

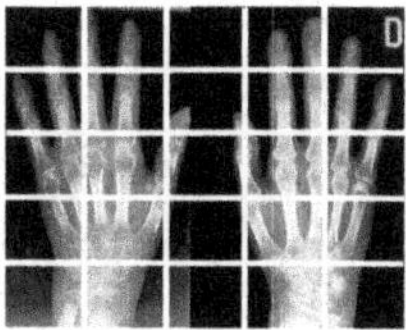

Fig. 2. This grid corresponds to the 25 patches extracted on this image. One histogram (96 bins) is extracted for each patch of the grid.

The visual similarity between two images i and j, $\delta_I(i,j)$, is computed as the average of pairwise L_2 distances between feature vectors extracted from patches.

$$\delta_I(i,j) = \frac{\sum_{k=1}^{n_p} L_2(fe(p_i^k), fe(p_j^k))}{n_p}$$

Our goal is, for any image i of the collection, to evaluate its visual similarity to all the other images j of the collection. Since $\delta_I(i,j) = \delta_I(j,i)$ and that $\delta_I(i,i) = 0$, the number of distances to compute is $(N^2/2) - N$. δ_I can be seen as disimilarity matrix. For our experiments, we have chosen L_2 as the Euclidian distance.

5 Concept Propagation

Concept propagation is based on the following hypothesis: two images that are visually similar share some common semantics. Indexing documents at a conceptual level enables to control the meaning of the propagation. Our model consists of a set of textual documents D, and a set of associated images I indexed by a set of concepts $\Delta(d) = \{c_i\}$. Each image in I is associated with at least one document in D. The image set I is projected into a space from which we only know a distance function δ_I between any image pair $\delta_I(i_1, i_2)$ from $I \times I$. The distance δ is related to a visual similarity. Each document d is connected to any other document through the image space. This produces a new function δ_D between two documents. Among possible choices for the building of δ_D, for our purpose, as our emphasis is the visual similarity, we propose to define $\delta_D(d_i, d_j)$ as the minimum distance δ_I between the set of images associated to d_i and the set associated to d_j. As each document is associated to a set of concepts, *concept propagation* consists of transferring concepts from a source document d_i to a target set of document $\tau(d_i)$ based on function $\delta_D(d_i, d_j)$. We define the target function $\tau(d)$ that selects a target set of documents and $\tau_k(d_i)$ as the k Nearest Neighbors of d_i according to δ_D. Finally, concept propagation P on document d is performed by producing a new concept set $\Delta_1(d)$ from the original one $\Delta_0(d)$:

$$\Delta_1(\tau_k(d)) = P(\Delta_0(\tau_k(d)), \Delta_0(d))$$

Propagation P can be performed several times to obtain a final indexing Δ_n after n iterations.

6 Experimentation

As our method is based on semantic indexing terms categories, we need an indexing method that enables filtering on semantic categories. We have chosen the Medical CLEF [1] test collection because documents of this collection are medical cases associating medical images and multi-lingual texts with a set of test queries.

We use UMLS meta-thesaurus which is structured as a semantic network composed of *semantic types* and of *semantic groups*. Here are some examples of semantic groups: Living Beings (LIVB), Diagnostic Procedures (PROC), Disorders (DISO), Anatomy (ANAT).

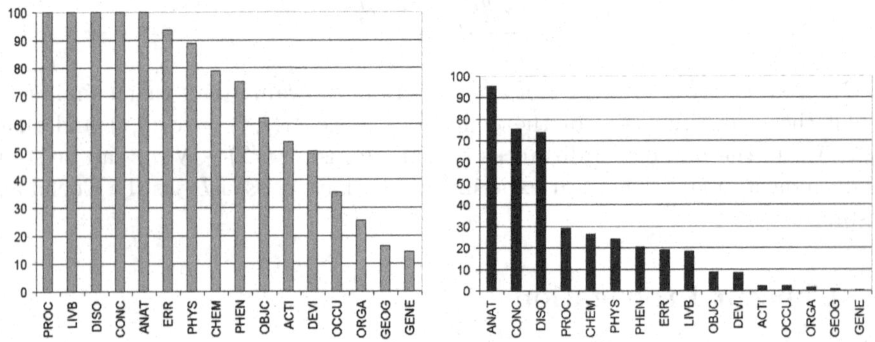

Fig. 3. Semantic group repartition on CasImage (left) and Peir (right) collections. CasImage and Peir are sub-collections of CLEFImage.

Fig. 3 shows the percentage of presence of a semantic group in the collection CasImage (left). Five main semantic groups appear in every document (Procedure, Living Beings, Disorders, Concepts & Ideas, Anatomy). The need for propagation of annotation is then rather low because this collection is already well annotated. The right side of Fig. 3 shows a much different repartition of the semantic groups. Only Anatomy is present in 95% of the documents. The goal of the propagation is to try to fill remaining 5% left. This task is rather difficult. The information on procedure, which includes the modality of the image, is much lower: Less that 30% of image have this information. Query distribution is different and emphasis Disorders, Procedures, Concepts and Anatomy. Our experimentations are focused on these semantic groups: due to the query distribution, we get more chance to measure a noticeable change when computing mean average precision using this query set. For testing concept propagation, we set up an indexing baseline (Fig. 4 and Fig. 5). We use the Log for term frequency weighting, inverse document frequency and cosine for matching distance. For this baseline, we obtain a mean average precision[2] of 0.1307.

[2] Computed using treceval program.

We have tested different propagation P with different k for the target function τ_k. The influence of k value is difficult to interpret because of its instability. We have experimentally fixed $k = 10$. The full propagation (noted FULL in Fig 4) just transfers every concept. Propagation using semantic group authorizes concept transfers only if no concept from this group exists in the target document. A more precise filtering (noted ST_ANAT and ST_PROC) performs at the semantic type level: in this case, only concepts of the semantic group are used. Concepts of this group are transferred only if no concept of the same semantic type is found in the target document.

Figure 4 shows propagation results for few semantic groups. The left figure shows the results for the first propagation, the right figure after 4 propagations. These results show a clear difference between this simple semantic filtering and no filtering: performing semantic filtering is *always* better than transferring all index information.

In order to examine closely the effect of propagation, we have degraded the original MetaMap indexing, by randomly deleting a given percentage of concepts belonging to the ANAT semantic groups. The goal is then to recover the lost information by the propagation mechanism described in section 5. Figure 5 shows

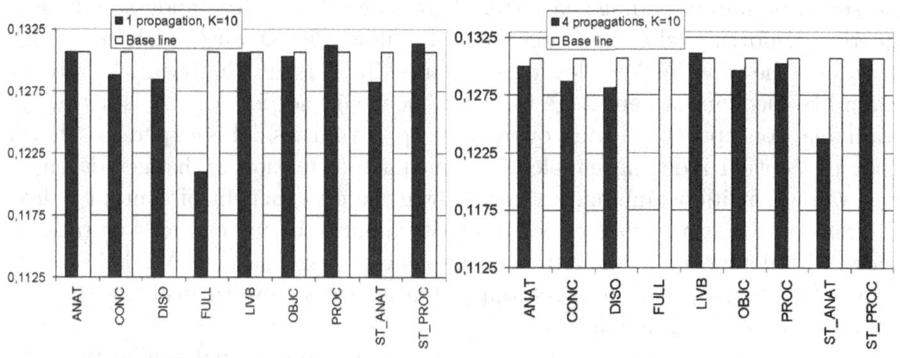

Fig. 4. Propagation 1 time (left), 4 times (right) for k=10

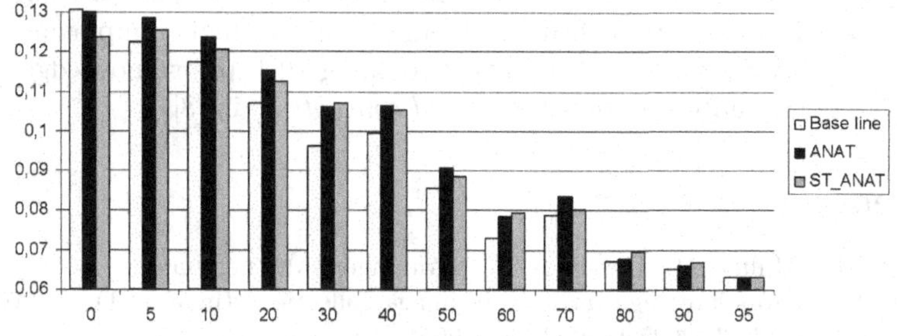

Fig. 5. Degradation of semantic group ANAT

results from 0% to 95% of anatomical concepts degradation. The MAP of the baseline also decreases when more degradation occurs. In Fig 5 we can see the evolution of the MAP for the baseline and ANAT and ST_ANAT extension when degradation percentage is increasing. The same $k = 10$ is chosen and the results are computed after 4 expansions (Δ_4 concepts set). Although we are not enable to fully recover removed information, we can notice that 5% of information lost affect the baseline but not really the extended version (for ANAT expansion only). Moreover, from 5% to 70% of information lost, our concept propagation always performs better than the baseline. From our point of view, this result clearly demonstrates the benefits on using a conceptual indexing and performing semantically controlled concepts propagation. If a large amount of information is lost (80% to 95%), no improvement is measured.

7 Conclusion and Future Works

Indexing for Information retrieval consists of associating annotations to documents to facilitate their access. User needs are usually highly semantic, and low level indexing (at signal level), is not precise enough to solve precise queries because of the well known semantic gap. To bridge this gap, we see no other possibilities than providing the system knowledge it needs to "understand" user queries. Unfortunately, collecting valuable knowledge at large scale is a huge task, and the way this knowledge can be used in IR is not that clear. In sub domain like medicine, where large valuable knowledge set like UMLS is nowadays available, conceptual indexing opens new opportunities. This paper has shown that conceptual indexing enables inter-media information exchanges. Even, if this work is at its beginning, we have shown the clear benefit of concept indexing and propagation using visual similarity, mainly in the case of low ratio of index completeness (by simulation using concept degradation). We believe our approach opens up new research opportunities to go toward bridging the gap from visual to semantic indexing.

One important improvement that should be made to visual similarity is the use of several categories of low-level features (e.g. texture features). The use of several types of features would enable adaptive propagation based on prior knowledge. For instance, geometric features are very important to characterize the anatomy. Therefore, these features should be given a higher importance when propagating anatomy-related concepts. Providing such prior knowledge can be done in a user-friendly way by using *visual concepts* as in [8].

References

1. Clough, P., Muller, H., Deselaers, T., Grubinger, M., Lehmanna, T., Jensen, J., Hersh, W.: Cross-language retrieval in image collections (imageclef). In: Working Notes for the CLEF 2005 Workshop. (2005)
2. Monay, F., Gatica-Perez, D.: On image auto-annotation with latent space models. In Rowe, L.A., Vin, H.M., Plagemann, T., Shenoy, P.J., Smith, J.R., eds.: ACM Multimedia, ACM (2003) 275–278

3. Li, J., Wang, J.Z.: Automatic linguistic indexing of pictures by a statistical modeling approach. IEEE Trans. Pattern Anal. Mach. Intell. **25** (2003) 1075–1088
4. Hare, J.S., Lewis, P.H.: Saliency-based models of image content and their application to auto-annotation by semantic propagation. In: Proceedings of the Second European Semantic Web Conference (ESWC2005), Heraklion, Crete (2005)
5. Iyengar, G., Duygulu, P., Feng, S., Ircing, P., Khudanpur, S.P., Klakow, D., Krause, M.R., Manmatha, R., Nock, H.J., Petkova, D., Pytlik, B., Virga, P.: Joint visual-text modeling for automatic retrieval of multimedia documents. In: MULTIMEDIA '05: Proceedings of the 13th annual ACM international conference on Multimedia, New York, NY, USA, ACM Press (2005) 21–30
6. Xu, J., Croft, W.: Improving the effectiveness of informational retrieval with local context analysis. ACM Transaction on Information System **18** (2000) 79–112
7. Aronson, A.R.: Effective mapping of biomedical text to the umls metathesaurus: the metamap program. Proc AMIA Symp (2001) 17–21
8. Maillot, N., Thonnat, M., Boucher, A.: Towards ontology-based cognitive vision. Mach. Vis. Appl. **16** (2004) 33–40

Query Structuring with Two-Stage Term Dependence in the Japanese Language

Koji Eguchi[1] and W. Bruce Croft[2]

[1] National Institute of Informatics, Tokyo 101-8430, Japan
eguchi@nii.ac.jp
[2] University of Massachusetts, Amherst, MA 01003-9264, USA
croft@cs.umass.edu

Abstract. We investigate the effectiveness of query structuring in the Japanese language by composing or decomposing compound words and phrases. Our method is based on a theoretical framework using Markov random fields. Our two-stage term dependence model captures both the global dependencies between query components explicitly delimited by separators in a query, and the local dependencies between constituents within a compound word when the compound word appears in a query component. We show that our model works well, particularly when using query structuring with compound words, through experiments using a 100-gigabyte web document collection mostly written in Japanese.

1 Introduction

Japanese text retrieval is required to handle several types of problems specific to the Japanese language, such as compound words and segmentation [1]. To treat these problems, word-based indexing is typically achieved by applying a morphological analyzer, and character-based indexing has also been investigated. Some researchers compared this kind of character-based indexing with word-based indexing, and found little difference between them in retrieval effectiveness (e.g., [1,2,3]). Some other researchers made use of supplemental phrase-based indexing in addition to word-based indexing for English (e.g., [4]) or for Japanese (e.g., [5]). However, we believe this kind of approaches is not appropriate for the languages, for instance Japanese, in which individual words are frequently composed into a long compound word and the formation of an endless variety of compound words is allowed.

Meanwhile, the structured query approach has been used to include more meaningful phrases in a proximity search query to improve retrieval effectiveness [6,7]. A few researchers have investigated this approach to retrieval for Japanese newspaper articles [1,3]; however, they emphasized formulating a query using n-grams and showed that this approach performed comparably in retrieval effectiveness with the word-based approach. We are not aware of any studies that have used structured queries to formulate queries reflecting Japanese compound words or phrases appropriately. Phrase-based queries are known to perform effectively, especially against large-scale and noisy text data such as typically appear

H.T. Ng et al. (Eds.): AIRS 2006, LNCS 4182, pp. 522–529, 2006.

on the Web [8,7]. Again, we have not seen any studies that used structured queries to effectively retrieve web documents written in Japanese.

In this paper, we use the structured query approach using word-based units to capture, in a query, compound words and more general phrases of the Japanese language. Our approach is based on a theoretical framework using Markov random fields [7]. We experiment using a large-scale web document collection mostly written in Japanese.

2 Retrieval Model and Term Dependence Model

'Indri' is a search engine platform that can handle large-scale document collections efficiently and effectively [9]. The retrieval model implemented in Indri combines the language modeling [10] and inference network [11] approaches to information retrieval. This model allows structured queries similar to those used in 'InQuery' [11] to be evaluated using language modeling estimates within the network. Because we focus on query formulation rather than retrieval models, we use Indri as a baseline platform for our experiments. We omit further details of Indri because of space limitations. See [9] for the details.

Metzler and Croft developed a general, formal framework for modeling term dependencies via Markov random fields (MRFs) [7], and showed that the model is very effective in a variety of retrieval situation using the Indri platform. MRFs are commonly used in statistical machine learning to model joint distributions succinctly. In [7], the joint distribution $P_\Lambda(Q, D)$ over queries Q and documents D, parameterized by Λ, was modeled using MRFs, and for ranking purposes the posterior $P_\Lambda(D|Q)$ was derived by the following ranking function, assuming a graph G that consists of a document node and query term nodes:

$$P_\Lambda(D|Q) \stackrel{rank}{=} \sum_{c \in C(G)} \lambda_c f(c) \tag{1}$$

where $Q = t_1...t_n$, $C(G)$ is the set of cliques in an MRF graph G, $f(c)$ is some real-valued feature function over clique values, and λ_c is the weight given to that particular feature function.

Full independence ('fi'), *sequential dependence* ('sd'), and *full dependence* ('fd') are assumed as three variants of the MRF model. The full independence variant makes the assumption that query terms are independent of each other. The sequential dependence variant assumes dependence between neighboring query terms, while the full dependence variant assumes that all query terms are in some way dependent on each other. To express these assumptions, the following specific ranking function was derived:

$$P_\Lambda(D|Q) \stackrel{rank}{=} \sum_{c \in T} \lambda_T f_T(c) + \sum_{c \in O} \lambda_O f_O(c) + \sum_{c \in O \cup U} \lambda_U f_U(c) \tag{2}$$

where T is defined as the set of 2-cliques involving a query term and a document D, O is the set of cliques containing the document node and two or more query terms that appear contiguously within the query, and U is the set of cliques containing the document node and two or more query terms appearing noncontiguously within the query.

3 Query Structuring with Two-Stage Term Dependence

In compound words that often appear for instance in Japanese, the dependencies of each constituent word are tighter than in more general phrases. Therefore, we consider that these term dependencies should be treated as global between query components that make up a whole query and as local within a compound word when the compound word appears in a query component. Metzler and Croft's term dependence model, which we summarized in the previous section, gives a theoretical framework for this study, but must be enhanced when we consider more complex dependencies as mentioned above. In this paper, we propose *two-stage term dependence model* that captures term dependencies both between query components in a query and between constituents within a compound word.

To achieve the model mentioned above, we extend the term dependence model given in Eq. (2), on the basis of Eq. (1), as follows:

$$P_\Lambda(D|Q) \stackrel{rank}{=} \sum_{c_q \in T(Q)} \lambda_T f_T(c_q) + \sum_{c_q \in O(Q)} \lambda_O f_O(c_q) + \sum_{c_q \in O(Q) \cup U(Q)} \lambda_U f_U(c_q) \quad (3)$$

where

$$f_T(c_q) = f'_T\Big(\sum_{q_k \in c_q} \sum_{c_t \in T(q_k)} \mu_T g_T(c_t)\Big)$$

$$f_O(c_q) = f'_O\Big(\sum_{q_k \in c_q} \sum_{c_t \in O(q_k)} \mu_O g_O(c_t)\Big)$$

$$f_U(c_q) = f'_U\Big(\sum_{q_k \in c_q} \sum_{c_t \in O(q_k) \cup U(q_k)} \mu_U g_U(c_t)\Big) \ . \quad (4)$$

Here, Q consists of query components $q_1 \cdots q_k \cdots q_m$, and each query component consists of individual terms $t_1 \cdots t_n$. $T(Q)$, $O(Q)$ and $U(Q)$ can be defined in the same manner as in Eq. (2) with the query components consisting of a whole query, while $T(q_k)$, $O(q_k)$ and $U(q_k)$ are defined with the individual terms consisting of a query component. The feature functions f'_T, f'_O and f'_U and another feature functions g_T, g_O and g_U can be given in the same manner as f_T, f_O and f_U that were defined in Section **2**, respectively. Hereafter, we assumed that the constraint $\lambda_T + \lambda_O + \lambda_U = 1$ was imposed independently of the query, and assumed $\mu_T = \mu_O = \mu_U = 1$ for simplicity. When Q consists of two or more query components and each of which has one term, Eq. (3) is equivalent to Eq. (2). The model given by Eq. (2) can be referred to as the *single-stage term dependence model.* When $f'_T(x) = f'_O(x) = f'_U(x) = x$ for any x, Eq. (3) represents dependencies only between constituent terms within each query component, which can be referred to as the *local term dependence model*; otherwise, Eq. (3) expresses the *two-stage term dependence model*.

According to Eq. (3), we assumed the following instances, considering special features of the Japanese language [12].

Two-stage term dependence models

(1) *glsd+* expresses the dependencies on the basis of the sequential dependence both between query components and between constituent terms within a query component, assuming dependence between neighboring elements. The beliefs (scores) for the resulting feature terms/phrases for each of f_T, f_O and f_U are combined as in Eq. (3).

(2) *glfd+* expresses the dependencies between query components on the basis of the full dependence, assuming all the query components are in some way dependent on each other. It expresses the dependencies between constituent terms within a query component on the basis of the sequential dependence.

Here in f_T, f_O and f_U, each compound word containing prefix/suffix word(s) is represented as an exact phrase and treated the same as the other words, on the basis of the empirical results reported in [12]. Let us take an example from the NTCIR-3 WEB topic set [13], which is written in Japanese. The title field of Topic 0015 was described as three query components, "オゾン層 オゾンホール 人体" (which mean 'ozone layer', 'ozone hole' and 'human body'). A morphological analyzer converted this to "オゾン" ('ozone' as a general noun) and "層" ('layer' as a suffix noun), "オゾン" ('ozone' as a general noun) and "ホール" ('hole' as a general noun), and "人体" ('human body' as a general noun). The following are Indri query expressions corresponding to Topic 0015, according to the glsd+ and glfd+ models, respectively:

```
 weight( λ_T #combine( #1( オゾン 層 ) オゾン ホール 人体 )
         λ_O #combine( #1( オゾン 層 ) #od2( オゾン ホール ) 人体 )
         λ_U #combine( #uwN_4( #1( オゾン 層 ) オゾン ホール )
                       #uwN_3( オゾン ホール 人体 ) ) )
#weight( λ_T #combine( #1( オゾン 層 ) オゾン ホール 人体 )
         λ_O #combine( #1( オゾン 層 ) #od2( オゾン ホール ) 人体 )
         λ_U #combine( #uwN_4( #1( オゾン 層 ) オゾン ホール )
                       #uwN_3( オゾン ホール 人体 )
                       #uwN_3( #1( オゾン 層 ) 人体 )
                       #uwN_5( #1( オゾン 層 ) オゾン ホール 人体 ) ) )
```

where #1(·) indicates exact phrase expressions; #od$M$(·) indicates phrase expressions in which the terms appear ordered, with at most $M-1$ terms between each; and #uwN_ℓ(·) indicates phrase expressions in which the specified terms appear unordered within a window of N_ℓ terms. N_ℓ is given by $(N_1 \times \ell)$ when ℓ terms appear in the window.

Local term dependence models

(3) *lsd+* indicates the glsd+ model with $f'_T(x) = f'_O(x) = f'_U(x) = x$ for any x, ignoring the dependencies between query components.

(4) *lfd+* indicates the glfd+ model with $f'_T(x) = f'_O(x) = f'_U(x) = x$ for any x, ignoring the dependencies between query components.

The following is an example query expression of 'lsd+' on Topic 0015.

```
#weight( λT #combine( #1(オゾン 層) オゾン ホール 人体 )
         λO #combine( #1(オゾン 層) #od2(オゾン ホール) 人体 )
         λU #combine( #1(オゾン 層) #uwN2(オゾン ホール) 人体 ) )
```

4 Experiments

4.1 Data and Experimental Setup

For experiments, we used a 100-gigabyte web document collection 'NW100G-01', which was used for the NTCIR-3 Web Retrieval Task ('NTCIR-3 WEB') [13] and for the NTCIR-5 WEB Task ('NTCIR-5 WEB') [14]. We used the topics and relevance judgment data of the NTCIR-3 WEB for training the model parameters[1]. We used the data set that was used in the NTCIR-5 WEB for testing[2]. All the topics were written in Japanese. The title field of each topic gives 1–3 query components that were suggested by the topic creator.

We used the texts that were extracted from and bundled with the NW100G-01 document collection. In these texts, all the HTML tags, comments, and explicitly declared scripts were removed. We segmented each document into words using the morphological analyzer 'MeCab version 0.81'[3]. We did not use the part-of-speech (POS) tagging function of the morphological analyzer for the documents, because the POS tagging function requires more time. We used Indri to make an index of the web documents in the NW100G-01 using these segmented texts. We used several types of stopwords in the querying phase, on the basis of the empirical results reported in [12].

In the experiments described in the following sections, we only used the terms specified in the title field. We performed morphological analysis using the 'MeCab' tool described above to segment each of the query component terms delimited by commas, and to add POS tags. Here, the POS tags are used to specify prefix and suffix words that appear in a query because, in the query structuring process, we make a distinction between compound words containing prefix and suffix words and other compound words, as described in Section **3**.

[1] For the training, we used the relevance judgment data based on the *page-unit document model* [13] included in the NTCIR-3 WEB test collection.

[2] We used the data set used for the *Query Term Expansion Subtask*. The topics were a subset of those created for the NTCIR-4 WEB [15]. The relevance judgments were additionally performed by extension of the relevance data of the NTCIR-4 WEB. The objectives of this paper are different from those of that task; however, the data set is suitable for our experiments.

[3] ⟨http://www.chasen.org/~taku/software/mecab/src/mecab-0.81.tar.gz⟩.

Table 1. Optimization results using a training data set

	AvgPrec$_a$	%increase	AvgPrec$_c$	%increase
base	0.1543	0.0000	0.1584	0.0000
lsd+	0.1624	5.2319	0.1749	10.4111
lfd+	0.1619	4.9120	0.1739	9.7744
glsd+	0.1640	6.2731	0.1776	12.0740
glfd+	0.1626	5.4140	0.1769	11.6788
naive-sd	0.1488	-3.5551	0.1472	-7.0743
naive-fd	0.1488	-3.5427	0.1473	-7.0496
ntcir-3	0.1506	-2.3774	0.1371	-13.4680

4.2 Experiments for Training

Using the NTCIR-3 WEB test collection, we optimized each of the models defined in Section **3**, changing each weight of λ_T, λ_O and λ_U from 0 to 1 in steps of 0.1, and changing the window size N for the unordered phrase feature as 2, 4, 8, 50 or ∞ times the number of words specified in the phrase expression. Additionally, we used $(\lambda_T, \lambda_O, \lambda_U) = (0.9, 0.05, 0.05)$ for each N value above. Note that stopword removal was only applied to the term feature f_T, not to the ordered/unordered phrase features f_O or f_U. The results of the optimization that maximized the mean average precision over all 47 topics ('AvgPrec$_a$') are shown in **Table 1**. This table includes the mean average precision over 23 topics that contain compound words in the title field as 'AvgPrec$_c$'. '%increase' was calculated on the basis of 'base', the result of retrieval not using query structuring. After optimization, the 'glsd+' model worked best when $(\lambda_T, \lambda_O, \lambda_U, N) = (0.9, 0.05, 0.05, \infty)$, while the 'glfd+' model worked best when $(\lambda_T, \lambda_O, \lambda_U, N) = (0.9, 0.05, 0.05, 50)$.

For comparison, we naively applied the single-stage term dependence model using either the sequential dependence or the full dependence variants defined in Section **2** to each of the query components delimited by commas in the title field of a topic, and combined the beliefs (scores) about the resulting structure expressions. We show the results of these as 'naive-sd' and 'naive-fd', respectively, in **Table 1**. These results suggest that Metzler and Croft's single-stage term dependence model must be enhanced to handle the more complex dependencies that appear in queries in the Japanese language. For reference, we also show the best results from NTCIR-3 WEB participation [13] ('ntcir-3') at the bottom of **Table 1**. This shows that even our baseline system worked well.

4.3 Experiments for Testing

For testing, we used the models optimized in the previous subsection. We used the relevance judgment data, for evaluation, that were provided by the organizers of the NTCIR-5 WEB task. The results are shown in **Table 2**. In this table, 'AvgPrec$_a$', 'AvgPrec$_c$' and 'AvgPrec$_o$' indicate the mean average precisions over all 35 topics, over the 22 topics that include compound words in the title field,

Table 2. Test results of phrase-based query structuring

	AvgPrec_a	%increase	AvgPrec_c	%increase	AvgPrec_o	%increase
base	0.1405	0.0000	0.1141	0.0000	0.1852	0.0000
lsd+	0.1521	8.2979	0.1326	16.2563	0.1852	0.0000
lfd+	0.1521	8.2389	0.1325	16.1407	0.1852	0.0000
glsd+	0.1503	6.9576	0.1313	15.1167	0.1823	-1.5496
glfd+	0.1588 *	13.0204	0.1400	22.6950	0.1906	2.9330

'*' indicates statistical significant improvement over 'base', 'lsd+', 'lfd+' and 'glsd+' where $p < 0.05$ with two-sided Wilcoxon signed-rank test.

and over the 13 topics that do not include the compound words, respectively. '%increase' was calculated on the basis of the result of retrieval not using query structuring ('base'). The results show that our two-stage term dependence models, especially the 'glfd+' model, gave 13% better performance than the baseline ('base'), which did not assume term dependence, and also better than the local term dependence models, 'lsd+' and 'lfd+'. The advantage of 'glfd+' over 'base', 'lsd+', 'lfd+' and 'glsd+' was statistically significant in average precision over all the topics. The results of 'AvgPrec_c' and 'AvgPrec_o' imply that our models work more effectively for queries expressed in compound words.

5 Conclusions

In this paper, we proposed the two-stage term dependence model, which was based on a theoretical framework using Markov random fields. Our two-stage term dependence model captures both the global dependence between query components explicitly delimited by separators in a query, and the local dependence between constituents within a compound word when the compound word appears in a query component. We found that our two-stage term dependence model worked significantly better than the baseline that did not assume term dependence at all, and better than using models that only assumed either global dependence or local dependence in the query. Our model is based on proximity search, which is typically supported by Indri [9] or InQuery [11].

We believe that our work is the first attempt to explicitly capture both long-range and short-range term dependencies. The two-stage term dependence model should be reasonable for other languages, if compound words or phrases can be specified in a query. Application to natural language-based queries, employing an automatic phrase detection technique, is worth pursuing as future work.

Acknowledgments

We thank Donald Metzler for valuable discussions and comments, and David Fisher for helpful technical assistance with Indri. This work was supported in part by the Overseas Research Scholars Program and the Grants-in-Aid for Scientific Research (#17680011 and #18650057) from the Ministry of Education,

Culture, Sports, Science and Technology, Japan, in part by the Telecommunications Advancement Foundation, Japan, and in part by the Center for Intelligent Information Retrieval. Any opinions, findings and conclusions or recommendations expressed in this material are those of the author(s) and do not necessarily reflect those of the sponsor.

References

1. Fujii, H., Croft, W.B.: A comparison of indexing techniques for Japanese text retrieval. In: Proceedings of the 16th Annual International ACM SIGIR Conference, Pittsburgh, Pennsylvania, USA (1993) 237–246
2. Chen, A., Gey, F.C.: Experiments on cross-language and patent retrieval at NTCIR-3 Workshop. In: Proceedings of the 3rd NTCIR Workshop, Tokyo, Japan (2002)
3. Moulinier, I., Molina-Salgado, H., Jackson, P.: Thomson Legal and Regulatory at NTCIR-3: Japanese, Chinese and English retrieval experiments. In: Proceedings of the 3rd NTCIR Workshop, Tokyo, Japan (2002)
4. Mitra, M., Buckley, C., Singhal, A., Cardie, C.: An analysis of statistical and syntactic phrases. In: Proceedings of RIAO-97. (1997) 200–214
5. Fujita, S.: Notes on phrasal indexing: JSCB evaluation experiments at NTCIR ad hoc. In: Proceedings of the First NTCIR Workshop, Tokyo, Japan (1999) 101–108
6. Croft, W.B., Turtle, H.R., Lewis, D.D.: The use of phrases and structured queries in information retrieval. In: Proceedings of the 14th Annual International ACM SIGIR Conference, Chicago, Illinois, USA (1991) 32–45
7. Metzler, D., Croft, W.B.: A markov random field model for term dependencies. In: Proceedings of the 28th Annual International ACM SIGIR Conference, Salvador, Brazil (2005) 472–479
8. Mishne, G., de Rijke, M.: Boosting web retrieval through query operations. In: Proceedings of the 27th European Conference on Information Retrieval Research, Santiago de Compostela, Spain (2005) 502–516
9. Metzler, D., Croft, W.B.: Combining the language model and inference network approaches to retrieval. Information Processing and Management **40**(5) (2004) 735–750
10. Croft, W.B., Lafferty, J., eds.: Language Modeling for Information Retrieval. Kluwer Academic Publishers (2003)
11. Turtle, H.R., Croft, W.B.: Evaluation of an inference network-based retrieval model. ACM Transactions on Information Systems **9**(3) (1991) 187–222
12. Eguchi, K.: NTCIR-5 query expansion experiments using term dependence models. In: Proceedings of the 5th NTCIR Workshop, Tokyo, Japan (2005)
13. Eguchi, K., Oyama, K., Ishida, E., Kando, N., Kuriyama, K.: Overview of the Web Retrieval Task at the Third NTCIR Workshop. In: Proceedings of the 3rd NTCIR Workshop, Tokyo, Japan (2003)
14. Yoshioka, M.: Overview of the NTCIR-5 WEB Query Expansion Task. In: Proceedings of the 5th NTCIR Workshop, Tokyo, Japan (2005)
15. Eguchi, K., Oyama, K., Aizawa, A., Ishikawa, H.: Overview of the Informational Retrieval Task at NTCIR-4 WEB. In: Proceedings of the 4th NTCIR Workshop, Tokyo, Japan (2004)

Automatic Expansion of Abbreviations in Chinese News Text

Guohong Fu[1], Kang-Kwong Luke[1], GuoDong Zhou[2], and Ruifeng Xu[3]

[1] Department of Linguistics, The University of Hong Kong, Hong Kong
[2] School of Computer Science and Technology, Suzhou University, China 215006
[3] Department of Systems Engineering and Engineering Management, The Chinese University of Hong Kong, Hong Kong
ghfu@hotmail.com, kkluke@hkusua.hku.hk, gdzhou@suda.edu.cn, rfxu@se.cuhk.edu.hk

Abstract. This paper presents an n-gram based approach to Chinese abbreviation expansion. In this study, we distinguish reduced abbreviations from non-reduced abbreviations that are created by elimination or generalization. For a reduced abbreviation, a mapping table is compiled to map each short-word in it to a set of long-words, and a bigram based Viterbi algorithm is thus applied to decode an appropriate combination of long-words as its full-form. For a non-reduced abbreviation, a dictionary of non-reduced abbreviation/full-form pairs is used to generate its expansion candidates, and a disambiguation technique is further employed to select a proper expansion based on bigram word segmentation. The evaluation on an abbreviation-expanded corpus built from the PKU corpus showed that the proposed system achieved a recall of 82.9% and a precision of 85.5% on average for different types of abbreviations in Chinese news text.

1 Introduction

Abbreviations (also referred to as short-forms of words or phrases) are widely used in current articles. Abbreviations form a special group of unknown words that mainly originate from technical terms and named entities. Consequently, expanding abbreviation to their original full-forms plays an important role in improving information extraction and information retrieval systems [1][2][3].

Over the past years, much progress has been achieved in English abbreviation resolution and various methods have been proposed for the identification and expansion of abbreviation in English, including rule-based methods[1], statistically - based methods [2] and machine learning methods [3][4]. However, the study of Chinese abbreviation expansion is still at its early stage. Only in recent years, have some studies been reported on the expansion of abbreviations in Chinese [5][6].

In this paper, we propose an n-gram language model (LM) based approach to Chinese abbreviation expansion. In this study, we distinguish reduced abbreviations from non-reduced abbreviations that are created by elimination or generalization, and apply different strategies to expand them to their respective full-forms. For a reduced abbreviation, a mapping table is first used to map each short-word in it to a set of

H.T. Ng et al. (Eds.): AIRS 2006, LNCS 4182, pp. 530–536, 2006.

long-words, and a bigram based Viterbi algorithm is then applied to decode a proper combination of long-words as its full-form. For a non-reduced abbreviation that is created by elimination or generalization, a dictionary of abbreviation/full-form pairs is applied to generate all its expansion candidates, and a disambiguation technique is further employed to select a proper expansion based on bigram word segmentation. Evaluation on an abbreviation-expanded corpus built from the Peking University (PKU) corpus shows that our system is effective for different Chinese abbreviations.

2 Abbreviations in Chinese News Text

In general, Chinese abbreviations are created using three major methods, namely reduction, elimination and generalization. Corresponding to these methods, there are three types of abbreviations in Chinese, namely reduced abbreviation, eliminated abbreviation and generalized abbreviation [6][7].

Given a full-form $F = f_1 f_2 \cdots f_m$ and its corresponding abbreviation $S = s_1 s_2 \cdots s_n$, let $f_i (1 \le i \le m)$ denote a constituent word of the full-form and $s_j (1 \le j \le n)$ denote one component of the relevant short-form, then the above three types of Chinese abbreviations can be formally redefined as follows:

If $n = m$ and s_i is the corresponding short-form of the constituent word f_i (namely for $i = 1$ to n, $s_i \in f_i$), then S is a reduced abbreviation. This means that each constituent word of a full-form should have remains in its short-form if it is created by reduction.

If $n < m$ and $\forall s_j \in F (1 \le j \le n)$, then S is an eliminated abbreviation. This implies that each component of an eliminated abbreviation should be a remaining part of its original full-form even though some parts of the full-form are eliminated during abbreviation.

If $n < m$ and $\exists s_j \notin F (1 \le j \le n)$, S is a generalized abbreviation. This means that some additional morphemes or words are usually needed to abbreviate an expression with generalization.

With the above formal definitions, we can distinguish reduced abbreviations from non-reduced abbreviations and deal with them in different ways. Here, non-reduced abbreviation is a general designation of the latter two types of abbreviations.

Table 1 presents a survey of Chinese abbreviations on the Peking University (PKU) corpus. The original PKU Corpus contains six month (Jan to Jun, 1998) of segmented and part-of-speech tagged news text from the People's Daily [8], in which the tag '*j*' is specified to label abbreviations. In this survey, the first two month is selected as the source data. Furthermore, all the explicitly-labeled abbreviations are manually paired with their respective full-forms.

It is observed from Table 1 that Chinese news text is not rich in abbreviations. Among a total of more than two million words, there are only about twenty thousand abbreviations. However, these abbreviations are widely distributed in different sentences. As can be seen in Table 1, more than 14% of sentences have abbreviation(s). In addition, about 60% of abbreviations are observed to be reduced abbreviations.

This demonstrates in a sense that reduction is the most popular method among the three methods for creating Chinese abbreviations.

Table 1. Number of abbreviations in the first two month of texts from the PKU corpus

Corpus	No. words	No. sentences	No. abbreviations Reduced (%)	Non-reduced (%)	Total	No. sentences with abbreviations
January	1.12M	47,288	6,505 (62.6%)	3,883 (37.4%)	10,388	6,729 (14.2%)
February	1.15M	48,095	6,655 (59.7%)	4,498 (40.3%)	11,153	7,137 (14.8%)

3 The Expansion System

Chinese abbreviations may be created either by the method of reduction or by the method of non-reduction (viz. elimination or generalization). However, we do not know exactly how a given abbreviation is created before expansion. To ensure any abbreviation can be expanded, we assume that a given abbreviation could be created the three methods and develop a statistically-based expansion system for Chinese.

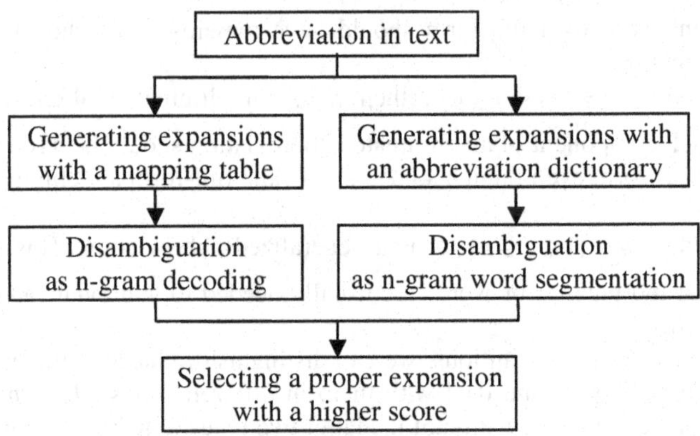

Fig. 1. Overview of the system for Chinese abbreviation expansion

Fig. 1 illustrates an overview of our system for expansion, which works in three steps: (1) Expanding as a reduced abbreviation: In this case, a given abbreviation is first assumed to be a reduced abbreviation. Then, a mapping table between short-words and long-words is applied to map each short-word within the abbreviation to a set of long-words. Finally, a decoding algorithm is employed to search an appropriate combination of long-words as its full-form based on n-gram language models (LMs). (2) Expanding as a non-reduced abbreviation: In this case, a given abbreviation is viewed as a non-reduced abbreviation. Then, a dictionary of non-reduced abbreviation /full-form pairs is used to generate a set of expansion candidates for it. Finally, an n-gram word segmentation is employed to select a proper expansion if any. (3) If the above two steps come out with two different expansions, a final disambiguation is used to make a choice between the two expansions by comparing their respective n-gram scores. The one with a higher score is chosen as the resulting expansion.

3.1 Expanding Reduced Abbreviations

3.1.1 Generating Expansions with a Mapping Table

The idea of generating expansion candidates using a mapping table is based on the characteristics of a reduced abbreviation: each character or word in a reduced abbreviation must match a word in its original full-form. In this study, characters or words consisting of an abbreviation are designated as short-words while the constituent words in a full-form are referred as long-words hereafter. Given a reduced abbreviation, if the matching long-words of each short-word in it are known, its expansion candidates can be thus determined by combing exhaustively the relevant long-words.

In this study, a mapping table like Table 2 is used to map each short-word in reduced abbreviations to a set of long-words. With this mapping table, expansion candidates for a reduced abbreviation can be generated as follows: First, the reduced abbreviation is segmented into a sequence of short-words using a mapping table and a dictionary of normal Chinese words. Then, each segmented short-word is mapped to a set of long-words by consulting the mapping table. All the generated long-words and their matching short-words are stored in a lattice structure. Obviously, any combination of the relevant long-words forms an expansion candidate.

Table 2. A mapping table between short-words and long-words

Short-words	Full-words	English translation
大	大会 l 大学 l 大学生 l 大型 l …	meeting l university l undergraduate l large-scale l …
联	联邦 l 联合会 l 联合国 l联盟 l …	federation l union l the United Nations l alliance l …
…	…	…

3.1.2 Disambiguation as N-Gram Decoding

The disambiguation of reduced abbreviations is actually a process of decoding in that the expansion candidates for a given reduced abbreviation are implicitly involved in a lattice of long-words. Obviously, each path from the beginning to the end of the lattice is a combination of these long-words that forms a potential expansion for the abbreviation. Consequently, the goal of disambiguation for a reduced abbreviation is to decode a best path from the lattice. In our system, an n-gram based Viterbi algorithm is applied to perform this task.

Given a reduced abbreviation S, let $F = f_1 f_2 \cdots f_n$ denote a certain combination of long-words generated with the above mapping table, L and R stand for the respective left and right context around the abbreviation. Expansion decoding aims to search a combination of long-words as the best path from the lattice that maximizes the n-gram probability, namely

$$\hat{F} = \arg\max_F P(LFR) = \arg\max_F P(Lf_1 f_2 \cdots f_n R) \approx \arg\max_F P(f_1 \mid L) \times \prod_{i=1,n} P(f_i \mid f_{i-N+1,i-1}) \times P(R \mid f_{n-N+2,n}) \quad (1)$$

Where $P(f_i \mid f_{i-N+1,i-1})$ denotes an n-gram probability and N denotes the number of contextual words. With a view to the problem of data sparseness, a bigram LM is employed in our system, namely $N = 2$. Let f_0 and f_{n+1} denote the respective words on the left and right of the abbreviation, we have $P(f_1 \mid L) = P(f_1 \mid f_0)$ and $P(R \mid f_{n-N+1,n}) = P(f_{n+1} \mid f_n)$. Thus, Equation (1) can be rewritten as

$$\hat{F} = \arg\max_F P(Lf_1 f_2 \cdots f_n R) \approx \arg\max_F \prod_{i=1}^{n+1} P(f_i \mid f_{i-1}) \quad (2)$$

3.2 Expanding Non-reduced Abbreviations

3.2.1 Generating Expansions with a Dictionary of Abbreviations

According to the definition in Section 3, a one-to-one mapping relationship no longer exists between constituent words of a non-reduced abbreviation and the component words within its full-form. The above mapping table is therefore not workable in generating expansions for a non-reduced abbreviation. In order to address this problem, a dictionary of abbreviation/full-form pairs is applied here, which maps each non-reduced abbreviation to a set of full-forms and can be manually compiled or automatically compiled from an abbreviation-expanded corpus. Table 3 presents some pairs of non-reduced abbreviations and their full-forms extracted from the abbreviation-expanded corpus in Table 1.

Table 3. A dictionary of non-reduced abbreviations

Short-forms	Full-forms	English translation
清华	清华大学	Tsinghua Univeristy
三通	通邮、通商、通航	the direct links in mail, transport and trade across the Taiwan Straits (viz. the Three Direct Links)
…	…	

3.2.2 Disambiguation as N-Gram Word Segmentation

In contrast to a reduced abbreviation whose expansion candidates are underlying in a lattice of long-words, a non-reduced abbreviation is directly mapped to a set of expansion candidates during expansion generation. Consequently, disambiguating a non-reduced abbreviation is to select a proper expansion from a set of candidates if any. In our system, this task is performed using bigram word segmentation [9]. The main idea is: each expansion candidate of a given abbreviation is segmented into a sequence of words using bigram LMs. The one whose segmentation has the maximum score will be identified as the most probable expansion.

4 Evaluation Results and Discussions

In evaluating our system, we conduct two experiments on the abbreviation-expanded corpora in Table 1, in which the first month is used for training while the second month is for open test. The results are summarized in Table 4 and Table 5, respectively.

Table 4. Evaluation results for different training data and LMs

Measures	Trained with unexpanded corpus		Trained with expanded corpus	
	Unigram	Bigram	Unigram	Bigram
Recall (%)	56.5	69.4	71.0	**82.9**
Precision (%)	65.9	77.5	75.1	**85.5**

Table 4 presents the evaluation results for different training data and n-gram LMs. Both the mapping table and the abbreviation dictionary used in this evaluation are extracted from the training corpus only. As can be seen from this table, our system is effective for a majority of abbreviations in Chinese. A recall of 82.9% and a precision of 85.5% are achieved on average for different abbreviations if a bigram LM trained with the expanded corpus is applied. Furthermore, an abbreviation-expanded corpus is of significance in developing a high-performance expansion system for Chinese. both recall and precision can be substantial improved by using an abbreviation-expanded corpus as the training data, in comparison with the original version of the corpus, in which all abbreviations remain unexpanded. Table 4 also shows that bigram LMs outperform unigram LMs, which proves that higher-order LMs are usually more powerful than lower-order LMs in Chinese abbreviation disambiguation.

Table 5. Evaluation results for different mapping tables and abbreviation dictionaries

The mapping table and the abbreviation dictionary are	Unigram LMs		Bigram LMs	
	Recall	Precision	Recall	Precision
from the training corpus only	71.0	75.1	82.9	85.5
from both the training corpus and the test data	76.9	80.1	87.3	88.7

Table 5 presents the results for different mapping tables and abbreviation dictionaries of different sizes. It is observed that the respective overall recall and precision are improved from 82.9% and 85.5% to 87.3% and 88.7% for bigram LMs if all the abbreviations in the test data are covered by the mapping table and the abbreviation dictionary. This indicates that a broad-coverage mapping table or abbreviation dictionary is of great value to Chinese abbreviation expansion. However, acquiring such types of knowledge bases is still a challenge because abbreviations are dynamically produced in Chinese text and their identification remains unresolved at present.

5 Conclusion

In this paper, we presented an n-gram based approach to Chinese abbreviation expansion. In order to generate expansion candidates for different Chinese abbreviations, we distinguish reduced abbreviations from non-reduced abbreviations. In our system, a given abbreviation is expanded as a reduced abbreviation and a non-reduced abbreviation, respectively. If two different expansions come out in the two cases, the one with higher score is selected as the resulting expansion. Evaluation on an abbreviation-expanded corpus built from the PKU corpus showed that our system achieved a recall of 82.9% and a precision of 85.5% on average for different abbreviations in Chinese news text. For future work, we hope to improve our system by exploring and acquiring dynamically more features for expansion disambiguation.

Acknowledgement

This study was supported in part by Hong Kong Research Grants Council Competitive Earmarked Research Grant 7271/03H. We also would like to thank the Institute of Computational Linguistics of the Peking University for their corpus.

References

1. Yu, H., G. Hripcsak, and C. Friedman: Mapping abbreviations to full-forms in biomedical articles. Journal of American Medical Information Association, Vol. 9, No.3 (2002) 262-272
2. Terada, A., T. Tokunaga, and H. Tanaka: Automatic expansion of abbreviations by using context and character information. Information Processing and Management, Vol.40, No.1 (2004) 31-45
3. Gaudan, S., H. Kirsch, and D. Rebholz-Schuhmann: Resolving abbreviations to their senses in Medline. Bioinformatics, Vol. 21, No.18, (2005) 3658-3664
4. Yu, Z., Y. Tsuruoka, and J. Tsujii: Automatic resolution of ambiguous abbreviations in biomedical texts using support vector machines and one sense per discourse hypothesis. In: Proceedings of the 26th ACM SIGIR, Toronto, Canada (2003) 57-62
5. Chang, J.-S., and Y.-T. Lai: A preliminary study on probabilistic models for Chinese abbreviations. In: Proceedings of the 3rd SIGHAN Workshop on Chinese Language Processing, Barcelona, Spain (2004) 9-16
6. Lee, H.-W.: A study of automatic expansion of Chinese abbreviations. MA Thesis, The University of Hong Kong, (2005)
7. Yin, Z.: Methodologies and principles of Chinese abbreviation formation. Language Teaching and Study, No.2 (1999) 73-82
8. Yu, S., H. Duan, S. Zhu, B. Swen, and B. Chang: Specification for corpus processing at Peking University: Word segmentation, POS tagging and phonetic notation. Journal of Chinese Language and Computing, Vol. 13, No.2 (2003) 121-158
9. Fu, G., and K.-K. Luke: Chinese unknown word identification using class-based LM. Lecture Notes in Artificial Intelligence (IJCNLP 2004), Vol.3248 (2005) 704-713

A Novel Ant-Based Clustering Approach for Document Clustering

Yulan He, Siu Cheung Hui, and Yongxiang Sim

School of Computer Engineering, Nanyang Technological University
Nanyang Avenue, Singapore 639798
{asylhe, asschui, S8137640I}@ntu.edu.sg

Abstract. Recently, much research has been proposed using nature inspired algorithms to perform complex machine learning tasks. Ant Colony Optimization (ACO) is one such algorithm based on swarm intelligence and is derived from a model inspired by the collective foraging behavior of ants. Taking advantage of the ACO in traits such as self-organization and robustness, this paper proposes a novel document clustering approach based on ACO. Unlike other ACO-based clustering approaches which are based on the same scenario that ants move around in a 2D grid and carry or drop objects to perform categorization. Our proposed ant-based clustering approach does not rely on a 2D grid structure. In addition, it can also generate optimal number of clusters without incorporating any other algorithms such as K-means or AHC. Experimental results on the subsets of 20 Newsgroup data show that the ant-based clustering approach outperforms the classical document clustering methods such as K-means and Agglomerate Hierarchical Clustering. It also achieves better results than those obtained using the Artificial Immune Network algorithm when tested in the same datasets.

1 Introduction

Nature inspired algorithms are problem solving techniques that attempt to simulate the occurrence of natural processes. Some of the natural processes that such algorithms are based on include the evolution of species [1,2], organization of insect colonies [3,4] and the working of immune systems [5,6]. Ant Colony Optimization (ACO) algorithm [3,4] belongs to the natural class of problem solving techniques which is initially inspired by the efficiency of real ants as they find their fastest path back to their nest when sourcing for food. An ant is able to find this path back due to the presence of pheromone deposited along the trail by either itself or other ants. An open loop feedback exists in this process as the chances of an ant taking a path increases with the amount of pheromone built up by other ants. This natural phenomenon has been applied to model the Traveling Salesman Problem (TSP) [3,4].

Early approaches in applying ACO to clustering [7,8,9] are to first partition the search area into grids. A population of ant-like agents then move around this 2D grid and carry or drop objects based on certain probabilities so as to

H.T. Ng et al. (Eds.): AIRS 2006, LNCS 4182, pp. 537–544, 2006.

categorize the objects. However, this may result in too many clusters as there might be objects left alone in the 2D grid and objects still carried by the ants when the algorithm stops. Therefore, Some other algorithms such as K-means are normally combined with ACO to minimize categorization errors [10,11,12]. More recently, variants of ant-based clustering have been proposed, such as using inhomogeneous population of ants which allow to skip several grid cells in one step [13], representing ants as data objects and allowing them to enter either the active state or the sleeping state on a 2D grid [14].

Existing approaches are all based on the same scenario that ants move around in a 2D grid and carry or drop objects to perform categorization. This paper proposes a novel ant-based clustering approach without relying on a 2D grid structure. In addition, it can also generate optimal number of clusters without incorporating any other algorithms such as K-means or AHC. When compared with both the classical document clustering algorithms such as K-means and AHC and the Artificial Immune Network (aiNet) based method [15], it shows improved performance when tested on the subsets of 20 Newsgroup data [16]. The rest of the paper is organized as follows. Section 2 briefly describes the Ant Colony Optimization (ACO) algorithm. The proposed ant-based clustering approach is discussed in Section 3. Experimental results are presented in Section 4. Finally, section 5 concludes the paper and outlines the possible future work.

2 Ant Colony Optimization

The first Ant Colony Optimization (ACO) algorithm has been applied to the Traveling Salesman Problem (TSP) [3,4]. Given a set of cities and the distances between them, the TSP is the problem of finding the shortest possible path which visits every city exactly once. More formally, it can be represented by a complete weighted graph $G = (N, E)$ where N is the set of nodes representing the cities and E is the set of edges. Each edge is assigned a value d_{ij} which is the distance between cities i and j. When applying the ACO algorithm to the TSP, a pheromone strength $\tau_{ij}(t)$ is associated to each edge (i, j), where $\tau_{ij}(t)$ is a numerical value which is modified during the execution of the algorithm and t is the iteration counter.

The skeleton of the ACO algorithm applied to the TSP is:

```
procedure ACO algorithm for TSPs
    set parameters, initialize pheromone trails
    while (termination condition not met) do
        Tour construction
        Pheromone update
    end
end ACO algorithm for TSPs
```

At first, each of the m ants is placed on a randomly chosen city. At each *Tour construction* step, ant k currently at city i, chooses to move to city j at the tth iteration based on the probability $P_{ij}^k(t)$ which is biased by the pheromone trail strength $\tau_{ij}(t)$ on the edge between city i and city j and a locally available heuristic information η_{ij}. Each ant is associated with a *tabu list* in which the current partial tour is stored, i.e. $tabu_k(s)$ stores a set of cities visited by ant k so far at time s. After all the ants have constructed their tours, *Pheromone update* is performed by allowing each ant to add pheromone on the edges it has visited. At the end of the iteration, the tabu list is emptied and each ant can choose an alternative path for the next cycle.

3 Ant-Based Algorithm to Document Clustering

In document clustering, the vector-space model is usually used to represent documents and documents are categorized into groups based on the similarity measure among them. For each document d_i in a collection $\mathcal{D}$, let $\mathcal{W}$ be the unique

```
1. Initialization.
   set the iteration counter t = 0
   For every edge (i, j), set an initial value τ_ij(t) for trail intensity and Δτ_ij = 0.
   Place m ants randomly on the n documents.
2. Set the tabu list index s = 1.
   for k = 1 to m do
        Place starting document of the kth ant in tabu_k(s)
   end for
3. Tour Construction.
   repeat until tabu list is full
        Set s = s + 1
        for k = 1 to m do
             Choose the document j to move to with probability P^k_ij(t)
             Move the kth ant to the document j
             Insert document j into the tabu list tabu_k(s)
        end for
   end repeat
4. Pheromone Update.
   for every edge (i, j) do
        Δτ_ij = Σ_{k=1}^{m} Δτ^k_ij
        compute τ_ij(t + 1) = (1 − ρ)τ_ij(t) + Δτ_ij
        set Δτ_ij = 0
   end for
5. Set t = t + 1
   if a stopping criteria is met, then
        print clustering results, stop
   else
        empty tabu lists, go to 2
   end if
```

Fig. 1. Ant-based document clustering algorithm

word items occurring in $\mathcal{D}$ and $M = |\mathcal{W}|$, then document d_i is represented by the vector $\boldsymbol{d_i} = (w_{i1}, w_{i2}, \cdots, w_{iM})$ where w_{ij} denotes the appearance of word w_j in document d_i which is normally weighted by *term frequency X inverse document frequency* (TFIDF).

Fig. 1 shows the ant-based document clustering algorithm. The design of the ant-based algorithm involves the specification of the following:

- $\tau_{ij}(t)$ represents the amount of pheromone associated with the document pair doc$_{ij}$ at iteration t. The initial amount of pheromone deposited at each path position is inversely proportional to the total number of documents which is defined by $\tau_{ij}(0) = \frac{1}{N}$ where N is the total number of documents in the collection $\mathcal{D}$.
  At every generation of the algorithm, τ_{ij} is updated by $\tau_{ij}(t+1) = (1 - \rho)\tau_{ij}(t) + \Delta\tau$ where $\rho \in (0, 1]$ determines the evaporation rate and the update of pheromone trail; $\Delta\tau$ is defined as the integrated similarity of a document with other documents within a cluster which is measured by:

$$\Delta\tau = \begin{cases} \sum_{j=1}^{N_i} [1 - \frac{\text{dist}(\boldsymbol{c_i}, \boldsymbol{d_j})}{\gamma}] & d_j \in c_i \\ 0 & otherwise \end{cases} \tag{1}$$

  where $\boldsymbol{c_i}$ is the centroid vector of the ith cluster, $\boldsymbol{d_j}$ is the jth document vector which belongs to cluster i, $\text{dist}(\boldsymbol{c_i}, \boldsymbol{d_j})$ is the distance between document d_j and the cluster centroid c_i, N_i stands for the number of documents which belongs to the ith cluster. The parameter γ is defined as swarm similarity coefficient and it affects the number of clusters as well as the convergence of the algorithm.
- η_{ij} is a problem-dependent heuristic function for the document pair doc$_{ij}$. It is defined as the Euclidean distance $\text{dist}(\boldsymbol{d_i}, \boldsymbol{d_j})$ between two documents d_i and d_j.
- Ant k moves from document i to document j at tth iteration by following probability $P_{ij}^k(t)$ defined by:

$$P_{ij}^k(t) = \begin{cases} \frac{[\tau_{ij}(t)]^\alpha \cdot [\eta_{ij}]^\beta}{\sum_{l \notin \text{tabu}_k(t)} [\tau_{il}(t)]^\alpha \cdot [\eta_{ij}]^\beta} & if\ \ j \notin \text{tabu}_k(t) \\ 0 & otherwise \end{cases} \tag{2}$$

  where $l \notin \text{tabu}_k(t)$ means l cannot be found in the tabu list of ant k at time t. In other words, l is a document that ant k has not visited yet. The parameters α and β control the bias on the pheromone trail or the problem-dependent heuristic function.
- Finally, a stopping criteria needs to be carefully set. It can either be a predefined maximum number of iterations or it can be the change in the average document distance to the cluster centroid between two successive iterations. The average document distance to the cluster centroid is defined as:

$$f = \frac{\sum_{i=1}^{N_C} \{ \frac{\sum_{j=1}^{N_i} \text{dist}(\boldsymbol{c_i}, \boldsymbol{d_j})}{N_i} \}}{N_C} \tag{3}$$

where $\boldsymbol{c_i}$ is the centroid vector of the ith cluster, $\boldsymbol{d_j}$ is the jth document vector which belongs to cluster i, $\text{dist}(\boldsymbol{c_i}, \boldsymbol{d_j})$ is the distance between document d_j and the cluster centroid c_i, N_i stands for the number of documents which belongs to the ith cluster. N_C stands for the total number of clusters.

Once the ant-based clustering algorithm has run successfully, a fully connected network of nodes will be formed. Each node represents a document, and every edge is associated with a certain level of pheromone intensity. The next step is essentially to break the linkages in order to generate clusters. Various methods can be applied such as minimum spanning trees. Here, the average pheromone strategy is used. The average pheromone of all the edges is first computed and then edges with pheromone intensity less than the average pheromone will be removed from the network and results in a partially connected graph. Nodes will then be separated by their connecting edges to form clusters.

4 Performance Evaluation

In this section, the performance of the proposed ant-based clustering algorithm is evaluated. The experimental setup is first explained followed by a comparative account of the results generated from different experiments conducted. This section also attempts to obtain the optimal parameters of ant-based clustering and reason intuitively its performance on clustering in comparison with other algorithms.

4.1 Experimental Setup

Experiments have been conducted on the 20 Newsgroup data set [16] which is in fact a benchmarking data set commonly used for experiments in text applications of machine learning techniques. A few combinations of subsets of documents are selected for experiments based on various degrees of difficulty. Table 1 lists the details of various subsets used. Each subset consists of 150 randomly chosen documents from various newsgroups. All newsgroup articles have their headers stripped and main body pre-processed only. Once transformed into term-document vectors, they are fed into the clustering engine.

The tunable parameters in the ant-based clustering algorithm include number of iterations i, number of ants m, rate of decay ρ, swarm similarity coefficient γ,

Table 1. Statistics on experimental data

Dataset	*Topics*	*No. of Docs Per Group*	*Total No. of Docs*
1	sci.crypt, sci.space	150, 150	300
2	sci.crypt, sci.electronics	150, 150	300
3	sci.space, rec.sport.baseball	150, 150	300
4	talk.politics.mideast, talk.religion.misc	150, 150	300

and the parameters α and β that control the bias on the pheromone trail. A set of experiments have been conducted and it was found that the optimal value or range of each parameter is $i = 100$, $14 \leq m \leq 19$, $0.1 \leq \rho \leq 0.3$, $0.4 \leq \gamma \leq 0.5$, $\alpha = 1$, and $\beta = 1$. From this section onwards, subsequent experiments will be carried out using these selected optimized values for the parameters.

4.2 Clustering Accuracy

This section evaluates the performance on clustering accuracy based on F-measure [17]. The 10-fold cross-validation technique is applied to each of the four sample data sets and the average F-measure score will be used as a comparative observation. The experimental results given in Table 2 shows that the ant-based clustering method performs much better than AHC and K-means in all four sample data sets. In AHC and K-means, information on the expected number of clusters must be supplied, but ACO is able to predict the cluster structure accurately by generating the exact number of clusters in each sample data set.

Table 2 also compares the published results from aiNet [15], a technique based on Artificial Immune System (AIS), with ant-based clustering using the identical sets of data. Belonging to the same class of nature inspired algorithms as ACO, AIS performs an evolutionary process on raw document data based on the immune network and affinity maturation principles. The proposed method uses Agglomerate Hierarchical Clustering (AHC) and K-means to construct antibodies and detect clusters. Also, Principal Component Analysis (PCA) is introduced for dimensionality reduction in a bid to improve clustering accuracy. From the published experimental results, none of the proposed AIS methods scored is close to the results of ant-based clustering in all four sample data sets. In fact, the results from the ant-based clustering algorithm have shows a relative improvement in performance over the aiNet_{pca}_K-means by 5% to 34% and a relative improvement over the aiNet_{pca}_AHC by 7% to 35%. Furthermore, it is observed that the ant-based approach is far more stable by producing consistent F-measure scores of higher than 0.8, while the aiNet scored varying F-measure results, ranging from 0.6 to 0.8.

Although both ACO and AIS belong to the same family of nature inspired algorithms, these two methods use entirely different approaches to model problem solving techniques. This results in a performance difference when applying both to the same problem domain. The evolution stage in ACO makes use of stochastic ants as decision tools for choosing a path to move based on pheromone trail intensity. The final network generated has edges of varying amounts of pheromone that represent the differences among documents in a collection. Such differences will allow easy partitioning of semantically similar documents by using the averaged pheromone level as a threshold for searching connected sub-graphs in the network. This entire evolving process models after the document clustering task almost perfectly.

On the other hand, using the AIS to model the same clustering task may not be an ideal case. The clonal selection theory is only capable of generating an

Table 2. Comparison of clustering accuracy of AHC, K-means, aiNet, and ant-based clustering

	F-Measure			
Method	*Subset 1*	*Subset 2*	*Subset 3*	*Subset 4*
AHC	0.665	0.654	0.700	0.631
K-means	0.794	0.580	0.513	0.624
aiNet_AHC	0.810	0.640	0.718	0.641
$aiNet_{pca}$_AHC	0.815	0.735	0.715	0.640
aiNet_K-means	0.807	0.628	0.630	0.639
$aiNet_{pca}$_K-means	0.836	0.661	0.631	0.646
Ant-Based	**0.874**	**0.811**	**0.803**	**0.865**

immune network based on affinity between antibodies and antigens (documents). Detection of clusters in this network requires assistance from other methods such as AHC or K-means, which may disrupt the biological ordering of antigens. Therefore, the performance of AIS is limited by the artificial clustering applied on its network.

5 Conclusions

This paper has proposed a novel ant-based clustering algorithm and its application to the unsupervised data clustering problem. Experimental results showed that the ant-based clustering method performs better than K-means and AHC by a wide margin. Moreover, the ant-based clustering method has achieved a higher degree of clustering accuracy than the Artificial Immune Network (aiNet) algorithm which also belongs to the same family of nature inspired algorithms.

In future work, it would be interesting to investigate the behavior of the ant-based algorithm using other sources of heuristic functions and pheromone update strategies. In addition, more intelligent methods of breaking linkages among documents can be devised to replace the existing average pheromone approach.

References

1. E. Yu and K.S. Sung. A genetic algorithm for a university weekly courses timetabling problem. *International Transactions in Operational Research*, 9(6):703–717, 2002.
2. E.K. Burke, D.G. Elliman, and R.F. Weare. A genetic algorithm based university timetabling system. In *Proceedings of the 2nd East-West International Conference on Computer Technologies in Education*, pages 35–40, Crimea, Ukraine, September 1994.
3. M. Dorigo, V. Maniezzo, and A. Colorni. Positive feedback as a search strategy. Technical report 91-016, Politecnico di milano, 1991. Dip. Elettronica.
4. M. Dorigo, V. Maniezzo, and A. Colorni. The ant system: Optimization by a colony of cooperating agents. *IEEE Transactions on Systems, Man, and Cybernetics - Part B*, 26(1):29–42, 1996.

5. L.N. de Castro and F.J. Von Zuben. Learning and optimization using the clonal selection principle. *IEEE Transactions on Evolutionary Computation, Special Issue on Artificial Immune Systems*, 6(3):239–251, 2002.
6. D. Dasgupta, Z. Ji, and F. Gonzlez. Artificial immune system (ais) research in the last five years. In *Proceedings of the International Conference on Evolutionary Computation Conference (CEC)*, Canbara, Australia, December 2003.
7. J. L. Deneubourg, S. Goss, N. Franks, A. Sendova-Franks, C. Detrain, and L. Chretien. The dynamics of collective sorting robot-like ants and ant-like robots. In *Proceedings of the first international conference on simulation of adaptive behavior on From animals to animats*, pages 356–363, Cambridge, MA, USA, 1990. MIT Press.
8. Lumer E. D. and Faieta B. Diversity and adaptation in populations of clustering ants. In Cli D., Husbands P., Meyer J., and Wilson S., editors, *Proceedings of the Third International Conference on Simulation of Adaptive Behaviour: From Animals to Animats 3*, pages 501–508, Cambridge, MA, 1994. MIT Press.
9. Kuntz P., Layzell P., and Snyers D. A colony of ant-like agents for partitioning in vlsi technology. In P. Husbands and I. Harvey, editors, *Proceedings of the Fourth European Conference on Artificial Life*, pages 417–424. MIT Press, 1997.
10. N. Monmarche. On data clustering with artificial ants. In Alex Alves Freitas, editor, *Data Mining with Evolutionary Algorithms: Research Directions*, pages 23–26, Orlando, Florida, 18 1999. AAAI Press.
11. B. Wu, Y. Zheng, S. Liu, and Z. Shi. Csim: a document clustering algorithm based on swarm intelligence. In *Proceedings of the 2002 congress on Evolutionary Computation*, Honolulu, USA, 2002.
12. Y. Peng, X. Hou, and S. Liu. The k-means clustering algorithm based on density and ant colony. In *IEEE International Conference in Neural Networks and Signal Processing*, Nanjing, China, December 2003.
13. Julia Handl and Bernd Meyer. Improved ant-based clustering and sorting. In *PPSN VII: Proceedings of the 7th International Conference on Parallel Problem Solving from Nature*, pages 913–923, London, UK, 2002. Springer-Verlag.
14. L. Chen, X. Xu, and Y. Chen. An adaptive ant colony clustering algorithm. In *Proceedings of the Third International Conference on Machine Learning and Cybernetics*, pages 1387–1392, Shanghai, China, August 2004.
15. Na Tang and V. Rao Vemuri. An artificial immune system approach to document clustering. In *Proceedings of the 2005 ACM symposium on Applied computing*, pages 918–922, New York, NY, USA, 2005. ACM Press.
16. *20 Newsgroups Data Set*, 2006. http://people.csail.mit.edu/jrennie/20Newsgroups/.
17. M. Steinbach, G. Karypis, and V. Kumar. A comparison of document clustering techniques. In *In KDD Workshop on Text Mining*, 2000.

Evaluating Scalability in Information Retrieval with Multigraded Relevance

Amélie Imafouo and Michel Beigbeder

Ecole Nationale Supérieure des Mines of Saint-Etienne
158 Cours Fauriel - 42023 Saint-Etienne, Cedex 2, France
{imafouo, beigbeder}@emse.fr

Abstract. For the user's point of view, in large environments, it can be desirable to have Information Retrieval Systems (IRS) that retrieve documents according to their relevance levels. Relevance levels have been studied in some previous Information Retrieval (IR) works while some others (few) IR research works tackled the questions of IRS effectiveness and collections size. These latter works used standard IR measures on collections of increasing size to analyze IRS effectiveness scalability. In this work, we bring together these two issues in IR (multigraded relevance and scalability) by designing some new metrics for evaluating the ability of IRS to rank documents according to their relevance levels when collection size increases.

1 Introduction

Nowadays, many factors support a growing production of information. A regular increase of 30% was noted between 1999 and 2002 in the information production [1]. The problem of accessing this mass of information comes under the field of domains like digital libraries and information retrieval but currently few works of these domains have taken into account the size effect in their approaches. The size of large collections (or web) coupled with and the ambiguity of user query make it difficult for search engines to return the most recent and relevant information in real-time. The need to learn more about they way collections size acts on retrieval effectiveness becomes increasingly pressing. In this work, we present works dealing with multigraded relevance and in the last part we present the metrics designed to evaluate the ability of IR systems to rank documents according to their relevance levels.

2 Multigraded Relevance Levels in IR

2.1 The Relevance as a Complex Cognitive and Multidimensional Concept

Relevance is the central concept for IR evaluation, usually considered as a binary notion. However, some research works showed that the relevance is a complex

H.T. Ng et al. (Eds.): AIRS 2006, LNCS 4182, pp. 545–552, 2006.

cognitive concept, that has many dimensions ([2], [3] , [4], [5]. Many different aspects of relevance have also been discussed by proposed definitions and classifications ([4], [6], [7], [8]). It is not an easy job to judge documents and give them a relevance level regarding a topic as many variables affect the relevance (*Rees et al.* [9]: about 40 variables, *Cuadra et al.* [10]: 38 variables). All these works and many others suggest that there is no single relevance (there are many relevances) and that relevance is a complex cognitive problem.

2.2 Multigraded Relevance in IR

In the user's point of view, it is desirable to have IRS that retrieve documents according to their relevance level [11]. IR evaluation methods should then credit (or at least recognize) IRS for their ability to retrieve highly relevant documents at the top of their results list, by taking into account various relevance levels of a document for a given query; they have been studied in some previous IR works (*Tang et al.* [12]: a seven-point scale, *Spink et al.* [13] used a three-point scale). Some test collections provide multigraded relevance assessments (TREC Web Track collection: three point scale [14], *INEX* collections: a multilevel scale, *NTCIR* evaluation campaign [15]). *Kekäläinen et al.* [11] used a four-points scale for relevance level : *highly relevant, fairly relevant, marginally relevant, not relevant.* Each of these relevance level has to be expressed by a numerical value for computing measures. One of the remaining question is the choice of these values and the semantic they should have. Their work also proposed *generalized non binary recall and precision*, that are extensions of standard binary recall and precision taking into account multiple relevance levels [11]. The *Discounted Cumulated Gain* and the *Cumulated Gain* are also proposed by the same authors in [16]. We present them using our formalism in section 3.2. *Sakai* [17] also proposed a measure based on of the *Cumulated Gain.*

Our conceptions meet those of *Kekäläinen et al.* [16] concerning the fact that multiple relevance levels should be taken into account when evaluating IRS. While information grows continuously, for the users lambda, one of the main issue for IRS will become to retrieve documents with highly relevance level at the top of the results list. We design metrics to allow the evaluation of this ability in IRS as collections size increase.

3 Protocols for Scalability Evaluation with Multigraded Relevance

Let C_1 and C_2 be two collections of different sizes such as $C_1 \subset C_2$ and S an IRS. The aim is to analyze how S behaves on each collection to determine if its effectiveness improves, remains the same or decreases when the collection size increases. Our measures are based on the comparisons of the relevance levels of the first documents in the results lists for the two collections.

3.1 Relevance Level *Importance*

For a given topic, we assume that a relevance level is given to every document regarding this topic. Let $\{rel_i\}$, $i = 1, \ldots, n$ be the set of possible relevance levels; two documents are *equivalent* if they have the same relevance level regarding this topic.

We define a total order relation on the set of the relevance levels noted $\succ$: $rel_i \succ rel_j$ when $i > j$. This total order relation gives the preference wished on retrieved documents but it gives no indication about the *importance* of a particular relevance level regarding the other relevance levels. However, it is the *importance* of a relevance level that characterizes the quality/quantity of information expected from a document of this relevance level. One may need to highly credit (resp penalize) retrieval systems that return the documents with the highest relevance level at the top (resp not at the top) of their results list: in this case, the highest relevance level must have a high *importance* (compared to the *importance* of the other relevance levels) when evaluating retrieval results. On the other side, some applications need to retain many documents of good relevance levels, the difference between a document of good relevance level and a document of high relevance level is not important. Thus, a function I that models the *importance* of relevance levels depends of the types of applications the IRS is designed for and is characterized by the following properties (a positive and increasing function):

- $I(rel_i) > 0$
- $I(rel_i) > I(rel_j)$ if $rel_i \succ rel_j$ i.e. $i > j$

The choice of the number of relevance levels and the attribution of numerical values to relevance levels is still an open problem in IR. *Kekäläinen* [18] used different empirical weighting schemes (see figure 1). Giving a numerical values of *importance* to relevance levels means nothing in the absolute; but in the relation with others relevance levels, these values can be associated to a semantics as we show it through the *gain function* (section 3.3).

	Highly relevant (HR)	Fairly Relevant (FR)	Marginally Relevant (MR)	Not Relevant (NR)
scheme 1	1	1	1	0
scheme 2	3	2	1	0
scheme 3	10	5	1	0
scheme 4	100	10	1	0

Fig. 1. Four schemes for assigning numerical values to relevance levels [18]. These values give the *importance* of the relevance level for us.

3.2 Cumulated Gain at a Given Rank

The *Cumulated Gain, CG* as proposed by *Kekäläinen et al.* [16], is computed at rank r by the sum of relevance levels of documents retrieved at any rank $k \leq r$:

$$\begin{cases} CG(1) = I(RelLevel(d_1) \\ CG(i) = CG(i-1) + I(RelLevel(d_i) \end{cases}$$

The *Discounted Cumulated Gain(DCG)* also computes relevance gains with a discount factor which is a decreasing function of the rank: the greater the rank, the smaller share of the document relevance level is added to the cumulated gain. This factor is needed to reduce progressively the impact of the gain of relevant information according to the rank (steep reduction with a function like the inverse of the rank $disf(k) = 1/k$ if the first documents are those we want to focalize on during the evaluation or less steeply with a function like the inverse of the log of the rank $disf(k) = 1/log_b(k)$ as in [16]). By averaging over a set of queries, the average performance of a particular IR method can be analyzed. Averaged vectors have the same length as the individual ones and each component i gives the average of the ith component in the individual vectors. The averaged vectors can directly be visualized as gain-by-rank graphs. The actual CG and DCG vectors are also compared to the best theoretically possible. We described the building of the best theoretically results list in section 3.4, as we re-use it for our metrics.

3.3 Information Gain Between Two Relevance Levels

For a given topic, in front of two documents with two different relevance levels, the same amount of relevant information is not expected from the two documents. It is interesting to quantify the relevant information gained (or lost) when moving from a relevance level to another, that is a function of the relevance levels: $Gain(rel_i, rel_j) = g(rel_i, rel_j)$, with these characteristics:

- $g(rel_i, rel_j) > g(rel_i, rel_k)$ if $rel_j \succ rel_k$ i.e. if $j > k$
- $g(rel_i, rel_j) < g(rel_k, rel_j)$ if $rel_i \succ rel_k$ i.e. if $i > k$
- $g(rel_i, rel_i) = 0$. There is neither a gain nor a loss of information when one stays on the same relevance level (even if one change the document because the documents of the same relevance level for a given topic are in the same equivalence class).

By deduction, we have: $g(rel_i, rel_j) < 0$ if $rel_i \succ rel_j$ i.e. if $i > j$.

Indeed if we have $rel_i \succ rel_j$, then this means that the quantity of relevant information contained in the document of relevance level rel_i is higher than the quantity of relevant information contained in a document of relevance level rel_j. Thus, when moving from a document of relevance level rel_i to a document of relevance level rel_j, one loses relevant information and so $g(rel_i, rel_j) < 0$.

In the same way, $g(rel_i, rel_j) > 0$ if $rel_j \succ rel_i$ i.e. if $i < j$

It is obvious that the gain function between two relevance levels depends on the *importance* associated to each of the relevance levels.

Thus $g(rel_i, rel_j) = h(I(rel_i), I(rel_j))$. An example of an h function is modelled by the mathematical distance (we can build a *distance* between different relevant levels, using their associated numerical value of *importance*).

For example (simple case) $d(rel_i, rel_j) = d(I(rel_i), I(rel_j))$. So we can choose:

$$\begin{cases} g(rel_i, rel_j) = -d(I(rel_i), I(rel_j))\, if\, rel_i \succ rel_j \\ g(rel_i, rel_j) = d(I(rel_i), I(rel_j))\, else \end{cases}$$

We respect all the properties of the function g.

3.4 Information Gain at a Rank When Collection Size Increases

We assume in this study that the measures proposed will be used to evaluate the effectiveness of a system on a collection that grows (from a first collection C_1 to a second collection C_2 with $C_1 \subseteq C_2$). When a collection C_1 grows by the addition of new documents and becomes a collection C_2, our assumption is that the effectiveness of a good retrieval system should at the worst case stay the same (from C_1 to C_2), whatever be (the relevance level of) the documents added. This effectiveness should not decrease, whatever be the documents added, as all the documents in C_2 were already in the collection C_1. And when new relevant documents are added, a good retrieval system effectiveness should stay the same or increase from C_1 to C_2.

For a given topic t, $d_k^t(C)$ is the document retrieved at rank k when the collection C is queried using the topic t. the information gain at rank k between the results lists of both collections is computed using the gain function as follows: $Move_k^t(C_1, C_2) = g(RelLevel(d_k^t(C_1)), RelLevel(d_k^t(C_2)))$

This Move expresses the relevant information gain (resp loss) at rank k when moving from C_1 results list for the topic t to C_2 results list for the topic t. We obtain a vector of weighted Moves by applying a discount factor $< disf(1) \times Move_1^t(C_1, C_2), \ldots, disf(N) \times Move_N^t(C_1, C_2) >$

Measure Type 1. There are two possibilities for using these vectors:

- For a given cut-off level N, either we sum the vectors' elements topic by topic to have a unique value for each topic. Thus we define the first metric as follows[1]:

$$Measure1_N^t(C_1, C_2) = \sum_{k=1}^{N} disf(k) \times Move_k^t(C_1, C_2)$$

- either we sum the weighted Moves rank by rank for all the topics and we obtain a single vector of N elements:

$$< disf(1) \times \textstyle\sum_t (Move_1^t(C_1, C_2)), \ldots, disf(N) \times \sum_t (Move_N^t(C_1, C_2)) >$$

  This sum-vector has the same size as vectors of weighted Moves for each topic; we can then visualize the vector as a gain/loss versus rank graph.

[1] For two collections C_i and C_j, this measure can only be computed on the set of topics t for which the IRS S provides a results list of N or more documents for both collections.

Thus, by querying an IRS on a set of collection $\{C_i\}$ such as $C_i \subset C_{i+1}$, we obtain information gains realized when collection size increases, and we can analyze the impact of collection size on the information gain. According to our assumptions, the measure $Measure1^t_N(C_1, C_2)$ should not be negative for a good retrieval system, as $C_1 \subset C_2$.

Measure Type 2. For a given collection C, the IRS S provide a result list *Retrieved*$^t(C)$ for a given topic t. Then we build a results list $Retrieved\_ideal^t_N(C)$ so-called *ideal* for this topic in the same way as [16].

Example: Consider $HR \succ FR \succ MR \succ NR$ the relevance levels of [16] (see table 1 and a topic t with 7 documents HR, 10 documents FR, 20 documents MR. We choose $N = 30$. The *ideal* results list of size 30 for topic t is as follows:

$$\underbrace{HR, \ldots, HR}_{7 times}, \underbrace{FR, \ldots, FR}_{10 times}, \underbrace{SR, \ldots, SR}_{13 times}$$

We can now build the weighted vectors of Moves between the results list for the collection C and the *ideal* results list:
$< disf(1) \times Move^t_1(C, Ideal\_C), \ldots, disf(N) \times Move^t_N(C, Ideal\_C)$.

As for the previous case, we have two possibilities of using these vectors for evaluation:

- At the cut-off level N and for the topic t we compute :

$$Measure2^t_N(C) = Measure1^t_N(Retrieved^t(C), Retrieved\_ideal^t_N(C))$$

  This measure expresses the information gain when moving from the collection C results list to an *ideal* results list. While the collection C size increases, we can then analyze the variation of its results list compared to an ideal results list.
- we sum the weighted vectors elements rank by rank for all the topics and we obtain a single vector of N elements; we can then visualize the vectors as a gain-versus-rank graph.

$$< disf(1) \times \sum_t (Move^t_1(C, Ideal\_C)), \ldots, disf(N) \times \sum_t (Move^t_N(C, Ideal\_C)) >$$

4 Discussions and Conclusions

In this work, we propose some metrics based on the notion of multigraded relevance levels for evaluating the way IRS scale. Their goal is to provide some information on the coherence between the ranking of documents retrieved by an IRS and the relevance levels of these documents as collection size increases. Some recent metrics in IR used a notion of relevance with multiple levels, e.g. the *Discounted Cumulated Gain* or the *Cumulated Gain.* For a given collection and an IRS, these metrics compute the relevant information gain obtained as one goes through the results list returned by an IRS on a given collection. Our metrics compute the relevant information gain obtained when a single IRS is used on a collection which grows. Thus we evaluate the ability of the IRS to rank the documents according to their relevance levels when collection size grows. All the metrics that use multigraded relevance need to associate a numerical value to each relevance level and this is still not well studied in IR: in this study, we formalize the (obvious) constraints linked to the attribution of numerical values to relevance levels through the *importance function* and the *gain function*.

We are now working on the relation between our metrics and the existing metrics that used multigraded relevance levels through some experiments.

References

1. Lyman, P., Varian, H.R., Swearingen, K., Charles, P., Good, N., Jordan, L.L., Pal, J.: How much informations 2003. http://www.sims.berkeley.edu/research/projects/how-much-info-2003/ (2003)
2. Mizzaro, S.: How many relevances in information retrieval? Interacting with Computers **10** (1998) 303–320
3. Barry, C.L.: User-de.ned relevance criteria: an exploratory study. Journal of the American Society for Information Science **45** (1994) 149Ű159
4. Saracevic, T.: Relevance: A review of and a framework for the thinking on the notion in information science. Journal of the American Society for Information Science **26** (1975) 321–343
5. Schamber L., E.M.B., Nilan, M.S.: A re-examination of relevance: toward a dynamic, situational definition. Information Processing and Management **26** (1990) 755Ű776
6. Wilson, P.: Situational relevance. Information Storage and Retrieval **9** (1973) 457–471
7. Cooper, W.S.: A definition of relevance for information retrieval. Information Storage and Retrieval (1971)
8. Cosijn, E., Ingwersen, P.: Dimensions of relevance. Information Processing and Management **36** (2000) 533Ű550
9. Rees, A.M., Schulz, D.G.: A field experimental approach to the study of relevance assessments in relation to document searching. 2 vols. Technical Report NSF Contract No. C-423, Center for Documentation and Communication Research, School of Library Science (1967)
10. Cuadra, C.A., Katter, R.V.: The relevance of relevance assessment. In: Proceedings of the American Documentation Institute. Volume 4., American Documentation Institute, Washington, DC (1967) 95–99
11. Kekäläinen, J., Järvelin, K.: Using graded relevance assessments in ir evaluation. Journal of the American Society for Information Science and Technology **53** (2002) 1120–1129
12. Tang, R., William M. Shaw, J., Vevea, J.L.: Towards the identification of the optimal number of relevance categories. Journal of the American Society for Information Science **50** (1999) 254–264
13. Spink, A., Greisdorf, H., Bateman, J.: From highly relevant to not relevant: examining different regions of relevance. Information Processing and Management: an International Journal **34** (1998) 599–621

14. Voorhees, E.M.: Evaluation by highly relevant documents. In: Proceedings of the 24th annual international ACM SIGIR Conference. (2001) 74–82
15. : Ntcir workshop 1: Proceedings of the first ntcir workshop on retrieval in japanese text retrieval and term recognition, tokyo, japan. In Kando, N., Nozue, T., eds.: NTCIR. (1999)
16. Järvelin, K., Kekäläinen, J.: Ir evaluation methods for retrieving highly relevant documents. In: Proceedings of the 23th annual international ACM SIGIR Conference. (2000) 41–48
17. Sakai, T.: Average gain ratio: A simple retrieval performance measure for evaluation with multiple relevance levels. In: SIGIR'03. (2003)
18. Kekäläinen, J.: Binary and graded relevance in ir evaluations -comparison of the effects on rankings of ir systems. Information Processing an Management **41** (2005)

Text Mining for Medical Documents Using a Hidden Markov Model

Hyeju Jang[1], Sa Kwang Song[2], and Sung Hyon Myaeng[1]

[1] Department of Computer Science, Information and Communications University, Daejeon, Korea
{hjjang, myaeng}@icu.ac.kr
[2] Electronics and Telecommunications Research Institute, Daejeon, Korea
smallj@etri.re.kr

Abstract. We propose a semantic tagger that provides high level concept information for phrases in clinical documents. It delineates such information from the statements written by doctors in patient records. The tagging, based on Hidden Markov Model (HMM), is performed on the documents that have been tagged with Unified Medical Language System (UMLS), Part-of-Speech (POS), and abbreviation tags. The result can be used to extract clinical knowledge that can support decision making or quality assurance of medical treatment.

1 Introduction

Patient records written by doctors are invaluable information especially in areas where experiences have great consequences. If doctors find useful information they need from patients' records readily, they can use it to deal with problems and treatments of current patients. That is, it can provide a support for medical decision making or for quality assurance of medical treatment.

In the treatment of chronic diseases, for example, the past records on the symptoms, therapies, or performances a patient has shown assist doctors to get a better understanding of different ways of controlling a disease of the current patient. As a result, they help their decisions for the direction of the next treatment.

Moreover, hospitals where medical records are kept in the computers are increasing nowadays. The growing availability of medical documents in a machine-readable form makes it possible to utilize the large quantity of medical information with linguistically and statistically motivated tools. Implicit knowledge embedded in a large medical corpus can be extracted by an automated means.

This paper describes a tagging system that yields high-level semantic tags for clinical documents in a medical information tracking system. The tags in this system are categories of information that phrases of medical records contain, such as *symptom, therapy,* and *performance*. They will allow the tracking system to retrieve past cases doctors want to know about a certain therapeutic method, for example. The tagging system uses existing medical terminological resources, and probabilistic Hidden Markov Models [1] for semantic annotation.

H.T. Ng et al. (Eds.): AIRS 2006, LNCS 4182, pp. 553 – 559, 2006.

The contributions of this research can be summarized in three aspects. First, from a practical point of view, it widens the possibility of helping doctors with the experiences and knowledge embedded in the past patient records. Second, from a technical point of view, it attempts to annotate clinical text on phrases semantically rather than syntactically, which are at higher level granularity than words that have been the target for most tagging work. Finally, it uses a special method to guess unknown phrases that don't appear in the training corpus for the robust tagging.

2 Related Works

The popular and conventional approach of part-of-speech (POS) tagging systems is to use a HMM model so as to find a most proper tag [2]. Some systems use a HMM with additional features. Julian Kupiec [3] and Dong Cutting et al. [4] described POS tagging systems, which have the concept of ambiguity class and equivalence class, respectively. Our system also adopted the equivalence class concept which group words into equivalence classes.

Tagging systems in the medical field have focused on the lexical level of syntactic and semantic tagging. Patrick Ruch [5] and Stephen B. Johnson [6] performed semantic tagging on terms lexically using the Unified Medical Language System (UMLS). On the other hand, Udo Hahn et al. [7] and Hans Paulussen [8] built POS taggers which categorized words syntactically.

There also have been the systems which extract information from the medical narratives [9, 10, 11]. Friedman [9, 10] defined six format types that characterize much of the information in the medical history sublanguage.

3 Methodology

The purpose of the tagging system is to annotate the clinical documents with semantic tags that can be used by a tracking system whose goal is to provide useful information to doctors. Our work is based on the list of questions doctors are interested in getting answers for, which was provided by Seoul National University Hospital (SNUH). Among them, we focused on the two questions: 'How can X be used in the treatment of Y?' and 'What are the performance characteristics of X in the setting of Y?' where X and Y can be substituted by {Medical Device, Biomedical or Dental Material, Food, Therapeutic or Preventive Procedure} and {Finding, Sign or Symptom, Disease or Syndrome}, respectively. Our tagging system assigns semantic tags to appropriate phrases so that the tracking system can answer those questions.

The semantic tags were chosen to answer the questions from the doctors in SNUH. While there are many interesting questions and therefore many tags to be used ultimately by a tracking system, we chose Symptom, Therapy, and Performance as the Target Semantic Tags (TST) for the current research. *Symptom* describes the state of a patient whereas *Therapy* means everything a medical expert performs for the patient, such as injection, operation, and examination. *Performance* means the effect or the result of a therapy and includes the results of some examinations or the change of a patient's status (e.g. getting better or getting worse).

TST in this research distinguish the tagging system unique because they represent higher level concepts. Unlike part-of-speech (POS) or UMLS semantic categories of a term, TST can be utilized by the application systems directly. In fact, TST was chosen for a particular application system in the first place. The categories of TST should be changed depending on the purpose of the application system, but the method we propose can be used in the same manner with an appropriate training corpus.

There can be different ways of assigning semantic tags to phrases. Our work is based on an observation that there is a specific sequence when people record something. For example, a description on a cause is followed by that of an effect. Events are usually described in their temporal order. We assumed that the narrative data in CDA documents has implicit rules about sequences.

In order to model the sequential aspect of the clinical documents, we opted for Hidden Markov Model (HMM). Unfortunately, we cannot fully use the grammar rules in our research because our corpus includes Korean and English words mixed. But with the idea that people tend to write things in a certain sequence, we chose to use HMM.

The system architecture for the semantic tagger using HMM is shown in Fig. 1. It is divided into two stages: training and tagging.

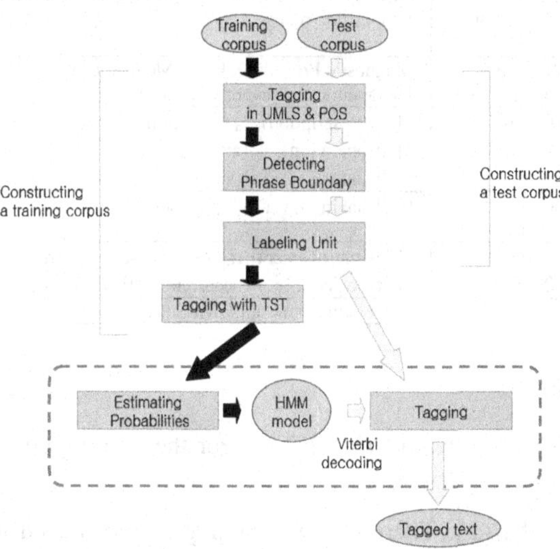

Fig. 1. The system architecture for the TST tagger

3.1 Common Part

1) Tagging in UMLS & POS: The corpus is first processed with UMLS tagging and POS tagging. The former is for classifying medical terms in their semantics whereas the latter is for understanding the syntactic role of words. Abbreviations in the corpus are processed based on the research in the same project.

2) Detecting Phrase Boundary: This is important because symptom, therapy, or performance in TST is described with a phrase or a whole sentence, not a word. This task is not as simple as that for other types of text since doctors usually don't write grammatically correct sentences. In addition, periods are used not only for indication of the end of a sentence but also for abbreviations, dates, floating point numbers and so on.

A phrase is defined to be a unit that ends with a predicate (i.e. a verb ending in Korean) and include a subject with some intervening words like function words and adverbs. Since doctors tend to write a subject like a lab test or medication in English and a predicate in English, a phrase tends to consist of both English and Korean words.

3) Labeling Phrase Units with Equivalence Class: Since there are many words occurring only once in the corpus, we place words into equivalence classes so that class labels are used in HMM (see Table 1 for the equivalence classes). Words are grouped into equivalence classes, and a phrase is expressed with the set of equivalence classes it contains. Fig. 2 shows how a phrase is transformed into an observance expressed with equivalence classes.

Table 1. Equivalence Classes for Words

UMLS tag for cause	Biomedical or Dental Material, Food
UMLS tag for disease or symptom	Finding, Sign or Symptom, Disease or Syndrome, Neoplastic Process
UMLS tag for therapy	Diagnostic Procedure, Food, Medical Device, Therapeutic or Preventive Procedure
Clue word for therapy	처방(prescription), 복용(administer medicine), 시행(operation), 후(after), 이후(later), 사용(use), 증량(increase), 수술(surgery), 중단(discontinue)
Clue word for symptom	발열(having fever), 관찰(observe)
Clue word for performance	호전(improvement), 감소(decrease), 상승(rise), 정상(normal), 발생(occurrence), 변화(change)
unknown	neither clue word nor UMLS tag

3.2 Training Part

1) Tagging with Target Semantic Tags (TST): For the training corpus, the tagging is done manually.

2) Estimating Probabilities: There are two training methods used in current tagging systems. One is to use the Baum-Welch algorithm [16] to train the probabilities, which does not require a tagged training corpus. Another advantage is that for different TST, the only thing we should do is just replace the corpus. The other method is to use tagged training data. This method counts frequencies of words/phrases/tags to estimate the probabilities required for a HMM model. The disadvantage of this method is that it needs a tagged training corpus whose quantity is enough to estimate the probabilities. Building a training corpus is a time-consuming and labor intensive work. Despite this disadvantage, we choose the second method because its accuracy is much higher than that of the Baum-Welch method. David Elworthy [13] and Bernard Merialdo [14] compared tagging performance using the Baum-Welch algorithm against the one using the tagged-training data, proving that using training data is much more effective.

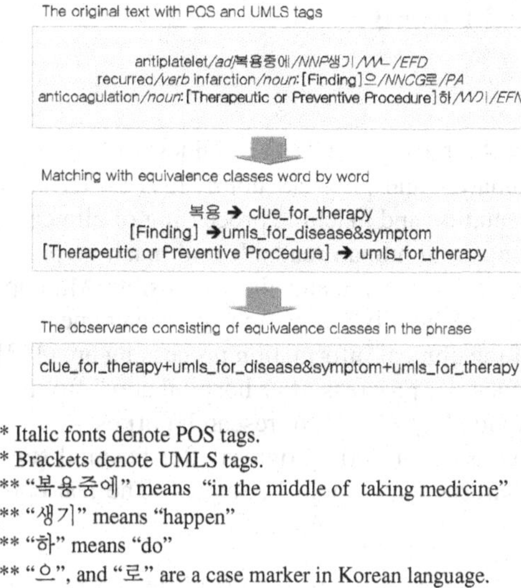
The original text with POS and UMLS tags
antiplatelet/adj복용중에/NNP생기/VV-/EFD
recurred/verb infarction/noun:[Finding]으/NNCG로/PA
anticoagulation/noun:[Therapeutic or Preventive Procedure]하/VV이/EFN
Matching with equivalence classes word by word
복용 → clue_for_therapy
[Finding] →umls_for_disease&symptom
[Therapeutic or Preventive Procedure] → umls_for_therapy
The observance consisting of equivalence classes in the phrase
clue_for_therapy+umls_for_disease&symptom+umls_for_therapy
* Italic fonts denote POS tags.
* Brackets denote UMLS tags.
** "복용중에" means "in the middle of taking medicine"
** "생기" means "happen"
** "하" means "do"
** "으", and "로" are a case marker in Korean language.

Fig. 2. The observance of a phrase with equivalence classes

3.3 Tagging Part

1) Tagging: The system finds a most probable tag sequence using the Viterbi algorithm [15] using the HMM model constructed in the training stage.

2) Tagging of unknown phrase units: phrases appearing in the test corpus are categorized into largely two groups. The first group is for a phrase with no component word known to the system and hence transformed to an equivalence class label. There is no clue in the phrase that can be used in predicting its meaning. Since the whole phrase is labeled as *unknown*, not a class label, its statistics can be gathered from the training corpus that contains many unknown phrases. The other group is for the unknown phrases that have some clues with the words comprising the phrase unit, which have their class labels. The reason why they are called *unknown* is because the particular combination of the class labels corresponding to the phrase is not simply available in the training corpus. We call such a clue combination, not sequence, a *pattern*. The probability of an unknown phrase can be estimated with the equivalence class labels although the unit itself is unknown (see Fig. 2 for an example).

When an unknown pattern appears as an observance, it is compared against the existing patterns so that the best pattern can be found, to which the unknown pattern can be transformed. That is, an unknown pattern is regarded as the best matching pattern. The pattern that matches best with the unknown pattern is chosen and its probability is the same as that of the selected pattern. The probability of that unknown pattern of observance is calculated using the probability of the most similar pattern. When more than one pattern is most similar, the probability of the unknown pattern becomes the average of the most similar patterns.

4 Experiments and Results

4.1 Data

The Clinical Document Architecture (CDA) provides a model for clinical documents such as discharge summaries and progress notes. It is an HL7 (Healthcare Level 7) standard for the representation and machine processing of clinical documents in a way that makes the documents both human readable and machine processable and guarantees preservation of the content by using the eXtensible Markup Language (XML) standard. It is a useful and intuitive approach to management of documents which make up a large part of the clinical information processing area [12].

We picked 300 sections of "progress after hospital stay" from the CDA documents as the target corpus provided by SNUH for research purposes.

The training corpus consists of 200 "progress after hospital stay" sections containing 1187 meaningful phrases that should be tagged. The test corpus is 100 sections with 601 phrases.

4.2 Performance

The level of accuracy of our system is calculated as the number of correct tags per the total number of tags. Fig.3 shows the comparison of the basic model with and without unknown phrase processing. Although the result of the system is not as good as expected, it is promising and undergoing further improvement. We suggest the direction of modification below.

- Increase the number of different equivalence classes. The number of tags corresponding to the equivalence classes is so small at this point that the transition probabilities are not very meaningful.
- Find better initial probabilities of a HMM model.
- Improve the unknown phrase guessing method.
- Get as much tagged text as we can afford.

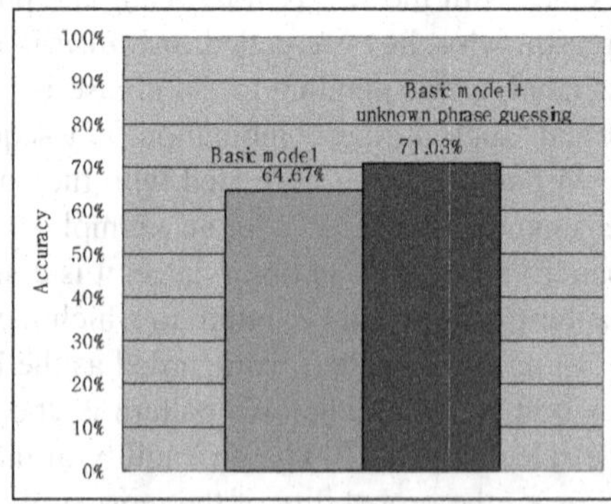

Fig. 3. The results of a basic HMM model and a HMM model with an unknown phrase guessing module

5 Conclusion

We showed a semantic tagger for medical documents using a HMM model. For future work, we are going to utilize symbols and numeric expressions to represent phrases better and to find a better way for matching equivalence classes. Moreover, we will design and compare against other methods such as Markov random fields, SVM, and so on.

Acknowledgement

This study was supported by a grant of the Korea Health 21 R&D Project, Ministry of Health & Welfare, Korea.

References

1. L.R.Rabiner et al., "An Introduction to Hidden Markov Models", IEEE ASSP Magazine, 1986
2. Linda Van Guilder, "Automated Part of Speech Tagging:A Brief Overview", Handout for LING361, 1995
3. Julian Kupiec, "Robust part-of-speech tagging using a hidden Markov model", Computer Speech and Language, pp. 225–242, 1992.
4. Doug Cutting et al., "A Practical Part-of-Speech Tagger", In Proceedings of the 3rd ACL, pp.133–140, 1992
5. Patrick Ruch, "MEDTAG: Tag-like Semantics for Medical Document Indexing", In Proceedings of AMIA'99, pp.35–42
6. Stephen B. Johnson, "A Semantic Lexicon for Medical Language Processing", JAMIA, 1999 May–Jun; 6(3): 205–218
7. Udo Hahn, "Tagging Medical Documents with High Accuracy", Pacific Rim International Conference on Artificial Intelligence Auckland, Newzealand , pp. 852–861, 2004
8. Hans Paulussen, "DILEMMA-2: A Lemmatizer-Tagger for Medical Abstracts", In Proceeings of ANLP, pp.141–146, 1992
9. Carol Friedman, "Automatic Structuring of Sublanguage Information," London: IEA, 1986, pp. 85–102.
10. Emile C. Chi et al., "Processing Free-text Input to Obtain a Database of Medical Information", In Proceedings of the 8th Annual ACM-SIGIR Conference, 1985
11. Udo Hahn, "Automatic Knowledge Acquisition from Medical Texts", In Proceedings of the 1996 AMIA Annual Fall Symposium, pp.383–387, 1996
12. What is CDA?: http://www.h17.org.au/CDA.htm#CDA
13. David Elworthy, "Does Baum-Welch Re-estimation Help Taggers?", Proceedings of the 27th ACL, 1989
14. Bernard Merialdo, "Tagging English Text with a Probabilistic Model", Computational Linguistics 20.2, pp155–172, 1994
15. A. J. Viterbi, "Error bounds for convolutional codes and an asymptotically optimal decoding algorithm", IEEE Transactions of Information Theory 13, pp 260–269, 1967
16. Baum, L, "An inequality and associated maximization technique in statistical estimation for probabilistic functions of a Markov process", Inequalities 3:1-8, 1972.

Multi-document Summarization Based on Unsupervised Clustering

Paul Ji

Center for Linguistics & Philology
University of Oxford
paul_dji@yahoo.co.uk

Abstract. In this paper, we propose a method for multi-document summarization based on unsupervised clustering. First, the main topics are determined by a MDL-based clustering strategy capable of inferring optimal cluster numbers. Then, the problem of multi-document summarization is formalized on the clusters using an entropy-based object function.

Keywords: multi-document summarization, clustering, entropy, optimization.

1 Introduction

Multi-document summarization has been a major concern in information processing due to multiple information overloads. Compared with single document summarization, multi-document summarization faces two difficulties. One is that given multiple documents for one event or relevant events, the information in the documents tends to overlap, which calls for effective analysis of the similarity and difference between contents of the documents. Another difficulty is that the structural information of documents cannot be readily used in the same way as in single document cases.

In general, the strategies for multi-document summarization fall in two categories, extraction-based and abstraction-based. An extraction-based strategy generally ranks sentences in some way and selects top-scoring sentences as summaries. An abstraction-based strategy involves some understanding of the contents in the documents, e.g., concepts and relations, etc., and then creates the summaries based on some generation techniques.

Under the extraction-based strategy, the main method is to define a criterion based on some superficial features, e.g., key words in documents and positions of sentences or formal representations of texts, e.g., conceptual graphs or tree structures, etc. [7], and then rank the sentences according to the criterion.

For the abstraction-based strategy, one main method is to produce some kind of semantic representation of the main contents in the documents, and then use some generation techniques to create summaries based on the semantic representation. One particular example is by information fusion [1], which tried to combine similar sentences across documents to create new sentences based on language generation technologies. Although this could model, to some extent, the human's efforts in summarization process, it faces some difficulties. One is that it relies on some external

H.T. Ng et al. (Eds.): AIRS 2006, LNCS 4182, pp. 560 – 566, 2006.

resources, e.g., dependency parsers, interpretation or generation rules, etc., which would limit its portability. Another difficulty is that when generating similar sentences, it needs manual intervention to set similarity thresholds or number of the clusters; however, this is not generally known in advance.

Another main method within this category is built on compression of sentences [6], in which longer sentences were converted to shorter ones based on parse tree transducers by keeping the main constituents and cutting down others. One difficulty of this approach is that it needs a full and effective parsing module for the languages involved. However, for languages other than English, such an effective full parsing system may be not available. Another problem is that the algorithm now focuses on sentence to sentence transductions, while for a multi-document summarization, a many to one sentence transducer is needed.

In this paper, we propose a method for multi-document summarization based on unsupervised clustering. In detail, the main topics are first determined by a MDL-based clustering strategy with capabilities of inferring optimal cluster numbers. Then, the problem of multi-document summarization is formalized using an entropy-based object function.

2 Motivation

Although sentence extraction method is not the usual way humans create summaries for documents, we cannot deny the fact that some sentences in the documents do represent some aspects of their contents to some extent. On the other hand, as contents increase on the Internet, speed will be a key factor for summarization. So, extraction-based summarization is still a promising candidate.

To ensure good coverage and avoid redundancy, the main themes should be determined in the multiple documents. To do that, some clustering methods have been proposed. Stein and Bagga et al. [11] grouped documents into clusters by clustering single document summaries, and then selected representative passages from each cluster to form a summary. Boros et al. [2] clustered sentences, with the cluster number predefined as 30, and the performance is not very good. Hardy et al. [4] selected representative passages from clusters of passages, and the system worked well in DUC 2001 with some parameter adjustments. Siddharthan et al. [10] clustered sentences with SimFinder [5] to generate summaries, in which a similarity threshold for clusters needed to be pre-specified, although the cluster number needed not.

One common problem among these clustering methods is that they need to specify the number of the clusters or the similarity threshold in advance. However, this number or threshold is not known before clustering in most cases. In this work, we try to use a MDL-based strategy [2] to automatically infer appropriate cluster numbers as well as members of the clusters.

After creating the clusters, the next problem is to select some sentences from the clusters. We adopt a global criterion for sentence selection, which is based on the overall performance of the summary in representing the whole document set. In this work, we adopt the pair-wise entropy of clusters to model the overall representativeness of a summary for a document set. In practice, we construct a vector for a candidate summary just like those for sentence clusters, add it to the group of the clusters and compute the entropy of the new cluster group. The optimal summary is the one, which results in the highest entropy of the new cluster group.

3 Sentence Clustering

We employ a Gaussian mixture model algorithm with a MDL criterion to estimate the optimal cluster number and the member of each cluster. The MDL criterion is given by 1).

$$-\log p(y \mid K, \theta) + \frac{1}{2} L \log(NM) \tag{1}$$

where y is the data points, K is the number of clusters, θ is the parameter set, N is the number of the data points, M is the number of the dimensions, and L is given by 2).

$$L = K(1 + M + \frac{M(M+1)}{2}) - 1 \tag{2}$$

For each K, an EM algorithm can be used to seek the solution, which contains K clusters as well as their members. The algorithm will be terminated until the change of the criterion is below a threshold. By comparing the values of *MDL* among all K, we can get K^*, which minimizes *MDL*.

4 Sentence Selection

After sentence clustering, we get n sentence clusters. Now the problem is to choose one representative sentence from each cluster, and they together form a summary. Let $v_1, v_2, \ldots, v_n$ be vectors for the n clusters respectively, $F(v_1, v_2, \ldots, v_n) = [v_1, v_2, \ldots, v_n]$ be the matrix for the n vectors, and $s_{i,j}$ be the cosine similarity between v_i and v_j, then the entropy for $F(v_1, v_2, \ldots, v_n)$ is defined as (3).

$$E(v_1, v_2, \ldots, v_n) = -\sum_{i=1}^{n}\sum_{j=1}^{n}(s_{i,j}\log s_{i,j} + (1 - s_{i,j})\log(1 - s_{i,j})) \tag{3}$$

Let v_0 be the vector for the summary, $F(v_0, v_1, \ldots, v_n)$ is the matrix for the n vectors plus the summary, then the entropy for $F(v_0, v_1, \ldots, v_n)$ is (4).

$$E(v_0, v_1, \ldots, v_n) = -\sum_{i=0}^{n}\sum_{j=0}^{n}(s_{i,j}\log s_{i,j} + (1 - s_{i,j})\log(1 - s_{i,j})) \tag{4}$$

Intuitively, $E(v_1, v_2, \ldots, v_n)$ and $E(v_0, v_1, \ldots, v_n)$ represents the perplexity of the cluster structure of $F(v_1, v_2, \ldots, v_n)$ and $F(v_0, v_1, \ldots, v_n)$ respectively. (5) is the increase of the perplexity when the summary vector is inserted.

$$E(v_0, v_1, \ldots, v_n) - E(v_1, v_2, \ldots, v_n) \tag{5}$$

Intuitively, a good summary, if having better coverage, tends to make (5) higher. With the number of the representative sentences fixed (the same as the number of the clusters), the increase of the coverage on one hand means the decrease of the redundancy on the other hand. So, if a summary having higher entropy difference as in (5), i.e, having higher entropy as in (4) (since (3) is fixed), it will ensure more coverage and less redundancy at the same time. Formally, the multi-summarization problem can be defined as in the following.

to seek a summary, whose vector v_0 maximizes (4)

For this optimization problem, we need to adopt some kind of search strategies. We choose two sentences nearest to the center of each cluster as the representative, and select one based on the measure (4). The search strategy is roughly deletion of one sentence at one time, unless the sentence is the only representative of one cluster, and each deletion maximizes the entropy of the remaining summary together with the n sentence clusters.

5 Experiments and Results

The data is document sets of DUC2004 [8] Task 2 run by National Institute of Standards and Technology (NIST). We chose ROUGE-1 as the evaluation measure for our results. Regarding the clustering configuration, we set the cluster number range as from 6 to 12, considering the size of the clusters.

Table 1 gives the comparison between the performance with cluster validation (i.e., automatically determining appropriate cluster numbers) and that without cluster validation (i.e., pre-specifying cluster numbers manually).

Table 1. Performance with/out cluster validation

Cluster No.	Validation	6	7	8	9	10	11	12
Scores	0.371	0.343	0.358	0.354	0.355	0.347	0.344	0.341

From Table 1, we can see that the performance with cluster validation is consistently better than that without cluster validation. The reason may be that the appropriate number of the clusters should be different for different sets of documents, and with the cluster number being fixed, it would not be ensured to produce the optimal cluster structure, which would affect the performance.

To confirm the reason, we checked the separability of the generated clusters based on two views. One view is from discrimination between clusters in general. To do that, we constructed a matrix $C \times C$, where C is the set of generated clusters. Fig. 1 shows the change of the average pair-wise entropy, $E(CC)$, which was normalized by the number of cluster pairs ($|C| \times |C|$) to ensure comparability.

Another view is from discrimination of probabilities of one sentence belonging to clusters. To do so, we constructed a matrix $S \times C$, where S and C are sets of sentences and clusters respectively. $p(s, c)$ is the probability of s falling in c, which is in fact the *cosine* similarity between vectors of s and c. Fig. 1 also shows the change of the average pair-wise entropy, $E(SC)$, for each cluster setting, which was normalized by the number of the sentence-cluster pairs to ensure comparability.

Fig.1 indicates that in both views, the normalized pair-wise entropy is minimized with cluster validation, which means that with the optimal cluster number automatically

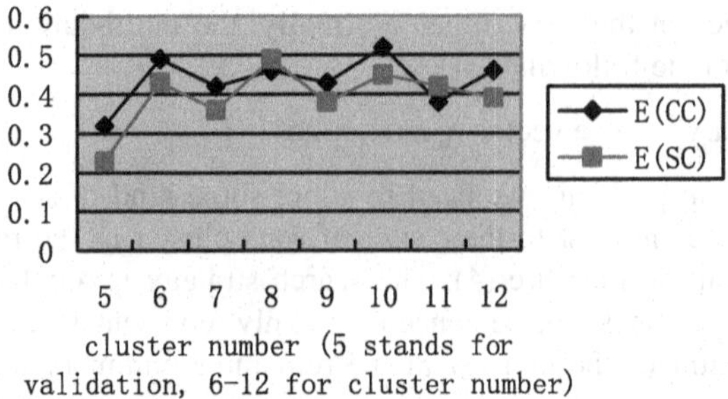

Fig. 1. Normalized average entropy *vs.* cluster number

determined, a clearer cluster structure would be acquired. In such cases, more representative sentences could be selected to form better summaries.

In addition to Rouge scores for evaluation, we can use the similarity between the extracted sentences and the cluster-based topics to evaluate the coverage. On the other hand, we can use the similarity between the extracted sentences to evaluate the redundancy. In this section, we use these two measures to compare our methods with supervised clustering (k is pre-defined and set as 7, due to the highest scores achieved among all pre-specified) and MEAD's ranking [9].

Regarding coverage, we took the highest similarity between sentences in a summary and a topic as the similarity between the summary and the topic, and chose the topics acquired with cluster validation as the ground-truth topics. Fig. 2 shows the change of the average summary-topic similarity applied to the generated summaries by the three methods, which shows that the unsupervised method acquired higher similarity between summaries and topics.

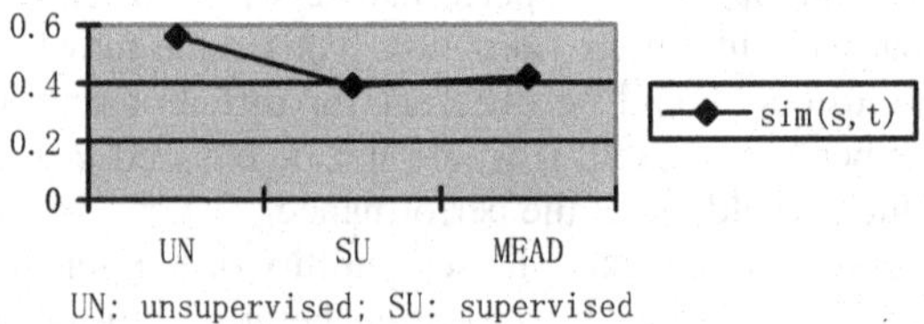

Fig. 2. Summary-Topic Similarity

Regarding redundancy, we chose representation entropy of a summary for evaluation. Intuitively, representation entropy of a summary captures the amount of the information embedded in the sentences. So, higher representation entropy corresponds to lower redundancy.

We constructed a $S \times S$ covariance matrix, where S is the set of the summary sentences. Let the eigenvalues be $x_1, x_2, \ldots, x_d$, where $d=|S|$, and $y_1, y_2, \ldots, y_d$ be the normalized eigenvalues:

12) $y_i = \frac{x_i}{\sum_{j=1}^{d} x_j}$

These normalized eigenvalues meet the properties of commonly used probabilities, so representational entropy of $S \times S$ (Hr) can be defined as usual, as in (5'). Fig. 3 shows the change of the representational entropy when applied to the generated summaries by the three methods.

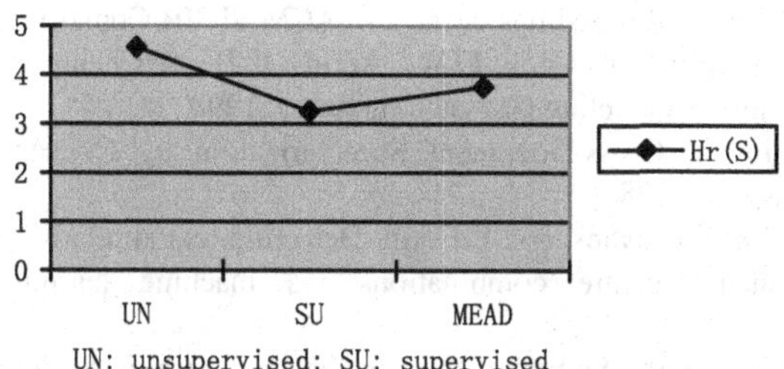

Fig. 3. Representational Entropy

The comparison indicates that the summary generated by the unsupervised clustering acquired the highest average representational entropy, which means the information is more widely distributed among the summary sentences.

6 Conclusions

In this paper, we propose an approach for multi-document summarization based on unsupervised clustering. For sentence clustering, we adopt a MDL-based strategy with capabilities of inferring optimal cluster numbers automatically. For sentence selection, we formalize it as an optimization problem using an entropy-based criterion.

As the first step for multi-document summarization, sentence clustering has an important influence on the overall performance. In fact, the sentences contain real feature words as well as some noise. To get better cluster structure, we need to reduce the noise. Together with automatic determination of optimal cluster numbers, feature selection would be a very interesting topic in future.

For multi-document summarization, we can see it as an optimization problem: to select a subset of sentences that meets some constraints. With sentences being clustered, one constraint can be created for the optimization problem: one sentence to be extracted per cluster. In this work, we formalize the overall optimization problem based on pair-wise entropy. There may be other ways to formalize the problems according to different views about multi-document summarization, e.g., a view based on coverage of the main concepts in the documents. To compare the performance across different views could be another interesting topic in future.

Another important future work is how to make the extracted summary more coherent, a main step we don't touch in this work. Time-based event analysis, conceptual focus and shift or anaphor co-reference would be future topics in this area.

References

1. R.Barzilay, K.R.McKeown, M.Elhadad 1999. Information fusion in the context of multi-document summarization, in Proceedings of the 37th ACL, 1999, Maryland.
2. E.Boros, P.B.Kantor and D.J.Neu. A Clustering Based Approach to Creating Multi-Document Summaries. Proceedings of the 24th ACM SIGIR Conference, LA, 2001.
3. Bouman, C.A., Shapiro, M., Cook, G.W., Atkins, C.B., & Cheng, H. Cluster: An unsupervsied algorithm for modeling Gaussian mixtures. 1998
4. H.Hardy, N.Shimizu. Cross-Document Summarization by Concept Classification. In SIGIR 2002, page 121-128.
5. V.Hatzivassiloglou, J.Klavans, and E.Eskin. Detecting text similarity over short passages: exploring linguistic feature combinations via machine learning. Proceedings of EMNLP'99.
6. K.Knight, D.Marcu, 2002. Summarization beyond sentence extraction: a probabilistic approach to sentence compression, *Artificial Intelligence*, 139(1), 2002.
7. W. Mann, S. Thompson. Rhetorical structure theory: towards a functional theory of text organization. *Text* 1988; 8 (3): 243–281.
8. P. Over, J. Yen, 2004. An Introduction to DUC2004: Intrinsic Evaluation of Generic New Text Summarization Systems, Proceedings of DUC 2004.
9. D.Radev, T.Allison, S.B.Goldensohn, J.Blitzer, A.Çelebi, S.Dimitrov, E.Drabek, A.Hakim, W.Lam, D.Liu, J.Otterbacher, H.Qi, H.Saggion. MEAD - a platform for multi-document multilingual text summarization. In Proceedings of LREC 2004, Lisbon, Portugal, May 2004.
10. A.Siddharthan, A.Nenkova and K.McKeown. Syntactic Simplication for Improving Content Selection in Multi-Document Summarization. In Proceeding of COLING 2004, Geneva, Switzerland.
11. G.C.Stein, A.Bagga, G.B.Wise. Multi-Document Summarization: Methodologies and Evaluations. In Conference TALN 2000, Lausanne, October 2000.

A Content-Based 3D Graphic Information Retrieval System

Yonghwan Kim and Soochan Hwang

Department of Computer Engineering, Hankuk Aviation University, 412-791, Korea
{yhkim93, schwang}@hau.ac.kr

Abstract. This paper presents a 3D graphic information retrieval system which supports a content-based retrieval for 3D. The user can pose a visual query involving various 3D graphic features such as inclusion of a given object, object's shape, descriptive information and spatial relations on the web interface. The data model underlying the retrieval system models 3D scenes using domain objects and their spatial relations. An XML-based data modeling language called 3DGML has been designed to support the data model. We discuss the data modeling technique and the retrieval system in detail.

1 Introduction

As more image and graphic data are used in applications, the methods that support content-based retrievals of 3D graphic information are desired. Most current 3D graphic systems focus on visualizing 3D images. They usually model a 3D image using lines and polygons with information on their placements in the space. One problem with this approach is that it is difficult to store semantics of 3D objects in a scene. The lack of such information makes it difficult to retrieve or manipulate a particular domain object separately from others.

We have developed a 3D graphic data model that supports a content-based retrieval for 3D scenes. This paper presents the data model and a web-based information retrieval system based on it. The 3D graphic data model, called 3DGML (3-Dimensional Graphical Markup language), allows the semantics of 3D objects to be incorporated into a 3D scene. The semantics implied in a 3D image include roles of component objects, semantic or spatial relationships among objects and composition hierarchies of objects.

In 3DGML, scenes are modeled as compositions of 3D graphic objects. A set of primitive 3D objects is used as building blocks for modeling 3D scenes. Larger 3D objects are defined through a composition of other objects. The user can search scenes using various 3D graphic features such as inclusion of a given object, object's shape, descriptive information and spatial relations of 3D objects they contain. The 3D graphic retrieval system presented in this paper was implemented using XML.

The remainder of this paper is organized as follows. The next chapter discusses previous research related to our work. Chapter 3 describes modeling 3D images with 3DGML. Chapter 4 presents the information retrieval system for 3D graphic information that we developed. Chapter 5 concludes the paper.

H.T. Ng et al. (Eds.): AIRS 2006, LNCS 4182, pp. 567 – 573, 2006.

2 Related Works

Research on 3D graphics has mainly concerned about the visualization of data to provide the user with a 3D feel [1, 2, 3]. Existing graphic systems treat a 3D object as a collection of lines and polygons rather than a unit of manipulation. It prevents them from supporting content-based retrievals or manipulations of 3D objects.

MPEG standard committees have been worked on the issues of modeling 3D images [4, 5]. MPEG-4 SNHC (Synthetic, Natural, and Hybrid Coding) aims to develop techniques for representing synthetic images with natural objects efficiently. 3D objects are represented by 3D meshes (surfaces), norm vectors, and their features such as color, texture, etc. However, modeling 3D objects as semantic units is not addressed by MPEG.

X3D (eXtensible 3D) is an Open Standards XML-enabled 3D file format and related access services for describing interactive 3D objects and worlds [6, 7]. It enables real-time communication of 3D data across all applications and network applications. X3D does not also consider the representation of semantics in 3D graphic modeling, which are used to retrieve 3D images from the database.

There have been some efforts to model the spatial relations of objects for 3D scenes. Xiong and Wang described a technique supporting similarity search for a chemical application [8]. They represent a 3D object as a 3D graph consisting of one or more substructures of connected subgraphs. Similarity of two objects is determined by comparing their substructures and edges. Gudivada and Jung proposed an algorithm for retrieving images of relevance based on similarity to user queries [9]. In their image representation scheme, an image is converted to an iconic image with human assistance. This method also determines the spatial relation of objects using the connectivity of graphs.

While the works discussed above considered geometrical similarity, they did not discuss modeling the semantics of 3D objects and querying based on the contents of 3D images. In order to address the problems discussed above a new data model is needed that treats 3D objects as first class objects in modeling. The new model should represent 3D data using semantic units rather than primitive geometrical objects.

3 Modeling 3D Graphic Data in 3DGML

We first present the 3D graphic data model with an example 3DGML document that models a 3D scene. A 3D scene is modeled using three types of components: 3D objects contained in the scene, spatial relations on the objects, and descriptors. A simple 3D object is modeled using basic objects called shapes. A *shape* object is a system-defined 3D graphic object such as box, cone, cylinder and sphere. An object of an arbitrary shape that is difficult to model with basic objects only is defined using one or more polygons. Such an object is called a *user-defined shape.* A complex 3D object is modeled as a composition of shapes, user-defined shapes, and other complex objects.

Every 3D object within a scene exists in the form of a *Gobject* (Graphic object). A Gobject is defined by extending an abstract object. An abstract object is a skeletal object that is used as a prototype of other objects. We called it *Aobject.* It is a template that does not physically exist in a scene. It specifies the shape of a 3D object and partially describes its appearance. Hence, the modeling of a scene typically involves defining abstract objects first. In many cases, abstract objects represent semantic units such as desk, chair, etc that are germane to an application domain.

Fig. 1 shows a 3D scene of furniture in an office that will be used in our discussion below. In Fig. 1, we show Gobject's id and Aobject's id in the parenthesis that the Gobject refers to.

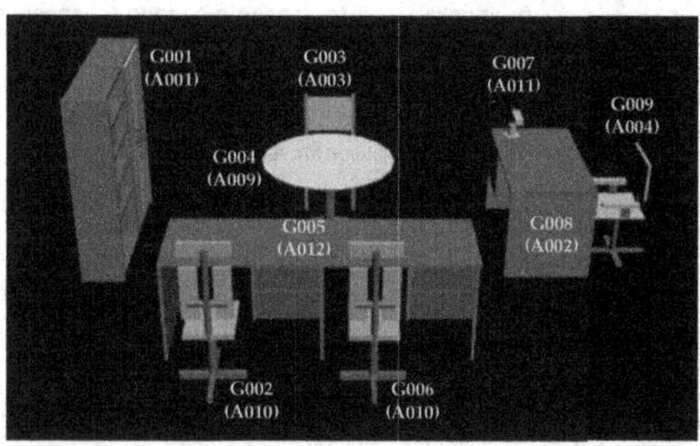

Fig. 1. The 3D scene of an office

Fig. 2 is a stripped version of a document that models the scene in 3DGML. The model in Fig. 2 defines the scene "OFFICE_01." It contains two major blocks: the *Definition* block that defines abstract objects used in the scene and the *Display* block that defines the actual contents of the scene.

The *Display* block contains the definitions of Gobjects G001 through G009 which model the objects labeled in Fig. 1, correspondingly. Object G001 is an example of a Gobject defined using A001, an abstract object, as its prototype. It is a concrete extension of A001 with a translation.

We now narrow our discussion to the modeling of bookcase G001. The 3DGML model shown in Fig. 2 defines bookcase G001 in a two-step process. It first defines an abstract object, A001 and defines G001 as a Gobject by declaring A001 to be its prototype and specifying transformation information needed to create a concrete graphic object to be inserted in the scene.

According to the definition of the abstract object A001, it consists of many basic shapes including a cylinder. The information defined by the shape object in the abstract definition still is not specific enough to be used in a scene as it barely defines the information needed by A001. This example defines the relative location and scaling factors within A001 and the default color for the objects contained in it. These values may need to be modified or complemented with further information to fit it in a specific scene. The definition of A001 contains one descriptor, which describes it as a bookcase. The *Contour* element of A001 contains a vector value of the contour object for A001.

The *Td_string* element defines the spatial relation for the objects contained in this scene. A 3D string can express the concepts such as A is located to the left of B, A is above B, and A is closer than B [10, 11]. A 3D string is a 3-tuple (u, v, w), where u, v, and w represent the 1D strings obtained when objects are projected to the X, Y, and

```
- <Scene>
    <Descriptor value="OFFICE_01" />
    <Td_String
      u="G001<G002<G003=G004=G005<G006<G007<G008<G009"
      v="G003=G009<G004=G005=G008<G002=G006<G001<G007"
      w="G003<G001<G007<G004=G008=G009<G005<G002=G006" />
  - <Definition>
    - <Aobject aid="A001">
        <Descriptor value="BOOKCASE" />
        <Contour value1="-0.2 0.9 -0.2" />
        <Td_String u="......" v="......" w="......" />
      - <Children>
        - <Transform translation="8.5 4.1 -16.2" rotation="0 0 0" scale="1 1 1">
          - <Shape shapeID="S004" color="0.337300 0.337300 0.337300">
              <Cylinder height="11.94" radius="0.9" />
            </Shape>
          </Transform>
        + <Transform translation= ... >
          <! ...... The definition of other shapes for Aobject A001 ......... >
      </Children>
    </Aobject>
  + <Aobject oid="A002">
      <! ...... The definitions of other Aobjects ........ ........ ........ ......... >
    </Definition>
  - <Display>
    - <Transform translation="-8.5 -4.14 16.25" rotation="0 0 0" scale="1 1 1">
        <Gobject gid="G001" refAid="A001" />
      </Transform>
    + <Transform translation="2.17 -4.14 16.33" rotation="0 0 0" scale="1 1 1">
      <! ...... The definitions of Gobjects G002 through G009...... ......... >
  </Display>
</Scene>
```

Fig. 2. The definition of the scene in Fig. 1

Z-axis, respectively. In Fig. 2, the *Td_String* element specifies the spatial relation of Gobjects G001 through G009. For example, the u value of the 3D string specifies the ordering of the nine objects with respect to the X-axis. "G001<G002" means that G001 is located to the left of G002. Since G003, G004, and G005 are on the same locations with respect to the X-axis and located to the left of G006, the 3D string is represented as "G003=G004=G005<G006."

4 A 3D Graphic Information Retrieval System

The implementation of the 3D graphic information retrieval system discussed so far is based on the XML technology. It has been implemented and runs with IIS(Internet Information Server) on the Windows NT platform. The XML parser was implemented in ASP using DOM API [12]. Parsed XML documents are stored in Oracle 9i. The system consists of the semantic editor, 3D object manager, the query coordinator and the database wrapper as shown in Fig. 3.

The user creates a 3D scene using a 3D graphic editor which generates a VRML file like 3D Max. Then, the semantic editor converts the VRML document to a 3DGML document. The database wrapper parses the generated 3DGML document and maps it to a relational database. The relational schema is defined to reflect the part of 3DGML structure that is used querying 3D images. The output generator converts the 3DGML document to a VRML document to display its 3D image on the Web browser.

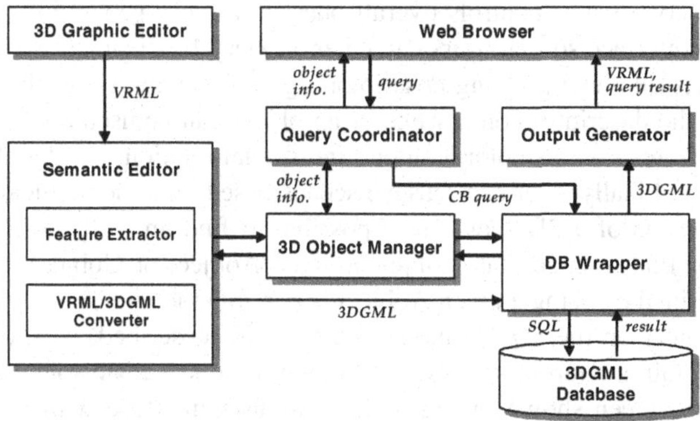

Fig. 3. An Overview of the 3D Graphic Database System

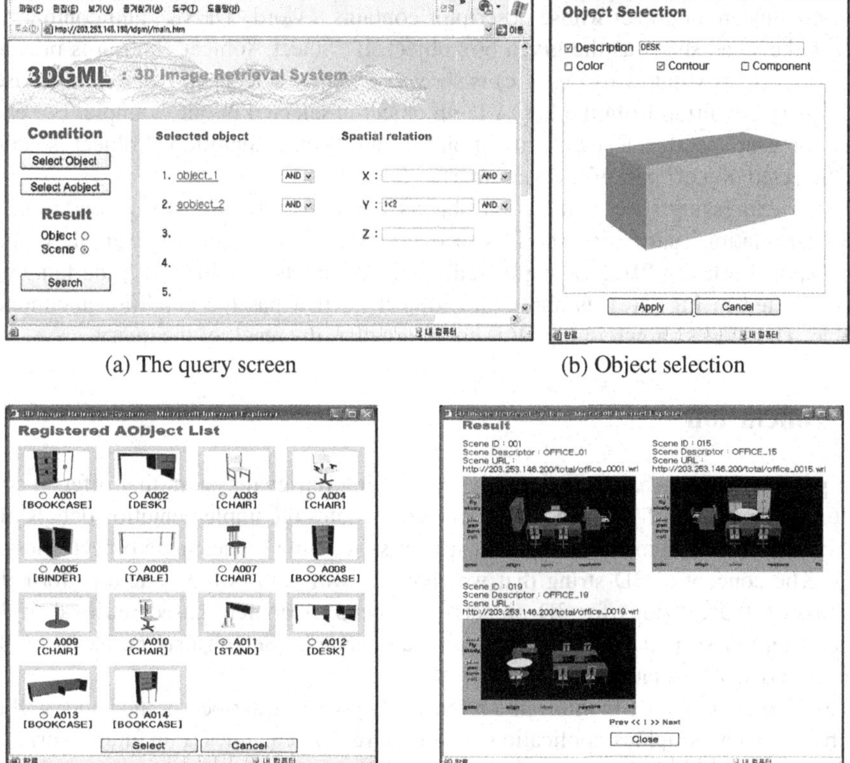

(a) The query screen (b) Object selection

(c) Aobject list (d) The result screen

Fig. 4. A sample query and the result

The query coordinator controls overall phases of query generation. It provides a graphical user interface so that users can enter a Query-By-Example style query. Our system supports queries involving descriptors, graphic features and the existence of objects. First, the description on a scene or an object can be used querying database. 3D graphic features such as color, contour and spatial relations can be also used as a query condition. Finally, user can retrieve scenes based on a hierarchical relationship among components of a 3D scene. It is possible to find an Aobject that contains a given shape or find a scene that contains a given Aobject or Gobject. Finding Gobjects that is defined by using a given Aobject is possible too.

A sample query on the 3D database system is now described, which retrieves the scenes that contain a lamp on a desk. Fig. 4 shows the screens displayed by the system. The query screen shown in Fig. 4(a) is the user interface with which the user enters queries.

The user specifies object selection conditions by pressing "Select Object" or "Select Aobject" buttons. When "Select Object" button is pressed, the object selection window of Fig. 4(b) is popped up for the expression of conditions. The conditions of Fig. 4(b) mean an object whose descriptor contains a word "DESK" and contour object is the same shape as the given box object. If "Select Aobject" button is pressed, the Aobject list window of Fig. 4(c) is shown. The user selects an Aobject to be used as a query condition from the list. A lamp object is selected in our example. For each time the user specifies a search condition for an object, a conditioned object is added to "Selected object" part of the user interface.

After expressing all conditions on objects, the user states 3D string conditions in "Spatial relation" part using the id's of conditioned objects. In our example of Fig. 4(a), spatial relation "1<2" is expressed, which means the "aobject_1", the lamp object selected in Fig. 4(c), is above the "object_1" that has the condition mentioned above. The scenes shown in Fig. 4(d) are returned as the result of the query.

5 Conclusion

We presented a web-based 3D graphic information retrieval system that offers a content-based retrieval of 3D scenes. Our model separates the implementation details of a 3D object from its semantic usage and supports modeling scenes in an object-oriented way. The concept of 3D string that we came up with allows the system to formally express spatial relations for 3D objects in a scene. A content-based retrieval of 3D objects on our system was described using an example. Search may be based on 3D shapes and spatial relations.

We expect that the 3D information retrieval systems described in the paper will be useful for many graphics applications that require 3D semantics. For future work, we are planning on providing more elaborate algorithm to represent the contour information of objects and the support of similarity query based on shapes of objects and scenes. We also consider the indexing methods that help fast searching of 3D graphic data using their features.

Acknowledgements

This work was supported by the Internet information Retrieval Research Center (IRC) in Hankuk Aviation University. IRC is a Regional Research Center of Gyeong-gi Province, designated by ITEP and Ministry of Commerce, Industry and Energy.

References

1. Jie, S.: Visualizing 3D Geographical Data with VRML. IEEE Proc. of the Int. on Computer Graphics (1998) 108–110
2. Weinhaus, F. M., Devarajan, V.: Texture Mapping 3D Models of Real-World Scenes. ACM Computing Survey, vol. 29, no. 4 (1997) 325–368
3. Hwang, S., Cho, S., Wang, T., Sheu, P. C.-Y.: A Fast 3D Visualization Methodology Using Characteristic Views of Objects. Int. J. of Software Eng. and Knowledge Eng., vol. 8, no. 1 (1998)
4. Huang, Q., Puri, A., Liu, Z.: Multimedia Search and Retrieval: New Concepts, System Implementation, and Application. IEEE Transactions on Circuits and System for Video Technology, vol. 10, no. 5 (2000) 679–692
5. Hunter, J.: MPEG-7 Behind the Scenes. D-Lib Magazine, vol. 5, no. 9 (1999).
6. Bray T., Paoli, J., Sperberg-McQueen, C. M.: Extensible Markup Language (XML) 1.0. http://www.w3.org/TR/1998/REC-xml-19980210 (1998)
7. Web 3D Consortium. X3D and Related Specifications. http://www.web3d.org/x3d/specification.html (2005)
8. Xiong, W., Wang, J. T. L.: Fast Similarity Search in Database of 3D Objects. IEEE Int. Conference on Tools with Artificial Intelligence (1998) 16–23
9. Gudivada, V. N., Jung, G. S.: Spatial Knowledge Representation and Retrieval in 3D Image Database. IEEE Pro. of the Int. Conference on Multimedia Computing and Systems (1995) 90–97
10. Hwang, J., Roh, J., Lee, K., Park, J., Hwang, S.: An XML-based 3-Dimensional Graphic Database System. The Human Society and the Internet. LNCS, Vol. 2105 (2001) 454–467
11. Chang, S. K., Shi, Q. Y., Yan, C. W.: Iconic Indexing by 2-D Strings. IEEE Transactions on Pattern Analysis and Machine Intelligence, vol. 9, no. 3 (1987)
12. W3C Consortium. Document Object Model (DOM). http://www.w3.org/DOM/ (1998)

Query Expansion for Contextual Question Using Genetic Algorithms

Yasutomo Kimura[1] and Kenji Araki[2]

[1] Otaru University of Commerce 3-5-21, Midori, Otaru, 047-8501 Japan
kimura@res.otaru-uc.ac.jp

[2] Graduate School of Information Science and Technology, Hokkaido University, Kita 14 Nishi 9, Kita-ku, Sapporo-shi, 060-0814 Japan
araki@media.eng.hokudai.ac.jp

Abstract. We propose a query expansion method using Genetic Algorithms(GA) in Japanese. Recently, question answering research focuses on contextual questions. Therefore a question answering system has to resolve contextual problems by using both previous questions and previous answers. This problem is largely related to query expansion because of the need to find new keywords. In the contextual processing, a query needs to find other suitable keywords from related resources. Although it is easy for a system to find related words, it is difficult to find a suitable combination of keywords. GA is better suited for a combination problem just like a knapsack problem. Therefore we apply GA to our contextual query expansion method. In the evaluation experiment, MRR was 0.2531 in 360 contextual questions. We confirm the MRR of our method is higher than that of the baseline. We illustrate our method and the experiment.

1 Introduction

Recently, question answering tasks focus on the problem of retrieving the appropriate answer rather than retrieving document lists (Ellen. M. Voorhees. et al. 2000). This task has recently been evaluated in TREC-QA (Ellen. M. Voorhees. 2004), NTCIR-QAC (Tsuneaki Kato et al. 2004) and so on. TREC Question answering was the first large-scale evaluation of domain independent question answering systems. QAC is a challenge encouraging the evaluation of question answering technology in Japanese. Such systems that answer isolated factoid questions are the most basic level of question answering technology. Question answering systems are used interactively to answer a series of related questions, whereas in the conventional setting, systems answer isolated questions one at a time. Question answering systems generally reply about location, organization, date and so on. Recently question-answering systems are able to correctly answer isolated question. Moreover, it is often said that QA tasks appear simple by not including the why-question types and the how-question types. However a series of related questions is more difficult than isolated questions. Question answering systems have to interpret a given question within the context of a specific dialogue. In the case of a series of

H.T. Ng et al. (Eds.): AIRS 2006, LNCS 4182, pp. 574–580, 2006.

related questions, QA systems have to include the context processing abilities of systems such as anaphora resolution and ellipses handling. While, Murata et al. have obtained high accuracy results by adding keywords instead of using anaphora resolution and ellipses handling, these methods sometimes can not find candidate answers using context processing (Masaki Murata et al. 2005). These methods have two problems. First, not every result includes a correct answer even if a system fills the ellipsis correctly. Secondly, a suitable query can not always be generated from the words that composed the question. To resolve these problems, we consider that a question answering system makes a query using related keywords. We also assume that the keyword exists near the IR results which are from related questions that have been previously retrieved and therefore context processing is not necessary. From here, the extraction of keywords would follow this process.

- To generate a query from the first related question which does not need context processing.
- To find sentences which include noun words which correspond to inputted words.
- To extract noun words from the extracted sentences.
- To make keyword candidates from the extracted words.

This process is similar to pseudo-relevance feedback. Pseudo-relevance feedback is generally used by query expansion. However in addition to keywords used in the query there exist many candidates. Moreover, we have to retrieve information to check for the best query. Therefore confirming the optimum keyword combination for a query takes much time. We propose a contextual question answering system using Genetic Algorithms because GA is suitable for finding the optimum combination 1989. GA is applied to make keyword combinations in contextual question answering methods.

2 Basic Idea

In natural language processing, GA has not been adapted much because it is difficult to configure a fitness function. For example with knapsack problems in GA 1989, the fitness function is simple. The knapsack problem is a problem in combinatorial optimization. This problem task gives a set of items, each with a cost and a value, then determines the number of each item to include in a collection so that the total cost is less than some given cost and the total value is as large as possible. Chen et al proposed GA method for Information Retrieval (Chen, Hsinchun et al. 1998). In a similar way, GA of our method uses a vocabulary from natural genetics. In our reasearch GA finds not the correct answer but a suitable query. The suitable query is a combination of keywords. In the case of isolated questions, a question answering system usually finds a suitable combination of keywords from the isolated question's words. On the other hand, a contextual question answering system has to add more related question words, previous answers and a high number of relevant words. This means a suitable query is selected by a combination of many keyword candidates.

In addition, GA is better suited for an optimization problem just like a knapsack problem. Therefore we apply GA to our contextual question answering system.

3 Process

3.1 Overview

In the case of isolated questions, a query is usually generated by a combination of question words. In the case of series type questions, the first question does not usually have to handle ellipses. From the second question which handles ellipses, our system generates a suitable query using GA. Figure 1 shows an overview of process.

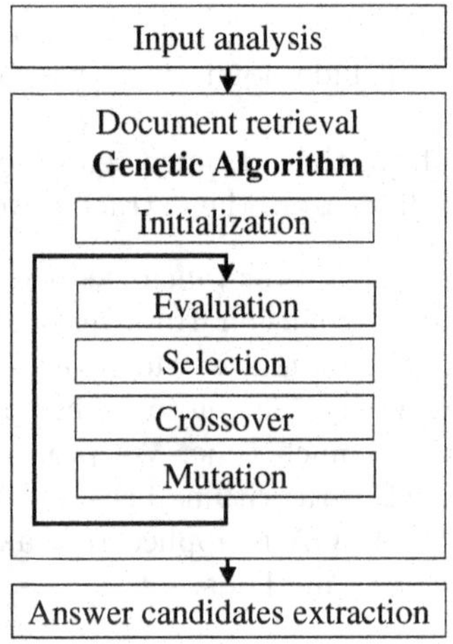

Fig. 1. Overview of GA process

Input analysis is performed extracting a question type and keywords. We focus on document retrieval and passage retrieval GA is applied. In the following sections, we will explain the details of our method.

4 Query Expansion Using GA

Related contextual question answering systems focus on anaphora and ellipsis resolution. However, a system sometimes can not extract correct documents by correct ellipsis resolution. In other words, even if a person makes a relevant query to question, search engines sometimes can not find any correct documents with the correct query. After the first question in a series question type, we consider that it is possible to make a suitable query from many related keywords just like neighbor words near question keywords in retrieved documents. Therefore our system tries combining related keywords using GA. In our GA, an answer format consists of keywords for searching. We consider that most suitable answers consists of words which are extracted from highly similar sentences with the question.

4.1 Initialization

Query candidates are as follows:

- Question's words
- Related words which compose a previous question.
- Related words which compose high similar sentence to the question.

A query randomly composes two or three keywords from keyword candidates because we made a preliminary experiment about number of keywords.

Initialization is as follows:

- To focus on questions without contextual processing.
- To generate a query from the question.
- To extract sentences which includes keywords of the query from retriever documents.
- To make a keyword candidates list from the extracted sentences.

4.2 Evaluation of Fitness Value

Fitness value means a measure for checking adaptability to the individual. In this paper, GA finds not the correct answer but a suitable query. In the case of finding a suitable query, we need information retrieval results for evaluation. If we could not confirm whether the query is the best query or not, we need to try all query patterns. A system has to check contextual questions many times compared with isolated questions. In our system, a fitness function mainly includes three elements which consist of keywords, related measure and possibility of document retrieval. First, our system randomly generates combinations of a few keywords. A combination of keywords is evaluated by sum of each keyword's IDF. Our system searches these queries in descending order until finding 10 successful queries.

Our system extracts related sentences using the keyword type. Keywords are divided into three types which consist of contextual keywords, question keywords and query keywords. A related evaluation function is shown as follows:

$$Fitness = Con + Inp + 0.5 \cdot Qry \tag{1}$$

$$Con = \frac{Sum\,of\,contextual\,keyword's\,IDF}{Number\,of\,contextual\,keywords} \tag{2}$$

$$Inp = \frac{Sum\,of\,question\,keyword's\,IDF}{Number\,of\,question\,keywords} \tag{3}$$

$$Qry = Sum\,of\,query\,keyword's\,IDF \tag{4}$$

4.3 Selection

A proportion of the existing population is selected to breed a new generation. Selection means to extract the best solutions in the existing population. In our method, the population consists of individual queries. Each query is evaluated by a fitness function. The important assumption of the selection is to obtain document candidates by the query.

4.4 Crossover

Crossover is a genetic operator used to vary the programming of chromosomes from one generation to the next. Two chromosomes which obtain high fitness value randomly are selected as parents. Our system generates new chromosomes(queries) by the crossover of two chromosomes.

4.5 Mutation

The purpose of mutation is to avoid local minima by preventing the population of chromosomes from becoming too similar to each other. Mutation means to generate new chromosome by changing randomly. Our system uses words which have never existed in tried queries in words of population and the random ratio is 2%.

4.6 Generation

A generation consists of selection, crossover and mutation. In each generation, the fitness of the whole population is evaluated. Individuality is evolving with generation. However it takes much time with trying many generations. Therefore, our system tries three generations.

5 Evaluation Experiment

In this experiment, we evaluate the contextual query expansion method by GA. the baseline is made up of keywords in a series and we try to two GA methods which consist of First Question GA(FQGA) method and Each Question GA(EQGA) method. FQGA generates query candidates based on retrieved documents from the first question of a series. EQGA generates query candidates based on retrieved documents from previous question.

A Question set consists of 360 questions in NTCIR5-QAC3 (Tsuneaki Kato et al. 2004). This question set consists of 50 series question set which includes 35 gathering series and 15 browsing series. Corpus are 2 years of newspaper articles from the Yomiuri newspaper and the Mainichi newspaper. The Mean Reciprocal Rank(MRR) is used as the evaluation measure (Ellen. M. Voorhees. et al. 2000). The top 20 answers for each instance were considered by the MRR. The result of MRR is shown in Table 1.

Table 1. Experiment

Method	MRR
Baseline	0.0316
First Question GA(FQGA)	0.2531
Each Question GA(EQGA)	0.1448

6 Consideration

This question set includes 35 gathering series and 15 browsing series. Gathering type questions are concerning a common global topic. For example, a gathering series shows as follows:

Q1 What genre does the "Harry Potter" series belong to?

Q2 Who is the author?

Q3 Who are the main characters in that series?

On the other hand, browsing type questions is more difficult than gathering type questions because each consecutive question shares a local context. For example, a browsing series shows as follows:

Q1 Where was Universal Studios of Japan constructed?

Q2 Which train station is the nearest?

Q3 Who was the actor that attended the ribbon-cutting ceremony on the opening day?

Q4 What is the name of the movie he features in that was released in the New Year season of 2001?

Q5 What is the name of the movie starring Kevin Costner released in the same season?

In this experiment, the FQGA method had the highest performance of the three methods. We confirmed the effectiveness of the FQGA method. Although the EQGA method was lower than the FQGA method, the EQGA method was better suited for browsing series. We found 11 questions that the EQGA method could correctly answer although other systems could not answer. In these questions, we confirmed that 7 questions were the gathering type.

7 Conclusion

In this paper, we described resolving contextual questions using GA. In contextual questions, we tried to handle ellipses by adding keywords in retrieved documents. Our system found a suitable combination of words in retrieved documents. In the experiment using NTCIR5-QAC3 data, we confirmed a high evaluation.

In future work, we will apply it to other QA.

References

Ellen. M. Voorhees.: Overview of the TREC 2004 Question Answering Track. (2004)

Ellen. M. Voorhees and Dawn M. Tice: Building a question answering test collection. In The 23 International ACM SIGIR Conference on Research and Development in Information Retrieval. **78** (2000) 200–207

Tsuneaki Kato, Jun'ichi Fukumoto, Fumito Masui and Noriko Kando.: Handling Information. Access Dialogue through QA Technologies – A novel challenge. HLT-NAACL2004 Workshop on Pragmatics of Question Answering. (2004) 70–77

Masaki Murata, Masao Utiyama and Hitoshi Isahara.: Japanese Question-Answering System Using Decreased Adding with Multiple Answers at NTCIR 5. NTCIR Workshop 5 Proceedings, (2005) 380–385 2005

D.E.Goldberg: Genetic Algorithms in Search, Optimization, and Machine Learning., Addison-Wesley (1989)

Tatsunori Mori and Shinpei Kawaguchi. : Answering Contextual Questions Based on the Cohesion with the Knowledge. In Proceedings of the Fifth NTCIR Workshop Meeting, (2005) 386–393.

Chen, Hsinchun and Shankaranarayanan, Ganesan and She, Linlin and Iyer, Anand A Machine Learning Approach to Inductive Query by Examples: An Experiment Using Relevance Feedback, ID3, Genetic Algorithms, and Simulated Annealing Journal of the American Society for Information Science **49(8)** (1998) 693–705.

Xu, Jinxi and Croft, W. Bruce Query Expansion Using Local and Global Document Analysis, Proceedings of the 19th Annual ACM-SIGIR Conference, (1996) 4–11.

Fine-Grained Named Entity Recognition Using Conditional Random Fields for Question Answering

Changki Lee, Yi-Gyu Hwang, Hyo-Jung Oh, Soojong Lim, Jeong Heo, Chung-Hee Lee, Hyeon-Jin Kim, Ji-Hyun Wang, and Myung-Gil Jang

Electronics and Telecommunications Research Institute (ETRI)
161 Gajeong-dong, Yuseong-gu, Daejeon,
305-350, Korea
{leeck, yghwang, ohj, isj, jeonghur, forever, jini, jhwang, mgjang}@etri.re.kr

Abstract. In many QA systems, fine-grained named entities are extracted by coarse-grained named entity recognizer and fine-grained named entity dictionary. In this paper, we describe a fine-grained Named Entity Recognition using Conditional Random Fields (CRFs) for question answering. We used CRFs to detect boundary of named entities and Maximum Entropy (ME) to classify named entity classes. Using the proposed approach, we could achieve an 83.2% precision, a 74.5% recall, and a 78.6% F1 for 147 fined-grained named entity types. Moreover, we reduced the training time to 27% without loss of performance compared to a baseline model. In the question answering, The QA system with passage retrieval and AIU archived about 26% improvement over QA with passage retrieval. The result demonstrated that our approach is effective for QA.

Keywords: Fine-Grained Named Entity Recognition, Conditional Random Fields, Question Answering.

1 Introduction

The Question-Answering (QA) system is an information retrieval system that finds an answer instead of finding a list of documents in response to a user's question. Passage extraction methods have been the most commonly used by many QA systems. In the passage extraction methods, sentences or passages that are regarded as the most relevant sentences or passages to the question are extracted and then answers are retrieved by using lexico-syntactic information or NLP techniques (especially named entity recognition).

Recent advances have brought some systems (such as BBN's IdentiFinder) to within 90% accuracy when classifying named entities into broad categories, such as person, organization, and location. These coarse categorizations are useful in many areas of natural language research, but more finely grained classification has additional advantages in QA [1, 2].

In many QA systems, fine-grained named entities are extracted by coarse-grained named entity recognizer and fine-grained named entity dictionary [6]. While much

H.T. Ng et al. (Eds.): AIRS 2006, LNCS 4182, pp. 581 – 587, 2006.

research has gone into the coarse categorization of named entities, we are not aware of much previous work using machine learning algorithms to perform more fine-grained classification.

Fleischman and Hovy describe a method for automatically classifying person instances into eight finer-grained subcategories [1]. They use a supervised learning method that considers the local context features and global semantic features. But a training data is highly skewed, because they use a simple bootstrapping method to generate the training data automatically. And they treat only person and location, but we treat all kinds of named entity types.

Mann explores the idea of a fine-grained proper noun ontology and its use in question answering [2]. He builds a proper noun ontology from unrestricted text. This automatically constructed ontology is then used on a question answering task. The disadvantage of this method is that its coverage is small, because he uses simple textual co-occurrence patterns.

In this paper, we describe a fine-grained Named Entity Recognition (NER) using Conditional Random Fields (CRFs) and its use in question answering.

2 Fine-Grained NER Using Conditional Random Fields

We define 147 fine-grained named entity types in consideration of user's asking points for finding answer candidates of a question answering system. They have 15 top levels and each top node consists of 2 or 4 layers. The base set of such types is; *person, study field, theory, artifacts, organization, location, civilization, date, time, quantity, event, animal, plant, material*, and *term*.

In many machine learning-based named entity recognitions (NERs), each name class N is subdivided into 2 sub-classes, i.e., B-N and I-N [3, 4]. Hence, there are total 295 classes (147 name classes * 2 sub-classes + 1 not-a-name class). In this case, CRFs can not be applied directly, because CRFs which have many classes are inefficient. To solve this problem, we break down the NE task in two parts; boundary detection using CRFs and NE classification using Maximum Entropy (ME).

Let $\mathbf{x} = \langle x_1, x_2, \ldots x_T \rangle$ be some observed input data sequences, such as a sequence of words of sentences in a document. Let $\mathbf{y}$ be a set of FSM states, each of which is associated with a name class and a boundary (i.e. B, I). Let $\mathbf{y} = \langle y_1, y_2, \ldots y_T \rangle$ be some sequences of states, (the values on T output nodes). We define the conditional probability of a state sequence $\mathbf{y}$ given an input sequence $\mathbf{x}$ as follows:

$$\begin{aligned} P(\mathbf{y} \mid \mathbf{x}) &= P(\mathbf{c}, \mathbf{b} \mid \mathbf{x}) = P(\mathbf{b} \mid \mathbf{x}) P(\mathbf{c} \mid \mathbf{b}, \mathbf{x}) \\ &\approx P(\mathbf{b} \mid \mathbf{x}) \prod_{i=1} P(c_i \mid b_i, x_i) \end{aligned} \quad (1)$$

where $\mathbf{b} = \langle b_1, b_2, \ldots b_T \rangle$ is a set of boundary states, $\mathbf{c} = \langle c_1, c_2, \ldots c_T \rangle$ is a set of name class states.

CRFs define the conditional probability of a boundary state sequence $\mathbf{b}$ given an input sequence $\mathbf{x}$ as follows:

$$P_{CRF}(\mathbf{b} \mid \mathbf{x}) = \frac{1}{Z(\mathbf{x})} \exp\left(\sum_{t=1}^{T} \sum_{k} \lambda_k f_k(b_{t-1}, b_t, \mathbf{x}, t) \right) \quad (2)$$

where $f_k(b_{t-1},b_t,\mathbf{x},t)$ is an arbitrary feature function over its arguments, and λ_k is a learned weight for each feature function. Higher λ weights make their corresponding FSM translations more likely.

We calculate the conditional probability of a state c_i given a named entity (b_i, x_i) extracted by boundary detector as follows:

$$P_{ME}(c_i \mid b_i, x_i) = \frac{1}{Z(b_i, x_i)} \exp\left(\sum_k \lambda_k f_k(c_i, b_i, x_i) \right) \tag{3}$$

We use the following features:

- Word feature—orthographical features of the (-2,-1,0,1,2) words.
- Suffix—suffixes which are contained in the current word among the entries in the suffix dictionary.
- POS tag—part-of-speech tag of the (-2,-1,0,1,2) words.
- NE dictionary—NE dictionary features.
- 15 character level regular expressions – such as [A-Z]*, [0-9]*, [A-Za-z0-9]*, …

The primary advantage of Conditional Random Fields (CRFs) over hidden Markov models is their conditional nature, resulting in the relaxation of independence assumptions required by HMMs. Additionally, CRFs avoid the label bias problem, a weakness exhibited by maximum entropy Markov models (MEMMs). CRFs outperform both MEMMs and HMMs on a number of real-world sequence labeling tasks [3, 4].

3 Experiment

3.1 Fine-Grained NER

The experiments for fine-grained named entity recognition (147 classes) were performed on our Korean fine-grained named entity data set. The data set consists of 6,000 documents tagged by human annotators. We used 5,500 documents as a training set and 500 documents as a test set, respectively. We trained the model by L-BFGS using our C++ implementation of CRFs and ME. We use a Gaussian prior of 1 (for ME) and 10 (for CRF). We perform 500 iterations for training.

Table 1. Sub-task of fine-grained NER (147 classes)

Sub-task	Method (# Class)	Training Time (second/iter.)	Pre. (%)	Rec. (%)	F1 (%)
Boundary detection	ME (3 classes)	**6.23**	87.4	**80.6**	83.9
Boundary detection	CRFs (3 classes)	24.14	**89.5**	79.7	**84.3**
NE classification	ME (147 classes)	18.36	83.7	84.2	84.0

Table 1 shows the performance of boundary detection and NE classification using CRF and ME. In boundary detection, we obtained the performance, 83.9% F1 and 84.3% F1 using ME and CRF respectively. In NE classification, we could achieve an 84.0% F1 using ME (we assume the boundary detection is perfect).

Table 2. Performance of fine-grained NER (147 classes)

Task	Method (# Class)	Training Time (second/iter.)	Pre. (%)	Rec. (%)	F1 (%)
NER-1 (baseline)	1 stage: ME (295 classes)	154.97	82.5	73.5	77.7
NER-2	1 stage: CRFs (295 classes)	12248.33	-	-	-
NER-3	2 stages: ME+ME (3+147 classes)	**24.59**	82.8	73.3	77.8
NER-4 (proposed)	2 stages:CRFs+ME (3+147 classes)	42.50	**83.2**	**74.5**	**78.6**

Table 2 shows the performance of fine-grained named entity recognition using CRF and ME. In named entity recognition, our proposed model NER-4 that is broken down in two parts (i.e., boundary detection using CRFs and NE classification using ME) obtained 78.6% F1. The baseline model NER-1 (ME-based) that is not broken down in two parts obtained 77.7% F1. We could not train NER-2 because training is too slow.

Table 2 also shows the training time (second per iteration) of fine-grained named entity recognition. The baseline model NER-1 using ME takes 155.0 seconds per iteration in training. Our proposed model NER-4 takes 42.5 seconds per iteration. So we reduced the training time to 27% without loss of performance.

3.2 QA with Fine-Grained NER

We performed another experiment for a question answering system to show the effect of the fine-grained named entity recognition. We used ETRI QA Test Set which consists of 402 pairs of question and answer in encyclopedia [5]. Our encyclopedia currently consists of 163,535 entries, 13 main categories, and 41 sub categories in Korean. For each question, the performance score is computed as the reciprocal answer rank(RAR) of the first correct answer.

Our fine-grained NER engine annotates with 147 answer types for each sentence and then the indexer generates Answer Index Unit (AIU) structures using the answer candidates (the fine-grained NE annotated words) and the content words which can be founded within a same sentence. We append distance information between answer candidates and content words to AIU structures [5].

Figure 1 contains as example of the data structure passed from the indexing module.

The answer processing module searches the relative answer candidates from index DB using question analysis and calculates the similarities and then extracts answers [5].

[Example sentence]
The Nightingale Award, the top honor in international nursing, was established at the International Red Cross in 1912 and is presented every two years (title: Nightingale)

[Fine-grained NER]
<The Nightingale Award:CV_PRIZE>, the top honor in international nursing, was established at **<the International Red Cross:OGG_SOCIETY>** in **<1912:DT_YEAR>** and is presented every **<two years:DT_DURATION>** (title: **<Nightingale:PS_NAME>**)

[Answer candidates]
The Nightingale Award:CV_PRIZE
the International Red Cross:OGG_SOCIETY
1912:DT_YEAR
two years:DT_DURATION
Nightingale:PS_NAME

[Sentence-based AIU structures]

AIU	(**The Nightingale Award:CV_PRIZE**, **the International Red Cross:OGG_SOCIETY**, distance info.), … , (**two year:DT_DURATION**, **Nightingale:PS_NAME**, distance info.), (**The Nightingale Award:CV_PRIZE**, top, distance info), (**The Nightingale Award:CV_PRIZE**, honor, distance info), …

Fig. 1. Example of index processing

Table 3. Performance of Question Answering

Task	MRAR
QA with passage retrieval	0.525
QA with passage retrieval and AIU (fine-grained NE)	**0.662**

Table 3 shows the performance of question answering system with or without Answer Index Unit (i.e. fine-grained named entity information). The QA system with passage retrieval and AIU achieved about 26% improvement over QA with passage retrieval.

4 Conclusion

In this paper, we describe a fine-grained Named Entity Recognition using Conditional Random Fields (CRFs) for question answering. We used CRFs to detect boundary of named entities and Maximum Entropy (ME) to classify named entity classes. Using the proposed approach, we could achieve an 83.2% precision, a 74.5% recall, and a 78.6% F1 for 147 fined-grained named entity types. Moreover, we reduced the training time to 27% without loss of performance compared to a baseline model. In the question answering, The QA system with passage retrieval and AIU (fine-grained NE) archived about 26% improvement over QA with passage retrieval. The result demonstrated that our approach is effective for QA.

References

[1] M. Fleischman and E. Hovy. *Fine grained classification of named entities*. COLING, 2002.
[2] G. Mann, *Fine-Grained Proper Noun Ontologies for Question Answering*, SemaNet'02: Building and Using Semantic Networks, 2002
[3] McCallum and W. Li. *Early results for named entity recognition with conditional random fields, feature induction and web-enhanced lexicons*, CoNLL, 2003.
[4] S. Fei, F. Pereira. *Shallow Parsing with Conditional Random Fields*, HLT & NAACL, 2003.
[5] H. Kim, J. Wang, C. Lee, C. Lee, M. Jang. *A LF based Answer Indexing Method for Encyclopedia Question Answering System*, AIRS, 2004.
[6] K. Han, H. Chung, S. Kim, Y. Song, J. Lee, H. Rim. *Korea University Question Answering System at TREC 2004*, TREC, 2004.

Appendix: Fine-Grained Named Entity Tag Set (147 classes)

1. Person	PS_NAME	PS_MYTH
2. Study Field	FD_OTHERS FD_SCIENCE FD_SOCIAL_SCIENCE	FD_MEDICINE FD_ART FD_PHILOSOPHY
3. Theory	TR_OTHERS TR_SCIENCE TR_TECHNOLOGY TR_SOCIAL_SCIENCE	TR_ART TR_PHILOSOPHY TR_MEDICINE
4. Artifacts	AF_CULTURAL_ASSET AF_BUILDING AF_MUSICAL_INSTRUMENT AF_ROAD AF_WEAPON AF_TRANSPORT AF_WORKS AFW_GEOGRAPHY AFW_MEDICAL_SCIENCE	AFW_RELIGION AFW_PHILOSOPHY AFW_ART AFWA_DANCE AFWA_MOVIE AFWA_LITERATURE AFWA_ART_CRAFT AFWA_THEATRICALS AFWA_MUSIC
5. Organization	OG_OTHERS OGG_ECONOMY OGG_EDUCATION OGG_MILITARY OGG_MEDIA OGG_SPORTS OGG_ART OGG_SOCIETY	OGG_MEDICINE OGG_RELIGION OGG_SCIENCE OGG_BUSINESS OGG_LIBRARY OGG_LAW OGG_POLITICS
6. Location	LC_OTHERS LCP_COUNTRY LCP_PROVINCE LCP_COUNTY LCP_CITY LCP_CAPITALCITY LCG_RIVER LCG_OCEAN	LCG_BAY LCG_MOUNTAIN LCG_ISLAND LCG_TOPOGRAPHY LCG_CONTINENT LC_TOUR LC_SPACE
7. Civilization	CV_NAME CV_TRIBE	CV_CLOTHING CV_POSITION

	CV_SPORTS CV_SPORTS_INST CV_POLICY CV_TAX CV_FUNDS CV_LANGUAGE CV_BUILDING_TYPE CV_FOOD CV_DRINK	CV_RELATION CV_OCCUPATION CV_CURRENCY CV_PRIZE CV_LAW, CVL_RIGHT, CVL_CRIME, CVL_PENALTY,
8. Date	DT_OTHERS DT_DURATION DT_DAY DT_MONTH	DT_YEAR DT_SEASON DT_GEOAGE DT_DYNASTY
9. Time	TI_OTHERS TI_DURATION TI_HOUR	TI_MINUTE TI_SECOND
10. Quantity	QT_OTHERS QT_AGE QT_SIZE QT_LENGTH QT_COUNT QT_MAN_COUNT QT_WEIGHT	QT_PERCENTAGE QT_SPEED QT_TEMPERATURE QT_VOLUME QT_ORDER QT_PRICE QT_PHONE
11. Event	EV_OTHERS EV_ACTIVITY EV_WAR_REVOLUTION	EV_SPORTS EV_FESTIVAL
12. Animal	AM_OTHERS AM_INSECT AM_BIRD AM_FISH AM_MAMMALIA	AM_AMPHIBIA AM_REPTILIA AM_TYPE AM_PART
13. Plant	PT_OTHERS PT_FRUIT PT_FLOWER PT_TREE	PT_GRASS PT_TYPE PT_PART
14. Material	MT_ELEMENT MT_METAL MT_ROCK	MT_CHEMICAL MTC_LIQUID MTC_GAS
15. Term	TM_COLOR TM_DIRECTION TM_CLIMATE TM_SHAPE TM_CELL_TISSUE	TMM_DISEASE TMM_DRUG TMI_HW TMI_SW

A Hybrid Model for Sentence Ordering in Extractive Multi-document Summarization

Dexi Liu[1,2], Zengchang Zhang[1], Yanxiang He[2], and Donghong Ji[3]

[1] School of Physics, Xiangfan University, Xiangfan 441053, P.R. China
[2] School of Computer, Wuhan University, Wuhan 430079, P.R. China
[3] Institute for Infocomm Research, Heng Mui Keng Terrace 119613, Singapore
dexiliu@gmail.com

Abstract. Ordering information is a critical task for multi-document summarization because it heavily influent the coherence of the generated summary. In this paper, we propose a hybrid model for sentence ordering in extractive multi-document summarization that combines four relations between sentences. This model regards sentence as vertex and combined relation as edge of a directed graph on which the approximately optimal ordering can be generated with PageRank analysis. Evaluation of our hybrid model shows a significant improvement of the ordering over strategies losing some relations and the results also indicate that this hybrid model is robust for articles with different genre.

1 Introduction

Automatic text summarization [1] that provides users with a condensed version of the original text, tries to release our reading burden, and most summarization today still relies on extraction of sentences from the original document [2]. In extractive summarization, a proper arrangement of these extracted sentences must be found if we want to generate a logical, coherent and readable summary. This issue is special in multi-document summarization. Sentence position in the original document, which yields a good clue to sentence arrangement for single-document summarization, is not enough for multi-document summarization because we must consider inter-document ordering at the same time [3].

Barzilay et al. [4] showed the impact of sentence ordering on readability of a summary and explored some strategies for sentence ordering in the context of multi-document summarization. Lapata [5] proposed another method to sentence ordering based on an unsupervised probabilistic model for text structuring that learns ordering constraints from a large corpus. However, the limitation of above-mentioned strategies is still obvious. Barzilay's strategy paid attention to chronological and topical relation whereas ignored sentence original ordering in the articles where summary comes from. Hence, if the articles are not event-based, the quality of summary will decrease because the temporal cue is invalid. As for Lapata's strategy, the probabilistic model of text structure is trained on a large corpus, so it performs badly when genre of the corpus and the article collection are mismatched.

H.T. Ng et al. (Eds.): AIRS 2006, LNCS 4182, pp. 588 – 592, 2006.

To overcome the limitation mentioned above, we propose a hybrid model for sentence ordering in extractive multi-document summarization that combines four relations between sentences - chronological relation, positional relation, topical relation and dependent relation. Our model regards sentence as vertex and combined relation as edge of a directed graph on which the approximately optimal ordering can be generated with PageRank analysis. Evaluation of our hybrid model shows a significant improvement of the ordering over strategies losing some relations and the results also indicate that this hybrid model is robust for articles with different genre.

2 Model

For the group of sentences extracted from the article collection, we construct a directed graph (we call it Precedence Graph). Let $G = (V, E)$ be a directed graph with the set of vertices V and set of directed edges E, where E is a subset of $V \times V$. For a given vertex v_i, let In(v_i) be the set of vertices that point to it (predecessors), and let Out(v_i) be the set of vertices that vertex v_i points. In our hybrid model, the Precedence Graph is constructed by adding a vertex for each sentence in the summary, and edges between vertices are established using sentence relations. Three of the pre-defined quantitative relations are integrated to precedent relation using linear model.

$$R_{i,j} = \lambda_P P_{i,j} + \lambda_D D_{i,j} + \lambda_C C_{i,j} \,. \tag{1}$$

where λ_P, λ_D, λ_C are the weight of positional relation $P_{i,j}$, dependent relation $D_{i,j}$ and chronological relation $C_{i,j}$ individually. If $R_{i,j}$>0, there exist a directed edge with value $R_{i,j}$ from v_i to v_j (see figure 1). In our case, we set λ_P=0.3, λ_D=0.4, λ_C=0.3 manually. By the way, figure 1 does not contain topical relation because it will be used only after the first vertex has been selected.

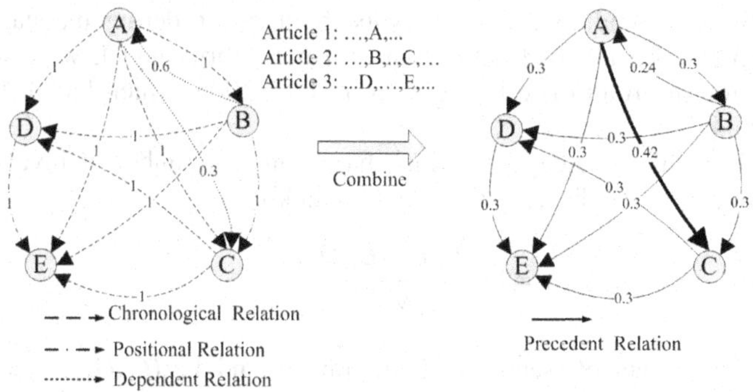

Fig. 1. An example for the hybrid model

Unfortunately, the next critical task of finding an optimal road according to the Precedence Graph is a NP-complete problem. Our model employs PageRank method [6], which is perhaps one of the most popular ranking algorithms, and was designed as a method for Web link analysis. PageRank method is a graph-based ranking algorithm to decide the importance of a vertex within a graph, by taking into account global information recursively computed from the entire graph, rather than relying only on local vertex-specific information. Score of vertex v_i is defined as:

$$S(v_i) = (1-d) + d * \sum_{v_j \in Out(v_i)} R_{i,j} \frac{S(v_j)}{\sum_{v_k \in In(v_j)} R_{k,j}} . \tag{2}$$

where d is a parameter set between 0 and 1 (we let d=0.1 in our experiments).

The next step following PageRank algorithm is the selection of vertex. We first select the vertex with highest score as the first sentence in summary. For the following vertices, not only the score itself but also the topical relation with immediately previous vertex should be taken into consideration. The succeed vertex of v_k should satisfy:

$$v_i = \arg\max_{v_{\hat{i}}}((1-\lambda_T)S(v_{\hat{i}}) + \lambda_T T_{k,\hat{i}}) . \tag{3}$$

where λ_T is the weight of topical relation $T_{i,j}$ (we let λ_T =0.4 in our experiments) . When the succeed vertex of v_k is picked up, vertex v_k and all edges linked with it are deleted from the Precedence Graph. This operator iterate until the graph is empty, and then an ordered summary is produced.

3 Experiments

We collected three groups of summaries generated by three different participants of task 2 in DUC2004 [7]. Three selected groups have high (id=65), fair (id=138), and low (id=111) ROUGE score [8] individually so that we can observe the degree of influence that our model affects on summarizer with different performance. For each group, we randomly selected 10 from 50 summaries produced by one participant. Five postgraduates were employed and everyone built one ordering manually for each summary. To test the adaptive performance of our hybrid model, we also randomly selected the fourth group of testing data from DUC2005, which has different genre with DUC2004.

We use Kendall's strategy [9], which based on the number of inversions in the rankings, to measure the distance between two ranking.

$$\tau = 1 - \frac{2 \times \mathrm{Inv}(O_i, O_j)}{N(N-1)/2} \tag{4}$$

where N is the number of sentences being ranked and $\mathrm{Inv}(O_i, O_j)$ is the number of interchanges of consecutive elements necessary to arrange them in their natural ordering (manual ordering in this case). If we think in terms of permutations, then τ can be interpreted as the minimum number of adjacent transpositions needed to bring one ordering to the other. We use the minimal one from 5 distance values corresponding to the 5 "ideal" orderings produced manually. Figure 2 shows the evaluation results.

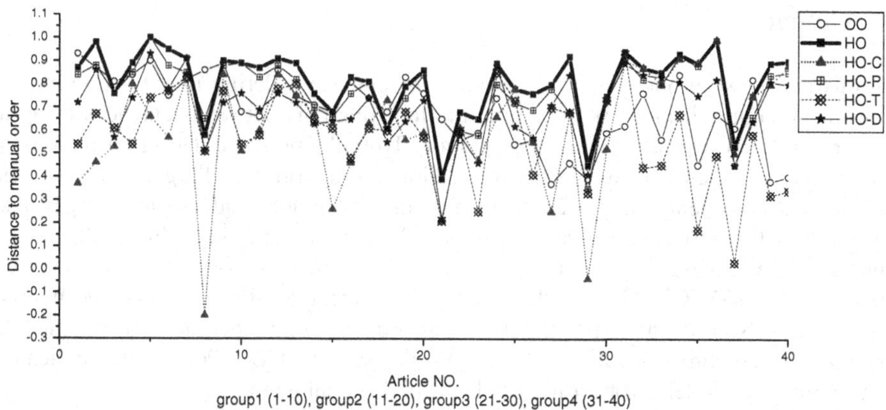

Fig. 2. Results and comparison of different models

As Figure 2 indicates, our hybrid ordering (HO) model yields more satisfactory result (as a whole) than any other model taking part in the comparison, including the strategy employed by the summarizer that produces the original ordering (OO). Although partially hybrid strategies such as HO-C (hybrid ordering model without chronological relation), HO-P (hybrid ordering model without positional relation), HO-T (hybrid ordering model without topical relation, and HO-D (hybrid ordering model without dependent relation) are all worse than HO, we found that different relation has significantly different influence on the ordering strategy performance. In context of DUC2004 dataset (group 1, 2, 3), HO-C performs worst and HO-T is slightly better than HO-C, whereas there aren't significant difference between HO-P, HO-D, and HO. In other word, the ordering strategy depend more on chronology and topical relation than on positional and dependent relation for DUC2004 dataset. After investigating the failure ordering of HO-C, we found that most of the articles have been used for task 2 in DUC2004 are event-based, so chronological cue is very important for ordering. Moreover, for the limitation of summary length, number of sentences in most summaries is less than 10 – the number of articles from which summary extracted, so the probability of two sentences coming from the same article is very low. Hence the importance of positional relation is not distinct.

On the contrary, HO-T and HO-P perform much worse than HO on DUC2005 dataset (group 4) while HO-D and HO-C perform closely to HO. Reason for this change is that most articles for DUC2005 are biography, where chronological cue is not so important for ordering. Although different relation plays different role in ordering summary with different genre, our hybrid model combining these relations together has robust performance. Figure 2 indicates clearly that the results of HO have no distinct change when article genre change from event-based to biography-based. Furthermore, after reordering, the improvement for "poor" summaries (group 3) is better than for "good" summaries (group 1) because the summaries generated by a good summarizer are more readable than by bad one.

4 Conclusions

In this paper, we propose a hybrid model for sentence ordering in extractive multi-document summarization. We combine the four relations between sentences using a linear model and we call it precedent relation. To find a proper ordering for the group of sentences extracted from an article collection, our hybrid model regards sentence as vertex and precedent relation as edge of a directed graph, and employs PageRank analysis method to generate an approximately optimal ordering. We evaluate the automatically generated orderings against manual orderings on the testing dataset extended from DUC2004 and DUC2005. Experiment results show that the hybrid model has a significant improvement compared with other partially hybrid model. Moreover, experiment results on DUC2004 dataset and DUC2005 dataset indicates that our hybrid model is robust for articles with different genre.

References

1. Mani, I.: Automatic Summarization. John Benjamins (2001)
2. Radev, Dragomir R., Hovy, Eduard H., McKeown, K.: Introduction to the Special Issue on Summarization. Computational Linguistics 28(4) (2002) 399-408
3. Okazaki, N., Matsuo, Y., Ishizuka M.: Coherent Arrangement of Sentences Extracted from Multiple Newspaper Articles. PRICAI (2004) 882-891
4. Barzilay, R., Elhadad, E., McKeown, K.: Inferring strategies for sentence ordering in multi-document summarization. Journal of Artificial Intelligence Research, 17 (2002) 35–55
5. Lapata, M.: Probabilistic text structuring: experiments with sentence ordering. In Proceedings of the 41st Meeting of the Association of Computational Linguistics (2003) 545–552
6. Brin, S., Page, L.: The anatomy of a large-scale hypertextual Web search engine. Computer Networks and ISDN Systems, 30 (1998) 1–7
7. Paul, O., James, Y.: An Introduction to DUC-2004. In Proceedings of the 4th Document Understanding Conference (DUC 2004). (2004)
8. Lin, C.Y., Hovy, E.: Automatic Evaluation of Summaries Using N-gram Co-Occurrence Statistics. In Proceedings of the Human Technology Conference (HLTNAACL-2003), Edmonton, Canada (2003)
9. Lebanon, G., Lafferty, J.: Combining rankings using conditional probability models on permutations. In Proceedings of the 19th International Conference on Machine Learning. Morgan Kaufmann Publishers, San Francisco, CA. (2002) 363-370

Automatic Query Type Identification Based on Click Through Information

Yiqun Liu[1], Min Zhang[1], Liyun Ru[2], and Shaoping Ma[1]

[1] State Key Lab of Intelligent Tech. & Sys., Tsinghua University, Beijing, China
liuyiqun03@mails.tsinghua.edu.cn
[2] Sogou Incorporation, Beijing, China
ruliyun@sohu-rd.com

Abstract. We report on a study that was undertaken to better identify users' goals behind web search queries by using click through data. Based on user logs which contain over 80 million queries and corresponding click through data, we found that query type identification benefits from click through data analysis; while anchor text information may not be so useful because it is only accessible for a small part (about 16%) of practical user queries. We also proposed two novel features extracted from click through data and a decision tree based classification algorithm for identifying user queries. Our experimental evaluation shows that this algorithm can correctly identify the goals for about 80% web search queries.[1]

1 Introduction

Web Search engine is currently one of the most important information access and management tools for WWW users. Most users interact with search engine using short queries which are composed of 4 words or even fewer. This phenomena of "short queries" has prevented search engines from finding users' information needs behind their queries.

With analysis into search engine user behavior, Broder [1] and Rose [2] independently found that search goals behind user queries can be informational, navigational or transactional (refered to as resource type by Rose). Further experiment results in TREC [3][4] showed that informational and navigational search results benefit from different kinds of evidences. Craswell [5] and Kraaij [6] found that anchor text and URL format offer improvement to content-only method for home page finding task, which covers a major part of navigational type queries. Bharat [7] proved that informational type searches may be improved using hyper link structure analysis. According to these researches, if query type can be identified for a given user query, retrieval algorithm can be adapted to this query type and search performance can be improved compared with a general purpose algorithm. That is why we should identify users' search goals behind their submitted queries.

[1] Supported by the Chinese National Key Foundation Research & Development Plan (2004CB318108), Natural Science Foundation (60223004, 60321002, 60303005, 60503064) and the Key Project of Chinese Ministry of Education (No. 104236).

H.T. Ng et al. (Eds.): AIRS 2006, LNCS 4182, pp. 593–600, 2006.

Query type identification can be performed by two means: Sometimes queries can be classified by simply looking at its content. "AIRS2006" is a navigational type query because the user probably wants to find the homepage of this conference; while "car accident" may be informational because the user seems to be looking for detailed information on "car accident". However, several queries can only be classified with the help of search context. For the query "information retrieval", it is impossible to guess without further information whether the user wants to locate the book written by CJ van Rijsbergen (navigational) or he wants to know something about IR (informational).

The remaining part of the paper is constructed as follows: Section 2 analyzes into search engine logs and studies the possibilities of using click through data and anchor text in query type prediction. Section 3 proposed two novel features extracted from click through data and developed a decision tree based classification algorithm. Experimental results of query classification are shown in Section 4. Finally come discussion and conclusion.

2 Analysis into Search Engine Logs

In order to verify reliability and scalability of our classification method, we obtained part of query logs from a widely-used Chinese search engine Sogou (www.sogou.com). The logs are collected from February 1st to February 28th in the year 2006. They contain 86538613 non-empty queries and 4345557 of them are unique. Query sessions are provided according to cookie information and there are totaly 26255952 sessions in these logs.

When we try to predict one user query's type, we can make use of click through data if and only if this query appears in past click through logs. In Figure 1, the category axis shows date when the logs are collected (all logs are collected in February 2006, so year/month information is omitted) and the value axis show the percentage of queries which have click through information.

We can see that new queries made up of 100% queries on the first day because no history information is available before that day. However, as time goes by the percentage of newly-appeared queries drops to about 10% (average data is 11.15% for the last 10 days). It means that click through data can be applied to classify about 90% queries for search engines.

According to previous works [8] and [9], if one web page shares the same anchor text as a query, the query is probably navigational type. In those works which use anchor text for query type identification, only those queries which match a certain number of anchor texts can be predicted using this evidence. So it is important to find how many percentage of web search queries have such matches with anchor texts.

We crawled over 202M Chinese web pages from the Web and extracted anchor text information from these pages. After reducing possible spams, noises and redundancies, the percentage of queries matching a certain number of anchor text is calculated. We found that the percentage of matching queries doesn't vary with time and there are less than 20% (16.24% on average) matching queries

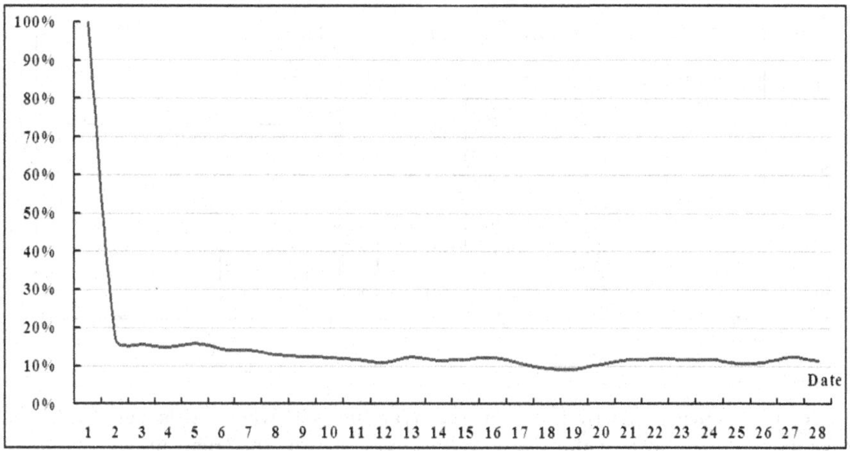

Fig. 1. Percentage of newly-appeared queries in query logs

each day. It means that anchor text evidence is only applicable for less than 20% queries in practical Web search environment.

We can conclude that click through data is suitable for query type identification for general purpose Web search engine. Anchor text evidence may be effective for a part of queries but it is not applicable for the major part. So our work is mainly focused in classification using click through evidence.

3 Query Type Identification Using Click Through Data

In this section, we propose two novel evidences extracted from click through data: *n Clicks Satisfied (nCS)* and *top n Results Satisfied (nRS)*. They can be used as features in our query type identification algorithm.

In order to find the differences between navigational and informational / transactional type queries, we developed a training set of queries which contains 153 navigational queries and 45 informational queries. These queries are randomly selected from query logs and manually classified by 3 assessors using voting to decide queries' categories.

3.1 N Clicks Satisfied (nCS) Evidence

N Clicks Satisfied (nCS) evidence is extracted from the number of user clicks for a particular query. It is based on the following assumption:

Assumption 1 (Less Effort Assumption): While performing a navigational type search request, user tend to click a small number of URLs in the result list.

Supposing one web search user has a navigational goal (CNN homepage, for example), he has a fixed search target in mind and would like to find just that target URL (www.cnn.com) and corresponding snippet in the result list. So it

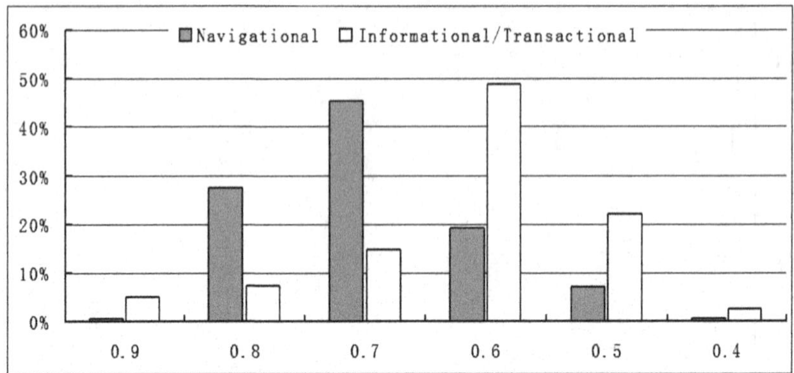

Fig. 2. nCS feature distribution in the training set when n is set to 2. Category axis shows nCS value and value axis shows the percentage of navigational/informational/transactional queries with a certain nCS value.

is impossible for him to click a number of URLs which are not the target page unless there exists cheating pages.

According to the *Less Effort Assumption*, we can judge a query type by the number of URLs which the user clicks. nCS feature is defined as:

$$nCS(Query\ \boldsymbol{q}) = \frac{\#(Session\ of\ \boldsymbol{q}\ that\ involves\ less\ than\ \boldsymbol{n}\ clicks)}{\#(Session\ of\ \boldsymbol{q})}. \qquad (1)$$

According Figure 2, navigational type queries have larger nCS than informational/transactional ones. Most navigational queries have a nCS larger than 0.7 while 70% informational/transactional queries' nCS is less than 0.7. It means this feature can separate a large part of navigational queries.

3.2 Top n Results Satisfied (nRS) Evidence

Top n Results Satisfied (nRS) evidence is extracted from the clicked URL's rank information. It is based on the following assumption:

Assumption 2 (Cover Page Assumption): While performing a navigational type search request, user tend to click only the first few URLs in the result list.

This assumption is related to the fact that navigational type queries have a much higher retrieval performance than informational/transactional ones. According to TREC web track and terabyte track experiments [3],[4] and [10] in the last few years, an ordinary IR system can return correct answers at the 1st ranking for 80% user queries.

According to the *Less Effort Assumption*, we can judge a query type by whether the user clicks other URLs besides the first $\boldsymbol{n}$ ones. Top n Results Satisfied (nRS) feature developed from this idea is defined as:

$$nRS(Query\ \boldsymbol{q}) = \frac{\#(Session\ of\ \boldsymbol{q}\ that\ involves\ clicks\ only\ on\ top\ \boldsymbol{n}\ results)}{\#(Session\ of\ \boldsymbol{q})}. \qquad (2)$$

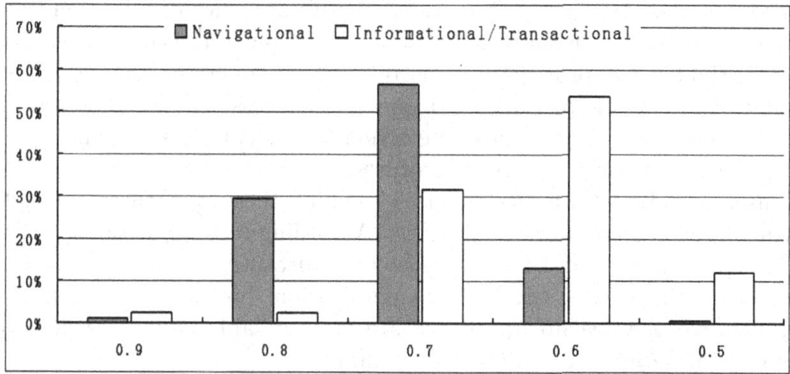

Fig. 3. nRS feature distribution in the training set when n is set to 5. Category axis shows nRS value and value axis shows the percentage of navigational/informational/transactional queries with a certain nRS value.

According to the nRS distribution shown in Figure 3, navigational type queries have larger nRS than informational/transactional ones. 80% navigational queries have a nRS value larger than 0.7 while about 70% informational/transactional queries' nRS is less than 0.7. It shows this feature can also be used to classify web search queries as well as nCS.

3.3 A Learning Based Identification Algorithm

Based on the two new features proposed in Section 4.1 and 4.2, we can separate informational/transactional queries from navigational ones. Besides these features, click distribution proposed by Lee [8] is also believed to be able to identify web search queries.

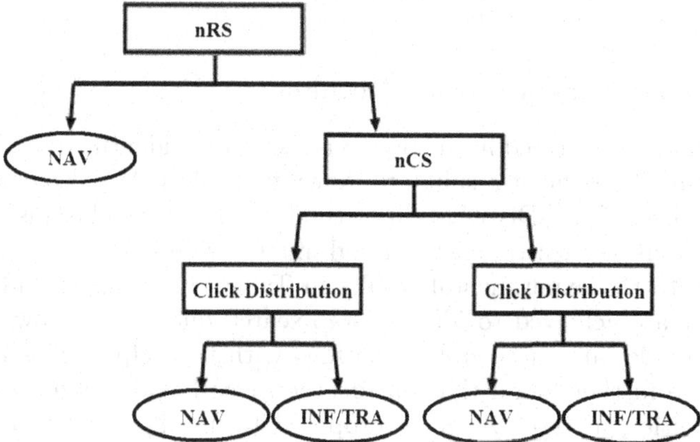

Fig. 4. A Query identification decision tree composed of click through features(INF:informational, TRA:transactional, NAV:navigational)

In order to combine these 3 features: nCS, nRS and click distribution to finish the query type identification task, we adopted a typical decision tree algorithm. It is a method for approximating discrete-valued functions that is robust of noisy data and capable of learning disjunctive expressions. We choose decision tree because it is usually the most effective and efficient classifier when we have a small number (3 features here) of features.

We used standard C4.5 algorithm to combine these 3 features and get the following decision tree shown in Figure 4. According to C4.5 algorithm, The effectiveness of features can be estimated by the distance away from the root. We can see that nRS is more effective in identification than nCS and click distribution. The two new features proposed are more reliable here according to the metric of information ratio in C4.5 algorithm.

4 Experiments and Discussions

4.1 Query Type Identification Test Set and Evaluation Measures

We developed a test set to verify the effectiveness of our identification algorithm. This test set is composed of 81 informational/transactional type queries and 152 navigational queries. The informational/transactional ones were obtained from a Chinese search engine contest organized by tianwang.com (part of the contest is specially designed to test the search engines' performance for informational/transactional queries) and the navigational ones are obtained from a widely-used Chinese web directory hao123.com (websites and their description are used as navigational type results and queries correspondingly). This test set is not assessed by the people who developed the training set in order to get rid of possible subjective noises.

We use traditional precision/recall framework to judge the effectiveness of the query type identification task. Precision and Recall values are calculated separately for two kinds of queries. They are also combined to F-measure value to judge the overall performance.

4.2 Query Type Identification Results

The test query set described in the previous section are identified by our decision tree algorithm. Experiment results are shown in Table 1. We also compared our method with Lee's Click-Distribution method [8] in Figure 5 because it is to our knowledge the most effective click-through information based method.

According to the experimental results in Table 1, precision and recall values over 80% are achieved to identify web search queries. It shows that most web search queries are successfully identified with the help of click through information. We further found that our decision tree based method outperforms Click-Distribution method which is proposed by Lee[8] both on the training set (improved by 21%) and on the test set(improved by 18%). Although Click-Distribution is quite effective for query type identification, by adding new features of nCS and nRS, better performance is achieved.

Table 1. Query type identification experimental results.(INF:informational, TRA:transactional, NAV:navigational)

	Training set			Test set		
	INF/TRA	NAV	Mixed	INF/TRA	NAV	Mixed
Precision	76.00%	91.07%	87.65%	73.74%	85.62%	81.49%
Recall	66.67%	90.71%	85.25%	72.84%	86.18%	81.54%
F-measure	0.71	0.91	0.86	0.73	0.85	0.81

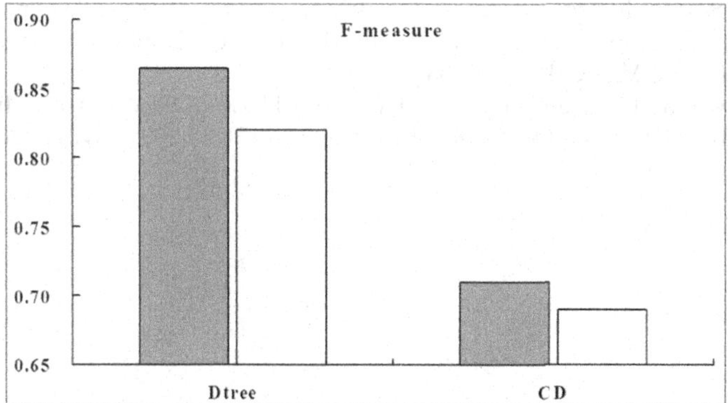

Fig. 5. F-measure values of our decision tree based method and the Click-Distribution based method, Dtree: decision tree based method, CD: Click-Distribution based method, Train/Test: experimental results based on the training/test set

5 Conclusions and Future Work

Given the vast amount of information on the World Wide Web, a typical short query of 1-3 words submitted to a search engine usually cannot offer enough information for the ranking algorithm to give a high quality result list. Using click through data to identify the user goals behind their queries may help search engine to better understand what users want so that more effective result ranking can be expected.

Future study will focus on following aspects: How well does these new features work together with evidences from the queries themselves? How should the traditional ranking models be adjusted for automatically identified queries? Is there any other proper query classification criterion in users' notion?

References

1. Andrei Broder: A taxonomy of web search. SIGIR Forum(2002), Volume 36(2):3-10.
2. Daniel E. Rose and Danny Levinson, Understanding User Goals in Web Search. In proceedings of the 13th World-Wide Web Conference, 2004.

3. N. Craswell, D. Hawking: Overview of the TREC-2002 web track. In The eleventh Text Retrieval Conference (TREC-2002), NIST, 2003.
4. N. Craswell, D. Hawking: Overview of the TREC-2003 web track. In the twelfth Text REtrieval Conference (TREC 2003), NIST, 2004.
5. N Craswell, D Hawking and S Robertson. Effective Site Finding using Link Anchor Information. In Proceedings of ACM SIGIR '01, 2001.
6. Kraaij W, Westerveld T, Hiemstra D. The importance of prior probabilities for entry page search. In Proceedings of ACM SIGIR '02, 2002.
7. Bharat K, Henzinger M. Improved algorithms for topic distillation in a hyperlinked environment. In Proceedings of ACM SIGIR '98, 1998.
8. Uichin Lee, Zhenyu Liu and Junghoo Cho, Automatic Identification of User Goals in Web Search, in proceedings of the 14th World-Wide Web Conference, 2005.
9. I. Kang and G. Kim. Query type classication for web document retrieval. In Proceedings of ACM SIGIR '03, 2003.
10. N. Craswell and D. Hawking. Overview of the TREC-2004 Web track. In the Thirteenth Text REtrieval Conference Proceedings (TREC 2004), NIST, 2005.

Japanese Question-Answering System for Contextual Questions Using Simple Connection Method, Decreased Adding with Multiple Answers, and Selection by Ratio

Masaki Murata, Masao Utiyama, and Hitoshi Isahara

National Institute of Information and Communications Technology,
3-5 Hikaridai, Seika-cho, Soraku-gun, Kyoto, 619-0289, Japan
{murata, mutiyama, isahara}@nict.go.jp
http://www.nict.go.jp/x/x161/member/murata/

Abstract. We participated in NTCIR QAC-1, QAC-2, and QAC-3, which were evaluation workshops for answering questions held by the National Institute of Informatics of Japan and studied question-answering systems for contextual questions. Contextual questions are defined as a series of questions with contexts. For example, the first question is "What is the capital of Japan?" and the one succeeding is related to the first such as "What was it called in ancient times?". Contextual question-answering can be considered interactive. This paper describes our system for contextual questions, which obtained the second best accuracy in QAC-1 and the best accuracy in both QAC-2 and QAC-3 for contextual question-answering. It is thus a high-performance system.

1 Introduction

Question-Answering Challenges (QAC) [3] handles interesting topics and contextual questions in particular. Contextual question answering can be considered to be interactive. Contextual questions are currently defined as a series of questions with contexts. For example, the first question is "What is the capital of Japan?" and the one succeeding is related to the first such as "What was it called in ancient times?". As we can determine the answers to these kinds of questions, we can automatically evaluate the answers that a question-answering system will output. Automatic evaluation is very important for scientific studies. We participated in NTCIR's QAC-1, QAC-2, and QAC-3, which were evaluation workshops for contextual question answering held by the National Institute of Informatics of Japan and studied question-answering systems for contextual questions. Our question-answering system used numerous effective methods to answer questions. We used a method of connecting question sentences to handle contextual questions. We also used a method of using multiple documents to obtain more accurate answers. We then used a special method of handling cases when multiple answers could be correct for a question [2].

H.T. Ng et al. (Eds.): AIRS 2006, LNCS 4182, pp. 601–607, 2006.

2 Our Question-Answering System

2.1 Prediction of Answer Type

The system applies manually defined heuristic rules to predict the answer type. There are 39 of these rules. Some of them are listed here:

1. When *dare* "who" occurs in a question, a person's name is given as the answer type.
2. When *itsu* "when" occurs in a question, a time expression is given as the answer type.
3. When *donokurai* "how many" occurs in a question, a numerical expression is given as the answer type.

2.2 Document Retrieval

Our system extracts terms from a question by using a morphological analyzer, ChaSen [1]. The analyzer first eliminates terms whose part of speech is a preposition or a similar type; it then retrieves by using the extracted terms. Our system extracts documents where a term occurs near each other.

2.3 Answer Detection

Our system first generates candidate expressions to detect the answer from the extracted documents. We initially used morpheme n-grams for the candidate expressions, but this approach generated too many candidates. Instead, we now use candidates only consisting of nouns, unknown words, and symbols. Moreover, we used the ChaSen analyzer to determine morphemes and their parts of speech.

Our approach to determining whether a candidate was a correct answer was to add score for the candidate, under the condition that it was near an extracted term, to score based on heuristic rules according to the answer type. The system then selected the candidates with the highest number of total points as the correct answers. (See our paper [2] for the detail.)

2.4 Use of Multiple Documents as Evidence with Decreased Weighting

Our system also use the method of decreased adding with multiple answers. Let us assume that the question, "What is the capital of Japan?", is input into a question-answering system, with the goal of obtaining the correct answer, "Tokyo". A typical question-answering system would output the candidate answers and scores listed in Table 1. These systems would also output document ID from which each candidate answer had been extracted. The system outputs the incorrect answer, "Kyoto", as the first answer for the example in Table 1. We developed a method of using multiple documents with decreased weighting as evidence. The method adds scores by assigning decreasing weights to them. Now suppose that we implement our method by multiplying the score of the i-th

Table 1. Candidate answers according to original scores, where "Tokyo" is correct answer

Rank	Candidate answer	Score	Document ID
1	Kyoto	3.3	926324
2	Tokyo	3.2	259312
3	Tokyo	2.8	451245
4	Tokyo	2.5	371922
5	Tokyo	2.4	221328
6	Beijing	2.3	113127
...	...	...	...

Table 2. Candidate answers obtained by adding scores with decreasing weights, where "Tokyo" is correct answer

Rank	Cand. ans.	Score	Document ID
1	Tokyo	4.3	259312, 451245, ...
2	Kyoto	3.3	926324
3	Beijing	2.3	113127
...	...	...	...

candidate by a factor of $0.3^{(i-1)}$[1] before adding up the scores. In this case, the score for "Tokyo" is 4.3 ($= 3.2 + 2.8 \times 0.3 + 2.5 \times 0.3^2 + 2.4 \times 0.3^3$), and we obtain the results in Table 2. As expected, "Tokyo" achieves the highest score.

2.5 Method for Cases Where Multiple Answers Can Be Correct

A question can have multiple answers in the QAC test collection. For example, when the question is "Which counties are permanent members of the United Nations Security Council?", there are five answers, "the United States of America", "the United Kingdom", "Russia", "France", and "China".

We proposed the *select-by-rate method* for cases where multiple answers can be correct. The *Select-by-rate method* outputs answers having a score more than a certain rate (rate for selection) of the highest score. For example, when we use 0.9 as the rate for selection and the highest score of answer candidates is 1100, we extract answer candidates with scores of more than 990 as the desired answers.

We also proposed the *select-one method*, that only extracts the top answer in every case. We also proposed the *select-two method*, that only extracts the top two answers in every case.

2.6 Method for Cases of Contextual Questions

Contextual questions in QAC are asked to answer questions. Contextual questions are defined as a series of questions with contexts. For example, the first

[1] The value of 0.3 was determined by preliminary experiments.

question is "What is the capital of Japan?" and the one succeeding is related to the first such as "What was it called in ancient times?".

We developed a method for cases of contextual questions, where we used the concatenation of all the questions from the first to the current question sentence in which an interrogative pronoun, adjective, or adverb in the first sentence had been changed to dummy symbols (e.g. "@@@") as the current question.

For example, when the previous question is "What is the capital of Japan?" and the current question is "What was it called in ancient times?", the question, "@@@ is the capital of Japan? What was it called in ancient times?" was constructed, and it was processed by question-answering systems. We refer to this method as *Conc1*.

We also used a method of adding answers for previous questions to the question itself. For example, when the previous question is "What is the capital of Japan?" and the current question is "What was it called in ancient times?", the answer "Tokyo" was extracted [,/from?] the question, "@@@ is the capital of Japan? Tokyo. What was it called in ancient times?" was constructed, and it was processed by question-answering systems. We refer to this method as *Conc2*.

3 Results in QACs 1, 2, and 3

3.1 QAC-1

Our results in the Task-3 (the tasks for contextual questions) of QAC-1 at the NTCIR-3 evaluation workshop are in Table 3. The task involved 20 sets of contextual questions each of which had two related questions. The average F-measure (MF) was used for evaluation. Our system obtained 0.17 for MF. Our team was in second place of the 7 teams who participated. The score for the best team was 0.19. Although our system was inferior to the best team's, another system of ours that changed the method for contextual questions (Conc1 → Conc2) after the workshop obtained an F-measure of 0.23, which was higher than that for the best team. We could not use a decreased method of adding using multiple documents as evidence in QAC-1, because we developed this after the QAC-1 workshop.

Table 3. Results in QAC-1

	Decreased adding method	For multiple answers	For contextual questions	MF
Submitted system	No use	Select-one	Conc1	0.167
System after workshop	No use	Select-one	Conc2	0.225

Table 4. Results in QAC-2

RunID	Decreased adding method	Rate in select-by-rate	For contextual qu.	MF
System 1	Use	0.95	Conc1	0.224
System 2	Use	0.97	Conc1	0.223

As the test collection for QAC-1 only had a total of 40 questions, the results were not valid.

3.2 QAC-2

Our results in the Task-3 (the tasks for contextual questions) of QAC-2 at the NTCIR-4 evaluation workshop are in Table 4. The task involved 36 sets of contextual questions each of which had five to ten related questions. There were a total of 251 questions. MF was used for evaluation. The RunIDs in the table indicate the ID numbers for our submitted systems. Our system obtained 0.22 for MF. Our team was in first place of the 7 teams who participated. The MF for the second best team was 0.20. Although there were many questions, the test collection was inconvenient as it could not be used for automatic evaluation or for experiments after the workshop.

3.3 QAC-3

Our results in QAC-3 at the NTCIR-5 evaluation workshop are in Table 5. Only a task for contextual questions was set. It involved 50 sets of contextual questions each of which had five to ten related questions. There were a total of 360 questions. MF was used for evaluation. The "Total", "First", and "Rest" in the table indicate the results for all the questions, those for the first questions in sets of contextual questions, and those for the remaining questions in the sets. Our system obtained 0.25 for MF. Our team was in first place of the 7 teams who participated. The MF for the second best team was 0.19. We found there were vast differences between our team and the second best team. The MF for "First" was higher than that for "Rest". This indicates that the first questions in the sets of contextual questions were relatively easy and the other questions

Table 5. Results in QAC-3

RunID	Decreased adding	Rate in select-by-rate	For contextual questions	MF		
				Total	First	Rest
System 1	Use	0.95	Conc1	0.236	0.403	0.209
System 2	Use	0.90	Conc1	0.250	0.450	0.218
System 3	Use	0.95	Conc2	0.208	0.403	0.177
After QAC-3	Use	0.85	Conc1	0.197	0.299	0.181
After QAC-3	No use	0.90	Conc1	0.174	0.236	0.165

Table 6. Results in QAC-3 (reference run 1 data)

RunID	Decreased adding	Rate in select-by-rate	For contextual questions	MF		
				Total	First	Rest
System 1	Use	0.95	—	0.305	0.403	0.289
System 2	Use	0.90	—	0.314	0.450	0.292
System 3	Use	0.95	—	0.305	0.403	0.289

Table 7. Results in QAC-3 (reference run 2 data)

RunID	Decreased adding	Rate in select-by-rate	For contextual questions	MF		
				Total	First	Rest
System 1	Use	0.95	No use	0.090	0.403	0.039
System 2	Use	0.90	No use	0.099	0.450	0.042
System 3	Use	0.95	No use	0.090	0.403	0.039

were difficult. To compare Systems 1 and 3, we found that System 3, which used Conc2 to add answers in the previous questions to the current question, obtained a lower MF than System 1, which did not use Conc2. We found that Conc2 was not effective in the QAC-3 test collection.

Reference run 1 data were given in QAC-3, and experiments using these data were conducted. Questions in the data were transformed into normal questions by supplementing omitted expressions (e.g., pronouns). For example, when the previous question was "What is the capital of Japan?" and the current question was "What was it called in ancient times?", the current question was transformed into "What was Tokyo called in ancient times?" and it was used to answer questions. We could estimate how difficult a contextual question was by comparing the experimental results for contextual questions with the results for the reference run data. The results are in Table 6. We found that the MF for the reference run data was higher than that for contextual questions. This meant that answering questions in the reference run data was easier than answering contextual questions. By comparing the results for "First" and "Rest" in the reference run data, we found that their MFs were very different and questions other than the first questions were more difficult than the first questions.

Our method for contextual questions was simple and connected to previous questions. It obtained 0.75 (= 0.218/0.292) of the MF for reference run 1 data.

Reference run 2 data were given in QAC-3, and experiments using these data were undertaken. Questions in the data exactly equaled the original question, but we could not use the previous questions. For example, when the previous question was "What is the capital of Japan?" and the current question was "What was it called in ancient times?", we could only use the current question "What was it called in ancient times?". The results are in Table 7. The MFs were very low. The MF was 0.099 (Table 7) when we did not use a simple connecting method. We obtained 0.14 (= 0.02/0.292) of the MF for reference run 1 data, while the simple connecting method obtained 0.75 (= 0.218/0.292) of the MF for reference run 1 data, indicating that our simple connecting method was effective.

The method of decreased adding obtained an MF of 0.250 and without it we obtained an MF of 0.197 (Table 5). This indicated that our method of decreased adding was effective.

The method of select-by-rate adding obtained the best MF when the rate for selection was 0.90.

As QAC-3 had numerous questions, and the test collection could be used for automatic evaluation and for experiments after the workshop, it was convenient and useful.

4 Conclusion

Our question-answering system used many effective methods to answer questions. We used a method of connecting question sentences to handle contextual questions. We used another method of using multiple documents to obtain more accurate answers. We used a special method for handling cases when multiple answers could be correct for a question. Our question-answering system obtained the second best accuracy in QAC-1 and the best accuracy in both QAC-2 and QAC-3 for contextual question answering. It is thus a high-performance system.

References

1. Yuji Matsumoto, Akira Kitauchi, Tatsuo Yamashita, Yoshitaka Hirano, Hiroshi Matsuda, and Masayuki Asahara. Japanese morphological analysis system ChaSen version 2.0 manual 2nd edition. 1999.
2. Masaki Murata, Masao Utiyama, and Hitoshi Isahara. Japanese question-answering system using decreased adding with multiple answers at ntcir 5. *Proceedings of the Fifth NTCIR Workshop*, 2005.
3. National Institute of Informatics. *Proceedings of the Third NTCIR Workshop (QAC)*. 2002.

Multi-document Summarization Using a Clustering-Based Hybrid Strategy

Yu Nie[1], Donghong Ji[1], Lingpeng Yang[1], Zhengyu Niu[1], and Tingting He[2]

[1] Institute for Infocomm Research, 21 Heng Mui Keng Terrace, Singapore 119613
{ynie, dhji, lpyang, zniu}@i2r.a-star.edu.sg
[2] Huazhong Normal University, 430079, Wuhan, China
the@mail.ccnu.edu.cn

Abstract. In this paper we propose a clustering-based hybrid approach for multi-document summarization which integrates sentence clustering, local recommendation and global search. For sentence clustering, we adopt a stability-based method which can determine the optimal cluster number automatically. We weight sentences with terms they contain for local sentence recommendation of each cluster. For global selection, we propose a global criterion to evaluate overall performance of a summary. Thus the sentences in the final summary are determined by not only the configuration of individual clusters but also the overall performance. This approach successfully gets top-level performance running on corpus of DUC04.

1 Introduction

Compared with single-document summarizations, information diversity and information redundancy are two main difficulties to be overcome in multi-document summarization tasks. The strategies for multi-document summarization roughly fall in three categories according to the way summaries are created.

The first kind of strategy ranks and extracts sentences from multi-documents similar with from single-document. However, compared with single-document approaches, the ranking here should consider not only sentence representative, but also mutual coverage between sentences to avoid redundancy. The difficulty for this approach is how to cover contribution of sentences to representative and their mutual difference simultaneously. MEAD [9] provides a two-phase ranking method which ranks sentences first and then adjusts summary by mutual information overlap of sentences. This method heavily relies on the initial selection of the first sentence, and the convergence of the adjusting process is questionable.

The second strategy is based on sentence or paragraph clustering. Sentences are extracted from clusters to form summaries. Intuitively, each cluster can be seen as a topic in the document collection, and selecting sentences from clusters helps the summary to get a good coverage of the topics and remove redundancy.

Many methods have been proposed on clustering-based multi-document summarization. Boros [3] clustered sentences to produce summaries with cluster number predefined. Hardy and Shimizu [4] selected representative passages from clusters of

H.T. Ng et al. (Eds.): AIRS 2006, LNCS 4182, pp. 608 – 614, 2006.

passages, and the system worked well in DUC01 with some parameter adjustments. Siddharthan et al. [11] and Blair-Goldensohn et al. [2] produced summaries by sentences clustered with a similarity threshold for clusters predefined.

The third strategy is based on information fusion [1], which tries to combine similar sentences across documents to create new sentences based on language generation technologies. Although this could model human's efforts in summarization process, it heavily relies on external resources (parsers, rules eg.). This limits its portability.

In this paper, we adopt the second strategy ---- sentence-clustering-based summarization.

The remainder of the paper is organized as follows. In section 2, we talk about the motivation of this work. In section 3, we explore the algorithm for sentence clustering. In section 4, we discuss about representative sentence selection from the clusters. In section 5, we present the experiments and results on DUC 2004 data. In section 6, we give out the conclusion.

2 Motivation

In this paper, we adopt the sentence-clustering-based strategy to determine main topics from source documents to reach good representative and avoid redundancy.

One immediate problem occurs that how many clusters are appropriate for sentences in source documents. Most existed clustering-based methods predefine the cluster number or similarity threshold in clustering process. However, the predefining depends on experience or guess, being hard to meet requirements from varies summarization tasks. To deal with this problem, we use a stability-based strategy [5] to automatically infer the cluster number.

Another problem is how to select representative sentences from clusters. Here, the search strategy is a key issue. In general, there are two kinds of search strategies. A local strategy tries to find a representative sentence for one cluster based on information of the cluster itself, while a global strategy tries to find the representative sentence based on the overall performance of the whole summary. Almost all clustering methods adopt only local strategies for representative sentence selection till now. In this paper, we propose a hybrid strategy to combine local and global search for sentence selection. Thus, sentences are selected not only observing individual clusters, but also observing their overall performance.

3 Sentence Clustering

Let v1 and v2 be feature vectors for sentence s1 and s2, their similarity is defined as the cosine distance.

$$\mathrm{sim}(s1, s2) = \cos(v1, v2) \tag{1}$$

To infer the appropriate cluster number, we adopt a stability method, which has been applied to other unsupervised learning problems [6][8][10]. Formally, let k be the cluster number, we need to find k which meets (2).

$$k = \arg\max_{k}\{criterion(k)\} \tag{2}$$

Here, the criterion is set up based on resampling based stability.

Let P^{μ} be a subset sampled from full sentence set in the document collection P with size $\alpha|P|$ (α is set as 0.9 in this paper), $C(C^{\mu})$ be $|P|\times|P|$($|P^{\mu}|\times|P^{\mu}|$) connectivity matrix based on the clustering result on $P(P^{\mu})$. Each entry $c_{ij}(c^{\mu}_{ij})$ of $C(C^{\mu})$ is calculated in the following way: if the entry pair p_i, $p_j(\in P(P^{\mu}))$ belong to the same cluster, then $c_{ij}(c^{\mu}_{ij})$ equals 1, otherwise, 0. Then the stability is defined in (3).

$$M(C^{\mu},C)=\frac{\sum_{i,j}1\{C^{\mu}_{i,j}=C_{i,j}=1, p_i\in P^{\mu}, p_j\in P^{\mu}\}}{\sum_{i,j}1\{C_{i,j}=1, p_i\in P^{\mu}, p_j\in P^{\mu}\}} \tag{3}$$

Intuitively, $M(C^{\mu}, C)$ denotes the consistency between the clustering results on C^{μ} and C. The assumption is that if k is actually the "natural" number of the clusters, then the clustering results on the subset P^{μ} generated by sampling should be similar to the clustering result on full sentence set P.

Obviously, the above function satisfies $0\leq M\leq 1$. It is noticed that $M(C^{\mu}, C)$ tends to decrease when increasing the value of k. Therefore to avoid the bias that small value of k is to be selected as cluster number, we use the cluster validity of a random predictor ρ_k to normalize $M(C^{\mu}, C)$. The random predictor ρ_k achieved the stability value by assigning uniformly drawn labels to objects, that is, splitting the data into k clusters randomly. Furthermore, for each k, we tried q times. So, the normalized object function can be defined as (4).

$$M_k^{norm}=\frac{1}{q}\sum_{i=1}^{q}M(C_k^{\mu_i},C_k)-\frac{1}{q}\sum_{i=1}^{q}M(C_{\rho_k}^{\mu_i},C_{\rho_k}) \tag{4}$$

Normalizing $M(C^{\mu}, C)$ by the stability of the random predictor can yield values independent of k (Levine et al. 2001).

After the number of optimal clusters is decided, k-means algorithm is implemented to cluster. Each output cluster is supposed to denote one topic in the document collection.

For the sake of running efficiency, we try to find the optimal cluster number k within the range 8~12 only in following experiments.

4 Representative Sentence Selection

Terms that are supposed to denote important concepts are extracted from texts (sentence cluster or the whole document collection) first. Then sentences are weighted based on terms they contain to select the representative sentence from each cluster.

4.1 Local Search Strategy

For local search strategies, we select the representative sentences based on the clusters themselves. The terms of a cluster are those extracted from the texts in the cluster. We use the terms extracted from the clusters, rather than those from the document collections, because they are supposed to contribute more in representative.

Equation (5) describes the weight scheme for a term in a cluster.

$$w_t = \frac{\log(1+tf) * \log(1+tl)}{\log(1+cf)} \tag{5}$$

Here *tf* is term frequency in the cluster, *tl* is the term length (number of words in the term, *cf* is cluster frequency (number of clusters which contain the term).

The weight of a sentence *s* is calculated as follows:

$$w_s = \sum_{t \in s} \frac{w_t}{\log(1+sl)} \tag{6}$$

Here *sl* is the length of the sentence (number of the words in the sentence).

4.2 Global Search Strategy

For global search strategy, sentences are selected according to their contribution to the performance of the whole summary. Thus a global criterion to evaluate the summary is necessary. For a summary *sm*, let *sml* be its length (word number in the summary) the criterion is defined as below.

$$\frac{\sum_{t \in sm} (\log(1+tf) * \log(1+tl))}{sml} \tag{7}$$

Terms here are extracted from the whole document set. *tf* is term frequency, *tl* is term length.

Intuitively, the criterion reflects the global term density of a summary. In general, we expect the summary to contain more terms, more longer terms, and as short as possible.

Assuming there are *n* clusters created, each cluster recommends *m* sentences, *n* sentences (one sentence per cluster) from *n*m* sentences are expected to maximize the global criterion.

For search heuristics, we adopt a strategy of deleting one sentence each time. A sentence, whose deletion maximizes the criterion among all sentences, is removed from the candidate sentence set in each step, until there is only one sentence for each cluster.

If there is a length limitation for the summarization task, further adjustment (adding or removing sentences) of the summary is necessary. The same global criterion is applied.

5 Experiments and Results

5.1 Data Set

We make experiments on document set of task2 of DUC04 with 50 sets of English TDT documents. Each set contains 10 documents. Four human model summaries are provided for each set for evaluation.

5.2 Evaluation Method

ROUGE [7] is used to automatically evaluation our experiment results. We follow the same requirement of DUC04 task2 ---- produce summaries of no more than 665 bytes from each set of 10 documents. All ROUGE evaluation configurations follow the configurations in DUC04: stop words included, Porter-stemmed, 4 human model summaries used, and the first 665 bytes cutoff.

5.3 Results

Comparison with DUC04 reports

Table 1 lists ROUGE scores of our summaries and of DUC04 runs.

We can find that our scores are much higher than the median, a bit lower than the best, in fact ranking the second in average among the task participants.

Table 1. Performance comparison (recommendation of top 3 sentences)

	ROUGE -1	ROUGE -L	ROUGE -W
Ours	0.377	0.384	0.132
DUC Best	0.382	0.389	0.134
DUC Median	0.343	0.355	0.121
DUC worst	0.242	0.276	0.094

Different Recommendations

To check effectiveness of the hybrid strategy, we test global search with different local recommendations, including recommendation of top 1, 2, 3 or all sentences from each cluster.

For comparison, we also list the scores of two purely global strategies, where global search is conducted directly on empty sentence set or the whole sentence set without sentence clustering respectively. For the former, one sentence is added each step based on the global criterion, and for the latter, one sentence is deleted each step based on the global criterion.

Table 2. Comparison between results of different stratedgies

Recommendations	ROUGE-1
1	0.369
2	0.371
3	**0.377**
all	0.370
Purely global-1	0.368
Purely global-2	0.365
CS-1	0.370
CS-2	0.361
CS-3	0.354

From table 2, we get several findings. One is that the hybrid strategy is better than the purely global search strategy, which indicates that the clustering and the local search provide good heuristics for better sentence selection.

Another finding is that the performance with 3 sentence selection is better than that with 2 sentence selection, which in turn is better than that with only the top sentence selected. This means that the sentences which form the optimal summary may be not necessarily the top sentences recommended by individual clusters.

The third finding is that with all sentences recommended, the score decreases, which seems to be contradictory with the fact that with larger search pool, a better solution tends to be found. However, note that better solutions depend on not only the search pool but also the appropriate search strategy, and with more sentences recommended, it would be more likely to reach local optimal, which may underlie the lower performance for global output.

At last, we find that our term based local recommendation performs better than centroid based recommendation. This prove the efficiency of the local recommendation stratedgy.

Validation Vs. Non-validation

To learn whether the automatic determination of the cluster number helps to improve the quality of the summary, remembering that the optimal cluster number was decided within range 8~12 by cluster validation, now we got 5 runs by denoting cluster number to 8~12 respectively. Results are listed in Table 3. We found that the run with cluster validation outperforms all these 5 runs. The reason might be that with fixed cluster number, the optimal clustering might be missed, which would affect the overall performance.

Table 3. Validation vs. Non-validation

	ROUGE-1	ROUGE-L	ROUGE-W
validation	0.377	0.384	0.132
Cluster-8	0.369	0.376	0.128
Cluster-9	0.370	0.374	0.127
Cluster-10	0.373	0.378	0.128
Cluster-11	0.370	0.379	0.129
Cluster-12	0.371	0.378	0.129

6 Conclusion

In this paper, we propose a clustering-based hybrid approach for multi-document summarization. It mainly consists of three steps: sentence clustering, local recommendation, and global selection. For sentence clustering, we adopt a stability based method to infer the optimal cluster number automatically. For local recommendation of sentences, we rank the sentences based on the terms they contain. For global selection of sentences, we propose a global criterion, and seek the optimal summaries based on this criterion.

We have some findings from experiments. First, clustering is useful to remove the redundancy, thus capable of outperforming the non-clustering method. Secondly, automatic cluster validation helps to improve the overall performance, because the optimal cluster numbers varies for different source documents. Furthermore, the combination of local and global outperforms separate search strategies. All this factors together make the clustering-based hybrid approach successfully get top-level performance.

References

1. Regina Barzilay, Kathleen R. McKeown, and Michael Elhadad. 1999. Information fusion in the context of multi-document summarization. ACL 1999, Maryland.
2. Sasha Blair-Goldensohn, David Evans. Columbia University at DUC 2004. In DUC 2004 Workshop, Boston, MA, 2004.
3. Endre Boros, Paul B. Kantor and David J. Neu. A Clustering Based Approach to Creating Multi-Document Summaries. DUC 2001 workshop.
4. Hilda Hardy, Nobuyuki Shimizu. Cross-Document Summarization by Concept Classification. In SIGIR 2002, page 121-128.
5. Lange, T., Braun, M., Roth, V., & Buhmann, J. M. (2002) Stability-Based Model Selection. Advances in Neural Information Processing Systems 15.
6. Levine,E. and Domany,E.. 2001. Resampling Method for Unsupervised Estimation of Clus-ter Calidity, Neural Computation, Vol.13, 2573-2593.
7. Chin-Yew Lin and Eduard Hovy. Automatic Evaluation of Summaries Using N-gram Co-Occurrence Statistics. In Proceedings of the Human Technology Conference (HLT-NAACL-2003), Edmonton, Canada.
8. Zhengyu Niu, Donghong Ji and Chew Lim Tan. 2004. Document Clustering Based on Clus-ter Validation, CIKM'04. November 8-13, 2004, Washington, DC, USA.
9. Dragomir Radev, Timothy Allison, Sasha Blair-Goldensohn, John Blitzer, Arda Çelebi, Stanko Dimitrov, Elliott Drabek, Ali Hakim, Wai Lam, Danyu Liu, Jahna Otterbacher, Hong Qi, Horacio Saggion, Simone Teufel, Michael Topper, Adam Winkel, and Zhang Zhu. MEAD - a platform for multidocument multilingual text summarization. In Proceedings of LREC 2004, Lisbon, Portugal, May 2004.
10. Volker Roth and Tilman Lange. Feature Selection in Clustering Problems, NIPS2003 workshop.
11. Advaith Siddharthan, Ani Nenkova and Kathleen McKeown. Syntactic Simplication for Improving Content Selection in Multi-Document Summarization. In Proceeding of COLING 2004, Geneva, Switzerland.

A Web User Preference Perception System Based on Fuzzy Data Mining Method

Wei-Shen Tai[1] and Chen-Tung Chen[2]

[1] Department of Information Management, National Yunlin University of Science and Technology, 123 University Road, Section 3, Douliou, Yunlin, Taiwan, R.O.C.
[2] Department of Information Management, National United University, 1 Lien Da Road, Kung-Ching Li, Miao Li, Taiwan, R.O.C.
g9423803@yuntech.edu.tw,
ctchen@nuu.edu.tw

Abstract. In a competitive environment, providing suitable information and products to meet customer requirements and improve customer satisfaction is one key factor to measure a company's competitiveness. In this paper, we propose a preference perception system by combining fuzzy set with data mining technology to detect the information preference of each user on a web-based environment. An experiment was implemented to demonstrate the feasibility and effectiveness of the proposed system in this study. It indicates that the proposed system can effectively perceive the change of information preference for users in a Web environment.

Keywords: Customer Satisfaction, Preference Perception, Data Mining, Fuzzy Set Theory, Web environment.

1 Introduction

Using the Internet to provide customers with personal products or services, an applicable solution is proposed to establish a low cost and high performance channel to accelerate interaction between businesses and customers [5]. As Internet users and information explosively grow, providing personal information or services to users actively has become an important issue for website owners who want to maintain customers' loyalty and satisfaction [15, 16]. On the other hand, business managers can understand the real requirement of each customer in accordance with the results of analyzing user behavior on the Web.

Over the years, data mining and data warehousing technologies have become powerful tools for businesses to predict complicated market trends. Data mining and data warehousing technologies can retrieve useful knowledge or information from an enormous database, then integrate, accumulate and translate it into a part of business wisdom [1]. In other words, data mining provides an effective solution for business managers to retrieve valuable information from a variety of data sources.

In this paper, we propose a personal-information preference perception model by combining data mining and fuzzy set theory to detect trends or changes in users'

H.T. Ng et al. (Eds.): AIRS 2006, LNCS 4182, pp. 615 – 624, 2006.

preference. It is helpful for business managers to understand customers' preferences and their changes. The organization of this paper is as follows. First, we introduce some literature including data mining and fuzzy set theory. Second, we propose a user preference perception model to detect the preference of users based on their browsing patterns. Third, an experiment is implemented to demonstrate the feasibility and effectiveness of the proposed model. Finally, some conclusions are presented at the end of this paper.

2 Data Mining and Fuzzy Set

As information technology grows, business managers can aggregate and store mass data more easily and efficiently. Nevertheless, managers also face a difficult work to analyze and interpret mass data in the same time. Therefore, data mining (DM) and knowledge discovery in database (KDD) were presented to help managers analyze and retrieve valuable information from mass data [3, 4].

2.1 Web Mining

Although users can acquire amounts of information from Internet sources, they also face a serious information overload problem. Up to now, browsers and search engines still can neither provide information to meet users' requirements exactly nor solve the information overload problem on the Internet effectively [11, 12]. Web mining is an effective tool to analyze users' browsing behavior and provide suitable information for users to meet their need, respectively [9].

Web usage mining obtains some important data from Web servers or proxy servers such as log files, user profiles, registration data, user sessions or transactions [9]. According to these data and mining processes, web usage mining provides a method to discover valuable information about users' intentions and to understand their browsing behaviors. However, the information preference of each user is often ambiguous and it is difficult to capture their intentions and behaviors. In other words, we can not define the information preference of each user directly. Therefore, fuzzy set theory is applied in this study to determine the preference degree of each user. In this paper, we use web usage mining to discover the browsing patterns of each user on the Web.

2.2 Fuzzy Set Theory

A fuzzy set $\tilde{A}$ in a universe of discourse X is characterized by a membership function $\mu_{\tilde{A}}(X)$ which associates with each element x in X a real number in the interval [0,1]. The function value $\mu_{\tilde{A}}(X)$ is termed the grade of membership of x in $\tilde{A}$ [8, 17].

Definition 2.1. A fuzzy set $\tilde{A}$ in the universe of discourse X is called a normal fuzzy set implying that $\exists x_i \in X$, $\mu_{\tilde{A}}(x_i) = 1$.

A fuzzy number $\tilde{n}$ is a fuzzy subset in the universe of discourse X whose membership function is both convex and normal (shown in Fig. 1).

Definition 2.2. The α-cut of a fuzzy number $\tilde{n}$ is defined as

$$\tilde{n}^{\alpha} = \{x_i : \mu_{\tilde{n}}(x_i) \geq \alpha, x_i \in X\}, \qquad (1)$$

where $\alpha \in [0,1]$.

The α-cut of fuzzy number $\tilde{n}$ is a non-empty bounded closed interval contained in X and it can be denoted by $\tilde{n}^{\alpha} = [n_l^{\alpha}, n_u^{\alpha}]$. The symbols n_l^{α} and n_u^{α} are the lower and upper bounds of the closed interval, respectively. Fig. 2 shows a fuzzy number $\tilde{n}$ with α-cut [7, 18].

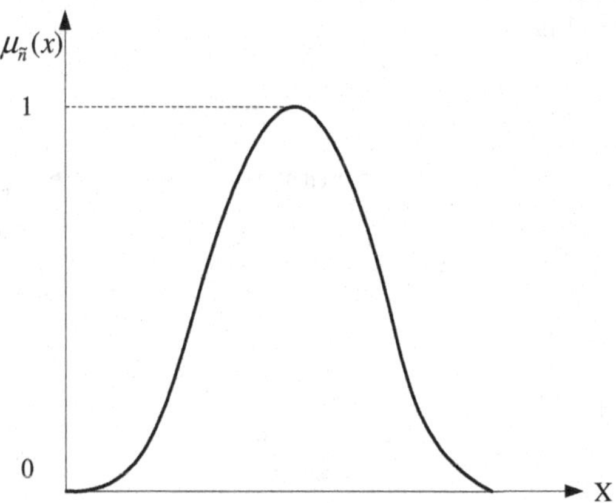

Fig. 1. A fuzzy number $\tilde{n}$

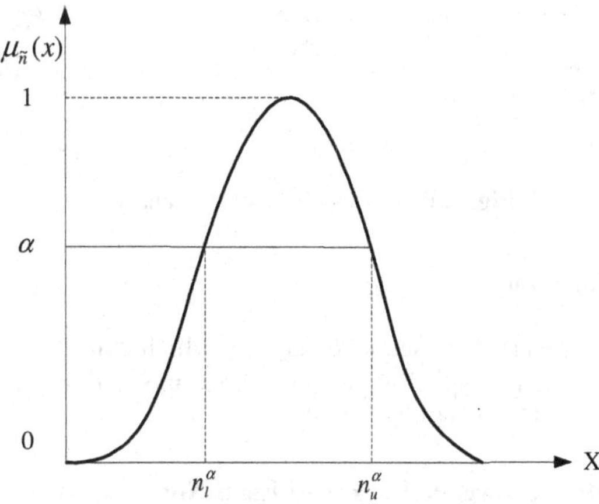

Fig. 2. A fuzzy number $\tilde{n}$ with α-cut

Up to now, fuzzy set theory has been applied in many areas such as fuzzy logic control [2], fuzzy decision-making [10], fuzzy expert system [6], and fuzzy information retrieve [13]. In this paper, we apply the concept of membership function to calculate the preference degree of each user based on his/her browsing time. It can help us to deal efficiently with the relationship between a user's browsing time and the length of content on the website.

3 Preference Perception Model

In this paper, we analyze the log data from web servers to obtain the surfing tracks and browsing time of each user on the website. The processes of preference analysis of users are shown in Fig. 3.

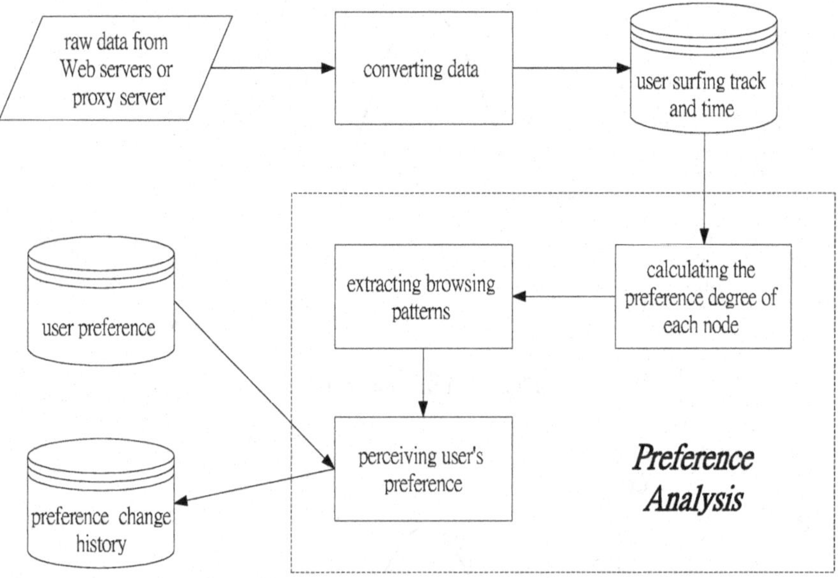

Fig. 3. Processes of preference analysis

3.1 Converting Log Data

On the Internet, the traversal subsequence is a set which consists of the browsing web page nodes of each user in a specific period. For example, T_1 is traversal subsequence and are denoted by T_1={Y, B, H, C, T, I, H, J}.

3.2 Calculating the Preference Degree of Each Node

In order to detect the information preference of each user, we must choose the interesting web page from his/her traversal subsequences. If the browsing time is too long, then the web page may be idled. Therefore, it is necessary to eliminate these nodes (web pages) whose browsing time exceeds the maximum threshold value. In this paper, the maximum threshold value is denoted by t_{max}. For a web page, we can compute the average browsing time (t_{avg}) of all visitors. If a visitor's browsing time exceeds the average browsing time for this web page, this visitor is very interested in this web page and his preference degree is regarded as 1. It means that this visitor is completely interested in the content of this web page. Conversely, if a visitor's browsing time for a web page is shorter, it means that the visitor has less interest in this web page. Therefore, the preference degree of the visitor for a web page can be determined in accordance with the browsing time. The preference degree of each visitor for a web page is calculated as follows:

$$P_{ki} = \begin{cases} \dfrac{t_{ki}}{t_{avg}}, & 0 \le t_{ki} < t_{avg} \\ 1, & t_{avg} \le t_{ki} \le t_{\max} \\ 0, & t_{\max} < t_{ki} \end{cases} \tag{2}$$

where, t_{ki} is the browsing time of the kth visitor in the ith web page, P_{ki} is the preference degree of the kth visitor for the ith web page. The preference degree can be represented by a membership function shown in Fig. 4.

According to the membership function of the preference degree,α-cut is applied to divide the web pages into two categories as follows:

(i) If $P_{ki} \ge \alpha$, we consider that the kth visitor is interested in the ith web page.
(ii) If $P_{ki} < \alpha$, we consider that the kth visitor has no interest in the ith web page.

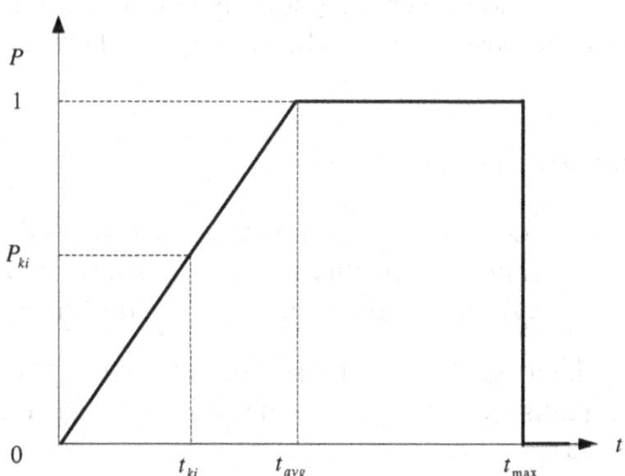

Fig. 4. The membership function of the preference degree

3.3 Perceiving Users' Preferences

According to the browsing patterns and the preference degree of each user in each category, the variation of a user's preferences will be perceived in the proposed model. In this study, the vector and the weighted value are applied to represent the preference degree of each user in different categories. The preference degree of each user in each category can be calculated as follows:

$$P_i = (w_{i1}^*, w_{i2}^* \cdots, w_{ik}^*) \tag{3}$$

where P_i is the preference degree of the ith user in each category, w_{ij} is the preference degree of the ith user in the jth category (j=1, 2,…,k).

Generally speaking, the importance degree of each category should be related to its frequency in a browsing pattern. If a category appears in less browsing patterns, it should be more important than those categories that appear in many browsing patterns at the same frequencies. According to this concept, we use the revised TFIDF (Term Frequency – Inverse Document Frequency) concept [14] to calculate the preference degree of each user in each category from browsing patterns. The formula is described as follows:

$$w_j^* = \sqrt{\frac{f_j}{n} \times \log\left(\frac{p_n}{p_j} + 1\right)} \tag{4}$$

where w_j^* is the preference degree of each user in the jth category, f_j is the appearance frequency of the jth category, n is the total number of appearance frequencies for each category in all browsing patterns, p_n is the total number of browsing patterns, p_j is the number of browsing patterns that contain the jth category.

4 System Design and Experiments

In this paper, a website providing personal news is built to demonstrate the feasibility and effectiveness of the proposed model. The system architecture and experiments are elaborated as follows:

4.1 System Architecture and Functions

In this study, a website is designed to demonstrate the proposed model with a Linux operating system. The architecture of this proposed system is shown in Fig. 5. The major components of this system and their functions are described as follows:

(1) Web server log database. When a registered member logs on the website, all of his/her surfing paths and browsing time are recorded and stored in the log database of the server.

(2) Converting log data. The system will convert the log data of each user into his/her traversal subsequence sets in accordance with his/her registered number, respectively.

(3) Eliminating the uninteresting nodes (web pages) of each user. At this stage, the system will divide these web pages into two categories, interesting and uninteresting nodes, in accordance with the preference for each node. Then, these uninteresting nodes for each user will be eliminated.

(4) Extracting browsing patterns. This proposed system will extract some applicable browsing patterns from users' traversal subsequence sets.

(5) Perceiving users' preferences. After a specific period, the proposed system will calculate the weight of each category and detect the variation in information preference of each user.
(6) Providing general or personal information. The website will provide the user personal information according to his/her preferences in the user profile database.
(7) Search engine. This search engine provides a friendly interface for the user to search information easily.
(8) User profile database. When a user registers in the web site, all the information about the user will be stored in the user profile database.
(9) Information database. Information is stored in the database according to category attribution. The system will update the information database every four hours.

4.2 Experiments and Result Analysis

In order to demonstrate the feasibility and effectiveness of the preference perception model, a website was built to make experiments at a University in Taiwan. The experimental processes are described as follows.

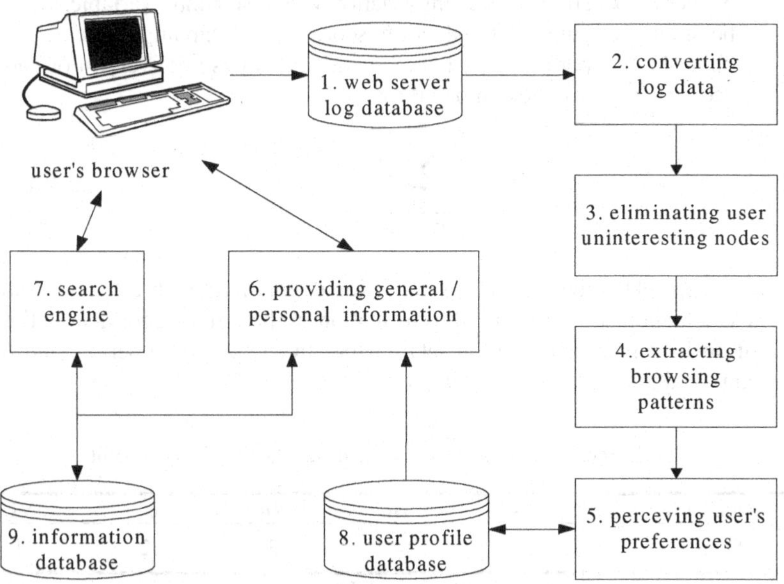

Fig. 5. The system architecture of the proposed system

4.2.1 Experimental Procedure

In this experiment, the participants are students at this University and the information content is news. The topical news categories are divided into "Stock and Security", "Society and Culture", "International", "Home and Living", "Mainland China", "Business and Economy", "Regional", "Government", "Entertainment", "Science and

Computer", "Recreation and Sports". The experimental procedure is described as follows:

Step 1. Member registration. All participants filled out their profiles such as name, password, email address and information preferences in the registration form.

Step 2. Providing personal news to each member. When a member logs into the website, he/she will get the personal news by category according to his/her information preferences from the user profile database.

Step 3. Recording browsing behavior of each member. The cookies which record surfed web pages, browsing time and timestamp are sent to the web server.

Step 4. Perceiving the variation of preferences of each member. According to the log data for three weeks, users' preferences are analyzed based on the user perception model proposed. Because the average reading speed of an adult is 200 words per minute approximately [5]. Therefore, the maximum reading time is set to 10 minutes (t_{max}=10) to filter out those web pages whose browsing time is abnormally long in this experiment. In addition, the parameters α and threshold of traversal sequence are 0.5 and 9, respectively.

Step 5. Satisfaction survey. After a member reads a piece of news, he/she can express his/her satisfaction for this information with a semantic variable, which can be converted into a satisfaction score by a mapping table (shown in Table 1). In a period, the satisfaction degree of push-delivery information in each category can be computed as:

$$S_{ci} = \frac{\sum_{t=1}^{T} S_{cti}}{T_{ci} * E_{\max}} \tag{5}$$

where S_{ci} ($0 \leqq S_{ci} \leqq 1$) is the ith user's satisfaction degree in the cth category, S_{cti} is the ith user's satisfaction score for the tth information in the c-th category, T_{ci} is the total number of push-delivery information sets for the cth category in a period, and E_{max} is the maximum value in the semantic table (in this case, E_{max}=5).

Table 1. Scores of semantic variables in satisfaction measurement

Semantic variable	VS	SA	NR	US	VU
Score	5	4	3	2	1

VS: Very satisfying, SA: Satisfying, NR: Normal, US: Unsatisfying, VU: Very unsatisfying.

4.2.2 Preference Analysis of User

The proposed system calculates the preference degree of each user according to the browsing time on the web page. For example, we can find 19 surfing paths for the 21st user (shown in Table 2). In a surfing path, *tid* denotes the number of categories, *did* denotes the number of documents. In the brace of a surfing path, the first number denotes the user's browsing time (unit: second), the second number denotes the degree of preference (between 0 and 1). For example, In the second web page (*tid* =9 and *did*=20483), the browsing time of the 21st user is 43 seconds and its average browsing time for all users is 57 seconds. Therefore, the preference degree of the 21st user for this web page is 0.754 (43/57).

Once the proposed system eliminates those nodes which degree of preference is lower than the threshold value (in this case, $\alpha = 0.5$). Then, the weight of the 21st user's preference in the second category is calculated as follows:

$$w_2^* = \sqrt{\frac{2}{17} \times \log\left(\frac{5}{1} + 1\right)} = 0.303$$

The preference degree of the 21st user for all categories can be denoted by $P_{21}^{w} = (w_1, w_2, w_3, w_4, w_5, w_6, w_7, w_8, w_9, w_{10}, w_{11})$ = (0, 0.303, 0, 0.214, 0, 0, 0.214, 0, 0.399, 0, 0.358). When the threshold of the preference degree is 0.2, the 21st user's preference will be denoted by P_{21} = (0, 1, 0, 1, 0, 0, 1, 0, 1, 0, 1).

Table 2. Surfing paths of the 21st user

No.	Surfing path
1	tid=9&did=20486(76/1)=>tid=9&did=20483(43/0.754) =>tid=11&did=20765(61/1)=>tid=11&did=20764(80/1) =>tid=11&did=20499(58/0.92)=>tid=11&did=20498(71/1) =>tid=3&did=20757(58/0.935)=>tid=11&did=21652(138/1)

4.2.3 Results Analysis

During the experiment for the proposed preference perception model, 16176 pieces of news and 8087 log records were retrieved. According to the new preference of each user, the proposed system provided new information or news to each user and calculated the satisfaction degree when he/she received the new information. According to the satisfaction measurement results, the average degree of satisfaction for the 23 members (their preferences had changed) is 72.4%. in the other words, we can prove that this proposed model can effectively and exactly detect the variation of information preference of each user on the Internet.

5 Conclusions

Because of the Internet's convenience and prevalence, more and more businesses provide information, services and products through it. If businesses can provide customers suitable services or products, they will improve not only customer satisfaction and loyalty but also competitiveness. Therefore, how to detect the information preference of users and provide suitable information to users are very important issues for business managers. In this paper, we proposed a personal-information preference perception system by combining fuzzy set with data mining technology in a web-based environment. According to the user preference perception model, a system for providing personal news is built to demonstrate the proposed model. As the result of the degree of satisfaction survey for all participants in this experiment, the satisfaction degree is up to 72 % for the users whose preference changed in the experimental period.

The user preference perception model proposed in this paper will enable businesses to provide more suitable information, services and products to meet their customers' requirements and allow both of them to enjoy the benefit of the Internet.

References

1. Berry, Michael J.A., Linoff, G., Data Mining Techniques: For Marketing, Sales, and Customer Support, New York: Wiley (1997).
2. Chen, C.S., Design of stable fuzzy control systems using Lyapunov's method in fuzzy hypercubes, Fuzzy Sets and Systems, 139 (2003) 95-110.
3. Fayyad, U., Uthurusamy, R., Data mining and knowledge discovery in databases, Communications of the ACM, 39 (1996) 24-26.
4. Fayyad, U., Uthurusamy, R., Evolving data mining into solutions for insights: Evolving data into mining solutions for insights, Communications of the ACM, 45 (2002) 28-31.
5. Grabe, W., Stoller, F.L., Reading for Academic Purpose: Guidelines for the ESL/EFL teacher, In Celce-Murcia, M. (Ed.), Teaching English as a second or foreign language: the 3rd ed. Boston: Heinle (2001) 187-204.
6. Hall, L.O., Rule chaining in fuzzy expert systems, IEEE Transactions on Fuzzy Systems, 9 (2001) 822-828.
7. Kaufmann, A., Gupta, M.M., Introduction to Fuzzy Arithmetic: Theory and Applications, New York: Van Nostrand Reinhold (1985).
8. Klir, G. J., Yuan, B., Fuzzy sets and fuzzy logic: theory and applications, Prentice Hall (1995).
9. Kosala, R., Blockeel, H., Web mining research: a survey, ACM SIGKDD Explorations Newsletter archive, 2 (2000) 1-15.
10. Lee, H.S., Optimal consensus of fuzzy opinions under group decision making environment, Fuzzy Sets and Systems, 132 (2002) 303-315.
11. Madria, S.K., Raymond, C., Bhowmick, S., and Mohania, M., Association rules for Web data mining in WHOWEDA, International Conference on Digital Libraries: Research and Practice (2000) 227-233.
12. Maes, P., Agents that reduce work and information overload, Communications of the ACM archive, 37 (1994) 30-40.
13. Miyamoto, S., Information clustering based on fuzzy multisets, Information Processing and Management, 39 (2003) 195-213.
14. Salton, G., Another look at automatic text-retrieval systems, Communications of the ACM, 29 (1986) 648-656.
15. Song, H.S., Kim, J. K., Kim, S.H., Mining the change of customer behavior in an internet shopping mall, Expert Systems with Applications, 21 (2001) 157-168.
16. Yen, Benjamin P.C., Kong, Robin C.W., Personalization of information access for electronic catalogs on the web, Electronic Commerce Research and Applications, 1 (2002) 20-40.
17. Zadeh, L.A., Fuzzy sets, Information and Control, 8 (1965) 338-353.
18. Zimmermann, H.J., Fuzzy Set Theory and its Applications (the 2nd ed.), Boston: Kluwer Academic Publishers (1991).

An Analysis on Topic Features and Difficulties Based on Web Navigational Retrieval Experiments

Masao Takaku[1], Keizo Oyama[2], and Akiko Aizawa[2]

[1] Research Organization of Information and Systems
[2] National Institute of Informatics
2-1-2 Hitotsubashi, Chiyoda-ku, Tokyo 101-8430, Japan
{masao, oyama, aizawa}@nii.ac.jp

Abstract. We analyze the relationship between topic features and difficulties in Web navigational retrieval tasks based on the experiments done on the NTCIR-5 Web test collection. Our analysis shows that the difficulties of a retrieval task are closely related to the specificity of the topic, and topics that are of some particular categories are more difficult than others. For example, a representative page of a company or an organization is easier on average to find than that of a person, a product, or an event. Our results show that adding metadata on a topic would potentially be useful for search engines to predict the difficulty of the task. Additionally, we show that the number of unique documents retrieved from different systems weakly correlates with the query's performance.

1 Introduction

We conducted the Web navigational retrieval subtask (Navi2) at the NTCIR-5 workshop[1]. Navi2 is a "known item search" task, a kind of navigational retrieval task, which seeks one or more representative pages of an entity including a product, a company, a person, an event, and a website. Throughout our experiments and those of participants in Navi2, typical information needs for a known item could be resolved within up to the 10th rank in the best four run results. However some needs suffered slightly poor performance in all systems.

Many information retrieval studies have been done on predicting topic difficulties [2][3][4][5][6]. In those studies and in ours, prediction of topic difficulty is seen to provide a hint for optimizing retrieval methods with specific features of a query.

Although many studies on predicting topic difficulties have been done, most of them have aimed at increasing effectiveness in ad-hoc retrieval situations. To our knowledge, however, few works have focused on predicting the performance of a query on the Web navigational retrieval. Jensen et al.[7] suggested a methodology for predicting a query difficulty from snippet texts in search result pages. For a navigational retrieval task, Oyama et al.[8] described the test collection used for the NTCIR-4 Web task and conducted preliminary analyses on the relationships between the difficulties of NTCIR-4 Web topics and the attributes of the topics.

H.T. Ng et al. (Eds.): AIRS 2006, LNCS 4182, pp. 625–632, 2006.

In this paper, we analyze results of NTCIR-5 Web navigational retrieval experiments, particularly to reveal relationships among topic difficulties, a topic's metadata, and query-corpus features.

2 Test Dataset

In our analysis, run submission results for the NTCIR-5 Web task[1] were used. We selected 14 run submissions, from the runs of five participant groups and organizers, regarding their groups and retrieval methods. Table 1 shows the selected run IDs and their retrieval methods for Navi2. In the table, "Content" represents a model based on content (fulltext) of pages, "Link" represents a hyperlink information-based model, and "Anchor" represents an anchor text-based model. We used these runs throughout our experiments.

Table 1. Selected runs from Navi-2 submissions

Retrieval methods	Selected runs
Content	JSWeb-3, OKSAT-Web-F-00, ORGREF-GC1, TNT-5
Anchor	K3100-12, ORGREF-R
Content+Anchor	OKSAT-Web-F-05, TNT-3
Content+Link	JSWeb-4, ORGREF-GC1-LF1
Anchor+Link	K3100-9
Content+Anchor+Link	OKSAT-Web-F-07, ORGREF-C20-P2, TNT-1

In Navi2, 400 topics were created and delivered to participants, and after relevance assessments there were 269 topics which had at least one highly relevant pages in the collection. We used these 269 topics for the evaluation and experiments described in this paper.

An example of a Navi2 topic is shown in Figure 1. Each topic has several metadata for its relevance judgment. All these metadata except for `<NUM>` were assigned by the assessor who created the topic. `<TITLE>` is a usual query form. `<TYPE>` means a level of TITLE's specificity, and three levels of specificity are assigned as follows: TYPE=1 means "a phrase in TITLE represents its target item", TYPE=2 means "two or more phrases in TITLE represent the item", and TYPE=3 means "one or more phrases in TITLE do not specifically represent the item." `<CATEGORY>` represents a genre of the target item. Eight categories are defined by the organizers. Table 2 shows defined categories. The first and second columns are the codes and the definitions of the categories, respectively. "`SPECIALTY`" specifies the searcher's knowledge level on the target item; the codes are defined as follows: "A" (searcher knows the item in detail), "B" (searcher knows its outline), "C" (searcher knows it to the extent the item can be identified among others), and "D" (searcher knows only its existence but knows very little about the item).

```
<TOPIC>
<NUM>1041</NUM>
<TYPE>1</TYPE><CATEGORY>B</CATEGORY>
<TITLE>UNESCO</TITLE>
<DESC>I want to visit to the homepage of UNESCO.</DESC>
<NARR>
<BACK>I would like to know about activities of UNESCO.</BACK>
<RELE>The top page of the National Federation of UNESCO
 Associations in Japan would be relevant.</RELE>
</NARR>
<USER SPECIALTY="C">Graduate-doctoral 1st year, female,
 5 years experience in searching</USER>
</TOPIC>
```

Fig. 1. Example of Navi2 topic (English translation)

Table 2. Example of 8 categories assigned to the topics

Code	Definition
A	Products / services
B	Company / organization
C	Persons
D	Facilities
E	Sights, historic spots, and natural things
F	Information resources
G	Online shops and online services
H	Events

3 Analysis Methods

Several analysis methods on topic features and topic difficulties were used.

We used standard reciprocal rank measure (RR) in the experiments related to topic difficulty, and we averaged it over fourteen selected runs on each topic. RR is an inverse of the rank of the top-ranked relevant document.

As topic features, we took TYPE, CATEGORY, SPECIALTY, and length of a query. For analysis of topic metadata, we grouped topics with the same metadata, and tested their differences with each other by using Student's t-test. When $p < 0.05$, we considered it as the significant difference. For length of a query, we used three variants of length of TITLE for analysis. These three variants were based on phrases (ql_{phrase}), words[1] (ql_{word}), and bytes (ql_{byte}).

Another kind of topic feature, Pool Size, was also tested. Pool Size is the number of unique documents in the pool, which was constructed from the selected run results in which pool depth were at most 100, a setting that is the same as

[1] We used the morphological analyzer MeCab for separating words in a Japanese query; See `http://mecab.sourceforge.jp/`

the default submission setting of Navi2. We compared these topic features with the following predictors derived from prior works:

- IDF_{min} is the minimum inverse document frequency (IDF) in a given query Q. For IDF, we used the same setting as in a prior work[5]:

$$IDF(w) = \frac{log_2(N + 0.5)/N_w}{log_2(N + 1)},$$

where w is a word[1] in Q, N_w is the number of documents in which the query term w appears, and N is the number of documents in the whole collection.
- Query Scope, proposed by He et al.[5]:

$$QScope = -log_2(N_Q/N),$$

where N_Q is the number of documents containing at least one of the query terms.
- AvICTF, proposed by Plachouras et al.[9]:

$$AvICTF = \frac{log_2 \prod_{w \in Q} token_{coll}/tf_{coll}}{|Q|},$$

where $token_{coll}$ means the number of tokens in the whole collection and tf_{coll} is the number of occurrences of a query term in the whole collection.
- Simplified Clarity Score, proposed by He et al.[5], which is based on the original notion of Cronen-Townsend et al.'s *clarity score*[4]:

$$SCS = \sum_{Q} P_{ml}(w|Q) \cdot log_2 \frac{P_{ml}(w|Q)}{P_{coll}(w)},$$

where $P_{ml}(w|Q)$ is given by qtf/ql and $P_{coll}(w)$ is given by $tf_{coll}/token_{coll}$.

We analyzed linear dependencies between query difficulty and each of these feature by using Pearson's correlation coefficient r.

4 Results

4.1 Topic Metadata

We grouped topics that had the same TYPE values and tested their average RR measures by using t-tests. The distribution of topic groups based on TYPE is given in Table 3. Columns 4–6 in the table show p-values of their differences; values in italics are significant.

The TYPE groups' difficulties are all statistically significant. The TYPE-1 group is the easiest group of topics, TYPE-2 is the second, and TYPE-3 is the most difficult group of topics. Next, we assigned a rank to each topic with TYPE-group difficulty, and computed rank correlations between the rank in the order

Table 3. Distribution of the TYPE groups

TYPE	# of topics	Average RR	1	2	3
1	145	0.31	-	*0.0000*	*0.0000*
2	98	0.18		-	*0.0245*
3	26	0.10			-

Table 4. Distribution of the CATEGORY groups

CATEGORY	# of topics	Average RR	B	E	D	F	G	H	A	C
B	60	0.31	-	0.546	0.397	0.077	*0.047*	*0.007*	*0.000*	*0.0008*
E	16	0.28		-	0.926	0.555	0.385	0.119	*0.032*	0.055
D	29	0.27			-	0.572	0.404	0.133	*0.032*	0.056
F	48	0.25				-	0.707	0.223	0.088	0.106
G	29	0.23					-	0.375	0.242	0.228
H	19	0.19						-	0.893	0.862
A	49	0.19							-	0.685
C	28	0.18								-

of TYPE-group and the rank in the order of average RR measure. Spearman's rank correlation ρ is 0.462, and Kendall's τ is 0.341.

The distribution of CATEGORY groups is shown in Table 4. Columns 4–11 in the table show p-values of their differences; values in italics are significant.

Category-B, which represents companies or organizations, is the easiest category, and E, D, and F follow. Significantly difficult groups compared with B are G, H, A, and C. Category-A, which represents products or services, is the second most difficult category in terms of average RR, and significantly easy groups compared with A are D, E, and B. Other combinations of groups did not show any statistical significance.

Table 5 shows the distribution of the topics' SPECIALTY. Columns 4–7 in the table show p-values of their differences. Differences among the SPECIALTY groups are slightly small, so no significance were observed.

Table 5. Distribution of the SPECIALTY groups

SPECIALTY	# of topics	Average RR	A	C	B	D
A	62	0.21	-	0.403	0.394	0.675
C	73	0.24		-	0.901	0.854
B	106	0.26			-	0.900
D	28	0.26				-

4.2 Query Length, Pool Size, and Query-Corpus Features

Table 6 shows the correlation coefficients of several predictors with the average RR. The upper value in each cell represents the correlation coefficient r (values in bold are significant) and the lower value in each cell represents its significance.

Table 6. Correlation coefficients of query-corpus features

	Avg.RR	ql_{phrase}	ql_{word}	ql_{byte}	IDF_{min}	QScope	AvICTF	SCS	PoolSize
Avg.RR	1	**-.438**	-.112	**-.256**	-.091	.112	-.077	.095	**-.342**
		.000	.067	.000	.137	.066	.209	.066	.000
ql_{phrase}		1	**.492**	**.643**	.054	**-.340**	**.231**	**-.304**	**.521**
			.000	.000	.377	.000	.000	.000	.000
ql_{word}			1	**.644**	.005	**-.727**	**.609**	**-.714**	**.153**
				.000	.930	.000	.000	.000	.012
ql_{byte}				1	.078	**-.378**	**.220**	**-.320**	**.132**
					.204	.000	.000	.000	.031
IDF_{min}					1	.035	-.047	.035	-.025
						.565	.446	.571	.687
QScope						1	**-.930**	**.957**	**-.444**
							.000	.000	.000
AvICTF							1	**-.988**	**.524**
								.000	.000
SCS								1	**-.486**
									.000
PoolSize									1

Two variants of query length, ql_{phrase} and ql_{word}, show negative correlation with the average RR; ql_{phrase} shows a relatively strong negative correlation. Note that the length of phrases in a query was limited to a maximum of 3 phrases during the process of the topic creation, and this variant is strongly correlated with TYPE metadata of topics, which was mentioned in the previous section; correlation between TYPE and phrase-based query length is 0.871.

Pool Size also shows weak negative correlations with average RR. Other features based on query-corpus statistics do not have any significant correlation with average RR.

5 Discussions

Our analysis for topics' metadata shows that specificity of the topics (TYPE) correlates with the average RR measure. This appears to be intuitive because this metadata directly indicates the topic's specificity.

Our analysis on genre of topics (CATEGORY) shows that a few groups of genres within topics have significant relationships to the difficulties and weakly correlate with the average RR measure. For instance, the topic group of a person, a product/service, or an event is more difficult than that of a company or an organization. Similar trends were observed in a prior experiment on the NTCIR-4 Web test collection[8], but the results were not based on statistical analysis. If a user can assign additional metadata for a query, such as a level of specificity (TYPE) or a genre (CATEGORY), on a Web search engine, it would be helpful for indicating the query's difficulty and thus prompting the search engine to adopt other retrieval methodologies. Such an application of making use of topic metadata remains as future work.

A user's level of knowledge on a topic does not affect the difficulty of the topic in our analysis. There are two possible interpretations for this result: 1) a user who is very familiar with a search target item does not always express a query in a way that gains good performance; or 2) assigning familiarity to a query would bias it in the process of topic creation. Further analysis on this point would be needed to better explain this result.

Comparison among features of pool size, query length, and query-corpus statistics shows that pool size has weak correlations with average RR, and that query-corpus features do not have any significant correlation with average RR. The correlation of the pool size with average RR appears to be slightly intuitive because of its nature. When the results of many retrieval systems agree, the number of unique documents returned is decreased. This suggests that an agreement in different systems' results on a query indicates easiness of the query in some way, and that systems do not always agree with one another in the case of difficult queries. If, for example, various kinds of search engines are available, the query difficulty could be estimated by using the agreement of the documents returned. Distributed retrieval engines or meta-search engines are such applications of realistic estimation, but even one search engine having some retrieval techniques within it could be applicable. Those applications of utilizing the agreement of different retrieval techniques remain as future work.

Topic features related to corpus statistics, such as QScope and SCS, do not correlate with average RR. We also calculated the correlation of the predictors based on the anchor text surrogate collection with average RR, but those experiments resulted in similar results. The reasons why those query-corpus features do not work well are not clear, and we will further investigate the relationships between query difficulty and those predictors in the future.

We also preliminarily investigated differences among retrieval methods, based on either fulltext, anchor text, or hyperlink information, but did not draw any clear conclusions. This remains to be examined in greater detail in the future.

References

1. Oyama, K., Takaku, M., Ishikawa, H., Aizawa, A., Yamana, H.: Overview of the NTCIR-5 WEB navigational retrieval subtask 2 (Navi-2). In Kando, N., Takaku, M., eds.: Proceedings of NTCIR-5 Workshop Meeting. (2005) 242–222
2. Carmel, D., Yom-Tov, E., Soboroff, I.: SIGIR workshop report: predicting query difficulty - methods and applications. SIGIR Forum **39** (2005) 25–28
3. Eguchi, K., Kuriyama, K., Kando, N.: Sensitivity of IR systems evaluation to topic difficulty. In: Proceedings of the 3rd International Conference on Language Resources and Evaluation (LREC 2002). Volume 2. (2002) 585–589
4. Cronen-Townsend, S., Zhou, Y., Croft, W.B.: Predicting query performance. In: SIGIR '02: Proceedings of the 25th Annual International ACM SIGIR Conference on Research and Development in Information Retrieval. (2002) 299–306

5. He, B., Ounis, I.: Inferring query performance using pre-retrieval predictors. In: Proceedings of the 11th International Conference of String Processing and Information Retrieval (SPIRE 2004). (2004) 43–54
6. Yom-Tov, E., Fine, S., Carmel, D., Darlow, A.: Learning to estimate query difficulty: including applications to missing content detection and distributed information retrieval. In: SIGIR '05: Proceedings of the 28th Annual International ACM SIGIR conference on Research and Development in Information Retrieval. (2005) 512–519
7. Jensen, E.C., Beitzel, S.M., Grossman, D., Frieder, O., Chowdhury, A.: Predicting query difficulty on the Web by learning visual clues. In: SIGIR '05: Proceedings of the 28th Annual International ACM SIGIR conference on Research and Development in Information Retrieval. (2005) 615–616
8. Oyama, K., Ishikawa, H., Eguchi, K., Aizawa, A.: The test collection for navigational retrieval on WWW data — design and characteristics. Progress in Informatics (2005) 59–73
9. Plachouras, V., He, B., Ounis, I.: University of Glasgow at TREC2004: Experiments in web, robust and terabyte tracks with Terrier. In: Proceedings of the 13th Text REtrieval Conference (TREC 2004). (2004)

Towards Automatic Domain Classification of Technical Terms: Estimating Domain Specificity of a Term Using the Web

Takehito Utsuro[1], Mitsuhiro Kida[2], Masatsugu Tonoike[3], and Satoshi Sato[4]

[1] Graduate School of Systems and Information Engineering, University of Tsukuba, 1-1-1, Tennodai, Tsukuba, 305-8573, Japan
[2] Nintendo Co.,Ltd., 11-1, Hokotate-cho, Kamitoba, Minami-ku, Kyoto-shi, 601-8116 Japan
[3] Graduate School of Informatics, Kyoto University, Yoshida-Honmachi, Sakyo-ku, Kyoto 606-8501, Japan
[4] Graduate School of Engineering, Nagoya University, Furo-cho, Chikusa-ku, Nagoya 464-8603, Japan

Abstract. This paper proposes a method of domain specificity estimation of technical terms using the Web. In the proposed method, it is assumed that, for a certain technical domain, a list of known technical terms of the domain is given. Technical documents of the domain are collected through the Web search engine, which are then used for generating a vector space model for the domain. The domain specificity of a target term is estimated according to the distribution of the domain of the sample pages of the target term. Experimental evaluation results show that the proposed method achieved mostly 90% precision/recall.

1 Introduction

Lexicons of technical terms are one of the most important language resources both for human use and for computational research areas such as information retrieval and natural language processing. Among various research issues regarding technical terms, full-/semi-automatic compilation of technical term lexicon is one of the central issues. In various research fields, novel technologies are invented every year, and related research areas around such novel technologies keep growing. Along with such invention of technologies, novel technical terms are created year by year. Considering such a situation, it requires a huge cost for manually compiling lexicons of technical terms for hundreds of thousands of technical domains. Therefore, it is inevitable to invent a technique of full-/semi-automatic compilation of technical term lexicons for various technical domains.

The whole task of compiling a technical term lexicon can be roughly decomposed into two sub-processes: (1) collecting candidates of technical terms of a technical domain, and, (2) judging whether each candidate is actually a technical term of the target technical domain. The technique of the first sub-process is

H.T. Ng et al. (Eds.): AIRS 2006, LNCS 4182, pp. 633–641, 2006.

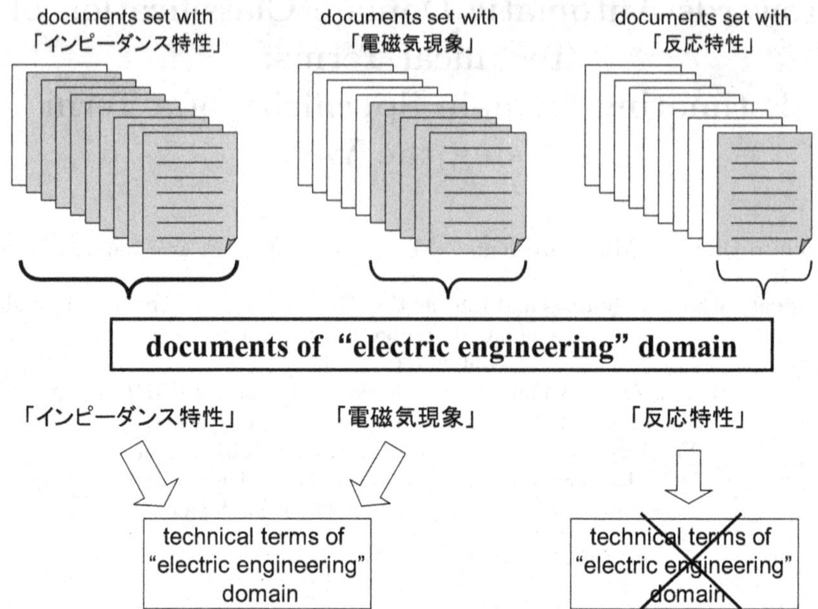

Fig. 1. Degree of specificity of a term based on the domain of the documents (Example terms: *impedance characteristic*, *electromagnetism*, and *response characteristic*)

closely related to research on automatic term recognition, and has been relatively well studied so far (e.g., [7]). On the other hand, the technique of the second sub-process has not been studied well so far. Exceptional cases are works such as [1,2], where their techniques are mainly based on the tendency of technical terms appearing in technical documents of limited domains rather than in documents of daily use such as newspaper and magazine articles. Although the underlying idea of those previous works is very interesting, those works are quite limited in that they require existence of certain amount of technical domain corpus. It is not practical for manually collecting technical domain corpus for hundreds of thousands of technical domains. Therefore, as for the second sub-process here, it is very important to invent a technique for automatically classifying the domain of a technical term.

Based on this observation, among several key issues regarding the second sub-process above, this paper mainly focuses on the issue of estimating the domain specificity of a term. In this paper, supposing that a target technical term and a technical domain are given, we propose a technique of automatically estimating the specificity of the target term with respect to the target domain. Here, the domain specificity of the term is judged among the following three levels: i) the term mostly appears in the target domain, ii) the term generally appears in the target domain as well as in other domains, iii) the term generally does not appear in the target domain.

The key idea of the proposed technique is as follows. In the proposed technique, we assume that sample technical terms of the target domain are available. Using such sample terms with search engine queries, we first collect a corpus of the target domain from the Web. In a similar way, we also collect sample pages that include the target term from the Web. Then, the similarities of the contents of the documents are measured between the corpus of the target domain and each of the sample pages that include the target term. Finally, the domain specificity of the target term is estimated according to the distribution of the domain of those sample pages.

Firgure 1 illustrates rough idea of this technique. Among the three example (Japanese) terms, the first term (*impedance characteristic*) mostly appears in the documents of the *"electric engineering"* domain on the Web. In the case of the second term (*electromagnetism*), about half of sample pages collected from the Web can be regarded as in the *"electric engineering"* domain, while the rest are not. On the other hand, in the case of the last term (*response characteristic*), only a few of the sample pages can be regarded as in the *"electric engineering"* domain. In our technique, such difference of the distribution can be easily identified, and the domain specificities of those three terms are estimated.

Experimental evaluation results show that the proposed method achieved mostly 90% precision/recall.

2 Domain Specificity Estimation of Technical Terms Using Documents Collected from the Web

2.1 Outline

As we introduced in the previous section, the underlying purpose of this paper is to invent a technique for automatically classifying the domain of a technical term. More specifically, this paper mainly focuses on the issue of estimating the domain specificity of a term t with respect to a domain C, supposing that the term t and the domain C are given.

Generally speaking, the coarsest-grained classification of domain specificity of a term is binary classification, namely, the class of terms that are used in a certain technical domain, vs. the class of terms that are *not* used in a certain technical domain. In this paper, we further classify the degree $g(t, C)$ of the domain specificity into the following three levels:

$$g(t,C)=\begin{cases} + \text{ (t mostly appears in the documents of the domain C.)} \\ \pm \text{ (t generally appears in the documents of the domain C as well as} \\ \quad \text{in those of the domains other than C.)} \\ - \text{ (t generally does not appear in the documents of the domain C.)} \end{cases}$$

(When we simply classify domain specificity of a term into two classes with the coarsest-grained binary classification above, we regard those with domain specificity '+' or '±' as those that are used in the domain, and those with domain specificity '−' as those that are *not* used in the domain.)

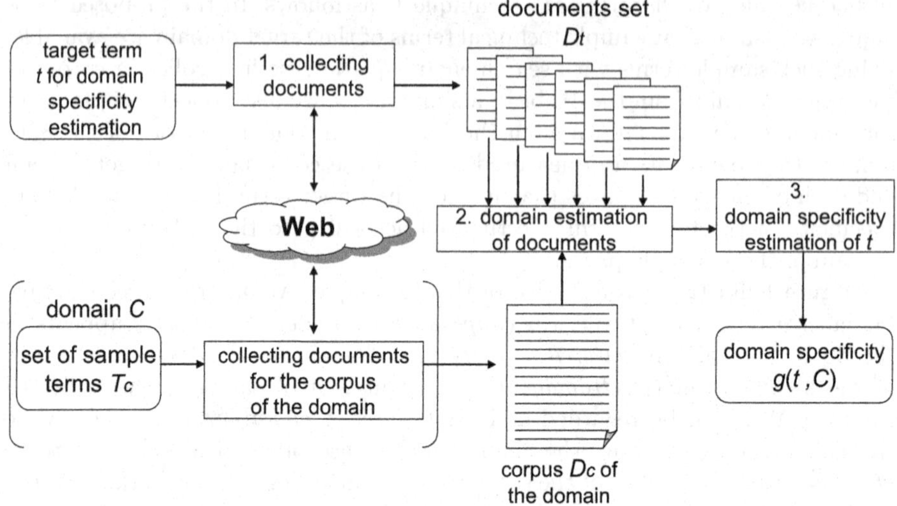

Fig. 2. Domain specificity estimation of terms based on Web documents

The input and output of the process of domain specificity estimation of a term t with respect to the domain C are given below:

input	target term t for domain specificity estimation, set T_C of sample terms of the domain C
output	domain specificity $g(t, C)$ of t with respect to C

The process of domain specificity estimation of a term is illustrated in Figure 2, where the whole process can be decomposed into two sub-processes: (a) that of constructing the corpus D_C of the domain C, and (b) that of estimating the specificity of a term t with respect to the domain C. In the process of domain specificity estimation, the domain of documents including the target term t is estimated, and the domain specificity of t is judged according to the distribution of the domains of the documents including t. The details of those two sub-processes are described in the followings.

2.2 Constructing the Corpus of the Domain

When constructing the corpus D_C of the domain C using the set T_C of sample terms of the domain C, first, for each term t in the set T_C, we collect into a set D_t the top 100 pages obtained from search engine queries that include the term t[1]. The search engine queries here are designed so that documents that describe the

[1] Related techniques for automatically constructing the corpus of the domain using the sample terms of the domain include those presented in [6,3]. We are planning to evaluate the performance of those related techniques and compare them with the one employed in this paper.

technical term t are ranked high. When constructing a corpus of the Japanese language, the search engine "goo"[2] is used. The specific queries that are used in this search engine are phrases with topic-marking postpositional particles such as "t-toha," "t-toiu," "t-wa," and an adnominal phrase "t-no," and "t."

Then, union of the sets D_t for each t is constructed and denoted as $D(T_C)$:

$$D(T_C) = \bigcup_{t \in T_C} D_t$$

Finally, in order to exclude noise texts from the set $D(T_C)$, the documents in the set $D(T_C)$ are ranked according to the number of sample terms (of the set T_C) that are included in each document. Through a preliminary experiment, we decided here that it is enough to keep top 500 documents, and regard them as the corpus D_C of the domain C.

2.3 Domain Specificity Estimation of Technical Terms

Given the corpus D_C of the domain C, domain specificity of a term t with respect to a domain C is estimated through the following three steps:

step 1. Collecting documents that include the term t from the Web, and constructing the set D_t of those documents.

step 2. For each document in the set D_t, estimating its domain by measuring similarity against the corpus D_C of the domain C. Then, given a certain lower bound L of document similarity, from D_t, extracting documents with large enough similarity values into a set $D_t(C, L)$.

step 3. Estimating the domain specificity $g(t, C)$ of t using the document set $D_t(C, L)$ constructed in the step 2.

Details of those three steps are given below:

Collecting Web Documents Including the Target Term. For each target term t, documents that include t are collected from the Web. According to a procedure that is similar to that of constructing the corpus of the domain C described in section 2.2, the top 100 pages obtained with search engine queries are collected into a set D_t.

Domain Estimation of Documents. For each document in the set D_t, its domain is estimated by measuring similarity against the corpus D_C of the domain C. Then, given a certain lower bound L of document similarity, documents with large enough similarity values are extracted from D_t into the set $D_t(C, L)$.

In the process of document similarity calculation, we simply employ a standard vector space model,[3] where a document vector is constructed, after removing

[2] http://www.goo.ne.jp/

[3] An alternative here is to apply supervised document classification techniques such as those based on machine learning technologies (e.g., [4]). Especially, since it is not easy to collect negative data here in the task of domain specificity estimation of a term, it seems very interesting to apply recently developed techniques without labeled negative data (e.g., [5]).

153 stop words, as a vector of frequencies of content words such as nouns and verbs in a document. Here, the corpus D_C of the domain C is regarded as a document d_C, and a document vector $dv(d_C)$ is constructed. For each document d_t in the set D_t, a document vector $dv(d_t)$ is also constructed. Then, the cosine similarity between $dv(d_t)$ and $dv(d_C)$ is calculated and is defined as the similarity $sim(d_t, D_C)$ between the document d_t and the corpus D_C of the domain C.

$$sim(d_t, D_C) = sim(d_t, d_C) = \cos(dv(d_t), dv(d_C)) = \frac{dv(d_t) \cdot dv(d_C)}{|dv(d_t)||dv(d_C)|}$$

Finally, suppose that a certain lower bound L of document similarity is given, documents d_t with the similarity value $sim(d_t, D_C)$ above or equal to L are regarded as those of the domain C, and are collected into a set $D_t(C, L)$:

$$D_t(C, L) = \{d_t | sim(d_t, D_C) \geq L\}$$

In experimental evaluation of section 3, the lower bound L is determined using a development term set for tuning various parameters of the whole process of estimating domain specificity of technical terms.

Domain Specificity Estimation of a Term. The domain specificity of the term t with respect to the domain C is estimated using the document sets D_t and $D_t(C, L)$. Here, this is done by simply calculating the following ratio r_L of the numbers of the documents within the two sets:

$$r_L = \frac{|D_t(C, L)|}{|D_t|}$$

Then, by introducing the two thresholds $a(\pm)$ and $a(+)$ for the ratio r_L, the specificity $g(t, C)$ of t is estimated with the following three levels:

$$g(t, C) = \begin{cases} + \ (a(+) \leq r_L) \\ \pm \ (a(\pm) \leq r_L < a(+)) \\ - \ (r_L < a(\pm)) \end{cases}$$

In experimental evaluation of section 3, as in the case of the lower bound L of the document similarity, the two thresholds $a(\pm)$ and $a(+)$ are also determined using the development term set mentioned above.

3 Experimental Evaluation

We evaluate the proposed method with five sample domains, namely, *"electric engineering"*, *"optics"*, *"aerospace engineering"*, *"nucleonics"*, and *"astronomy"*. For each domain C of those five domains, the set T_C of sample (Japanese) terms is constructed by randomly selecting 100 terms[4] from an existing (Japanese) lexicon

[4] Through a preliminary experiment, we conclude that it is not necessary to start with the set T_C of sample terms which has more than 100 sample terms. The number of minimum requirement for the size of T_C varies according to domains.

Table 1. Number of terms for experimental evaluation

	number of terms for each degree of domain specificity					
	development set T_{dev}			evaluation set T_{eval}		
	+	±	−	+	±	−
electric engineering	43	14	43	48	20	32
optics	35	15	50	40	24	36
aerospace engineering	39	10	51	36	24	40
nucleonics	22	24	54	34	28	38
astronomy	41	12	47	35	15	50

Table 2. Precision/recall of domain specificity estimation with threshold $a(\pm)$

	L	$a(\pm)$	development set T_{dev}		evaluation set T_{eval}	
			precision	recall	precision	recall
electric engineering	0.2	0.4	0.96(54/56)	0.95(54/57)	0.95(59/62)	0.87(59/68)
optics	0.2	0.4	0.94(49/52)	0.98(49/50)	1.00(60/60)	0.94(60/64)
aerospace engineering	0.2	0.4	0.94(42/44)	0.86(42/49)	0.79(54/68)	0.90(54/60)
nucleonics	0.25	0.2	0.92(36/39)	0.78(36/46)	0.95(60/63)	0.97(60/62)
astronomy	0.15	0.4	0.96(51/53)	0.96(51/53)	0.86(48/56)	0.96(48/50)

of technical terms for human use. For each of the five domains, we then manually constructed the development term set T_{dev} and the evaluation term set T_{eva}, each of which has 100 terms (those with frequency more than or equal to five, and hits of the search engine within 100 $\sim$ 10,000.), respectively. For each of the domain specificity '+', '±', and '−', Table 1 lists the number of terms of the class. In our experimental evaluation, the development term set T_{dev} is used for tuning the lower bound L of the document similarity, as well as the two thresholds $a(\pm)$ and $a(+)$, where those parameter values are determined so as to maximizing F score ($\alpha = 0.75$). Here, we chose the value of the weight α as 0.75, since we prefer precision to recall in our application such as automatic technical term collection, where automatically collected technical terms are recursively utilized as seed technical terms in the later steps of a bootstrapping process.

$$\text{F score} = \frac{1}{\alpha \frac{1}{\text{precision}} + (1-\alpha)\frac{1}{\text{recall}}}$$

Parameter values as well as experimental evaluation results are summarized in the Tables 2 and 3. Generally speaking, the task of discriminating the terms with domain specificity '+' against the rest is much harder than that of discriminating those with '+' and '±' against those with '−'. In the Table 3, especially, the results with the domains *"aerospace engineering"* and *"nucleonics"* are with lower precision values than other domains. Each of these two domains has closely related another domain (e.g., *"military"* for *"aerospace engineering"*, and *"radiation medicine"* for *"nucleonics"*), where technical terms of these two domains

Table 3. Precision/recall of domain specificity estimation with threshold $a(+)$

	L	$a(+)$	development set T_{dev}		evaluation set T_{eval}	
			precision	recall	precision	recall
electric engineering	0.2	0.7	0.97(32/33)	0.74(32/43)	0.92(24/26)	0.50(24/48)
optics	0.2	0.7	0.83(20/24)	0.57(20/35)	0.82(23/28)	0.58(23/40)
aerospace engineering	0.2	0.5	0.90(28/31)	0.72(28/39)	0.53(27/51)	0.75(27/36)
nucleonics	0.25	0.3	0.55(18/33)	0.82(18/22)	0.57(32/56)	0.94(32/34)
astronomy	0.15	0.7	0.89(34/38)	0.83(34/41)	0.87(33/38)	0.94(33/35)

tend to appear also in the documents of such closely related domains. Most of the errors are caused due to the existence of those close domains.

4 Concluding Remarks

This paper proposes a method of domain specificity estimation of technical terms using the Web. In the proposed method, it is assumed that, for a certain technical domain, a list of known technical terms of the domain is given. Technical documents of the domain are collected through the Web search engine, which

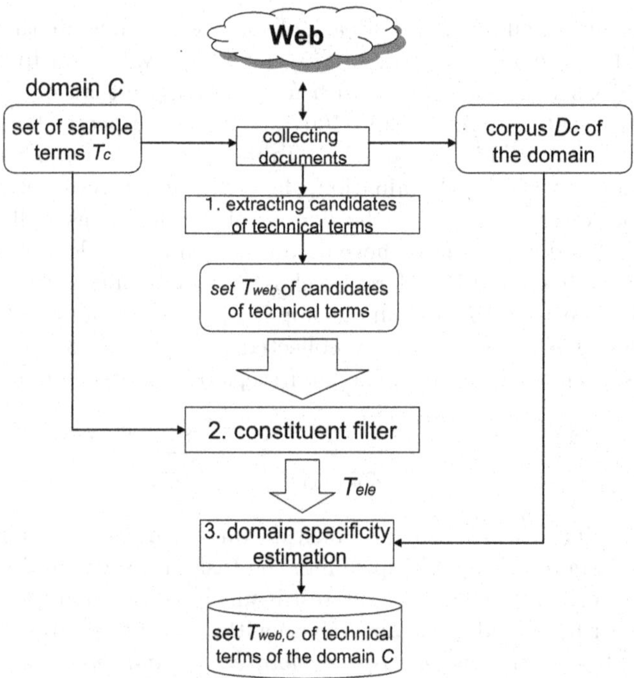

Fig. 3. Web based domain specificity estimation of terms not included in existing lexicons

are then used for generating a vector space model for the domain. The domain specificity of a target term is estimated according to the distribution of the domain of the sample pages of the target term. Experimental evaluation results showed that the proposed method achieved mostly 90% precision/recall.

Related techniques on domain specificity estimation of a term include the one based on straightforward application of similarity calculation of language models between the domain corpus and the target term. Although this technique seems mathematically well-defined, it is weak in that it assumes a single language model per a target term. Especially, when a target term appears in the documents of more than one domain, the technique proposed in this paper seems to be advantageous, since it independently estimates the domain of each individual document including the target term.

Figure 3 illustrates a further application of the proposed technique of domain specificity estimation of technical terms. Here, from the corpus D_C of the domain C, candidates of technical terms that are not included in any of existing lexicons of technical terms for human use are collected. Then, after excluding terms which do not share constituent nouns against the sample terms of the given set T_C, the domain specificity of the remaining terms are automatically estimated. In the result of experimental evaluation, out of the 1,000 candidates of technical terms per a domain, we discovered about 100 $\sim$ 200 novel technical terms that are not included in any of existing lexicons of technical terms. This result clearly supports the effectiveness of the proposed technique for the purpose of full-/semi-automatic compilation of technical term lexicons.

References

1. T. M. Chung. A corpus comparison approach for terminology extraction. *Terminology*, 9(2):221–246, 2004.
2. P. Drouin. Term extraction using non-technical corpora as a point of leverage. *Terminology*, 9(1):99–117, 2003.
3. C.-C. Huang, K.-M. Lin, and L.-F. Chien. Automatic training corpora acquisition through Web mining. In *Proceedings of IEEE/WIC/ACM International Conference on Web Intelligence*, pages 193–199, 2005.
4. T. Joachims. *Learning to Classify Text Using Support Vector Machines: Methods, Theory, and Algorithms.* Springer-Verlag, 2002.
5. B. Liu, Y. Dai, X. Li, W. S. Lee, and P. S. Yu. Building text classifiers using positive and unlabeled examples. In *Proceedings of the 3rd IEEE International Conference on Data Mining*, pages 179–186, 2003.
6. B. Liu, X. Li, W. S. Lee, and P. S. Yu. Text classification by labeling words. In *Proceedings of the 19th AAAI*, pages 425–430, 2004.
7. H. Nakagawa and T. Mori. Automatic term recognition based on statistics of compound nouns and their components. *Terminology*, 9(2):201–219, 2003.

Evaluating Score Normalization Methods in Data Fusion

Shengli Wu[1], Fabio Crestani[2], and Yaxin Bi[1]

[1] School of Computing and Mathematics
University of Ulster, Northern Ireland, UK
{s.wu1, y.bi}@ulster.ac.uk
[2] Department of Computer and Information Sciences
University of Strathclyde, Glasgow, UK
f.crestani@cis.strath.ac.uk

Abstract. In data fusion, score normalization is a step to make scores, which are obtained from different component systems for all documents, comparable to each other. It is an indispensable step for effective data fusion algorithms such as CombSum and CombMNZ to combine them. In this paper, we evaluate four linear score normalization methods, namely the fitting method, Zero-one, Sum, and ZMUV, through extensive experiments. The experimental results show that the fitting method and Zero-one appear to be the two leading methods.

1 Introduction

Data fusion in information retrieval has been investigated by many researchers and has been a well-established technique. The essential idea is to more accurately estimate the relevance of all the retrieved documents for a given query via combining these retrieved documents from multiple information retrieval systems into a single list. It can be used as an alternative for implementing an effective information retrieval system, it can also be useful in the WWW environment as a meta-search engine to merge resultant documents from different Web search engines.

For data fusion algorithms, several factors need to be considered: (a) performance of all component results; (b) correlation among all component results; (c) available information associated with retrieved documents. A lot of research work (e.g., in [1,6,7,9,10]) has been done on these topics.

For a component result, it comprises a ranked list of documents. Sometimes a score is associated with each document to indicate the estimated relevance of the document to the information need. Various experiments (for example, in [2,8]) have confirmed that rankings are not as informative as scores, therefore, we should use score information preferentially if it is possible. In this paper we assume that all component systems provide scores for retrieved documents.

Different component systems may use very different policies to apply scores to retrieved documents. This difference can be both on range and distribution. Therefore, It is necessary to normalize scores for effective data fusion algorithms such as CombSum and CombMNZ to combine them.

H.T. Ng et al. (Eds.): AIRS 2006, LNCS 4182, pp. 642–648, 2006.

Fox and Shaw designed a series of data fusion algorithms including CombSum and CombMNZ [3]. CombSum sets the score of each document in the fused result to the sum of the scores obtained by the individual result, while in CombMNZ the score of each document is obtained by multiplying this sum by the number of results which have non-zero scores. Lee [4] suggested a linear transformation method for score normalization. All normalized scores are in the range of [0,1], with the minimal score mapped to 0 and the maximal score to 1. This method is referred to as Zero-one later in this paper. Actually this can be improved since the top-ranked documents are not always relevant and bottom-ranked documents are not always irrelevant. A smaller range $[a,b]$ $(0 < a < b < 1)$ would be a better solution for score normalization. This method is referred to as the fitting method in this paper.

Montague and Aslam [5] suggested two other linear transformation methods Sum and ZMUV (Zero-Mean and Unit-Variance). In Sum, the minimal score is mapped to 0 and the sum of all scores in the result to 1. In ZMUV, the average of all scores is mapped to 0 and their variance to 1.

In this paper, our major goal is to evaluate the four linear normalization methods (Zero-one, the fitting method, Sum and ZMUV). However, some of our experimental results are useful for us to better understand score normalization in general.

2 Effect of Different Score Normalization Methods to Data Fusion

In this section we report an experiment which investigates the effect of different score normalization methods on data fusion. For a group of results, we combine them using CombSum and CombMNZ several times, each time with a different score normalization method. Then we compare the performance of the fused results from the same data fusion method but different score normalization methods. For the fitting method, [0.06, 0.6] is arbitrarily chosen in all cases. Some other values may provides better performance. ZMUV cannot be properly used for CombMNZ because about half of the scores in every result are negative. Instead, as in Montague and Aslam's experiment [5], we used ZMUV2, a variation of ZMUV, which added 2 to every score normalized with ZMUV.

We used 4 groups of results, which were subsets of results submitted to TREC 7 (ad hoc track), 8 (ad hoc track), 9 (web track) and 2001 (web track) [7]. All selected results include 1000 documents for every query and have a mean average precision of 0.15 or over. In each group, we randomly select 3-10 systems for a data fusion run with CombSum and CombMNZ, and 200 runs were tested for any given number of results.

Mean average precision was used for the evaluation. Figures 1-2 show the experimental result.

The experimental results of all four score normalization methods are very consistent. Also all score normalization methods have a very similar performance for both data fusion methods CombSum and CombMNZ. We can observe that

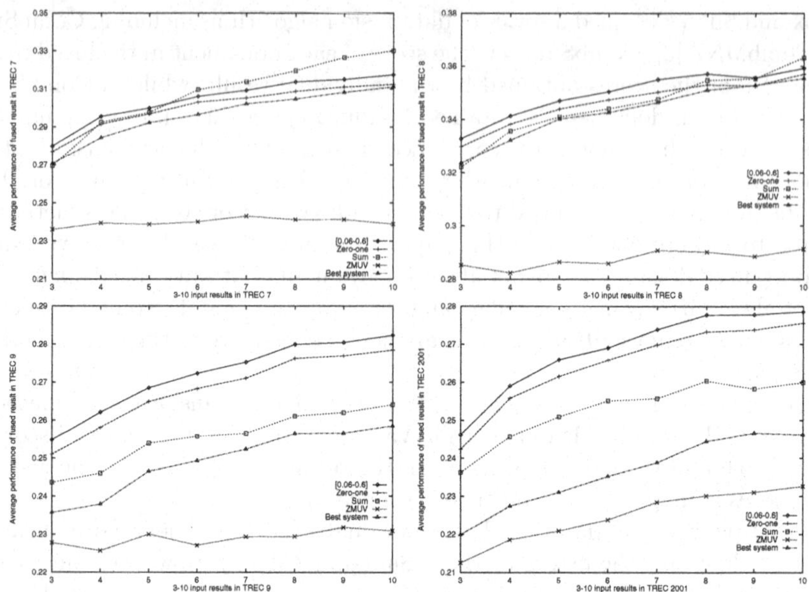

Fig. 1. Mean average precision of CombSum in TREC 7-2001 with different normalization methods

ZMUV2 is always the worst in all cases. The fitting method [0.06-0.6] is better than the Zero-one method using CombSum, but the Zero-one method is better than the fitting method [0.06-0.6] using CombMNZ. The fitting method and the Zero-one method perform better than Sum in most cases. In a few cases, Sum is a little better. Therefore, it suggests that the Zero-one method and the fitting method are very likely the best normalization method and ZMUV is the worst. The claim that both Sum and ZMUV are significantly better than Zero-one [5] cannot be supported in our experiment. According to our analysis in Section 2, Sum can lead to good fusing results when the (different) weights assigned to all input results fit their performances.

Another observation is: using the fitting method, Zero-one and Sum normalization methods, the fused results are better than the best component system on average, it confirms that data fusion is an effective technique to improve retrieval performance.

3 Distribution of Relevant Documents in Different Score Ranges

Actually, all these four score normalization methods (Zero-one, the fitting method, Sum, and ZMUV) are linear transformation methods. That is to say, it is straightforward to define a "general" schema to include all of them. We need to set up two pairs of ending points - the maximal score and the minimal score for raw scores

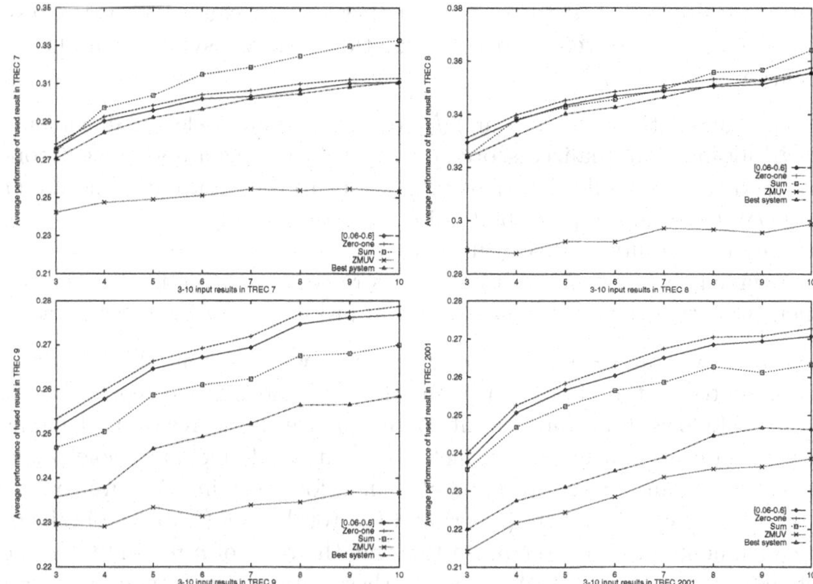

Fig. 2. Mean average precision of CombMNZ in TREC 7-2001 with different normalization methods

and normalized scores. We use *max* and *min* to denote the maximal and minimal normalized score, and *r_max* and *r_min* to denote the maximal and minimal raw score respectively. Please note that *r_max* is not necessarily the maximal raw score, and *r_min* is not necessarily the minimal raw score obtained by documents in the result. But instead we should guarantee that any raw score is between *r_max* and *r_min*. Then for any raw score *r_s*, we use the following formula to normalize it:

$$s = min + \frac{r\_s - r\_min}{r\_max - r\_min}(max - min)$$

If only considering one result, then there is no difference at all using any of the four methods. Any normalized score in any method can be linearly transformed into a corresponding score in another method. For example, suppose s_{zmuv} is a normalized score obtained with ZMUV, we can transform it into a normalized score for Zero-one by $s_{zero-one}{=}(s_{zmuv} - min_{zmuv})/(max_{zmuv} - min_{zmuv})$, where max_{zmuv} and min_{zmuv} denotes the normalized maximal score and the normalized minimal score with ZMUV respectively.

In data fusion, we need multiple results at the same time. The difference among these four normalization methods is that Sum and ZMUV normalize scores in different results into different ranges. The key issue is: can this bring any benefit to data fusion?

The goal of score normalization is to make scores in different results comparable. Ideally, if there is a linear relation between score and probability of relevance, it is a perfect condition for data fusion algorithms such as CombSum.

More specifically, we need to meet the following two conditions for n results r_i $(i = 1, 2, ..., n)$ retrieved from n information retrieval systems for the same information need:

- r is any one of the results, d_1 and d_2 are two randomly selected documents in r. d_1 obtains a normalized score of s_1, and d_2 obtains a normalized score of s_2, then the probability of d_1 being relevant to the information need equals to s_1/s_2 times of the probability of d_2 being relevant;
- for any two results r_1 and r_2 in n results, d_1 is a document in r_1 and d_2 is a document in r_2. If d_1's normalized score equals to d_2's normalized score, then these two documents have an equal probability of being relevant.

Next we report an experiment to examine to which extent the two requirements presented in the above can be met by those normalized scores. The procedure is as follows: first, for a result, we normalize the scores of its documents using a given score normalization method. Next we divide the whole range of scores into a certain number of intervals. Then for every interval, we count the number of relevant documents (*num*) and the total score (*t_score*) obtained by all the documents in that interval. In this way the ratio of *num* and *t_score* can be defined for every interval. We compare these ratios to investigate if the score normalization method is a good method or not.

We are not aimed at any particular component system here. A more effective way to understand the scoring mechanism of a particular information retrieval system is to investigate the scoring algorithm used in that system, not just analyse some query results produced by it. Rather than that, we try to find out some collective behaviour of a relatively large number of information retrieval systems in an open environment such as TREC. In such a situation, to analyse these query results becomes a sensible solution. There are a few reasons for this: Firstly, if the number of systems is large, it is a tedious task to understand the implementation mechanisms of all these systems; secondly, it is not possible if the technique used in some of the systems are not published; thirdly, the collective behaviour of these systems may not be clear even we know each of them well.

As in Section 2, we used the same four groups of results. We normalised all these results using the fitting method [0.06-0.6], Zero-one, Sum and ZMUV1 (a variation of ZMUV, which added 1 to every score normalized with ZMUV) over 50 queries. Then for all normalized results with a particular method, we divided them into 20 groups. Each group is corresponding to a interval and the scores of all documents in that group is located in the same interval.

For the fitting method, we divide [0.06,0.6] into 20 equal intervals [0.06,0.087), [0.087,0.114),..., [0.573,0.6]. For Zero-one, we divide [0,1] into 20 equal intervals [0,0.05), [0.05,0.1),..., [0.95,1]. For Sum, we divide [0,1] into 20 intervals [0,0.00005), [0.00005,0.0001), ..., [0.00095,1], since all normalized scores in Sum are very small and most of them are smaller than 0.001. For ZMUV1, 20 intervals (-∞, 0.1), [0.1,0.2),..., [1.9,+∞) are defined. For each group, we calculate its ratio of *num* and *t_score*. All results are shown in Figure 3. The ideal situation is that the curve is parallel to the horizontal axis, which means a linear relation between score and document's relevance and CombSum is a perfect method for

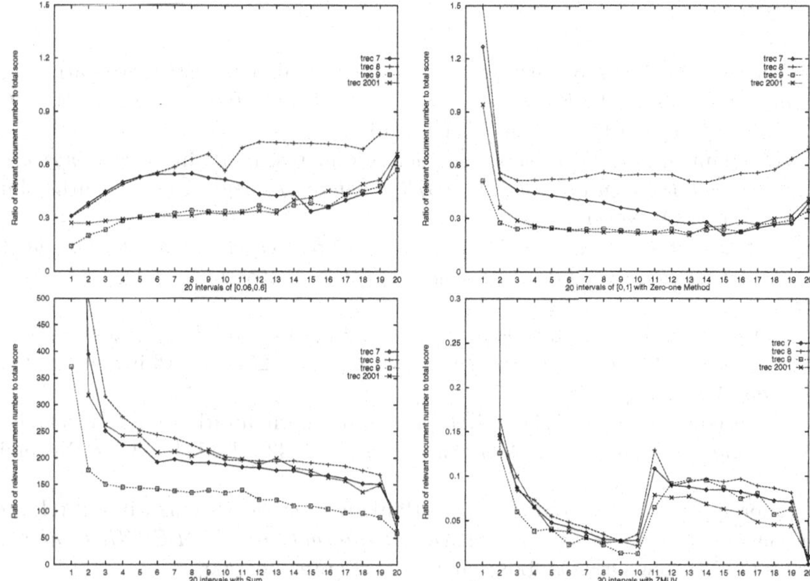

Fig. 3. Score distribution in TREC 7-2001 with different normalization methods

data fusion. It demonstrates that the fitting method is the best since all the curves generated are the most flat. For all other score normalization methods, their corresponding curves are quite flat as well, which suggests that a linear function is a good option to describe the relation between score and document's relevance.

For Zero-one, Sum and ZMUV1, we observe that very often their curves fly into the sky in Intervals 1 and 2. This is understandable since all the normalized scores with Zero-one and Sum are approaching 0 in Intervals 1 and 2. It is not the case for their raw scores and there are a few relevant documents in these two intervals. Therefore the curves are deformed. The situation is even worse for ZMUV1 since negative and positive scores coexist in Interval 1.

4 Conclusion

In this paper, we have evaluated four linear score normalization methods, namely the fitting method, Zero-one, Sum and ZMUV. Comparison analysis and extensive experimentation has been carried out to investigate them. On average, The fitting method and the Zero-one method appear to be the two leading methods. More specifically, the fitting method is more favourable to CombSum, while the Zero-one method is more favourable to CombMNZ. Sum is not as good as the fitting method and Zero-one but outperforms them occasionally. ZMUV always performs very badly and it does not seem to be a proper score normalization method.

References

1. E. Amitay, D. Carmel, R. Lempel, and A. Soffer. Scaling ir-system evaluation using term relevance sets. In *Proceedings of the 27th Annual International ACM SIGIR Conference*, pages 10–17, Sheffield, UK, July 2004.
2. J. A. Aslam and M. Montague. Models for metasearch. In *Proceedings of the 24th Annual International ACM SIGIR Conference*, pages 24–37, Gaithersburg, Maryland, USA, November 2003.
3. E. A. Fox and J. Shaw. Combination of multiple searches. In *The Second Text REtrieval Conference (TREC-2)*, pages 243–252, Gaitherburg, MD, USA, August 1994.
4. J. H. Lee. Analysis of multiple evidence combination. In *Proceedings of the 20th Annual International ACM SIGIR Conference*, pages 267–275, Philadelphia, Pennsylvania, USA, July 1997.
5. M. Montague and J. A. Aslam. Relevance score normalization for metasearch. In *Proceedings of ACM CIKM Conference*, pages 427–433, Berkeley, USA, November 2001.
6. I. Soboroff, C. Nicholas, and P. Cahan. Ranking retrieval systems without relevance judgments. In *Proceedings of 24th Annual International ACM SIGIR Conference*, pages 66–73, New Orleans, Louisiana, USA, September 2001.
7. TREC. http://trec.nist.gov/.
8. S. Wu and F. Crestani. Data fusion with estimated weights. In *Proceedings of the 2002 ACM CIKM International Conference on Information and Knowledge Management*, pages 648–651, McLean, VA, USA, November 2002.
9. S. Wu and F. Crestani. Methods for ranking information retrieval systems without relevance judgments. In *Proceedings of the 2003 ACM Symposium on applied computing (SAC)*, pages 811–816, Melbourne, Florida, USA, March 2003.
10. S. Wu and S. McClean. Performance prediction of data fusion for information retrieval. *Information Processing & Management*, 42(4):899–915, July 2006.

WIDIT: Integrated Approach to HARD Topic Search

Kiduk Yang, Ning Yu, Hui Zhang, Shahrier Akram, and Ivan Record

School of Library and Information Science, Indiana University, Bloomington, Indiana 47405, U.S.A.
{kiyang, nyu, hz3, sakram, irecord}@indiana.edu

1 Introduction

Web Information Discovery Tool (WIDIT) Laboratory at the Indiana University School of Library, whose basic approach to combine multiple methods as well as to leverage multiple sources of evidence, participated in 2005 Text Retrieval Conference's Hard track (HARD-2005) to investigate methods of effectively dealing with HARD topics by exploring a variety of query expansion strategies, the results of which were combined via an automatic fusion optimization process. We hypothesized that the "difficulty" of topics is often due to the lack of appropriate query terms and/or misguided emphasis on non-pivotal query terms by the system. Thus, our first-tier solution was to devise a wide range of query expansion methods that can not only enrich the query with useful term additions but also identify important query terms. Our automatic query expansion included such techniques as noun phrase extraction, synonym identification, definition term extraction, keyword extraction by overlapping sliding window, and Web query expansion. The results of automatic expansion were used in soliciting user feedback, which was utilized in a post-retrieval reranking process. The paper describes our participation in HARD-2005 and is organized as follows. Section 2 gives an overview of HARD track, section 3 describes the WIDIT approach to HARD-2005, and section 4 discusses the results and implications, followed by the concluding remarks in section 5.

2 HARD-2005 Overview

The goal of the TREC's HARD track in 2005 was to achieve "high accuracy retrieval from documents" (i.e. improved retrieval performance at top ranks) against a set of difficult topics using targeted interactions with the searcher[1]. The document collection used in HARD-2005 is the AQUAINT corpus from Linguistic Data Consortium (LDC), and the topic set for HARD-2005 consisted of 50 "difficult" topics, which were selected from a pool of past TREC topics that most systems performed poorly on. HARD participants were first to submit for each topic an automated retrieval result (i.e., Baseline Run) along with an HTML form with which to collect user feedback (i.e., Clarification Form). The Clarification Forms (CF) were then filled out by TREC assessors and the results were sent back to TREC participants to be utilized to improve retrieval performance in the subsequent submissions (Final Run).

[1] See TREC 2005 HARD Track Guidelines at http://ciir.cs.umass.edu/research/hard/guidelines.html

H.T. Ng et al. (Eds.): AIRS 2006, LNCS 4182, pp. 649–658, 2006.

The main question in HARD-2005, with its "difficult" topic set and CF mechanism, is how user feedback can improve retrieval performance of difficult topics. At a first glance, one may be inclined to suggest as a solution *relevance feedback*, which is a well-known user feedback mechanism in IR (Rocchio, 1971; Salton & Buckley, 1990). Since difficult topics tend to retrieve non-relevant documents at top ranks, however, it will be difficult for the user to evaluate enough retrieval results in a short time period to find relevant documents with which to deploy relevance feedback.

3 WIDIT Approach to HARD-2005

We hypothesized that the "difficulty" of topics is often due to the lack of appropriate query terms and/or misguided emphasis on non-pivotal query terms by the system. Thus, our first-tier solution was to devise a wide range of query expansion methods that can not only enrich the query with useful term additions but also identify important query terms. Our automatic query expansion included such techniques as noun phrase extraction, synonym identification, definition term extraction, keyword extraction by overlapping sliding window, and Web query expansion. The results of automatic expansion were used in soliciting user feedback, which was utilized in a post-retrieval reranking process.

For synonym identification, we integrated a sense disambiguation module into WIDIT's synset identification module so that best synonym set can be selected according to the term context. To reduce the noise from word definitions, we applied the overlapping sliding window (OSW) method to multiple definitions harvested from web and extracted the overlapping terms. To extract noun phrase, we combined the results of multiple NLP taggers as well as applying the OSW method. OSW method was also applied to topic fields to identify important topic terms. The Web query expansion method was a slight modification of the PIRC approach (Grunfeld et al., 2004; Kwok et al., 2005).

To produce the optimum baseline results, we merged various combinations of query formulation results and the query expansion results using a weighted sum fusion formula. The fusion weights were determined using previous year's Robust data to train the system via an automatic fusion optimization process, where best performing systems in selected categories (e.g., short query, top 10 systems, etc.) were combined using average precision as fusion weights until the performance gain fell below a threshold.

We viewed the clarification form as both manual query expansion and relevance feedback mechanism. Our clarification form included query term synonyms, noun phrase, and best sentences from top documents of the baseline result. Since difficult topics tend to produce few relevant documents in top ranks, we clustered the results and selected the best sentence from each cluster to include in the CF. In addition to expanding the query with user-selected terms from the clarification form, we also utilized the user's best sentence selection by boosting the rank of the documents in which selected sentences occurred.

WIDIT HARD system consists of five main modules: indexing, retrieval, fusion (i.e. result merging), reranking, and query expansion modules. The overview of WIDIT HARD system architecture is displayed in Figure 1.

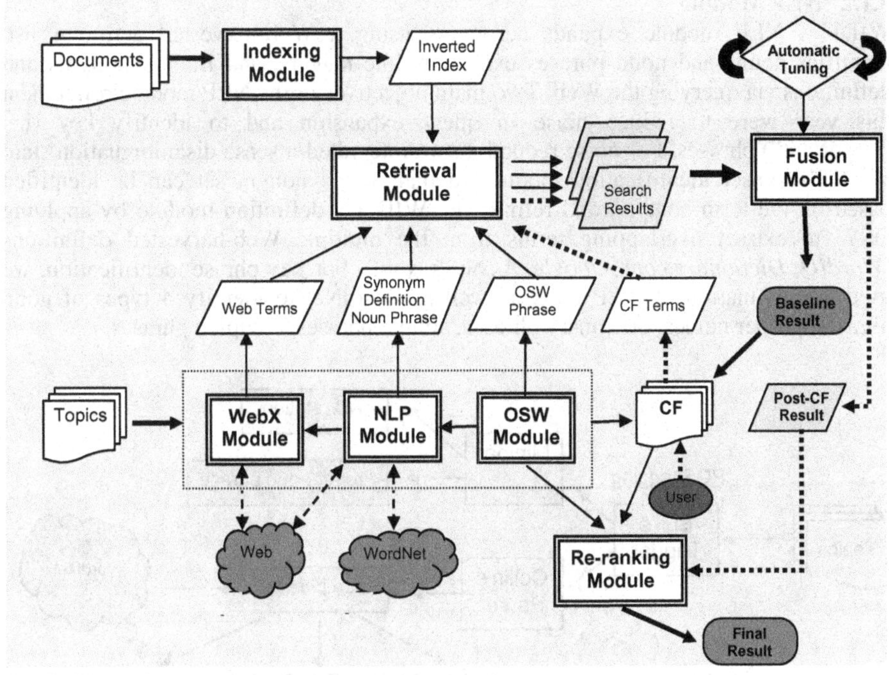

Fig. 1. WIDIT HARD System Architecture

3.1 Query Expansion

The query expansion module consists of three submodules: Web expansion (WebX) module expands the query with terms from Google search results; NLP module finds synonyms and definitions of query terms and identifies nouns and noun phrases in the query; Overlapping Sliding Window (OSW) module extract key phrases from the query.

3.1.1 OSW Module

The main function of OSW module is to identify important phrases. OSW method, which is based on the assumption that phrases appearing in multiple fields/sources tend to be important, works as follows:

i. Define window size and the number or maximum words allowed between words.
ii. Slide window from the first word in the source field/source. For each of the phrase it catches, look for the same/similar phrase in the search fields/sources.
iii. Produce the OSW phrases
iv. Change source field/source and repeat step 1 to 3 till all the fields/sources have been used.

OSW method was applied to topic descriptions and query term definitions to identify key phrases.

3.1.2 NLP Module

WIDIT's NLP module expands acronyms using a Web-harvested acronym list, identifies nouns and noun phrases using multiple taggers, and finds synonyms and definitions via querying the Web. Two main objectives of the NLP module refinement this year were to reduce noise in query expansion and to identify key (i.e. "important") phrases. For noise reduction, we integrated a sense disambiguation[2] into WIDIT's synset identification module so that best synonym set can be identified based on the term context, and refined the WIDIT's definition module by applying OSW to extract overlapping terms from the multiple Web-harvested definitions (*WordIQ*, *Dictionary.com*, *Google*, *Answers.com*). For key phrase identification, we used a combination of NLP tools as well as WordNet to identify 4 types of noun phrases: proper names, dictionary phrases, simple phrases, complex phrases.

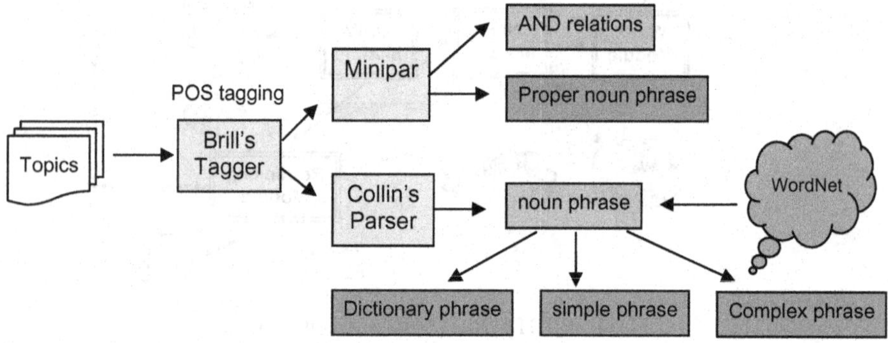

Fig. 2. Noun Phrase Identification Diagram

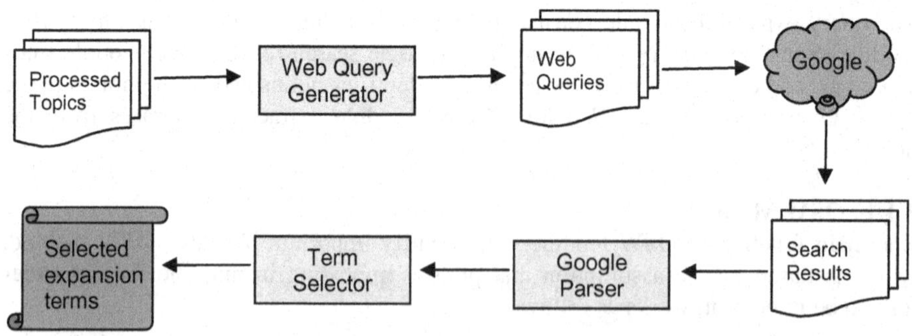

Fig. 3. WebX Module Architecture

3.1.3 WebX Module

The PIRCS group has demonstrated that web expansion is an "effective" method for improving the performance of weak (i.e. difficult) topics (Grunfeld et al., 2004; Kwok et al., 2005). WebX module, which is based on the PIRC approach, expands the query

[2] WordNet word sense disambiguation software developed by the Natural Language Processing Group at the University of Minnesota, Duluth.

with related terms harvested from Google search results. WebX module consists of Web query construction, Web search, search result parsing and term selection (Figure 3). The Web query generator constructs Web queries by selecting up to 5 most salient terms from the processed HARD topics (i.e., stopped and stemmed text, nouns, phrases). The queries are then sent to Google, and subsequent search results (the snippets and the body texts) are parsed to extract up to 60 terms per topic to be used as query expansion terms.

3.2 Fusion

The fusion module combines the multiple sets of search results after retrieval time. In our earlier study (Yang, 2002b), similarity merge approach proved ineffective when combining content- and link-based results, so we devised three variations of the weighted sum fusion formula, which were shown to be more effective in combining fusion components that are dissimilar (Yang, 2002a). Equation (4) describes the simple *Weight Sum* (WS) formula, which sums the normalized system scores multiplied by system contribution weights.:

$$FS_{WS} = \Sigma(w_i * NS_i), \quad (4)$$

where: FS = fusion score of a document,
w_i = weight of system i,
NS_i = normalized score of a document by system i,
= $(S_i - S_{min}) / (S_{max} - S_{min})$

One of the main challenges in using the weighted fusion formula lies in determination of the optimum weights for each system (w_i). Last year, we devised a novel man-machine hybrid approach called the *Dynamic Tuning* to tune the fusion formula (Yang, Yu, & Lee, 2005; Yang & Yu, 2005). This year, we devised another alternative fusion weight determination method called *Automatic Fusion Optimization by Category* (AFOC). AFOC involves iterations of fusion runs (i.e., result merging), where best performing systems in selected categories (e.g., short query, top 10 systems, etc.) are combined using average precision as fusion weights until the performance gain falls below a threshold. The current AFOC implementation does not guarantee true optimization since the process will stop when a local optimum is encountered. Figure 4 illustrates the automatic fusion optimization process.

3.3 Clairfication Form

The main objective of our CF design strategy was to obtain accurate feedback that validates and supplements the system efforts for dealing with difficult topics (e.g. query expansion). Consequently, our Clarification Forms served as manual query expansion and relevance feedback mechanism, which included such components as the selection from candidate expansion terms and phrase, the validation of BoolAnd relations. In addition to displaying important phrases and best sentences from top 200 retrieved documents[3], our CF included synonym sets, definition terms, and query term relations with the use of JavaScript to make the interaction more friendly and

[3] Since weak topics tend to retrieve non-relevant documents at top ranks, we clustered the top 200 documents and selected the best sentence from each cluster.

efficient. The CF terms selected by the user was used to create a CF-expanded query. Phrases, BoolAnd terms and relevant documents identified in CF were used by the reranking module to boosts the ranks of documents with important phrases and relevant documents identified by the user.

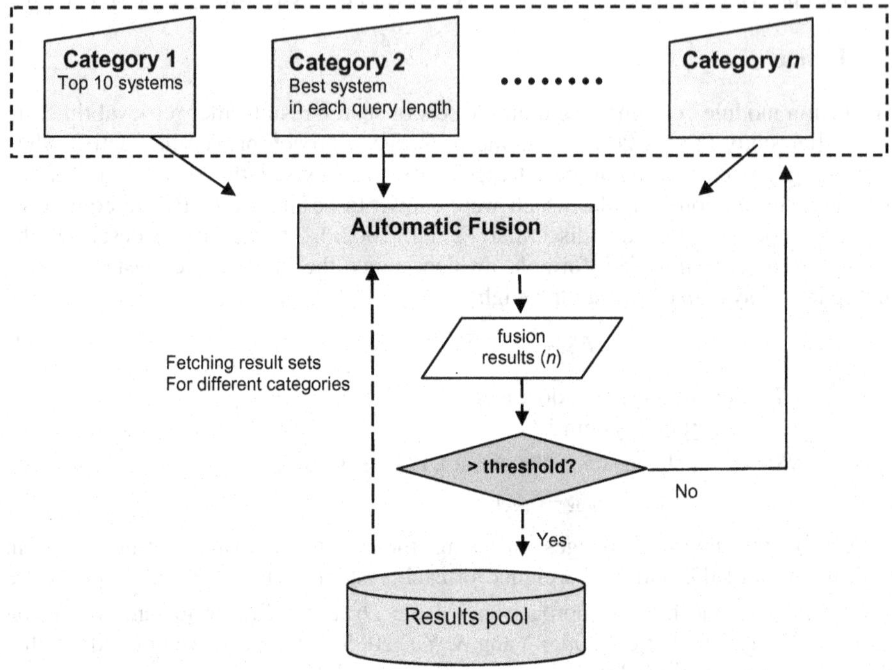

Fig. 4. Automatic Fusion Optimization by Category

3.4 Reranking

The objective of reranking is to float low ranking relevant documents to the top ranks based on post-retrieval analysis of reranking factors. After identifying reranking factors such as *OSW terms*, *CF terms*, and *CF-reldocs*, which are relevant documents identified in CF form, we computed the reranking factor scores (*rf_sc*) for top *k* documents and boosted the ranks of documents with *rf_sc* above a threshold score above a fixed rank R using the following formula:

$$\text{doc\_score} = rf\_sc + \text{doc\_score@rankR} \tag{5}$$

Although reranking does not retrieve any new relevant documents (i.e. no recall improvement), it can produce high precision improvement via post-retrieval compensation (e.g. phrase matching) or force rank-boosting to accommodate trusted information (e.g. *CF-reldocs*).

4 Results

Web query expansion (*WebX*) was the most effective method of all the query expansion methods. Figure 5, which shows Web query expansion results by query length, plots the retrieval performance gain (indicated by the bar above the zero line) or loss (indicated by the bar below the zero line) of various *WebX* methods over non-expansion query results. As can be seen in the figure, *WebX* showed most gain in performance for short queries (i.e. title) but had an adverse effect for longer queries (i.e. description) except when using the rotating window approach (blue, green, and yellow bars).

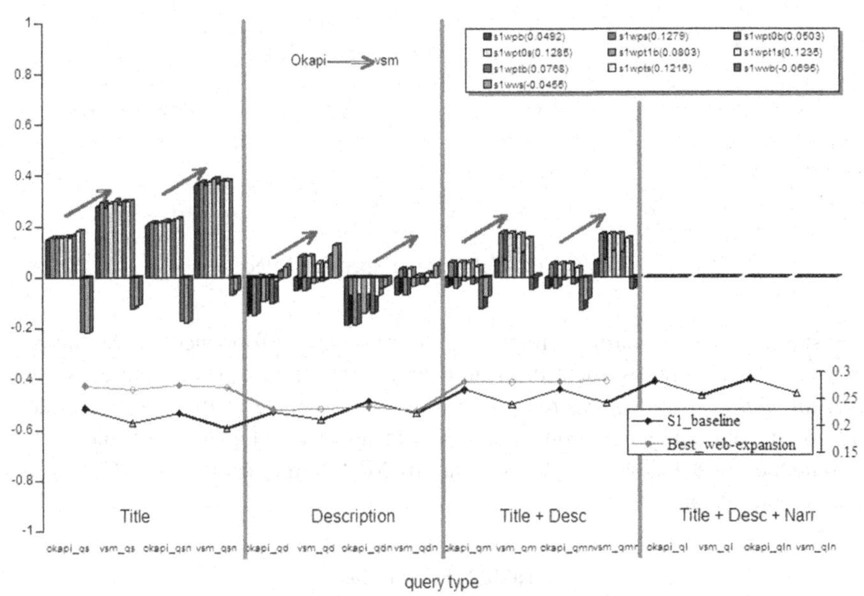

Fig. 5. Web Query Expansion Effect

Among the non-*WebX* query expansion methods, *Proper Noun Phrases*, *OSW Phrases*, and *CF Terms* helped retrieval performance for longer queries, although the rate of performance gains fall much below *WebX* methods (Figure 6). It is interesting to note that the effect of query expansion is influenced by the query length in an opposite manner between *WebX* and non-*WebX* methods. Without query expansion, longer queries usually outperform the shorter queries. With query expansion, however, query length has opposite effect on *WebX* and non-*WebX* methods (i.e., WebX methods works well with short queries, whereas non-WebX methods works better with longer queries). The composite effects of query expansion and query length suggest that *WebX* should be applied to short queries, which contain less noise that can be exaggerated by Web expansion, and non- *WebX* should be applied to longer queries, which contain more information that query expansion methods can leverage.

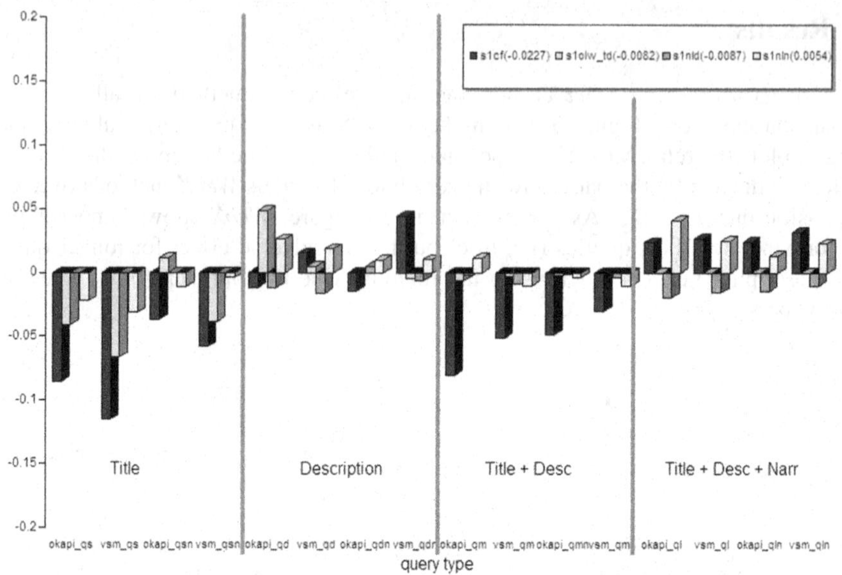

Fig. 6. Non-Web Query Expansion Effect

Fusion (i.e. result merging) improved the retrieval performance across the board with almost 50% improvement in mean average precision for short queries, showing that *Automatic Fusion Optimization by Category* is a viable method to streamline the process of combining numerous result sets in an efficient manner (Table 1). We attribute the lower fusion performance gain by MRP to the fact that *AFOC* used MAP in tuning the fusion formula.

Table 1. Fusion Effect

	Mean Avg. Precision	Mean R-Precision
Baseline Title Run	0.1694	0.2416
Baseline Description Run	0.1698	0.2395
Baseline Fusion Run	0.2324	0.2961
Final Title Run	0.2513 (+48%)	0.3020 (+25%)
Final Description Run	0.2062 (+21%)	0.3020 (+10%)
Final Fusion Run	0.2918 (+25%)	0.3442 (+16%)

Figure 7 shows the effect of reranking factors. The main reranking factors in Figure 7 are *OSW phrases* (*O* in run labels: e.g., RODXX), *CF terms* (C in run labels: e.g., RCDXX), and *CF-reldocs* (D in run labels: e.g., RODXX, RCDXX). Examination of top reranking systems suggests that CF-relevant documents has the most positive effect on retrieval, followed by *OSW* and *CF Terms*.

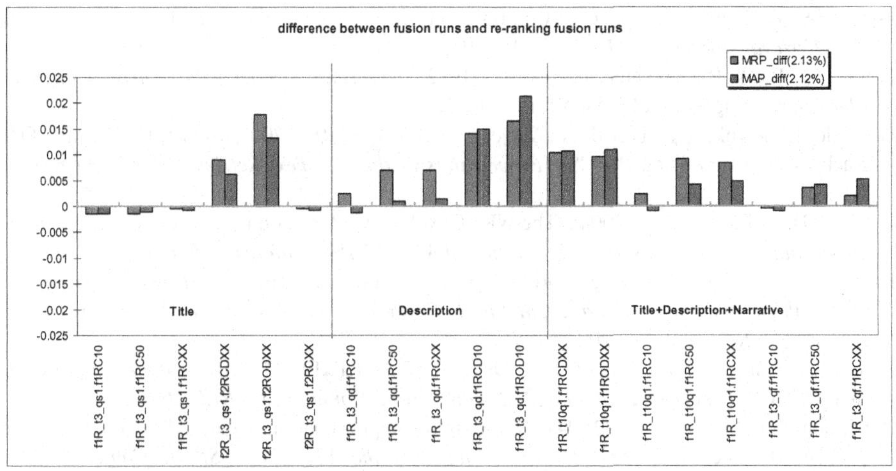

Fig. 7. Reranking Effect

5 Concluding Remarks

We investigated an integrated approach to HARD topic search. To address two major characteristics of difficult topics, we first devised a variety of automatic query expansion methods designed specifically to identify important query terms and phrases as well as to discover new query terms. The automatic query expansion methods, which generated candidate query expansion terms while improving baseline search results, seeded the content of the clarification form, where the system approach to HARD topic search was validated by the user and fed back into the system to further improve the retrieval performance. In keeping with the WIDIT philosophy of fusion, which is capture in the statement "the whole is greater than sum of its parts", we examined the effects of fusion throughout our investigation and found combining of not only data (e.g., short and long queries) and methods (e.g., query expansion methods) but also the system and the user (e.g., construction and utilization of the clarification form) to be quite beneficial. Furthermore, the study revealed the different behaviors by Web-based and non-Web-based query expansion methods when compounded with query length, which suggests the desirability of a flexible system that dynamically adapt its strategies to given situations over a static system. Finally, we devised an effective automatic fusion optimization process that can be deployed in situations where there are too many fusion components to be combined and tuned manually.

References

Buckley, C., Salton, G., & Allan, J., & Singhal, A. (1995). Automatic query expansion using SMART: TREC 3. *Proceeding of the 3rd Text Rerieval Conference (TREC-3)*, 1-19.

Buckley, C., Singhal, A., & Mitra, M. (1997). Using query zoning and correlation within SMART: TREC 5. *Proceeding of the 5th Text REtrieval Conference (TREC-5)*, 105-118.

Fox, E. A., & Shaw, J. A. (1995). Combination of multiple searches. *Proceeding of the3rd Text Rerieval Conference (TREC-3)*, 105-108.

Frakes, W. B., & Baeza-Yates, R. (Eds.). (1992). *Information retrieval: Data structures & algorithms*. Englewood Cliffs, NJ: Prentice Hall.

Grunfeld, L., Kwok, K.L., Dinstl, N., & Deng, P. (2004). TREC 2003 Robust, HARD, and QA track experiments using PIRCS. *Proceedings of the 12th Text Retrieval Conference*, 510-521.

Harman, D. & Buckley, C. (2004). The NRRC Reliable Information Access (RIA) workshop. *Proceedings of the 27th Annual International ACM SIGIR Conference*, 528-529.

Krovetz, R. (1993). Viewing morphology as an inference process. *Proceedings of the Sixteenth Annual International ACM SIGIR Conference on Research and Development in Information Retrieval*, 191-203.

Kwok, K.L., Grunfeld, L., Sun, H.L., & Deng, P. (2005) TREC2004 robust track experiments using PIRCS. *Proceedings of the* 13th *Text REtrieval Conference (TREC 2004).*

Robertson, S. E. & Walker, S. (1994). Some simple approximations to the 2-Poisson model for probabilistic weighted retrieval. *Proceedings of the* 17th *ACM SIGIR Conference on Research and Development in Information Retrieval*, 232-241

Rocchio, J. J., Jr. (1971). Relevance feedback in information retrieval. In G. Salton (Ed.), *The Smart System-- experments in automatic document processing*, 313-323. Englewood Cliffs, NJ: Prentice-Hall, Inc.

Salton, G., & Buckley, C. (1990). Improving retrieval performance by relevance feedback. *Journal of the American Society for Information Science, 41*, 288-297.

Singhal, A., Buckley, C., & Mitra, M. (1996). Pivoted document length normalization. *Proceedings of the ACM SIGIR Conference on Research and Development in Information Retrieval*, 21-29.

Yang, K. (2002a). Combining Text-, Link-, and Classification-based Retrieval Methods to Enhance Information Discovery on the Web. (*Doctoral Dissertation*. University of North Carolina).

Yang, K. (2002b). Combining Text- and Link-based Retrieval Methods for Web IR. *Proceedings of the 10th Text Rerieval Conference (TREC2001)*, 609-618.

Yang, K., & Yu, N. (2005). WIDIT: Fusion-based Approach to Web Search Optimization. *Asian Information Retrieval Symposium 2005.*

Yang, K, Yu, N., & Lee, Y (2005). Dynamic Tuning for Fusion: Harnessing Human Intelligence to Optimize System Performance. *Proceedings of the 9th World Multi-Conference on Systemics, Cybernetics and Informatics.*

Automatic Query Expansion Using Data Manifold

Lingpeng Yang[1], Donghong Ji[1], Yu Nie[1], and Tingting He[2]

[1] Institute for Infocomm Research, 21 Heng Mui Keng Terrace, Singapore 119613
{lpyang, dhji, ynie}@i2r.a-star.edu.sg
[2] Huazhong Normal University, 430079, Wuhan, China
tthe@mail.ccnu.edu.cn

Abstract. This paper proposes an automatic query expansion method that combines document re-ranking and standard Rocchio's relevance feedback. The document re-ranking method ranks the top retrieved documents based on the intrinsic manifold structure collectively revealed by a great amount of data. This is done by using a semi-supervised learning algorithm to integrate pseudo relevant documents with documents to be re-ranked. Given an initial ranked list of retrieved documents, the document re-ranking approach picks a set of documents from the top ones (including query itself) as pseudo relevant documents. In this way, the intrinsic relationship of all the retrieved documents to be re-ranked with the pseudo relevant documents (pseudo irrelevant documents are missing) can be determined via a semi-supervised learning algorithm. Finally, all the retrieved documents can be re-ranked according to above relationship. Evaluation on benchmark corpora show that the approach can achieve much better performance than standard Rocchio's relevance feedback and performance better than other related approaches.

1 Introduction

Automatic query expansion (QE) has been shown to produce good retrieval performance [1, 2, 4, 7] in information retrieval (IR). Many methods have been proposed to improve the effectiveness of automatic query expansion, including term clustering, statistical factor analysis, and document re-ranking [3, 8, 9].

In this paper, we propose a method for automatic query expansion that combines document re-ranking (DR) with Rocchio's classical formula. We first reorder the positions of top initially retrieved documents, and then use standard Rocchio's relevance feedback to do query expansion.

The document re-ranking method using the intrinsic manifold structure of data is derived from semi-supervised learning [10]. The ranking problem is viewed as a special case of semi-supervised learning, in which only positive documents (e.g., query itself) are available while negative documents (irrelevant documents) are missing.

This algorithm first form a weighted network on data, and assign a positive ranking score to each relevant documents and zero to the remaining documents that are to be ranked. All documents then spread their ranking scores to their nearby neighbors via the weighted network. The spread process is repeated until a global stable state is achieved, and all documents are ranked according to their final ranking scores.

H.T. Ng et al. (Eds.): AIRS 2006, LNCS 4182, pp. 659 – 665, 2006.

The rest of this paper is organized as the following. In Section 2, we describe the re-ranking algorithm and automatic query expansion method in detail. Section 3 will give out the experimental results. Finally we present the conclusion in Section 4.

2 Algorithm

2.1 Document Re-ranking

Given a query q and M ranked retrieved documents acquired from the initial retrieval, since we do not know the exact relevant documents, we simply pick top R ones as the pseudo relevant documents (query q itself is always considered as one pseudo relevant document).

In this way, given a query q and M ranked retrieved documents, we can have two necessary sets of data: M documents to be re-ranked, R pseudo relevant documents. As a result, document re-ranking can be achieved by ranking the M documents according to their ranking scores with the R pseudo relevant documents via a semi-supervised algorithm as shown in Figure 1.

Following are some notations for the document reranking algorithm:

- o q : the query
- o $\{r_j\}$ ($1 \le j \le R$): the R pseudo relevant documents
- o $\{m_j\}$ ($1 \le j \le M$): the initial M retrieved documents to be re-ranked
- o $X = \{x_i\}$ ($1 \le i \le R+M$) refers to the union set of the above two categories of documents in the above order, i.e. x_i ($1 \le i \le R$) represents the R pseudo relevant documents $\{r_j\}$($1 \le j \le R$), and x_i ($R+1 \le i \le R+M$) represents the initial M retrieved documents $\{m_j\}$ ($1 \le j \le M$) to be re-ranked.
- o $F = \{f_i\}^T$ ($1 \le i \le R+M$) is a ranking function which assigns each document x_i a ranking value f_i. If we have prior knowledge about the confidences of relevant documents, then we can assign different ranking scores to the relevant documents proportional to their respective confidences.
- o $y = \{y_i\}^T$ ($1 \le i \le R+M$) is a vector where $y_i = 1$ if x_i is a (pseudo) relevant document, and $y_i = 0$ otherwise.

In the document re-ranking algorithm, the manifold structure in X is represented as a connected graph. The connected graph is constructed as following: Sort distance (e.g., Euclidean distance) d_{ij} ($1 \le i, j \le R+M$) between document x_i and x_j in ascending order, and then repeat connecting the two documents with an edge according the order until a connected graph is obtained.

The ranking scores of any vertex in the graph are spread to nearby vertices through weighted edges until a global stable state is achieved. Here, each vertex corresponds to a document, and the edge between any two documents x_i and x_j is weighted by w_{ij} to measure their similarity. Here w_{ij} is defined as follows: $w_{ij} = \exp(-d_{ij}^2/2\sigma^2)$ if $i \ne j$ and there is an edge linking x_i and x_j, and $w_{ii} = 0$ because there are no loops in the graph ($1 \le i,j \le R+M$), where σ is a covariance factor to control the weight W_{ij}. Finally, the $(R+M) \times (R+M)$ affinity matrix $W = [w_{ij}]$ is symmetrically normalized by $S = D^{-1/2}WD^{-1/2}$ in which D is the diagonal matrix with (i, i)-element equal to the sum of the i-th row of W.

Input: q: query;
M: the set/the number of retrieved documents to be re-ranked;
R: the set/the number of pseudo relevant documents;
Algorithm: DocumentReranking(q, M, R)
BEGIN
Set the iteration index t=0
BEGIN DO Loop
Spread ranking score by

$$F(t+1) = \alpha S F(t) + (1-\alpha) y \tag{1}$$

where α is a parameter in [0, 1);
END DO Loop when F converges;
Rank document x_i ($R+1 \leq i \leq R+M$) according its ranking scores f_i (largest ranked first).
END

Fig. 1. algorithm of document re-ranking

The parameter α in algorithm specifies the relative contributions to the ranking scores from neighbors and the initial ranking scores.

This algorithm has been shown to converge to a closed form

$$f^* = \beta(I - \alpha S)^{-1} y \tag{2}$$

where $\beta = 1 - \alpha$ [9]. We can use this closed form to compute the ranking scores of points directly if the data scale is not very large.

2.2 Automatic Query Expansion

After document re-ranking, we use Rocchio's retrieval feedback [5], as improved by Salton and Buckley [6]. Assuming that the relevant documents are the top r documents retrieved and that the information about the non-relevant documents is absent, Rocchio's formula adapted to the retrieval feedback setting becomes:

$$W_{t,Q_{exp}} = \alpha \cdot W_{t,Q_{un\,exp}} + \frac{\beta}{r} \cdot \sum_{k=1}^{r} w_{t,k} \tag{3}$$

Where $W_{t,\,Qexp}$ is the weight assigned to term t after query expansion, and $W_{t,Qunexp}$ and $w_{t,k}$ are the weights of term t in the unexpanded query and in the pseudo-relevant documents, respectively, according to a weighting scheme applied to the whole collection.

3 Experiments and Evaluation

We used the NTCIR-3 CLIR Chinese SLIR subcollection as test data and the 42 D-run query topics as queries. We used the closed form expression (2), where parameter α is fixed as 0.99 and σ is as 1.25. Bi-grams were used as index units and vector space model as retrieval model.

We used the NTCIR's relaxed relevance judgment and rigid relevance judgment to measure the precision of retrieved documents. Relaxed relevance judgments consider highly relevant, relevant, and partially relevant documents, while rigid relevance judgments only consider highly relevant and relevant documents. We used Mean Average Precision (MAP) to measure the overall retrieval performance.

The first experiment tested the effectiveness of the ranking algorithm using only the query itself as pseudo relevant document (*R*=1). Table 2 lists the MAP values for re-ranking different number of top retrieved documents. In Table 2, the first column indicates the number of top documents to be re-ranked, INI refers to the initial retrieval result, +x% denotes the improvement against the baseline [INI], and MAP(relaxed)/MAP(rigid) represent MAP values on relaxed /rigid relevance measure.

From Table 1, we can see that document re-ranking improves MAP(rigid) and MAP(relaxed) by 9.8%-16.1% and 10.4%-12.9% respectively. We also conducted the paired t-tests. Table 1 lists their significance marks for each re-ranking setting. In the following, * correspond to p-values less or equal 0.05, which means significant difference. From Table 1, we can see that the improvement is significant for both MAP(rigid) and MAP(relaxed). This implies that the re-ranking can improve the performance of top retrieved documents significantly.

Table 1. MAPs of DR (*R*=1) and Initial Retrieval

	MAP(rigid)	MAP(relaxed)
INI	0.1688	0.2197
M=20	0.1853 +9.8%*	0.2428 +10.5%*
M=25	0.1867 +10.6%*	0.2426 +10.4%*
M=30	0.1916 +13.5%*	0.2434 +10.8%*
M=35	0.1917 +13.6%*	0.2436 +10.9%*
M=40	0.1891 +12.0%*	0.2429 +10.5%*
M=50	0.1888 +11.8%*	0.2428 +10.5%*
M=60	0.1885 +11.7%*	0.2427 +10.4%*
M=70	0.191 +13.2%*	0.2438 +11.0%*
M=80	0.1955 +15.8%*	0.2448 +11.4%*
M=90	0.1937 +14.8%*	0.2449 +11.5%*
M=100	0.1960 +16.1%*	0.2480 +12.9%*

From the closed form expression (2), we can see that the re-ranking of the documents depends completely on their intrinsic data structure. In other words, the initial ranking has no influence in the re-ranking. This is in contradictory with the traditional view that top documents in initial retrieval are generally important in IR. To incorporate the roles of initial retrieval, the pseudo relevant documents could be expanded from the top documents, and then the closed form expression could be used for the new relevant document sets. In the experiment, we simply considered top 10 documents as pseudo relevant documents and used them for re-ranking. Table 2 lists the MAP values for re-ranking different number of top retrieved documents by the new pseudo relevant document set.

Table 2. MAPs of DR: top 10 Documents as Pseudo Relevant Documents

	MAP(rigid)	MAP(relaxed)
INI	0.1688	0.2197
M=20	0.1973 +16.9%*	0.2617 +19.1%*
M=25	0.1956 +15.9%	0.256 +16.5%
M=30	0.1987 +17.1%	0.2544 +15.8%
M=35	0.1975 +14.8%	0.2575 +17.2%*
M=40	0.1941 +15%	0.2560 +16.5%
M=50	0.1946 +15.3%*	0.2564 +16.7%
M=60	0.1990 +17.9%	0.2570 +17%*
M=70	0.1994 +18.1%*	0.2575 +17.2%*
M=80	0.2118 +25.5%*	0.2595 +18.1%*
M=90	0.2129 +26.1%*	0.2625 +19.5%*
M=100	0.2149 +27.3%*	0.2656 +20.9%*

From Table 2, we can see that incorporation of top 10 documents as pseudo relevant documents produced better results than only original query topic q as pseudo relevant document. This implies that initial retrieval helps the pure manifold-based re-ranking.

To see whether the MAP improvement against the original query q as pseudo relevant document, Table 2 lists the t-test significance marks. From Table 2, we can see that the improvement against the original query as pseudo relevant document was significant in most cases, especially when more documents were used for re-ranking.

Now that document re-ranking can improve the performance of initial search, it should help query expansion. To confirm it, we combined the re-ranking with standard Rocchio's relevance feedback. That is, we first re-ranked top 100 retrieved documents, and then applied standard Rocchio's relevance feedback on re-ranked top documents. In the experiment, we selected 200 bi-grams from the top N (N=15, 20, 25 or 30) retrieved documents to do query expansion.

Table 3 gives the MAP values for standard QE (Rocchio) and extended QE (re-ranking + Rocchio), where the first column indicate the number of the top documents used for query expansion.

Table 3. Standard QE and Extended QE

	Standard QE		Extended QE	
	MAP(rigid)	MAP(relaxed)	MAP(rigid)	MAP(relaxed)
INI	0.1688	0.2197	0.1688	0.2197
N=15	0.2196 +30.1%	0.2836 +29.1%	0.2491 +47.6%*	0.3118 +41.9%*
N=20	0.2229 +32.0%	0.2853 +29.9%	0.2513 +48.9%*	0.3127 +43%*
N=25	0.2216 +31.3%	0.2843 +29.4%	0.2520 +49.3%*	0.3166 +44.1%*
N=30	0.2208 +30.8%	0.2839 +29.2%	0.2518 +49.2%*	0.3157 +43.7%*

From Table 3, we can see that using Rocchio's standard QE improved MAPs by 29.1%-32% against initial search, while using extended QE achieved better results with improvement by 41.9%-49.3%. This indicates that the re-ranking helps query expansion to improve the performance.

To see whether the MAP difference between extended QE and standard QE is significant, Table 3 also lists the significance marks, which suggests that extended QE significantly outperforms standard QE.

To further check the effectiveness of extended QE, we compare our method with Mitra's query expansion [4], which uses Maximal Marginal Relevance module to adjust term correlation to re-order retrieved documents before standard query expansion. Table 4 gives the MAP values for Mitra's method and extended QE, where the first column indicate the number of the top documents re-ranked before doing standard query expansion. From Table 4, when *M* is small (*M*<40), our method gets comparable results than Mitra's method; when *M* is larger (*M*>=40), our method gets better results.

Table 4. Comparison with Mitra's Query Expansion method (*N*=20)

	Mitra's QE		Extended QE	
	MAP(rigid)	MAP(relaxed)	MAP(rigid)	MAP(relaxed)
INI	0.1688	0.2197	0.1688	0.2197
M=30	0.2378 +40.8%	0.3011 +37.1%	0.2376 +40.7%˜	0.3004 +36.8%˜
M=40	0.2341 +38.6%	0.2989 +36.0%	0.2479 +46.9%*	0.3098 +41.0%*
M=50	0.2315 +37.1%	0.2937 +33.6%	0.2501 +48.2%*	0.3133 +42.6%*

4 Conclusions

In this paper, we propose an automatic query expansion method that combines document re-ranking and standard Rocchio's relevance feedback. The document re-ranking method is based on semi-supervised learning algorithm, which integrates top retrieved documents with the query in learning process by representing them as vertices in a weighted graph and iteratively propagating the ranking information from any vertex to nearby vertices until this process converges.

Our analysis and experimental results demonstrates the potential of this manifold learning based algorithm in information retrieval. The extended QE achieves significant improvement over standard Rocchio's relevance feedback, and the re-ranking method itself gets significant improvement against the initial retrieval.

References

1. Carpineto C. Demori, R., Romano, G. Bigi B. An Information-Theoretic Approach to Automatic Query Expansion. ACM Transactions on Information Systems, Vol. 19, No. 1, 2001, Pages 1-27.
2. Crouch, C., Crouch, D., Chen, Q., Holtz, S. Improving the Retrieval Effectiveness of Very Short Queries, Information Processing and Management, 38(2002).
3. Kurland O., Lee L. PageRank without Hyper-links: Structural Re-ranking using Links Induced by Language models. In the Proceedings of the 28th Annual International ACM SIGIR Conference on Research and Development in Information Retrieval. 2005.

4. Mitra M., Singhal A., Buckley C. Improving Automatic Query Expansion. In the proceedings of the 21th Annual International ACM SIGIR Conference on Research and Development in Information Retrieval. 1998.
5. Rocchio, J. Relevance Feedback in Information Retrieval. In the SMART retrieval system – Experiments in Automatic Query Expansion, G.Salton, Ed., Prentice Hall, Englewood Cliffs, N.J. 1971.
6. Salton, G., Buckley, C. Improving retrieval performance by relevance feedback. Journal of the American Society of Information Science. 41, 288-297. 1990.
7. Xu, J., Croft, B. Improving the Effectiveness of Information Retrieval with Local Context Analysis. ACM Transactions on Information Systems. 18, 1, 79-112. 2000.
8. Yang L.P., Ji D.H., Leong M.K. Chinese Document Re-ranking Based on Term Distribution and Maximal Marginal Relevance. In proceeding of the Second Asia Information Retrieval Symposium. LNCS 3689, pp. 299-311. 2005.
9. Zhang B.Y, .Li H., Liu Y., Ji L., Xi W., Fan W., Chen Z., Ma W. Improving Search Results using Affinity Graph. In the Proceedings of the 28th Annual International ACM SIGIR Confer-ence on Research and Development in Information Retrieval. 2005.
10. Zhou, D.Y., Weston J., Gretton A., Bousquet O., Schölkopf B. Ranking on Data Manifolds. Advances in Neural Information Processing Systems 16, 169-176. 2004.

An Empirical Comparison of Translation Disambiguation Techniques for Chinese–English Cross-Language Information Retrieval

Ying Zhang, Phil Vines, and Justin Zobel

School of Computer Science and Information Technology, RMIT University, GPO Box 2476V, Melbourne, Australia

Abstract. Disambiguation techniques are typically employed to reduce translation errors introduced during query translation in cross-lingual information retrieval. Previous work has used several techniques — based on term similarity, term co-occurrence, and language modelling. However, the previous experiments were conducted on different data sets, and thus the relative merits of each technique is presently unclear. The goal of this work is to compare the effectiveness of these techniques on the same Chinese–English data sets. Our results show that despite the different underlying models and formulae used, the aggregated results are comparable. However, there is wide variation in the translation of individual queries, suggesting that there is scope for further improvement.

1 Introduction

When using simple dictionary translations without addressing the problem of translation ambiguity, the effectiveness of cross-language information retrieval (CLIR) can be 60% lower than that of monolingual retrieval [1]. Translation ambiguity stems from the fact that many words do not have a unique translation, and sometimes the alternate translations have very different meanings. This problem is particularly severe in view of the observed tendency of web users to enter short queries; it is difficult for even a human to reliably determine the intended meaning from the available context. The dictionary-based translation approaches are prone to error due to the likelihood of selecting the wrong translation of a query term among alternatives provided by the dictionary. Techniques, that use statistics obtained from training corpora have been proposed to reduce the ambiguity and errors introduced during query translation. However, use of different data sets and language pairs has meant that it has not been possible to draw clear conclusions about the relative merits of the different disambiguation techniques. By using the same data sets, we aim to compare and understand the merits of the different disambiguation techniques.

2 Background

Amongst the various approaches to CLIR, translation of the query language to that of the document language has been most commonly used, and researchers

H.T. Ng et al. (Eds.): AIRS 2006, LNCS 4182, pp. 666–672, 2006.

have applied dictionary-based translation methods with success. Web queries are generally short, and thus sophisticated natural language processing techniques can flounder due to lack of context. However, any dictionary-based technique must address the fact that many words have multiple translations. This disambiguation phase is crucial to the translation result. Interestingly, researchers have used different disambiguation techniques utilizing statistics obtained from the test collection, all seemingly with good results. These have included using term similarity [2], co-occurrence statistics in the target document collection [1,3,4,5,6], and probabilistic methods based on a language model [7,8].

As these previous experiments were generally carried out on different test collections, it is unclear whether any particular approach is superior. The goal of this work is to compare the effectiveness of these techniques on the same data sets. We first review the different approaches that have been used.

2.1 Term Similarity

Adriani [2] proposed a disambiguation technique based on the concept of statistical term similarity. A term-similarity matrix was built using the statistical term-distribution parameters obtained from the corpus terms. The term similarity is obtained using the *Dice similarity coefficient* (DSC) — a term association measure commonly used in clustering. The term similarity value between term x and y, $\mathrm{SIM}(x, y)$, is calculated as follows:

$$\mathrm{SIM}(x,y) = 2\sum_{i=1}^{n}(w'_{xi} \times w'_{yi}) \Big/ \left(\sum_{i=1}^{n} w_{xi}^2 + \sum_{i=1}^{n} w_{yi}^2\right)$$

where

w_{xi} = the weight of term x in document i
w_{yi} = the weight of term y in document i
w'_{xi} = w_{xi} if document i also contains term y, or 0 otherwise
w'_{yi} = w_{yi} if document i also contains term x, or 0 otherwise
n = the number of documents in the collection

The term weight w_{xi} of term x in document i is computed using the standard `tf.idf` weighting formula. This algorithm computes the sum of maximum similarity values between each candidate translation of a term and the translations of all other terms in the query. They used one document as the window of co-occurrence. For each query term, the translation with the highest sum is selected as its translation. The results of their Indonesian–English and English–Indonesian CLIR experiments demonstrated the effectiveness of this disambiguation technique. Gao at al. [3] employed a similar approximate algorithm for choosing optimal translations in English–Chinese CLIR.

2.2 Term Co-occurrence

Ballesteros and Croft [1] used co-occurrence statistics obtained from the target corpus for translation disambiguation. Their hypothesis is that the correct translations of query terms should co-occur in target language documents and incorrect translations should tend not to co-occur. Other studies including [4,5,6] also used similar approaches to select the best translation(s). Building on the work [1,4,6], Gao et al. [3] observed that the correlation between two terms is stronger when the distance between them is shorter. They extended the previous co-occurrence model by incorporating a distance factor $D(x,y) = e^{-\alpha(\mathrm{Dis}(x,y)-1)}$. The mutual information (MI) between term x and y, $\mathrm{MI}(x,y)$, is calculated as follows:

$$\mathrm{MI}(x,y) = \log\left(\frac{f_w(x,y)}{f_x f_y} + 1\right) \times D(x,y)$$

where $f_w(x,y)$ is the frequency with which x and y co-occur within a window size of w in the collection; f_x is the collection frequency of x; and f_y is the collection frequency of y. $D(x,y)$ decreases exponentially when the distance between two terms x and y increases, where α is the decay rate, which is empirically set as 0.8 ; and $\mathrm{Dis}(x,y)$ is the average distance between x and y in the collection. They experimented on the TREC-9 Chinese collection and showed that the addition of the distance factor leads to substantial improvements over the basic co-occurrence model.

2.3 Language Modelling

Given a query s_1, s_2, s_3 ..., s_n, each translation candidate set T is a sequence of words t_1, t_2, t_3 ..., t_n. In previous work [8], we used a probability model $P(T) = P(t_1, t_2, t_3 \,..., \, t_n)$ to estimate the maximum likelihood of each sequence of possible translations and selected the translation set T with the highest probability value $P(T)$ among all possible translation sets. Our disambiguation technique is based on a bigram *hidden Markov model* (HMM), such models have been used widely for probabilistic modelling of sequence data:

$$P(t_1, t_2, t_3 ... \, t_n) = P(t_1) \prod_{a=2}^{n} P(t_a | \, t_{a-1})$$

To compute the probability of a sequence of words, we need to calculate the quantities $P(t)$, the probability of word t, and $P(t|t')$, the probability of t in the context of t', as follows:

$$P(t) = \frac{f(t)}{N}\ , \ P(t|t') = \frac{P_w(t,t')}{\sum_{t''} P_w(t'',t')}$$

where $f(t)$ is the collection frequency of term t, N is the number of terms in the document collection, and $P_w(t,t')$ is the probability of term t' occurring after term t within a window of size w. The zero-frequency problem arises in the context of probabilistic language models, when the model encounters an event in a context in which it has not been seen before. Smoothing provides a way to estimate and assign the probability to that unseen event. We use the following absolute discounting and interpolation formula, which applies the smoothing method proposed by [7]. In this method,

$$P(t|t') = \max\left\{\frac{f_w(t,t') - \beta}{N}, 0\right\} + \beta P(t)P(t')$$

where $f_w(t,t')$ is the frequency of term t' occurring after term t within a window size w. [7] successfully used this formula to compute the frequency of term t' and t within a text window of fixed size through an order-free bigram language model in Italian–English CLIR. The absolute discounting term β is equal to the estimate proposed by [9]:

$$\beta = \frac{n_1}{n_1 + 2n_2}$$

where n_k is the number of terms with collection frequency k.

3 Experiments

To test the effectiveness of Chinese–English CLIR using the different techniques described in Section 2, we conducted a set of experiments using the data sets of NTCIR 4 and 5. A Chinese topic contains four parts: *title*, *description*, *narrative*, and *key words* relevant to the whole topic. We chose to use the titles of the Chinese topics only as queries for two reasons: first, ambiguity problem is often resolved implicitly when queries are long enough (the additional words provide sufficient context to resolve confusion) but is still a critical problem when queries are short [10]; second, web queries are often short, and the average length of the titles approximates the web queries.

Our experiments consist of four runs: a mono-lingual reference in RUN_{mono}, the term similarity technique in RUN_{ts}, the term co-occurrence technique in RUN_{tc}, and our technique based on a HMM in RUN_{lm}. Our experiments used the Lemur IR system[1] developed by the Computer Science Department at the University of Massachusetts and the School of Computer Science at Carnegie Mellon University.

The relevance judgements provided by NTCIR are at two levels — strictly relevant documents known as *rigid relevance*, and documents that are likely to be relevant, known as *relaxed relevance*. In this paper, we used the rigid relevance judgements provide by NTCIR to report our results.

4 Results and Analysis

A comparison of the characteristics of several translation disambiguation techniques is tabulated in Table 1. The term similarity technique uses both `tf` and `idf`, and selects the best translation of each query term from multiple candidates by comparing their statistical associations with the candidate translations of all other query terms within each document. The term co-occurrence technique makes use of `tf` and of the frequency of terms co-occurring within a window size of w in the collection. In addition, a distance factor is incorporated to discriminate strong and weak term correlation. Our language modelling technique uses

[1] http://www.lemurproject.org/

Table 1. A comparison of the characteristics of different translation disambiguation techniques

	RUN$_{ts}$	**RUN**$_{tc}$	**RUN**$_{lm}$
Model	Term similarity (DSC)	Term co-occurrence (MI)	Language modelling (Bigram HMM)
Co-occurrence frequency	×	√	√
Term distance	×	√	×
Term order	×	×	√
`tf.idf` weighting	√	`tf` only	`tf` only
Window size	A document	A sentence	4 terms
Translation selection	The translations of all other query terms are considered.		Only the translations of the immediately preceding query term are considered.

Table 2. Chinese–English CLIR results using different disambiguation techniques

	NTCIR-5 data set				**NTCIR-4 data set**			
	RUN$_{mono}$	RUN$_{ts}$	RUN$_{tc}$	RUN$_{lm}$	RUN$_{mono}$	RUN$_{ts}$	RUN$_{tc}$	RUN$_{lm}$
MAP	0.363	0.312	0.316	0.312	0.242	0.199	0.198	0.200
% Mono	—	85.9	87.0	85.9	—	81.9	81.8	82.5
Recall	0.812	0.776	0.779	0.765	0.612	0.521	0.545	0.529
P@10	0.476	0.408	0.414	0.380	0.403	0.331	0.326	0.335
R-precision	0.527	0.318	0.322	0.312	0.289	0.229	0.225	0.240

`tf` and the frequency of terms co-occurring within a window size of w together with an absolute discounting model for smoothing. When using a bigram HMM language model, only the co-occurrences of the candidate translations of adjacent query terms are considered. Interestingly the language modelling disambiguation technique increases in complexity in accordance with the product of the number of possible translations for each query term, and thus becomes impractical for longer queries. Each of the techniques uses different models, formulae and parameters; nonetheless each achieved comparable results across multiple data sets, as shown in Table 2.

We used a *t-test* to analyze the statistical significance of our results. These results suggest that there is no significant difference in the overall effectiveness on the data sets tested. However, analysis of the individual queries reveals a different story. Of the 93 NTCIR 4 and 5 queries with ambiguity (at least one query term has multiple translations), the three different disambiguation techniques produced the same translation in only 19% of the queries (10 out 50 NTCIR-4

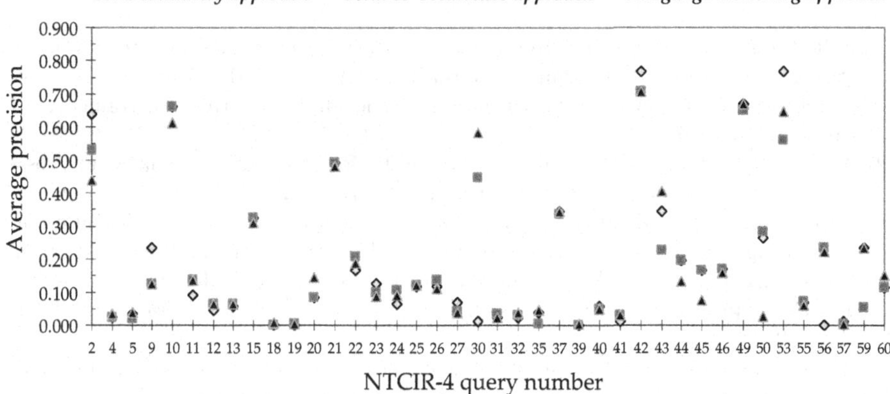

Fig. 1. An sample of different average precision values generated by different disambiguation techniques for individual queries

queries and 8 out of 43 NTCIR-5 queries), and manual inspection showed that these were indeed correct translations. In the other 81% of queries with differing translations, there was often significant difference in the average precision results for each of the approaches. An indicative sample of the differing results for the 50 NTCIR-4 ambiguous queries is shown in Figure 1. In some cases there were only minor differences in average precision. For example, in query 40, the Chinese term "问题" was variously translated as "problem" and "issue". This term has a low weight and the other more significant terms — "足球"(soccer), "世界杯"(world cup), and "门票"(ticket) — were correctly translated. However in other cases, one or more technique has effectively failed, as a result of selecting a totally inappropriate translation. For example, in query 30 "动物复制技术", the term similarity approach translated the query as "animal copy skill", whereas the other methods correctly translated the query as "animal clone technology".

5 Conclusion

This research has shown that, superficially, all these translation disambiguation methods are comparable (no significant difference) when averaged across a query set. However, at the individual query level, each of the techniques frequently produce differing results. This means that it may be possible to develop a new approach that combines the best elements of these current methods. At the very least, it should be possible to use a "combination of evidence" approach [11] to improve the overall translation quality. Further, when all methods produce the same translation, we can have a higher degree of confidence in the correctness, and differing translations could act as a trigger for further analysis.

References

1. Ballesteros, L., Croft, W.B.: Resolving ambiguity for cross-language retrieval. In: Proceedings of the 21st Annual International ACM SIGIR Conference on Research and Development in Information Retrieval, Melbourne, Australia, ACM Press (1998) 64–71
2. Adriani, M.: Using statistical term similarity for sense disambiguationin cross-language information retrieval. Information Retrieval **2** (2000) 71–82
3. Gao, J., Zhou, M., Nie, J.Y., He, H., Chen, W.: Resolving query translation ambiguity using a decaying co-occurrence model and syntactic dependence relations. In: Proceedings of the 25th Annual International ACM SIGIR Conference on Research and Development in Information Retrieval, Tampere, Finland, ACM Press (2002) 183–190
4. Hiemstra, D., de Jong, F.: Disambiguation strategies for cross-language information retrieval. In: Proceedings of the 3rd European Conference on Research and Advanced Technology for Digital Libraries, Paris, France, Springer-Verlag (1999) 274–293
5. Kwok, K.L.: Exploiting a chinese-english bilingual wordlist for english-chinese cross language information retrieval. In: Proceedings of the 5th International Workshop on Information Retrieval with Asian Languages, Hong Kong, China, ACM Press (2000) 173–179
6. Jang, M.G., Myaeng, S.H., Park, S.Y.: Using mutual information to resolve query translation ambiguities and query term weighting. In: Proceedings of the 37th Annual Meeting of the Association for Computational Linguistics on Computational Linguistics, College Park, Maryland, Association for Computational Linguistics (1999) 223–229
7. Federico, M., Bertoldi, N.: Statistical cross-language information retrieval using n-best query translations. In: Proceedings of the 25th Annual International ACM SIGIR Conference on Research and Development in Information Retrieval, Tampere, Finland, ACM Press (2002) 167–174
8. Zhang, Y., Vines, P., Zobel, J.: Chinese oov translation and post-translation query expansion in chinese–english cross-lingual information retrieval. ACM Transactions on Asian Language Information Processing **4** (2005) 57–77
9. Ney, H., Essen, U., Kneser, R.: On structuring probabilistic dependences in stochastic language modelling. Computer Speech and Language **8** (1994) 1–38
10. Sanderson, M.: Word sense disambiguation and information retrieval. In: Proceedings of the 17th Annual International ACM SIGIR Conference on Research and Development in Information Retrieval, Dublin, Ireland, ACM Press (1994) 142–151
11. Smets, P.: The combination of evidence in the transferable belief model. IEEE Transaction on Pattern Analysis and Machine Intelligence **12** (1990) 447–458

Web Mining for Lexical Context-Specific Paraphrasing

Shiqi Zhao, Ting Liu, Xincheng Yuan, Sheng Li, and Yu Zhang

Information Retrieval Laboratory, School of Computer Science and Technology, Box 321, Harbin Institute of Technology, Harbin, P.R. China, 150001
{zhaosq, tliu, xcyuan, lis, yzhang}@ir.hit.edu.cn

Abstract. In most applications of paraphrasing, contextual information should be considered since a word may have different paraphrases in different contexts. This paper presents a method that automatically acquires lexical context-specific paraphrases from the web. The method includes two main stages, candidate paraphrase extraction and paraphrase validation. Evaluations were conducted on a news title corpus whereby the context-specific paraphrasing method was compared with the Chinese synonymous thesaurus. Results show that the precision of our method is above 60% and the recall is above 55%, which outperforms the thesaurus significantly.

Keywords: context-specific paraphrasing, web mining.

1 Introduction

Lexical paraphrasing aims to acquire paraphrases of words, which is elementary but very important in many NLP applications. For instance, in Question Answering (QA), paraphrases should be detected in question and answer sentences so that the exact answers can be pinpointed. In automatic evaluation of Machine Translation (MT), lexical paraphrases need to be recognized in order to evaluate the systems' translation results more accurately. In Information Extraction (IE), paraphrases of the words in IE patterns should be identified so as to extract the required information from texts.

Two broad approaches to lexical paraphrasing have dominated the literature. One approach acquires paraphrases from dictionaries, such as WordNet in English [2], [6] and Tongyici Cilin in Chinese [7]. The other approach collects lexical paraphrases from monolingual or bilingual corpora. Lin identified words with similar meaning by measuring the similarity of the contextual words [8]. Barzilay and McKeown extracted paraphrases from a corpus of multiple English translations of the same source text [3]. Bannard and Callison-Burch derived paraphrases using bilingual parallel corpora [1]. Wu and Zhou extracted lexical paraphrases with multiple resources, including a monolingual dictionary, a bilingual corpus, and a large monolingual corpus [9].

These methods facilitate the acquisition of paraphrases. However, none of them specify the contexts in which the derived paraphrases can be adapted. Recently, topic adaptation for paraphrasing has been researched. For example, Kaji and Kurohashi selected lexical paraphrases according to different topics [5]. However, the topics are limited and predefined rather than any given context.

H.T. Ng et al. (Eds.): AIRS 2006, LNCS 4182, pp. 673 – 679, 2006.

This paper addresses the problem of context-specific paraphrasing. Here, a specific context means a sentence in which a word occurs. A new web mining method is presented to extract lexical context-specific paraphrases. In our method, if a word occurs in different sentences, different paraphrases should be extracted according to each sentence.

2 Method

2.1 Candidate Paraphrase Extraction

Two stages are included: candidate paraphrase extraction and paraphrase validation. The method for candidate paraphrase extraction is based on two principles. The first is authors on the web create information independently, thus their "vocabularies" vary greatly [4]. In other words, if a concept is widely discussed on the web, then various expressions (lexical paraphrases) will be found. The other principle is that lexical paraphrases play similar syntactic roles in sentences, which indicates that paraphrases of a given word w in sentence S can be derived by extracting words whose syntactic roles are similar with w. Three main steps are included in candidate paraphrase extraction:

Step1: Query S on the web and retrieve similar sentences. In this step, the sentence S is searched on the web using Baidu. From the retrieved snippets, sentences whose similarities with S exceed a predefined threshold T_{CE} are retained as candidate sentences. Word overlapping rate (WOR) is used here for computing the similarity between S and any candidate sentence S':

$$WOR(S, S') = \frac{|WS(S) \cap WS(S')|}{\max(|WS(S)|, |WS(S')|)} \tag{1}$$

where "*WS(.)*" denotes the set of words in a sentence. "|.|" denotes the cardinality of a set.

Step2: Extract candidates according to syntactic similarity. In this step, sentence S and all the candidate sentences are first parsed by a Chinese dependency parser. In a dependency result, two words and their dependency relation are represented as a triple. For example, "<他, SBV, 喜┌ >" is a triple. The criterion shown in Fig. 1 is used for extracting candidate paraphrases.

Step3: Filter candidates using ECilin. HIT IR-Lab Tongyici Cilin (Extended) (ECilin for short), [1] is utilized here for filtering incorrect candidates. ECilin is a Chinese synonym dictionary. In ECilin, each word has a sense code and all sense codes are organized into a hierarchy that contains five levels. At the first level, words are classified into 12 classes, while at the fifth level thousands of classes are formed with synonyms put in the same class. In this stage, the first level of ECilin is used for candidate filtering. In detail, if w' and w are not in the same class defined in ECilin's first level, then w' is filtered. In addition, if a candidate is a stopword, then it is filtered.

---

[1] ECilin is an extended version of TongyiciCilin, which was developed by Information Retrieval Lab of Harbin Institute of Technology (http://www.ir-lab.org/).

Given:

S: original sentence;
S': candidate sentence;
$DT(S)$: dependency tree of S;
$DT(S')$: dependency tree of S';
$<w_1, rel, w_2>$: a triple in $DT(S)$;
$<w_1', rel', w_2'>$: a triple in $DT(S')$.

Criterion:

If $rel=rel'$ and $w_2=w_2'$, then w_1' is extracted as a candidate paraphrase of w_1.
If $w_1=w_1'$ and $rel=rel'$, then w_2' is extracted as a candidate paraphrase of w_2.

Fig. 1. Criterion for candidate paraphrase extraction

2.2 Paraphrase Validation

Though the obtained candidates are filtered, there still remain a lot of incorrect candidates. Therefore, a method for validating candidate paraphrases is necessary. Let $w_1, \ldots, w_n$ be n candidate paraphrases of word w in sentence S. We generate n new sentences $S_1, \ldots, S_n$ first, in which S_i ($1 \le i \le n$) is generated by replacing w in S with candidate w_i. Intuitively, if searching S and S_i using a search engine can retrieve similar results (snippets), then one can say that S and S_i are similar in meaning. Accordingly, word w and w_i are similar in this specific context.

2.2.1 Assumption for Paraphrase Validation

For word w in sentence S (and each candidate w_i in S_i), a pseudo document $PD_S(w)$ ($PD_{Si}(w_i)$) is constructed. First, Search S (S_i) in Baidu, and obtain top 100 retrieved snippets. Then, sentences containing w (w_i) are extracted from the snippets. $PD_S(w)$ ($PD_{Si}(w_i)$) is constructed using these extracted sentences. In this work, paraphrases are validated based on the following assumption:

Assumption: Given w in S, and a candidate paraphrase w_i in S_i, if the similarity between their pseudo documents $PD_S(w)$ and $PD_{Si}(w_i)$ exceeds a predefined threshold T, then w_i is validated as w's paraphrase within the specific sentence S.

2.2.2 Similarity Measurements for Pseudo Documents

According to the assumption, a similarity measurement is needed for computing similarities between pseudo documents. Here, two different similarity measurements are investigated: VSM-based similarity and syntactic similarity.

VSM-based similarity (VSMSim): Given two pseudo documents $PD_{S1}(w_1)$ and $PD_{S2}(w_2)$. In VSM, they are represented as vectors V_1 and V_2, in which the weight of each word is calculated using a *tf·itf* heuristic:

$$tf \cdot itf(w, PD) = tf(w, PD) \times \log \frac{\max(tf(w', C_{CD}))}{tf(w, C_{CD})} \quad (2)$$

where $tf(w, PD)$ denotes the term frequency of word w in pseudo document PD. $tf(w, C_{CD})$ is w's term frequency counted on a China Daily Corpus (C_{CD}). $max(tf(w', C_{CD}))$

is the largest term frequency obtained on the corpus. The VSM-based similarity is calculated as the cosine similarity between V_1 and V_2:

$$VSMSim(PD_{S1}(w_1), PD_{S2}(w_2)) = \cos(V_1, V_2) = \frac{V_1 \bullet V_2}{\|V_1\|\|V_2\|} \tag{3}$$

where " $\bullet$ " denotes inner product. " $\|\cdot\|$ " denotes the length of a vector.

Syntactic similarity (SYNSim): In order to compute syntactic similarity, $PD_{S1}(w_1)$ and $PD_{S2}(w_2)$ are first parsed using the dependency parser described above. The syntactic similarity of pseudo documents is calculated with the same method as described in [8], as is rewritten in Equation (4). The similarity is calculated through the surrounding contextual words which have dependency relations with the investigated words according to the parsing results.

$$SYNSim(PD_{S1}(w_1), PD_{S2}(w_2)) = \frac{\sum_{(rel,w)\in T(w_1)\cap T(w_2)} (I(w_1, rel, w) + I(w_2, rel, w))}{\sum_{(rel,w)\in T(w_1)} I(w_1, rel, w) + \sum_{(rel,w)\in T(w_2)} I(w_2, rel, w)} \tag{4}$$

where $T(w_i)$ denotes the set of words that have the dependency relation *rel* with w_i.

$$I(w_i, rel, w_j) = \log \frac{p(w_i, rel, w_j)}{p(w_i \mid rel) p(w_j \mid rel) p(rel)} \tag{5}$$

3 Results and Analysis

3.1 Data and Metrics

In the experiments, a corpus of news titles is chosen as test data. That is mainly because in many applications, such as QA, IE, and multi-document summarization, the words and sentences to be paraphrased are usually from news articles. The news titles are collected from "sina news (http://news.sina.com.cn/)". All titles in the "important news" section from March 15, 2006 to April 5, 2006 are downloaded. 257 titles are left after removing duplications.

The metrics are precision, recall, and f-measure. Let $M_1, \ldots, M_T$ be T paraphrasing methods to be compared. N is the number of sentences in test data. n_i is the number of words in the *i-th* sentence that can be paraphrased by all the T methods. nt_{ij} is the number of acquired paraphrases for the *j-th* paraphrased word in the *i-th* sentence using method M_t $(1 \le t \le T)$. mt_{ij} is the number of correct paraphrases (judged manually) in the nt_{ij} paraphrases. Precision of method M_t is defined as:

$$precision(M_t) = \frac{\sum_{i=1}^{N} \sum_{j=1}^{n_i} \frac{mt_{ij}}{nt_{ij}}}{\sum_{i=1}^{N} n_i} \quad (1 \le t \le T) \tag{6}$$

Recall is difficult to calculate since it is impossible to enumerate all paraphrases that a word has within a context. Therefore, an approximate approach is used to calculate recall of each method. Specifically, for the *j-th* paraphrased word in the *i-th* sentence, all its correct paraphrases acquired by the *T* methods are put together (with duplication removed). Let m_{ij} be the total number. Then we assume that m_{ij} is the number of paraphrases that the word can really have within this specific sentence. Recall of method M_t is defined as:

$$recall(M_t) = \frac{\sum_{i=1}^{N}\sum_{j=1}^{n_i}\frac{mt_{ij}}{m_{ij}}}{\sum_{i=1}^{N} n_i} \quad (1 \le t \le T) \tag{7}$$

The f-measure of method M_t is defined as:

$$f-measure(M_t) = \frac{2 \times precision(M_t) \times recall(M_t)}{precision(M_t) + recall(M_t)} \quad (1 \le t \le T) \tag{8}$$

3.2 Experimental Results

In the experiments, four methods are completed and compared, including: (1) M_{ECilin}: the method that extracts paraphrases using the fifth level (synonyms) of ECilin. (2) $M_{CSP\text{-}Candi}$: the context-specific paraphrasing (CSP) method that extracts candidate paraphrases as described in Section 2.1 without paraphrase validation. (3) $M_{CSP\text{-}VSM}$: CSP method using VSMSim in paraphrase validation. (4) $M_{CSP\text{-}SYN}$: CSP method using SYNSim in paraphrase validation.

Three thresholds are used in the methods: (1) T_{CE}: threshold for candidate extraction; (2) T_{VSM}: threshold for VSMSim in validation; (3) T_{SYN}: threshold for SYNSim in validation. In the experiments, these thresholds are empirically set 0.30, 0.60, and 0.08 respectively. The comparing results are shown in Table 1:

Table 1. Comparing results of four methods

Method	Precision (%)	Recall (%)	F-measure (%)
M_{ECilin}	12.66	44.27	19.69
$M_{CSP\text{-}Candi}$	48.94	63.41	55.24
$M_{CSP\text{-}VSM}$	63.34	56.06	59.48
$M_{CSP\text{-}SYN}$	66.10	55.28	60.21

3.3 Analysis

3.3.1 Comparison with Method Using ECilin

As can be seen from Table 1, all the three CSP methods, i.e. $M_{CSP\text{-}Candi}$, $M_{CSP\text{-}VSM}$, and $M_{CSP\text{-}SYN}$ outperform M_{ECilin} significantly in precision, recall, and f-measure. Specifically, precision of M_{ECilin} is quite low, which shows that most synonyms defined in ECilin are not paraphrases in specific contexts. For example, the Chinese words "逝世(die)" and "毙命(die)" are synonyms. However, they can never be used in the same context, as the former expresses the death of a personage while the latter is usually used to express the death of an evil person. In contrast, in the CSP methods, these kinds of synonyms cannot be extracted as paraphrases, which makes precision much higher. On the other hand, recall of M_{ECilin} is also much lower than that of the CSP methods, which demonstrates that paraphrases in specific contexts are not necessarily synonyms.

3.3.2 Evaluation of Paraphrase Validation

In this section, the effectiveness of paraphrase validation stage is analyzed. It can be seen from Table 1 that both $M_{CSP\text{-}VSM}$ and $M_{CSP\text{-}SYN}$ outperform $M_{CSP\text{-}Candi}$ greatly in precision, which suggests that the validation methods using distinct similarity measurements are both effective in filtering incorrect candidates. At the same time, it can be found that recall decreases after the validation stage. The decrease indicates that some correct paraphrases are filtered in the validation by mistake. Nevertheless, the increases in f-measure demonstrate the effectiveness of paraphrase validation.

3.3.3 Comparison of Similarity Measurements

This section compares $M_{CSP\text{-}VSM}$ and $M_{CSP\text{-}SYN}$ so as to evaluate the two similarity measurements. It can be seen from Table 1 that $M_{CSP\text{-}VSM}$ and $M_{CSP\text{-}SYN}$ produce similar results. However, $M_{CSP\text{-}SYN}$ is better in precision while $M_{CSP\text{-}VSM}$ is better in recall. The reason why $M_{CSP\text{-}SYN}$ achieves a higher precision is that syntactic information is helpful in filtering incorrect candidates. For example, the sentence "巴林 客轮 沉没 48 人 遇难 (Tourist boat sinks off Bahrain, at least 48 died)" is from our test data. For the word "客轮(tourist boat)", "海岸(coast)" is extracted as a candidate paraphrase mistakenly. In $M_{CSP\text{-}VSM}$, this incorrect candidate cannot be filtered in validation, since their PDs share a lot of identical words, which makes them quite similar when represented as vectors in VSM. Nevertheless, these two words play different syntactic functions in sentences and have dependency relations with quite different words in PDs. Therefore, in the validation of $M_{CSP\text{-}SYN}$, this incorrect candidate can be easily filtered.

4 Conclusions

This paper proposes a web mining method to automatically acquire context-specific paraphrases. There are three main contributions. First, this work focuses on the problem of context-specific paraphrasing, which has seldom been addressed before. Second, a novel two-stage web mining method is presented. Third, a novel assumption is introduced in paraphrase validation and two different similarity measurements are investigated.

For the presented CSP methods, $M_{CSP\text{-}VSP}$ and $M_{CSP\text{-}SYN}$, precisions are 63.34% and 66.10%, and recalls are 56.06% and 55.28% respectively. The results significantly outperform the method using ECilin.

In the future work, paraphrase validation will be improved. Especially, different similarity measurements will be combined so as to get an optimal compromise of precision and recall.

Acknowledgments. This research was supported by National Natural Science Foundation of China (60435020, 60575042, 60503072). We thank Wanxiang Che and Weigang Li for useful discussions and their valuable comments on this paper.

References

1. Bannard, C., Callison-Burch, C.: Paraphrasing with Bilingual Parallel Corpora. In Proceedings of the 43rd Annual Meeting of the Association for Computational Linguistics (ACL2005), pages 597-604, 2005.
2. Barzilay, R., Elhadad, M.: Using Lexical Chains for Text Summarization. In Proceeding of the Intelligent Scalable Text Summarization Workshop (ISTS'97), pages 10-17, 1997.
3. Barzilay, R., McKeown, K. R.: Extracting Paraphrases from a Parallel Corpus. In Proceedings of the 39th Annual Meeting of the Association for Computational Linguistics (ACL2001), pages 50-57, 2001.
4. Cui, H., Wen, J.-R., Nie, J.-Y., Ma, W.-Y.: Probabilistic Query Expansion Using Query Logs. In Proceedings of Eleventh International World Wide Web Conference (WWW2002), pages 325-332, 2002.
5. Kaji, N., Kurohashi, S.: Lexical Choice via Topic Adaptation for Paraphrasing Written Language to Spoken Language. In Proceedings of the Second International Joint Conference on Natural Language Processing (IJCNLP2005), pages 981-992, 2005.
6. Langkilde, I., Knight, K.: Generation that Exploits Corpus-based Statistical Knowledge. In Proceedings of the 17th International Conference on Computational Linguistics and the 36th Annual Meeting of the Association for Computational Linguistics (COLING-ACL), pages 704-710, 1998.
7. Li, W., Liu, T., Zhang, Y., Li, S., He, W.: Automated Generalization of Phrasal Paraphrases from the Web. In Proceedings of the Third International Workshop on Paraphrasing (IWP2005), pages 49-56, 2005.
8. Lin, D.: Automatic Retrieval and Clustering of Similar Words. In Proceedings of the 17th International Conference on Computational Linguistics and the 36th Annual Meeting of the Association for Computational Linguistics (COLING-ACL), pages 768-774, 1998.
9. Wu, H., Zhou, M.: Optimizing Synonym Extraction Using Monolingual and Bilingual Resources. In Proceedings of the Second International Workshop on Paraphrasing (IWP2003), pages 72-79, 2003.

Author Index

Lecture Notes in Computer Science

For information about Vols. 1–4110

please contact your bookseller or Springer

Vol. 4228: D.E. Lightfoot, C.A. Szyperski (Eds.), Modular Programming Languages. X, 415 pages. 2006.

Vol. 4224: E. Corchado, H. Yin, V. Botti, C. Fyfe (Eds.), Intelligent, Data Engineering and, Automated Learning – IDEAL 2006. XXVII, 1447 pages. 2006.

Vol. 4219: D. Zamboni, C. Kruegel (Eds.), Recent Advances in Intrusion Detection. XII, 323 pages. 2006.

Vol. 4217: P. Cuenca, L. Orozco-Barbosa (Eds.), Personal Wireless Communications. XV, 532 pages. 2006.

Vol. 4213: J. Fürnkranz, T. Scheffer, M. Spiliopoulou (Eds.), Knowledge Discovery in Databases: PKDD 2006. XXII, 660 pages. 2006. (Sublibrary LNAI).

Vol. 4212: J. Fürnkranz, T. Scheffer, M. Spiliopoulou (Eds.), Machine Learning: ECML 2006. XXIII, 851 pages. 2006. (Sublibrary LNAI).

Vol. 4208: M. Gerndt, D. Kranzlmüller (Eds.), High Performance Computing and Communications. XXII, 938 pages. 2006.

Vol. 4206: P. Dourish, A. Friday (Eds.), UbiComp 2006: Ubiquitous Computing. XIX, 526 pages. 2006.

Vol. 4205: G. Bourque, N. El-Mabrouk (Eds.), Comparative Genomics. X, 231 pages. 2006. (Sublibrary LNBI).

Vol. 4202: E. Asarin, P. Bouyer (Eds.), Formal Modeling and Analysis of Timed Systems. XI, 369 pages. 2006.

Vol. 4201: Y. Sakakibara, S. Kobayashi, K. Sato, T. Nishino, E. Tomita (Eds.), Grammatical Inference: Algorithms and Applications. XII, 359 pages. 2006. (Sublibrary LNAI).

Vol. 4197: M. Raubal, H.J. Miller, A.U. Frank, M.F. Goodchild (Eds.), Geographic, Information Science. XIII, 419 pages. 2006.

Vol. 4196: K. Fischer, E. André, I.J. Timm, N. Zhong (Eds.), Multiagent System Technologies. X, 185 pages. 2006. (Sublibrary LNAI).

Vol. 4194: V.G. Ganzha, E.W. Mayr, E.V. Vorozhtsov (Eds.), Computer Algebra in Scientific Computing. XI, 313 pages. 2006.

Vol. 4193: T.P. Runarsson, H.-G. Beyer, E. Burke, J.J. Merelo-Guervós, L. D. Whitley, X. Yao (Eds.), Parallel Problem Solving from Nature - PPSN IX. XIX, 1061 pages. 2006.

Vol. 4192: B. Mohr, J.L. Träff, J. Worringen, J. Dongarra (Eds.), Recent Advances in Parallel Virtual Machine and Message Passing Interface. XVI, 414 pages. 2006.

Vol. 4189: D. Gollmann, J. Meier, A. Sabelfeld (Eds.), Computer Security – ESORICS 2006. XI, 548 pages. 2006.

Vol. 4188: P. Sojka, I. Kopeček, K. Pala (Eds.), Text, Speech and Dialogue. XIV, 721 pages. 2006. (Sublibrary LNAI).

Vol. 4187: J.J. Alferes, J. Bailey, W. May, U. Schwertel (Eds.), Principles and Practice of Semantic Web Reasoning. XI, 277 pages. 2006.

Vol. 4186: C. Jesshope, C. Egan (Eds.), Advances in Computer Systems Architecture. XIV, 605 pages. 2006.

Vol. 4185: R. Mizoguchi, Z. Shi, F. Giunchiglia (Eds.), The Semantic Web – ASWC 2006. XX, 778 pages. 2006.

Vol. 4184: M. Bravetti, M. Núñez, G. Zavattaro (Eds.), Web Services and Formal Methods. X, 289 pages. 2006.

Vol. 4183: J. Euzenat, J. Domingue (Eds.), Artificial Intelligence: Methodology, Systems, and Applications. XIII, 291 pages. 2006. (Sublibrary LNAI).

Vol. 4182: H.T. Ng, M.-K. Leong, M.-Y. Kan, D. Ji (Eds.), Information Retrieval Technology. XVI, 684 pages. 2006.

Vol. 4180: M. Kohlhase, OMDoc – An Open Markup Format for Mathematical Documents [version 1.2]. XIX, 428 pages. 2006. (Sublibrary LNAI).

Vol. 4179: J. Blanc-Talon, W. Philips, D. Popescu, P. Scheunders (Eds.), Advanced Concepts for Intelligent Vision Systems. XXIV, 1224 pages. 2006.

Vol. 4178: A. Corradini, H. Ehrig, U. Montanari, L. Ribeiro, G. Rozenberg (Eds.), Graph Transformations. XII, 473 pages. 2006.

Vol. 4176: S.K. Katsikas, J. Lopez, M. Backes, S. Gritzalis, B. Preneel (Eds.), Information Security. XIV, 548 pages. 2006.

Vol. 4175: P. Bücher, B.M.E. Moret (Eds.), Algorithms in Bioinformatics. XII, 402 pages. 2006. (Sublibrary LNBI).

Vol. 4174: K. Franke, K.-R. Müller, B. Nickolay, R. Schäfer (Eds.), Pattern Recognition. XX, 773 pages. 2006.

Vol. 4173: S. El Yacoubi, B. Chopard, S. Bandini (Eds.), Cellular Automata. XV, 734 pages. 2006.

Vol. 4172: J. Gonzalo, C. Thanos, M. F. Verdejo, R.C. Carrasco (Eds.), Research and Advanced Technology for Digital Libraries. XVII, 569 pages. 2006.

Vol. 4169: H.L. Bodlaender, M.A. Langston (Eds.), Parameterized and Exact Computation. XI, 279 pages. 2006.

Vol. 4168: Y. Azar, T. Erlebach (Eds.), Algorithms – ESA 2006. XVIII, 843 pages. 2006.

Vol. 4167: S. Dolev (Ed.), Distributed Computing. XV, 576 pages. 2006.

Vol. 4165: W. Jonker, M. Petković (Eds.), Secure, Data Management. X, 185 pages. 2006.

Vol. 4163: H. Bersini, J. Carneiro (Eds.), Artificial Immune Systems. XII, 460 pages. 2006.

Vol. 4162: R. Královič, P. Urzyczyn (Eds.), Mathematical Foundations of Computer Science 2006. XV, 814 pages. 2006.

Vol. 4161: R. Harper, M. Rauterberg, M. Combetto (Eds.), Entertainment Computing - ICEC 2006. XXVII, 417 pages. 2006.

Vol. 4160: M. Fisher, W.v.d. Hoek, B. Konev, A. Lisitsa (Eds.), Logics in Artificial Intelligence. XII, 516 pages. 2006. (Sublibrary LNAI).

Vol. 4159: J. Ma, H. Jin, L.T. Yang, J.J.-P. Tsai (Eds.), Ubiquitous Intelligence and Computing. XXII, 1190 pages. 2006.

Vol. 4158: L.T. Yang, H. Jin, J. Ma, T. Ungerer (Eds.), Autonomic and Trusted Computing. XIV, 613 pages. 2006.

Vol. 4156: S. Amer-Yahia, Z. Bellahsène, E. Hunt, R. Unland, J.X. Yu (Eds.), Database and XML Technologies. IX, 123 pages. 2006.

Vol. 4155: O. Stock, M. Schaerf (Eds.), Reasoning, Action and Interaction in AI Theories and Systems. XVIII, 343 pages. 2006. (Sublibrary LNAI).

Vol. 4154: Y.A. Dimitriadis, I. Zigurs, E. Gómez-Sánchez (Eds.), Groupware: Design, Implementation, and Use. XIV, 438 pages. 2006.

Vol. 4153: N. Zheng, X. Jiang, X. Lan (Eds.), Advances in Machine Vision, Image Processing, and Pattern Analysis. XIII, 506 pages. 2006.

Vol. 4152: Y. Manolopoulos, J. Pokorný, T. Sellis (Eds.), Advances in Databases and Information Systems. XV, 448 pages. 2006.

Vol. 4151: A. Iglesias, N. Takayama (Eds.), Mathematical Software - ICMS 2006. XVII, 452 pages. 2006.

Vol. 4150: M. Dorigo, L.M. Gambardella, M. Birattari, A. Martinoli, R. Poli, T. Stützle (Eds.), Ant Colony Optimization and Swarm Intelligence. XVI, 526 pages. 2006.

Vol. 4149: M. Klusch, M. Rovatsos, T.R. Payne (Eds.), Cooperative Information Agents X. XII, 477 pages. 2006. (Sublibrary LNAI).

Vol. 4148: J. Vounckx, N. Azemard, P. Maurine (Eds.), Integrated Circuit and System Design. XVI, 677 pages. 2006.

Vol. 4146: J.C. Rajapakse, L. Wong, R. Acharya (Eds.), Pattern Recognition in Bioinformatics. XIV, 186 pages. 2006. (Sublibrary LNBI).

Vol. 4144: T. Ball, R.B. Jones (Eds.), Computer Aided Verification. XV, 564 pages. 2006.

Vol. 4142: A. Campilho, M. Kamel (Eds.), Image Analysis and Recognition, Part II. XXVII, 923 pages. 2006.

Vol. 4141: A. Campilho, M. Kamel (Eds.), Image Analysis and Recognition, Part I. XXVIII, 939 pages. 2006.

Vol. 4139: T. Salakoski, F. Ginter, S. Pyysalo, T. Pahikkala, Advances in Natural Language Processing. XVI, 771 pages. 2006. (Sublibrary LNAI).

Vol. 4138: X. Cheng, W. Li, T. Znati (Eds.), Wireless Algorithms, Systems, and Applications. XVI, 709 pages. 2006.

Vol. 4137: C. Baier, H. Hermanns (Eds.), CONCUR 2006 – Concurrency Theory. XIII, 525 pages. 2006.

Vol. 4136: R.A. Schmidt (Ed.), Relations and Kleene Algebra in Computer Science. XI, 433 pages. 2006.

Vol. 4135: C.S. Calude, M.J. Dinneen, G. Păun, G. Rozenberg, S. Stepney (Eds.), Unconventional Computation. X, 267 pages. 2006.

Vol. 4134: K. Yi (Ed.), Static Analysis. XIII, 443 pages. 2006.

Vol. 4133: J. Gratch, M. Young, R. Aylett, D. Ballin, P. Olivier (Eds.), Intelligent Virtual Agents. XIV, 472 pages. 2006. (Sublibrary LNAI).

Vol. 4132: S. Kollias, A. Stafylopatis, W. Duch, E. Oja (Eds.), Artificial Neural Networks – ICANN 2006, Part II. XXXIV, 1028 pages. 2006.

Vol. 4131: S. Kollias, A. Stafylopatis, W. Duch, E. Oja (Eds.), Artificial Neural Networks – ICANN 2006, Part I. XXXIV, 1008 pages. 2006.

Vol. 4130: U. Furbach, N. Shankar (Eds.), Automated Reasoning. XV, 680 pages. 2006. (Sublibrary LNAI).

Vol. 4129: D. McGookin, S. Brewster (Eds.), Haptic and Audio Interaction Design. XII, 167 pages. 2006.

Vol. 4128: W.E. Nagel, W.V. Walter, W. Lehner (Eds.), Euro-Par 2006 Parallel Processing. XXXIII, 1221 pages. 2006.

Vol. 4127: E. Damiani, P. Liu (Eds.), Data and Applications Security XX. X, 319 pages. 2006.

Vol. 4126: P. Barahona, F. Bry, E. Franconi, N. Henze, U. Sattler, Reasoning Web. XII, 269 pages. 2006.

Vol. 4124: H. de Meer, J.P. G. Sterbenz (Eds.), Self-Organizing Systems. XIV, 261 pages. 2006.

Vol. 4121: A. Biere, C.P. Gomes (Eds.), Theory and Applications of Satisfiability Testing - SAT 2006. XII, 438 pages. 2006.

Vol. 4120: J. Calmet, T. Ida, D. Wang (Eds.), Artificial Intelligence and Symbolic Computation. XIII, 269 pages. 2006. (Sublibrary LNAI).

Vol. 4119: C. Dony, J.L. Knudsen, A. Romanovsky, A. Tripathi (Eds.), Advanced Topics in Exception Handling Components. X, 302 pages. 2006.

Vol. 4117: C. Dwork (Ed.), Advances in Cryptology - CRYPTO 2006. XIII, 621 pages. 2006.

Vol. 4116: R. De Prisco, M. Yung (Eds.), Security and Cryptography for Networks. XI, 366 pages. 2006.

Vol. 4115: D.-S. Huang, K. Li, G.W. Irwin (Eds.), Computational Intelligence and Bioinformatics, Part III. XXI, 803 pages. 2006. (Sublibrary LNBI).

Vol. 4114: D.-S. Huang, K. Li, G.W. Irwin (Eds.), Computational Intelligence, Part II. XXVII, 1337 pages. 2006. (Sublibrary LNAI).

Vol. 4113: D.-S. Huang, K. Li, G.W. Irwin (Eds.), Intelligent Computing, Part I. XXVII, 1331 pages. 2006.

Vol. 4112: D.Z. Chen, D. T. Lee (Eds.), Computing and Combinatorics. XIV, 528 pages. 2006.

Vol. 4111: F.S. de Boer, M.M. Bonsangue, S. Graf, W.-P. de Roever (Eds.), Formal Methods for Components and Objects. VIII, 447 pages. 2006.